Minerals

Their Constitution and O
Second edition

D0086007

The new edition of this popular textbook once again provides an indispensable guide for the next generation of mineralogists.

Minerals is an authoritative and comprehensive study of modern mineralogy, designed for use on one- or two-semester courses, for undergraduate and graduate students in the fields of geology, materials science, and environmental science.

This second edition has been thoughtfully reorganized, making it more accessible to students, whilst still being suitable for an advanced mineralogy course. Fully updated and revised, important additions include expanded introductions to many chapters, a new introductory chapter on crystal chemistry, revised figures, and an extended color plates section containing beautiful color photographs. Text boxes include historical background and case studies to engage students, and end-of-chapter questions help them reinforce concepts. With new online resources provided to support learning and teaching, including laboratory exercises, PowerPoint slides, useful web links, and mineral identification tables, this is a sound investment for students and a valuable reference for researchers, collectors, and anyone interested in minerals.

Hans-Rudolf Wenk is Professor of the Graduate School in the Department of Earth and Planetary Science at the University of California, Berkeley. Since joining the Berkeley faculty, he has been engaged in teaching and research, covering a wide field of mineralogy, from feldspars to carbonates, metamorphic rocks to shales, and from the Earth's surface to the inner core. His particular focus has been on microstructures, investigated using electron microscopy and synchrotron X-rays.

Andrey Bulakh is Professor in the Department of Mineralogy at St. Petersburg State University. He is a specialist in mineralogy, geochemistry, and the origin of alkaline rocks and carbonatites. More recently, he has studied the history of ornamental stones in architecture. He has written several books that are widely used at Russian universities, and was a long-time member of the Commission on New Minerals, Nomenclature and Classification of the International Mineralogical Association.

Minerals

Their Constitution and Origin

Second Edition

Hans-Rudolf Wenk

University of California, Berkeley, USA

Andrey Bulakh

St. Petersburg State University, Russia

CAMBRIDGE
UNIVERSITY PRESS

University Printing House, Cambridge CB2 8BS, United Kingdom

Cambridge University Press is part of the University of Cambridge.

It furthers the University's mission by disseminating knowledge in the pursuit of education, learning and research at the highest international levels of excellence.

www.cambridge.org
Information on this title: www.cambridge.org/wenk

© Hans-Rudolf Wenk and Andrey Bulakh 2016

First published 2004
Second edition 2016
Reprinted 2017

Printed in the United Kingdom by TJ International Ltd Padstow Cornwall.

A catalog record for this publication is available from the British Library

Library of Congress Cataloging in Publication data
Names: Wenk, Hans–Rudolf, 1941– author. | Bulakh, A. G. (Andrey Glebovich), author.
Title: Minerals : their constitution and origin / Hans-Rudolf Wenk, Andrey Bulakh.
Description: Second edition. | Cambridge : Cambridge University Press, 2016. | Includes bibliographical references and index.
Identifiers: LCCN 2015042018| ISBN 9781107106260 (hardback : alk. paper) | ISBN 9781107514041 (pbk. : alk. paper)
Subjects: LCSH: Mineralogy.
Classification: LCC QE363.2 .W46 2016 | DDC 549–dc23 LC record available at http://lccn.loc.gov/2015042018

ISBN 978-1-107-10626-0 Hardback
ISBN 978-1-107-51404-1 Paperback

Additional resources for this publication at www.cambridge.org/wenk

....

Contents

Color plates section is found between pp. 314 and 315

Preface

Minerals: Their Constitution and Origin is an introduction to mineralogy for undergraduate and graduate students in the fields of geology, materials science, and environmental science. It has been designed as a textbook for use in a one- or two-semester course but gives students a broader view and covers all aspects of mineralogy in a modern and integrated way. It provides detailed references to important publications on principles of crystallography and mineralogy. The book is not only descriptive but for interested readers derives basic principles such as aspects of symmetry theory, background on stereographic projection, X-ray diffraction and thermodynamics, based on general background from mathematics, physics and chemistry. The overall goal is to emphasize concepts and to minimize nomenclature. The text includes appendices covering identification of hand specimens and optical properties. With the broad approach, the book is not only a textbook for students but also a reference for teachers, researchers, collectors and anyone interested in minerals. The book is written in a modular fashion that permits instructors to select or omit some parts, depending on the level of the course, without compromising the continuity.

Today mineralogy is not just part of a geology curriculum. The importance of mineralogy has broadened to a wide variety of disciplines, from igneous petrology to soils science, from archaeology to cement engineering, from materials science to structural geology. Our book provides an alternative to existing texts by focusing more tightly on concepts, at the expense of completeness, and by integrating geological processes and applications more closely with the discussions of systematic mineralogy.

The book is divided into six parts:

Part I deals with general concepts of crystal chemistry, bonding, chemical formulas, mineral classification and hand specimen identification.

Part II introduces concepts of symmetry expressed in the morphology and structure of crystals. It then explores defects in crystal structures and the diversity of features observed during crystal growth.

Part III centers on the physics of minerals. First it shows how to use X-ray diffraction to determine the structural features introduced in Part II. A chapter on physical properties is optional but is significant for modern mineral physics and geophysics. We introduce optical properties and the use of the petrographic microscope because most mineralogists need to have this background before mineral systems are discussed in detail. The chapter on mineral identification with a microscope relies on access to relevant laboratory equipment. If there is no such access to microscopes, or if a separate course in optical mineralogy is available, chapters on optical mineralogy can be skipped. Part III concludes with a discussion of advanced analytical techniques, introducing equipment that may be encountered in modern mineralogical research laboratories.

Part IV discusses the wide range of mineral formation. It also provides some background in thermodynamics for understanding mineral equilibria in geological environments and phase transformations. Later chapters include applications of thermodynamics to sedimentary, hydrothermal, metamorphic, and igneous processes to demonstrate its relevancy.

Part V is a systematic treatment of mineral groups and about 200 of the most important minerals. Each chapter combines mineral characteristics with a discussion of a mineral-forming environment particularly linked to this mineral group, and information about mineral origin and mineral-forming processes. Part V starts with the most common minerals in the crust, quartz and feldspars, and ends with an overview of rare organic minerals like mellite.

Part VI on applied mineralogy deals with topics such as metal deposits, gems, cement, and human health, and explores how minerals form in the universe and were active components at each stage of the evolution of the Earth. We now have a much better understanding of minerals in the deep Earth, thanks largely to progress in seismology and experiments at ultra-high pressure and temperature. This part is largely independent from the rest of the book and these chapters can be used as reading assignments and form good starting points for term projects. The chapters should illustrate to students that mineralogy is not just complicated formulas, strange names,

Miller indices, and point-groups, but has practical significance.

Appendices contain determinative tables and important technical terms are defined in a glossary.

There are many excellent mineralogy textbooks, ranging from the early Niggli (1920) monograph (which still contains much of the information that is needed), to modern books such as Hibbard (2002), Klein and Dutrow (2007), Nesse (2011) and Klein and Philpotts (2012). Our book has a different emphasis. The goal is to be selective in including material rather than all inclusive, yet trying to remain quantitative, scientifically sound, and avoiding superficiality. It is well known that many students are frightened of mathematical expressions. We are using some equations here and there, but they can be skipped, without losing the thread, if students do not have the necessary background. But since most geology programs require mathematics and physics courses, it seems only reasonable to show students that this material is useful and to show some quantitative relationships; for example, how trigonometry can be used to calculate interfacial angles; how X-ray diffraction patterns are linked to lattice parameters; basic thermodynamics to understand a boundary in a phase diagram; simple linear algebra to appreciate why a second-rank tensor, such as the optical indicatrix, has the shape of an ellipsoid; or how complex numbers can be used to add waves more easily analytically than graphically to obtain diffraction intensities. We also have not shied away from referring to important references, including the classic studies of von Laue (1913) on X-ray diffraction, van't Hoff (1912) on the geochemistry of salt deposits, Bowen (1915) on experimental petrology, and also recent discoveries such as the structure of the lower mantle (e.g. Lekic et al., 2012), mineral identification on Mars (e.g. Bish et al., 2013) or isotope analyses at the atomic scale (e.g. Valley et al., 2014). This provides links to follow up on details about some of the milestones in mineralogy for readers who are interested.

The origin of this book goes back to 1993, when Dasha Sinitsyna, a student from (then) Leningrad, visited Berkeley on an exchange program and brought a little red book on mineralogy, written by her professor, Andrey Bulakh (1989), which caught Rudy Wenk's attention because it was an inspiring brief introduction to mineralogy. Over the following years we established further contact, in part through the

exchange of another student, Anton Chakmouradyan. After reciprocal visits to St Petersburg and Berkeley, sponsored by the University of California Education Abroad Program, the authors decided to attempt to produce an English mineralogy book, in the spirit of the Russian version but expanded it considerably.

The different backgrounds of the authors guarantee a broad view: Andrey Bulakh is a specialist on alkaline rocks and minerals and geochemistry and has written several books that are widely used at Russian universities, including the latest (Bulakh, 2011). Rudy Wenk's earlier research focused on metamorphic rocks, deformation fabrics, and investigations of microstructures in feldspars and carbonates with electron microscopes. More recently it has emphasized minerals at high pressure, stress and temperature with aspects such as anisotropy in the deep Earth (see http://eps.berkeley.edu/people/hans-rudolf-wenk). Both have taught introductory mineralogy at major universities for a long time. In this book we have tried to unite our expertise.

The first edition was published 12 years ago. Why have we prepared a new edition? The basic concepts of mineralogy and crystallography have not changed and a lot still relies on investigations with the petrographic microscope, introduced almost 200 years ago, and X-ray diffraction, celebrating in 2014 its hundredth anniversary with the UNESCO Year of Crystallography. But in 12 years a few things have happened: 1000 new minerals were added to the 4000 in 2002, but none of those are the subject of the book. Important is the shift in mineral production. South Africa is no longer the leading supplier of diamonds and China has become by far the main producer of steel. Particularly it manufactures a whopping 60% of the world's cement.

In 12 years the internet has also made profound changes. If you want to know the density of olivine, or the price of gold, you no longer go to a library but to Wikipedia. The new edition takes this into account by referring not only to books for "Further reading" but also recommends webpages with important mineral information. Appendices on mineral properties are provided not only in printed format but as digital files as well. And we added digital materials that may be useful for instructors: PowerPoint files from which teachers can select slides, and sample laboratory exercises based on a one-semester Berkeley mineralogy course. A Kindle edition is also available.

Compared with the first edition we have reorganized the content to make it easier to use for teaching. Part I starts with crystal chemistry and connects students with what they learnt in chemistry lectures, then links it to mineral classification which is mainly based on chemical composition, and introduces hand-specimen identification to bring students early in contact with actual minerals. With such a background it makes it easier to advance in Part II to the more abstract but important concepts of symmetry principles as well as graphic representations of crystal forms such as the stereographic projection. In Part V we have added a brief chapter on organic minerals, though rare but very interesting to make readers aware of different types of bonding and crystal structures.

The book has benefited from the help of many colleagues. Some generously contributed illustrations, others reviewed parts of the manuscript and provided valuable input in discussions. Foremost our thanks go to students who, over many years, taught us what for them is important in mineralogy, made us appreciate the difficult subjects, and guided us to topics of most interest.

There are different acknowledgements: One is to obtain permissions from publishers which we appreciate greatly and will give details in the next section.

Another is to appreciate people who have become friends. In alphabetical order we acknowledge first those who share some of their outstanding images, Joszef Arnoth (Naturhistorisches Museum, Basel, e.g., Arnoth, 1986); Mark Bailey (Asbestos TEM Laboratory, Berkeley); Regine Buxtorf (Basel); John Christensen (LBNL, Berkeley); Frank de Wit (Terhorst Nl, www.strahlen.org); Ken Finger (UCMP, Berkeley); Andreas Freund (ESRF, Grenoble); Walter Gabriel (Münchenstein-CH); John Grimsich (Berkeley CA); Gustaaf Hallegraeff (IMAS, University of Tasmania, Hobart); Henry Hänni (University of Basel-CH); Gregory Ivanyuk (Apatiti); Ray Joesten (University of Connecticut); Deborah Kelley (University of Washington, Seattle); Steven Kesler (University of Michigan, Ann Arbor); E.C. Klatt (Mercer University, Florida); Maya Kopylova (University of British Columbia Vancouver); George Kourounis (Toronto, http://www.stormchaser.ca/Stormchaser.html); Rob Lavinski (Richardson TX, The Arkenstone, http://www.irocks.com); Wayne and Dona Leicht (Laguna Beach CA, http://www.Kristalle.com); Wendy Mao (Stanford University CA); Andreas Massanek (Technical University of

Freiberg, e.g., Hofmann and Massanek, 1998); Remo Maurizio (Vicosoprano, e.g., Bedogné et al., 1995); Olaf Medenbach (Ruhr University, Bochum, e.g., Medenbach and Wilk, 1986; Medenbach and Medenbach, 2001); Terry Mitchell (Los Alamos National Laboratory); Hans-Ude Nissen (ETH Zurich); Janet Oldak (UCSC, Los Angeles CA); Michael Queen (Carlsbad); Colin Robinson (University of Leeds, UK); Thomas Schüpbach (Digital Studio, Ipsach-CH); Jeffrey Scovil (Phoenix, scovilphotography.com; e.g., Pough, 1996; Scovil, 1996); Tim Teague (UC Berkeley); Erica and Harold Van Pelt (Los Angeles, e.g., Keller, 1990; Sofianides and Harlow, 1990); Gustaaf Van Tendeloo (University of Antwerp); Mark Thompson (Lunenburg MA; Mark Thompson Information Service, http://mtinfopage.com/minerals.htm); Alexander Van Driessche (Vrije Universiteit, Brussel); Max Weibel (ETH Zurich, e.g., Weibel, 1973); Dan and Diana Weinrich (Weinrich Minerals, Inc., Grover MO, http://danweinrich.com); Elizabeth Wenk (Macquarie University, Sydney); Roland Wessicken (ETH Zurich); and a collective of the Museum of Geology at Beijing (e.g., Gao Zhen-xi, 1980). All are enthusiastic mineralogists. Two of them even have a mineral named after them: medenbachite and lavinskiite. Two appear in photographs as scale: Frank de Wit next to a giant halite crystal (Figure 10.12b) and George Kourounis in a sulfur fumarole (Plate 19d).

For other figures, data, reviews of book chapters, and comments we appreciate help from Jill Banfield (UC Berkeley), Dmitriy Belakovskiy (Fersman Museum, Moscow), David Bish (University of Indiana), David Blake (NASA, Ames CA), Douglas Bock (CSIRO, Sydney), Evelyn Denzin (Athens, Georgia), Robert Downs (University of Arizona, Tucson), Edward Garnero (Arizona State University, Tempe), A. Filippenko (UC Berkeley), Valeriy Ivanikov, Catherine McCammon (Bayerisches Geoinstitut), Wolfgang Müller (T.U. Darmstadt), Barbara Romanowicz (UC Berkeley), Masha Sitnikova (BRD, Hannover), Alan Stern (SwRI, Boulder CO); Roman Vasin (JINR, Dubna), and Sergei and Vladimir Krivovichev, Igor Pekov, S. Petov, Eugeniy Treivus, and Anatoly Zolotarev (St. Petersburg State University). Rudy Wenk is particularly indebted to his wife Julia for her meticulous review that discovered many embarrassing errors.

To interact with these people has been the most rewarding experience and we appreciate their enthusiasm and support. It is like a great family of mineral enthusiasts which started with Theophrastus and Pliny, included Stensen, Goethe (a dedicated mineral collector; visit his house in Weimar), and then of course Dana who created systematic mineralogy.

Last but not least we are grateful to the staff at Cambridge University Press for their dedication and patience. Of course blame for all remaining deficiencies, omissions, and errors in content rests with the authors and we continue to appreciate the input from readers who suggest corrections. We can make corrections in future printings and do not have to wait for another edition; so do not hesitate to contact us to report errors and make suggestions. We dedicate the book to our wives, Julia Wenk and Victoria Kondratieva.

H.-R. Wenk
A. Bulakh

Figure credits

For more details see also captions and references.

The authors are grateful to the following publishers, institutions, and individuals for permission to reproduce material:

Cambridge University Press

Figure 16.31: Putnis, A. (1992). *Introduction to Mineral Sciences*. Cambridge University Press, Cambridge. Fig. 4.35.

Figure 38.3: Fowler, C. M. R. (2005). *The Solid Earth. An Introduction to Global Geophysics*, 2nd edn. Cambridge University Press, Cambridge. Fig. 8.17.

Dover Publications

Figure 7.8b: Haeckel, E. (1904). *Kunstformen der Natur*. Bibliografische Institut, Leipzig. English translation (1974): *Art Forms in Nature*, Dover Publications, New York.

Figure 10.10a,b: Bentley, W. A. and Humphreys, W. J. (1962). *Snow Crystals*. Paperback edn. Dover Publications, New York.

Elsevier
Books

Figure 1.5a: Baikow, V. E. (1967). *Manufacture and Refining of Raw Cane Sugar*. Elsevier, Amsterdam. Fig. 17.1. Copyright 1967 with permission from Elsevier.

Figure 10.8a: Verma, A. R. (1953). *Crystal Growth and Dislocations*. Academic Press, New York. Fig. 56. Copyright 1953 with permission from Elsevier.

Figure 12.19: Dillon, F. J. (1963). Domains and domain walls. In *Magnetism, Volume 3*, ed. G. T. Rado and H. Suhl, pp. 415–464. Academic Press, New York. Fig. 13. Copyright 1963 with permission from Elsevier.

Journals

Figure 17.7b: Barber, D. J. and Wenk, H.-R. (1979). On geological aspects of calcite microstructure. *Tectonophysics*, **54**, 45–60. Fig. 9. Copyright 1979 with permission from Elsevier.

Figure 33.6: Cann, J. R., Strens, M. R. and Rice, A. (1985). A simple magma-driven thermal balance model for the formation of volcanogenic massive sulfides. *Earth Planet. Sci. Lett.*, **76**, 123–134. Fig. 4. Copyright 1985 with permission from Elsevier.

Figure 37.6: Daulton, T. L., Eisenhour, D. D., Bernatowicz, T. J., Lewis, R. S. and Buseck, P. R. (1996). Genesis of presolar diamonds: comparative high-resolution transmission electron microscope study of meteoritic and terrestrial nano-diamonds. *Geochim. Cosmochim. Acta*, **60**, 4853–4872. Fig. 7. Copyright 1996 with permission from Elsevier.

Figure 37.7: Wood, J. A. and Hashimoto, A. (1993). Mineral equilibrium in fractionated nebular systems. *Geochim. Cosmochim. Acta*, **57**, 2377–2388. Fig. 5. Copyright 1993 with permission from Elsevier.

International Union of Crystallography

Figure 7.7c: MacGillavry, C. H. (1976). *Symmetry Aspects of M. C. Escher's Periodic Drawings*, 2nd edn. International Union of Crystallography. Fig. 10.

Figure 11.3a,b: Ewald, P. P. (1962). *Fifty Years of X-ray Diffraction*. International Union of Crystallography, Utrecht. Figs. 4.4-1 and 4.4-3.

Figure 32.6: Gerkin, R. E., Lundstedt, A. P. and Reppart, W. J. (1984). Structure of fluorene $C_{13}H_{10}$, at 159K. *Acta Cryst. C*, **40**, 1892–1894. Fig. 2.

Figure 32.9: Swaminathan, S., Craven, B. M. and McMullan, R. K. (1984). The crystal structure and molecular thermal motion of urea at 12, 60 and 123 K from neutron diffraction. *Acta Cryst. B*, **40**, 300–306. Fig. 1.

Geological Society of America

Figures 9.6 and 29.12: Page, R. H. and Wenk, H.-R. (1979). Phyllosilicate alteration of plagioclase studied by transmission electron microscopy. *Geology*, **7**, 393–397. Figs. 2C and 2F.

Iraq

Figure 7.7a: Mallowan, M. E. L. and Cruikshank, R. J. (1933). Excavations at Tel Arpachiyah. *Iraq*, **2**, 1–178.

Maxwell Museum of Anthropology, University of New Mexico

Figure 7.7b: Courtesy of the Maxwell Museum of Anthropology, University of New Mexico. Photographer: C. Baudoin.

Mineralogical Association of Canada

Figure 32.5: Echigo, T. and Kimata, M. (2010). Crystal chemistry and genesis of organic minerals: A review of oxalate and polycyclic aromatic hydrocarbon minerals. *Can. Mineral.*, **48**, 1329–1358. Fig. 13, p. 1343.

Figure 32.8: Echigo, T. and Kimata, M. (2010). Crystal chemistry and genesis of organic minerals: a review of oxalate and polycyclic aromatic hydrocarbon minerals. *Can. Mineral.*, **48**, 1329–1358. Fig. 8, p. 1339.

Figure 36.3: Wicks, F. J., Kjoller, K. and Henderson, G. S. (1992). Imaging the hydroxyl surface of lizardite at atomic resolution with the atomic force microscope. *Can. Mineral.*, **30**, 83–91. Fig. 8, p. 90.

Figure 36.5: Semkin, R. G. and Kramer, J. R. (1976). Sediment geochemistry of Sudbury area lakes. *Can. Mineral.*, **14**, 73–90. Fig. 13, p. 83.

Mineralogical Society of America

Figure 4.2: Sriramadas, A. (1957). Diagrams for the correlation of unit cell edges and refractive indices with the chemical composition of garnets. *Am. Mineral.*, **42**, 294–298. Fig. 1, p. 295.

Figures 9.15 and 16.2: Meisheng, Hu, Wenk, H.-R. and Sinitsyna, D. (1992). Microstructures in natural perovskites. *Am. Mineral.*, **77**, 359–373. Fig. 10, p. 367; Fig. 15, p. 370.

Figure 10.12a,b: Rickwood, P. C. (1981). The largest crystals. *Am. Mineral.*, **66**, 885–907. Figs. 10 and 11, p. 903.

Figure 16.28: Bischoff, W. D., Sharma, S. K. and MacKenzie, F. T. (1985). Carbonate ion disorder in synthetic and biogenic magnesian calcites: a Raman spectral study. *Am. Mineral.*, **70**, 581–589. Fig. 2, p. 583.

Figure 16.32: McKeown, D. A. and Post, D. A. (2001). Characterization of manganese oxide mineralogy in rock varnish and dendrites using X-ray absorption spectroscopy. *Am. Mineral.*, **86**, 701–713. Fig. 2, p. 39.

Figure 16.35: Phillips, B. L. (2000). NMR spectroscopy of phase transitions in minerals. In *Transformation Processes in Minerals*. Reviews in Mineralogy, vol. 39, Mineralogical Society of America, pp. 203–240. p. 223.

Figure 16.37: McCammon, C. A. (2000). Insights into phase transformations from Mössbauer spectroscopy. In *Transformation Processes in Minerals*. Reviews in Mineralogy, vol. 39, Mineralogical Society of America, pp. 241–257. p. 251.

Figure 21.13: Steiger, R. H. and Hart, S. R. (1967). The microcline–orthoclase transition within a contact aureole. *Am. Mineral.*, **52**, 87–116. Fig. 2, p. 91.

Figure 21.23: Barron, L. M. (1972). Thermodynamic multicomponent silicate equilibrium phase calculations. *Am. Mineral.*, **57**, 809–823.

Figure 25.9: Devouard, B., Posfai, M., Hua, X., Bazylinski, D. A., Frankel, R. B. and Buseck, P. R. (1997). Magnetite from magnetotactic bacteria: size distributions and twinning. *Am. Mineral.*, **83**, 1387–1398. Fig. 12, p. 389.

Figure 30.12: Veblen, D. R. and Buseck, P. (1980). Microstructures and reaction mechanisms in biopyriboles. *Am. Mineral.*, **65**, 599–623. Fig. 5, p. 608.

Figure 30.25: Greenwood, H. J. (1967). Wollastonite: stability in $H_2O–CO_2$ mixtures and occurrence in a contact-metamorphic aureole near Salmo, British Columbia, Canada. *Am. Mineral.*, **52**, 1669–1680.

Figure 36.2: Guthrie, G. D. and Mossman, B. T. (1993). *Health Effects of Mineral Dust*. Reviews in Mineralogy, vol. 28. p. 1 and p. 231. RA1231.M55 H424 Public Health.

Mineralogical Magazine

Figure 30.20: Joesten, R. (1986). The role of magmatic reaction, diffusion and annealing in the evolution of coronitic microstructure in troctolitic gabbro from Risör, Norway. *Mineral. Mag.*, **50**, 441–467.

Oxford University Press

Figure 36.1: Stanton, M. F., Layard, M., Tegeris, A., Miller, E., May, M., Morgan, E. and Smith, A. (1981). Relation of particle dimension to carcinogenicity in amphibole asbestoses and other fibrous materials. *J. Natl. Cancer Inst.*, **67**, 965–975, by permission of Oxford University Press.

Royal Geological Society of Cornwall

Figure 26.12: Hosking, K. F. G. (1951). Primary ore deposition in Cornwall. *Trans. Roy. Geol. Soc. Cornwall*, **18**, 309–356.

Schweizerische Mineralogische und Petrographische Gesellschaft

Figure 28.10: Wenk, H.-R., Wenk, E. and Wallace, J. H. (1974). Metamorphic mineral assemblages in pelitic rocks of the Bergell Alps. *Schweiz. Mineral. Petrog. Mitt.*, **54**, 507–554.

Figure 30.26: Trommsdorff, V. (1966). Progressive Metamorphose kieseliger Karbonatgesteine in den Zentralalpen zwischen Bernina und Simplon. *Schweiz. Mineral. Petrog. Mitt.*, **46**, 431–460.

Schweizer Strahler

Figures 18.8 and 18.9: Mullis, J. (1991). Bergkristall. *Schweizer Strahler*, **9**, 127–161. Figs. 2 and 5.

Society of Economic Geologists

Figure 18.2: Fournier, R. O. (1985). The behavior of silica in hydrothermal solutions. In *Geology and Geochemistry of Epithermal Systems*, ed. B. R. Berger and P. M. Bethke. Reviews in Geology, vol. 2, pp. 63–79. Society of Economists and Geologists, Chelsea, MI.

Figure 26.10: Sillitoe, R. H. (1973). The tops and bottoms of porphyry copper deposits. *Econ. Geol.*, **68**, 799–815. Fig. 1, p. 800.

Figure 33.7: Carr, H. W., Groves, D. I. and Cawthorne, R. G. (1994). The importance of synmagmatic deformation in the formation of Merensky Reef potholes in the Bushveld complex. *Econ. Geol.*, **89**, 1398–1410. Fig. 1, p. 1399, Fig. 2, p. 1402.

Figure 36.7: Smith, K. S. and Huyck, H. L. O. (1999). An overview of the abundance, relative mobility, bioavailability, and human toxicity of metals. In *The Environmental Geochemistry of Mineral Deposits, Volume A*, ed. G. S. Plumlee and M. J. Logsdon, pp. 29–70. Society of Economic Geologists, Littleton, CO.

Society of Glass Technology

Figure 18.4: Lee, R. W. (1964). On the role of hydroxyl on the diffusion of hydrogen in fused silica. *Phys. Chem. Glasses*, **5**, 35–43.

Springer-Verlag
Books

Figure 9.13: Lally, J. S., Heuer, A. H. and Nord, G. L. (1976). Precipitation in the ilmenite-hematite system. In *Electron Microscopy in Mineralogy*, ed. H.-R. Wenk, pp. 214–219. Springer-Verlag, Berlin. Fig. 1a. Copyright Springer-Verlag 1976.

Figure 18.1: Hoefs, J. (1987). *Stable Isotope Geochemistry*, 3rd edn. Springer-Verlag, Berlin. Fig. 31. Copyright Springer-Verlag 1987.

Figure 22.7: Mitchell, R. H. (1986), *Kimberlites: Mineralogy, Geochemistry and Petrology*. Plenum Press, New York. Fig. 3.1.

Figures 20.8 and 21.17: Champness, P. and Lorimer, G. (1976). Exsolution in silicates. In *Electron Microscopy in Mineralogy*, ed. H.-R. Wenk, pp. 174–204. Springer-Verlag, Berlin. Figs. 9 and 14. Copyright Springer-Verlag 1976.

Figures 21.16 and 21.18: Smith, J. V. and Brown, W. L. (1988). *Feldspar Mineralogy*. Springer-Verlag, Berlin. Figs. 1.4 and 2.4. Copyright Springer-Verlag 1988.

Figure 29.20: Pédro, G. (1997). Clay minerals in weathered rock materials and in soils. In *Soils and Sediments. Mineralogy and Geochemistry*, ed. H. Paquet and N. Clauer, pp. 1–20. Springer-Verlag, Berlin. Fig. 3. Copyright Springer-Verlag 1997.

Figure 30.7: Wenk, H.-R. (ed.) (1976). *Electron Microscopy in Mineralogy*. Springer-Verlag, Berlin. p. 2. Copyright Springer-Verlag 1976.

Figure 30.14: Liebau, F. (1985). *Structural Chemistry of Silicates: Structure, Bonding, Classification*. Springer-Verlag, Berlin. Fig. 10.3. Copyright Springer-Verlag 1985.

Figure 30.23: Winkler, H. G. F. (1979). *Petrogenesis of Metamorphic Rocks*, 5th edn. Springer-Verlag, Berlin. Fig. 5.6. Copyright Springer-Verlag 1979.

Figures 31.6a and 31.6c: Gottardi, G. and Galli, E. (1985). *Natural Zeolites*. Springer-Verlag, Berlin. Figs. 2.1E and 6.1. Copyright Springer-Verlag 1985.

Figures 37.10 and 37.11: Ringwood, A. E. (1979). *Origin of the Earth and Moon*. Springer-Verlag,

Berlin. Figs. 2.1 and 12.1. Copyright Springer-Verlag 1979.

Journals

Figures 9.5 and 17.7b: Barber, D. J., Heard, H. C. and Wenk, H.-R. (1981). Deformation of dolomite single crystals from 20–800C. *Phys. Chem. Miner.*, **7**, 271–286. Fig. 3a,b. Copyright Springer-Verlag 1981.

Figures 16.3 and 12.10: Wenk, H.-R., Hu, M., Lindsey, T. and Morris, W. (1991). Superstructures in ankerite and calcite. *Phys. Chem. Mineral.*, **17**, 527–539. Fig. 3c. Copyright Springer-Verlag 1991.

Figures 16.30 and 12.29: Aines, R. D., Kirby, S. H. and Rossman, G. R. (1984). Hydrogen speciation in synthetic quartz. *Phys. Chem. Mineral.*, **11**, 204–212. Fig. 1. Copyright Springer-Verlag 1984.

Figures 19.15b and 21.15b: Müller, W. F., Wenk, H. R. and Thomas G. (1972). Structural variations in anorthites. *Contrib. Mineral. Petrol.*, **34**, 304–314. Copyright Springer-Verlag 1976.

Taylor and Francis Group

Figure 17.7a: Westmacott, K. H., Barnes, R. S. and Smallman, R. E. (1962). The observation of dislocation "climb" source. *Phil. Mag.*, **7** (ser. 8), 1585–1613. Fig. 2.

University of Chicago Press

Figure 21.24: Bowen, N. L. and Tuttle, O. F. (1950). The system $NaAlSi_3O_8$–$KAlSi_3O_8$–H_2O. *J. Geol.*, **58**, 498–511. Fig. 3.

Figure 24.6: Goldsmith, J. R. and Heard, H. C. (1961). Subsolidus phase relations in the system $CaCO_3$–$MgCO_3$. *J. Geol.*, **69**, 45–74. Fig. 4.

Wepf & Co. AG Verlag

Figure 10.15: Stalder *et al.* (1973). Tafel 17c (Figure 5.30). Republished 1998 as *Mineralienlexikon der Schweiz*, Wepf, Basel.

John Wiley & Sons
Books

Figure 26.13: Evans, A. M. (1993). *Ore Geology and Industrial Minerals: An Introduction*, 3rd edn. Blackwell, Oxford. Fig. 4.13.

Figures 2.21 and 34.5: Klein, C. (2002). *Mineral Science*, 22nd edn. Wiley, New York. Copyright 2002 John Wiley & Sons. This material is used by permission of John Wiley & Sons, Inc.

Figure 16.29: Nakamoto, K. (1997). *Infrared and Raman Spectra of Inorganic and Coordination Compounds, Part A: Theory and Applications in Inorganic Chemistry*, 5th edn. Wiley, New York. Copyright 1997 John Wiley & Sons. This material is used by permission of John Wiley & Sons, Inc.

Journals

Figure 12.13: Morris, G. B., Raitt, R. W. and Shor, G. G. (1969). Velocity anisotropy and delay time maps of the mantle near Hawaii. *J. Geophys. Res.*, **74**, 4300–4316. Fig. 12.

Figure 16.9: Vasin, R., Wenk, H.-R., Kanitpanyacharoen, W., Matthies, S. and Wirth, R. (2013). Anisotropy of Kimmeridge shale. *J. Geophys. Res.*, **118**, 1–26, doi:10.1002/jgrb.50259. Fig. 1.

Figure 16.39: McCammon, C. A. (1995). Mössbauer spectroscopy of minerals. In *Mineral Physics and Crystallography. A Handbook of Physical Constants*. American Geophysical Union, Washington, DC, pp. 332–347. Fig. 2.

Figure 38.1: Anderson, D. L. and Hart, R. S. (1976). An earth model based on free oscillations and body waves. *J. Geophys. Res.*, **81**, 1461–1475.

Figure 38.6: Shen, G., Mao, H.-K., Hemley, R. J., Duffy, T. S. and Rivers, M. L. (1998). Melting and crystal structures of iron at high pressures and temperatures. *Geophys. Res. Lett.*, **25**, 373–376. Fig. 3.

Yale University Press

Figure 24.12: Garrels, R. M., Thompson, M. E. and Siever, R. (1960). Stability of some carbonates at 25 °C and one atmosphere total pressure. *Am. J. Sci.*, **258**, 402–418. Fig. 5.

Praise for the first edition

"I think this book represents a sound undergraduate investment – a textbook that an undergraduate could visit and revisit throughout their degree program, to remind them of the basics and, by following up the references, to provide a deeper understanding of the subjects covered."

Chemistry World

"... the book provides a good coverage of minerals, with clear diagrams and photographs to supplement the text. ... there is much of value in this book. ... the text is clear; and deeper treatments can be skipped, while still gaining knowledge of the wider range of mineralogy."

OUGS Newsletter

"Wenk and Bulakh's *Minerals* is both authoritative and accessible, providing a thorough grounding in many aspects of modern mineralogy in a first-rate text."

New Scientist

"... this is a refreshing new mineral textbook and is a wonderful resource to freshen up an undergraduate course. Every lecturer who teaches mineralogy and every Earth Sciences library should get a copy. ... Very highly recommended."

Geological Magazine

Part I | Minerals as chemical compounds

1 | Subject and history of mineralogy

Along with mathematics and astronomy, mineralogy is among the oldest branches of science. In this introductory chapter we explore the roots and evolution of mineralogy, with a first book written by Theophrastus in 300 BC to over 20 Nobel Prizes awarded for research involving crystals. We will discuss the different directions in mineralogy and provide a summary of the organization of the book.

1.1 What is mineralogy?

The answer to the question posed above may seem obvious: mineralogy is the study of minerals. From introductory geology classes you may know that all rocks and ores consist of minerals. For instance, quartz, biotite, and feldspar are the main minerals of granites; and hematite and magnetite are the major minerals of iron ores. At one point mineralogy was well defined as dealing with those naturally occurring elementary building blocks of the Earth that are chemically and structurally homogeneous. This simple definition of a mineral has changed over time. As the definition of "mineral" has become more vague the boundaries of "mineralogy" have opened and increasingly overlap other sciences.

In this book we take a broad view of mineralogy. Minerals are naturally occurring, macroscopically homogeneous chemical compounds with regular crystal structures. Traditionally also included are homogeneous compounds that do not have a regular structure such as opal (a colloidal solid), natural liquid mercury, and amorphous mineral products formed by radioactive decay, known as *metamict* minerals. Rocks, ores, and mineral deposits, which are studied in petrology and geochemistry, will also be discussed in order to emphasize the geological processes that are of central interest to all who study Earth materials.

Other materials are more peripheral but nevertheless have similar properties and obey the same laws as the minerals mentioned above. For example, ice is mainly the object of glaciology, planetary and soils science. Apatite (a major constituent of bones and teeth), carbonates (which form the skeletons of mollusks), and oxalates and urates (of which human kidney stones are composed) are studied in medicine and in a specific branch of mineralogy called "biomineralogy". Crystals growing in concrete, known as *cement minerals*, are not natural products but they are of enormous economic importance. Other artificial compounds with a crystal structure occur as a result of industrial transformations in natural conditions. They may form because of chemical alteration of buried waste products or by means of interaction between the soils and contaminated groundwater. Modern environmental geology, hydrology, and soils science are concerned with these mineral-like materials and study them with methods similar to those employed in mineralogy.

Mineralogy is broadening its scope and now overlaps other disciplines in a way that was not envisioned a few years ago. For example, newly discovered high-temperature superconductors have a structure related to the mineral perovskite and often possess a morphology resembling clays. We will include a discussion of some of the new subjects of mineralogy in this text, but our focus is on those natural, chemically and structurally homogeneous substances that form due to geological and biogeological processes in the Earth. In particular, we concentrate on rock- and ore-forming minerals, which are a small subset of the almost 5000 mineral species that are currently known.

1.2 History

Mineralogy and crystallography are old branches of science. For example, crystals, with their regular

3

morphology, are known to have fascinated Ancient Greek philosophers. Indeed, the name crystal derives from the Greek *krystallos* (meaning "ice") and was applied to quartz, since the Ancient Greeks thought that this mineral was water that had crystallized at high pressure deep inside the Earth. Accordingly, the German term *Bergkristall* (meaning "mountain crystal"), a synonym for quartz, has survived to this day. Note that the term crystal is also used for glass with a brilliant reflection. Such glass is not related to minerals and, in fact, is not even crystalline. In this section we will discuss a few highlights in the early history of mineralogy. Those interested in more details will find a great deal of information in Burke (1966) and Groth (1926).

The Greek philosopher Theophrastus (300 BC), a pupil of Aristotle, wrote the first known book on minerals, entitled *On Stones* (see Caley and Richards, 1956). Similar to the focus of this text, Theophrastus' book was greatly concerned with the origin of minerals. It begins: "Of those substances formed in the ground, some are made of water and some of earth. Metals obtained by mining come from water; stones, including the precious kinds, come from earth." As we will see later, this statement, although highly simplified, does contain some truth. Some minerals precipitate from solutions, others crystallize from a melt. The name *mineral* originates from Latin and relates to materials that are excavated in mines (*mina* is the Latin word for "mine", and *minare* is Latin for "to mine"). Pliny the Elder, who was killed in the eruption of Vesuvius in August AD 79, summarized the knowledge on minerals at his time, describing over 30 minerals in his *Historia Naturalis* (see Lenz, 1861), among them galena, molybdenite, chalcopyrite, beryl, and augite. The mineralogy of Pliny emphasized minerals that were of economic interest, giving descriptions and discussing their occurrence and usage. For example, he writes about diamond: "More expensive than all other gems and any other human possessions are diamonds (*adamas*). They are only known to kings, and even among them only to a few. Indian diamonds resemble quartz (*krystallos*). They are transparent, have a regular form with smooth surfaces and are the size of a hazelnut."

It was nearly 1500 years later, with the publication of *De Re Metallica* in 1556 by the German mining engineer and physician Georg Bauer (known as

Agricola) from Freiberg, Saxony (Figure 1.1a), that mineralogy emerged as a science. Succeeding centuries brought important advances, one of which was the discovery by Niels Stensen (Nicolas Steno) in 1669 (Figure 1.1b) that angles between the regular faces of crystals are always the same, in spite of differences in shape or size. Probably inspired by the observation of a regular cleavage in calcite by the Swedish mineralogist Torbern Bergmann (1773), in 1784 the French scientist René J. Haüy (Figure 1.1c) interpreted this law of constancy of interfacial angles. He suggested that all crystals are composed of elementary building blocks, which he called "integral molecules". The building blocks later became known as "unit cells". A macroscopic crystal can be thought of as a three-dimensional periodic array of such unit cells. The regularity of faces and interfacial angles is obvious, for example in a cubic crystal of halite or in an octahedron of magnetite.

During the eighteenth and nineteenth centuries, most of the important minerals were described. A. G. Werner (Figure 1.1d), a mining geologist in Freiberg, proposed a chemical classification of minerals (e.g. Hoffmann, 1789). This classification was later refined by Swedish chemist J. J. Berzelius and is still in use today. It became the basis for the first comprehensive textbooks in mineralogy, such as those by Haüy (1801), Dana (1837), Breithaupt (1849), and Groth (1904). Mathematicians, among them J. F. C. Hessell (1830), A. Bravais (1850), E. S. Fedorow (1885, 1892), and A. Schoenflies (1891), investigated the possible symmetries of a material with a regular morphology and periodic internal structure, as suggested by Haüy, and developed a system to classify minerals according to symmetry. At that time there was considerable debate among chemists and mineralogists about whether the concept of internal structure (or "lattice") really applied or whether, instead, crystals were continuous. Johann Joseph Prechtl (1810) advanced the concept of stacking of spherical particles and hypothesized that "atoms" had no form in the liquid state but took spherical form during solidification. The sketches of William Hyde Wollaston (1813; Figure 1.2a) illustrate different arrangements of spheres that are the basis of simple metal structures, as we will see in Chapter 2. While Wollaston thought that all atoms were of the same size, William Barlow (1897) suggested that the size may be characteristic of elements and that

Fig. 1.1 Four pioneers of modern mineralogy. (a) Georg Bauer (Agricola), (b) Niels Stensen (Nicolas Steno), (c) René J. Haüy, and (d) Abraham Gottlob Werner.

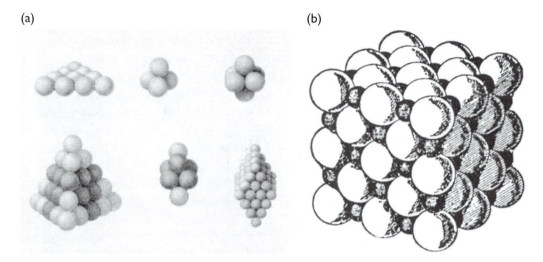

Fig. 1.2 Early models of crystal structures. (a) Wollaston (1813) assumed that crystals consisted of close-packed spheres (different gray shades indicate different elements). (b) Barlow (1897) refined the model by assuming different sizes of atoms and correctly predicting the structure of halite.

a crystal structure of binary compounds could be obtained as close-packing of such spheres. He correctly predicted the structure of halite (Figure 1.2b).

Before 1800 most research on minerals and crystals was based on visual observations and measurements of angles between crystal faces with mechanical goniometers. This changed in 1809 when Wollaston developed a reflecting goniometer that allowed much more accurate measurements. It was perfected in 1893 with a two-circle goniometer (Figure 1.3a), independently by E. S. Fedorow in St. Petersburg, S. Czapski in Jena, and V. Goldschmidt in Heidelberg who published the fundamental "Kristallographische Winkeltabellen" (Goldschmidt, 1897) based on goniometer measurements (see, e.g., Burchard, 1998).

Another essential instrument is the petrographic microscope, which is still the most important tool used to identify minerals in rocks (Chapters 13 and 14). Its development started in 1828 with a microscope designed by W. Nicol that could analyze mineral fragments with polarized light and was advanced over many years. The instrument proposed by Rosenbusch (1876) and built with R. Fuess is basically the foundation of modern petrographic microscopes (Figure 1.3b), without which mineralogic–petrologic research would be unthinkable (e.g., Medenbach, 2014).

The theories about the atomic structure of crystals based on crystal morphology and symmetry were highly speculative and the issue was only resolved in 1912, when Max von Laue and his coworkers in Munich irradiated crystals with X-rays and observed diffraction, proving that crystals have indeed a lattice structure and that X-rays are waves (Chapter 11). At this point mineralogy became an experimental science and expanded considerably, as will be detailed in Part III. In 1914 William Lawrence Bragg published the first crystal structure determination, which described the detailed atomic arrangements in halite. A modern textbook of mineralogy incorporating the new advances in structural investigations was written by Paul Niggli (1920). If you look through it, you will discover many figures that are still reproduced in textbooks today. X-ray diffraction was the favored analytical technique at that time. Much later, electron microscopes and spectrometers became important tools. High-temperature and high-pressure techniques eventually became available to produce minerals in the laboratory under any conditions found in the Earth and beyond. A significant part of this research on minerals was carried out by physicists and chemists, and the boundaries of mineralogy became more and more blurred.

Mineralogy is established both as an independent science and as a support discipline for many other branches of science (Figure 1.4). Table 1.1 lists some famous mineralogists who have made outstanding contributions to science. The relevance of mineralogy–crystallography, in the context of scientific endeavor, is also highlighted by the unusual number of Nobel Prizes in Physics, Chemistry, and even Medicine that have been awarded for achievements related to

(a) 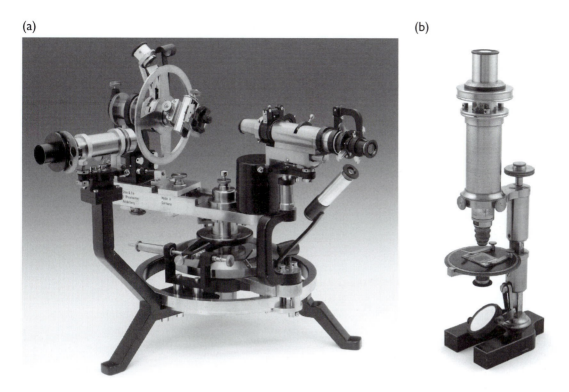 (b)

Fig. 1.3 (a) Two-circle Stoe goniometer developed by Goldschmidt to measure angular relationships between crystal faces (courtesy O. Medenbach). (b) The first "modern" petrographic microscope built by Fuess and Rosenbusch in 1876 (courtesy O. Medenbach and T. Mappes).

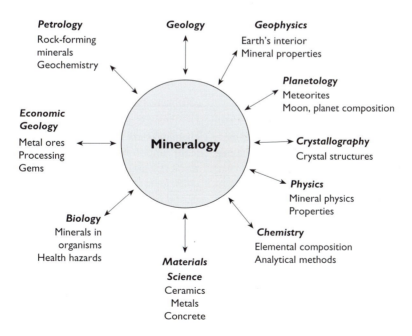

Fig. 1.4 Minerals are the core of mineralogy. But mineralogy is of interest to many different disciplines.

Table 1.1 Some famous mineralogists (not including living mineralogists)

Name, date	Country	Contribution
Georg Bauer (Agricola) 1494–1555	Germany	Detailed description of minerals
Niels Stensen (Nicolas Steno) 1638–1686	Denmark	Law of interfacial angles
Torbern O. Bergman 1735–1784	Sweden	Cleavage of calcite
René Just Haüy 1743–1822	France	Concept of unit-parallelepipeds
Abraham G. Werner 1750–1817	Germany	Origin and properties of minerals
Lorentz Pansner 1777–1851	Germany/Russia	Hardness and density of minerals
Johan J. Berzelius 1779–1848	Sweden	Chemical mineralogy
Johann A. Breithaupt 1791–1873	Germany	Density of minerals, parageneses
Eilhard Mitscherlich 1794–1863	Germany	Isomorphism and polymorphism
Johann F.C. Hessel 1796–1872	Germany	Point-group symmetry
Auguste Bravais 1811–1863	France	Lattice types
James Dwight Dana 1813–1895	USA	Systematic mineralogy
Nicolai Koksharoff 1818–1892	Russia	Goniometry of crystals
Harry Rosenbusch 1836–1914	Germany	Optical mineralogy
Gustav Tschermak 1836–1927	Austria	Silicate structures
Paul von Groth 1843–1927	Germany	Chemical crystallography
Ephgraph S. Fedorow 1853–1919	Russia	Space-group symmetry
Artur Schoenflies 1853–1928	Germany	Space-group symmetry
Viktor Goldschmidt 1853–1933	Germany	Geometry of crystals
Penti Eskola 1883–1964	Finland	Igneous minerals
Alexander Fersman 1883–1945	Russia	Mineral-forming processes
Norman L. Bowen 1887–1956	USA	Experimental petrology
Paul Niggli 1888–1953	Switzerland	Mineral-forming geological processes
Viktor M. Goldschmidt 1888–1947	Norway/Germany	Crystal chemistry, geochemistry
William L. Bragg 1890–1971	Great Britain	Crystal structure of minerals
Nicolai Belov 1891–1982	Russia	Mineral structures
Paul Ramdohr 1890–1985	Germany	Ore minerals
Cecil E. Tilley 1894–1973	Great Britain	Igneous and metamorphic minerals
Tom F. W. Barth 1899–1971	Norway	Petrology
Francis J. Turner 1904–1985	New Zealand/USA	Metamorphic minerals and deformation
Dmitry P. Grigoriev 1909–2003	Russia	Crystal growth

crystallography and research methods used in mineralogy, beginning with Röntgen (1901), von Laue (1914), and the Braggs (1915) for X-rays and more recently to Brockhouse and Shull for neutron diffraction (1994), to Curt, Kroto, and Smalley for the structure of fullerenes (1996), and to Geim and Novoselov for the structure of graphene (2010).

1.3 Major directions of investigation

Mineralogy is concerned with the characterization of properties and chemical composition of minerals and the study of the conditions of their formation. Since minerals are substances that concentrate certain chemical elements (such as metals), they are economically important and are studied to define mineral resources and exploration techniques. Any mineralogist must be able to identify minerals in order to search for them in the field and to investigate mineral samples with the most important laboratory techniques. Mineralogy is also the science that relates naturally occurring substances of crystalline structure to the more basic sciences of crystallography, chemistry, materials science, and solid-state physics. Mineralogy borrows from these disciplines information about the atomic structure, bonding characteristics, chemical stability, and growth processes of various compounds, and adopts various analytical methods. In return, mineralogy often provides answers to puzzling features in complex human-made products.

We can distinguish several major directions in mineralogical studies. Sometimes they are independent, but often they overlap and cannot be separated. Some investigations may be classified as "basic mineralogy", whereas others can be considered as "applied mineralogy"; both types of investigation include experimentation and theory. It is in part this diversity that makes mineralogy such a fascinating topic.

There are several major branches of *basic mineralogy*:

- Crystal chemistry of minerals (composition, structure, and bonding)
- Physical properties of minerals (e.g., density, optical properties, color)
- Studies of mineral formation including:
  - General principles of crystal growth
  - Geological processes on the surface and in the interior of the Earth
  - Chemical reactions, and the influence of temperature and pressure
- Relationships between mineral structure, chemical composition, properties, crystal habit, and the conditions under which minerals form

The most important directions of *applied mineralogy* are:

- Mineral identification, determination of morphology, composition, and properties
- Exploration mineralogy and ore deposits
- Industrial mineralogy (cement minerals and zeolites are examples)
- Gemology
- Mineralogical aspects of material science and solid-state physics (many ceramic products have mineral equivalents)
- Biomineralogy
- Minerals as health hazards (e.g., asbestos)

This book is organized into six parts that cover these various fields. Part I deals with the chemistry and elementary structure of crystals. The chemical composition of minerals has become the basis for classification. In order to connect readers early with macroscopic properties, we will also have a section on basic mineral identification of hand specimens. An intrinsic feature of crystals is symmetry, and it will be discussed in some detail in Part II. Symmetry is expressed in the external morphology of crystals as

well as in the arrangement of atoms in the crystal structure. Part III introduces methods used for the physical investigation of minerals, ranging from X-ray diffraction to optical microscopy and advanced techniques such as spectroscopy and electron microscopy. Of the physical properties, the optical properties are most important for petrologists, who study minerals in rocks by means of thin sections with the petrographic microscope. Part IV deals with the diversity of minerals and explores the general conditions and processes of mineral formation, with a discussion of thermodynamic principles that govern the chemical reactions. Part V contains a systematic survey of the most important minerals, including their structure, diagnostic properties, geological occurrence, and industrial use. We also highlight in this part some of the major geological processes of mineral formation in sedimentary, igneous, and metamorphic environments. Part VI introduces applied mineralogy – outlining the major branches, from mineral resources to cement minerals – and the major methods of investigation. Included in this part are chapters on gemology and the health aspects of minerals. The concluding two chapters provide an overview and review of the distribution of minerals in the universe, the solar system, and the Earth. Four appendices follow the main part of the book: the first two may be used for mineral identification from hand specimens, the third and fourth for identification with the petrographic microscope.

1.4 Some preliminary advice

Mineralogy is not an easy field for a novice to enter. Not only are you confronted with many new concepts, there are also new names to absorb and you need to develop your own judgment to distinguish the crucial from the optional. Memorizing mineral names is not the most inspiring aspect of mineralogy, but it is useful to learn the most important minerals and their general composition, just as it is useful in language studies to learn the important words. In addition to reading books on the subject, you have to become practically acquainted by working with actual mineral specimens. The lectures in a mineralogy course need to be complemented with a laboratory that uses hand specimens and introduces laboratory techniques such as the petrographic microscope and X-ray diffraction. A good start is to visit museum collections that display spectacular and aesthetically beautiful mineral

(a)

(b)

Fig. 1.5 (a) Crystal of sucrose ($C_{12}H_{22}O_{11}$) with regular flat faces. Sugar is not a mineral because it is industrially manufactured (from Baikow, 1967). (b) Crystal of halite (NaCl) from Fulda, Germany (courtesy O. Medenbach). The morphology displays cubic building blocks that suggest an internal structure with cubic unit cells as suggested by Haüy.

samples. If you have the opportunity, begin collecting some of your own samples in the field, buy some at flea markets and bring them to class. For beginners, get some specimens first from pegmatites, skarns, and hydrothermal deposits because these predominantly coarse and well-developed crystals can be easily recognized.

In the process of these practical exercises begin to use a notebook in which you enter mineral descriptions. It is enough to allot just half a page to a mineral or a page to a deposit. Relate minerals to their geological occurrence and classify them into minerals observed in igneous rocks (ultrabasic, basic, acid, alkaline, and volcanic rocks), pegmatites (mica-bearing, quartz-bearing, etc.), skarns (developing at a contact of igneous rock with limestone or dolomite), deposits of the weathering crusts (laterites, bauxites, and oxidized ores), chemical (from real and colloidal solutions), biogenic, and diagenetic deposits (alteration of sediments during compaction and burial). Describe in detail which features of major minerals you have observed with your own eyes. Slowly integrate your knowledge so that each mineral becomes an entity with a name, a chemical fingerprint (for some remember the formula, for complex ones remember at least a list of major elements), distinguishing features, the geological system in which they occur, and, for many, an industrial application (ranging from metal ore to food additive). As you familiarize yourself with mineralogy, minerals will become much more than dry names. They will emerge as multifaceted building blocks that help us understand the processes that govern the Earth.

1.5 Definition of crystal and mineral

Before we enter the field let us try to define *crystal* and *mineral*. This is a first attempt which will become clearer as we gain more background:

- A *crystal* is a homogeneous chemical compound with a regular and periodic arrangement of atoms. Examples are halite, "salt" (NaCl), and quartz (SiO_2). But crystals are not restricted to minerals: they make up most solid matter such as sugar (Figure 1.5a), cellulose, metals, bones, and even DNA.
- A *mineral* is a chemical compound that forms by a geological process. Figure 1.5b shows a crystal of halite where the cubic morphology suggests an internal structure with cubic unit cells. Most minerals are crystalline.

1.6 Summary

Mineralogy dates back to antiquity but evolved as a main branch of modern science in the eighteenth century. Much of our present knowledge about minerals was attained by research done in the twentieth century with modern analytical techniques such as X-ray diffraction. Minerals are defined as homogeneous chemical compounds that form by a geological process. Most minerals are crystals with a periodic arrangement of atoms.

Test your knowledge

1. What is a mineral and what is a crystal?
2. Name some principal objects and applications of mineralogy?

3. Which branches of science and engineering are most closely related to mineralogy?
4. What are the major directions of modern mineralogical research?
5. Find a mineral in your daily environment.

Further reading

Blackburn, W. H. and Dennen, W. H. (1994). *Principles of Mineralogy*, 2nd edn. Brown Publishing, Dubuque, IA.

Hibbard, M. J. (2002). *Mineralogy. A Geologist's Point of View*. Wiley, New York.

Klein, C. and Dutrow, B. (2007). *Manual of Mineral Science*, 23rd edn. Wiley, New York.

Klein, C. and Philpotts, A. (2012). *Earth Materials: Introduction to Mineralogy and Petrology*. Cambridge University Press, Cambridge.

Nesse, W. D. (2011). *Introduction to Mineralogy*, 2nd edn. Oxford University Press, New York.

2 | Elements, bonding, simple structures, and ionic radii

The closest field to mineralogy is chemistry. Originally, basic concepts of chemistry emerged based on mineral properties, and today much of mineralogy relies directly on chemical principles such as elemental composition and bonding. Mineralogy could not be discussed without the periodic table as a background. In this chapter we review some of the major principles of chemistry and illustrate how they apply to minerals, such as metallic bonding in gold (Au), ionic bonding in halite (NaCl), and covalent bonding in diamond (C). You will become aware how features of atoms link to properties of minerals. It is assumed that you have some knowledge of chemistry.

2.1 Chemical elements

Many mineral properties are closely related to the underlying chemical properties of constituent atoms and molecules. Let us start, therefore, by reviewing some fundamental chemistry. The basic building unit of a crystal is the atom. Atoms are composed of a very small *nucleus* containing positively charged *protons* and neutral *neutrons*. Nuclei range in diameter from 1.75 fm for hydrogen to about 15 fm for uranium (1 fm = 1 femtometer = 10^{-15} meters). Negatively charged electrons surround the nucleus and are distributed over a much larger volume. For an isolated atom electrons are distributed over roughly a spherical space with a diameter of 1–3 Å (1 Å = 1 ångström = 10^{-10}

meters = 0.1 nanometers (nm)), i.e., 10^{5} times larger than the nucleus (Figure 2.1a). Electrons are responsible mainly for the chemical behavior of atoms and for bonding, which combines atoms to form larger molecules and crystals. While the nucleus with protons and neutrons is very small, it contributes most of the mass to atoms (proton ~ 1.67×10^{-27} kg, neutron ~ 1.67×10^{-27} kg, electron ~ 9.1×10^{-31} kg).

Depending on the number of protons, atoms form different *elements* with distinct chemical properties. Each has an abbreviation such as H for hydrogen or Si for silicon. As of January 2016, 118 elements have been confirmed. The atomic number of an element is the number of protons found in an atom of that

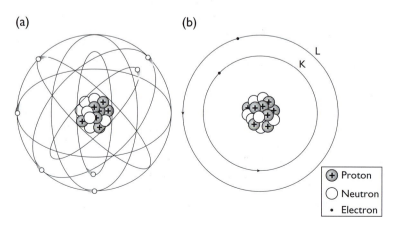

(a) (b)

⊕	Proton
◯	Neutron
•	Electron

Fig. 2.1 Generalized models of atomic structure. (a) Within an atom a small nucleus consisting of protons and neutrons is surrounded by an electron "cloud". (The size of the nucleus relative to the electron cloud is greatly exaggerated.) (b) A more detailed view of the Bohr model of the atom reveals that electrons are arranged in shells (K, L, M, etc.).

element. It is also equal to the number of electrons when the atom is in a neutral state. The number of neutrons can vary. For example, H can have 1 proton and 0 neutrons in ^{1}H, and is then called hydrogen, or it can have 1 proton and 1 neutron in ^{2}H, also known as deuterium, or even 2 neutrons in ^{3}H, known as tritium. The isotope ^{3}H is not stable and decays over time. We will discuss this so-called radioactive decay later.

Elements are represented in the *periodic table* (Figure 2.2), introduced in 1869 by Dmitri Mendeleev, and elements are placed into rows and columns that arrange atoms with specific electronic configurations. In this table the full names of elements, as well as their abbreviated symbols are given, together with atomic number (the number of protons or electrons, respectively) and average atomic weight (defined as a dimensionless number corresponding to 1/12 of the weight of ^{12}C, carbon composed of 6 protons and 6 neutrons) which depends on the isotopic composition.

In the simplified view of the atomic structure formulated by Niels Bohr, electrons are arranged in shells, labeled K, L, M, etc. (Figure 2.1b). Only the K-shell is occupied for elements in the first row of the periodic table, the K- and L-shells are occupied for second-row elements, and the K-, L-, and M-shells are occupied for third-row elements. In a qualitative way, with increasing diameter, shells occupy more space and can accommodate more electrons: 2, 8, and 18 for K, L, and M, respectively. This simple shell model was expanded by L. V. de Broglie, who demonstrated that electrons do not only have particle properties, with positions that can be defined in space and time, but that they also behave like waves, and are distributed in space in a much more complicated way than on a single orbit. It was Erwin Schrödinger who unified these views in a new model of the atom, based on quantum mechanics. The *Schrödinger equation*, proposed in 1926 and still in wide use today, relates the probability of finding an electron at a specific place and a given time to the mass and potential energy of the particle. On the basis of this model it was concluded that electrons in shells do not follow a circular orbit, but are present with a certain probability and are arranged in *orbitals*. Up to two electrons can occupy an orbital, each with opposite spin. The distribution of electrons in orbitals has different geometries: *s*-orbitals are spherical (Figure 2.3a), *p*-orbitals are directional along three main orthogonal axes (Figure 2.3b), and *d*-orbitals are of more complicated

shapes (Figure 2.3c). There is one *s*-orbital in each shell, and there can be up to three *p*-orbitals and up to five *d*-orbitals, each occupied by up to two electrons. The K-shell has only *s*-orbitals, the L-shell *s*- and *p*-orbitals, the M-shell *s*-, *p*-, and *d*-orbitals, etc.

Shells and orbitals are filled in a regular fashion to maintain the lowest energy. We illustrate the orbital-filling process for the element silicon (Si), which has 14 electrons in the neutral state. Two electrons are in the *s*-orbital of the first (K) shell, and this orbital is denoted as $1s^2$ (the superscript 2 indicating that there are two electrons). The K-shell can only have *s*-orbitals and is therefore full. The next two electrons are in the *s*-orbital of the second (L) shell, $2s^2$, and six electrons are in the *p*-orbitals of the second shell, two along each axis ($2p_x^2, 2p_y^2, 2p_z^2$). This L-shell has only *s*- and *p*-orbitals and is also full. The remaining four electrons are in the third shell, with two in the *s*-orbital $3s^2$, and two in the *p*-orbitals $3p_x$ and $3p_y$. (In principle, the two electrons in the 3*p*-orbitals could both be accommodated in the $3p_x^2$-orbital, but this would require a higher energy than $3p_x$ and $3p_y$ do.) The electron distribution of Si can be expressed compactly as

$$1s^2 \qquad 2s^2 \quad 2p_x^2 \quad 2p_y^2 \quad 2p_z^2 \qquad 3s^2 \quad 3p_x \quad 3p_y$$

$$\underset{\text{K}}{\uparrow\downarrow} \qquad \underbrace{\uparrow\downarrow \quad \uparrow\downarrow \quad \uparrow\downarrow \quad \uparrow\downarrow}_{\text{L}} \qquad \underbrace{\uparrow\downarrow \quad \uparrow \quad \uparrow}_{\text{M}}$$

Arrows below the orbitals indicate schematically the electrons and their spins.

With increasing atomic number in each shell, first *s*-, then *p*-, and finally *d*-orbitals are filled. In the fourth row of the periodic table (elements K through Kr) first *s*-electrons of the fourth shell are filled and only then are *d*-orbitals of the third shell filled (elements Sc through Zn). Between Ga and Kr, *p*-orbitals of the fourth shell are completed. This is illustrated for elements up to Kr in Table 2.1. As we have mentioned, each orbital can accommodate two electrons. If there are single electrons in an orbital, these so-called *valence electrons* are easily lost, for example in chemical reactions, owing to low energy barriers. In the case of the silicon atom described above, there are two such valence electrons, $3p_x$ and $3p_y$. The highest stability of an electron configuration is achieved if an outer electron shell is completely filled, which is the case for elements in the last column of the periodic table; these are known as the *noble* or the *inert* gases. Inert gas elements resist chemical reactions and bonding with other elements. Elements of the first

Group

1	2	3	4	5	6	7	8	9	10	11	12	13	14	15	16	17	18
1 H 1.008 Hydrogen ±1																	2 He 4.003 Helium
3 Li 6.941 Lithium 1	4 Be 9.012 Beryllium 2											5 B 10.811 Boron 3	6 C 12.011 Carbon ±4	7 N 14.007 Nitrogen ±3	8 O 16.00 Oxygen −2	9 F 18.998 Fluorine −1	10 Ne 20.180 Neon
11 Na 22.990 Sodium 1	12 Mg 24.305 Magnesium 2											13 Al 26.982 Aluminum +3	14 Si 28.086 Silicon 4	15 P 30.974 Phosphorus 5	16 S 32.066 Sulfur −2,6	17 Cl 35.453 Chlorine ±1	18 Ar 39.948 Argon
19 K 39.098 Potassium 1	20 Ca 40.078 Calcium 2	21 Sc 44.956 Scandium 3	22 Ti 47.867 Titanium 4	23 V 50.942 Vanadium 5	24 Cr 51.996 Chromium 3,6	25 Mn 54.938 Manganese 2,4,7	26 Fe 55.85 Iron 2,3	27 Co 58.933 Cobalt 2,3	28 Ni 58.693 Nickel 2	29 Cu 63.546 Copper 2	30 Zn 65.39 Zinc 2	31 Ga 69.723 Gallium 3	32 Ge 72.61 Germanium 2,4	33 As 74.922 Arsenic ±3,5	34 Se 78.96 Selenium ±2,4	35 Br 79.904 Bromine ±1	36 Kr 83.80 Krypton 2
37 Rb 85.468 Rubidium 1	38 Sr 87.62 Strontium 2	39 Y 88.906 Yttrium 3	40 Zr 91.224 Zirconium 4	41 Nb 92.906 Niobium 5	42 Mo 95.94 Molybdenum 4,6	43 Tc (98) Technetium 4,7	44 Ru 101.97 Ruthenium 3,4	45 Rh 102.906 Rhodium 3	46 Pd 106.42 Palladium 2,4	47 Ag 107.868 Silver 1	48 Cd 112.411 Cadmium 2	49 In 114.818 Indium 3	50 Sn 118.710 Tin ±4	51 Sb 121.760 Antimony ±3	52 Te 127.60 Tellurium ±2,4	53 I 126.904 Iodine ±1	54 Xe 131.29 Xenon 2
55 Cs 132.905 Caesium 1	56 Ba 137.327 Barium 2	57 La→ 138.906 Lanthanum 3	72 Hf 178.49 Hafnium 4	73 Ta 180.948 Tantalum 5	74 W 183.84 Tungsten 4,6	75 Re 186.207 Rhenium 4,7	76 Os 190.23 Osmium 4	77 Ir 192.217 Iridium 4	78 Pt 195.078 Platinum 2,4	79 Au 196.967 Gold 3	80 Hg 200.59 Mercury 1,3	81 Tl 204.383 Thallium 1,3	82 Pb 207.2 Lead 2,4	83 Bi 208.980 Bismuth 3	84 Po (209) Polonium ±2,4	85 At (210) Astatine ±1	86 Rn (222) Radon 2
87 Fr (223) Francium 1	88 Ra (226) Radium 2	89 Ac⇉ (227) Actinium 3	104 Rf (261) Rutherfordium 4	105 Db (262) Dubnium 5	106 Sg (263) Seaborgium 6	107 Bh (262) Bohrium 7	108 Hs (265) Hassium	109 Mt (266) Meitnerium	110 Ds (271) Darmstadtium	111 Rg (272) Roentgenium	112 Cn (285) Copernicium	113 Uut (284) Ununtrium	114 Fl (289) Flerovium	115 Uup (289) Ununpentium	116 Lv (293) Livermorium	117 Uus (294) Ununseptium	118 Uuo (294) Ununoctium

Key:
8 — Atomic number
O — Symbol
16.00 — Atomic weight
Oxygen — Name
−2 — Common oxidation state

Lanthanide series →

58 Ce 140.116 Cerium 3,4	59 Pr 140.908 Praseodymium 3	60 Nd 144.24 Neodymium 3	61 Pm (145) Promethium 3	62 Sm 150.36 Samarium 3	63 Eu 151.964 Europium 3	64 Gd 157.25 Gadolinium 3	65 Tb 158.925 Terbium 3	66 Dy 162.50 Dysprosium 3	67 Ho 164.930 Holmium 3	68 Er 167.26 Erbium 3	69 Tm 168.934 Thulium 3	70 Yb 173.04 Ytterbium 3	71 Lu 174.967 Lutetium 3

Actinide series ⇉

90 Th 232.038 Thorium 4	91 Pa 231.036 Protactinium 5	92 U 238.029 Uranium 4	93 Np (237) Neptunium 5	94 Pu (244) Plutonium 4	95 Am (243) Americium 3	96 Cm (247) Curium 3	97 Bk (247) Berkelium 3	98 Cf (251) Californium 3	99 Es (252) Einsteinium 3	100 Fm (257) Fermium 3	101 Md (258) Mendelevium 3	102 No (259) Nobelium 2,3	103 Lr (262) Lawrencium 3

Fig. 2.2 Periodic table of chemical elements. The most abundant elements (by weight percentage) in the Earth's crust are indicated by light shading. Elements that do not occur naturally are highlighted by dark shading.

(a)

(b)

(c)

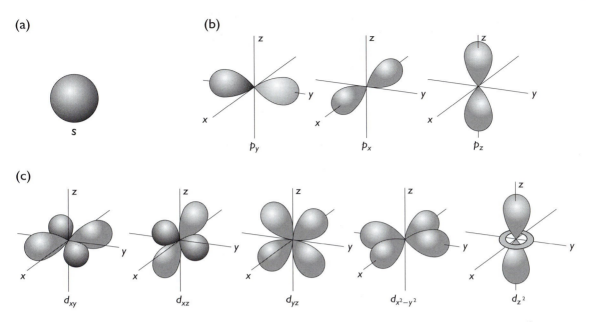

Fig. 2.3 Within a shell, electrons are arranged in orbitals. (a) The *s*-orbitals have a spherical geometry, (b) *p*-orbitals are directional along the principal axes, and (c) *d*-orbitals display more complicated distributions.

column (Group 1) have one electron more than a filled outer shell, whereas elements of the next-to-last column (Group 17) are one electron short of having a filled outer shell. The importance of these electron configurations will become more apparent when we discuss bonding.

Of the 118 known elements, 94 occur naturally on the Earth, while the others have been produced only in the laboratory (dark shading in Figure 2.2). The distribution of elements is by no means uniform, nor is there a simple relationship between the elemental distribution and placement within the periodic table. Figure 2.4 illustrates the abundance of the major elements in the Earth's crust and in the Earth as a whole. The crustal abundances are clearly best defined by direct observations. Whole Earth abundances are based on estimates inferred from meteoritic evidence and from physical properties of the Earth's interior. As we can see, oxygen, silicon, and aluminum are by far the most common elements in the Earth's crust and the major components of common rocks. The most common minerals are therefore compounds of oxygen, silicon, and aluminum and, to a lesser extent, of magnesium, iron, titanium, calcium, potassium, sodium, and phosphorus. Examples are quartz (SiO_2), feldspar ($CaAl_2Si_2O_8$), and olivine (Mg_2SiO_4). If we look at the whole Earth, magnesium is a major component of the lower mantle and iron dominates in

the core, and the abundances of these elements are significantly higher than in the crust. If we were to look at the elemental abundances of the whole solar system, hydrogen, helium, and carbon would dominate. These light elements have largely been lost during the condensation of the inner planets, but ice (H_2O), methane (CH_4), and nitrogen are still the main components in the outer planets.

In this book, therefore, we will deal mainly with compounds that are combinations of elements from the first three or four rows of the periodic table (Figure 2.2). There are some important exceptions, such as gold (in native gold), lead (in galena, PbS), and barium (in barite, $BaSO_4$), all in the sixth row, that will be included.

2.2 Bonding

In the solid state, atoms are closely surrounded by neighboring atoms. The forces that bind atoms together are electrical in nature. Their type and intensity are largely responsible for the physical and chemical properties of crystals. These electrical forces between atoms are referred to as bonds. Depending on the electronic structure of an atom and its nearest neighbor, the types of bonding forces can vary. In general, in the solid state at low temperature, where thermal vibration is only moderate, four types of

Table 2.1 Electron configurations of atoms from hydrogen to krypton

	Shell K									
Element	1s	2s	2p	3s	3p	3d	4s	4p	4d	4f
1. H	1									
2. He	2									
3. Li	2	1								
4. Be	2	2								
5. B	2	2	1							
6. C	2	2	2							
7. N	2	2	3							
8. O	2	2	4							
9. F	2	2	5							
10. Ne	2	2	6							
11. Na	2	2	6	1						
12. Mg	2	2	6	2						
13. Al	2	2	6	2	1					
14. Si	2	2	6	2	2					
15. P	2	2	6	2	3					
16. S	2	2	6	2	4					
17. Cl	2	2	6	2	5					
18. Ar	2	2	6	2	6					
19. K	2	2	6	2	6		1			
20. Ca	2	2	6	2	6		2			
21. Sc	2	2	6	2	6	1	2			
22. Ti	2	2	6	2	6	2	2			
23. V	2	2	6	2	6	3	2			
24. Cr	2	2	6	2	6	4	1			
25. Mn	2	2	6	2	6	5	2			
26. Fe	2	2	6	2	6	6	2			
27. Co	2	2	6	2	6	7	2			
28. Ni	2	2	6	2	6	8	2			
29. Cu	2	2	6	2	6	10	1			
30. Zn	2	2	6	2	6	10	2			
31. Ga	2	2	6	2	6	10	2	1		
32. Ge	2	2	6	2	6	10	2	2		
33. As	2	2	6	2	6	10	2	3		
34. Se	2	2	6	2	6	10	2	4		
35. Br	2	2	6	2	6	10	2	5		
36. Kr	2	2	6	2	6	10	2	6		

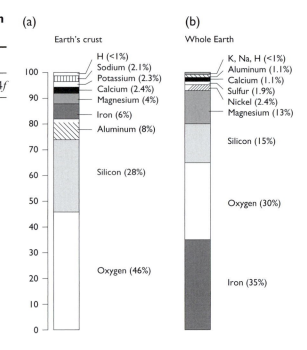

Fig. 2.4 Abundance of elements by weight percentage (weight%) (a) in the Earth's crust, (b) in the whole Earth.

neutral atoms; it arises because the electron distribution in atoms is not uniform. *Ionic bonding* relies on electrostatic attraction between atoms of different charge, where electrons have been removed (positive charge) or added (negative charge). As we will discuss below, such charged atoms are called ions. In *covalent bonding*, single electrons are shared between two atoms in a common orbital. In reality all these bonding forces are active in a crystal, but in different minerals some forces can dominate. For example, in halite (NaCl) bonding is largely ionic, whereas in diamond (C) it is covalent. Bonding can be mixed, and different types of bonding may exist between different atoms in a mineral structure. For example, the important Si–O bond involves covalent and ionic forces. In the case of graphite (C) some bonds are largely covalent, whereas others are of the van der Waals type. A very schematic triangular representation with examples of mineral species for different types of bonding (excluding van der Waals bonding) is shown in Figure 2.5.

Metallic bonding

The solid phases of three-quarters of all elements are *metals*, with mineral representatives including gold,

bonding are distinguished, which we will discuss briefly. In *metallic bonding* some outer electrons have been removed from the atoms and move freely within the structure. The attractive force between the positively charged atoms and the negatively charged *electron cloud* holds such structures together. The *van der Waals bond* is a weak overall attraction between

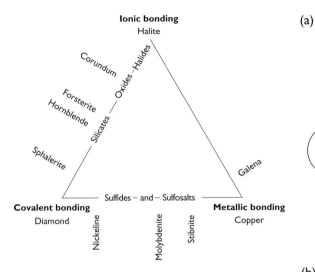

Fig. 2.5 Triangular representation of ionic, covalent, and metallic bonding with some mineral representatives.

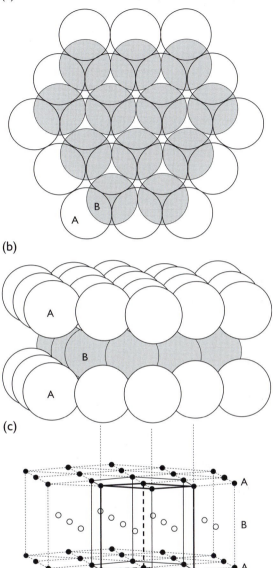

silver, copper, iron, and platinum. In metals, some valence electrons are shared with the whole structure, and move relatively freely. The image of an *electron gas* between atoms with a positive charge is often used. Bonding is achieved between the positively charged atoms and the electron gas. The mobile electrons are the reason for the high electrical and thermal conductivity of metals. The electrical conductivity is expressed in Ohm's law. (The current I is proportional to the applied voltage V and inversely proportional to the resistance R: $I = V/R$.) The ratio of electrical to thermal conductivity is identical for all metals because both conductivities depend on the movements of the free electrons.

The electrostatic attraction of atoms in a metal causes each atom to surround itself with as many neighboring atoms as is geometrically possible. Since atoms have a more or less spherical shape, we obtain a regular geometrical arrangement if we place them as closely packed as possible. Arranging spheres of equal size on a plane provides a layer with a hexagonal pattern (open circles in Figure 2.6a). This first layer we label A. Going into the third dimension, the closest stacking is achieved if we put atoms into depressions in the first layer (shaded circles in Figure 2.6a). We find that the shaded circles form a second hexagonal layer B which is displaced relative to the first layer. The third layer of atoms we put into depressions of the B layer, directly over the first layer. Since this layer is exactly over the first layer, we label it A again, and

Fig. 2.6 Hexagonal close-packing of spheres. (a) Single layer displaying a hexagonal pattern (open circles, A), with a second layer B (shaded) positioned in the depressions of the first layer. (b) Perspective view of the stacking of A and B layers (the third layer lies exactly over the first layer with an AB–AB stacking. (c) Representation of atoms as small circles and outline of the unit cell for this structure (bold) in a hexagonal prism.

Figure 2.6b shows a three-dimensional view of this stacking ABA, which can be repeated to infinity in a macroscopic crystal (ABABAB . . .). The arrangement of atoms is known as a *crystal structure*, and, in this particular case, the arrangement is close-packed because there is a minimum of open space between atoms. The array of atoms displays a hexagonal pattern if viewed from above (Figure 2.6a), a property we will later refer to as *symmetry* (Part II). This arrangement is known as *hexagonal close-packed structure*, often abbreviated as *hcp*. Metals such as beryllium, titanium, zinc, and zirconium crystallize in this structure type at ambient conditions.

There is an alternative way of stacking close-packed layers. Instead of placing the third layer exactly over the first layer, we can position it over the depressions in the first layer that were not used for the second layer. Layer C in Figure 2.7a shows this (shaded spheres). In this case, only the fourth layer stacks over the first layer and the sequence repeats. As Figure 2.7b illustrates, the stacking ABCA looks like a pyramid, and not only the base but each of the three sides of the pyramid is a close-packed layer. In Figure 2.7c a more extensive sequence of this stacking is shown (ABCABC . . .) and the pyramid is tilted so that it fits into a cube. The close-packed layers are perpendicular to the body diagonals of the cube. A small cube is shown at the lower right corner, with centers of atoms in the corners and centers of faces. This structure type with ABC stacking is known as cubic close-packed, or *face-centered cubic (fcc)*. It is a very important structure type and elements such as aluminum, copper, and gold have such an atomic arrangement.

In both stacking patterns each atom has 12 closest neighbors, six in the same plane, and three above and three below (see, e.g., Figure 2.6a). We say that the *coordination number* (CN) is 12. Both structures are equally economical in occupying space and represent the densest packing of spheres (74% of the volume is occupied). There are other ways to stack close-packed layers, but these two types of stacking represent by far the two most important metal structures.

Rather than using representations of crystal structures as stacks or clusters of atoms as in Figures 2.6b and 2.7c, crystallographers prefer a more transparent system, similar to Haüy's concept of elementary building blocks that repeat in three dimensions, alluded to in Chapter 1. These building blocks are called *unit cells*, and we will discuss them systematically in Chapter 7, introducing just the basics here. Unit cells are polyhedra with three pairs of parallel faces that repeat periodically in three dimensions. To visualize these polyhedra more easily, we show atoms at a smaller-than-actual size relative to their interatomic

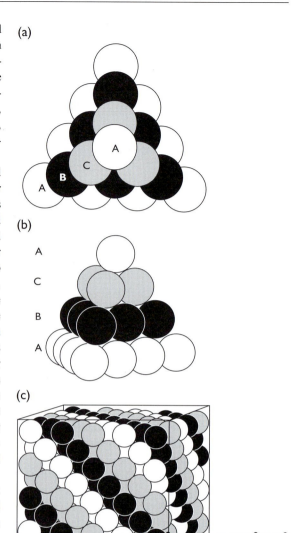

Fig. 2.7 Cubic close-packing of spheres. (a) Plan view of a stacking in which the third layer (C) lies over the alternate depressions of the first layer, resulting in an ABCABC stacking. (b) Perspective view of this stacking pyramid (tetrahedron). Notice that, on all four faces of the pyramid, atoms are close-packed. (c) The structure is extended and then tilted to fit into a cube. A small cube (lower right corner) identifies the unit cell, with centers of atoms on corners and face centers.

distances, leaving space between them. This allows us to better see the unit cell interior. Doing so, it is quite easy to identify a unit polyhedron in the ABAB stacking (Figure 2.6c). The unit cell of the hcp structure, which we can repeat by translation to cover all

(a)   (b)   (c)

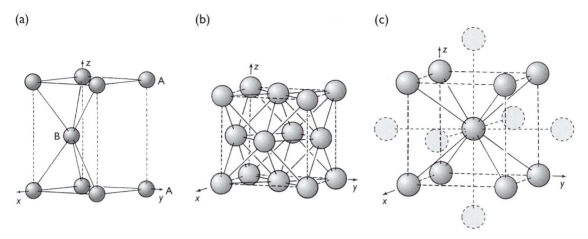

Fig. 2.8 Simple structures in metals with the unit cell indicated by dashed lines. (a) Unit cell for hexagonal close-packed (hcp) stacking ABAB. (b) Cubic unit cell for stacking ABCABC, with the close-packed layers perpendicular to the body diagonal of the cube. The unit cell is face-centered cubic (fcc). (c) Body-centered cubic (bcc) arrangement. In these representations the size of the atoms is reduced, relative to their distance, to better visualize the internal atomic arrangement. Atoms are shaded, solid lines connect closest neighbors. In the case of bcc (c), second-closest neighbors are shown (light shading) and connected by dotted lines.

atoms in the crystal, is highlighted in Figure 2.8a. It consists of rhomboids at the base and at the top, with rectangles as sides. Atoms are at the corners (A) and one is inside the polyhedron (B), above one of the triangles in the A layer. In the case of the fcc structure, a cubic unit cell is chosen as outlined in Figure 2.7c, lower right, and highlighted in Figure 2.8b. In this representation we can immediately see atoms at the corners and the centers of faces of the cube. Go to Box 2.1 and try some three-dimensional visualizations of these structures.

Since interatomic distances are similar, there is only a relatively small difference in energy between the fcc and hcp structures for a given metal. However, the mechanical properties of metals crystallizing in the two structures are very different. Hcp metals have only a single set of close-packed layers, whereas fcc metals have four equivalent sets (corresponding to the four body diagonals). The close-packed layers act as slip planes during deformation. Fcc metals such as copper, aluminum, gold, and silver with many equivalent slip planes are ductile, whereas hcp metals such as beryllium, zinc, titanium, and zirconium are much more brittle and difficult to deform. This difference in deformation properties is significant for technological applications.

There is a third simple structure type in metals in which atoms are arranged in the corners and the center of a cubic unit cell (Figure 2.8c). In this *body-centered*

cubic (bcc) structure, each atom has eight closest neighbors and six only slightly more remote neighbors (outside the unit cube and indicated by lighter shading). This also provides a fairly dense packing with 68% of the volume occupied, as compared to 74% each for the fcc and hcp structures. Representatives of bcc structures include α-iron (steel and a rare natural mineral, kamacite, which occurs mainly in iron meteorites) and tungsten.

Ionic bonding

An atom becomes ionized if it attains an inert gas configuration by loss or gain of one or more electrons. Halite (NaCl) is a good mineral example for ionic bonding. Sodium loses an electron to form the neon configuration: Na $(1s^2\,2s^2\,2p^6\,3s^1) - e^- \rightarrow Na^+(1s^2\,2s^2\,2p^6)$, and chlorine gains an electron to have the argon configuration: Cl $(1s^2\,2s^2\,2p^6\,3s^2\,3p^5) + e^- \rightarrow Cl^-$ $(1s^2\,2s^2\,2p^6\,3s^2\,3p^6)$. The ion with the positive charge is called a *cation*, that with the negative charge an *anion*. *Ionic bonding* is basically due to the electrostatic attraction between oppositely charged ions and therefore subject to Coulomb's law, which describes the relationship between the attractive or repulsive force F, point charges q_1 and q_2, and their distance d (Figure 2.9a):

$$F \approx (q_1 \times q_2)/d^2 \tag{2.1}$$

Box 2.1 Analytical focus: Three-dimensional visualizations of crystal structures

We have introduced some simple crystal structures of metals and will later discuss some much more complicated structures of many minerals. All these structures are three-dimensional (3D), yet representations in books are only two-dimensional projections. With the help of computing facilities, the 3D space becomes much more accessible and we suggest to readers to try this for the simple metal structures and later for more complex mineral structures. One way is to use the Crystallographic Open Database (Graziulis *et al.*, 2009; www.crystallography.net) and we look here first at the fcc structure of gold.

In this software start with "Search"; select Au as element and 1, 1 as number of distinct elements. "Send" will access the list of structures that are available in the database. We pick the second one and click on COD ID: 9008463, which will display a projection of the structure in Jmol, with one atom in the corner of the unit cell. (You need to install Java on your computer.) Then right-click and under "Symmetry" choose "Reload Polyhedra", which will expand the structure to all atoms in the unit cell. Using your left mouse button you can rotate the structure and view it from different directions. The square now becomes a cube. Try this and become familiar with the 3D view. Then do the same for hexagonal close-packing (selecting Zr as element) and the cubic body-centered structure (selecting Fe as element).

There are other software sources to visualize crystal structures and available on the internet. Four examples are:

Crystalviewer provided by the Publisher (www.cambridge.org/earthmaterials/crystalviewer).
CrystalMaker (http://www.crystalmaker.com)
JMOL (http://jmol.sourceforge.net)
VESTA Visualization for Electronic and Structural Analysis (http://jp-minerals.org/vesta/en/)

Be aware that with internet sources details change as programs and computing facilities are updated.

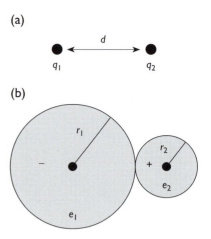

Fig. 2.9 Coulomb's law. (a) Point charges, and (b) application to ions that are in contact.

If we consider a very simple ionic structure where ions are spherical and in contact and only take nearest-neighbor interactions into account (Figure 2.9b), this can be rephrased as

$$F \approx \left(e_1^+ \times e_2^-\right)/(r_1 + r_2)^2 \qquad (2.2)$$

where e_1 and e_2 are the ionic charges of cation and anion, and r_1 and r_2 are the radii of the corresponding ions. We can use this simplified formulation to understand some features of ionic structures. The strength of the attraction F correlates with the melting point. Thus, for sodium halides with a 1^+ charge, the melting point decreases with increasing interatomic distance $(r_1 + r_2)$:

Compound	Interatomic distance (Å)	Melting point (°C)
NaF	2.31	988
NaCl	2.79	801
NaBr	2.94	740
NaI	3.18	660

On the other hand, for two crystals of the same structure and similar interatomic distance, the melting point increases with the ionic charge:

Compound	Charge	Interatomic distance (Å)	Melting point (°C)
NaF	1	2.31	988
CaO	2	2.40	2570

(a)   (b)   (c)

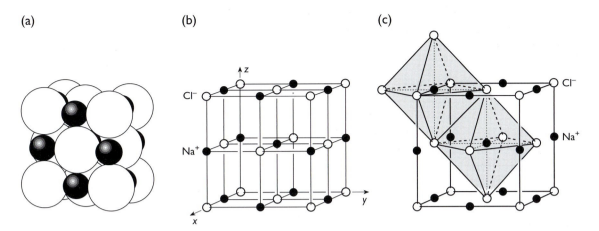

Fig. 2.10 (a) Structure of halite (NaCl) with alternating Na^+ (black) and Cl^- (white). (b) A cubic unit cell is outlined with Cl^- in the corners in a representation with reduced sizes of atoms. The structure can be viewed as a combination of two fcc structures that are translated. (c) Each Na^+ is surrounded by six Cl^- and we can display this relationship by drawing an octahedron (shaded), known as a coordination polyhedron.

Among minerals, ionic bonding is prevalent, such as in halite, with alternating Na^+ and Cl^- that are in contact (Figure 2.10a). Again it is easier for visualization purposes to draw atoms as small spheres with exaggerated interatomic spacing. In so doing we recognize a cube-shaped unit cell with Cl^- in the corners (Figure 2.10b). Each Na^+ is surrounded by six Cl^- (Figure 2.10c, indicated by dotted lines) and vice versa. We will discuss this structure in more detail later in this chapter. Ionic structures, at least of simple binary compounds, typically display fairly high symmetry. Again, try the 3D visualization of the halite structure with the method described in Box 2.1.

Covalent bonding

Whereas ions lose or gain electrons to achieve a more stable configuration, covalently bonded atoms share electrons to fill partially occupied orbitals. For example, two chlorine atoms combine to form a Cl_2 molecule:

$$\begin{array}{cc} \overset{\bullet\bullet}{\underset{\bullet\bullet}{\textbf{:}\,Cl\,\overset{\bullet}{}}} & \overset{\circ\circ}{\underset{\circ\circ}{\,Cl\,\overset{\circ}{}\textbf{:}}} \end{array} \rightleftharpoons \begin{array}{cc} \overset{\bullet\bullet}{\underset{\bullet\bullet}{\textbf{:}\,Cl\,}} & \overset{\circ\circ}{\underset{\circ\circ}{\,\textbf{:}\,Cl\,\textbf{:}}} \end{array}$$

The covalent bond is directional between two atoms. The directionality is particularly emphasized for p-orbitals with p_x, p_y, and p_z aligned more or less

along orthogonal axes. A good example is water (H_2O). The electron configuration of oxygen is

$$1s^2\ 2s^2\ 2p_x^2\ 2p_y\ 2p_z$$

The two unpaired p-orbitals can bond with hydrogen atoms. Since p-orbitals are at right angles to one another, O–H bonds should also be at right angles (Figure 2.11a). This is not strictly true, however, because hydrogen and oxygen atoms are slightly ionized and thus an electrostatic repulsion between the slightly positive hydrogen atoms causes the angle to open to 104.5° (Figure 2.11b).

Perhaps the best mineral example for covalent bonding is diamond. The electronic configuration of an isolated carbon atom is

$$1s^2\ 2s^2\ 2p_x\ 2p_y$$

(a)   (b)

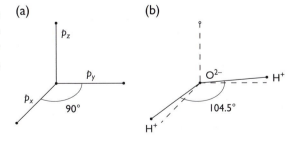

Fig. 2.11 Molecular bonding in H_2O. (a) Ideal p_x, p_y, and p_z oxygen bonds at right angles to one another. (b) Angular distortion to 104.5° due to electrostatic repulsion of the two hydrogen atoms that are slightly ionized.

In diamond crystals this electron configuration is complicated by the influences from neighboring atoms that cause electrons in 2s-orbitals and p-orbitals to change places, resulting in four unpaired orbitals with overlapping energies in what is called hybridization:

$$1s^2\ 2s\ 2p_x\ 2p_y\ 2p_z$$

In order for such a hybridized carbon atom to be stable, it must be bonded with four neighbors, sharing their orbitals. This is illustrated with a simple two-dimensional scheme (each arrow represents an electron):

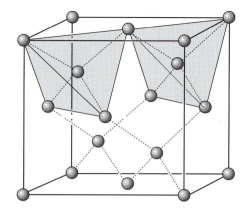

Fig. 2.12 Cubic structure of diamond, which can also be viewed as fcc, but with additional carbon atoms in alternate eighth cubes. Bonds between neighboring carbon atoms are shown by dotted lines. Each carbon atom is surrounded by four neighboring carbon atoms, outlining a tetrahedron (shaded). Tetrahedra are connected over corners.

In diamond this bonding is not achieved in a planar configuration as in the sketch above, but rather in three dimensions, where each carbon atom is surrounded by four neighbors (Figure 2.12, dotted lines) in the form of a tetrahedron (Figure 2.12, shaded). Tetrahedra are linked with each other over corners and form a three-dimensional framework. The diamond structure has a cubic unit cell. In general, however, covalent structures, owing to their directionality, are of low symmetry, and there is no tendency towards close-packing.

Van der Waals bonding

Van der Waals bonding is produced by a weak attraction between atoms. A slightly irregular distribution of electrons in the outermost shell causes charge fluctuations. This is due to the very dynamic nature of the electron distribution, with locally changing charge densities. It causes a random and very brief formation of dipoles between two atoms. Inert gases crystallize with van der Waals bonding, with similar close-packed structures as for metals. In minerals, van der Waals bonding is only important for specific bonds, such as between layers in the silicate mineral talc, or carbon layers in graphite, and between rings of eight sulfur atoms in native sulfur. The symmetry

of these minerals is generally low because of the shape of molecules.

In these simple models, van der Waals bonding is the weakest type of bonding, followed by metallic and ionic. Covalent bonding is strongest, and this is expressed in the hardness of crystals (Table 2.2). Covalent bonding is directional, resulting in a low symmetry arrangement. For more detailed treatments on bonding, one should consult textbooks on chemistry and solid-state physics. For example, in metallic bonding the wave nature of electrons and the fact that electrons are not really free but are constrained by quantum considerations are addressed by the Bloch theory.

2.3 Ionic radii

It is useful to introduce the concept of electron density as the probability of finding an electron in a volume unit. The electron density distribution of a separated ion or atom with inert gas characteristics (i.e., corresponding to elements in the last column of the periodic table; see Figure 2.2) has approximately spherical symmetry. Also, for many purposes, atoms in crystals with metallic and ionic bonding can be approximated by spheres. It is difficult to assign an absolute size to these spheres, since the overall electron density

Table 2.2 Summary of characteristic properties of different types of bonding, with mineral examples

Property	Bond type			
	Metallic	Ionic	Covalent	Van der Waals
Bond strength	Variable	Strong	Very strong	Weak
Structure	Nondirectional	Nondirectional	Directional	Nondirectional
Symmetry	High symmetry	High symmetry	Low symmetry	Low symmetry
Hardness	Variable	Strong	Very strong	Weak
Melting point	Variable	Variable	High	Low
Electrical	Good conductors	Poor conductors	Insulators	Insulators
Mineral examples	Copper (Cu)	Halite (NaCl)	Diamond (C)	Sulfur (S) (weak bond only)

decreases gradually with increasing distance from the nucleus. However, between ions in a crystal structure such as that of halite (Figure 2.10a) there is a balance between attractive and repulsive forces. This is shown schematically in Figure 2.13. Coulomb attraction discussed earlier is inversely proportional to the square of the distance between the center of the ions (dashed curve). This attraction is countered by a strong repulsion as electron shells of anion and cation begin to overlap, known as Born repulsion, as well as repulsion

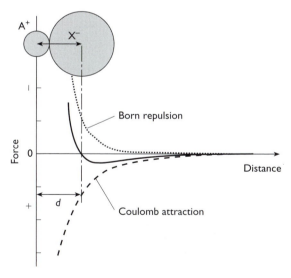

Fig. 2.13 Coulomb attraction between oppositely charged ions (dashed line) and Born repulsion due to overlaps of electron shells (dotted line) produce a combined curve (solid line) with balanced forces at a stable equilibrium distance.

between positively charged nuclei (dotted curve). The net force is a difference between the two (solid curve), and it is balanced at a certain equilibrium separation. Such a model for ionic crystal structures that assumes ions are hard spheres of fixed radius and in contact is extremely useful. The interatomic distance d between a cation A^+ and an anion X^- is the sum of the radii r_A^+ and r_X^- of the two ions.

The sum of two radii can be derived from the atomic positions that are determined experimentally by X-ray diffraction methods (Chapter 11). If the absolute radius of only one ion is known, then the radii of all ions can be readily calculated. On the basis of the optical determinations of the radii of F^- (1.33 Å) and O^{2-} (1.40 Å) by J. A. Wasastjerna in 1923, and on the first structure determinations from X-ray diffraction experiments by W. L. Bragg (1914), V. M. Goldschmidt (1926) derived ionic radii of many elements. Table 2.3 illustrates his findings for some alkali halides crystallizing in the halite structure (Figure 2.10). His results agree closely with modern lattice parameter determinations. Obviously, the model of hard spheres is better for ionic, metallic, or van der Waals structures than for covalent ones, where binding forces are directed and the electron density distribution is nonspherical.

On the basis of our present understanding of atomic structure, the calculations of radii have been refined, and more accurate atomic dimensions are now available. The size of an ion is determined primarily by its electronic configuration, its ionic state and the number of anions surrounding a cation (coordination number).

Table 2.3 Measured interatomic distances and derived ionic radii (in italics) for some alkali halides. The radii of F⁻ (underlined) are assumed to be known (Shannon and Prewitt, 1969).

Halide AX	$r_A^+ + r_X^-$ (Å)	r_A^+ (Å)	r_X^- (Å)
CsF	3.00	*1.70*	1.33
RbF	2.85	*1.49*	1.33
KF	2.71	*1.38*	1.33
NaF	2.35	*1.02*	1.33
LiF	2.07	*0.74*	1.33
RbCl	3.33	1.49	*1.81*
KCl	3.19	1.38	1.81
NaCl	2.83	1.02	1.81
LiCl	2.55	0.74	1.81

1. For a given element the ionic radius decreases with increasing positive charge and increases with increasing negative charge. For example, the atomic and ionic radii for lead, iron, and silicon are summarized below.

Species[a]	Radius (Å)	Species	Radius (Å)	Species	Radius (Å)
Pb^{4-}	2.15	Fe^0	1.26	Si^{4-}	1.98
Pb^0	1.74	Fe^{2+}	0.77	Si^0	1.18
Pb^{2+}	1.18	Fe^{3+}	0.65	Si^{4+}	0.40
Pb^{4+}	0.78				

[a] Superscript zero denotes the neutral state.

2. Radii of elements in the same vertical column (group) in the periodic table with identical ionic charge increase in size with increasing atomic number (Z), for example:

		Li	Na	K	Rb	Cs
Z		3	11	19	37	55
Charge	0	1.56	1.86	2.23	2.36	2.55
Charge	1+	0.74	1.02	1.38	1.49	1.70

However, the increase in radius is not proportional to Z.

3. The radius of ions with the same electronic configuration, but an increasing positive charge, decreases. For example, for ions with a $1s^2\, 2s^2\, 2p^6$ configuration:

Ions	O^{2-}	F^-	Na^+	Mg^{2+}	Al^{3+}	Si^{4+}
Ionic radius (Å)	1.40	1.33	1.02	0.72	0.53	0.40

The relative sizes of some ions with the same electron configuration are also illustrated in Figure 2.14.

4. Anions are generally larger than cations. In particular, the most common anions (O^{2-} = 1.40 Å, Cl^- = 1.81 Å) are among the largest ions in minerals. Therefore, in ionic compounds most of the volume is occupied by anions, and their arrangement primarily determines many of the simple ionic structures.

There are several other factors influencing the size of the ion. For example, contributions from covalent bonding may distort the spherical symmetry of an ion. Also, the number of neighboring anions that surround a cation (coordination) affects the radius, as in the case of Ca^{2+}:

Neighbors	Radius (Å)
6	1.00
8	1.12
12	1.35

The influence of the coordination number on ionic radii is illustrated for some examples in Figure 2.15.

2.4 Radius ratio and coordination polyhedra

How is the size of an ion related to the crystal structure? If the model of touching spheres holds, intuitively we would expect a large cation to have more anions as adjacent neighbors than a small cation has. The number of closest neighbors of opposite charge around an ion (the coordination number, or CN) can range from 2 to 12 in ionic structures. The surrounding ions lie on corners of a more or less regular polyhedron, the so-called *coordination polyhedron*. Highly symmetrical polyhedra (tetrahedron with four corners and four faces, CN = 4; octahedron with 6 corners and 8 faces, CN = 6; cube with 8 corners and 6 faces, CN = 8; a more irregular polyhedron, called a cuboctahedron with 12 corners and 8 triangular and 6 square

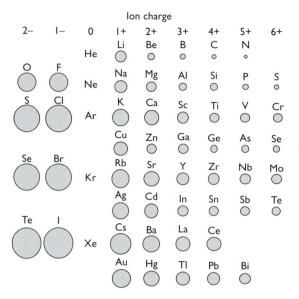

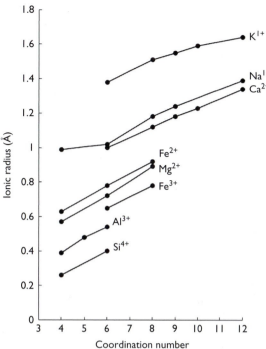

Fig. 2.14 Relative sizes of some ions with the same electron configuration.

Fig. 2.15 Variation of ionic radii with coordination number for some elements (data from Shannon and Prewitt, 1969).

faces, CN = 12) are most common (Figure 2.16c–f). An octahedral coordination polyhedron in the structure of halite is outlined in Figure 2.10c. In some cases the coordination can be planar (a triangle, CN = 3, as in Figure 2.16b, or a square) or even linear (Figure 2.16a, CN = 2).

In a polyhedron, an ideal close-packing of spheres in which the larger anions are in contact and are touched by the cation can, for a given CN, be achieved only for a specific ratio of the ionic radii. We first consider two simple, two-dimensional cases. In the first case a compound AX with a fairly large cation A^+ is surrounded by four anions X^- of equal size (Figure 2.17a, left). We keep the size of the anions constant and reduce the size of the cation. The anions come closer to the cation centers and, therefore, bonding in the crystal structure becomes stronger. This is because the energy to separate the crystal into ionic species depends primarily on Coulomb attraction between cations and anions, which increases with decreasing distance (equation 2.1). At a certain cation radius the anions touch each other (Figure 2.17a, middle). A further decrease in cation radius does not cause the configuration to shrink; instead, the cation "rattles" around inside the polyhedron (Figure 2.17a, right) and the electrostatic component of the lattice energy remains constant, which is illustrated schematically in Figure 2.18 (CN = 4). In the second case the

cation is surrounded by three anions in a triangular configuration (Figure 2.17b, left). As we reduce the radius of the cation to the size when anions were touching in the square configuration (Figure 2.17b, middle), anions still do not touch. Only at a much smaller radius do anions and cations touch (Figure 2.17b, right). Therefore, for triangular coordination the lattice energy decreases to a much smaller value of cation radius (Figure 2.18, CN = 3), and for small cations the triangular configuration is therefore advantageous and allows for closer packing with reduced lattice energy.

The ideal cation:anion radius ratio $(r_A:r_X)$, when both cations and anions are touching, can be calculated from geometrical considerations (Box 2.2).

The same method can be used to determine the critical radius ratios for three-dimensional coordination polyhedra. For 12-fold coordination, corresponding to simple close-packing, the radius ratio $r_A:r_X$ is 1.0. Table 2.4 lists the most common coordination polyhedra in minerals (3, 4, 6, 8, and 12), ideal stability ranges, and some examples. There are cases of 5-, 7-, and 10-fold coordination in more

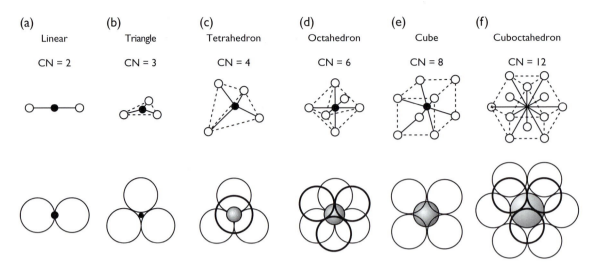

Fig. 2.16 Common coordination polyhedra in crystal structures: (a) linear (CN = 2); (b) triangle (CN = 3); (c) tetrahedron (CN = 4); (d) octahedron (CN = 6); (e) cube (CN = 8); (f) cuboctahedron (CN = 12).

complicated structures. The polyhedra in these cases are distorted, and it is more difficult to determine ideal stability ranges. As you can see there are some exceptions, particularly close to the ideal limits. Exceptions generally indicate that bonding is not purely ionic or ions are not ideally spherical.

In most mineral structures, the cations are smaller than the anions and there are also often more anions than cations (e.g., olivine (Mg_2SiO_4), kyanite (Al_2SiO_5), pyrite (FeS_2)). The anions therefore dominate the total volume and arrange themselves in a close-packed

structure to minimize the energy, just as atoms do in hcp and fcc metals (see Figures 2.6 and 2.7).

In a close-packed arrangement of spheres, there are two types of polyhedral interstices (spaces) between the spheres. Figure 2.20 illustrates two layers, a bottom layer with open circles (A) and a top layer with closed circles (B), both representing the same sort of anion, for example O^{2-}. Between these atoms we can place a tetrahedron T (a triangle as base and an apex on top) and an octahedron O that lies on a

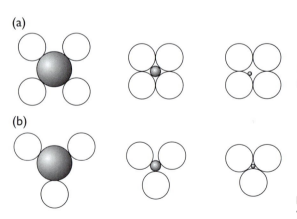

Fig. 2.17 Two-dimensional illustration of the influence of radius ratio on packing in ionic structures. The anion radius (white circles) remains the same, whereas the cation radius (shaded circles) is reduced from left to right. (a) Square coordination (ideal packing is obtained with a radius ratio of 0.41). (b) Triangular coordination (ideal packing is obtained with a radius ratio of 0.15).

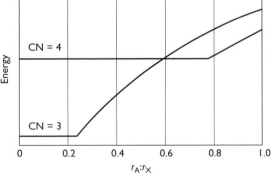

Fig. 2.18 Two-dimensional ionic packing (cf. Fig. 2.17). The electrostatic component of the lattice energy (energy it takes to separate a crystal into the free ionic species) decreases with closer proximity of ions. (CN: coordination number.) When anions are in contact, a further decrease in cation radius does not cause a further decrease in energy and a different coordination is preferred. The energy values are negative, with zero on top, and the scale is arbitrary.

Box 2.2 Additional information: Ideal radius ratio for two-dimensional geometry

(a)

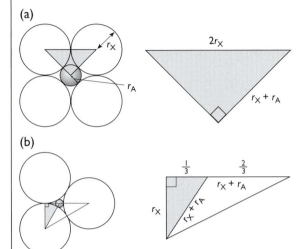

Fig. 2.19 Derivation of the ideal radius ratio for (a) square and (b) triangular coordination. Compare with Figure 2.17. On the left side is the geometry for ideal packing, on the right side is the enlarged triangle that is used for the derivation of the radius ratio; r_X is the radius of the anion and r_A that of the cation.

(b)

Square coordination (Figure 2.19a)

We consider a right triangle with two equal sides and the centers of two anions and the cation as corners. The hypotenuse has a length of $2r_X$ and the side adjacent to the right angle of $r_A + r_X$. Applying Pythagoras' theorem we obtain

$$2(r_A + r_X)^2 = 4r_X^2 \tag{2.3}$$

Taking the square root this reduces to

$$r_A + r_X = \sqrt{2}r_X \tag{2.4}$$

and solving this equation for r_A we obtain

$$r_A = (\sqrt{2} - 1)r_X = 0.414r_X \tag{2.5}$$

Triangular coordination (Figure 2.19b)

We consider a right triangle with two corners in the centers of anions, and the third corner halfway between two anions. The center of the cation lies on the longer side (at ⅓ and ⅔ distance from both corners). It outlines a smaller right triangle (shaded) with sides corresponding to $1r_X$, $\frac{1}{2}(r_A + r_X)$ and $(r_A + r_X)$ (hypotenuse). Again we apply Pythagoras' theorem to this smaller triangle.

$$(r_A + r_X)^2 = \left[\frac{1}{2}(r_A + r_X)\right]^2 + r_X^2 \tag{2.6}$$

Rearranging and taking the square root we obtain

$$\sqrt{3}/2(r_A + r_X) = r_X \tag{2.7}$$

and solving for r_A

$$r_A = (2/\sqrt{3} - 1)r_X = 0.155r_X \tag{2.8}$$

For idealized two-dimensional structures, the 3-fold coordination is stable approximately between a radius ratio r_A:r_X of 0.155 (ideal radius ratio for CN = 3) and 0.414 (ideal radius ratio for CN = 4).

Table 2.4 Common coordination polyhedra (CN) with ionic radii and radius ratios for cation–anion pairs of particular minerals

CN A–X	Mineral example	r_A (Å)	r_X (Å)	$r_A{:}r_X$	Ideal limit
3 Triangle					0.155
C–O	Calcite, $CaCO_3$	Small	1.36	—	
B–O	Borax, $Na_2(B_4O_5(OH)_4) \cdot 8H_2O$	0.02	1.36	0.015	
4 Tetrahedron					0.225
S–O	Barite, $BaSO_4$	0.12	1.40	0.09	
Si–O	Quartz, SiO_2	0.40	1.40	0.28	
Al–O	Orthoclase, $KAlSi_3O_8$	0.41	1.40	0.29	
Zn–S	Sphalerite, ZnS	0.60	1.84	0.33	
6 Octahedron					0.414
Ti–O	Rutile, TiO_2	0.61	1.40	0.44	
Fe^{3+}–O	Hematite, Fe_2O_3	0.65	1.40	0.46	
Mg–O	Diopside, $CaMgSi_2O_6$	0.72	1.40	0.51	
Na–Cl	Halite, NaCl	1.02	1.81	0.56	
Pb–S	Galena, PbS	1.18	1.84	0.64	
8 Cube					0.732
Ca–F	Fluorite, CaF_2	1.12	1.33	0.84	
Cs–Cl	CsCl (not a mineral)	1.70	1.81	0.94	
12 Cuboctahedron with 12 corners					1.000
K–O	Muscovite, $KAl_2Si_3AlO_{10}(OH)_2$	1.60	1.40	1.14	
Ca–O	Perovskite, $CaTiO_3$	1.35	1.40	0.96	

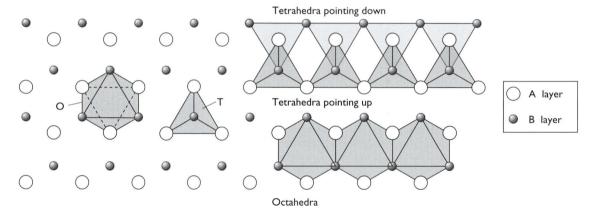

Tetrahedra pointing down

Tetrahedra pointing up

Octahedra

A layer

B layer

Fig. 2.20 Tetrahedral (T) and octahedral (O) interstices in close-packed structures. Two layers of the structure are shown. Open circles are at the bottom (A layer) and shaded circles (B layer) are in the indentations above the A layer. Shading outlines two polyhedra with atoms in the corners. O, octahedron; T, tetrahedron. On the right side of the figure is a row of upward and downward pointing tetrahedral interstices (on top) and a row of octahedral interstices (bottom).

triangular face (the lower triangle in the A layer is rotated against the top triangle in the B layer). The right side of Figure 2.20 shows on top all the tetrahedral interstices, some pointing up (darker shading), and some pointed down (lighter shading). At the bottom is a row of octahedral interstices. This illustrates that there are twice as many tetrahedral interstices as there are spheres, and the same number of octahedral interstices as there are spheres. The tetrahedral interstice (CN = 4) is smaller than the octahedral interstice (CN = 6) and can accommodate small cations such as silicon, beryllium, and zinc, whereas larger cations (iron, magnesium, and calcium) prefer octahedral interstices. In general, only a fraction of the interstices are occupied. For example, in the silicate olivine, oxygen ions have a more or less hcp arrangement, silicon ions are in a tetrahedral arrangement, and magnesium and iron are in octahedral interstices. Also in kyanite, with a structure that in detail is very complicated, oxygen ions form a close-packing arrangement, silicon ions are in tetrahedral interstices, and aluminum ions are in octahedral interstices. In halite, chloride ions are in a cubic close-packed arrangement, and sodium ions occupy octahedral interstices (Figure 2.10c). We will get to know a large number of crystals that fit this general pattern of close-packing of anions and cations in tetrahedral and octahedral interstices.

2.5 Some general rules concerning ionic structures

In the previous sections we have explored characteristic features of ionic structures. Electrostatic attraction between anions and cations, and the size of the ions, determine many features of crystal structures. The ionic bond relies on charge balance, even on a local scale, and deviations from local charge balance render a structure unstable. The influence of electrostatic attraction and repulsion on crystal structures, and the effects of charge balance/imbalance are summarized in rules first formulated by Linus Pauling in 1929. We review three of them below. Whenever a rule for ionic structures is violated, it is an indication that the structure is not truly ionic.

1. A coordination polyhedron of anions is formed about each cation. The cation–anion distance is determined by the sum of the ionic radii. The coordination number of the cation depends on the radius ratio. This is an expression of the rigid sphere concept with spheres in contact.
2. The *electrostatic valency principle* expresses that the electrostatic charges should be balanced between closest neighbors. This can be evaluated quantitatively by introducing the concept of the total strength of the valency bond p, defined as the ratio of cation charge z to the coordination number CN and representing the number of electrons per bond:

$$p = z/CN \tag{2.9}$$

In a stable coordinated structure, the sum of the bond strengths for all bonds that reach an anion from all the neighboring cations is equal to the absolute value of the charge y of the anion:

$$\sum p = \sum (z/CN) = |y| \tag{2.10}$$

An example is halite (Figure 2.10c), where Na^+ is surrounded by 6 Cl^- with a charge of 1 ($p = 1/6$). Each Cl^- has 6 Na^+ as next neighbors and summing over all these six bonds ($6 \times 1/6 = 1$), we obtain the charge of the anion.

3. The existence of edges, and particularly of faces, common to two anion polyhedra in a structure decreases its stability. This effect is large for cations with high valence and small coordination numbers. If this rule is violated (e.g., tetrahedra in Figure 2.21a–c), cations come into close proximity, which is an unstable situation due to Coulomb repulsion. Compared to corner-sharing tetrahedra (100%, Figure 2.21a), the cation distance for edge-sharing tetrahedra is reduced to 58% (Figure 2.21b), and for face-sharing tetrahedra to 33% (Figure 2.21c). For more highly coordinated polyhedra (e.g., octahedra in Figure 2.21d–f), this problem is less critical. For edge-sharing octahedra (Figure 2.21e), the cation distance is 71% as compared to that for corner-sharing octahedra (Figure 2.21d), and edge sharing often occurs.

In quartz (SiO_2), with a tetrahedral structure, tetrahedra only share corners. In rutile (TiO_2), with an octahedral structure, octahedra share two edges.

2.6 Summary

This chapter reviews some basic concepts of chemistry, such as the structure of atoms and the periodic table, that are a basis for any discussion of minerals. Then different types of bonding are introduced with typical crystal structure examples. Ionic structures are explained in some detail with concepts such as ionic radii, coordination polyhedra, and the classic Pauling rules. Remember:

• Metallic, ionic, covalent bonding
• Hexagonal and cubic close-packing of spheres in metal structures

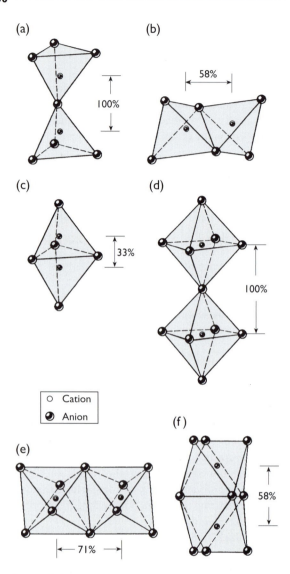

(a)

(b)

58%

100%

(c)

33%

(d)

100%

O Cation

◑ Anion

(f)

(e)

58%

71%

Fig. 2.21 In a stable structure, tetrahedra share (a) a corner, but not an edge (b), or a face (c) because this would bring cations into close proximity (cation–cation distances, relative to the corner-sharing arrangement, are indicated). For octahedra (d–f) this restriction is less critical because cation–cation distances are larger (after Klein, 2002).

- Ionic radius/radius ratio
- Coordination number defining the number of anions surrounding a cation

- Coordination polyhedra such as tetrahedra in SiO_2, octahedra in MgO, and cubes in CaO
- Interstices in close-packed structures (octahedral, tetrahedral)

Test your knowledge

1. Derive electron configurations for elements such as K, Cl, Al with the help of the periodic table (see Figure 2.2).
2. Which types of bonding are prevalent in minerals? Give specific examples.
3. Give examples of cations bonded to oxygen in octahedral coordination and tetrahedral coordination. Name some mineral examples.
4. Take a set of ping-pong balls and construct several close-packed layers. Put two layers on top of each other and then identify octahedral and tetrahedral interstices. Next, stack four layers to produce hexagonal close-packing. Take three layers and produce cubic close-packing and identify the fcc unit cell.
5. What is the ideal coordination number if cations and anions have equal size and are touching? Can you name an example of such a cation–anion pair?
6. Why do SiO_4 tetrahedra in silicates never share edges and faces?

Further reading

Evans, R. C. (1964). *An Introduction to Crystal Chemistry*. Cambridge University Press, Cambridge.

Harrison, W. A. (2011). *Solid State Theory*. Dover Publications, New York.

Kittel, C. (2004). *Introduction to Solid State Physics*, 8th edn. Wiley, New York.

Pauling, L. (1960). *The Nature of the Chemical Bond and the Structure of Molecules and Crystals*, 3rd edn. Cornell University Press, Ithaca, NY.

Putnis, A. (1992). *Introduction to Mineral Sciences*. Cambridge University Press, Cambridge.

Smart, L. E. and Moore, E. A. (2012). *Solid State Chemistry: An Introduction*, 4th edn. CRC Press, Boca Raton, FL

3 | Isomorphism, solid solutions, and polymorphism

In the previous chapter we have seen how the size of elements depends on their position in the periodic table. If atoms or ions have similar sizes and electron configurations, they often have very similar structures. This is called *isomorphism* and is of particular interest in mineralogy, with carbonate minerals such as magnesite $MgCO_3$, siderite $FeCO_3$, smithsonite $ZnCO_3$, rhodochrosite $MnCO_3$, and calcite $CaCO_3$ all with the same rhombohedral crystal structure, in spite of the elemental differences. In some minerals, especially at higher temperatures, there is a continuous compositional range, called *solid solution*, such as in olivine, ranging from forsterite Mg_2SiO_4 to fayalite Fe_2SiO_4. By contrast, SiO_2 may form different structures at high temperature (e.g., cristobalite), high pressure (e.g., coesite), and intermediate conditions (quartz). Such *polymorphism* can be used to determine conditions under which rocks formed.

3.1 Isomorphism and solid solutions

Early mineralogists such as Nicolas Steno and René J. Haüy established that minerals of different composition have a different morphology and that these morphological differences can be used to distinguish minerals. In 1821 the German chemist Eilhard Mitscherlich made the unexpected observation that compounds of different chemical composition can have a very similar morphology and called this phenomenon *isomorphism*. Based on what we now know about crystal chemistry (see Chapter 2), we can interpret this result in terms of crystal structures – that is, in a crystal structure, atoms or ions can be replaced by others of similar size without changing the structure type.

Carbonates, in particular, provide good examples of minerals with distinct chemical compositions that nonetheless have almost identical crystal forms and the same angles, as well as very similar rhombohedral crystal structures:

Magnesite	$MgCO_3$	$r_{Mg^{2+}} = 0.72$ Å
Siderite	$FeCO_3$	$r_{Fe^{2+}} = 0.77$ Å
Smithsonite	$ZnCO_3$	$r_{Zn^{2+}} = 0.75$ Å
Rhodochrosite	$MnCO_3$	$r_{Mn^{2+}} = 0.82$ Å
Calcite	$CaCO_3$	$r_{Ca^{2+}} = 1.0$ Å (low pressure)

If the radius r of the substituting cation is larger than 1 Å, then the structure changes to a different type with an orthorhombic structure:

Aragonite	$CaCO_3$	$r_{Ca^{2+}} = 1.00$ Å (high pressure)
Strontianite	$SrCO_3$	$r_{Sr^{2+}} = 1.16$ Å
Witherite	$BaCO_3$	$r_{Ba^{2+}} = 1.36$ Å

Notice that the radius of Ca^{2+} is close to 1 Å and $CaCO_3$ can exist in either the rhombohedral structure as calcite or the orthorhombic structure as aragonite, depending on external conditions.

A special type of isomorphism is when atoms or cations replace one another in arbitrary amounts. Such crystals are called *solid solutions* or *mixed crystals*. A good example of a solid solution is olivine, with similar ionic radii of magnesium and iron. The pure chemical components, called *end members*, are:

Forsterite	Mg_2SiO_4	$r_{Mg^{2+}} = 0.72$ Å
Fayalite	Fe_2SiO_4	$r_{Fe^{2+}} = 0.77$ Å

Most natural olivines, for example in basalts or peridotites, have intermediate compositions. There are many solid solutions in minerals and we will discuss them later. For now let us simply mention pyroxenes and feldspars as examples. In pyroxenes there are solid solutions between magnesium and iron, as in olivine:

| Diopside | $CaMgSi_2O_6$ | $r_{Mg^{2+}} = 0.72$ Å |
| Hedenbergite | $CaFeSi_2O_6$ | $r_{Fe^{2+}} = 0.77$ Å |

and also between calcium and magnesium:

| Diopside | $CaMgSi_2O_6$ | $r_{Ca^{2+}} = 1.00$ Å |
| Enstatite | $Mg_2Si_2O_6$ | $r_{Mg^{2+}} = 0.72$ Å |

Feldspar is a collective name for minerals that contain various proportions of the following end members:

Anorthite	$CaAl_2Si_2O_8$	$r_{Ca^{2+}} = 1.00$ Å
Albite	$NaAlSi_3O_8$	$r_{Na^+} = 1.02$ Å
Orthoclase	$KAlSi_3O_8$	$r_{K^+} = 1.38$ Å
Celsian	$BaAl_2Si_2O_8$	$r_{Ba^{2+}} = 1.36$ Å

There are complications when a divalent alkali ion (e.g., Ca^{2+}) is replaced by a univalent earth alkali ion (e.g., Na^+), as in the solid solution anorthite–albite (plagioclase), because this creates a charge imbalance. In this case, a coupled substitution of Si^{4+} for Al^{3+}, which are both of similar size ($r_{Si^{4+}} = 0.26$ Å, $r_{Al^{3+}} = 0.39$ Å, for coordination number 4), has to occur to maintain electrostatic neutrality and we can write a general formula $(Na_xCa_{1-x})Si_2Al(Si_xAl_{1-x})O_8$.

Solid solutions between magnesium and iron (as in olivine and diopside) occur at all temperatures. Mixing of magnesium and calcium (in pyroxenes), and potassium and sodium (in feldspars), on the other hand, occur only at high temperature, since there is a considerable size difference between substituting ions. This will be discussed in Chapter 21.

In a mixed crystal, substitution by atoms of different sizes causes changes in lattice parameters, as is illustrated in Figure 3.1 for some face-centered cubic (fcc) metals. The relationship between solid-solution composition and lattice parameter a is fairly linear if there is no interaction between substituting atoms. (This correspondence is known as *Vegard's rule.*) However, if there is additional attraction or repulsion between like and unlike atoms, one observes deviations from the linear relationship that are most significant for intermediate compositions.

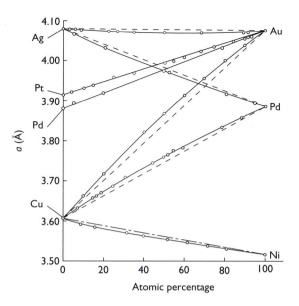

Fig. 3.1 Solid solutions between some cubic metals and the influence of the chemical composition on the lattice parameter. Solid lines connect experimental data whereas dashed lines are straight lines for reference.

Finally, we note that there also exist compounds that have identical crystal structures but that do not mix chemically. Examples of such *isostructural* crystals are halite (NaCl), galena (PbS), metacinnabar (HgS), and periclase (MgO). The structures of these compounds are identical owing to the very similar cation:anion radius ratios that the compounds display. But electron configurations of atoms and chemical bonding in each of these pairs of minerals are different.

3.2 Polymorphism and phase transitions

With the example of carbonates, we demonstrated above that the structure of a chemical compound does not change if we substitute different ions of similar size. However, structures of the same chemical compound may be different under different physical conditions. For example, $CaCO_3$ crystallizes as calcite at low pressure and as aragonite at high pressure. If a compound exists with different crystal structures, depending on external conditions, this is known as *polymorphism* and was first discovered by Mitscherlich (1820). Examples of polymorphic minerals are numerous. Polymorphism is a very useful property for petrologists. In particular, it allows one to estimate the temperature and pressure conditions during the crystallization of rocks because

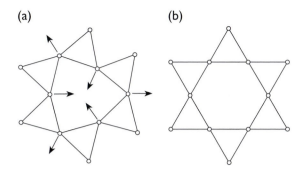

(a)   (b)

Fig. 3.2 Schematic projection along the c-axis of the structure of (a) trigonal low-temperature quartz (α) and (b) hexagonal high-temperature quartz (β). Circles are representative of SiO_4^{4-} tetrahedra that extend in the real three-dimensional structure of quartz as spirals along the c-axis (cf. Figure 21.5).

the original minerals formed during crystallization are often preserved. One example is aragonite, which forms in high-pressure subduction environments and still exists, for example, in rocks of the Coast Ranges of California, USA.

Another example of polymorphism is SiO_2. At low temperature and low pressure this chemical compound crystallizes as quartz. A schematic view of the crystal structure, projected along the c-axis, displays an arrangement of triangles with *trigonal* symmetry (Figure 3.2a). (Circles represent silicon atoms that are tetrahedrally coordinated by oxygens.) This variety of quartz is called low-temperature or α-quartz. Above a certain temperature, the arrangement of tetrahedra becomes more symmetrical and we recognize a 6-fold

symmetry axis (*hexagonal*) (Figure 3.2b). This polymorph of SiO_2 is called high-temperature or β-quartz. At even higher temperature, SiO_2 exists in yet a different structure, called cristobalite. All these SiO_2 polymorphs have in common that four oxygen atoms, outlining a tetrahedron, surround silicon and these tetrahedra are linked over corners to a three-dimensional framework. However, at very high pressure, SiO_2 changes its coordination number and silicon is in octahedral coordination as in stishovite, found at meteor impact sites.

When a compound changes its structure, for example by heating or under pressure, it undergoes a *phase transition*. Phase transitions are special isochemical cases of the more general class of phase transformations. In a very general way, when pressure is applied, a mineral may transform to a structure with denser packing. If the temperature is raised, thermal vibrations become more pronounced. Atoms may lose their identity and the structure may become disordered. Changes may be very minor, such as between α- and β-quartz, where there is a slight distortion of the structure (Figure 3.2); or they may be substantial, as in the transition of diamond to graphite, which involves a change in bonding type, and a total rearrangement. M. J. Buerger (1951) classified phase transitions into three types, reconstructive, order–disorder, and displacive, based on the structural changes that occur (Table 3.1). This is schematically illustrated in Figure 3.3 for a hypothetical two-dimensional ionic structure with square coordination polygons.

Table 3.1 Structural classification of phase transitions by M. J. Buerger

	Reconstructive (with diffusion)			Order–disorder (with diffusion)	Displacive (no diffusion)
Examples					
	C	SiO_2	SiO_2	$KAlSi_3O_8$	SiO_2
	Graphite	Quartz	Quartz	Microcline	α-(low) quartz
	Diamond	Stishovite	Cristobalite	Sanidine	β-(high) quartz
Structural change					
	Bonding	Closest neighbors	Second-closest neighbors		
	Structure of parent and daughter different			Close structural relationship	
Kinetics					
Reactions		Slow		Intermediate	Rapid
Products		Quenchable		Quenchable	Reversible

(a)

(b)

(c)

(d)

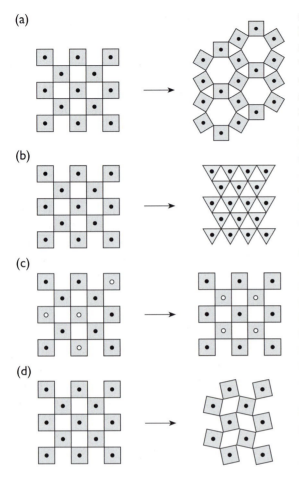

Fig. 3.3 Different types of phase transformations illustrated in two-dimensional structural sketches. Polygons represent coordination polyhedra with a cation in the center. (a) Reconstructive phase transitions in second coordination that rearrange the polygons but leave them intact. (b) Reconstructive phase transition in first coordination that breaks nearest-neighbor bonds and transforms squares to triangles. (c) Ordering transformation with a rearrangement of two types of cation. (d) Displacive transformation, resulting in a slight distortion of the square array.

Reconstructive transitions

These transitions require the breaking of bonds and a subsequent internal rearrangement, resulting in a new structure that may be quite different from the initial one. In Figure 3.3a coordination polyhedra remain intact, but they are differently arranged and linked. In the original structure they are in four-membered rings; in the transformed structure they are in three-

and six-membered rings. Since the closest neighborhood, i.e., the environment of cations by surrounding anions, remains intact, Buerger classified those transitions as *secondary coordination* transitions. Mineral examples are quartz–cristobalite (SiO_2), calcite–aragonite ($CaCO_3$), sphalerite–wurtzite (ZnS), and rutile–anatase (TiO_2). Changes in *primary coordination* involve breakage of closest neighbor bonds. In Figure 3.3b the square coordination changes to a triangular coordination, requiring breakage of anion–cation bonds. A good example for such transitions is quartz (coordination number (Si) = 4) transforming at high pressure to stishovite (coordination number (Si) = 6). The most severe reconstructive transitions occur when there is a change in the type of bonding, such as in the transition from hexagonal metallic graphite to cubic covalent diamond. Depending on the structural differences, reconstructive transitions may involve large changes in internal lattice energy and volume. Since bonds are broken, the kinetics are generally sluggish, and phases can be preserved if they are rapidly quenched. For example, the diamond phase of carbon is preserved for billions of years in kimberlites, even though diamond is not stable at the pressure and temperature conditions found on the surface of the Earth. In reconstructive transitions, atoms have to become mobile within the structure with local breakage of bonds due to thermal activation. This process is called diffusion.

Order–disorder transitions

These transitions apply to structures where two different atomic species can occupy the same lattice sites. At high temperature two atomic species (black and white circles in Figure 3.3c) may be distributed randomly in the coordination polyhedra (left side). Upon cooling the distribution becomes regular, with black and white species alternating (right side). Such a distribution is said to be "ordered". Figure 3.4 illustrates, with the distribution of black and white squares, that there may be a whole range of ordering patterns. In Figure 3.4b the distribution is random; in Figure 3.4a it is more ordered than random, with a tendency for alternation of black and white squares, and in Figure 3.4c it is less ordered than random, with the development of clusters of equal kind.

A simple example to illustrate ordering is the Cu–Au system. At high temperature Cu and Au atoms are distributed randomly over the sites of an fcc lattice,

(a) (b) (c)

Fig. 3.4 Intermediate states of ordering, illustrated with the distribution of black and white squares of equal numbers to illustrate (a) more ordered than random with a tendency for alternation, (b) random, and (c) less ordered than random with a tendency for clustering.

with an equal probability for Cu and Au to be on any site (Figure 3.5a). At these high-temperature conditions Au–Cu forms a solid solution and any amount of Cu can substitute for Au, causing a change in the lattice parameter a (Figure 3.1), which can be used to determine the composition. At low temperature, the atoms are ordered and the ordering pattern depends on the composition. For a composition Au_3Cu, Cu occupies the corners of the fcc lattice and Au is at the face centers (Figure 3.5b). The structure is still cubic, but no longer face-centered (the atoms on the centers of faces are different from those at the corners of the unit cell). For a composition AuCu, a layered tetragonal structure forms (Figure 3.5c). This structure has only a single 4-fold symmetry axis, compared to three axes in the cubic case. An ordered structure is always characterized by a reduction in symmetry. In the case of the disordered structure, a translation ½ along x and ½ along y brings atoms to coincidence. This is no longer the case in the ordered structure.

Feldspars are good mineral examples for ordering. The high-temperature monoclinic potassium feldspar

sanidine ($KAlSi_3O_8$), found in volcanic rocks, shows a disordered distribution of aluminum and silicon over tetrahedra in the structure, whereas in the low-temperature triclinic feldspar *microcline* ($KAlSi_3O_8$) aluminum and silicon are ordered. Since ordering requires diffusion, the transition is sluggish and sanidine is preserved in rapidly cooled volcanic rocks.

Displacive transitions

These transitions involve only a distortion of the lattice, and do not require the breakage of bonds. In Figure 3.3d, the regular arrangement of squares changes to a distorted arrangement by slightly rotating adjacent squares in opposite directions. The individual square is not changed, only the angles between neighboring squares. The transition from α-(low-temperature) to β-(high-temperature) quartz (Figure 3.2) is typical of a displacive transition. Relative to hexagonal β-quartz, oxygen atoms are slightly displaced in trigonal α-quartz (indicated by arrows in Figure 3.2a). The lattice energies of the parent and daughter phases in displacive transitions are very similar. Also, since no diffusion is required, displacive transitions are rapid and cannot be quenched. In the case of quartz, the high-temperature phase cannot be preserved. All quartz at ambient conditions is α-quartz and all quartz above 573 °C is β-quartz.

As our discussion above indicates, polymorphic phase transitions are usually discussed in terms of temperature and pressure changes and we will look at this in a more quantitative manner in Chapter 19. In some cases, however, an additional factor is the applied shear stress: for example, in the technologically important transformation in steel of α-Fe (bcc, ferrite)

(a) (b) (c)

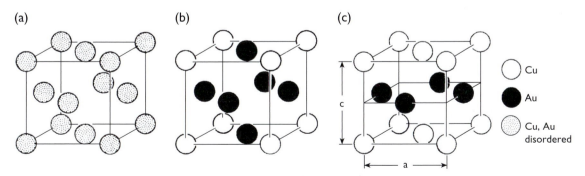

Fig. 3.5 Superstructures in the system Au–Cu. (a) Disordered fcc lattice. (b) Cubic primitive lattice in Au_3Cu. (c) Tetragonal lattice in Au–Cu.

to metastable tetragonal martensite; such stress-induced transitions are called *martensitic*.

3.3 Summary

In Chapter 3 we have considered some very simple cases relating bonding type and chemical composition to crystal structures. Particularly in minerals this is not so simple because atoms of similar size and electronic properties can substitute without changing the structure (isomorphism) and this results in solid solutions. This usually only occurs at higher temperatures. By contrast, a simple compound such as SiO_2 may crystallize in different structures depending on temperature–pressure conditions (e.g., quartz, cristobalite, coesite, stishovite). This is called polymorphism and is very important because of the wide range of conditions in the Earth. There are different types of phase transitions, relating different structures: displacive transformations are due to a slight distortion, order–disorder transformations require diffusion, and reconstructive transformations show not much resemblance between the two phases.

Test your knowledge

1. Give examples of atom pairs that form solid solutions in metals and ionic/covalent compounds.
2. Name two minerals that are isostructural but not isomorphic.
3. Carbonates can be used to illustrate both isomorphism and polymorphism. Give two mineral examples for each.
4. Explain the displacive transformation between α- and β-quartz.
5. Review the kinetics of the different types of phase transitions.

Further reading

Barrett, C. S. and Massalski, T. B. (1980). *Structure of Metals*, 3rd edn. Pergamon Press, Oxford.

Christian, J. W. (1981). *The Theory of Phase Transformations in Metals and Alloys: An Advanced Textbook in Physical Metallurgy*, 2nd edn. Pergamon Press, Oxford.

Porter, D. A., Easterling, K. E. and Sherif, M. (2009). *Phase Transformations in Metals and Alloys*, 3rd edn. CRC Press, Boca Raton. FL.

4 | Chemical formulas of minerals

Having gone over bonding and phase transitions, the elemental composition of compounds emerges as a critical property and, indeed, all minerals are associated with a characteristic chemical formula. This chapter introduces how to obtain formulas from chemical analyses. There are simple formulas such as NaCl for halite and more complex formulas for minerals with solid solutions such as plagioclase $(Ca_{1-x}Na_x)((Al_{1-x}Si_x)AlSi_2O_8)$. We also introduce graphical representations of multicomponent systems.

4.1 Ideal formulas

The chemical formula is an essential attribute of a mineral. There are many ways to write chemical formulas. So-called ideal formulas such as Al_2O_3 for corundum, $CaCO_3$ for calcite, Mg_2SiO_4 for forsterite, ZnS for sphalerite, FeS_2 for pyrite, or NaCl for halite correspond with compositions of chemically pure minerals.

Parentheses or brackets are used to highlight some structural properties. For example, $CaCO_3$ could be written as $Ca(CO_3)$ and Mg_2SiO_4 as $Mg_2(SiO_4)$ to indicate structural units. Parentheses include anion or other complexes in structures of minerals. More complicated examples are silicates where it is necessary to identify the tetrahedral structural units as in $Na(AlSi_3O_8)$ for albite, $Na_8(AlSiO_4)_6Cl_2$ for sodalite, or $Ca_{19}Al_{10}Mg_3(SiO_4)_{10}(Si_2O_7)_4(OH)_{10}$ for vesuvianite.

All chemical formulas are charge balanced. For calcite the 2^+ charge of calcium is balanced by the 2^- charge of the anion group (CO_3). For sodalite the 8^+ charge of sodium ions is balanced by the sum of the 6^- charge of six anion groups $(AlSiO_4)$ and the 2^- charge of two chlorine ions.

4.2 Empirical formulas

Many minerals have variable chemical compositions because of isomorphic substitution of chemical elements in their structures. That is the reason why the same mineral can have different compositions and therefore different formulas in different natural occurrences.

For example, calcite from two rock types can be described with formulas such as $(Ca_{0.9}Mn_{0.1})_{\Sigma=1.0}(CO_3)$ and $(Ca_{0.8}Mg_{0.2})_{\Sigma=1.0}(CO_3)$, and olivine with $(Mg_{1.6}Fe_{0.4})_{\Sigma=2.0}(SiO_4)$ and $(Mg_{1.4}Fe_{0.4}Mn_{0.2})_{\Sigma=2.0}(SiO_4)$. Again, formulas are charge balanced. For calcite the 2^+ charges of $(Ca_{0.9}Mn_{0.1})_{\Sigma=1.0}$ and $(Ca_{0.8}Mg_{0.8})_{\Sigma=1.0}$ are balanced by the 2^- charge of the anion (CO_3).

Formulas are derived from quantitative chemical analyses, and that is why they are often called empirical. For oxides such as perovskite and ilmenite and many other oxygen compounds, chemical analyses are often expressed as weight% of oxide components (e.g., second column in Table 4.1 and Table 4.2). Why are oxide components listed? It follows an old tradition of analytical chemistry to represent chemical compositions of such substances with combinations of simple oxides. In this way, perovskite is equivalent with $CaO \cdot TiO_2$ and ilmenite with $FeO \cdot TiO_2$.

Any chemical analysis contains minor errors, and the sum of components often does not add exactly to 100%. We now introduce a method by which formulas are calculated from analytical data.

4.3 Calculation of chemical formulas from weight percentage of oxides

Calculation of a chemical formula from weight% oxides is straightforward for stoichiometric compounds and is easily extended to solid solutions. The process becomes more difficult, however, if atomic vacancies are present in such minerals as amphiboles and sheet

Table 4.1 Calculation of chemical formula of perovskite from weight% oxides

Chemical component	Weight%	Molecular weight	Molecular proportion	Formula
CaO	41.25	56.08	0.736	1
TiO_2	58.75	79.90	0.736	1
Total	100.00			

Table 4.2 Calculation of chemical formula of ilmenite from weight% oxides

Chemical components	Weight%	Molecular weight	Molecular proportions	Atomic proportions		Numbers of cations
				Cations	Oxygens	
TiO_2	53.80	79.90	0.6733	0.6733	1.3466	1.000
MgO	2.72	40.32	0.0675	0.0675	0.0675	0.100 ⎫
FeO	38.70	71.85	0.5386	0.5386	0.5386	0.800 ⎬ 1.0
MnO	4.77	70.93	0.0672	0.0672	0.0672	0.100 ⎭
Total	99.99				2.0199	

silicates, if ions have variable oxidation states, or if a mineral contains components that were not analyzed.

We will first illustrate the procedure for stoichiometric perovskite and then extend it to ilmenite with a solid solution, and a plagioclase feldspar. For more complicated cases consult the literature (e.g., Droop, 1987; Bulakh and Zussman, 1994). Many computer programs are available for calculating mineral formulas.

In our first example, we will calculate the formula of perovskite from the weight% values of oxides (second column of Table 4.1). To do so, we divide the weight% value of each oxide by its molecular weight (third column) to obtain molecular proportions (fourth column). After normalizing the resulting proportions to clear fractions (i.e., dividing by 0.736), we obtain the formula 1 CaO + 1 TiO_2 = $CaTiO_3$ (fifth column).

Another example, the trigonal oxide mineral ilmenite, may contain some substitutions of Mg and Mn for Fe. A typical analysis is given in Table 4.2. Again, we obtain molecular proportions by dividing weight% values by the molecular weight. We then separate the molecular proportions into atomic proportions of cations and oxygens. From crystal chemistry we know that the ideal formula of ilmenite is $FeTiO_3$. Thus the atomic proportions for the sample analyzed in Table 4.2 can be normalized such that the sum of oxygen atoms is three. We multiply

the atomic proportions of cations by 3/2.0199 = 1.485 to obtain the corresponding number of cations (last column). Using the cation proportions in Table 4.2, we can write the formula for this particular ilmenite as $(Fe_{0.80}Mg_{0.10}Mn_{0.10})_{\Sigma=1.00}Ti_{1.00}O_{3.00}$. Let us check charge balance. The charges 2^+ of $(Fe_{0.80}Mg_{0.10}Mn_{0.10})$ and 4^+ of Ti are fully compensated by the charge 6^- of three oxygen anions.

Our third example is a plagioclase feldspar with an extensive solid solution of Ca, Na, and K. Again, we obtain molecular and atomic proportions from weight% values (Table 4.3). Since an ideal formula of feldspar such as albite is $NaAlSi_3O_8$, we normalize the atomic proportions to eight oxygen atoms (last column) and write the formula as $(Ca_{0.14}Na_{0.79}K_{0.07})_{\Sigma=1.00}(Si_{2.85}Al_{1.15})_{\Sigma=4.00}O_8$. Note that the tetrahedral cations (Si, Al) add up to four, and the large cations (Ca, Na, and K) sum approximately to 1. The charge 1.14^+ of $(Ca_{0.14}Na_{0.79}K_{0.07})$ and 14.85^+ of $(Si_{2.85}Al_{1.15})$ are compensated with the charge 16^- of eight oxygen anions (the discrepancy 0.01^- is connected with the precision of the calculations or analysis).

4.4 Simplified formulas

In many handbooks, formulas such as (Zn,Fe)S, (Zn, Fe,Mn)S and (Zn,Cd)S for sphalerite, (Fe,Mg,Mn)

Table 4.3 Calculation of chemical formula of plagioclase from weight% oxides

Chemical components	Weight%	Molecular weight	Molecular proportions	Atomic proportions		Numbers of cations (8 oxygen atoms)
				Cations	Oxygens	
SiO_2	64.60	60.09	1.075	1.075	2.150	2.85 ⎫
Al_2O_3	22.04	101.96	0.216	0.432	0.648	1.15 ⎬ 4.0
CaO	2.94	56.08	0.052	0.052	0.052	0.14 ⎫
Na_2O	9.28	61.98	0.150	0.300	0.150	0.79 ⎬ 1.0
K_2O	1.27	94.20	0.013	0.026	0.013	0.07 ⎭
Total	100.13				3.013	

$(Ti,Fe)O_3$ for ilmenite, $(Mg,Fe,Mn)_2SiO_4$ for forsterite are used. Isomorphic chemical elements are joined in parentheses. They are named simplified formulas because numbers of atoms are not specified.

For example, diopside can be described both with the ideal formula $CaMg(Si_2O_6)$ and with some simplified formulas such as $Ca(Mg,Fe,Mn)Si_2O_6$, (Ca,Na) $(Mg,Fe)Si_2O_6$, $(Ca,Na)(Mg,Cr)(Si,Al)_2O_6$, depending on the emphasis. None of these expressions are wrong, but in this textbook we will follow two basic rules.

1. We prefer an *ideal formula* and thus use in most cases the formulas of end members. We describe diopside as $CaMgSi_2O_6$ and sphalerite as ZnS, even though these ideal compositions rarely exist in nature.
2. A formula has to be charge balanced.

Hollandite is a common oxide mineral in manganese ores. It is frequently described with a simplified formula $Ba(Mn^{2+}, Mn^{4+})_8O_{16}$ and a substitution Mn^{2+} $\rightleftharpoons Mn^{4+}$. Only if we write $BaMn^{2+}Mn^{4+}_7 O_{16}$ are charges balanced, with a fixed $Mn^{2+}:Mn^{4+}$ ratio. Intermediate substitutions are either balanced by anion vacancies or by more complicated substitutions.

Garnets are chemically rather complex minerals. They have a common formula $R^1_3R^2_2(SiO_4)_3$ (R^1 and R^2 representing two different cations). Garnets may be white, green, yellow, brown, raspberry-red, pink, or black in color, with considerable variations in density, hardness, and refractive index. This variation can be explained in terms of fluctuations in chemical composition due to isomorphism.

4.5 How to use ternary diagrams

Triangular (or ternary) diagrams are often used for the representation of mineral compositions with multiple components and we will explain this for an idealized system. The chemical compositions of minerals with solid solutions between several major components are often represented in linear (binary solid solution A–B; Figure 4.1a), triangular (ternary solid solution A–B–C; Figure 4.1b), or tetrahedral diagrams (quaternary solid solution A–B–C–D; Figure 4.1c). In most cases the chemical concentrations are expressed by mole percentage (mol%).

We focus here on triangular, or ternary, diagrams because there are many examples among minerals and rocks that are composed largely of three components. Let a mineral system be described by A_a, B_b, and C_c, with a, b, and c being fractions or mol% values of components A, B, and C, respectively. The numbers a, b, and c add up to 1 or 100%. A triangle is divided with a grid, composed of three sets of parallel lines (Figure 4.2). Corners labeled A, B, and C represent end-member compositions of each component. If a composition is $A_{50}B_{50}$, it plots halfway between A and B (point X). Compositions that contain all three components, such as $A_{70}B_{20}C_{10}$, plot in the interior of the triangle and we find the point by counting the gridlines (point Y). Only two components are needed to find the point as an intersection of two gridlines. The third component is redundant, assuming that the three components are normalized to 100%.

As Figure 4.3a shows, there is a broad range of compositions between pyrope, almandine, and spessartine. Often only the formulas of the chemically pure substances are ascribed to mineral species. For example, almandine is assigned the formula Fe_3Al_2 $(SiO_4)_3$, but in reality its composition ranges between three components and $(Fe,Mg,Mn)_3Al_2(SiO_4)_3$ would

(a)

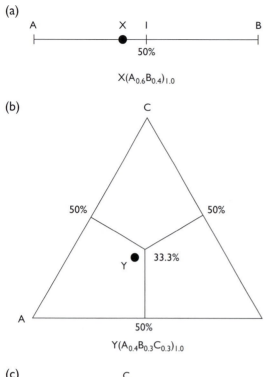

(b)

(c)

Fig. 4.1 Compositional variations of solid solutions can be represented in (a) linear, (b) triangular, or (c) tetrahedral variation diagrams. Mineral names are usually assigned by dividing fields halfway between end members. In the case of the mineral with the composition of the large dot (X, Y, and Z), the mineral name is A.

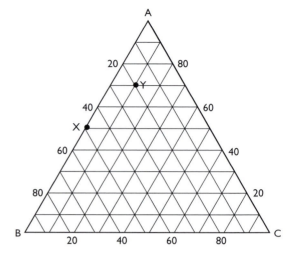

Fig. 4.2 Representation of minerals with compositions $A_{50}B_{50}$ (X) and $A_{70}B_{20}C_{10}$ (Y) in the A–B–C ternary diagram.

be a more correct formula. The order in which the cations are listed in empirical formulas indicates their relative abundance. The expression (Fe,Mg,Mn) adds up to 1. We plot in Figure 4.3a the compositions of known natural garnets on a triangular diagram of three "end-member" garnets (almandine $R^1 = Fe^{2+}$, pyrope $R^1 = Mg$, and spessartine $R^1 = Mn^{2+}$, for all $R^2 = Al$).

With ionic substitutions, the garnet structure remains more or less unchanged. However, unit cell parameters may increase or decrease, depending on the size of substituting ions, the bond type, and its strength. This is illustrated in the ternary diagram in Figure 4.3b which displays how the lattice parameter a and refractive index n correlate with the proportion of Fe^{2+}, Mg, and Mn in garnets.

Ca, Mg, Mn, and Fe carbonates (calcite ($CaCO_3$), magnesite ($MgCO_3$), rhodochrosite ($MnCO_3$), and siderite ($FeCO_3$), respectively) are another example of a mineral series with compositional variations. All four minerals have variable compositions (Figure 4.4), but, contrary to garnets, most natural carbonates cluster close to end-member compositions. While carbonates with intermediate compositions, such as dolomite ($CaMg(CO_3)_2$), huntite ($CaMg_3(CO_3)_4$), ankerite

(a)

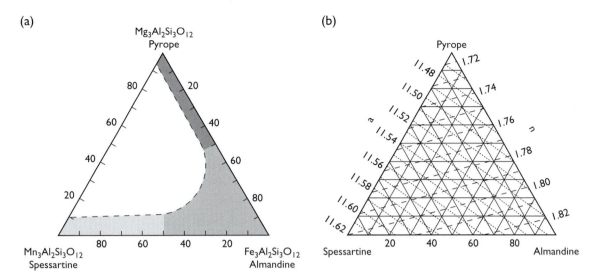

Fig. 4.3 Ternary diagram illustrating compositional variations in the garnets almandine, pyrope, and spessartine. (a) Names assigned to specific compositional ranges. The shaded range illustrates garnet compositions observed in natural rocks. (b) Variations of the lattice parameter a in ångströms (dashed lines) and refractive index n (dotted lines) with composition (data from Sriramadas, 1957).

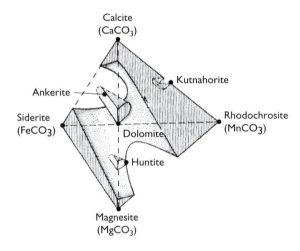

Fig. 4.4 Tetrahedron representing compositions of rhombohedral carbonates (end members $CaCO_3$, $MgCO_3$, $FeCO_3$, $MnCO_3$, outlining fields of observed compositions (dashed regions) and mineral names.

$(CaFe(CO_3)_2)$, and kutnahorite $(CaMn(CO_3)_2)$, do exist, the structure of these minerals is different from the carbonate end members listed earlier and separate names are assigned.

4.6 Summary

If you are a chemist, this chapter may appear trivial, but for geologists it is not so obvious to connect chemical formulas to chemical analyses which are often expressed in weight percentage of oxides, both in data from experimental facilities and in published documents. This is even more difficult if the mineral is a solid solution with an intermediate composition. Graphical representations are frequently used. For mineralogy, ternary diagrams are most important and you will encounter many in this book.

5 | Chemical classification and names of minerals

Minerals, as has become evident by now, have a well-defined chemical composition and it is indeed this chemical composition that has become the basic principle used to classify minerals. The over 5000 minerals known today are divided into 14 chemical groups of which silica compounds, so-called silicates, are the most abundant and significant in rocks. In this book we will discuss about 200 minerals in some detail.

5.1 Minerals, mineral species, and mineral varieties

In the first chapter we defined a mineral as a *naturally occurring solid with well-defined chemistry* and crystal structure that is formed by geological processes. Note that substances formed by human intervention (e.g., compounds forming in cement, products of interaction between seawater and metallurgical slag, or products of coal combustion) are not regarded as minerals. A certain mineral may exist with different morphologies, showing different properties and slight variations in its internal structure. Because of isomorphism and solid solutions, the chemical composition of a mineral may also fluctuate. Each particular mineral is therefore a sort of "individual", much as individual plants or animals within a species differ from each other. Biologists introduced the term *species* to collect individuals with similar characteristics, basing their definition largely on morphological factors. Similarly, the term *mineral species* has been introduced to include natural crystals with similar structural and chemical properties. In order to assign a mineral name to a chemical compound, it has to be analyzed in detail. Minerals either formed on Earth or were brought to Earth, e.g., from the Moon or planet Mars, or by meteorites.

5.2 Chemical classification of minerals

In 1789 A. G. Werner from Freiberg, Germany, proposed a chemical classification for minerals which

was emphasized in 1814 by the Swedish chemist J. Berzelius, claiming that mineralogy was part of chemistry. James W. Dana successfully used this system in his well-known *System of Mineralogy* (1837, and in the much enlarged fifth edition of 1868). Dana distinguished five mineral *classes*: (a) native elements, (b) sulfuric and arsenic compounds, (c) halides, (d) oxides, and (e) organic substances. This classification reflects the state of progress in quantitative chemical analyses during the nineteenth century.

In the twentieth century, beginning with the work of Lawrence Bragg, crystal structures could finally be determined and structural classifications were proposed. For example, such chemically diverse compounds as NaCl (halite), PbS (galena), and MgO (periclase) have the same crystal structure due to their similar cation:anion ratio. Structural classifications have since become conventional in solid-state physics.

In modern mineralogy a compromise was reached (Table 5.1). Overall, a chemical classification is most useful. However, as silicate structures were determined, it became obvious that minerals of similar composition, such as forsterite (Mg_2SiO_4), enstatite ($MgSiO_3$), and talc ($Mg_3Si_4O_{10}(OH)_2$), have entirely different properties, and thus a structural classification is more natural for these silicates. We emphasize that all classifications are arbitrary and are a product of the human mind, which seeks to order and simplify. Classifications are needed, but keep in mind that they are also limited because nature does not always follow

Table 5.1 General classification of minerals used in this book

The International Mineralogical Association (IMA) proposes a hierarchical scheme (Mills et al., 2009), based on Strunz and Nickel (2001). Groups in square brackets are omitted from the discussion.

Nickel–Strunz classes	Generalized formulas	IMA classes and °subclasses with *our additions	Chapter	Examples
1: Native elements	X	*Native metals and semimetals	22	Gold, Au
	$X_mY_nZ_k\cdots$	*[Intermetallics]		[Auricupride, Cu_3Au]
	X	*Native nonmetals	22	Sulfur, S
*[Carbides, nitrides, phosphides]	X_mC_n			[Moissanite, SiC]
*[Silicides]	X_mSi_n			[Hapkeite, Fe_2Si]
2: Sulfides, and sulfosalts	X_mS_n	Sulfides, [selenides, tellurides], arsenides, [antimonides, bismuthides]	26	Pyrite, FeS_2
	$X_m(Y_nS_k)_l$	Sulfosalts, [sulfarsenites, sulfantimonites, sulfbismuthites, etc.]	26	Tetrahedrite, $Cu_{12}(SbS_3)_4S$
3: Halogenides	X_mCl_n	Fluorides	23	Fluorite, CaF_2
		Chlorides	23	Halite, NaCl
		[Iodides, bromides]		[Iodargyrite, AgI]
4: Oxides	$X_mY_nO_k$	Oxides	27	Corundum, Al_2O_3
	$X_m(OH)_n$	Hydroxides	27	Brucite, $Mg(OH)_2$
	$X_m(SO_3)_n$	[Arsenites (including antimonites, bismuthites, sulfites, selenites, and tellurites)]		[Sidpietersite, $Pb_4(S_2O_3)O_2(OH)_2$]
5: Carbonates and nitrates	$X_m(CO_3)_n$	Carbonates	24	Calcite, $CaCO_3$
	$X_m(NO_3)_n$	Nitrates	24	Niter, KNO_3
6: Borates	$X_m(BO_3)_n$	°Nesoborates	24	Ludwigite, $Mg_2Fe(BO_3)O_2$
		[°Soroborates]		[Szaibelyite, $Mg_2((B_2O_5(OH))(OH)]$
		°Cycloborates	24	Borax, $Na_2(B_4O_5(OH)_4)\cdot8H_2O$
		°Inoborates	24	Kernite $Na_2(B_4O_6(OH)_2)\cdot3H_2O$
		[°Phylloborates]		[Johachidolite, $CaAl(B_3O_7)$]
		[°Tectoborates]		[Boracite, $Mg_3(B_4B_3O_{12})OCl$]
*[Iodates]	$X_m(IO_3)_n$			[Lautarite, $Ca(IO_3)_2$]

Table 5.1 (cont.)

Nickel–Strunz classes	Generalized formulas	IMA classes and °subclasses with *our additions		Chapter	Examples
7: Sulfates, selenates, and tellurates	$X_m(SO_4)_n$	Sulfates, [selenates], [tellurates]		25	Gypsum, $CaSO_4 \cdot 2H_2O$
	$X_m(CrO_4)_n$	[Chromates]			[Crocoite, $Pb(CrO_4)$]
		[Molybdates]			[Powellite, $Ca(MoO_4)$]
	$X_m(WO_4)_n$	Tungstates		25	Scheelite, $Ca(WO_4)$
8: Phosphates, arsenates, and vanadates	$X_m(PO_4)_n$	Phosphates		25	Apatite, $Ca_5(PO_4)_3(OH)$
	$X_m(AsO_4)_n$	Arsenates		25	Annabergite, $Ni_3(AsO_4)2 \cdot 8H_2O$
	$X_m(VO_4)_n$	Vanadates		25	Vanadinite, $Pb_5(VO_4)_3Cl$
9: Silicates and germanates	$Y_lX_m(SiO_4)_n$		°09.A: neso- *Ortho	28	Forsterite, $Mg_2(SiO_4)$
	$Y_lX_m(Si_2O_7)_n$		°09.B: soro-	28	Lawsonite $CaAl_2(OH)_2(Si_2O_7) \cdot H_2O$
	$Y_lX_m(Si_6O_{18})_n$		°09.C: cyclo- *Ring	28	Beryl, $Be_3Al_2(Si_6O_{18})$
	$Y_lX_m(Si_2O_6)_m$,		°09.D: ino- *Chain	30	Diopside, $CaMg(Si_2O_6)$
	$Y_lX_m(Si_4O_{11})_n$				Tremolite, $Ca_2Mg_5(Si_4O_{11})_2(OH)_2$
	$Y_lX_m(Si_2O_5)_n(OH)_k$		°09.E: phyllo- *Sheet	29	Talc, $Mg_3(Si_4O_{10})(OH)_2$
	$Y_lX_m((Al+Si)_nO_{2n})$		°09.F: tecto- *Framework	31	Albite, $Na(AlSi_3O_8)$
			°09.G: tecto-with zeolitic H_2O	31	Natrolite, $Na_2(Al_2Si_3O_{10}) \cdot 2H_2O$
10: Organic compounds	$M_mC_nH_kO_l$	°10.A. Salt of organic acids			Dashkovaite, $Mg(HCOO)_2 \cdot 2H_2O$
	C_mH_n	°10.B. Hydrocarbons			Evenkite, $C_{23}H_{48}$
	$C_nH_kO_lN$	*Oxy-nitrocompounds			Urea, $CO(NH_2)_2$
	$M_mC_nH_{kl}N$	*Cyclic nitrocompounds			Abelsonite, $NiC_{31}H_{32}N_4$

rules. For example, is the evaporite mineral hanksite $(KNa_2Cl(CO_3)_2(SO_4)_9)$ a chloride, a carbonate, or a sulfate? Is sillimanite (Al_2SiO_5), with its fibrous habit and tetrahedral chains, best viewed as a chain or as an orthosilicate? The most impressive examples of minerals that break down classification barriers are biopyriboles, intermediate between sheet silicates (*biotite*), *py*roxenes, and amph*iboles* (Thompson, 1978).

Our knowledge of the various mineral divisions and classes has mushroomed with the development of new analytical techniques. Consider how the number of known minerals has evolved over the last 200 years (Figure 5.1):

- Just as background: Plinius described about 30 minerals in his *Historia Naturalis* (~AD 70).
- At the end of the eighteenth century about 100 mineral species (the major rock-forming minerals, components of ores, and economically important minerals) were known; they include many of the minerals that are discussed in this book.
- In 1915, just before the discovery of X-ray diffraction, about 1000 minerals were known, largely owing to advances in chemistry and through the use of the petrographic microscope.
- Between 1915 and 1960, another 1000 or so new minerals were discovered, bringing the total to 2000.

Most of these minerals were identified by combined means of X-ray diffraction and chemical analysis.

- Since 1960, with the advent of electron microprobes and electron microscopy, another 3000 minerals have been added to the list of known minerals, and the number will continue to increase as new methods of characterization at a very small scale become available. In fact, over half of the minerals were discovered after 1980.

The number changes constantly because many newly discovered minerals have not yet been studied sufficiently to be accepted as "mineral species". In addition, several older minerals are being eliminated because they are today considered to represent just a chemical or structural variety of another mineral or are synonymous with older names (e.g., amethyst is a variety of quartz, emerald is a chromium-bearing beryl, titanite is a newer name for sphene). So-called *polytypes*, characterized by a different stacking of structural units without changes in the chemical composition, are structural varieties of the same mineral species, as in the case of micas.

Compounds with different crystal structures represent different minerals, even if the chemical composition is the same, as is the case with polymorphs. For example, silicon dioxides quartz, cristobalite,

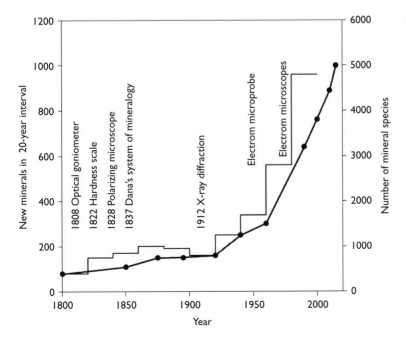

Fig. 5.1 Number of known mineral species and the number of newly discovered minerals in 20-year intervals.

tridymite, coesite, and stishovite are different mineral species.

In contrast, we apply the term "chemical varieties" to those representatives of a mineral species that deviate slightly from an accepted chemical formula. The accepted formula generally corresponds to a formula for the chemically pure substance. "Structural varieties" differ in some structural details.

The most common minerals are those that contain the most abundant elements (Table 5.2; see also Chapter 2). If we order the elements according to their abundance in the Earth's crust, we obtain the following sequence (atomic percentage in parentheses): O (53.4), H (17.3), Si (16.1), Al (4.8), Na (1.8), Mg (1.7), Ca (1.4), Fe (1.3), K (1.0), C (0.51), Ti (0.22), Cl (0.10), F (0.07) (Figure 2.4). This sequence generally correlates with the number of minerals that contain these elements, but there are exceptions. For example, many rare elements such as arsenic, phosphorus, lead, and copper are found in a large number of minerals. The number of minerals containing a certain element depends on this element's chemical activity (high activity produces more minerals), its tendency to substitute for other elements (generally isomorphism decreases the probability of an element occurring in many minerals), and the uneven elemental distribution in different rocks.

The element hydrogen plays an important role in minerals. Hydrogen is present in more than half of all minerals, and it may occur in the form of H^+, OH^-, H_3O^+, or H_2O. *Protons* (i.e., H^+) are rarely present, and, if present, the proton is always surrounded by a pair of oxygen atoms. An example is diaspore ($HAlO_2$), whose structure consists of close-packed oxygens, with aluminum occupying octahedral interstices and H^+ located between two oxygen atoms. In comparison, the *hydroxide* ion (OH^-) is quite common, examples including sheet silicates such as muscovite ($KAl_2(AlSi_3O_{10})(OH)_2$), gibbsite ($Al(OH)_3$), and malachite ($Cu_2(CO_3)(OH)_2$). In size, the hydroxide ion is similar to the oxygen atom and substitutes for it. By contrast, H_3O^+ is much larger and can substitute for K^+. Finally, *molecular* H_2O can exist in a mineral structure at specific atomic positions such as in gypsum ($CaSO_4 \cdot 2H_2O$). When the mineral is heated, H_2O molecules are released and the structure breaks down. In zeolites and clay minerals, molecular water is located in large interconnected "cavities" or "tunnels" (0.5–1.0 nm in diameter) within their structure, which explains why water is not firmly bound in these structures and its content may vary. During heating, water is expelled from these minerals without changing their structure greatly and can later be reintroduced.

Of the 5000 known mineral species about 200 are common rock-forming minerals (Deer *et al.*, 2013) and our book will focus on those (Part V). There are about 20 important gems (Chapter 34), though many of those go under different names, such as the mineral corundum with gem varieties blue sapphire and red ruby, or beryl with blue aquamarine, green emerald, and pink morganite. About 50 minerals are significant for metal ore production, such as goethite, hematite, and magnetite for iron, alunite and bauxite for aluminum, braunite, psilomelane, and pyrolusite for manganese, bornite and chalcopyrite for copper, and nickeline and pentlandite for nickel (Chapter 33).

Table 5.2 The numbers of minerals containing specific chemical elements as major components

Range	Minerals (actual number)
>1000	O (3929), H (2700), Si (1420), Ca (1130)
<1000–300	S (978), Al (959), Fe (920), Na (850), Cu (588), P (559), As (536), Mg (547), Mn (481), Pb (462), C (384), K (371), F (314), Cl (319)
<300–100	Ti (293), B (257), U (227), Sb (221), Zn (202), Bi (206), Ba (206), V (191), REE (190), Te (159), Ag (145), Ni (119), Nb (108), Se (108), Zr (108), Be (108), Sr (106), Li (106), Y (101)
<100–30	Sn (92), Sn (88), N (88), Hg (87), Cr (80), Pd (61), Co (56), Ta (52), Mo (52), Tl (47), Th (44), W (43)
<30	Pt (28), Cd (27), Ge (27), Au (26), Ir (25), I (22), Cs (22), Rh (12), In (11), Sc (11), Br (7), Ga (5), Ru (5), Os (4), Rb (3), Re (2)

Source: Krivovichev and Charykova, 2014.

Ten minerals form by biological processes, such as apatite in bones, calcite and aragonite in sea shells, and magnetite in magnetotactic bacteria. By far the majority of minerals have a regular crystal structure. An exception is amorphous opal.

5.3 Mineral names

Some mineral names used have been given by an investigator to honor a famous mineralogist (e.g., *haüyne* for René J. Haüy) or an illustrious colleague (*howieite* for R. A. Howie). There is even a mineral *wenkite*, not named after the first author of this book but his father, by one of his students who found it. Rules dictate that you cannot name a mineral after yourself. Other names refer to localities where the mineral was first discovered (e.g., *tremolite* for Val Tremola in Switzerland, *pigeonite* for Pigeon Cove in Minnesota). *Bulachite* is not named after the second author, but after a town in the Black Forest in Germany. Some names indicate the chemical composition of a mineral (e.g., *calcite* for calcium, *sodalite* for sodium). Certain mineral names are linked to properties (e.g., *orthoclase* has right (orthogonal) angles between cleavages, *albite* comes from the Latin word *alba*, meaning "white"), or the mode of a mineral's occurrence (*monazite* for the tendency of this mineral to form isolated crystals). Presently a commission of the International Mineralogical Association (IMA) must approve by vote the name of every newly discovered mineral, and criteria to establish new minerals are based mainly on diffraction spectra and chemical analyses.

5.4 Summary

In this short chapter we introduce important concepts that will be the basis and follow us through the rest of the book. A chemical classification is used to divide minerals into 10 groups. Of those, silicates are by far the most important group. We show how the number of minerals has increased from ~100 in 1900 to 5000 today. A new mineral has to be found in a natural locality and the uniqueness has to be approved by the International Mineralogical Association. Within a mineral species there may be structural and chemical varieties. Solid solutions (e.g., in garnets and

feldspars) are most significant but also trace elements can change the appearance of minerals significantly (e.g., corundum, ruby, sapphire). In this book we cover only a small selection of the most important minerals, with the emphasis placed on rock-forming minerals. For more systematic treatments, we refer the reader to comprehensive reference books.

Test your knowledge

1. What is the difference between a mineral species and a mineral variety? Give examples.
2. Mineral names have different origins: find four examples for each, a mineral named after a person, after a locality, and after a property.
3. Some solid solutions have been introduced, with variable composition. Plot, in a ternary representation, a feldspar of composition anorthite (An) 55%, albite (Ab) 40%, and orthoclase (Or) 5%.
4. Most minerals are classified according to their chemical composition; which are the major groups? Silicates are an exception; why is this so?
5. Without using books or notes, prepare a list of minerals that you know right now.

Further reading

Back, M. (2014). *Fleischer's Glossary of Mineral Species*, 11th edn. The Mineralogical Record Inc., Tuscon, AZ.

Clark, A. (1993). *M. Hey's Mineral Index*, 3rd edn. Chapman & Hall, London.

Gaines, R. V., Skinner, C. W., Foord, E. E., Mason, B. and Rosenzweig, A. (1997). *Dana's New Mineralogy: The System of Mineralogy of J. D. Dana and E. S. Dana*, 3 vols. Wiley, New York.

Krivovichev, V. G. (2008). *Mineralogical Glossary (Mineralogicheskiy slovar)*. St. Petersburg State University, St. Petersburg. In Russian, with English index of minerals.

Mitchell, R. S. (1979). *Mineral Names: What Do They Mean?* Van Nostrand Reinhold Co., New York.

Ramdohr, P. (1969). *The Ore Minerals and their Intergrowths.* Pergamon Press, Oxford.

Strunz, H. (1982). *Mineralogische Tabellen.* Akademische Verlagsgesellschaft, Leipzig.

Strunz, H. and Nickel, E. H. (2001). *Strunz Mineralogical Tables*, 9th edn. Schweizerbart, Stuttgart.

Important internet links: webpages with mineral information (see also links at the end of Chapter 6)

Crystallography Open Database COD. (www.crystallography.net)

Crystal Structure Database (B. Downs and P. Heese). Mineralogical Society of America, Washington, DC. (www.minsocam.org/MSA/Crystal Database .html)

ICDD, Crystal Data Identification File, International Center for Diffraction Data, Newtown Square, PA, USA: over 182 500 entries. (www.icdd.com)

ICSD, Inorganic Crystal Structure Database, FIZ Karlsruhe, Germany: complete structural information for inorganic compounds, including minerals, over 53 000 entries. (www.fiz-karls ruhe.de)

6 | Mineral identification of hand specimens

With 5000 mineral species, you probably wonder about how you can indentify a mineral that you find in the field or in a collection. While we will go into much more detail later in the book (Chapter 11 about X-ray diffraction, Chapters 13 and 14 about optical properties, and Part V about individual mineral groups), it seems appropriate to introduce simple hand specimen identification early, based on such properties as morphology, color, hardness, and density. This way you get an intuitive feeling for mineral identities.

6.1 Different scales

All geologists must be able to identify – more or less by inspection – most of the common rock-forming minerals and certain important accessory and ore minerals in rocks of all kinds. Such ability is developed largely through practice and experience, involving repeated observations of characteristic simple physical properties. Some of these observations can be made directly, with the naked eye. However, because many of the mineral grains in rocks are small, commonly less than 1 mm in diameter, a high-quality hand lens (magnification 5× or 10×) is an indispensable tool for routine field and laboratory observations.

Most hand specimen identification is based only upon the state of aggregation and on simple physical properties that can be determined by inspection or by some rapid and easily performed nondestructive tests. Many common minerals can be identified reasonably accurately in this fashion, even in the field. Note, however, that most of these properties are qualitative descriptions and often vary within a mineral species. In other cases, two distinct minerals may have very similar crystals and physical properties (e.g., proustite/pyrargyrite, quartz/phenakite), and can be told apart only after a detailed examination of their optical characteristics, chemical composition, or X-ray diffraction patterns.

6.2 State of aggregation (including crystallographic form and habit)

Some crystals, particularly those that have grown to large size in veins or cavities in otherwise finer grained rocks, have characteristic shapes and dimensions that reflect the processes by which they have been deposited. Others occur in amorphous-looking masses with characteristic surface textures. Highly subjective and somewhat fanciful terms are used to describe the state of aggregation of mineral bodies and the visible shapes of individual crystals.

Most minerals occur as small, uniformly sized grains, making up the polycrystalline aggregates we call *rocks*. With a hand lens the geologist can generally determine the approximate shapes and sizes of such grains and from these make tentative or firm identifications of the minerals present.

Of particular help in identification are any crystal faces that may be developed on the grains. Terms used to describe the degree of development of faces are:

- *Euhedral*: Grains fully enclosed by recognizable crystal faces.
- *Subhedral*: Grains partly enclosed by recognizable crystal faces.
- *Anhedral*: Grains with no visible crystal faces.

Careful study of the angles between crystal faces on euhedral and subhedral grains can be a guide to the symmetry of the crystals present and thus to their identity. Of equal importance in this respect are the relative dimensions of grains, especially those showing crystal faces. Three main types of morphology for mineral grains can be recognized, as follows:

1. *Granular* or *equant*: Grains are more or less equidimensional or spheroidal. This shape is common

in crystals of the isometric system such as garnet, which commonly crystallizes as dodecahedra. However, anhedral equant grains may crystallize in any system. Quartz, for example, is usually anhedral and equant in quartz-bearing rocks such as granite, quartzite, or sandstone, although its crystal symmetry is trigonal. In vesicles, quartz has the characteristic prismatic-rhombohedral morphology.

2. *Tabular* or *platy*: Crystals have two, more or less equal dimensions and one significantly shorter dimension. This shape is most typical of crystals with a sheet-like arrangements of atoms. Mica and chlorite are good examples of rock-forming minerals that commonly adopt this habit, generally in metamorphic rocks, where a preferred orientation of the grains contributes to the texture called slaty cleavage or schistosity (also known as foliation). Graphite and molybdenite are examples of ore minerals with similar structures and habits. The tabular habit is rarely found in cubic crystals, but it can arise in most other crystal systems.

3. *Prismatic, acicular*, or *fibrous*: Crystals are more or less rod shaped, with one long dimension and two roughly equal shorter dimensions. This habit is typical of crystals with one principal axis of symmetry (3-, 4-, or 6-fold) bounded by prisms, sets of identical crystal faces parallel to the long dimension of the grain. Trigonal and hexagonal crystals typically have 3, 6, or 12 faces, while tetragonal crystals typically have 4 or 8 faces. However, under specific growth conditions, such crystals become elongated perpendicular to the principal axis of symmetry. Quartz, for example, when growing into a cavity, will often form almost perfect hexagonal (six-sided) prisms; in other cases, it might form spherulites consisting of fibers (this variety is called chalcedony).

6.3 Color, streak, and luster

The way in which an incident light beam interacts with a mineral is expressed by three different properties:

1. *Color*: The color of a mineral as directly observed can be so characteristic as to be an important aid in identification, but it can also be misleading. Most minerals, even the rock-forming silicates in which isomorphous series are present, can show a bewildering variety of colors. In general, color by itself is insufficient to permit identification, but it can be useful when taken together with other properties.

There are many examples and we will only point out a few of them. If you have seen a sample of yellow sulfur (S) (Plate 1a; note that plates are all collected in the center of the book), red cinnabar (HgS) (Plate 1b), green malachite $Cu_2CO_3(OH)_2$ (Plate 1c), or blue turquoise $CuAl_6(PO_4)_4(OH)_8$ $4H_2O$ (Plate 1d) only once, you will easily recognize these minerals based on their color. In other cases color can be misleading. Corundum (Al_2O_3) occurs as white/gray crystals (Plate 1e) or as red ruby (Plate 1f) or blue sapphire (Plate 1g), depending on trace elements in the crystal structure. The same is true for beryl ($Be_3Al_2(SiO_3)_6$), which can be blue as aquamarine, green as emerald, or pink as morganite. Also quartz (SiO_2) can display a wide range of colors. Often it is clear (Plate 2a), but can be brown as smoky quartz (Plate 2b), violet as amethyst (Plate 2c), pink as rose quartz, or yellow as citrine. Color will be discussed in some detail in Chapter 15.

2. *Streak*: Many minerals show a characteristic color when reduced to a fine powder, regardless of the color they show in a bulk specimen. An example is hematite, which can range from red to metallic gray in hand specimens but always produces a dark red-brown powder on grinding. This property is called *streak* because the simplest way to produce a fine powder from most minerals is to scratch a sharp edge of a specimen across a rough ceramic plate (a streak plate), leaving a trail of dust-like powder. A colorless mineral, or one that is harder than the streak plate, scratches the plate to leave a trail of white powder. Most rock-forming silicates are harder than a streak plate, and the streak is of little help in their identification. Streak is most useful and diagnostic in the study of opaque ore minerals (e.g., sulfides and oxides), particularly those with metallic or submetallic luster (see below). Some of these ore minerals have a metallic or shining streak, whereas others have a nonmetallic streak. Following is a list of some common opaque minerals and the color of their streaks:

3. *Luster*: The term luster is used to describe the character of the light reflected from the surface of

Metallic streak	
Gold-yellow	gold
Silver-white	silver, arsenic, bismuth
Copper-red	copper
Grayish-white	platinum

Nonmetallic streak	
Black	pyrolusite, graphite, covellite, ilmenite, magnetite
Greenish-black	chalcopyrite, pyrite
Brownish-black	pyrite, marcasite
Gray-black	chalcocite, bornite (pale), galena, pyrrhotite, covellite, marcasite, arsenopyrite (dark)
Gray	antimony, graphite, stibnite, molybdenite (bluish to greenish)
Brown	sphalerite (pale to colorless), rutile (pale)
Brownish-red	cuprite (shining), hematite, manganite
Brownish-yellow	goethite
Red	cinnabar, hematite (dark)
Orange-red	realgar
Yellow	orpiment (pale)
Green	malachite (pale)
Blue	azurite (pale), lazurite

a mineral and depends on the refractive index. Terms in common use refer the luster of a mineral to that of some common material (e.g., *metallic* luster, *waxy* luster, *earthy* luster, and so on). The main division is into metallic and nonmetallic and the latter can have many different expressions. The luster of some minerals is as follows:

(a) *Metallic/submetallic*

Metallic: silver, mercury (liquid), bismuth, galena, molybdenite, stibnite, chalcocite, graphite, covellite, bornite, copper, pyrrhotite, chalcopyrite, pyrite, marcasite, gold.

Submetallic (semi-metallic): hematite, ilmenite, rutile, pyrolusite, manganite, goethite, wolframite, magnetite.

(b) *Nonmetallic*

Adamantine (diamond-like): cassiterite, zircon, sphalerite, diamond, scheelite, realgar, cinnabar, cuprite, wulfenite.

Resinous (oily, greasy, waxy): apatite, nepheline, halite, gypsum, serpentine, talc, sulfur, orpiment, sodalite, chalcedony.

Vitreous (glassy): quartz, opal, amphibole, pyroxene, olivine, feldspar, barite, celestite, anhydrite, beryl, garnet, tourmaline, dolomite, calcite, fluorite, spinel, cordierite, kyanite, epidote, apatite, topaz. Many other rock-forming minerals are in this group.

Earthy: graphite, goethite, limonite, clay minerals, anglesite, magnesite, hematite, chlorite.

Minerals with a perfect cleavage may exhibit a conspicuous *pearly* luster, which is usually seen only on the surface of cleavage. Note that some minerals can occur with a variety of lusters or show different luster on crystal faces and broken surfaces.

6.4 Mechanical properties

With little more equipment than a hammer and a pocket knife, a geologist can examine some of the mechanical properties of a mineral. The most obvious and easily determined of these are described below.

Hardness

Hardness is defined loosely as resistance to indentation or abrasion. Quantitative tests for hardness can be made under a microscope using a diamond indenter and are an important part of the mineralogy of opaque ore minerals. The average geologist studies merely relative hardness, expressed by the resistance offered by a smooth surface of a mineral to scratching by a sharp edge on a material of known hardness. Around 1800 the German mineralogist Friedrich Mohs devised a relative scale that compares the hardness of some standard minerals (Table 6.1) and this scale is still used universally.

The Mohs scale is nonlinear as compared with an absolute scale such as the Vickers indentation hardness (Figure 6.1), but each mineral on the Mohs scale will scratch minerals with a lower number and can be scratched by minerals with a higher number. For example, most rock-forming silicates are 6 or harder on the Mohs scale, and thus will scratch minerals of hardness 5 or less. The position of an unknown mineral on the scale can be established by such scratching tests. Sets of these minerals (excluding diamond) are commercially available at modest cost, but, in

Table 6.1 Mohs hardness scale

Scale no.	Mineral
1	Talc
2	Gypsum
3	Calcite
4	Fluorite
5	Apatite
6	Orthoclase
7	Quartz
8	Topaz
9	Corundum
10	Diamond

practice, a rough determination of hardness is generally made using the simpler scale below:

Material	Hardness
Thumbnail	about 2
Copper penny	3 to 4
Blade of pocket knife, window glass	5 to 6
Quartz or streak plate	6 to 7

When testing the hardness of a mineral, the following practices should be observed:

- First, examine the surface to be scratched with a hand lens before the test to detect the presence of old scratches.
- After the test, brush away any powder and use the hand lens on both surfaces to confirm the presence of a scratch and on which surface it lies.
- Make certain that the material is really scratched and not just granulated on pre-existing open fractures or cleavage surfaces (see below).

After the test, a few important points should be borne in mind when interpreting the results:

- Remember that minerals of similar hardness can simultaneously scratch each other.
- A crystal may be slightly harder on one face than on another, and harder in one direction than in another on the same face.

Cleavage and fracture

Most minerals break when struck sharply with a hammer and the nature of the surface so produced

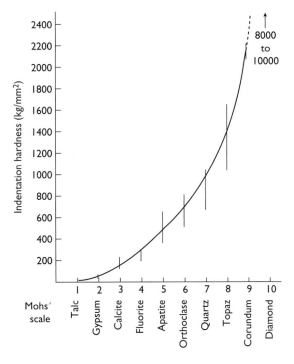

Fig. 6.1 Relationship between the Mohs relative hardness scale and the Vickers indentation hardness (see Tabor, 1954).

can be a diagnostic property. Two types of failure surfaces are recognized:

1. *Cleavage*: Some crystals break in one (Figure 6.2a) or more smooth plane surfaces (Figure 6.2b–e) whose orientation is determined by the regular atomic structure of the crystal. Such surfaces, called cleavages, occur along planes of weak atomic cohesion, and they reflect the internal symmetry of the crystal structure in much the same way as do crystal faces. Although cleavage surfaces superficially resemble crystal faces, they can be generally distinguished by the fact that they occur in parallel families spaced on the atomic scale. The crystal can be broken at any point along a cleavage plane, whereas most crystals do not necessarily break in the direction of a crystal face. Also, when observed through a hand lens, most cleavages display a "stepped" surface where the actual plane of breakage shifts from one level to another in the crystal (Figure 6.2c,e). If a mineral has a single excellent cleavage (graphite, molybdenite, and mica are examples), flakes are produced (Figure 6.2a). If it has two excellent cleavages, the mineral

(a)

(b)

(c)

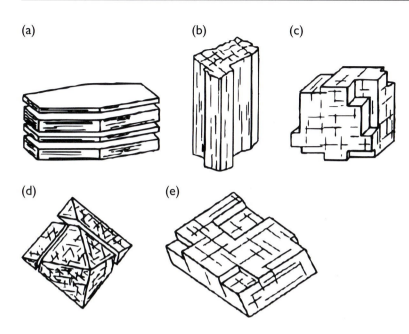

(d)

(e)

Fig. 6.2 Examples of cleavage in minerals. (a) Single cleavage causing a crystal to break up into flakes as in mica. (b) Two dominant cleavages produce fibrous or prismatic fragments, as in amphiboles. (c) Three cleavages at right angles, as in halite, produce cubic fragments. (d) Octahedral cleavage in fluorite. (e) Symmetrical cleavage in trigonal calcite. The 3-fold symmetry axis is vertical.

breaks into fibers (sillimanite and wollastonite are examples, Figure 6.2b).

Many minerals cleave in more than one plane. If the cleavages are of the same kind of form (i.e., they are related by symmetry operations of the crystal structure), the angles between them can be a guide to the symmetry and should be estimated. For example, halite has cleavage planes parallel to the cube faces at right angles (Figure 6.2c) and fluorite has an octahedral cleavage (Figure 6.2d). Calcite has a cleavage consisting of three identical planes inclined at about 70° to one another, symmetrically arranged about the 3-fold rotation axis of the crystal (Figures 6.2e and 6.3). This pattern can be seen clearly on a cleavage rhomb of Iceland spar (a glass-clear variety of calcite). Albite, on the other hand, has two cleavage planes that are unrelated by symmetry operations. For this reason the two albite cleavages are not identical, with one being more highly developed than the other and they are at a slightly oblique angle.

Cleavage surfaces should not be confused with other planar surfaces of breakage found on some crystals, called *partings*. These are discretely spaced surfaces separating parts of a crystal in which the structure differs in orientation (e.g., twin-lamellae), and surfaces saturated with inclusions (e.g., exsolution lamellae or fluid inclusions), or weakened by stress (e.g., in metamorphic quartz). The latter

Fig. 6.3 Calcite crystal from Mt. Baldwin, California, USA, displaying rhombohedral cleavage steps. Width 100 mm.

types of surface are not always developed on a given mineral and, unlike cleavage surfaces, are not pervasive on the atomic scale.

2. *Fracture*: Many mineral crystals (e.g., quartz) show only poorly defined cleavage or none at all. When such crystals are struck they break on generally irregularly oriented curved surfaces decided more by the stress distribution in the crystal at the time of rupture than by the atomic structure of the mineral. Such arbitrary surfaces of breakage are termed fractures and are classified on the basis

of their general appearance. *Irregular* fracture is typical of minerals without cleavage such as apatite. A special type is *conchoidal* fracture, with smooth curved surfaces as in quartz and opal. *Stepped* fracture occurs in minerals with several cleavages, for example halite, galena, and feldspars. *Splintery* fracture is observed in fibrous minerals such as actinolite and anthophyllite.

6.5 Density and specific gravity

Density is the mass of a unit volume of a material, generally expressed in grams per cubic centimeter (see also Section 12.4). The *specific gravity* is the ratio of the weight of a body and the weight of an equal volume of water, a dimensionless quantity. For practical purposes, density and specific gravity can be taken as numerically equal. Mineral specimens with widely different densities and similar sizes can be distinguished easily by "heft", i.e., by the feeling of weight in a specimen bounced lightly in the hand. Try to gauge relative heft with specimens of known density and acquire a feeling for minerals with low (1.5–2.5 g/cm^3), intermediate (2.5–3.0 g/cm^3), and high (above 3.0 g/cm^3) densities. As may be expected, the density is high for close-packed structures as in metals, sulfides, and oxides, and low for open polyhedral structures as in borates, sulfates, carbonates, and silicates.

6.6 Other properties

Certain minerals have obvious and diagnostic properties not covered in the above discussion. Examples are:

Feel: Talc and serpentine feel slippery or "soapy".

Taste: Water-soluble minerals such as halite have unmistakable tastes when touched lightly on the tongue.

Odor: Some freshly broken or lightly heated minerals emit characteristic odors. This property is especially true of some sulfur- and arsenic-bearing minerals.

Magnetic properties: A few minerals – notably magnetite, and some varieties of ilmenite and pyrrhotite – are strongly magnetic. This property can be tested with a hand magnet.

Fluorescence: In ultraviolet light many minerals fluoresce (named after the mineral fluorite) visibly with characteristic colors (see the discussion in Chapter 15 and also Table 15.2). Note that some minerals continue to emit light *after* the exposure to ultraviolet rays; this property is called phosphorescence.

Radioactivity: Minerals can be weakly or strongly radioactive. Radioactivity can be measured with a Geiger counter.

Change of color: Some minerals (e.g., colored varieties of quartz and sodalite) may temporarily or permanently change their color when exposed to radiation, i.e., heat, X-rays, photons, etc.

Tarnish: Some opaque minerals (such as sulfides) react with atmospheric moisture, oxygen, and carbon dioxide to become coated with a thin film of brightly colored secondary minerals (e.g., bornite).

Color effects associated with microstructures: Several of these effects such as iridescence associated with specific mineral types will be discussed in Chapter 15.

Chemical composition: In some cases simple chemical tests can be performed in the field (e.g., calcite reacts with diluted hydrochloric acid, dolomite with concentrated hydrochloric acid). One can also use alizarin red to distinguish between calcite and dolomite. It reacts to form a red film on the surface of calcite rocks and gives a white film on dolomite.

6.7 Associations of minerals

Most minerals are intimately associated with one another in rocks. Many rocks – particularly those formed by igneous and metamorphic processes, including primary ore deposits – are thought to be more or less in chemical equilibrium at the time of formation, and thus coexisting minerals or associations of minerals are in well-defined chemical relations to one another. Sometimes the identity of an unknown mineral can be closely determined by noting which other minerals occur with it. Whenever you examine a specimen, try to determine all the minerals present and find the origin of the assemblage. For example, galena is often associated with sphalerite, quartz, and calcite in hydrothermal veins.

6.8 Some directions for practical mineral identification

Developing one's skills in identifying minerals in hand specimens can be learned only by conscientious practice and careful observation. The more time you spend examining the minerals and reading about them in reference books, the easier it will be for you to identify them when you encounter them in the field, laboratory, or on a collector's shelf.

Many minerals can be recognized by particular combinations of easily observed or measured physical properties, and it is wise to follow some standard procedures. The most useful and diagnostic of these properties are color, luster, streak, hardness, density (or specific gravity), fracture/cleavage, state of aggregation (including crystallographic forms, if present), and association (other minerals present), as described above. It is useful to acquaint yourself with a mineral set in a reference collection for practicing identification. Study a mineral, read about its principal properties, and try to recognize these properties in reference specimens. You may then try to improve your skills by diagnosing unknown minerals. In addition, try to memorize the chemical formulas of the more common minerals, or at least remember which elements are present. Through keen observation and much experience and training, you will develop your visual memory. Such practice will also help you to find your own methods of mineral identification.

In routine identification, the first step is usually to classify a mineral according to luster into two major groups, metallic and nonmetallic (Figure 6.4, and determinative tables in Appendix 1). The next step is to classify minerals in each of these main groups according to hardness and color (Figures 6.5 and 6.6, with important mineral examples for native elements, halides, phosphates, sulfates, sulfides, sulfosalts, oxides, and carbonates). Finally, one should apply all other diagnostic properties and confirm the identification.

For silicates the procedure is different. It is best to study them in the following order: (1) ortho- and ring silicates, (2) chain silicates, (3) sheet silicates, and (4) framework silicates. *Orthosilicates* are often found in well-developed crystals of various colors and have high hardness (≥ 6). *Chain silicates* rarely form ideal crystals. More often they occur as prismatic grains that have a cleavage along the elongation direction. Their color is usually green of various intensities (to black varieties) and shades. However, spodumene, tremolite, and wollastonite are colorless, rhodonite is pink, and glaucophane is grayish-blue. The hardness of all the chain silicates is approximately the same, ranging from 5 to 6. *Sheet silicates* are distinguished

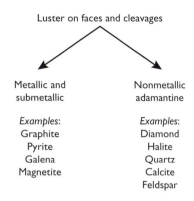

Fig. 6.4 Main division of minerals into those with metallic and nonmetallic luster (with some examples).

from all other silicates by the platy shape of their crystals and by their excellent basal cleavage. The sheet silicates differ in color (most typically, they are green, brown, colorless, or pink), and have a low hardness ranging from 1 (talc) to 3 (muscovite). *Framework silicates* usually are colorless or slightly colored. They have a vitreous luster and medium hardness (about 5–6).

In the process of these practical exercises, begin to use a notebook in which you enter mineral descriptions. It is enough to allot just half a page to a mineral or a page to a deposit. Relate minerals to their geological occurrence and classify them into minerals observed in igneous rocks (ultrabasic, basic, acid, alkaline volcanic rocks), pegmatites (mica-bearing, quartz-bearing, etc.), skarns (developing at a contact of igneous rock with limestone or dolomite), deposits of the weathering crusts (laterites, bauxites, and oxidized ores), chemical (from real and colloidal solutions), biogenic, and diagenetic deposits (alteration of sediments during compaction and burial). Describe in detail which features of major minerals you have observed with your own eyes. Slowly integrate your knowledge so that each mineral becomes an entity with a name, a chemical fingerprint (for some remember the formula, for complex ones remember at least a list of major elements), distinguishing features, the geological system in which they occur, and, for many, an industrial application (ranging from metal ore to food additive). As you familiarize yourself with mineralogy, minerals will become much more than dry names. They will emerge as multifaceted building blocks that help us to understand the processes that govern the Earth.

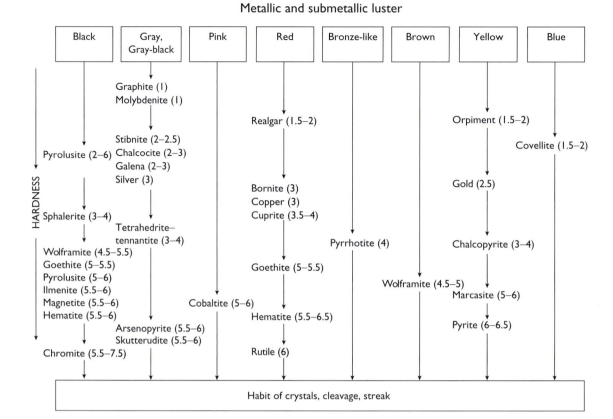

Fig. 6.5 Classification of metallic minerals according to hardness and color, with important examples. (Hardness is given in parentheses.)

6.9 Summary

This chapter is a first stab at mineral identification. It emphasizes macroscopic samples (hand specimens) that are found in the field. Preliminary identification is done with minimal instrumentation such as a hand lens, a pocket knife, and a hand magnet. Later in the book more advanced laboratory techniques will be introduced, most importantly the petrographic microscope (Chapters 13 and 14) and X-ray diffraction (Chapter 11). In hand specimens, become familiar with the following properties:

- Morphology: euhedral, subhedral, anhedral
- Habit: equant, platy, prismatic
- Cleavage: angle between cleavage planes
- Color, streak, luster
- Density, heft
- Ten minerals on the Mohs hardness scale (finger nail, knife blade)

Test your knowledge

For this section you need to get some hands-on experience with minerals. Unless you are reading this book as part of a mineralogy class with a laboratory, here are some suggestions:

1. Visit your closest mineralogy collection, for instance, at a university or a museum. Look at the samples in view of the characteristic features described in this chapter, such as color, morphology, and aggregation.
2. Some of the most interesting mineral specimens are displayed at various gem and mineral shows around the world; a few notable examples are the Tucson Gem and Mineral Show in Arizona (February), Saint-Marie aux Mines Mineral, Gem and Fossil Show in Alsace, France (June), and the Munich Mineral Show in Germany (October). From the local calendar of cultural events or from

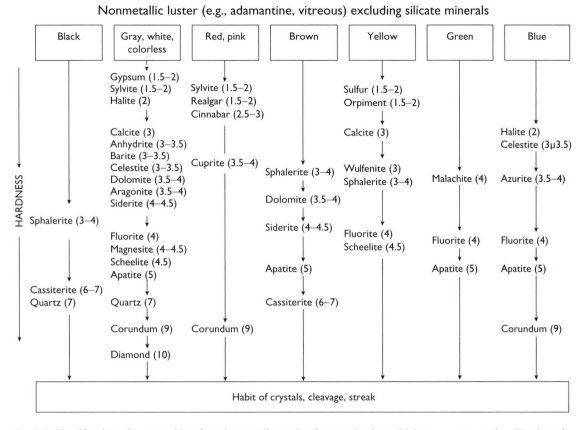

Fig. 6.6 Classification of nonmetallic minerals according to hardness and color, with important examples. (Hardness is given in parentheses.)

internet resources, find out whether there is an annual mineral show in your area.

3. Try to obtain some of the commonest minerals, either in the field, at a mineral show, or at a flea market. Halite (table salt), calcite, feldspar, quartz, magnetite, pyrite, and mica are good candidates for beginners. Perform some of the tests on them described above, such as hardness, cleavage, heft, and streak (use an old electrical fuse as a streak plate).

4. Join some of the local mineral societies that arrange fieldtrips.

Further reading

Bishop, A. C., Woolley, A. R. and Hamilton, W. R. (1999). *Cambridge Guide to Minerals, Rocks and Fossils*, 2nd edn. Cambridge University Press, Cambridge.

Hurlbut, C. S. and Sharp, W. E. (1998). *Dana's Minerals and How to Study Them*, 4th edn. Wiley, New York.

Johnsen, O. (2002). *Minerals of the World. Princeton Field Guides*. Princeton University Press, Princeton, NJ.

Medenbach, O. and Medenbach, U. (2001). *Mineralien: Erkennen und Bestimmen*. Steinbach's Naturführer. Mosaikverlag, Steinbach, Germany.

Medenbach, O. and Wilk, H. (1986). *The Magic of Minerals*. Springer-Verlag, Berlin.

Pellant, C. (2002). *Rocks and Minerals*. Smithsonian Handbooks, Washington, DC.

Polk, P. (2012). *Collecting Rocks, Gems and Minerals: Identification, Values and Lapidary Uses*, 2nd edn. Krause Publications, Iola, WI.

Pough, F. H. (1996). *A Field Guide to Rocks and Minerals*, 5th edn. Houghton Mifflin, New York.

Prinz, M., Harlow, G. and Peters, J. (1978). *Simon Schuster's Guide to Rocks and Minerals*. Simon and Schuster, New York.

Schumann, W. (2008). *Minerals of the World*, 2nd edn. Sterling Publishing, New York.

Important internet links: webpages with mineral properties, images, and identification

Mineral galleries, minerals for sale, Dan and Diana Weinrich (www.danweinrich.com)

Mineral galleries, minerals for sale, Arkenstone (www.irocks.com)

Mineral exhibits and events in France (www.leregne mineral.fr)

Mineral properties, images, and links to mineral dealers (www.mindat.org)

Mineral magazine, Germany (www.mineralien-welt .de)

Mineralogical Record, Journal (www.minrec.org)

Mineralogy database with properties and images (www.webmineral.com)

www.geosystems.no

www.lapis.org

www.minbook.com

www.mineralnews.com

www.rocksandminerals.org

Part II | Symmetry expressed in crystal structures and morphology

7 | The concept of a lattice and description of crystal structures

Minerals originally fascinated scientists because of their regular morphology and symmetry. This was the basis for the discovery of the lattice, i.e., a periodic repetition of units in three dimensions. Unit cells and the 14 different lattice types will be introduced as well as rational indices to describe lattice planes (*hkl*) and lattice directions [*uvw*]. Parameters describing the crystal structure will be defined.

7.1 Discovery of the lattice

In Chapter 2 we saw how interatomic bonding forces determine the internal structure of crystals. For example, in the case of close-packing in a metal, atoms repeat periodically in three dimensions. A translation of an atom by an interatomic distance superposes it on the next atom. In ionic and covalent structures, the atomic arrangement is more complicated and ion groups or molecules repeat, rather than individual atoms. A regular internal crystal structure was proposed in the eighteenth century, based on some unique macroscopic properties of crystals. The concept of a periodic crystal structure was developed from observations of the plane faces that are observed on freely growing crystals, the characteristic angles between faces, and the regular cleavage that is observed in many minerals. Only much later, in the twentieth century, was it determined that this regular and periodic internal structure was due to the regular bonding forces between atoms.

In 1669, Nicolas Steno discovered that the angles between corresponding faces of quartz crystals are always the same, irrespective of the actual size of the faces. At that time science moved at a slow pace. Over 50 years later, in 1723, Michael A. Cappeller observed that each mineral species has a characteristic set of interfacial angles (these angles can be measured with a protractor) and proposed a *law of constant interfacial angles* for minerals in general (Figure 7.1). For example, a magnetite crystal may occur as a perfect octahedron (Figure 7.2a) or as

distorted octahedra (Figure 7.2b,c). In either case the angles between faces are identical. In 1773 Torbern Bergmann studied the regular cleavage of calcite. If a calcite crystal is crushed, it breaks into small fragments that take the shape of little rhombohedra. If one of these little rhombohedra is crushed again, it

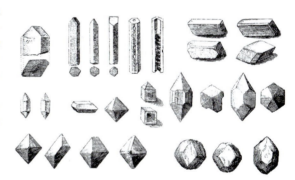

Fig. 7.1 Crystals with different morphology but equal angles (from Cappeller, 1723), among them quartz and calcite.

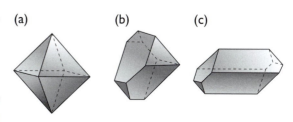

Fig. 7.2 Regular (a) and distorted octahedra (b,c). All of the corresponding faces are parallel, but their sizes vary.

forms a set of even smaller rhombohedra of microscopic scale. Similar observations were made for
halite, except that the small fragments were not
rhombohedra but rather cubes. Bergmann supposed
that these small fragments might be the building
blocks of crystals. A few years later, in 1784, René
J. Haüy came up with an ingenious theory to explain
both the growth morphology and the regular cleavage of crystals. He proposed that all crystals are built
up from elementary parallelepipeds (Figure 7.3a),
filling space without gaps (a parallelepiped is a polyhedron consisting of three pairs of parallel faces).
This model explains, for example, the morphology
of a dodecahedron as faces bordering stacks of cubic
parallelepipeds (Figure 7.3a) and we are very much
reminded of Haüy's concept from a modern scanning
electron microscope image of the growth pattern in a
europium–tellurium alloy (Figure 7.3b). Haüy
claimed that all external crystal faces are such planes
bordering the stack of parallelepipeds. This concept
remained a hypothesis for over 100 years, but today
we know that it is basically correct. Parallelepipeds
are idealized units in the crystal structure which we
called *unit cells* in Chapter 2. Each cell may contain a
single atom (Figure 7.4a), or a group of different
atoms and ions (Figure 7.4b). The macroscopic crystal is then an assembly of the elementary cells,
repeating periodically in three dimensions.

In order to better understand the array of parallelepipeds, it is useful to abstract each cell and its content
by a point. The crystal can then be idealized by a
three-dimensional periodic array of points (Figure
7.4c). The law of interfacial angles now becomes obvious: crystal faces are discrete planes containing points
of this array, and crystal edges are lines of points.
Angles are therefore determined by the geometry
of the array and are not arbitrary. Ludwig August
Seeber, in 1824, introduced the term *lattice* for this
regular array of points. A lattice is periodic in three
dimensions and has the property that the environment
of each point is identical and in the same orientation.
If you could place yourself in a lattice point and
observe all of the neighboring points, the perspective
would be the same no matter which point you picked.
The unit cell is the parallelepiped formed by eight
lattice points.

To fully describe the internal structure of a crystal,
two pieces of information are needed. First, one must
characterize the geometry of the unit cell and therefore the lattice. Second, one must identify the content

(a)

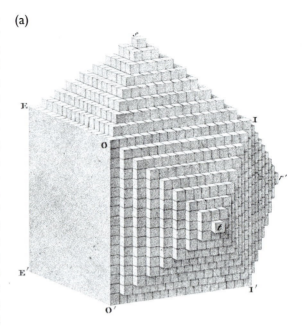

(b)

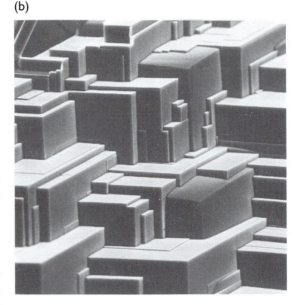

Fig. 7.3 (a) Haüy's (1801) notion that crystals such as
cubic halite (NaCl) are built up of elementary
parallelepipeds. In this original image, the
parallelepipeds are cubic cells that can produce a
macroscopic dodecahedron. (b) A modern
scanning electron microscope image of growth in a
europium–tellurium alloy illustrates Haüy's concept,
although the growth pattern is less regular than in
Haüy's idealized figure (courtesy R. Wessicken). The
width of the image is 0.15 mm.

(a)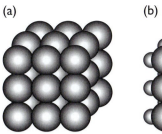

(b)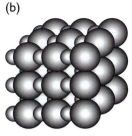

(c)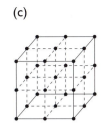

Fig. 7.4 (a) A cubic crystal with repeating atoms, (b) with repeating groups of different atoms, and (c) with an idealized periodic structure that represents the crystal by a simple array of points.

of the unit cell, i.e., the type and position of the atoms or ions. The atomic arrangement in the unit cell is called the *crystal structure*. The macroscopic crystal is then obtained by a periodic repetition of the unit cell through translation.

7.2 Symmetry considerations

Early mineralogists such as Steno and Haüy noted an extraordinary feature of crystals, namely their distinctive symmetry. While symmetry is a universal principle (see Box 7.1), it is most succinctly expressed in the morphology and structure of crystals, and, in fact, symmetry theory was developed largely by crystallographers. We refer to an object as being symmetrical if it can be moved in some way and yet appear exactly as before. For example, in Figure 7.5a a hand is shown. A single hand is asymmetrical and represents the symmetry *motif* for the remaining parts of the figure. In Figure 7.5b the hand is repeated by translation, forming an array of three identical (or *congruent*) hands. We call this pattern with a set of three hands symmetrical because if we translate the first hand by a vector *t*, it superposes exactly on the second hand. *Translation*, symbol *t*, is a symmetry operation producing a simple shift of the motif.

(a)   (b)   (c)   (d)

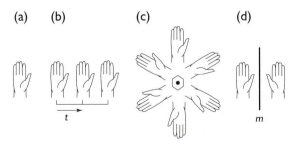

Fig. 7.5 (a) A hand as a symmetry motif. (b) A symmetrical array of three hands repeated by translation. (c) Six hands repeated by 60° rotations. (d) A pair of hands repeated by mirror reflection. This produces a left hand from a right hand.

Figure 7.5c shows a ring of six identical hands. In this case applying 60° rotations around an axis perpendicular to the ring (and the page) generates the set of six hands. The seventh rotation superposes the next hand exactly on the first one. *Rotation* is another symmetry operation, and a *rotation axis* is the corresponding symmetry element. In Figure 7.5c we have a 6-fold rotation, the rotation angle (ϕ) is 60°, and the symbol is a hexagon. There are other rotation angles, as we will soon learn.

Yet another way to repeat an object is shown in Figure 7.5d. In this case the pair of hands is related by a mirror reflection. The symmetry operation is called *mirror reflection*, symbol *m*, and the corresponding symmetry element a *mirror plane*. Contrary to the pattern with translated or rotated hands, mirror reflection does not produce an identical repetition. Each element of the symmetric motif on the right-hand side is an equal distance from the mirror plane on the left-hand side, in this case producing a left hand from a right hand. Such a repetition is called *enantiomorphous*, to distinguish it from a congruent repetition in the cases of translation and rotation.

Another important aspect of symmetrical patterns is that the final pattern is formed by a continued repetition until the initial motif is itself reproduced. For translation, this property is obvious because the translation operation always repeats the motif. In the case of mirror reflection, this return to the original motif is accomplished by the initial reflection from left to right and then back again from right to left. In the case of rotation we must return to the original hand after a finite number of incremental rotations. This makes it necessary that the rotation angle ϕ be an integral submultiple of a complete rotation, that is, $\phi = 360°/n$, where $n = 1, 2, 3, \ldots$ For example, a 4-fold rotation ($n = 4$) has a rotation angle $\phi = 90°$.

Let us now return to the crystal structure of halite, examining it in terms of its symmetry elements (Figure 7.6). We easily find 4-fold rotation axes

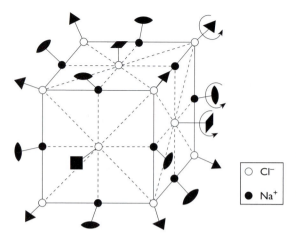

Fig. 7.6 The crystal structure of halite (NaCl) with mirror planes (represented by dashed lines) and rotation axes added. Symbols for rotation axes are discussed in the text. The solid lines outline the unit cell.

perpendicular to the faces of the cube (■), but there are also 3-fold rotation axes along the body diagonals (▲). In addition, there are 2-fold rotation axes (◖) along edge diagonals and a whole set of mirror planes (dashed lines). Finally, there is translation, implicitly dictated by the alternating sodium and chloride ions in the crystal structure. The cause of this symmetrical array lies in the balance of bonding forces (Chapter 2). We will return to a more systematic discussion of symmetry in crystals in the next chapter, but for now we will apply some of these considerations to our discussion of lattices.

7.3 The unit cell as the basic building block of a crystal

At the beginning of the chapter, Haüy's concept of a unit parallelepiped, called a unit cell, was introduced. There are a number of constraints one must consider in choosing a unit cell. In Chapter 2 we discussed some simple crystal structures such as halite (NaCl), which contains regularly alternating chloride and sodium ions. Let us make a list of general considerations for unit cells and how these apply to the NaCl example:

1. *A unit cell must be large enough to contain an integral number of formula units.* Thus, the unit cell of halite must contain at least one sodium and one chloride ion.
2. *Each corner of a unit cell must be identical, with an identical environment.* If this were not the case we could not repeat the unit cell by translation to form a periodic crystal. If we place Cl^- in one corner of the unit cell of halite, Cl^- must occupy all corners.
3. *A unit cell must express the symmetry of the atomic relationship.* In the case of the NaCl structure, we have seen above that the symmetry is expressed in rotation axes and mirror planes (Figure 7.6). A unit cell in the shape of a cube, with either Cl^- or Na^+ in the corners, best expresses this symmetry.

Unit cells are classified based on their symmetry. The three 4-fold rotation axes at right angles in the NaCl unit cell identify the unit cell as *cubic* because the unit parallelepiped has the shape of a cube (Figure 7.6). (Also characteristic for the symmetry of a cube are

---

Box 7.1 Background: Symmetry in art and nature

While symmetry is most conspicuous and is clearly expressed in crystals, symmetry is omnipresent. Take, for example, the prehistoric polychrome plate from Mesopotamia in Figure 7.7a, showing a 4-fold rotation axis and mirror planes. This archetypal symmetrical pattern was used much later in pottery of early Christianity and perfected in Islamic art. While cultures in the Western world on the whole preferred mirror symmetry, Native Americans almost exclusively used rotational symmetry elements, as displayed in a pre-Columbian Mimbres bowl from New Mexico (Figure 7.7b). The artist who is best known for the use of symmetry in graphics is M. C. Escher, whose drawings include not only rotational and mirror symmetries but also translational symmetries and periodic arrangements of motifs having many similarities to crystal structures (Figure 7.7c). Greek temples emphasize translation by the repetition of pillars; Romanesque churches have almost perfect mirror symmetry. Symmetry is not restricted to pottery, paintings, or architecture. Johann Sebastian Bach also used it extensively in his compositions. Throughout Bach's *Art of the Fugue*, for example, we find translations and mirror reflections that were intentionally applied to achieve artistic effects (Figure 7.7d).

Box 7.1 (cont.)

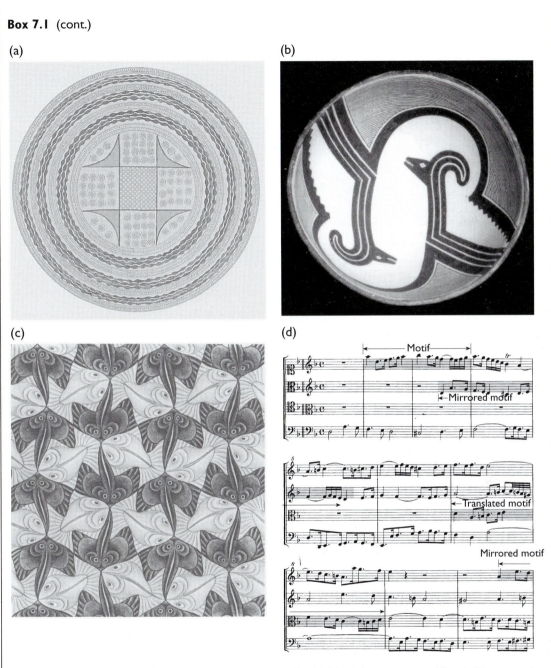

Fig. 7.7 Symmetry in art. (a) Polychrome plate from Arpachiyah with Maltese square, prehistoric Mesopotamia, Halaf Period, 7000–5000 BC (from Mallowan and Cruikshank, 1933). (b) Prehistoric Mimbres bowl from southern New Mexico, with bighorn sheep displaying 2-fold rotational symmetry, AD 1000 (courtesy Maxwell Museum of Anthropology, University of New Mexico, Albuquerque, New Mexico; see also Brody, 1980). (Photo: C. Baudoin.) (c) Complex rotational and translational symmetries in the graphic art of M. C. Escher (from MacGillavry, 1976). (d) Translational and mirror symmetries in the *Art of Fugue* by Johann Sebastian Bach (*Contrapunctus 6 a 4 in Stylo Francese*). The basic motif is identified.

Continued

Box 7.1 (cont.)

(a)

(c)

(b)

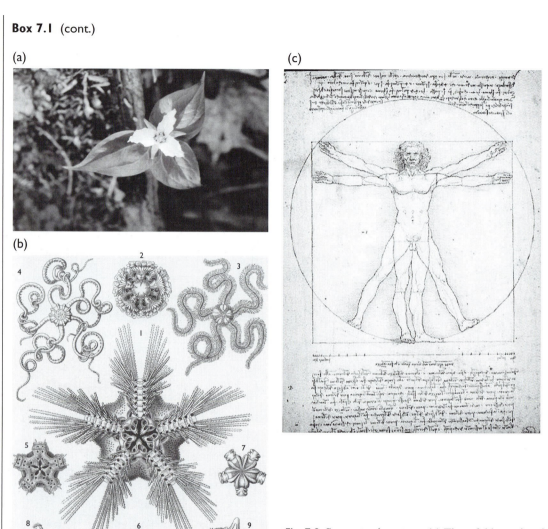

Fig. 7.8 Symmetry in nature. (a) Three-fold rotational symmetry in leaves and flowers of a *Trillium* lily (photograph E. H. Wenk). (b) Five-fold rotational symmetry in starfish (from Haeckel, 1904). (c) Mirror symmetry in the human body (drawing by Leonardo da Vinci).

In the natural world, symmetry is present not only in minerals but also in living organisms. Examples include rotation axes in plants, such as a 3-fold axis in a *Trillium* lily (Figure 7.8a), or a 5-fold axis in a starfish (Figure 7.8b), and mirror planes in vertebrates (Figure 7.8c). Interestingly in the last two cases the symmetry is only external (the heart being on the left side, and the right and left sides of the brain differing in the case of vertebrates). Carl van Linné (Linnaeus) used symmetry as a guiding principle for his classification of plants.

four 3-fold axes in the directions of the body diagonals.) This is the highest possible symmetry. Such unit cells have equal sides and right angles between edges.

On the other end of the symmetry spectrum are *triclinic* unit cells, where adjoining edges are of different lengths and are at oblique angles to one another (Figure 7.9a). There is a whole set of

(a)

(b)

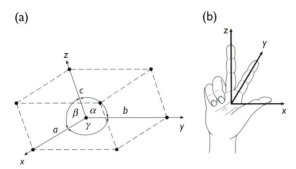

Fig. 7.9 (a) Unit cell with lattice parameters a, b, c, α, β, γ, and crystallographic axes x, y, z that define a right-handed coordinate system. (b) Definition of a right-handed coordinate system.

intermediate symmetries, which is considered in the next section.

In order to describe the shape of a unit cell, we need either three vectors, $\boldsymbol{a}$, $\boldsymbol{b}$, and $\boldsymbol{c}$, or six numbers consisting of three cell lengths, a, b, c (measured in ångströms or nanometers, where 10 Å = 1 nm) and three angles α, β, γ. The angle α is between the edges containing cell lengths b and c, β is between a and c, and γ is between a and b. These six numbers are called *lattice parameters* (Figure 7.9a). The edges of the unit cell provide a convenient coordinate system to describe the geometry of a crystal with axes x, y, and z. The axes x, y, z are chosen such that they form a right-handed coordinate system (Figure 7.9b). (Note that axes x, y, z are directions, while a, b, c are cell lengths. But mineralogists are not entirely consistent with this nomenclature and, as in this book, in some contexts the x-axis is called the a-axis, etc.) The axes x, y, z define a coordinate system that is, in general, not rectangular.

There is an infinite range of shapes and sizes of unit cells due to particular values of lattice parameters a, b, c, α, β, γ, but there are only a very limited number of lattices with different types of symmetry. The reason for this lack of diversity is that the periodic structure of the lattice is only compatible with a few possible rotations. The two-dimensional coverage of a surface with different polygons in Figure 7.10 illustrates that a two-dimensional unit cell can only be a parallelogram (with a 2-fold rotation axis perpendicular to it), a rectangle (with a 2-fold axis and mirror planes), a triangle (with a 3-fold axis), a square (with a 4-fold axis), and a hexagon (with a 6-fold axis), because otherwise space could not be filled uniformly and there would be gaps. In the case of pentagons and

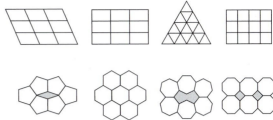

Fig. 7.10 Coverage of a surface with polygons. Note that gaps (shaded areas) exist within pentagons, heptagons, and octagons.

heptagons, the coverage is not periodic, i.e., it does not repeat the motif by translation, while in the case of octagons, the pattern is periodic but has gaps.

Without too much difficulty, the types of possible rotation axes in crystals can be derived geometrically. Consider a lattice plane and, perpendicular to it, an n-fold rotation axis (Figure 7.11a). Symmetry requires that after a rotation of angle $\phi = 360°/n$, all points of the rotated lattice plane coincide with points on the original lattice plane, and that after n rotations the lattice plane is again in the starting position. Now consider a line of points $P_{-2} P_{-1} P_0 P_1 P_2 \ldots$ in the lattice plane with points spaced by a distance a (Figure 7.11b) and apply the symmetry rotation by an angle $\phi = 360°/n$ in the counterclockwise direction, which repeats the line as $P'_{-2} P'_{-1} P'_0 P'_1 P'_2 \ldots$ The line continues to repeat after each rotational increment ϕ. (These lines are not plotted in Figure 7.11b.) Just before rotating back to the initial line again (rotation step $n - 1$), we have a line $P''_{-2} P''_{-1} P''_0 P''_1 P''_2 \ldots$ that is at an angle of $-\phi$ to the initial line. $P'_1 P''_{-1}$ are two lattice points defining a lattice line that is parallel to the original line. In order to satisfy the lattice condition, the distance $P'_1 P''_{-1}$ has to be an integer multiple of the unit cell distance a. In the right triangle $P_0 P'_1 X$ we calculate

$$\cos \phi = \frac{N \times \dfrac{a}{2}}{a} = N/2 \qquad (7.1)$$

where N is an integer, and, since $|\cos \phi| \leq 1$, we find the following solutions for ϕ:

N	−2	−1	0	1	2
$\cos \phi$	−1	−½	0	½	1
ϕ	180°	120°	90°	60°	0° = 360°
n-fold	2	3	4	6	1

(a)   (b)

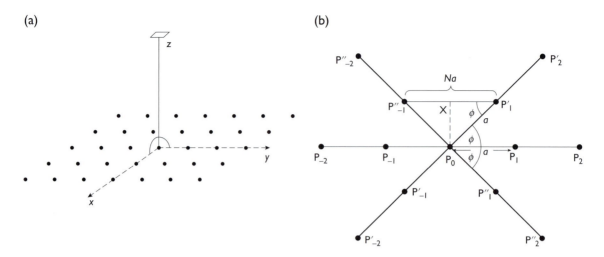

Fig. 7.11 (a) A rotation axis perpendicular to a lattice plane x, y. (b) Derivation of possible rotation angles for lattices. If lattice points along line P_i P_{-i} are rotated, new points P'_i P'_{-i} and P''_i P''_{-i} are produced.

This means that only 1-, 2-, 3-, 4-, and 6-fold rotation axes can occur in crystals. A lattice does not allow for axes with $n = 5, 7, 8$, or higher, as in Figure 7.11b where the angle ϕ is 45° and the array of points is not a lattice. A 1-fold rotation axis means no symmetry, since any object is brought to coincidence after a full 360° rotation.

This derivation for a two-dimensional lattice plane holds for three-dimensional lattices as well. Three-dimensional lattices are simply stacks of identical lattice planes, parallel to each other, with none or some displacement of corresponding points when viewed from above the planes. Consider, for example, the symmetrically different arrangements shown in Figure 7.12. In Figure 7.12a we have a layer with oblique rows of lattice points (closed circles). A single layer always has a 2-fold axis perpendicular to it. But if we add a second layer (open circles), this 2-fold rotation disappears, unless the second layer is exactly above the first (Figure 7.12b), or halfway between lattice points of the first layer (Figure 7.12c). Similarly a layer with a square pattern of lattice points maintains its 4-fold axis only if the second layer is exactly above the first (Figure 7.12d) or is displaced by half a translation (Figure 7.12e). For hexagonal layers, as in a close-packed structure, the 6-fold axis is maintained if the second layer is exactly above the first (Figure 7.12f), but degenerates into a 3-fold axis if a second layer is above the centers of alternating triangles of the first layer (open circles), and a third layer is above the centers of the remaining triangles (crosses in Figure 7.12g).

Thus far we have generated seven symmetrically distinct lattice types with just a single rotation axis perpendicular to a lattice layer: 1-, 2-, 3-, 4-, and 6-fold. If we combine a 2-fold rotation with a mirror plane (Figure 7.12h), we obtain a rectangular array of points. This symmetry is maintained if the next layer is exactly above the first (Figure 7.12h), or diagonally displaced (Figure 7.12i). The same symmetry also applies to a pattern with a rhombus-shaped array with the second layer above the first (Figure 7.12j), or translated (Figure 7.12k). We have just added four more lattices with a 2-fold axis and a parallel mirror plane to our set of seven.

There is one more special type, which can be derived from the square pattern: if we place the second layer above the first, just as in Figure 7.12d, but at the same distance between layers as the distance x between lattice points within a layer, a cubic array results that has three 4-fold axes, with one perpendicular to each face of the cube (Figure 7.12l). This symmetry, with three 4-fold axes, is maintained if the second layer is displaced to the center of each square and at a height $x/2$ (Figure 7.12m). In fact, you may recognize here the bcc structure introduced in Chapter 2 (Figure 2.8c). If we take again a square layer and position it diagonally displaced, but at height $x/\sqrt{2}$ above the first layers, another cubic lattice is generated (Figure 7.12n). This time it is the fcc lattice (Figure 2.8b).

We have just derived, more intuitively than rigorously, the 14 different lattice types that crystals can take.

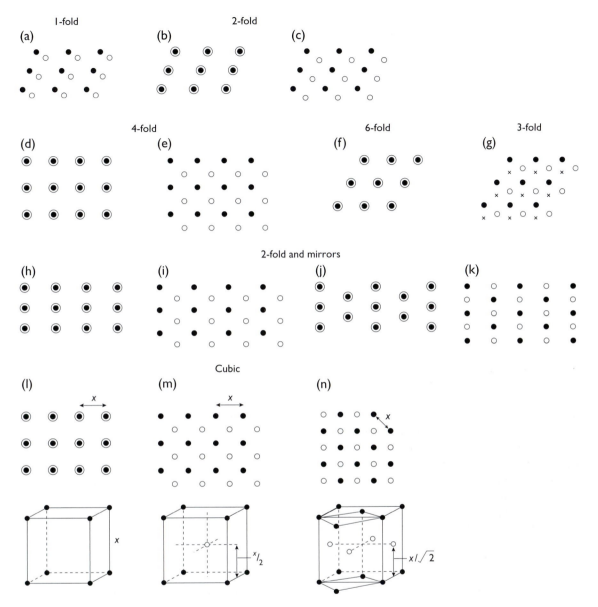

Fig. 7.12 Derivation of 14 symmetrically different lattice types, produced from a two-dimensional lattice layer (solid circles) by adding a second layer (open circles): (a) has no symmetry, (b,c) have a 2-fold rotation axis, (d,e) have a 4-fold rotation axis, (f) has a 6-fold rotation axis, (g) has a 3-fold rotation axis (with three alternating layers), (h–k) have 2-fold rotation axes and mirror planes, and (l–n) have a cubic symmetry. The first layer is solid circles, the second layer is open circles, and, where different from the first, the third layer is crosses.

Each type has a different translational, rotational, or mirror symmetry. These 14 fundamental lattices were first derived by the French mineralogist Auguste Bravais (1850) and are named after him as *Bravais lattices*.

The next task is to define an appropriate unit cell in the lattices. We have already mentioned above that the unit cell must display the symmetry of the atomic

relationship. There are some other conventions for choosing unit cells and lattice parameters (see Box 7.2). These conventions may appear to be trivial details, but they are important in quantitatively describing crystal structures and their physical properties, and in interpreting information about crystals obtained from reference books.

Box 7.2 Additional information: Some conventions for choosing a unit cell

- Symmetry is the primary criterion for choosing a unit cell. The highest possible symmetry is preferable. This is illustrated for a two-dimensional case, where a unit cell is a parallelogram. In the array of points in Figure 7.13a, we pick a cell that conforms to the mirror plane of the pattern (shaded), rather than an oblique cell. In this particular case the unit cell has a point in the center, in addition to the points in the corners, and is said to be *centered*. A unit cell with points only in the corners is called *primitive*.
- Smaller cells are preferred over larger cells and thus, in two dimensions, a small parallelogram with lattice points only in the corners is better than a larger parallelogram that has additional lattice points (Figure 7.13b), unless the symmetry criterion requires such a choice (Figure 7.13a).
- A cell that is the least distorted and its angles closest to 90° is preferred, mainly for reasons of easier visualization (Figure 7.13c).
- In assigning vectors *a*, *b*, *c* to the cell edges, the axes *x*, *y*, *z* must form a right-handed coordinate system (Figure 7.9b).
- Vector *c* (*z*-axis) is along the axis that has the highest rotational symmetry (3-, 4-, and 6-fold axis). An exception is a lattice with a single 2-fold axis (called *monoclinic*), in which case the 2-fold axis is generally chosen along *b* (the *y*-axis) ("second setting").
- In general, cell lengths are ordered such that $b > a > c$ (although for minerals many exceptions exist).
- Angles α, β, if not 90°, are obtuse (>90°) between the positive ends of axes and as close as possible to 90° to have least distortion (Figure 7.13c).

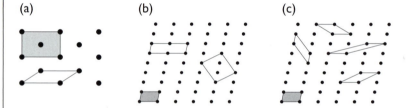

(a)　　　　(b)　　　　(c)

Fig. 7.13 Choices of unit cells for a two-dimensional lattice. (a) A unit cell that displays the rectangular mirror symmetry is preferred. (b) If this does not reduce the symmetry, choose a unit cell with the smallest volume (shaded). (c) Of the different choices, the one with least distortion (angles closest to 90°) is generally preferred.

Unit cells in the 14 Bravais lattices are shown in Figure 7.14. The symmetry of each is different, as is expressed in angles between the axes and relative lengths of the axes (we will explain this in more detail below). Note that seven of the unit cells have lattice points only in the corners. They constitute the seven *primitive lattices* designated by a symbol *P*. The other seven Bravais lattices have the same unit cells as a corresponding primitive cell, but in addition have either lattice points in the center of the cell and are called body-centered (symbol *I*, for "inside"), or a lattice point at the center of one pair of faces or of all pairs of faces and are called face-centered (symbols *C* for centering of a face cut by the *z*-axis, and *F* for centering of all faces). The special rhombohedral cell, which is a variant of the stacking of hexagonal lattice layers (Figure 7.12g), is assigned a symbol *R*.

There are restrictions about adding points to primitive lattices, as is illustrated in Figure 7.15 for the two-dimensional case. In general, adding points (open circles in Figure 7.15a) is not compatible with a lattice, because the two sets of points (open and solid circles) have a different arrangement of closest neighbors. However, if points are added in the center of a face (in two dimensions; see Figure 7.15b) it does not destroy the lattice character. After adding points to a point array, one must determine whether the

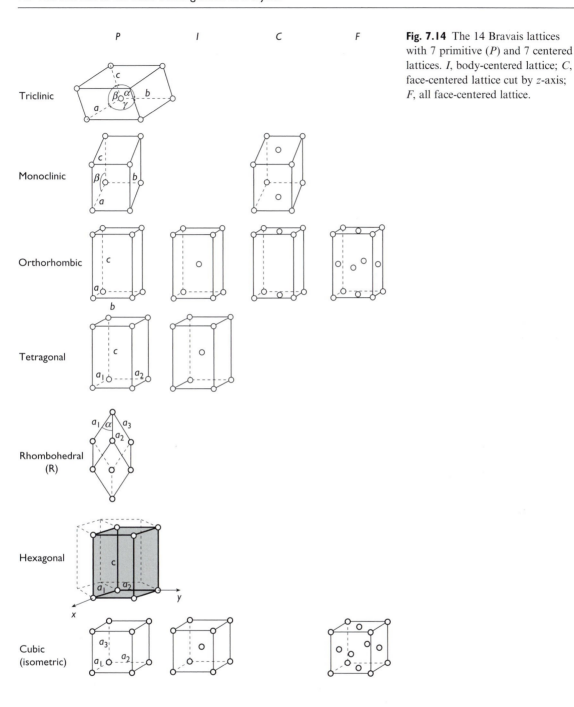

Fig. 7.14 The 14 Bravais lattices with 7 primitive (*P*) and 7 centered lattices. *I*, body-centered lattice; *C*, face-centered lattice cut by *z*-axis; *F*, all face-centered lattice.

symmetry has changed and whether the new array is really different. In the case of the rectangular unit cell (Figure 7.15b), the symmetry is maintained (two mirror planes) and the pattern is different. In the primitive cell, closest neighbors are at right angles (Figure 7.15a); in the face-centered cell, they are at oblique angles (Figure 7.15b). If points are added in the center of a hexagonal unit cell (open circles in Figure 7.15c), the hexagonal symmetry is destroyed and there is no longer a 6-fold axis. Adding a face center to a square unit cell produces a lattice, but this new lattice can be interpreted as a primitive square lattice with a smaller unit cell and axes at 45° (Figure 7.15d). Such a choice is preferred.

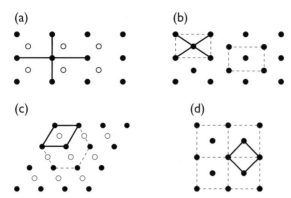

Fig. 7.15 Centering of lattices. (a) Adding arbitrary points (open circles) to a rectangular lattice actually destroys the lattice. (b) Points can be added in the center of a face of a rectangular lattice without destroying the lattice character. (c) Centering of a hexagonal lattice destroys the hexagonal symmetry. (d) Centering the face of a square lattice is equivalent to a primitive square lattice with a smaller unit cell. Open circles denote additional lattice points that are not allowed.

Let us now look more closely at the geometry of the unit cell: there are seven shapes that are distinctly different, and these cell shapes are the basis for the most fundamental classification of crystals into seven systems (Figure 7.14, left column). The most general shape of the unit cell is a parallelepiped with all angles and edge lengths arbitrary. Since all three axes are inclined, it is called the *triclinic system* with no mirror planes or rotation axes. In the *monoclinic system* only one axis is inclined and the other two are at right angles. In the *orthorhombic system* all axes are at right angles, but the cell lengths are arbitrary and nonequal. In the *tetragonal system* all axes are at right angles, two cell lengths are equal (sometimes called a_1 and a_2 to indicate their equivalence) and the third is arbitrary. The *rhombohedral system* has a unit cell in the

shape of a rhombohedron with equal cell lengths and arbitrary but equal angles. This parallelepiped has a 3-fold axis in one body diagonal. The *hexagonal system* has two axes at 120° and the third perpendicular. Two sides are equal in length (often called a_1 and a_2). Finally, the most special unit cell is a cube with all angles equal to 90° and all cell edges of equal length (a_1, a_2, and a_3). This is called the *cubic system* (also known as *isometric*). The lattice parameters and rotational axes for the seven crystal systems are summarized in Table 7.1.

7.4 Representation of lattice lines and planes with rational indices

Macroscopic features, such as faces and edges on a crystal, as well as microscopic structures, such as lattice planes and lattice directions, are essential to describe crystals and we need to have a simple system to describe them. It turns out that the unit cell is a very convenient coordinate reference frame to represent lattice lines and planes. In the following discussion we replace the stack of unit cells by their corners and represent the internal crystal structure by a point lattice. Thus each point of a lattice is, by definition, identical, and, since the lattice extends to infinity in all dimensions, the choice of an origin for the lattice coordinate system is arbitrary.

In a lattice we can identify linear features called directions or lines, and planar features called planes. Figure 7.16a illustrates a *lattice line* as a straight line passing through any two lattice points, and *lattice planes* as planes passing through any three lattice points. A lattice plane contains an infinite number of coplanar lattice lines. A crystal often displays a set of *planar faces* (Figure 7.16b), defining a polyhedron. Faces correspond to lattice planes, and the intersection of two faces defines an edge, corresponding to a

Table 7.1 The seven crystal systems

System	Lattice parameters		Parallelepiped	Rotation axes
Triclinic	$a \neq b \neq c$	$\alpha \neq \beta \neq \gamma \neq 90°$	Parallelohedron	No symmetry
Monoclinic	$a \neq b \neq c$	$\alpha = \gamma = 90°, \beta \neq 90°$	Prism	One 2-fold
Orthorhombic	$a \neq b \neq c$	$\alpha = \beta = \gamma = 90°$	Rectangular prism	Three 2-fold
Tetragonal	$a = b \neq c$	$\alpha = \beta = \gamma = 90°$	Square prism	One 4-fold, two 2-fold
Rhombohedral	$a = b = c$	$\alpha = \beta = \gamma \neq 90°$	Rhombohedron	One 3-fold, one 2-fold
Hexagonal	$a = b \neq c$	$\alpha = \beta = 90°, \gamma = 120°$	Rhombic prism	One 6-fold, two 2-fold
Cubic	$a = b = c$	$\alpha = \beta = \gamma = 90°$	Cube	Three 4-fold, four 3-fold, six 2-fold

(a)

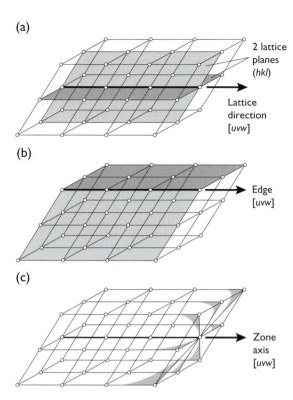

2 lattice
planes
(*hkl*)

Lattice
direction
[*uvw*]

(b)

Edge
[*uvw*]

(c)

Zone
axis
[*uvw*]

Fig. 7.16 (a) A lattice array and indicated on it a lattice
line and two intersecting lattice planes (shaded). (b) Two
crystal faces (shaded) define an edge. (c) A set of
intersecting lattice planes defines a zone axis.

lattice line. If several lattice planes share the same
edge, they are said to be cozonal and the common
edge is called a *zone axis* (Figure 7.16c). For many
crystallographic applications we are interested only in
the orientation of a lattice plane, or a lattice line,
rather than in its particular position. A stack of parallel planes or lines can then be viewed as being equivalent, and we can therefore shift (or translate) planes
and lines arbitrarily, including through the origin. To
do so, however, we must first develop a notational
system that efficiently describes lattice lines and lattice
planes.

Lattice directions such as *r* in Figure 7.17a can be
identified relative to the crystal lattice with axes *x*, *y*,
and *z*. A first step is to translate the line through the
arbitrary origin (O). Then we can describe the line
with the vector equation, $r = ua + vb + wc$, where *r*
is the vector from the origin to the closest lattice point
on the line; *a*, *b*, *c* are unit vectors along the three
crystal axes and outline the unit cell. As an example,
lattice line [132] is shown in Figure 7.17a. The values

u, *v*, *w* are integers, called *direction* or *zone axis*
indices, that are identified by putting them inside
brackets [*uvw*]. Using zone axis indices the symbol
[100] would denote the *x*-axis, having only the single
vector component *a* (Figure 7.17b). Other examples
shown are [111], [012], and [011]. When the direction,
rather than the length, of a lattice line is of interest, it
is customary to reduce zone indices by dividing them
by a common denominator of the nonzero indices; for
example, [022] becomes [011], and [396] becomes
[132]. Geometrically this means that the index corresponds to the first lattice point (from the origin) that
the direction intersects. A bar above an index
indicates that the corresponding vector component is
negative, for example [$\bar{1}$00] points in the negative
x-direction (read: bar-one zero zero).

Similar indices can also be used to specify lattice
planes, although this process is a bit more complicated. First, we center our macroscopic crystal in the
x, *y*, *z* coordinate system such that the center of the
crystal is at the origin (Figure 7.18a). Then we extend
each face so that it intersects all three axes, which is
shown for faces A and B in Figure 7.18a. We now
translate the plane until it intersects the three axes at
three different lattice points. For face A this is at
2, 2, 1, and for face B at 1, 1, 2 (Figure 7.18b). The
axis intersection points are simple multiples of the
unit cell lengths, i.e., *ma*, *nb*, and *oc*. The "axis
intercept" integers *m*, *n*, and *o* are called *Weiss*
indices (Weiss, 1819) and are used to describe the
orientation of a lattice plane with respect to the
x, *y*, *z* axes. A bar above an index number indicates
that the lattice plane intersects a crystal axis in the
negative direction.

W. H. Miller (1839) proposed using reciprocal
values of the axis intercepts *mno*, normalized to be
integers, *hkl*, to specify a plane. The conversion to
reciprocal values becomes particularly relevant when
lattice planes are replaced by their normals, or *poles*,
as is routinely done in many constructions. These new
but equivalent indices are known as *Miller indices*,
and the symbol (*hkl*) (in parentheses) distinguishes
lattice planes from lattice directions [*uvw*] (in
brackets). Miller indices (*hkl*) are obtained by taking
the reciprocal, or inverting, the axis intercepts *mno*
and then multiplying each value by the lowest
common multiple of the three denominators. For
example, a plane with Miller indices (211) has axis
intercepts ½*a*, 1*b*, 1*c*. Similarly, a plane with indices
(111) corresponds to one that intersects each axis at

(a)

(b)

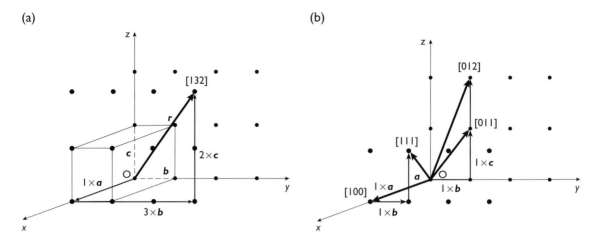

Fig. 7.17 Lattice lines and their specification by zone indices. (a) The lattice direction [132] for the line represented by vector *r* is shown in an orthorhombic crystal. (b) Representations are shown for lattice directions [100], [111], [012], and [011] (heavy arrows).

(a)

(b)

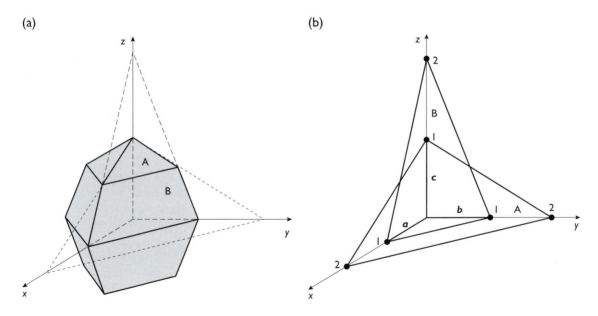

Fig. 7.18 Specification of a lattice plane by axis intercepts. (a) An orthorhombic crystal with faces A and B is centered in the *x*, *y*, *z* coordinate system. The faces are extended to intersect the axes (dashed lines). (b) After translation, the axis intercepts of the faces are at lattice points 2, 2, 1 for face A and 1, 1, 2 for face B.

the first lattice point on the axis, i.e., at 1*a*, 1*b*, and 1*c*. A zero value in Miller indices indicates that the plane *does not* intersect an axis; rather it is parallel to it. For example (100) is a lattice plane that is parallel to the *y*- and *z*-axes. As with zone indices, a bar above a Miller index value indicates that the plane intersects an axis on the negative side.

Let us follow the procedure to determine Miller indices of lattice planes A and B in Figure 7.18b. For plane A we carry out the following steps:

- The Weiss indices, *mno*, of face A are 2:2:1.
- We then take the reciprocal of each ratio value, obtaining ½, ½, 1.

- Next we find the lowest common multiple of the denominators of the reciprocal values. In this case, the lowest common multiple is 2.
- Multiplying by 2 to clear fractions, we obtain 1:1:2.
- The values of this ratio are then enclosed in parentheses, with no commas separating values (unless one of the values has two digits and ambiguity may arise). Thus the Miller indices for lattice plane A are (112).

For the steeper lattice plane B, we move in the reverse order, converting the plane's Miller indices, which can then be used to plot the plane. Plane B has Miller indices (221). How can we visualize its orientation?

- The first step is to convert Miller indices to Weiss axis intercept indices. We take the reciprocal values ½, ½, 1.
- We find the lowest common multiple of the denominators of these reciprocal values. This multiple is 2.
- Multiplying by 2 to clear fractions provides 1, 1, 2. These are the Weiss indices.
- We multiply the Weiss indices by the axis lengths and find the axis intercepts to be 1a, 1b, and 2c. Plotting these, we can construct face B (Figure 7.18b).

It is important to distinguish between Miller indices for lattice planes (hkl) and direction indices for lattice lines [uvw]. Only for orthogonal unit cells is the plane (100) perpendicular to the direction [100], and only for cubic crystals is the general plane (hkl) perpendicular to the lattice direction [$u = h$, $v = k$, $w = l$]. To visualize lattice planes from their Miller indices, it is also important to remember that an index 0 indicates that the plane is parallel to the corresponding axis and does *not* intersect it. Some examples of common lattice planes and their general Miller indices are shown in Figure 7.19, taken from the classic book of Paul Niggli (1920).

Note that all of these indices are integers, no matter what the symmetry of the crystal is or what the actual lattice parameters are. This elegant description, however, relies on the unit cell as a reference system, and thus knowledge about the lattice parameters is necessary if we need to obtain angular relationships. To illustrate this point, consider the following example:

Determine the angle between lattice planes (001) and (112) for a tetragonal crystal with a = 5 Å, b = 5 Å, and c = 8 Å.

- First we convert the Miller indices (001) to Weiss indices. The result is ∞ ∞ 1. The plane (001) does not intersect the x- and y-axes, i.e., it is parallel to the x- and y-axes and intersects the z-axis at OC = 1c. For simplicity we also translate it and draw it as the plane xy in Figure 7.20a (OAB), intersecting z at 0 (since only angular relationships are of concern, parallel translations of planes are always allowed).
- Next we convert (112) to Weiss indices obtaining 221. In order to draw the plane we have to obtain the axis intercepts, multiplying the Weiss indices by the corresponding cell lengths. The intercept along the x-axis is OA = 2a = 10 Å, along the y-axis it is OB = 2b = 10 Å, and along the z-axis it is OC = 1c = 8 Å.

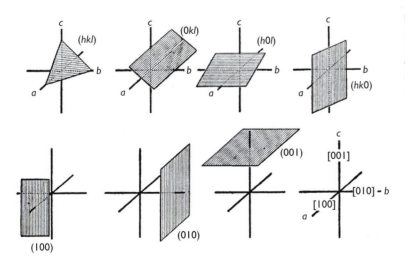

Fig. 7.19 Some frequently observed lattice planes and their Miller indices (hkl) for orthorhombic crystals. The bottom right diagram shows the lattice direction indices for the three lattice axes (from Niggli, 1920).

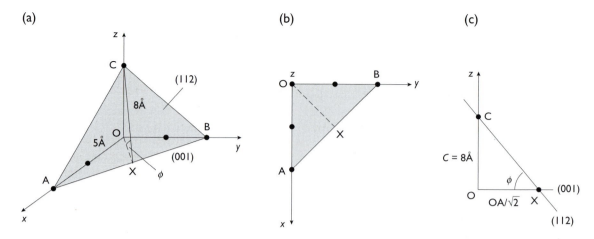

Fig. 7.20 Derivation of the interfacial angle ϕ between lattice planes (001) and (112) of a tetragonal crystal (in this case measured on the inside of the crystal). (a) Three-dimensional view of the planes. (b) Projection along the z-axis, illustrating the trace of the lattice plane (112) on (001). (c) A section containing the z-axis and displaying the true angle ϕ.

- Now we can plot the lattice plane ABC in the three-dimensional sketch shown in Figure 7.20a. We also draw the trace of the lattice plane (112) on (001), projected along the z-axis, in Figure 7.20b.
- The true angle ϕ between the two planes can be measured in the triangle OXC (Figure 7.20c).
- We obtain OX as $OA/\sqrt{2} = 7.07$ Å (Figure 7.20b).
- In triangle OXC we then have $\tan \phi = OC/OX = 8$ Å$/7.07$ Å $= 1.13$, from which we get the final answer, $\phi = 48°$.

Similar calculations can be done for other unit cells. Of course the geometry (and trigonometry) becomes more complicated for unit cells with lower symmetry. Miller and direction indices are applicable to all symmetries. However, in the case of hexagonal crystals, often a modified four-index system is used to account for the fact that there are three equivalent axes (a_1, a_2, and a_3) perpendicular to the z-axis (Box 7.3).

7.5 Relations between lattice planes and lattice lines

A lattice line is the intersection of two lattice planes (Figure 7.16a). Therefore we should be able to derive the corresponding direction indices $[uvw]$ for this line from the Miller indices of the two intersecting planes $(h_1k_1l_1)$ and $(h_2k_2l_2)$. The equations for two planes through the origin are given by

$$h_1x + k_1y + l_1z = 0 \tag{7.2a}$$

$$h_2x + k_2y + l_2z = 0 \tag{7.2b}$$

where x, y, z are the coordinates for all points on these planes. For the intersecting line, both equations must be satisfied, and here the ratio $x : y : z$ corresponds to the direction symbols $u : v : w$. Since we only need the ratios, the two equations with three unknowns can be solved:

$$(k_1l_2 - k_2l_1) : (l_1h_2 - l_2h_1) : (h_1k_2 - h_2k_1)$$

There is a simple recipe as an aid for remembering this:

- Write out the Miller indices twice for both planes, (320) and (112); then cut off the first and last columns.

- Cross-multiply as indicated by the arrows; going from top to bottom is positive, from bottom to top (dashed) is negative. For our specific example,

$$(2 \times 2 - 1 \times 0) \quad (0 \times 1 - 2 \times 3) \quad (3 \times 1 - 1 \times 2)$$
$$u = 4 \qquad v = -6 \qquad w = 1$$

- Renormalize by dividing by the common denominator if necessary.

Box 7.3 Additional information: Miller–Bravais indices in the hexagonal system

Hexagonal crystals are a somewhat special case. They have a unique c-axis, but perpendicular to it are three equivalent axes a_1, a_2, and a_3 (Figure 7.21). There is no reason to give preference to one of them. Accordingly, a coordinate system with four axes is often used, one along the 6-fold rotation axis (c) and three axes perpendicular to it (a_1, a_2, and a_3), separated by angles of $120°$. The a_3-axis is not independent of a_1 and a_2 and can, in principle, be omitted as we have done in Figure 7.14. In fact, as the following discussion demonstrates, it is rather awkward to use four indices for a three-dimensional space, yet it is done conventionally for crystal forms, and mineralogists have to be aware of it.

In the hexagonal system Miller indices are extended to Miller–Bravais indices ($hkil$) to better highlight the hexagonal symmetry. Figure 7.21a is a two-dimensional representation of a hexagonal lattice, showing only one lattice plane. The third dimension (c) is irrelevant in this discussion. Remember to convert Miller to Weiss to obtain axis intercepts. From this figure we derive that $i = -(h + k)$, for example the Miller plane (110) (axis intercepts 1100) is equivalent to the Miller–Bravais plane ($11\bar{2}0$), intersecting a_3 at $-\frac{1}{2}$, and the Miller plane ($\bar{1}20$) (axis intercepts $\bar{1}\frac{1}{2}00$) is equivalent to the Miller–Bravais plane ($\bar{1}230$).

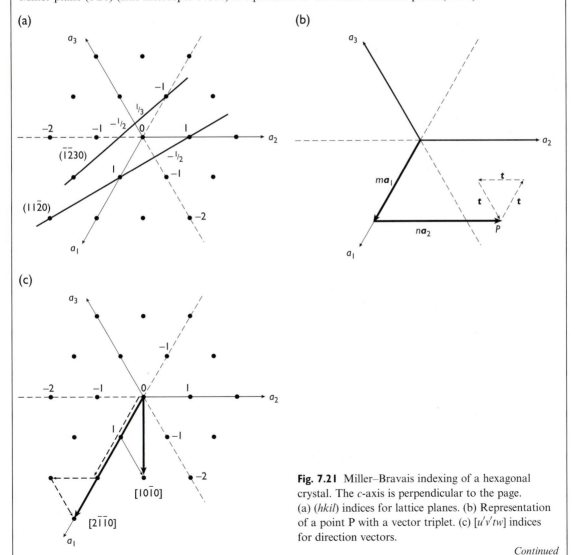

(a)

(b)

(c)

Fig. 7.21 Miller–Bravais indexing of a hexagonal crystal. The c-axis is perpendicular to the page. (a) ($hkil$) indices for lattice planes. (b) Representation of a point P with a vector triplet. (c) $[u'v'tw]$ indices for direction vectors.

Continued

Box 7.3 (cont.)

Likewise, four index direction symbols $[u'v'tw]$ can be used, where $t = -(u' + v')$ or $u' + v' + t = 0$. While the conversion from three- to four-index lattice plane symbols is easy (simply adding $i = -(h + k)$), the conversion is more difficult for vector symbols. The description of a three-dimensional vector as a combination of three vectors, $r = ua_1 + va_1 + wc$, is unique, but there are an infinite number of ways to describe a vector as a combination of four vectors, three of which are coplanar, $r = u'a_1 + v'a_2 + ta_3 + w\,c$. However, only one combination satisfies the same rule applicable to Miller–Bravais plane indices, namely $t = -(u' + v')$. J. D. H. Donnay (1947) derived the conversion. Take point P in the plane a_1, a_2, a_3 of Figure 7.21b that is defined by vectors $ma_1 + na_2 + 0a_3$. The figure illustrates that we can increase or decrease each vector by the same number t, since the sum of three vectors so introduced is equal to zero. The point P can therefore be represented by coordinates $m - t, n - t, -t$, in which t is any number, positive or negative. In particular t may be chosen equal to $(m + n)/3$, such that the sum of the three coordinates $u' = m - (m + n)/3$, $v' = n - (m + n)/3$, and $t = -(m + n)/3$ becomes equal to zero. The direction symbol $[uv0w]$ may therefore be written as $[u'v'tw]$ with $t = -(u' + v')$. Examples $[10\overline{1}0]$ and $[2\overline{1}\overline{1}0]$ are shown on Figure 7.21c.

To convert from the four-index notation to the three-index notation, do the following:

In the case of a face symbol $(hkil)$, simply omit the third index and obtain (hkl), for example $(\overline{2}\overline{1}30)$ becomes $(\overline{2}\overline{1}0)$.

For a direction symbol $[u'v'tw]$, subtract t from the first three indices and obtain $[u' - t, v' - t, w]$, for example $[2\overline{1}\overline{1}0]$ becomes $[300]$, or normalized to $[100]$ corresponding to the a_1 axis.

Exactly the same cross-multiplication procedure can be used to find the Miller indices of the lattice plane given by two lattice lines $[u_1v_1w_1]$ and $[u_2v_2w_2]$ in that plane.

During crystal growth, unit cells are added in all three dimensions, according to Haüy's picture (Figure 7.3a). Some surfaces are more stable and grow more slowly, and those faces become large and determine the exterior morphology. In other directions, crystals grow fast, resulting in small faces, such as the tips of a prism or a needle. As is illustrated in Figure 7.22, lattice planes with simple Miller indices such as (010), (100) and (110) contain a high density of lattice points per surface element and have large interplanar spacings. Such faces are most commonly observed on crystal polyhedra. Victor Goldschmidt determined statistically that lattice planes with indices $(h_1k_1l_1)$ and $(h_2k_2l_2)$ are more common than those with indices $(h_1 + h_2, k_1 + k_2, l_1 + l_2)$. This can be illustrated for a collection of 21 common cubic minerals. Faces (001) and (111) are observed in all 21, whereas faces (225) are observed only in four. In the diagram below, lines from $(h_1k_1l_1)$ and $(h_2k_2l_2)$ terminate at $(h_1 + h_2, k_1 + k_2, l_1 + l_2)$, with the number of instances in which face $(h_1 + h_2, k_1 + k_2, l_1 + l_2)$ was observed among the 21 cubic minerals shown in bold.

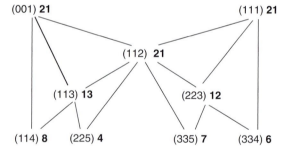

7.6 Crystal structure

In the previous section we illustrated the convenience of the unit cell framework to describe lattice directions and lattice planes without requiring information about absolute lengths and angles. The same unit cell system can be applied to describe the position of atoms within the unit cell. Figure 7.23 shows three atoms, A, B, and C, in a unit cell and their specification with a vector sum, $r = xa + yb + zc$. This representation is analogous to that of a lattice direction, except that the atom is, in general, not a lattice point and its coordinates x, y, z are therefore not integers as with $[uvw]$.

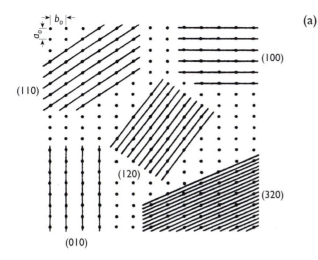

(010)

Fig. 7.22 Two-dimensional lattice, with lattice planes that show different packing by lattice points. Lattice planes with simple Miller indices such as (010) and (100) have the highest point density and the largest interplanar spacing.

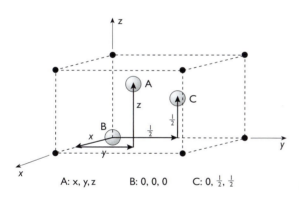

A: x, y, z   B: 0, 0, 0   C: 0, $\frac{1}{2}$, $\frac{1}{2}$

Fig. 7.23 Specification of an atom's position within the unit cell by fractional coordinates x, y, z, illustrated for three atoms A, B, and C. B and C are in special positions; A is in a general position.

The atomic coordinates x, y, z are given in fractions of a unit cell length and range between 0 and <1. If a coordinate is 1, it is counted as belonging to the next unit cell and reset to 0. Atomic coordinates can be complicated fractions, as is the case for atom A in Figure 7.23. In that case the atom is stated to be in a *general position* and fractional coordinates are labeled x, y, z. Coordinates can also be zero, as in the case for atom B, which is exactly in the corner of the unit cell (coordinates 0, 0, 0); or they can be simple fractions, as for atom C, which is in the center of the yz-face and has coordinates 0, ½, ½. Atomic

(a)

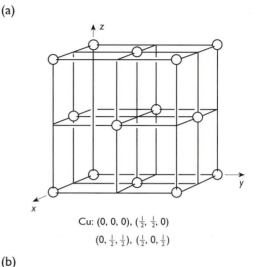

Cu: (0, 0, 0), ($\frac{1}{2}$, $\frac{1}{2}$, 0)

(0, $\frac{1}{2}$, $\frac{1}{2}$), ($\frac{1}{2}$, 0, $\frac{1}{2}$)

(b)

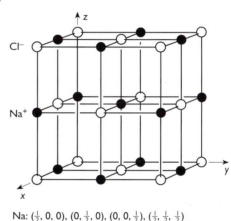

Na: ($\frac{1}{2}$, 0, 0), (0, $\frac{1}{2}$, 0), (0, 0, $\frac{1}{2}$), ($\frac{1}{2}$, $\frac{1}{2}$, $\frac{1}{2}$)

Cl: (0, 0, 0), ($\frac{1}{2}$, $\frac{1}{2}$, 0), (0, $\frac{1}{2}$, $\frac{1}{2}$), ($\frac{1}{2}$, 0, $\frac{1}{2}$)

(c)

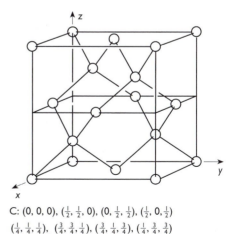

C: (0, 0, 0), ($\frac{1}{2}$, $\frac{1}{2}$, 0), (0, $\frac{1}{2}$, $\frac{1}{2}$), ($\frac{1}{2}$, 0, $\frac{1}{2}$)

($\frac{1}{4}$, $\frac{1}{4}$, $\frac{1}{4}$), ($\frac{3}{4}$, $\frac{3}{4}$, $\frac{1}{4}$), ($\frac{3}{4}$, $\frac{1}{4}$, $\frac{3}{4}$), ($\frac{1}{4}$, $\frac{3}{4}$, $\frac{3}{4}$)

Fig. 7.24 Unit cells, atomic positions, and atomic coordinates for (a) copper (fcc), (b) halite, and (c) diamond. All of these crystals have a cubic unit cell.

positions with simple coordinates (such as 0, 0, 0) or relationships between coordinates due to symmetry (such as 0, ½, ½ or *x*, *x*, *x*) are called *special positions*. Atoms in the unit cell may be related by symmetry and such symmetry-related atoms are called *equivalent positions*. The most important symmetry relationship is if for every atom at coordinates *x*, *y*, *z* there is an equivalent atom at coordinates $-x$, $-y$, $-z$. This represents an "inversion" in the origin. Crystal structures with such corresponding atoms are called *centrosymmetrical*. A majority of minerals are centrosymmetrical; an exception is quartz. Most natural organic crystals, such as DNA and sugar, are noncentric.

Conventionally, atoms are labeled according to their elemental identity, for example O for oxygen. If there is more than one atom present of the same element and at different coordinates, a number is assigned (such as O1 and O2). Atomic coordinates of all crystalline materials are available in reference books and databases such as the COD (www.crystallography.net) discussed in Box 2.1.

In Figure 7.24 we review three crystal structures that were discussed in Chapter 2: copper (fcc) (Figure 7.24a), halite (Figure 7.24b), and diamond (Figure 7.24c). All have a cubic unit cell, but unit cells are occupied by different atoms in different positions, and corresponding atomic coordinates are given below the unit cell images. In the case of these structures, all coordinates are simple fractions, but this is not always the case.

7.7 Summary

This chapter provides a brief introduction to symmetry concepts in general and then uses symmetry to define unit cells and lattices with periodic arrangements. Unit cells are described by six lattice parameters. Rational indices are used to define lattice planes and lattice directions. Within a unit cell atomic positions are defined by fractional atomic coordinates (crystal structure). Remember these important concepts:

- Crystal symmetry
- Unit cell parameters *a*, *b*, *c*, *α*, *β*, *γ*
- Seven crystal systems
- Fourteen Bravais lattices
- Lattice directions and direction indices [*uvw*]
- Lattice planes and Miller indices (*hkl*)
- Miller–Bravais indices for hexagonal crystals (*hkil*), $i = -(h + k)$
- Atomic coordinates *x*, *y*, *z*

Test your knowledge

1. Find all the symmetry planes and rotation axes in a cube.
2. Derive the angle between a (100) and a (111) face of a cubic mineral.
3. Which directions ([*uvw*] indices) are defined between faces (213) and ($\bar{4}$32)?
4. List the atomic coordinates of atoms in a hexagonal close-packed structure. For the unit cell see Figure 2.8a.
5. Make a copy of Figure 7.7c, which is a two-dimensional periodic pattern. Add on it rotation axes, glide planes (glide lines), and translations. Identify a suitable unit cell.
6. In a book with illustrations of classical art, find symmetry elements in the artistic subject matter.

Further reading

Boisen, M. and Gibbs, G. V. (eds.) (1990). *Mathematical Crystallography*. Reviews in Mineralogy, vol. 15, revised. Mineralogical Society of America, Washington, DC.

Buerger, M. J. (1956). *Elementary Crystallography*. Wiley, New York.

Hahn, T. (ed.) (2006). *International Tables for Crystallography. Volume A: Space-group symmetry*. International Union of Crystallography.

Phillips, F. C. (1963). *An Introduction to Crystallography*. Longmans, London.

Weyl, H. (1989). *Symmetry*. Princeton University Press, Princeton, NJ.

8 | Crystal symmetries: point-groups and space-groups

The lattice geometry discussed in the previous chapter restricts the possible combinations of symmetry elements that can exist in crystals. There are 32 possible combinations of rotations, mirror planes, and inversions that are expressed in the crystal morphology (point-groups) and 230 combinations of rotations, mirror planes, inversions, and translations that are expressed in the crystal structure (space-groups). The chapter also introduces efficient graphical representations of crystal forms. Spherical projections have wide applications in structural geology and materials science, well beyond mineralogy.

8.1 Introduction

In the previous chapter we have become familiar with the extraordinary regularity of the internal structure of crystals. The local balancing of bonding forces between atoms leads to a periodic repetition of elementary units. We have seen that these unit cells and the corresponding lattice arrays are diagnostic for specific minerals and have classified them according to their symmetry. Symmetry emerged as a central feature of minerals and crystals. In Chapter 7, we recognized seven crystal systems, ranging from the highest cubic symmetry to the lowest triclinic symmetry. In this chapter we will look at symmetry more formally, particularly in view of the possible symmetries that are present in the external morphology of crystals.

A characteristic feature of crystals is "directionality": specific directions in crystals are inherently different, and these differences are implicit in the lattice structure. Take, for example, a cubic crystal of galena (Figure 8.1a,e): if we view it along a crystal axis (e.g., [100]), we observe a 4-fold symmetry (Figure 8.1b); if the crystal is viewed along a body diagonal (e.g., [111]), it displays 3-fold symmetry (Figure 8.1c); if the cube is viewed along the bisectrix of two faces (e.g., [110]), two different mirror planes are observed (Figure 8.1d). If the growth velocity were *isotropic*, or equal in all directions, crystals would occur as spheres. Instead they display a regular morphology with planar surfaces. The largest faces are indicative of

directions in which the growth velocity is slowest. A material displaying directionality is called *anisotropic*.

Only rarely, however, does the internal structure of a crystal differ in all possible directions. Take again the cubic crystal of Figure 8.1. If the crystal viewed along [111] is turned 120°, it is indistinguishable from the original setting. If parts of a crystal are identical in different directions, we say that they are related by symmetry. Symmetry is fundamental to anisotropic crystals. In order to understand and visualize symmetry and anisotropy, we must first learn how to represent directional properties, which we do in the next section.

8.2 Spherical representations of morphology

Crystals are three-dimensional. So far we have represented crystal structures and lattices by perspective sketches, but on these plots it is difficult to see detailed geometrical relationships. For example, what is the angle between a [100] direction and a [111] direction in a cubic crystal? It would be useful to have a way to capture the quantitative geometrical relationships in crystals and allow us to measure geometrical features. A method that projects the crystal onto a sphere has proven to be very useful in this regard.

The procedure is as follows: take a crystal with well-developed planar faces and place it in the center

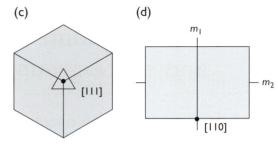

(a)

[111]

[100]

[110]

(b)

[100]

(c)

[111]

(d)

m_1

m_2

[110]

(e)

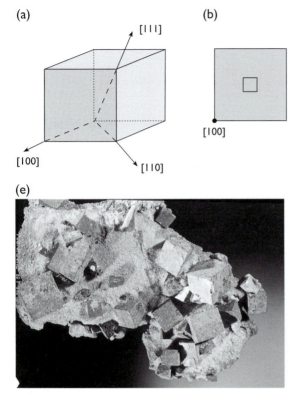

Fig. 8.1 (a) A cube of galena (PbS) looks very different when viewed in different directions. (b) In a view along [100] it appears as a square in projection, (c) looking along [111] the cube appears as a regular hexagon, and (d) looking along [110] the projection is a rectangle. (e) This is illustrated in an actual mineral sample with several crystals of galena from Tristate District, USA (40 mm × 67 mm) (courtesy O. Medenbach).

of a large sphere. In Figure 8.2a this is shown for a cubic crystal, with large cube faces of the type (100), and smaller faces for an octahedron of the type (111) and dodecahedron of the type (110) with all symmetrically related faces. Then construct *normals* (perpendicular lines) to each face, going through the center of the sphere and intersecting the surface of the sphere. Each point of intersection is called a *pole*, and in Figure 8.2a Miller indices are assigned. Thus we can represent crystal faces (or lattice planes) (*hkl*) by points on a spherical surface.

In Figure 8.2b we have again a sphere and on it an arbitrary point P representing a pole to the shaded lattice plane (*hkl*). If we define a coordinate system such as north and south poles (N and S), as well as an east-point E on the equator, then we can specify the point P by two spherical angular coordinates: a *polar angle ρ* (55°), measured from the north pole, and an *azimuth φ* (210°), measured from some arbitrary origin E (Figure 8.2b) in a counterclockwise direction along the equator. This is directly analogous to geographical coordinates on the Earth (Figure 8.2c), except that geographers prefer to use the latitude (distance from the equator) rather than the polar angle (or colatitude)

to define a location. Geographical longitude corresponds to the azimuth. In geography one uses the North Pole (the northern end of the Earth's rotational axis) and the meridian that passes through Greenwich, UK, as a reference coordinate system. (The equator is then simply the great circle perpendicular to the Earth's rotational axis.) Latitudes range from 90° north to 90° south of the equator, and longitudes from 180° east to 180° west of Greenwich. New York, for example, has a latitude 41° north (corresponding to a polar angle of 49°) and a longitude of 74° west. For crystals it is customary to use the *z*-axis [001] as the north pole, and the normal to the lattice plane (010), which lies on the equator, for all crystal systems, as the origin of the azimuth (Figure 8.2b). For crystals the azimuth φ is measured counterclockwise from the normal to (010) when viewing in the [001] direction. In this system (used in this chapter) the spherical coordinates for some lattice plane poles of a cubic crystal are: $φ = 0°$, $ρ = 0°$ for (001), $φ = 270°$, $ρ = 90°$ for (100), $φ = 0°$, $ρ = 45°$ for (011), and $φ = 90°$, $ρ = 135°$ for $(\overline{1}0\overline{1})$ (Figure 8.2a).

A face or lattice plane can be represented two ways on a sphere: either we construct the normal P, as described above, and obtain a point, or else we

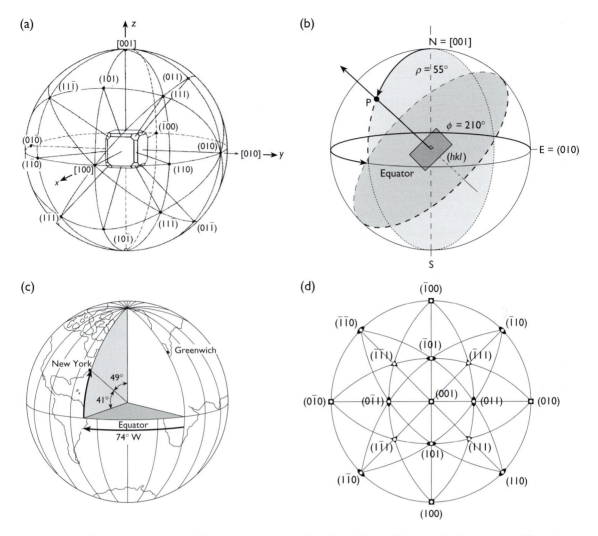

Fig. 8.2 Graphical representation of directions by means of a sphere. (a) A cubic crystal with a variety of faces is placed inside a sphere, and normals to its faces are constructed. Each normal intersects the sphere in a point, called a pole. (b) A lattice plane (*hkl*) (shaded) is translated through the center of the sphere. It intersects the sphere along a great circle (shown dashed). If we construct the normal to the lattice plane, it intersects the sphere at point P. The location of point P on the sphere is specified by spherical coordinates, an azimuth ϕ and a polar angle ρ. The azimuth is measured counterclockwise from the E-point, while the polar angle is measured from the north pole. (c) These coordinates are similar to geographical coordinates used to specify the location of a place on the Earth. Geographical longitude corresponds to the azimuth and is measured from the Greenwich meridian; geographical latitude is measured from the equator, rather than from the pole. (d) Stereographic projection of all the upper hemisphere poles of the cubic crystal in (a).

translate the plane until it passes through the center of the sphere and then map the intersection of the plane (or extension of a face) with the surface. The intersection (or trace) is a *great circle*, i.e., a circle on the sphere's surface having the same radius as the sphere, and is shown on Figure 8.2b by a dashed line with the plane shaded. Both methods uniquely describe the orientation of the plane relative to given coordinates.

Similarly lattice directions, or zone axes [*uvw*] are first translated until they go through the center of the crystal (and correspondingly the center of the sphere). We intersect them with the surface of the sphere and represent them as a point. The lattice direction [001] has coordinates $\phi = 0°$, $\rho = 0°$.

Stereographic projection

A sphere as shown in Figure 8.2 is three-dimensional and to represent it on paper it needs to be projected. There are various projections for different applications. Each projection from three to two dimensions distorts the sphere in some ways. For crystal geometry the stereographic projection proved to be very useful. In stereographic projection, points on the sphere with radius R are projected on to points on the sphere's equatorial plane. The principle of stereographic projection is simple: the sphere's surface is viewed from the south pole (point S) and point P_1 (at coordinates $\phi = 330°$, $\rho = 60°$) on the sphere is projected to point P'_1 on the equatorial plane where the line SP_1 intersects the equatorial plane (Figure 8.3a). The perspective view of the sphere in Figure 8.3a illustrates the projection of a circle on the sphere passing through P_1 onto the equatorial plane. The exact geometrical relationships can be seen by looking at a section through the sphere that contains the meridian on which the

point P_1, the south pole (point S), and the north pole (point N) lie (Figure 8.3b). If the radius of the sphere is R, the distance $d = OP'_1$ is $R \tan(\rho/2)$. Points on the lower hemisphere (such as P_2) project outside the equator (P'_2), but such projections are generally avoided. Instead, the lower hemisphere is viewed from the north pole, and the projection points (such as P''_2 in Figure 8.3b) are assigned different symbols (open circles) to distinguish them from upper hemisphere projections. We now construct a circle to represent the equatorial plane (Figure 8.3c), representing the shaded circle in Figure 8.3a. On it we mark the origin E (east point) for counting the azimuth (i.e., the pole of (010)). Then we measure the azimuth of P_1 ($\phi = 330°$ counterclockwise from E), and draw a line from N to P_1^0 in Figure 8.3c. The projection P'_1 is plotted at a distance $d = R \tan(\rho/2)$ from the center (N).

Wulff net and constructions with stereographic projection

The plotting of projection points, as explained above, is simplified if we use a template of a projected

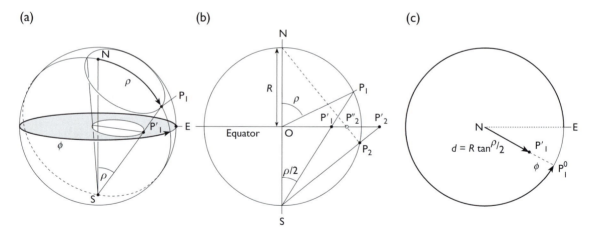

Fig. 8.3 Mechanics of stereographic projection. (a) Geometrical relationships displayed in a perspective drawing, illustrating the projection of a point P_1 on the sphere onto a point P_1 on the equatorial plane (shaded). S is south pole and N is north pole, E is the E-point. Also shown is the projection of a circle on the sphere onto the equatorial plane. (b) Section of the meridian that contains the point P_1 (at coordinates $\phi = 330°$, $\rho = 60°$), and the poles N and S, illustrating the geometry of stereographic projection. The point P_1 is viewed from the south pole S, and P'_1 is obtained as the intersection with the equatorial plane ($OP'_1 = R \tan(\rho/2)$). R is the radius of the sphere. Also shown is the projection of point P_2 on the southern (lower) hemisphere that projects to a point P'_2 outside the sphere. To avoid this problem, P'_2 is generally plotted as a point inside the sphere by projecting from the north pole N (point P''_2 on the equatorial plane). To distinguish upper and lower hemisphere projections, lower hemisphere projections such as P''_2 are characterized with an open circle. (c) Plot of the equatorial plane (shaded in (a)), and construction of the projection P'_1 (NP'_1 in (c) is OP'_1 in (b)).

coordinate system, called a *Wulff net*, named after the Russian crystallographer George V. Wulff (1863–1925). Using the Wulff net we do not need a protractor or calculator to plot points. The Wulff net is a coordinate system on a sphere with circles of equal longitude (great circles) and circles of equal latitude (small circles) projected onto a plane by means of stereographic projection. However, rather than projecting it with the north pole in the center (Figure 8.4a), the sphere is rotated so that the north pole is on top (Figure 8.4b). This net represents an auxiliary coordinate system to perform constructions, not the actual coordinate system that is used for representing poles (with azimuth and polar angles). Usually nets have a 10 cm radius and circles are drawn in 2° intervals. Such a net can be downloaded from the website for this book (www.cambridge.org/wenk). Figure 8.4 shows a Wulff net with circles at 10° intervals. The azimuth on this net should be numbered on the outer (primitive) circle, beginning at the bottom and advancing clockwise (Figure 8.4b shows these numbers from 0 to 35 in 10° increments). Also mark the E-point. Since these nets are used a great deal, it is customary to mount them on a piece of cardboard

(Figure 8.4c). Constructions are not actually done on the net but on a sheet of tracing paper that is superposed on the net. By applying a thumbtack (T) from the back, the tracing paper can be rotated around the center of the Wulff net (Figure 8.4c). Alternatively you can pin down the tracing paper with a sharp pencil point and spin it around the center.

We introduce below the procedures of constructions using the Wulff net with three examples.

1. *Plot a pole P_1 with spherical coordinates $\phi = 330°$, $\rho = 60°$*
   - Place a sheet of tracing paper over the Wulff net and mark the 0° azimuth and the center.
   - Rotate your tracing paper over the center of the fixed Wulff net (shown in light gray lines in Figure 8.5a) until your zero mark on the tracing paper is at 33 (i.e., 330°). (All construction marks on the tracing paper are shown in black.)
   - Plot the projection of pole P_1 by measuring 60° on the radius from the center towards the E-point (on the Wulff net) (Figure 8.5a). The pole is perpendicular to a lattice plane (*hkl*) and we can draw the intersection of that lattice plane

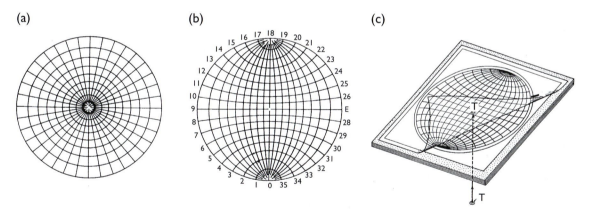

(a) (b) (c)

Fig. 8.4 (a) Coordinate system on a sphere with circles of equal longitude (great circles, appearing as straight lines in this view) and circles of equal latitude (small circles), projected in stereographic projection. Angular intervals between circles are 10°. (b) The same coordinate system, but rotated so that the north pole is at the top. This projection of the coordinate system is called a Wulff net and is used for constructions. Angular labels are applied to the outer circle. Note that numbering on this auxiliary net starts at the bottom and progresses clockwise, which produces a clockwise representation with the azimuth starting at the E-point, when plotted on tracing paper. (c) In practice, the Wulff net is usually mounted on a piece of cardboard and a thumbtack T is applied, so that a superimposed sheet of tracing paper can be rotated about the center. Perspective views such as those shown in Figure 8.2a–c display patterns on a sphere. In effect, we have projected the three-dimensional crystal structure onto a spherical surface. However, in order to achieve useful constructions, we need to project this spherical surface onto a plane. There are several standard methods for projecting a spherical surface on to a plane. We describe here the two most important ones: *stereographic projection* is used mainly in crystallography and materials science; *equal-area projection* is used in structural geology. Both are closely related.

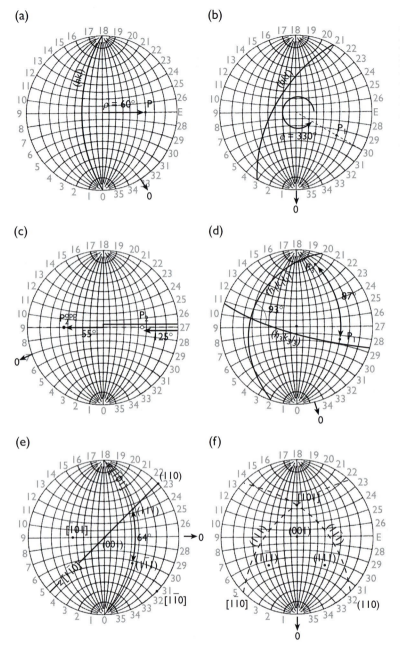

Fig. 8.5 Procedures for constructions in stereographic projection with the help of a Wulff net. The Wulff net is shown with light gray shades. Marks on the tracing paper are in black. (a) Plotting of pole P_1 at coordinates $\phi = 330°$ and $\rho = 60°$ by rotating the tracing paper until the zero mark coincides with 33 (330°) on the Wulff net and plotting the point at a distance 60° from the center. Also shown is the trace of the plane that appears as a great circle. (b) After plotting pole and trace, the tracing paper is rotated back into its original position to show the proper disposition. (c) Plotting of pole P_2 on the lower hemisphere at coordinates $\phi = 70°$ and $\rho = 125°$. We can either plot the opposite end (P_2^{opp}) or project from the north pole (P_2) and assign a different symbol (open circle). (d) Method of measuring an angle between two poles, P_1 and P_3, by placing both on a great circle and counting the angles. Traces of the corresponding faces are also shown. (e) Construction of lattice plane poles (001), (110), (111), ($1\bar{1}1$) and zone axes $[1\bar{1}0]$ and $[\bar{1}01]$ of a cubic crystal. Traces of zone circles $z[1\bar{1}0]$ and $z[\bar{1}01]$ are also shown. (f) The tracing paper with the construction in (e) is rotated back into its original position. Dashed lines are great circles representing traces of lattice planes (111) and ($1\bar{1}1$) with an intersection that determines the zone axis $[\bar{1}01]$. (Compare with Figure 8.2d, where the pattern is completed by applying cubic symmetry.)

(the so-called trace) with the sphere, assuming that the plane goes through the center of the sphere. The plane is perpendicular to the pole and the intersection is a great circle. We draw this great circle on the tracing paper (labeled *hkl*), measuring 90° from P_1 on the horizontal diameter (Figure 8.5a). Try to visualize this plane, which is fairly steeply inclined to the equatorial plane.

• The labeling of the Wulff net appears contrary to our previous definition that the azimuth is counted from the E-point counterclockwise. However, this is only an auxiliary labeling, which you can verify by rotating the tracing paper back, until the zero mark coincides with the zero mark on the Wulff net (Figure 8.5b) and you realize that the azimuthal angle is indeed counted 330° counterclockwise from the E-point and the plot is identical with the one of point P_1 done without the Wulff net (Figure 8.3c).

- If ρ is greater than 90° (e.g., pole P_2 $\phi = 70°$, $\rho = 125°$), the pole will correspond to a spherical point in the lower hemisphere, and when projected onto the equatorial plane it would plot outside the peripheral circle (Figure 8.3b). If we use a Wulff net that does not extend beyond the peripheral circle, we have two choices: we can plot the opposite end of the normal P_2^{opp} that re-enters at $\phi + 180°$, $\rho = 180° - \rho$ (i.e., at coordinates $\phi = 250°$, $\rho = 55°$) (Figure 8.5c). Alternatively, we continue along the radius and re-enter the peripheral circle on the lower hemisphere (35°) and plot P_2, indicating with an open circle symbol that the pole is on the lower hemisphere. This corresponds to a projection from the north pole (Figure 8.3b). Some caution is advised: many constructions on the Wulff net cannot be done by simply combining poles on upper and lower hemispheres!

2. *Determine the angle between two crystal faces P_1 and P_3. The coordinates of the poles of these faces are $\phi = 330°$, $\rho = 60°$ for P_1 and $\phi = 60°$, $\rho = 80°$ for P_3. (The internal angle between two faces is the supplement of the angle between the poles.)*
   - P_1 and the trace of the face $(h_1k_1l_1)$ are already plotted. We plot P_3 and the trace $(h_3k_3l_3)$ the same way from its ϕ and ρ values.
   - We now rotate the tracing paper above the Wulff net until both poles, P_1 and P_3, lie on a great circle (Figure 8.5d). (The zero arrow on the tracing paper lies over the 35 = 350° mark on the Wulff net.) We can read the angular distance between the two points by counting the divisions on the great circle. In this case the angle between the poles is 87°, therefore the angle between the two faces is 93°.

3. *We now return to the cubic crystal of Figure 8.2a with poles on faces (100), (110), (111), etc., and want to plot them in stereographic projection. We also would like to determine the angle between (001) and (111), and between (111) and $(1\bar{1}1)$.*
   - The plane (001) is parallel to x and y and, for crystals belonging to the cubic system, the pole is parallel to z and has spherical coordinates $\phi = 0°$, $\rho = 0°$; similarly (110) has spherical coordinates $\phi = 45°$, $\rho = 90°$. These two poles are easily plotted (Figure 8.5e).
   - Coordinates of (111) are more difficult to establish. We could do this with trigonometry

(Figure 7.18) but it is easier to apply zonal relationships and construct its position. On the crystal in Figure 8.2a we see that faces (001), (111), and (110) are cozonal, i.e., they share the common direction $[1\bar{1}0]$. (Try to calculate the zone from pairs of faces according to the (hkl)–$[uvw]$ cross-multiplication rule in Chapter 7.) The poles therefore lie on the same great circle (zone circle), perpendicular to $[1\bar{1}0]$. Face (111) is also cozonal with (010) and (101), with a zone circle perpendicular to $[\bar{1}01]$. It is therefore at the intersection of the two zone circles. Coordinates of $[1\bar{1}0]$ are $\phi = 225°$, $\rho = 90°$ and of $[\bar{1}01]$ are $\phi = 90°$, $\rho = 45°$.

- We plot $[1\bar{1}0]$ and $[\bar{1}01]$ and construct for each point the great circle perpendicular to it, using the Wulff net and plotting its trace. The zone circle perpendicular to $[1\bar{1}0]$ is vertical and the trace is therefore a straight line. The zone circle perpendicular to $[\bar{1}01]$ is inclined and we have to use the Wulff net to plot its trace (Figure 8.5e). The pole of (111) is obtained as the intersection of the two zone circles.

- Next we rotate the tracing paper back into its starting position (Figure 8.5f) and can read off the spherical coordinates of (111) as: $\phi = 325°$, $\rho = 53°$. Also, we can complete adding all other poles of the crystal in Figure 8.2a by symmetry (Figure 8.2d). $(1\bar{1}1)$ is shown in Figure 8.5f.

- Finally we need to determine the angles between poles (001) and (111), and (111) and $(1\bar{1}1)$. This is done by rotating the tracing paper to bring pairs on a great circle of the Wulff net. For example in Figure 8.5e we can read the interfacial angle between (111) and $(1\bar{1}1)$ as 64°. (Note that in this construction the interfacial angle is defined as the angle between the two poles.) Similarly the angle between (001) and (111) poles is determined as 53°.

- We have established a relationship between lattice plane poles (hkl) and zones $[uvw]$. In Figure 8.5e, the intersection of two zone circles (solid lines) $z[110]$ and $z[\bar{1}01]$ determines the pole (111). In Figure 8.5f we have drawn the traces of planes (111) and $(1\bar{1}1)$ (dashed lines) and their intersection determines the zone axis $[\bar{1}01]$. This construction, as well as the corresponding cross-multiplication rule, applies to all crystal symmetries. However, for example for triclinic symmetry, the pole to (001) and the direction [001] do not coincide!

When we project a three-dimensional object onto a two-dimensional plane, some information is lost or distorted. For example, one characteristic of stereographic projections is that all circles on the sphere, including small circles, appear as circles in projection (Figure 8.3a). Therefore all lines in a Wulff net are circles (Figure 8.4b). However, the geometrical center of a projected circle, when drawn with a compass, does not, in general, coincide with the projection point of the "true" center established on the sphere. As a further example of projection distortions, we note that, while equal angles between two directions have the same distance in projection, whether close to the north pole (center) or close to the equator (peripheral circle), equal areas do not. For example, a $10° \times 10°$ segment is much smaller near the north pole than near the equator (Figure 8.4b). This is a handicap for many applications and, therefore, another projection method was developed which we will discuss in the next section.

Equal-area projection

The stereographic projection is commonly used when angular relationships are important. For other applications, however, the equal-area projection is preferred. This latter method works as follows: using again the meridian section that contains the direction P of interest, we find this time the projection P' by rotating P around the north pole N onto a horizontal plane through the north pole (Figure 8.6a). The distance NP is identical with the distance NP' and, by applying trigonometry to the right triangle SNP (Figure 8.6a), we find $NP = 2R \sin(\rho/2)$, where R is the radius of the sphere. Point E on the equator projects as E' with a distance $r = NE' = \sqrt{2}R$. Contrary to stereographic projection, the radius of the equatorial circle is not the same as the radius of the sphere, and values for distances NP' need to be renormalized accordingly. As with the stereographic projection method, points on the lower hemisphere project outside the equator. The south pole would project as a circle at a distance $2R$. (In the stereographic projection method, the south pole would project into infinity.) As was the case in the stereographic projection method, poles on the lower hemisphere are generally projected by rotating around the south pole, instead of the north pole, and marked with different symbols (open circles), so that all poles are inside the equatorial circle (peripheral circle).

An auxiliary net for equal-area constructions is also available. It is called the equal-area or Schmidt net named after the structural geologist Walter Schmidt (Figure 8.6b) and can be used in a manner analogous to that of the Wulff net (Figure 8.4b). All the

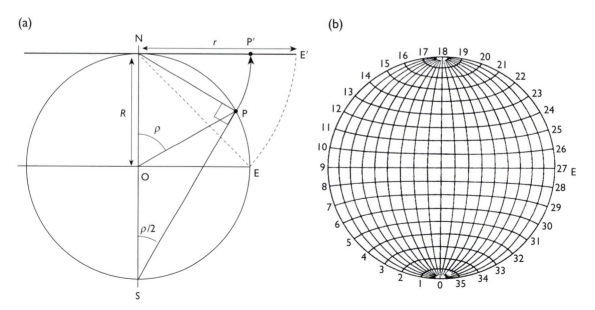

Fig. 8.6 Equal-area projection. (a) Geometric relationships for equal-area projection (analogous to those for stereographic projection in Figure 8.3b). (b) The equal-area or Schmidt net is a projection of an auxiliary coordinate system, equivalent to the Wulff net for stereographic projection (see Figure 8.4b).

constructions illustrated above can be done the same way with a Schmidt net. The only difference between the two methods is the distance of a pole from the center. For equal-area projection, corresponding poles are closer to the peripheral circle than for stereographic projection. A Schmidt net with 10 cm radius can be downloaded from the website for this book.

Unlike stereographic projection, in equal-area projection, apart from the peripheral circle, neither great circles nor small circles on the sphere appear as circles in projection and cannot be drawn with a compass. An angle on the Schmidt net is larger in the center than at the periphery. However, areas of regions with identical angular dimensions on the sphere are equal in projection. For example a $10° \times 10°$ domain in the center and near the E-point in Figure 8.6b are both of the same area. This equal-area feature is important when point densities need to be represented and statistically evaluated.

Typical applications of equal-area projection are in structural geology, where strikes and dips of foliation or bedding planes, as well as bearings and plunges of fold axes, are plotted. Equal-area projection is also used in representations of the preferred orientation of crystals in rocks. Figure 8.7a illustrates the direction pattern of [100] of antigorite crystals in a deformed serpentinite. Each crystal is represented by a pole and, in the case illustrated, the poles cluster in the center. In Figure 8.7b pole densities are contoured, and the pattern expresses pole densities per 1% area. If poles are randomly oriented, pole densities are everywhere

the same (1). If there is preferred orientation some directions have much higher pole densities (in this case 12). Such diagrams with contoured pole densities are called *pole figures*.

It is important to be able to visualize the three-dimensionality of a spherical projection and intuitively to see points in projection as directions in space. This visualization skill takes some practice, and the easiest way to become comfortable with such projections is by using the analogy to geographical representations of the Earth (Figure 8.2c). In these circular diagrams, always visualize a sphere extending over the circle.

8.3 Point-group symmetry

Stereographic projection and symmetry

In the previous section we discussed how to represent three-dimensional crystal directions and lattice planes by two-dimensional projections and applied these techniques to a cubic crystal. Note that in making such a projection, we lose information concerning the size of faces and the length of edges of a crystal. While size of faces can vary greatly, interfacial angles are strictly constant as stated in Steno's *law of constancy of interfacial angles*. Therefore, a spherical method is an ideal way to assess the geometry of a crystal.

The crystal morphology can be displayed as a perspective drawing such as that of a fluorite crystal with a combination of cubic (e.g., 100), octahedral (e.g., 111), and dodecahedral (e.g., 110) faces (Figure 8.2a).

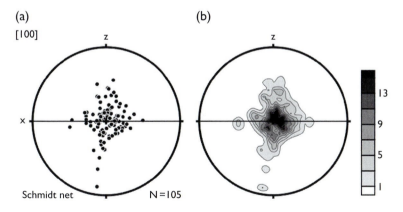

(a)
[100]

(b)

Schmidt net   N = 105

Fig. 8.7 (a) A pole distribution (*c*-axes of antigorite in a deformed serpentinite from Japan) in equal-area projection. The *c*-axes are plotted relative to sample coordinates; Z is the pole to the schistosity, Y (in center) is the lineation direction. Poles were measured with a universal stage on a petrographic microscope. (b) Pole densities are contoured and a pole figure is produced that represents pole densities in multiples of a random distribution (see also Soda and Wenk, 2014).

This method of display is ineffective, however, when it comes to an assessment of angular relationships. Instead, stereographic projection can be used and we have illustrated this in the previous section. For the complicated perspective drawing in Figure 8.2a we obtain a highly symmetrical pattern in projection (Figure 8.2d). With the help of a Wulff net, we can determine all interfacial angles.

We can also plot symmetry elements on a stereogram, including rotation axes and mirror planes. For example, Figure 8.8a illustrates, by means of a perspective drawing, the rotation axes that are present in a cubic crystal. As was done in Chapter 7, we use a square symbol to denote a 4-fold axis, a triangle to indicate a 3-fold axis, and a lens-shaped symbol to identify a 2-fold axis (see Figure 7.6). We transfer these rotation axes into the stereogram (Figure 8.8b), using the method outlined earlier. Dashed lines are entered for reference.

On this very symmetrical diagram we next add an arbitrary pole to a lattice plane (star) and repeat this pole by applying the rotational symmetry operations. It is easy to see how the pole repeats itself four times, rotating around the a_3-axis (Figures 8.8b and 8.9a). We follow a small circle (dotted line) and repeat the point after a 90° rotational interval. Miller indices change during this rotation and we obtain from (hkl) (1), $(\bar{k}hl)$ (2), $(\bar{h}\bar{k}l)$ (3), and $(k\bar{h}l)$ (4). Next we take each of the four generated poles and rotate them around the a_1-axis (Figure 8.9b). From the previously generated four poles (now labeled with larger symbols) we obtain a total of $4 \times 4 = 16$ poles (12 of which are new and have Miller indices assigned), with 8 located on the upper hemisphere (closed circles) and 8 on the lower hemisphere (open circles). Again, during this rotation, the poles follow small circle paths (dashed lines), which we can verify with the Wulff net (Figure 8.4b). Lastly we apply the a_2 rotation to all 16 poles (Figure 8.9c). Doing this, we note that the new positions of many poles coincide with already generated poles. At the end we have added 8 new poles (smaller symbols), for a total of 24.

On this pole pattern (Figure 8.9c and summarized in Figure 8.8b) we notice that 3-fold rotations and 2-fold rotations are already implicit and do not generate new poles, i.e., we did not need them to generate the 24 poles. The 24 poles represent a polyhedron that is typical of this combination of rotation axes (Figure 8.8c). Crystallographers call this particular polyhedron a pentagon-trioctahedron, which resembles an octahedron, but with each triangular face of the octahedron divided into three pentagons. Such a polyhedron in which all faces are related by symmetry is called a *crystal form*. Each face of a form has a Miller index "hkl" with the numerical values being permutations of h, k, and l, both positive and negative, as shown in Figure 8.9. All poles on the lower hemisphere have a negative l. We give the collection of all (hkl)s produced by symmetry the symbol $\{hkl\}$ (parentheses). Note that (hkl) is the symbol for a

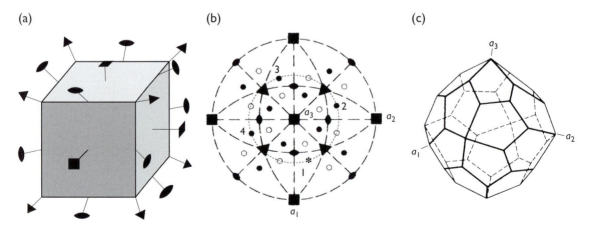

(a) (b) (c)

Fig. 8.8 (a) Perspective drawing of a cubic crystal with rotation axes indicated (compare with Figure 7.6). (b) Projection of the rotation axes in stereographic projection (same symbols are used as in (a)). Also indicated is a pole to a lattice plane (star) and its 24 repetitions due to the rotational symmetry. Dashed lines are for reference to better recognize the cubic symmetry. Closed symbols are poles on upper hemisphere; open symbols are poles on lower hemisphere. (c) The resulting poles can be interpreted as faces of a polyhedron, in this case a pentagon-trioctahedron.

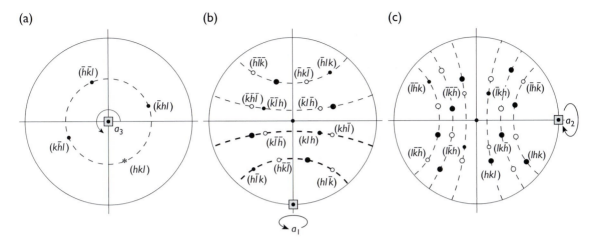

Fig. 8.9 Consecutive rotations about the three 4-fold rotation axes, generating from one pole (*hkl*), marked by star, symmetrically equivalent poles. (a) Rotation about a_3 produces four poles that all lie on a small circle (dashed line). (b) Rotation about a_1 produces a total of 16 poles that are distributed over small circles (dashed lines). Old poles are indicated by larger symbols. (c) Rotation about a_2 produces an additional eight poles. Compare the final pattern with Figure 8.8b. Closed circles are on upper hemisphere; open circles are on lower hemisphere.

particular face, whereas {*hkl*} is the symbol for all faces of a form. Similarly, crystal directions [*uvw*] are multiplied by symmetry, and to express the whole set of symmetrically equivalent directions we use the symbol ⟨*uvw*⟩. Thus the symbol ⟨100⟩ for a cubic crystal implies [100], [010], [001], [$\bar{1}$00], [0$\bar{1}$0], and [00$\bar{1}$].

Symmetry operations revisited

We have already touched upon symmetry considerations in Chapter 7, where we introduced three symmetry operations: translation, rotation, and mirror reflection. We now discuss these operations in more detail.

Translation is the most basic symmetry operation present in all crystals owing to their periodic lattice structure (see Figure 7.5b). Yet translation is not expressed directly in the symmetry of the external forms because the translation distances are very small (on the order of a few ångströms) and cannot be seen directly. We have noted earlier, however, that translational symmetry (i.e., the lattice character of crystals) imposes limitations on the types of rotation axes that are expressed in the external crystal forms. Thus we will exclude translation from the discussion of symmetries of crystal forms for now, but return to it at the end of this chapter.

Rotations and mirror reflections are expressed in crystal forms. A cubic crystal, for example, has quite

a large number of symmetry elements, whereas a triclinic crystal has very few. There are a limited number of possible combinations of rotations and mirror reflections in crystals. In 1830, Johann F. C. Hessell determined that there are only 32 different such combinations, and these combinations are now called *symmetry classes* or *point-groups*. To each such combination a symbol is assigned. Physicists commonly use the older Schoenflies symbols that are more readily applied in group theory, but crystallographers prefer the newer International or Hermann–Mauguin symbols, which are easier to visualize. Both symbol systems are listed in Table 8.1. We will now explore possible point-group symmetries, without going through a rigorous derivation. (Those interested in a more systematic discussion should consult Buerger, 1978, pp. 23–68.)

By means of *rotation* (Hermann–Mauguin symbol *n* = 1, 2, 3, 4, 6) around an *n*-fold symmetry axis, a crystal comes to coincidence after an angular rotation of $\phi = 360°/n$. After *n* such rotations the crystal is again in the starting position. Only 1-, 2-, 3-, 4-, and 6-fold rotation axes can occur in crystals (as noted in Chapter 7, the lattice structure of crystals does not allow for axes with *n* = 5, 7, 8, or higher). The 1-fold rotation is trivial, because every object is brought to coincidence after a 360° rotation. Rotation axes are indicated by symbols on stereoplots, as we have done in Figure 8.8b for a cubic crystal. In Figure 8.10a we

Table 8.1 Crystal systems and point-groups (symmetry classes)

System	Point-group			Abridged	Schoenflies	Multiplicity[a]
Lattice parameters	Hermann–Mauguin (complete) direction of symmetry element					
Triclinic	1			1	C_1	1
$a, b, c; \alpha, \beta, \gamma$	$\bar{1}$			$\bar{1}$	C_i	2^a
Monoclinic	$[010] = y$					
$a, b, c; \beta$	m			m	C_s	2
	2			2	C_2	2
	$2/m$			$2/m$	C_{2h}	4^a
Orthorhombic	$[100] = x$	$[010] = y$	$[001] = z$			
a, b, c	m	m	2	$mm2$	C_{2v}	4
	2	2	2	222	D_2	4
	$2/m$	$2/m$	$2/m$	mmm	D_{2h}	8^a
Tetragonal	$[001] = z$	$\langle 100 \rangle = x, y$	$\langle 110 \rangle$			
a, c	4			4	C_4	4
	$\bar{4}$			$\bar{4}$	S_4	4
	$4/m$			$4/m$	C_{4h}	8^a
	$\bar{4}$	2	m	$\bar{4}2m$	D_{2d}	8
	4	m	m	$4mm$	C_{4v}	8
	4	2	2	422	D_4	8
	$4/m$	$2/m$	$2/m$	$4/mmm$	D_{4h}	16^a
Trigonal	z	x, y, u				
a, c	$[0001]$	$\langle 2\bar{1}\bar{1}0 \rangle$	$\langle 10\bar{1}0 \rangle$			
	3			3	C_3	3
	$\bar{3}$			$\bar{3}$	C_{3i}	6^a
	3	m		$3m$	C_{3v}	6
	3	2		32	D_3	6
	$\bar{3}$	$2/m$		$\bar{3}m$	D_{3d}	12
Hexagonal	6			6	C_6	6
a, c	$\bar{6}$			$\bar{6} = 3/m$	C_{3h}	6
	$6/m$			$6/m$	C_{6h}	12^a
	$\bar{6}$	m	2	$\bar{6}m2$	D_{3h}	12^a
	6	m	m	$6mm$	C_{6v}	12
	6	2	2	622	D_6	12
	$6/m$	$2/m$	$2/m$	$6/mmm$	D_{6h}	24^a
Cubic	$\langle 100 \rangle; x, y, z$	$\langle 111 \rangle$	$\langle 110 \rangle$			
a	2	3		23	T	12
	$2/m$	$\bar{3}$		$m\bar{3}$	T_h	24^a
	$\bar{4}$	3	m	$\bar{4}3m$	T_d	24
	4	3	2	432	O	24
	$4/m$	$\bar{3}$	$2/m$	$m\bar{3}m$	O_h	48^a

[a] Centric (Laue group).

take a simple case with a single 2-fold rotation axis that generates two identical motifs (e.g., the number 5). After applying the 2-fold rotation two times we return to the starting point.

A *mirror reflection* (symbol m) produces a mirror image, and the operation needs to be applied twice to reproduce the original object. A mirror plane is indicated by a solid line (great circle) in a stereoplot

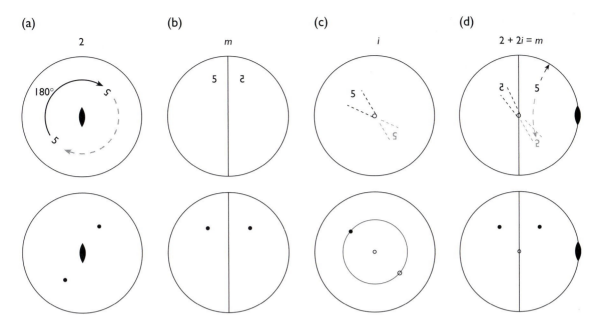

Fig. 8.10 Stereograms illustrating the repetition by (a) rotation, (b) mirror reflection, (c) inversion, and (d) a combination of 2-fold rotation and inversion, which is equivalent to a mirror plane normal to the 2-fold axis. (Top) Repetition of a "5" symbol is used to illustrate a change in handedness during mirror reflection and inversion. Marks on lower hemisphere are in light gray, those on upper hemisphere in black. (Bottom) Same as top, but the "5" motif is now replaced by a point. Closed circles are on the upper hemisphere; open circles are on the lower hemisphere.

(Figure 8.10b). Recall that a mirror reflection does not produce an identical repetition of an object, but rather an enantiomorphic repetition, creating a left-handed object from a right-handed object (Figure 7.5d).

There is a fourth symmetry operation called *inversion* (symbol *i*), and the corresponding symmetry element is called an inversion center, indicated by a small open circle in the center of the stereoplot. In Figure 8.10c (top) a motif on the upper hemisphere is transformed into a motif on the lower hemisphere by an inversion in the center of the sphere. As with mirror reflection, the inversion operation produces a mirror image and a change of handedness of the object. Many crystal structures possess one or more inversion centers, i.e., points in the unit cell from which all atoms have equivalents at the opposite end. (For example, in the structure of halite (Figure 7.24b), one inversion center is in the corner of the unit cell.) This internal inversion also applies to the external morphology, with crystal faces inverted on opposite ends of the inversion center.

In the bottom row of Figure 8.10 we have replaced the number 5 motif by a point, or pole, and we will use this representation in the discussion that follows. Be aware, however, that poles do not tell us directly whether the symmetry operation changes the handedness. Remember that translations and rotations keep the handedness, while mirror reflections and inversions change the handedness.

Symmetry operations can also be combined. We have already seen in Figure 8.8a an example of a cubic crystal with three 4-fold, four 3-fold, and six 2-fold axes. Let us now look at a simple combination of a 2-fold rotation and an inversion (Figure 8.10d). In Figure 8.10d (top) we apply the two operations in consecutive steps: first we rotate the number 5 around the 2-fold axis (which we have put into the equatorial plane) and obtain a 5 on the lower hemisphere after a 180° rotation. We then invert the number 5 and obtain a final 5 with changed handedness on the upper hemisphere. Particularly in the representation with poles, and leaving out the intermediate step (Figure 8.10d, bottom), we recognize a mirror plane perpendicular to the 2-fold axis. This is the same as the pattern in Figure 8.10b, bottom, illustrating that a 2-fold rotation–inversion is equivalent to a mirror plane perpendicular to the 2-fold axis. Thus we do not need to consider both mirror reflection and inversion in the

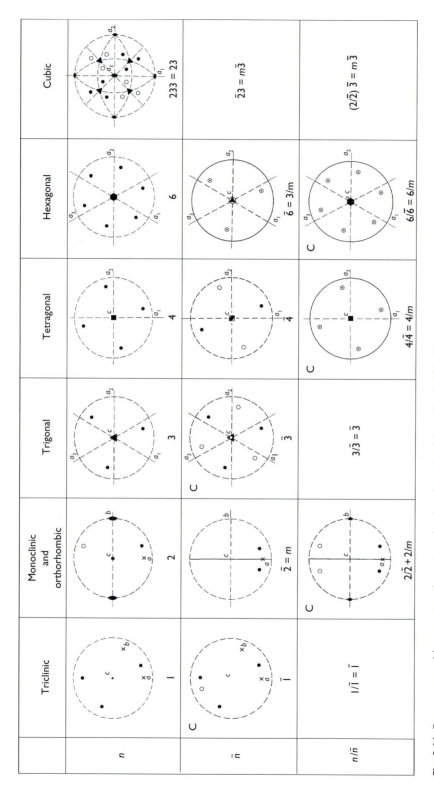

Fig. 8.11 Stereograms with symmetry elements and poles of a general form for the 32 point-groups.

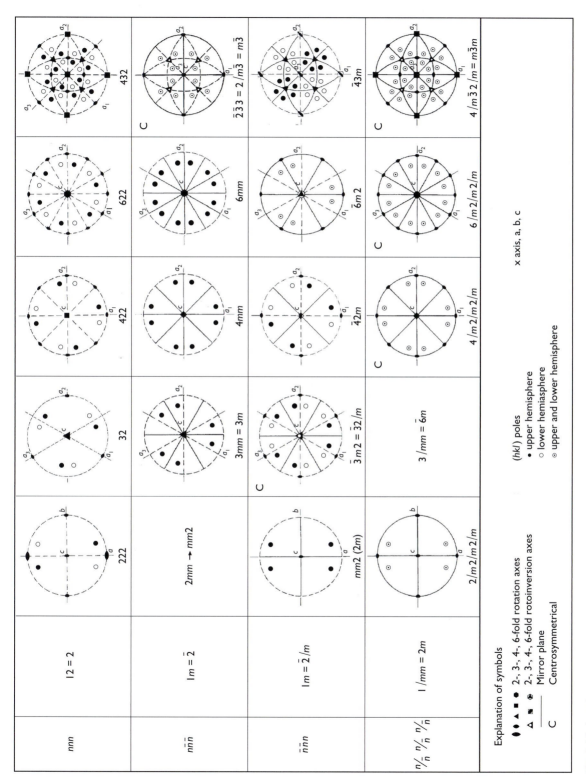

Fig. 8.11 (*cont.*)

derivation of possible combinations of symmetry elements in crystals.

We have combined (in Figure 8.10d) a 2-fold rotation with an inversion; similarly we can combine other rotation axes with an inversion and obtain so-called *rotoinversion axes* (general symbol $\bar{n}$). They are plotted with the same polygonal symbol as rotation axes in stereoplots, but with an open circle in the center. Some rotoinversion axes produce mirror planes perpendicular to the rotation axis, as in the case of the 2-fold rotoinversion, others do not. Rotation axes are called *proper* if they are simple rotations and *improper* if they are rotoinversions.

Thirty-two point-groups

The 32 point-groups are simply combinations of these symmetry elements, and the different possibilities are best explored with stereoplots. First, we consider point-groups with a single rotation axis, the so-called *monaxial point-groups*, which are illustrated in the first three rows of Figure 8.11, except for the last column to which we will return later. If the rotation axis is proper, there are five possible groups, with Hermann–Mauguin symbols 1 (no symmetry), 2, 3, 4, and 6. Stereoplots for these five groups (top row of Figure 8.11) display rotation axes and repetitions of a general pole, i.e., a pole to a lattice plane (*hkl*), that is in no special position relative to the symmetry elements, as with that illustrated in Figure 8.8. Next we have the groups with a single improper axis $\bar{n}$, i.e., a combined rotation and inversion. There are again five such groups $(\bar{1}, \bar{2}, \bar{3}, \bar{4}, \bar{6})$. The Wulff net can be used to construct the pattern for repetitions of a general pole, illustrated in the second row of Figure 8.11. Group $\bar{2}$ is generally shown as *m* because, as we have seen, a mirror plane is equivalent to a 2-fold improper rotation. Similarly, group $\bar{6}$ is equivalent to $3/m$, where the symbol "*/m*" denotes that a mirror plane is perpendicular to the rotation axis. Proper and improper rotations may be combined (general symbol $n/\bar{n}$). These groups are displayed in the third row of Figure 8.11. With combined proper–improper rotations we apply first the proper rotation to a pole and then the improper rotation to all poles that have been generated. The results illustrate that not all the patterns are new. For example $1/\bar{1} = \bar{1}$ and $3/\bar{3} = \bar{3}$. Verify this by performing all the symmetry operations. In the remaining cases, with even-fold rotation axes we produce a pattern with a mirror plane *m* perpendicular

to the rotation axis, and the symbols are $2/m$ (for $2/\bar{2}$), $4/m$ (for $4/\bar{4}$), and $6/m$ (for $6/\bar{6}$). In total, there are 13 monaxial point-groups with a single rotation axis.

A crystal may contain rotation axes in different directions, defining the *polyaxial point-groups*. Again, the lattice imposes restrictions on the directions in which axes may be present. If there is more than one axis, there have to be at least three, because after a rotation about a first axis n_1 that transforms a pole P_1 to P_2, and a second rotation n_2 that transforms P_2 into P_3, there has to be a third rotation n_3 that returns P_3 into P_1 (shown in stereographic projection in Figure 8.12). The general symbol for a polyaxial point-group is n_1 n_2 n_3. The simplest case is 222 in the orthorhombic system with three 2-fold axes at right angles to each other (Figure 8.13a). Next is 322, generally abbreviated to 32, with a 3-fold axis and two 2-fold axes perpendicular to it at an angle of 60° to each other (Figure 8.13b). (The third 2-fold axis in the stereoplot of Figure 8.13b is produced by the 3-fold rotation axis.) Point-group 422 has two types of 2-fold axes at 90° to the 4-fold axis and an angle of 45° between them (Figure 8.13c). (Again the 4-fold rotation produces additional symmetrical 2-fold axes.) In these tetragonal crystals the two types of 2-fold axes are 100 and 110. Point-group 622 has two types of 2-fold axes at 90° to the 6-fold axis and an angle of 30° between them (Figure 8.13d). In these hexagonal crystals the two types of 2-fold axes are along $a = \langle 2\bar{1}\bar{1}0 \rangle$ and $\langle 10\bar{1}0 \rangle$.

In the cubic system there are two types of polyaxial groups with axes along $\langle 100 \rangle$, $\langle 111 \rangle$, and $\langle 110 \rangle$ of the cube (Figure 8.13e,f). In the first type, 233 (or simply 23, and not to be confused with the trigonal point-group 32), 2-fold axes are along $\langle 100 \rangle$ and corresponding 3-fold axes are $\langle 111 \rangle$ and $\langle 1\bar{1}1 \rangle$ (Figure 8.13e). Finally in 432, the highest cubic symmetry, there are three 4-fold axes along $\langle 100 \rangle$, four 3-fold axes along $\langle 111 \rangle$, and six 2-fold axes along $\langle 110 \rangle$ (Figure 8.13f). This is the case we have already explored earlier (see Figures 8.8 and 8.9). Thus there are six polyaxial groups with only proper rotations (shown in the fourth row of Figure 8.11 and 23 in the first row).

As with monaxial groups, polyaxial groups can also have proper (P) and improper (I) rotations; however, after applying three rotations, identity must result. This means that if the first rotation is proper and the second one is improper, then the third one has to be improper to generate from the now left-handed object the right-handed one that we started out with. Thus,

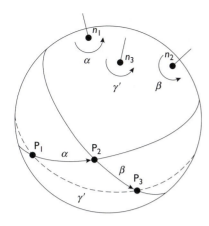

Fig. 8.12 In a crystal with three rotation axes, a rotation around n_1 transforms point P_1 to P_2. A second rotation around n_2 transforms point P_2 to P_3. There must be a third rotation n_3 that returns point P_3 to P_1.

only combinations PPP, IIP, IPI, and PII are possible; in contrast, combinations such as III and IPP would produce a mirror image after three rotations. Without going into details, we just state that there are nine such combinations: $mm2$ ($2m$), $3mm$ (called $3m$), $\bar{3}m2$ (called $\bar{3}2/m$), $4mm$, $\bar{4}2m$, $6mm$, $\bar{6}m2$, $2\bar{3}3$ (called $2/m\bar{3}$ or $m\bar{3}$), and $\bar{4}3m$ (rows five and six of Figure 8.11).

Finally, we can always add a center of symmetry to a polyaxial point-group, effectively combining proper and improper rotations, i.e., $n_1/\bar{n}_1\ n_2/\bar{n}_2\ n_3/\bar{n}_3$. This yields another four new point-groups: $2/m\ 2/m\ 2/m$, $4/m\ 2/m\ 2/m$, $6/m\ 2/m\ 2/m$, and $4/m\ 3/\bar{3}\ 2/m$ (called $4/m\ \bar{3}\ 2/m$ or $m\ \bar{3}\ m$) (row seven of Figure 8.11).

With these 19 polyaxial point-groups and 13 monaxial point-groups we come up with a total of 32 point-groups, representing all possible combinations of rotations, inversions, and mirror reflections in crystals. Review the stereoplots in Figure 8.11 and confirm how poles are repeated by symmetry operations.

Hermann–Mauguin symbols, as introduced above, are relatively easy to read, particularly in an unabbreviated form. The general symbol is $n_1/\bar{n}_1\ n_2/\bar{n}_2\ n_3/\bar{n}_3$ and stands for three n-fold axes in different directions. All other rotation axes that may be present in the crystal are symmetrically equivalent to one of these three axes. There may be a single axis (monaxial), or there may be three axes (polyaxial). Table 8.1 indicates in which crystallographic direction $[uvw]$ the axes are located in each point-group. The symbol $n/\bar{n}$ means that a proper axis is combined with an improper axis and both are parallel. Since $\bar{2}$ is equivalent to a mirror plane perpendicular to $\bar{2}$, the symbol $n/\bar{2} = n/m$ indicates that a mirror plane is perpendicular to an n-fold axis, 222 signifies that there are three different 2-fold axes in different directions (from Table 8.1 we identify the directions as [100], [010], and [001]). The symbol $2mm$ denotes that two mirror planes are parallel to a 2-fold axis, etc.

The full symbol is often abbreviated; for example, instead of $4/m\ 2/m\ 2/m$ one writes $4/mmm$ because not all symmetry operations are necessary to produce all repetitions. This means that with the symmetry elements in the abbreviated symbol we can generate all additional symmetry elements. We have seen how a

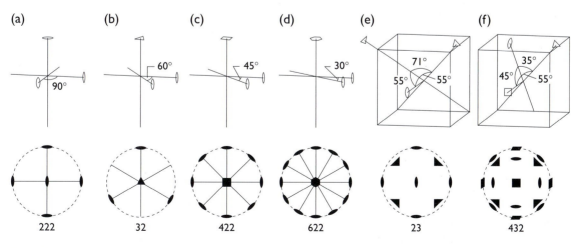

Fig. 8.13 Disposition of rotation axes in polyaxial point-groups. On top are perspective sketches and below are corresponding stereograms.

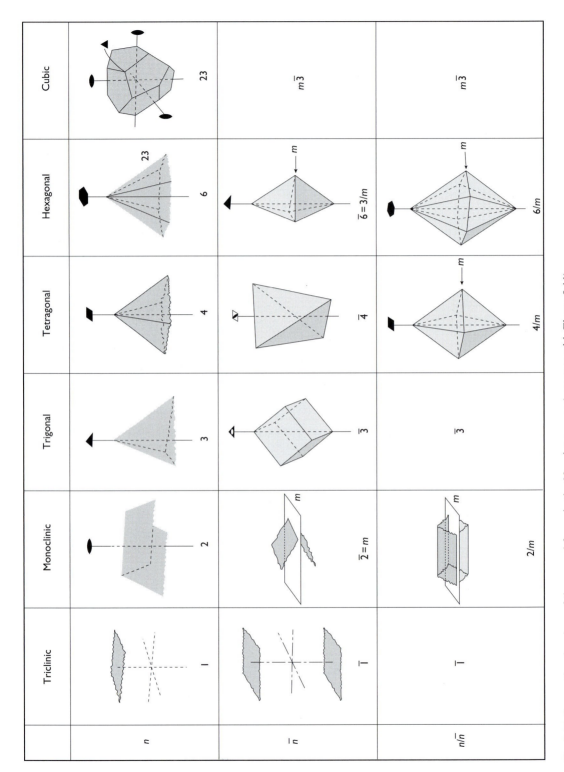

Fig. 8.14 Perspective drawings of the general forms in the 32 point-groups (compare with Figure 8.11).

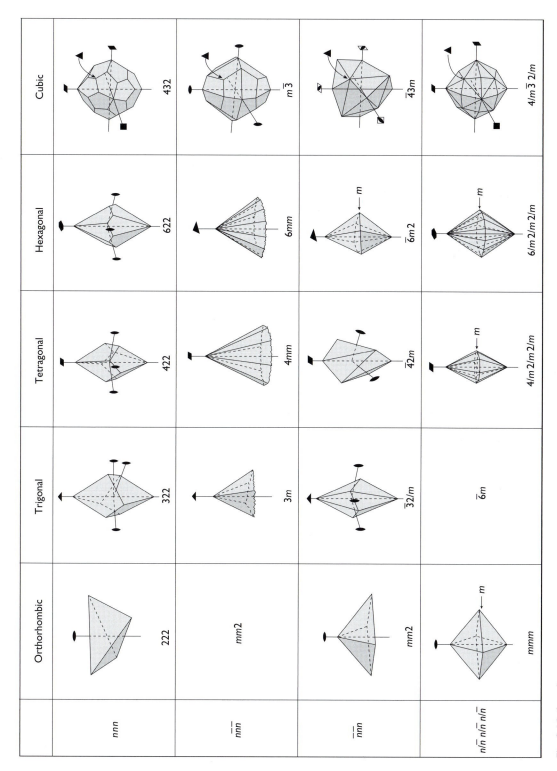

Fig. 8.14 (cont.)

general pole is multiplied through symmetry operations (e.g., in Figure 8.9). This multiplicity is also indicated in Table 8.1 and ranges from 1 (in 1) to 48 in $m\,\bar{3}\,m$. Of the 32 point-groups, 11 have only proper rotations, without mirror planes or an inversion center. As we will see in Chapter 12, some physical properties, such as optical activity, are observed only in crystals from these *enantiomorphic* point-groups. Also, there are 11 point-groups with a center of symmetry (indicated by superscript *a* in the multiplicity column of Table 8.1). In centrosymmetrical point-groups there is for every crystal face (*hkl*) or crystal direction [*uvw*] an equivalent face $(\bar{h}\,\bar{k}\,\bar{l})$ and an equivalent direction $[\bar{u}\,\bar{v}\,\bar{w}]$.

8.4 Crystallographic forms

The symmetry of a crystal is expressed in its morphology. In Figure 8.11, symmetry operations generated a set of poles. Each of these poles is representative of a lattice plane or a crystal face, and the collection of poles represents a polyhedron or form. In some cases this "crystallographic polyhedron" is not a usual geometrical polyhedron for which the whole surface is covered with faces. The form {*hkl*}, for example {123}, in point-group 1 is a single plane. There is no symmetry operation that generates an equivalent face. In other cases the form is a normal polyhedron, such as {123} in point-group 432, which is a pentagon-trioctahedron with 24 faces (Figure 8.8c). If the faces of a form do not cover the whole surface, we call it an *open form*, contrary to a regular *closed form*. For open forms, several forms have to be combined to cover the whole surface of a crystal. In the case of point-group 1, a combination of {100}, {010}, {001}, {$\bar{1}$00}, {0$\bar{1}$0}, {00$\bar{1}$}, each consisting of a single face, would cover the surface.

Every point-group has a characteristic form. Figure 8.14 illustrates, in perspective drawings, all the forms corresponding to the poles in Figure 8.11. These forms are called *general forms*, because they consist of faces with no special relationship, either to crystallographic axes or to the symmetry elements (rotation axes and mirror planes). They have indices {*hkl*} with *h*, *k*, and *l* representing arbitrary numbers without relationship (e.g., {125}). Try to relate the forms and faces in Figure 8.14 to the stereoplots in Figure 8.11. *Special forms*, on the other hand, have special relations to crystallographic axes and symmetry elements and are therefore repeated less often

than the faces in the general form. Miller indices are related (e.g., {111}) or special numbers (e.g., {100}).

We illustrate general and special forms, and corresponding polyhedra, for point-group 4/*m* 2/*m* 2/*m* (Figure 8.15). The general form {*hkl*} has eight poles on the upper hemisphere and eight poles on the lower hemisphere (Figure 8.15a). This corresponds to a pyramid with eight faces on top and eight on the bottom, and crystallographers call it a *ditetragonal bipyramid*. It is called a bipyramid because it has a top and a bottom, and it is called ditetragonal because the 4-fold symmetry is split into eight faces by mirror planes. Symmetrically equivalent (*hkl*)s are labeled in Figure 8.15a. There are several special forms (Figure 8.15b). If the pole is at (001), it is not repeated by the 4-fold rotation. The only symmetry operation that generates an additional pole is the mirror plane perpendicular to the 4-fold axis, producing (00$\bar{1}$). Thus the form {001} consists of only two poles, corresponding to two parallel faces, called a *pinacoid*. If the pole lies on a mirror plane with Miller indices (*h*0*l*) or (*hhl*), the corresponding form is a *tetragonal bipyramid* with only four faces on the top and on the bottom. If the pole lies on the equator, it represents a face parallel to the *z*-axis and the Miller index *l* is zero. The form {*hk*0} is a prism with eight parallel faces (*ditetragonal prism*), while the forms {100} and {110} are prisms with four parallel faces (*tetragonal prisms*).

Miller indices of special forms either contain zeroes or have two or more indices that are the same; for example {100}, {*hhl*}, and {*h*0*l*}. The multiplicity of special forms is lower than that of the general form, since poles that lie on a rotation axis or on a mirror plane are not repeated by the symmetry operation. Each symmetry class (except triclinic) can have a variety of forms, depending upon the relationship between the orientations of lattice planes and symmetry elements. While the general form is unique for a point-group, a general form in one point-group may be a special form in another. For example, we have seen that a tetragonal bipyramid is a special form in 4/*m* 2/*m* 2/*m*, but it is a general form in 4/*m*.

Also, depending on the point-group, a form with the same Miller indices may be a *closed polyhedron*, such as a bipyramid {*hkl*} in point-group 4/*m* (Figure 8.16a), or may be *open* on one side, such as a pyramid {*hkl*} in point-group 4 (Figure 8.16b). The surface of a freely growing crystal may consist of a single form, such as {110} in cubic garnet

(a)

(b)

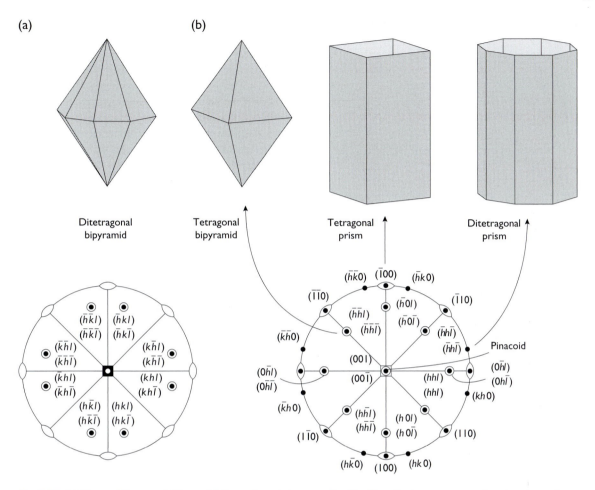

Fig. 8.15 (a) General form and (b) special forms in point-group 4/m 2/m 2/m. On top are perspective drawings of forms, below it corresponding stereograms in which Miller indices of poles are marked.

(a)

(b)

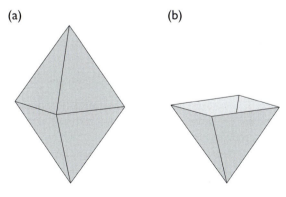

Fig. 8.16 The form {hkl} can be (a) a tetragonal bipyramid (closed form) in point-group 4/m, or (b) a tetragonal pyramid (open form) in point-group 4.

(Figure 8.17a) and {210} in pyrite (Figure 8.17b), or may represent a combination of forms, such as {10$\bar{1}$0}, {10$\bar{1}$1}, {01$\bar{1}$1}, and {51$\bar{6}$1} in trigonal quartz (Figure 8.17c). Special forms with simple Miller indices are more commonly observed than general forms because, as we have already noted in Chapter 7, lattice planes with simple Miller indices are more closely packed with lattice points, usually have a lower surface energy, and are more stable. There is no need to remember the names of all the crystal forms. Nevertheless, everyone should be familiar with the commonest polyhedra and be able to decipher their symmetry (Box 8.1).

The best-known polyhedra are the five regular polyhedra (also known as Platonic solids) with indistinguishable faces, edges, and corners (Table 8.2,

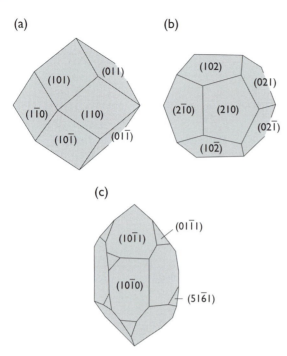

(a)

(011)
(101)
(1̄10) (110)
(101̄) (011̄)

(b)

(102)
(021)
(2̄10) (210)
(021̄)
(102̄)

(c)

(011̄1)
(101̄1)
(101̄0)
(51̄6̄1)

Fig. 8.17 (a) Cubic garnet (($Fe,Ca,Al)Si_3O_{12}$) often crystallizes with a single closed form such as a rhombic dodecahedron {110}. (b) Pyrite (FeS_2) frequently occurs as a pentagon-dodecahedron {210}. (c) Quartz (SiO_2) generally displays a combination of open forms such as a hexagonal prism {101̄0}, trigonal pyramids {101̄1}, {011̄1}, and others.

Figure 8.18). The faces are simple regular polygons, i.e., triangles, squares, and pentagons. As is the case for all other polyhedra, Leonhard Euler's theorem, which relates the number of faces (f), corners (c), and edges (e),

applies: $c + f = e + 2$. Note, however, that only cubic {100}, octahedral {111}, and tetrahedral {111} forms are present in crystals. This situation arises because dodecahedra and icosahedra imply the presence of 5-fold rotation axes, which, as we have previously observed, are not allowed. These polyhedra are incompatible with a lattice structure and cannot be described with rational Miller indices. Minerals of the cubic system may display other polyhedra with 12 faces. For example, pyrite often crystallizes as a so-called pentagon-dodecahedron {210} (Figure 8.17b). However, contrary to the Platonic pentagon-dodecahedron (Figure 8.18d), the pyrite pentagon-dodecahedron is not regular: a closer look reveals that one side of each pentagon face is longer than the others.

Prisms are forms consisting of a group of cozonal faces (with poles lying on a great circle) repeated to form an open-ended tube. Prisms can be produced by a variety of symmetry operations, including the combination $2/m$ (Figure 8.19a), but the commonest types of prism (Figure 8.19b) are generated by a repetition of a face that is parallel to a rotation axis by a 3-fold, 4-fold, or 6-fold rotation (trigonal, tetragonal, and hexagonal prisms, respectively). *Pyramids* are a group of faces inclined to a rotation axis. Figure 8.18c, for example, illustrates a trigonal pyramid with a 3-fold rotation axis. If there is a mirror plane perpendicular to the rotation axis, a pyramid on top is repeated at the bottom (bipyramid in Figure 8.19d). An interesting form is the *rhombohedron* (Figure 8.19e), in which a trigonal pyramid on top is repeated at the bottom not by a mirror plane, but rather by an inversion. The rhombohedron is a typical form observed in calcite. The faces of a rhombohedron are rhombuses. An important open form is the *base* with Miller indices (001) or (0001) in the hexagonal system. It consists of a single plane also known as *basal* plane.

8.5 Some comments on space-groups

The 32 point-groups describe the symmetry of the external crystal morphology as combinations of rotations, mirror reflections, and inversions. At the atomic level another symmetry operation, *translation*, is significant. Translation has already been introduced as the basic principle of the lattice: every atom repeats after a unit cell translation along each of the three crystallographic axes. The German mathematician Arthur M. Schoenflies (1891) and the Russian mineralogist Ephgraph S. von Fedorow (1885) determined

---

Box 8.1 Additional information: Important crystal forms to remember

Cubic

Cube	{100}	Figure 8.18c
Octahedron	{111}	Figure 8.18b
Tetrahedron	{111}	Figure 8.18a
Rhombic dodecahedron	{110}	Figure 8.17a

Lower symmetry

Prism	Figure 8.19a,b
Pyramid	Figure 8.19c
Bipyramid	Figure 8.19d
Rhombohedron	Figure 8.19e

Table 8.2 The five regular polyhedra

Polyhedron	Polygon	Faces	Corners	Edges	Figure
Tetrahedron	Triangle	4	4	6	Figure 8.18a
Octahedron	Triangle	8	6	12	Figure 8.18b
Cube	Square	6	8	12	Figure 8.18c
Dodecahedron	Pentagon	12	20	30	Figure 8.18d
Icosahedron	Triangle	20	12	30	Figure 8.18e

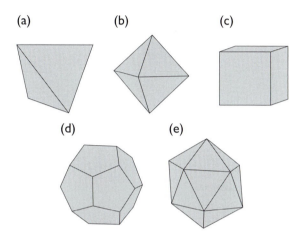

Fig. 8.18 Regular (Platonic) polyhedra: (a) tetrahedron, (b) octahedron, (c) cube (hexahedron), (d) dodecahedron, (e) icosahedron. All are composed of regular polygons (see Table 8.2).

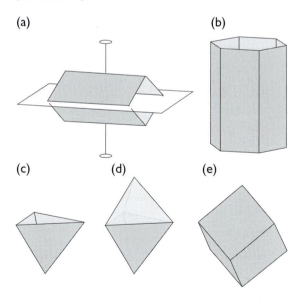

Fig. 8.19 Some important crystal forms: (a) monoclinic prism, (b) hexagonal prism, (c) trigonal pyramid, (d) trigonal bipyramid, (e) rhombohedron.

independently that there can be only 230 different combinations of rotation, mirror reflection, inversion, and translation in crystals with a lattice structure. These combinations are called *space-groups*. The structure of any crystal belongs to one of these 230 groups. Interestingly, the derivation of possible structural symmetries occurred 25 years before the discovery of X-ray diffraction, which confirmed that crystals had indeed a lattice structure. We do not review space-groups in any depth here and just introduce a few characteristic features with an example. Anyone interested in details of space-groups and their symmetry elements should consult the *International Tables for Crystallography*, *Volume A* (Hahn, 2006).

Translation may be combined with rotation and mirror reflection. Take, for example, the structure of the tetragonal mineral anatase (TiO_2), in which six oxygen atoms in octahedral coordination surround titanium. Some of the titanium atoms define the corners of the unit cell (Figure 8.20a). For simplicity only the coordination octahedra are shown. We recognize two mirror planes parallel to the z-axis, which are labeled m in Figure 8.20b (a pair of symmetrically equivalent octahedra is highlighted by darker shading). Also obvious in the anatase structure is a $\bar{4}$ rotoinversion axis parallel to the z-axis that goes through the center of the unit cell (Figure 8.20c). Again a set of symmetrically equivalent octahedra is indicated by shading. The 180° rotation is easy to see, but there is also a 90° rotation, combined with inversion in the center. However, there is more symmetry in this structure than is apparent at first glance.

Perpendicular to the z-axis, at $z = 3/8c$, is a plane on which octahedra are mirrored and can be brought to coincidence if translated half the unit cell distance along a, resulting in a glide component $\frac{1}{2}a$, where a is the lattice vector in the x-direction (Figure 8.20d). Such a mirror-translation plane is called a *glide plane*, and is given the symbol a rather than m, to indicate a glide component of $\frac{1}{2}a$. Glide planes with diagonal

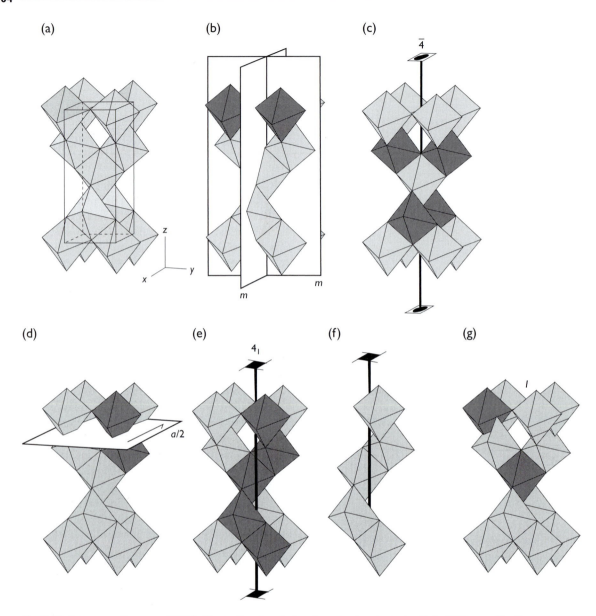

Fig. 8.20 Structure of anatase (TiO$_2$), illustrating translational symmetry elements. Only TiO$_6$ coordination polyhedra are shown. (a) Coordination polyhedra and outline of the unit cell. (b) Conventional mirror planes (*m*) parallel to the *z*-axis. (c) Conventional 4-fold rotoinversion axis. (d) Glide plane (*a*) perpendicular to the *z*-axis. (e) A 4-fold screw axis (4$_1$). (f) Repetition of a single octahedron to highlight the screw character. (g) Repetition of octahedra by a body-centering translation (*I*). Symmetrically equivalent pairs of octahedra are indicated with darker shading (courtesy S. Krivovichev).

glide components have symbols *n* and *d*, depending on the glide component. There is no *n* glide plane in anatase; a *d* glide plane exists, perpendicular to [110] but is difficult to see and is not shown in Figure 8.20. Figure 8.21 compares atomic repetitions for a (010) mirror plane and a *c* glide plane.

Parallel to the *z*-axis, but in the center of the quarter square of side length ½*a*, is an axis that repeats octahedra by a rotation, combined with a translation. This produces a screwlike repetition as indicated for one octahedron in the structure with darker shading (Figure 8.20e) and the assembly of symmetrically

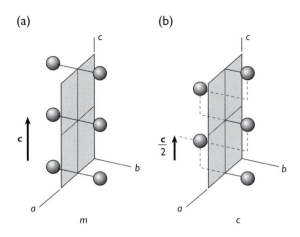

Fig. 8.21 Repetitions of atoms by mirror planes: (a) conventional mirror plane (m), (b) glide plane c with a glide component $c/2$.

equivalent octahedra is shown separately in Figure 8.20f to highlight the screw character. Such an axis that combines rotation with translation is called a *screw axis*. The symbol used to represent this axis is 4_1 to distinguish it from the pure rotation axis 4. The subscript 1 indicates a translation of one-quarter of the unit cell dimension in the direction of the rotation axis. A 4-fold screw axis with a translation of half the unit cell, not present in anatase, would have a symbol 4_2. After four of the tetragonal rotations-translations, we are back at the start (i.e., at the identical point of the next unit cell). Figure 8.22 compares atomic repetitions for a 4-fold rotation axis 4 (Figure 8.22a), a 4-fold rotoinversion axis $\bar{4}$ (Figure 8.22b), and a 4-fold screw axis 4_1 (Figure 8.22c).

In anatase there is yet another translational symmetry element. For each atom at position x, y, z there is an equivalent atom at $x + \dfrac{1}{2}$, $y + \dfrac{1}{2}$, $z + \dfrac{1}{2}$, i.e., a translation of half the body diagonal, bringing the center of the unit cell to coincidence with a corner. This is highlighted in Figure 8.20g for the octahedron centered around the origin. A symbol I is used to designate the cell as body-centered. Anatase has thus a body-centered rather than a primitive unit cell.

The space-group description of the crystal structure has to take all these translational symmetries into account and therefore the space-group symbol is more complex than the point-group symbol. The point-group of anatase is $4/m\ 2/m\ 2/m$ (abbreviated $4/mmm$, Table 8.1). The space-group symbol is $I\ 4_1/a\ 2/m\ 2/d$ (or abbreviated $I\ 4_1/amd$). We read from it that the unit cell is body-centered (I), the 4-fold axis is a screw rather than a pure rotation axis (4_1), the 2-fold axes are simple rotation axes (2). Of the mirror planes perpendicular to the rotation axes, planes perpendicular to [001], the 4_1 axis, at $z = 3/8$ are a glide planes, with a glide component $\frac{1}{2}a$; planes perpendicular to [100] at $x = 0$ and $\frac{1}{2}$ are normal m mirror planes; and planes perpendicular to [110] are diagonal glide planes with a glide component $\frac{1}{2}(-a + b + c)$.

Rather than using stereoplots to represent the symmetry elements of the crystal structure, a representation that outlines the unit cell is standard. Such plots have been tabulated for all 230 space-groups (Hahn, 2006). Figure 8.23 is an example for space-group $I\ 4_1/amd$. Two diagrams are used, both displaying a z-projection of the tetragonal unit cell. The first one

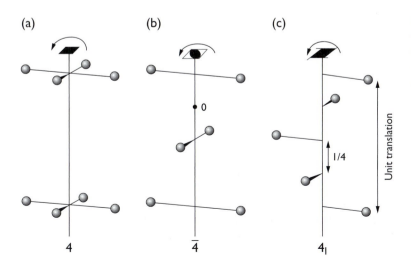

Fig. 8.22 Repetition of atoms by 4-fold rotations: (a) a simple rotation axis 4, (b) a rotoinversion axis $\bar{4}$ (0 indicates an inversion center), and (c) a screw axis 4_1.

(a)   (b)

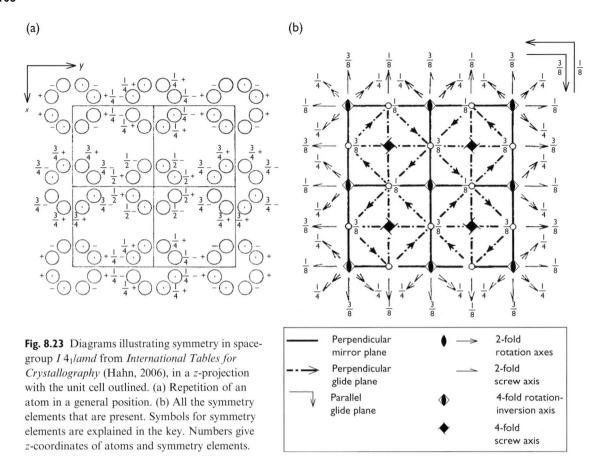

Fig. 8.23 Diagrams illustrating symmetry in space-group $I\,4_1/amd$ from *International Tables for Crystallography* (Hahn, 2006), in a z-projection with the unit cell outlined. (a) Repetition of an atom in a general position. (b) All the symmetry elements that are present. Symbols for symmetry elements are explained in the key. Numbers give z-coordinates of atoms and symmetry elements.

—— Perpendicular mirror plane	● →	2-fold rotation axes
- · - → Perpendicular glide plane	⟶	2-fold screw axis
Parallel glide plane	◆	4-fold rotation-inversion axis
	◆	4-fold screw axis

(Figure 8.23a) shows the unit cell with a general atom at coordinates x, y, z repeating due to symmetry (the symbols $+$, $-$, $\frac{1}{2}+$, $\frac{1}{2}+$ give the z-displacements). As you can count, there are 32 equivalent positions in the unit cell for an atom at general coordinates (in anatase, atoms are all at special coordinates and this number is reduced). The second diagram (Figure 8.23b) shows all symmetry elements. Open circles are inversion centers with the z-coordinates indicated. Regular arrows and squares with lens symbols are rotation axes, while single arrows and squares with wings are screw axes. The d glide plane (dot-dashed) has a $\frac{1}{2}(-\boldsymbol{a} + \boldsymbol{b})$ glide component parallel to the arrow (in the projection plane) and $\frac{1}{2}\boldsymbol{c}$ perpendicular to the projection plane, resulting in a total glide of $\frac{1}{2}(-\boldsymbol{a} + \boldsymbol{b} + \boldsymbol{c})$. The number gives the z-levels of the axes.

8.6 Summary

Crystals are three-dimensional, therefore they are difficult to visualize, analyze, and represent. The chapter

introduces first projections that represent directions mapped on a sphere to a plane, in analogy to geographical projections of a sector on a spherical Earth to a planar map. The two most important projections for crystals are stereographic and equal-area projection. Then the 32 point-groups are derived that describe the crystal morphology and 230 space-groups are introduced that classify the different symmetries of crystal structures. *Point-group symmetry* is a combination of rotation axes, mirror reflections, and inversions. There are monaxial and polyaxial point-groups. By including translation as an additional symmetry element, *space-group symmetry* has more symmetry elements: screw axes in addition to rotation axes, and glide planes in addition to mirror planes.

Test your knowledge

1. Which are the basic symmetry operations that are expressed in the morphology of a euhedral crystal?
2. Take a sheet of tracing paper and a stereonet. Plot the poles to faces $P_1\ \phi = 10°$, $\rho = 40°$, and

P_2 $\phi = 290°$, $\rho = 75°$. Determine the angle between the poles (corresponding to the interfacial angle).

3. Prove, by construction on a stereonet, that a mirror reflection is equivalent to a combination of 2-fold rotation and inversion.

4. Point-group $mm2$ has several symmetry elements. Plot the symmetry elements in stereographic projection, then choose a pole 1 ($\phi = 20°$, $\rho = 30°$) and generate all symmetrically equivalent poles. Assuming that pole 1 has Miller indices (123), what are the Miller indices of all the symmetrically equivalent poles?

5. The eight poles, which you have constructed in the previous question, correspond to a polyhedron. Sketch this polyhedron in a perspective drawing and try to visualize it.

6. Prisms, pyramids, and bipyramids are the most important noncubic crystal forms. Sketch those polyhedra for a crystal with tetragonal symmetry.

7. Octahedron and tetrahedron both have the symbol {111}. Explain the differences between the two on the basis of symmetry.

8. Look at the minerals in Plates 3a-d, 23a,d, 25f, 26a,b, 28e, 29c, 31e, 32g and list those where you can identify specific crystal forms, particularly cubes and octahedra.

Further reading

Borchard-Ott, W. (2011). *Crystallography: An Introduction*, 3rd edn. Springer.

Buerger, M. J. (1971). *Introduction to Crystal Geometry*. McGraw-Hill, New York.

Klein, C. and Dutrow, B. (2007). *Manual of Mineral Science*, 23rd edn. Wiley, New York.

Ladd, M. (2014). *Symmetry of Crystals and Molecules*. Oxford University Press, Oxford.

Phillips, F. C. (1977). *An Introduction to Crystallography*, 4th edn. Longman, London.

Rousseau, J.-J. (1998). *Basic Crystallography*. Wiley, New York.

Sands, D. E. (1994). *Introduction to Crystallography*. Dover Publications, New York.

9 | Crystalline defects

The concept that crystals are compounds with a periodic three-dimensional structure can explain macroscopic morphology and even atomic arrangements. On a closer look one finds that this periodicity is interrupted, perhaps in one out of 1000 unit cells. These interruptions of periodicity are known as crystal defects. They can be the result of growth, phase transitions, and deformation. This chapter describes the most important defects such as vacancies, dislocations, stacking faults, and twins.

9.1 Types of defects

We have introduced crystals as materials with periodic repetitions of atoms in three dimensions. This is very idealized and most real crystals have lattice defects that are introduced during growth, phase transformations, or deformation. *Point defects* are single "mistakes" in the regular and periodic positioning of atoms. They can take the form of a missing atom or *vacancy* (V), or an *interstitial* atom (I) introduced between normal lattice sites (Figure 9.1a). As individuals, point defects are difficult to observe, but, when numerous, they affect macroscopic properties (e.g., chemical composition and electrical resistivity). With *line defects* or *dislocations*, the perfect lattice is disrupted and displaced along a line labeled D which extends perpendicular to the sketch (Figure 9.1b). *Planar defects* are characterized by displacement over a whole plane of atoms such as in the so-called *stacking fault* (SF), again extending perpendicular to the sketch (Figure 9.1c). Other planar defects are *twins* where the orientation flips across a lattice plane. Linear and planar defects cause substantial local distortions of the crystal structure. The strain produced by these distortions can best be imaged with the transmission electron microscope (TEM) because electrons passing through the distorted crystal are deflected (see Chapter 16).

9.2 Point defects

The main influence of a point defect extends over only a few atomic diameters. The defect may be the result of a single atom (vacancy or interstitial; Figure 9.2a) or, more rarely, a cluster of a few. In ionic crystals, the formation of a vacancy involves a local readjustment of charge to maintain neutrality in the crystal as a whole. If a positive ion vacancy is compensated for by a negative ion vacancy, such a pair is called a Schottky defect (Figure 9.2b). Alternatively, a positive vacancy may be compensated by a positive interstitial nearby (Frenkel defect, Figure 9.2c).

9.3 Dislocations (line defects)

When single crystals are subjected to shearing stresses, plastic deformation occurs on one or more sets of defined lattice planes (hkl) known as *slip planes* (see Chapter 17). The slip planes are sheared in a specific lattice direction $[uvw]$, which is the *slip direction*. Large discrepancies between the observed strengths of crystals and their theoretical strengths, calculated on the assumption that slip occurred instantaneously across the slip planes, led E. Orowan (1934), M. Polanyi (1934), G. I. Taylor (1934), and others to introduce the concept of dislocations. Figure 9.3 shows a slip plane, over part of which slip has occurred. This slippage has transported the upper part of the crystal over the lower part by one unit cell in the direction of the vector b, so that the lattice perfection is restored across the slipped area except along the curved *dislocation line* extending from point A to point B. The displacement vector b is called the *Burgers vector*. At A, where the dislocation line is perpendicular to b, the lattice distortion is effectively caused by an extra vertical plane on top. Such a distortion is called an *edge dislocation* (Figure 9.4a). At B, where the dislocation line is parallel to *b*, the lattice planes perpendicular to it are distorted and form a continuous helix, so that the distortion is a *screw* (Figure 9.4b). It is apparent that the character of a dislocation and of the strain field around it

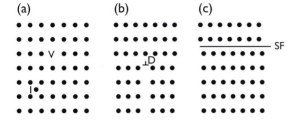

Fig. 9.1 Defects can be classified by the way they interrupt an ideal lattice. (a) With point defects a lattice point is missing (vacancy, V) or an additional lattice point is present (interstitial, I). (b) If an extra lattice plane is inserted in part of the crystal, this plane produces a line defect or dislocation (D). (The dislocation is noted by the upside-down T symbol.) (c) Part of the lattice may be displaced across a plane, producing a planar defect. In this case the defect is a stacking fault (SF). Dislocation and stacking fault extend perpendicular to the sketches.

changes as its orientation changes with respect to *b*. In general a dislocation will have edge and screw components. Slip of the whole upper part of the crystal in Figure 9.3 over the lower part occurs when the dislocation glides across the slip plane. Ultimately all the bonds across the slip plane have been broken, but the process occurs only in the vicinity of the dislocation so that relatively low stresses are required to cause slip and changes in crystal shape.

Taylor (1934) inferred the presence of dislocations mainly because of the weakness of many crystals when subjected to a stress. Without dislocations crystals would be extremely strong because many bonds would need to be broken at once. His model has much later been confirmed by direct imaging of dislocations with the transmission electron microscope (TEM) (see Chapter 16). Since the electronic structure is distorted along the dislocation core, this distortion leads to strong contrast when electrons pass through a crystal, and dislocations appear as lines (Figure 9.5). Dislocations are originally introduced during crystal growth at a density of about 10^6 dislocations/cm^2, corresponding about to 1 dislocation every 10 μm. They multiply during deformation and can reach densities as high as 10^{14} dislocations/cm^2 (as will be discussed in Chapter 17).

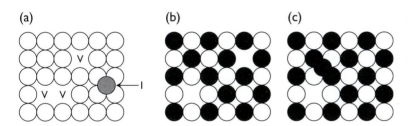

Fig. 9.2 Point defects in crystals. (a) Vacancies (V) and interstitial (I). (b) Schottky defects in ionic crystals with vacant cations (black) and anions (white) to balance charge. (c) Frenkel defects in ionic crystals with a cation vacancy balancing a cation interstitial (black).

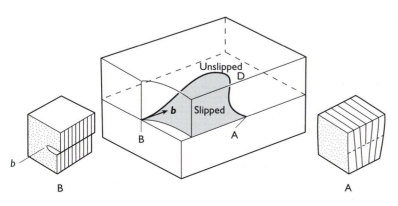

Fig. 9.3 A dislocation is produced when the upper part of a crystal slips over the lower part in the direction *b*. The geometry is different if the displacement (Burgers vector *b*) is perpendicular (A) or parallel (B) to the dislocation line D.

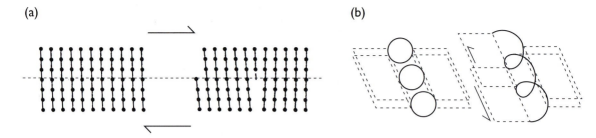

Fig. 9.4 Enlarged image of the structural distortion of a crystal during shear deformation: (a) an edge dislocation and (b) a screw dislocation.

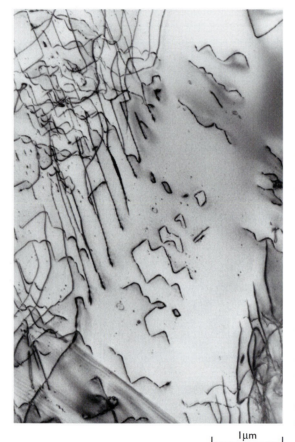

1 μm

Fig. 9.5 TEM image of dislocations in experimentally deformed dolomite (courtesy D. J. Barber; see also Barber *et al.*, 1981).

9.4 Planar defects during growth

Stacking faults

In Chapter 2 we discussed the structure of metals with a regular stacking of hexagonal close-packed layers (Figures 2.6 and 2.7). As a crystal grows it adds

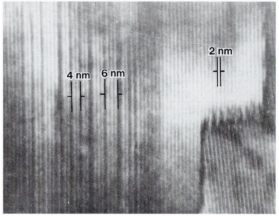

Fig. 9.6 High-resolution TEM image of lattice planes with stacking faults in sericite. On the right side is a grain boundary between two misoriented crystals (Page and Wenk, 1979).

another layer of atoms over the depressions in the first layer. When it comes to the third layer there are two options, adding atoms over the atoms in the first layer or adding atoms over the second type of depressions in the first layer, thus producing either hexagonal close-packing as in zinc or cubic close-packing as in copper. Energy differences are relatively small since the closest neighbor environment is the same. We mentioned stacking of the type A–B–A–B... versus A–B–C–A–B–C... There is a fair probability that the growing crystal may make a mistake, resulting in a sequence such as A–B–A–B–A–B–C–A–B–A, interrupting the periodic pattern. This produces a stacking fault, which can be geometrically described by a translation to restore the ideal crystal.

Stacking faults are present in many minerals, but particularly pervasive in sheet silicates (Chapter 29) where they interrupt the layered structure, as in polytypes of muscovite (Figure 9.6).

Twinning

During regular growth one lattice layer after the next is added to a crystal face. The new layer has the same structure and composition and is in the same orientation as the previous one (Figure 9.7a). However, there is a certain chance that a new layer is added in the "wrong" orientation, still maintaining continuity and coherence (Figure 9.7b). As with stacking faults, this change in orientation increases the internal energy only very slightly, because only second-closest neighbors across the interface are no longer in equilibrium positions. The final result is a bicrystal with identical structure and composition on either side of the planar defect, but with different orientations. Such an intergrowth, called a *twin*, is usually recognized by the fact that there are re-entrant angles that generally do not exist in single crystals. In a euhedral single crystal, all interfacial angles are larger than 180° when measured on the outside of a crystal (Figure 9.7a), but the twin in Figure 9.7b shows a re-entrant angle β that is smaller than 180°.

In general, twins are defined as intergrowths in which the two parts (twin and host) share a lattice plane (*twin plane*), or a lattice direction (*twin axis*). The relative orientation of the two parts of the crystal is well defined, either as a mirror reflection across the twin plane or as a rotation around the twin axis. Both the twin plane and the twin axis can be described with

rational indices. However, neither can be a symmetry element in the point-group of the crystal, for otherwise the twin operation would not create a new orientation.

Host and twin may join along a rational crystal plane, and such twins are called *contact twins*. In reflection twins, this plane can, but need not, be the twin plane. The plane of intergrowth is called the *composition plane*. A good example in which the twin and composition planes are coincident is a simple twin on the basal plane of calcite, shown in an actual sample and an idealized sketch in Figure 9.8a. Other examples of contact twins are the swallow-tail twin in monoclinic gypsum, where (100) is both the twin plane and the composition plane (Figure 9.9a). The multiple so-called "Albite law" twins in plagioclase feldspar, often occurring in a lamellar pattern, are also of this type (Figure 9.9b). The twin plane is (010).

Other twins are complex, and often patchy intergrowths of two crystals exist. In such *penetration twins* there is generally no obvious contact plane between twin and host, just an irregular surface, yet here too there is a crystallographically defined twin plane or twin axis relating the two parts. A good example is trigonal cinnabar with [0001] as twin axis (180° rotation), shown in Figure 9.8b. Other examples of penetration twins are [110] twins in fluorite (Figure 9.9c), [001] twins in pyrite that produce the "iron cross" morphology for dodecahedral habit (Figure 9.9d), cross-shaped twins in staurolite with {031} as twin plane (Figure 9.9e), and [001] twins in orthoclase ("Carlsbad law") (Figure 9.9f).

The relative orientation of host and twin is best visualized with stereographic projection and this is illustrated for quartz in Box 9.1.

While twins frequently form during crystal growth, they can also result from phase transformations or develop during deformation (see Chapter 17). In calcite {01-18} is a deformation twin and Plate 2d shows how twin lamellae look in a thin section viewed along the c-axis. Some commonly observed twin laws are summarized in Table 9.1. More examples will be discussed with specific minerals in Part V.

9.5 Planar defects during phase transformations

As we have noted in our discussion of polymorphism and solid solutions (Chapter 3), many minerals undergo phase transformations during cooling. At

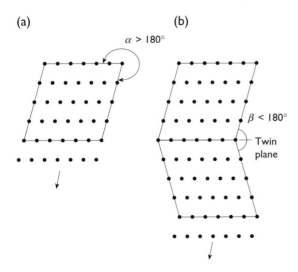

Fig. 9.7 Origin of twinning as a growth defect. (a) Growth of a single crystal with addition of lattice planes. α denotes the exterior interfacial angle. (b) Formation of a twin if a lattice plane is added in the "wrong" orientation, producing a re-entrant angle β.

(a)

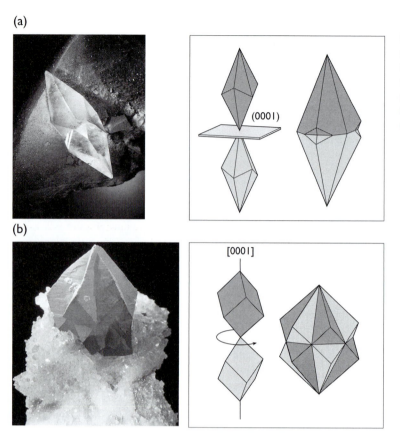

Fig. 9.8 Examples of actual twins in minerals (left side) and the schematic interpretation of the morphology (right side). (a) Contact twin of calcite, (b) penetration twin of cinnabar (courtesy O. Medenbach).

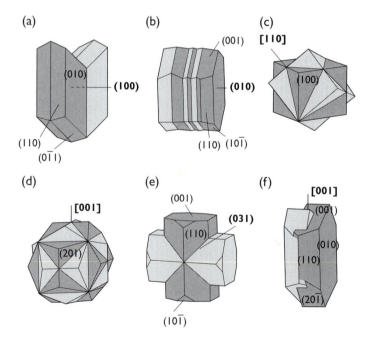

Fig. 9.9 Some typical twins in minerals: (a) (100) contact twin in gypsum ("swallow-tail"), (b) lamellar [010] contact twins in triclinic plagioclase feldspars ("Albite law"), (c) [110] twin in fluorite with cubic morphology, (d) [001] twin in dodecahedral (210) pyrite (so-called "iron cross"), (e) cross-shaped (031) twins in staurolite, (f) [001] penetration twin in orthoclase ("Carlsbad law"). Twin and host with different shading. Twin operation (mirror plane or rotation axis) marked with bold letters.

Box 9.1 Additional information: Stereographic projection and twinning in quartz

Low-temperature α-quartz (SiO_2) is trigonal and crystallizes in point-group 32. This point-group has no center of symmetry and no mirror planes. Figure 9.10a displays a stereoplot both of symmetry elements in quartz and of poles of the unit rhombohedron $\{10\bar{1}1\}$. If we produce an intergrowth in which two crystals are related by a 2-fold rotation on [0001], we obtain the pattern shown in Figure 9.10b. This intergrowth, called a *Dauphiné twin*, has effectively a $2 \times 3 = 6$-fold axis overall, but not locally. The two intergrown crystals are related by a rotation and therefore host and twin are both either right or left handed. A sketch of a Dauphiné penetration twin is added in Figure 9.10e, with the twin indicated by shading. The composition surfaces are generally irregular. In another intergrowth of quartz, the two crystals are related by a mirror reflection on $\{11\bar{2}0\}$ (Figure 9.10c). In such *Brazil twins*, projections of poles of host and twin coincide and the overall pattern remains trigonal. But the mirror reflection transforms the handedness. If the host is left handed, the twin is right handed. In this case twinning has effectively added a center of symmetry. Also here the twins are often intergrown with an irregular interface (Figure 9.10f).

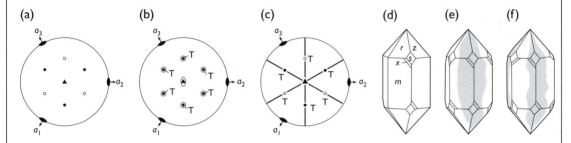

Fig. 9.10 Twinning in quartz. (a)–(c) Stereographic projections of quartz, illustrating symmetry elements and projection of the poles of the $\{10\bar{1}1\}$ rhombohedron (r). The mirror planes are marked with heavy lines. (d)–(f) Sketches of corresponding crystals with faces $m = \{10\bar{1}0\}, r = \{10\bar{1}1\}, z = \{01\bar{1}1\}, s = \{11\bar{2}1\}$, and $x = \{5\bar{1}\bar{6}1\}$. (a,d) Single crystal, (b,e) Dauphiné twin, (c,f) Brazil twin. The twin is indicated by T in the stereograms and by shading in the crystal sketches.

high temperature, they are often structurally disordered and chemically homogeneous. Upon cooling, they may order themselves to a structure with a lower symmetry (e.g., from hexagonal to trigonal quartz, or from monoclinic sanidine to triclinic microcline). Alternatively, a mineral that is homogeneous at high temperature may separate into domains of different composition at lower temperature by a process called *exsolution*. The substitution may involve all atoms in the structure, such as in Au–Cu, or only a certain set such as K and Na in alkali feldspars $(K,Na)AlSi_3O_8$.

This is illustrated schematically for a solid solution with atoms A and B in Figure 9.11. At high temperature, A and B are distributed randomly, i.e., on every site in the structure there is an equal probability of

having either A or B (Figure 9.11, top). A unit cell is outlined by shading. When this system is cooled, two different processes may take place by diffusion and rearrangement of atomic species.

If like atoms attract each other, there is a tendency for a solid solution to undergo a phase separation by *exsolution*, resulting in domains of different composition of A and B (Figure 9.11, bottom left). The boundary between the compositionally different domains within a macroscopic crystal is called an *interphase interface* and is often planar to minimize the lattice misfit between the domains. Since A and B are atoms of different size, lattice parameters a_A and a_B in the domains will also be different. Frequently planar interfaces repeat in a macroscopic crystal, giving rise to an

Table 9.1 Commonly observed twin laws in minerals

Mineral	Twin plane (*m*)	Twin axis (2)	Name of law	Common origin
Quartz		[0001]	Dauphiné	Growth, deformation, transformation
Quartz	$\{11\bar{2}0\}$		Brazil	Growth
Calcite		[0001]		Growth
Calcite	(0001)			Growth
Calcite	$\{01\bar{1}8\}$			Deformation
Aragonite	$\{110\}$			Growth
Gypsum	(100)		Swallow-tail	Growth
Pyrite		$\langle 001 \rangle$	Iron cross	Growth
Cinnabar		[0001]		Growth
Magnetite	$\{111\}$		Spinel	Growth
Perovskite	$\{121\}$			Growth, transformation
Staurolite	$\{031\}$			Growth
Sanidine		[001]	Carlsbad	Growth
Microcline	(010)		Albite	Transformation, growth, deformation
Microcline		[010]	Pericline	Transformation, growth, deformation

array of exsolution lamellae that are often submicroscopic. Calcium and magnesium in pyroxenes frequently undergo exsolution.

If unlike atoms attract, the result is an ordered arrangement of A and B, with the two atomic species alternating (Figure 9.11, bottom right). In this case, the unit cell (shaded) is larger to account for the

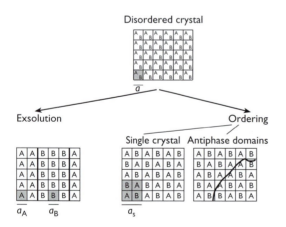

Fig. 9.11 Distribution of atoms A and B in a macroscopic crystal. At high temperature (top), atoms A and B are disordered. At low temperature (bottom), exsolution occurs if like atoms attract each other (bottom left), and ordering takes place if unlike atoms attract each other (bottom right). During exsolution and/or ordering, domains may form that differ in composition and/or ordering pattern.

periodic repeat of A and B with a lattice parameter a_s. Aluminum and silicon in feldspars are examples of elements with a tendency for ordering.

If exsolution occurs, the two phases may have very similar lattice parameters. In such a case the inclusion of A in B is strictly *coherent* (Figure 9.12a). If there is a slight difference and inclusions are small, lattices may still be coherent but with strain across the interface (Figure 9.12b). For larger inclusions, dislocations accommodate the strain (D in Figure 9.12c). An example is shown in Figure 9.13, where hematite (Fe_2O_3) inclusions in ilmenite ($FeTiO_3$) are decorated with a regular array of such "misfit" dislocations. If the structures of A and B are very different, the intergrowth becomes *incoherent* (Figure 9.12d).

We now look at some aspects of *ordering*. As we have seen, the rearrangement of atoms from a disordered to an ordered state is often accompanied by a reduction in symmetry. This loss in symmetry can be due either to compositional ordering (such as in the Al–Si distribution in feldspars), or positional ordering (such as the geometry of silicon tetrahedra in α- and β-quartz). Once ordering commences from two independent nucleation sites in a disordered crystal (Figure 9.14a), the ordered regions or domains continue to expand until they impinge on each other (Figure 9.14b). The domains will either coalesce into a single domain (this occurs when the atomic sequences at the interfaces are perfectly "in step"), or will be separated by a boundary. An ordered crystal

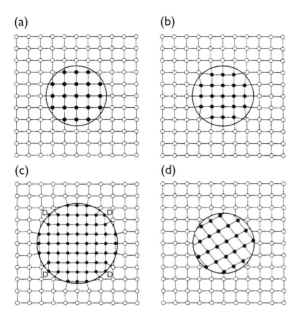

Fig. 9.12 (a) During exsolution the two phases (black and white circles) may have very similar lattice parameters, and thus an inclusion of the black phase in the white phase is coherent. (b) Generally lattice parameters are slightly different, resulting in a lattice strain across the interface. (c) If the inclusion is larger, this strain is accommodated by periodic dislocations (D). (d) If the structures of the black and white phases are very different the intergrowth becomes incoherent.

Fig. 9.13 TEM image showing small lenticular inclusions of hematite in ilmenite from Bancroft, Ontario, with misfit dislocations decorating the boundary between hematite and ilmenite regions. The larger dark lines are normal dislocations in ilmenite (from Lally *et al.*, 1976).

can consist of one domain or of many domains. The number and size of the domains are dependent upon the crystal's cooling history. Unlike exsolution domains, ordering domains all have the same chemical composition (Figure 9.11, bottom right).

Boundaries that separate ordered domains can be classified according to the structural change across the boundary. If the lattices in the two domains can be brought to coincidence by a translation, the boundaries between the lattices are called *antiphase boundaries* or APBs. Note that "antiphase" has nothing to do with antimatter and the name relates to the wave description of crystal structures. APBs require the presence of an ordered superstructure (Figure 9.11, bottom right).

The types of defect that develop during a phase transformation depend on the structural changes, particularly the change in space-group symmetry. For example, in the system Au_3Cu translational symmetry is lost by going from an fcc structure to a primitive cubic structure and this results in APBs. On the other

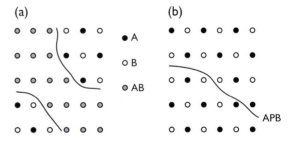

Fig. 9.14 When different regions in a disordered crystal start to order independently from nucleation sites and ordered regions grow (a), these regions may form a single crystal when they coalesce or (b) produce an antiphase boundary (APB) with two regions "out of phase".

hand, transformation of hexagonal β-quartz to trigonal α-quartz involves a loss in rotational symmetry and may produce twinning.

The mineral perovskite ($CaTiO_3$) is an example where ordering transformations can produce both

(a) (b)

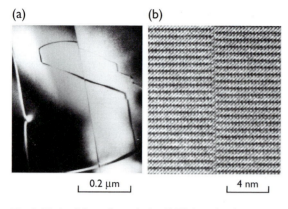

| 0.2 μm | | 4 nm |

Fig. 9.15 Antiphase boundaries (APBs) and twin boundaries in perovskite due to phase transformations. (a) Low-magnification TEM image of curved APB crossing a twin boundary. (b) High-resolution micrograph of an APB where the translational offset of lattice planes across the boundary is easily visible (from Hu *et al.*, 1992).

twin boundaries and APBs. Figure 9.15a shows a TEM image of a curved APB crossing a straight twin boundary. The high-resolution image of the same APB displays the translational offset of lattice planes across the antiphase boundary (Figure 9.15b). Planar defects may be produced during phase transformations, but they can also form during growth and even during deformation.

9.6 Quasicrystals

Quasicrystals violate the conventional rules of crystallography because their structure is not strictly periodic. We have mentioned earlier that icosahedra are not compatible with lattice symmetries, especially 5-fold rotations. About a hundred substances of this type are currently known and they are a focus of new technologies because of their unusual properties. So far only one type of quasicrystal has been identified in nature. Icosahedrite is a fine-grained Cu–Zn–Fe–Al alloy deposited on the surface of a volcanic rock from Kamchatka, Siberia (Bindi *et al.*, 2011).

9.7 Radiation defects and radioactive decay

The defects discussed above have a well-defined geometry. Also, the density of the defects is relatively low, so that overall the lattice remains intact. Radiation,

however, can inflict more serious damage to the crystal structure. Take, for example, a small inclusion of zircon ($ZrSiO_4$) in a biotite ($K(Mg,Fe)_3\ Si_3(Al,Fe) O_{10}(OH)_2$) crystal. Zircon generally contains traces of uranium and thorium, which emit γ-radiation. Over geological time, this radiation alters the crystal structure of biotite. One result of these alterations is the development of halos in biotite, with different colors that surround the zircon inclusions, which can be readily observed with a petrographic microscope (Plate 2e,f, left side of the image). We introduce here for the first time thin sections and microscope images. They will be discussed in detail in Chapters 13 and 14. Plate 2e is an image with plane polarized light which illustrates color effects. Plate 2f is an image with crossed polarizers which is produced by interference and emphasizes anisotropy of optical properties. Similar halos are observed around inclusions of monazite ($CePO_4$) in cordierite ($Al_3(Mg,Fe^{2+})_2Si_5AlO_{18}$) (Plate 2e,f, right side of image). More generally, the lattice structures of uranium- and thorium-containing minerals are often completely destroyed over time in a process known as *metamictization*.

Some isotopes of elements are subject to such decay and are called radioactive isotopes; for example, $^{40}_{19}K$ which transforms to $^{40}_{18}Ar$ through electron capture:

$$^{40}_{19}K + e^- \rightarrow {}^{40}_{18}Ar + \gamma$$

where γ is high-energy electromagnetic radiation that is emitted as the nucleus drops into a less-excited state. The decay is not spontaneous but occurs with a probability over a time period. It is measured by "half-life", the time it takes for half of the unstable isotopes to decay. In the case of $^{40}_{19}K$ the half-life is 1.28×10^9 years. By determining the isotopic composition of a mineral (e.g., $^{40}_{19}K$ and $^{40}_{18}Ar$ in microcline feldspar) its age can be estimated.

Another reaction is

$$^{87}_{37}Rb \rightarrow {}^{87}_{38}Sr + \beta$$

where β is radiation in the form of electrons, converting neutrons into protons. The half-life of $^{87}_{37}Rb$ is $\sim 5 \times 10^{11}$ years.

More complex radioactive decay occurs in chains, such as $^{238}_{92}U$ decaying first into $^{234}_{90}Th$ and ultimately into $^{206}_{82}Pb$. The goal of isotope geochemistry is to estimate formation conditions and ages of rocks from the isotopic composition of minerals.

9.8 Summary

While the overall structure of crystals is periodic, there are nevertheless defects that interrupt the regular stacking and are significant for properties such as chemical composition and strength. Defects are classified into point defects (vacancies and interstitials), line defects (edge and screw dislocations), and planar defects (interphase interfaces, antiphase boundaries, stacking faults, and twins). Defects are introduced during growth, phase transformations, and deformation. Planar and line defects are described with Miller and zone indices that relate their geometry to the unit cell. In radioactive minerals radiation defects are produced that ultimately destroy the crystal structure of the mineral and its surroundings (metamictization).

Test your knowledge

1. Give examples of atom pairs that form solid solutions in minerals (review Chapter 3).
2. Carbonates can be used to illustrate both isomorphism and polymorphism. Give two mineral examples for each.
3. Explain how antiphase boundaries form.
4. Describe the lattice deformation around an edge dislocation.
5. Give important examples of twins in specific minerals.
6. Give an example of a mineral where metamictization is observed.

Further reading

Barrett, C. S. and Massalski, T. B. (1980). *Structure of Metals*, 3rd edn. Pergamon Press, Oxford.

Hirth, J. P. and Lothe, J. (1982). *Theory of Dislocations*, 2nd edn. Wiley, New York.

Hull, D. and Bacon, D. J. (2011). *Introduction to Dislocations*, 5th edn. Butterworth-Heinemann, Oxford.

Kelley, A. A. and Knowles, K. M. (2012). *Crystallography and Crystal Defects*, 2nd edn. Wiley, New York.

Klassen-Neklyudova, M. V. (1964). *Mechanical Twinning of Crystals*. (Translated by J. E. S. Bradley.) Consultants Bureau, New York.

10 | Crystal growth and aggregation

Chapter 8 introduced crystal forms compatible with the 32 point-group symmetries. Ideal morphologies are rarely observed, and here complexities of growth are described. The concept of crystal habit, growth effects such as epitaxy, chemical zoning, porphyroblasts, dendrites, and different types of aggregation are introduced.

10.1 Crystal habit

Crystals have extraordinary properties of internal structure and external morphology. In several chapters we have described these features in detail, with emphasis on symmetry. In this chapter we will take a broader look at the morphology of minerals. Unfortunately, most of the time, we do not observe ideal symmetrical polyhedra, as described in Chapter 8. We will explore some of the complexities of crystal growth and their morphologies.

The external appearance of a crystal, its combination of crystal forms, and the relative development of these forms are collectively called the *crystal habit*. Even though minerals often do not occur as regular

polyhedra, there is nevertheless a characteristic shape to many minerals and it is used in mineral identification (see Chapter 6). There are several terms to describe this appearance. Crystals can be *equiaxed* or *equant* (similarly developed in all three dimensions; Figure 10.1a–c). This can be in the cubic system such as pyrite (Figure 10.1a), or sphalerite (Figure 10.1b), but equiaxed crystals also exist in other systems such as trigonal hematite (Figure 10.1c). Crystals can be *elongated* if one dimension dominates (Figure 10.1d–g), or *flattened* if one dimension is suppressed (Figure 10.1h–j). *Columnar*, *prismatic*, *acicular*, *fibrous*, and *hair-like* are terms used to distinguish different elongated crystal types, depending on how much

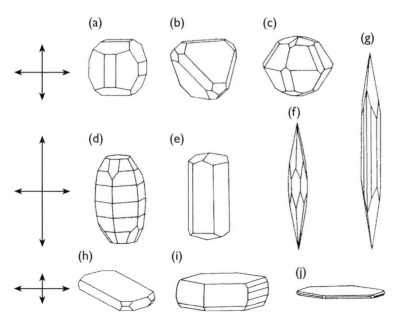

(a) (b) (c) (g) (f) (d) (e) (h) (i) (j)

Fig. 10.1 Crystal habit: (a–c) equant, (d–g) elongated, and (h–j) flattened crystals. (a) Pyrite with dodecahedron and cube; (b) sphalerite with tetrahedron dominating; (c) equiaxed hematite; (d) barrel-shaped corundum; (e) prismatic calcite; (f) acicular hematite; (g) acicular stibnite; (h) tabular orthoclase; (i) platy muscovite; (j) platy hematite.

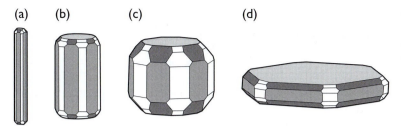

Fig. 10.2 Different habits in a hexagonal mineral with prismatic, rhombohedral, and basal forms. (a) Acicular, (b) prismatic, (c) equiaxed, (d) platy. Different shadings are applied to equivalent forms.

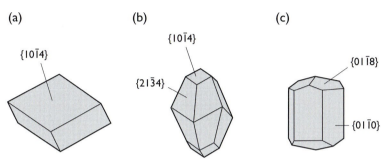

Fig. 10.3 Euhedral crystals of calcite: (a) rhombohedron $\{10\bar{1}4\}$; (b) combination of $\{21\bar{3}4\}$ (scalenohedron) and rhombohedron $\{10\bar{1}4\}$; (c) combination of prism $\{10\bar{1}0\}$ and rhombohedron $\{01\bar{1}8\}$.

one-dimensional growth dominates. *Tabular* and *platy* are terms used to describe flattened crystals.

In prismatic crystals one may recognize a 3-, 4-, or 6-fold rotation axis parallel to the elongation. Elongated and flattened crystals are rarely observed in the cubic system but may occur in any other system. The same mineral may occur with an equant, acicular, or fibrous habit, depending upon the conditions under which it grows. An example is hematite, which can be equant (Figure 10.1c), acicular (Figure 10.1f), or platy (Figure 10.1j). Figure 10.2 shows a hexagonal mineral with several forms (two prisms, several pyramids, and a base) and the corresponding morphology for fibrous, prismatic, equiaxed, and platy habit. The variation in relative size of corresponding faces is indicated by different shading patterns. For acicular growth the basal form is barely developed, while it is large for platy habit.

Qualitative terms that describe the habit have been used in assigning names to many minerals, often in a Greek translation. For example, scapolite (an aluminosilicate) obtained its name due to its prismatic habit after the Greek *skapos* (meaning "pillar"), acmite (a variety of pyroxene) has been named according to its acicular shape after the Greek *acme* (meaning "spike"), and sanidine (one of the feldspars) reflects its tabular crystal habit after the Greek *sanis* (for "board").

Sometimes it is possible to recognize individual forms. Halite generally displays a cube $\{100\}$ (see Figure 1.5b, Plate 3a), while magnetite shows an octahedron $\{111\}$ (Plate 3b). For other minerals crystal forms are more variable. Pyrite may occur as a cube $\{100\}$ (Plate 3c) or as a pentagon-dodecahedron $\{210\}$ (Plate 3d, Figure 8.17b). Calcite may display a rhombohedron $\{10\bar{1}4\}$ (Figure 10.3a), or a combination of the form $\{21\bar{3}4\}$, a so-called scalenohedron, and the rhombohedron $\{10\bar{1}4\}$ (Figure 10.3b). Sometimes calcite develops a prism $\{10\bar{1}0\}$, capped by a rhombohedron $\{01\bar{1}8\}$ (Figure 10.3c). Quartz, growing in free space, is almost always prismatic, with a prism $\{10\bar{1}0\}$, a large positive rhombohedron $\{10\bar{1}1\}$, and a smaller negative rhombohedron $\{01\bar{1}1\}$ (Figure 8.17c, and the book cover).

Striations on faces form if a crystal grows in a solution under changing chemical concentration, either decreasing or increasing. In pyrite the distinctly striated cube face is a combination of dominant $\{100\}$ with subsidiary pentagon-dodecahedral growth $\{210\}$. The striations on the cube faces correspond to the intersection of the cube faces with faces of the pentagon-dodecahedron (Figure 10.4a and Plates 3c, d). These striations in pyrite are evidence that pyrite crystallizes in point-group $m\bar{3}$, which does not have a 4-fold axis perpendicular to the cube face. Similarly, striations on the prism face $\{10\bar{1}0\}$ of quartz result

(a)

(b)

Fig. 10.4 (a) Striations on cube faces of pyrite from the Kassandra Peninsula in Macedonia due to a growth that combines cube (major) and pentagon-dodecahedron (minor) (courtesy J. Arnoth). Width 8 cm. (b) Striations on the prism faces of quartz with Tessin habit from Val Bedretto, Central Alps, due to a combination of prism (major) and rhombohedron (minor) (courtesy T. Schuepbach). Height of crystal 8 cm.

from a combination of dominant prismatic and subsidiary rhombohedral $\{10\bar{1}1\}$ growth (Figure 10.4b).

10.2 Nucleation and growth

Crystal nucleation and growth generally begin in an environment when the concentrations of the elements of a crystal reach a certain level of supersaturation in the environment. This level can be reached by evaporation of a solvent as, for example, water in desert lakes and marine lagoons, with crystallization of evaporite minerals such as halite, sylvite, and gypsum. A temperature decrease may also initiate crystallization: examples are ice from liquid water and igneous minerals in cooling magmas. Finally, crystallization may be initiated by chemical reactions. An example is metamorphic schist that forms when mudstone, composed of clays, recrystallizes at elevated temperatures and pressures and new minerals crystallize, such as mica, feldspar, quartz, and garnet. These environments are very diverse and many factors are involved, with no satisfactory answers; e.g., why calcite sometimes forms with the morphology of a rhombohedron and sometimes with prismatic habit (Figure 10.3).

First a crystal needs to nucleate. If some precursor crystals are available, new crystals may start growing on surfaces of substrates, as for example rutile (TiO_2) on hematite (Fe_2O_3) (Plate 3e). Newly formed minerals may share a lattice plane and the substrate face is used as a template. Such minerals are said to be in an *epitaxial* relationship. Substrates do not need to be the same crystal but have some structural similarities. Epitaxy is of great industrial importance, for example in the manufacturing of thin crystal films with favorable orientations. High-temperature superconducting oxides are difficult to grow as single crystals, but they readily precipitate on crystals of corundum (sapphire), because of the similarity in structure.

In other cases nucleation and orientated growth are not controlled by the structure of the substrate but rather by the orientation relative to the surface, with selected growth in the fastest direction perpendicular to the surface. Quartz crystals in fissures usually grow with *c*-axes perpendicular to the free surface (e.g., Plate 2a,b).

Since the crystal structure is anisotropic, growth is also direction dependent and this is the reason why a characteristic crystal morphology develops. This is illustrated in the sketch of Figure 10.5. The starting

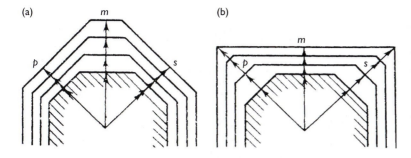

(a)   *m*

p   *s*

(b)   *m*

p   *s*

Fig. 10.5 Sketch to illustrate the effect of relative growth velocities on the dominance of faces. Faces with the slowest growth velocities dominate the morphology. (a) Slower growing faces *p* and *s* begin to dominate face *m*. (b) Faces *p* and *s* are dominated by slower growing face *m*.

point is the shaded crystal. In Figure 10.5a the direction perpendicular to *m* grows fastest and in Figure 10.5b the directions perpendicular to *p* and *s* grow fastest. The final morphology is determined by the slowest growing directions.

This can be complicated by competition between different crystals, as for quartz growing in an open druse, with preferential growth parallel to the *c*-axis. In Figure 10.6 crystals are originally oriented fairly randomly on a surface (I). As they grow, crystallites impinge on neighboring grains which limits their growth (II). Only those crystals that have a growth vector oriented towards the free space of a cavity continue to grow and determine the ultimate orientation selection (III).

Examined closely, even euhedral crystals often do not display ideally planar surfaces, but instead may contain striations, indentations, and conical protuberances. Crystal growth is locally heterogeneous, and the assumption that consecutive layers of lattice planes are regularly added is highly simplified. In detail, dislocations (see Chapter 9) are a critical ingredient of crystal growth (Frank, 1949). Figure 10.7a illustrates the lattice structure around a screw dislocation with a step in the lattice plane. In such steps it is most easy to add new unit cells, and growth preferentially proceeds from screw dislocations in a spiral fashion (Figure 10.7b). Growth spirals were observed for silicon carbide (Figure 10.8a) and twisted crystals have been documented in quartz (Figure 10.8b) (see, e.g., Shtukenberg *et al.*, 2014).

Growth velocities of minerals vary greatly. Very fast growth (cm/hour) may occur during temperature reduction in a saturated saline lake, such as crystallization of borax in Searles Lake, California, USA (Smith, 1979). Very slow velocities (~0.05 μm/1000 years) have been suggested for growth of dolomite in

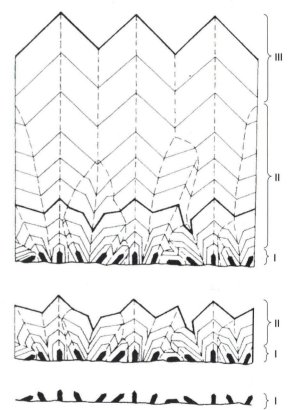

III

II

I

II

I

I

Fig. 10.6 Sketch illustrating changes in morphology as amethyst grows in a druse. Originally the crystal orientation may be fairly random (I), at a later stage crystals impinge (II), but ultimately only those crystals with a growth direction towards the free space dominate (III) (after Grigor'ev, 1965).

mudstones of nearby Deep Spring Lake (Peterson *et al.*, 1963). In cooling magmas of Kilauea volcano in Hawaii, growth rates of plagioclase of 15–30 μm/ year have been documented (Cashman and Marsh, 1988).

(a)

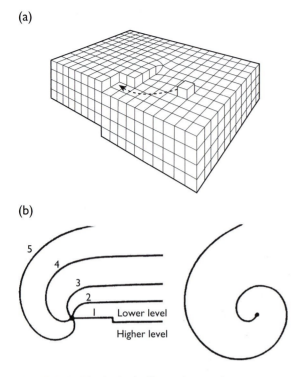

(b)

Fig. 10.7 (a) The lattice is distorted around a screw dislocation, producing steps on the surface. (b) During growth, unit cells are added, leading to spirals in the growth pattern.

Probably most information is available about the growth of calcite in stalactites of limestone caves (e.g., Fairchild and Baker, 2012). Growth rates range from 1 to 2000 μm/year, depending on climate (temperature and rainfall) as well as acidity of overlying strata.

Skeletal crystals grow from highly supersaturated solutions, or melts, when diffusion of the atoms that compose the crystal is limited. Locations with a high specific surface, i.e., edges and corners that are surrounded by the liquid phase, grow faster than faces (Figure 10.9a). The result is that edges are protruding and faces are set back, as in the halite crystal in Figure 10.9b or the quartz in Plate 3f. Skeletal growth is also prevalent in undercooled viscous volcanic melts and when volcanic glass crystallizes (*devitrification*) and diffusion is very sluggish.

Dendrites are an extreme case of prevalent apex and edge growth, but irregular material diffusion causes additional nucleation of new branches. The branching is subject to chance and can be described as a *fractal* phenomenon. The unique crystals resemble a branching plant, hence the name "dendrites". Dendritic growth is typical in native copper (Plate 4a), and in pyrolusite (MnO_2) and other manganese oxides, precipitating on fractures in limestone (Plate 4b).

(a)

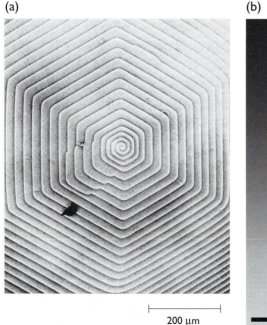

200 μm

(b)

5 mm

Fig. 10.8 (a) Growth spirals of the (0001) plane of hexagonal silicon carbide (carborundum) observed with phase contrast microscopy. Step height is 16.5 nm (from Verma, 1953). (b) Growth spirals in quartz (from Zorz, 2009).

(a)

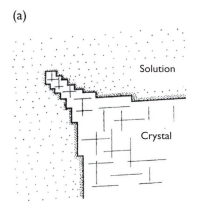

Solution

Crystal

(b)

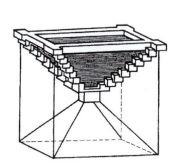

Fig. 10.9 (a) Section through a skeletal halite (NaCl) crystal immersed in a solution, illustrating that the tip of an edge has more surrounding solution than a flat face, and can therefore grow more rapidly. Such growth can also occur if impurities decorate faces. (b) Skeletal growth in halite with protruding edges and reset faces.

(a)

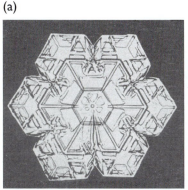

(b)

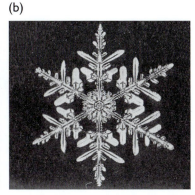

Fig. 10.10 Photographs of (a) tabular and (b) dendritic snowflakes (H_2O) (from Bentley and Humphreys, 1962).

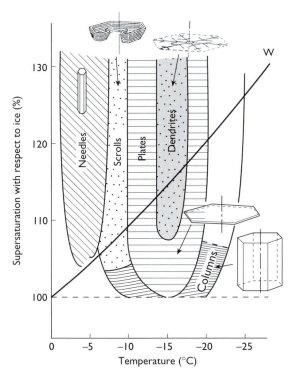

Tourmaline from Namibia exemplifies a case where symmetry is expressed in the skeletal morphology. Tourmaline (point-group $3m$) has no center of symmetry and the two ends of the [0001] axis are structurally different. Crystals generally grow with a prismatic morphology parallel to [0001]. A closer look at the positive end of the crystal displays regular growth faces (Plate 4c), while the negative end shows extensive skeletal features with indentations, also with trigonal symmetry but very different from the positive end (Plate 4d).

The growth morphology of crystals varies with physical conditions. This has been studied extensively for snow, where the habit ranges from acicular needles to plates (Figure 10.10a) and dendrites (Figure 10.10b), all with a characteristic hexagonal shape. The variation in snow crystal morphology is a function of both supersaturation and temperature (Figure 10.11).

Fig. 10.11 Morphology of snow crystals as a function of supersaturation and temperature. The line W gives the saturation vapor pressure with respect to supercooled water (data from Nakaya, 1954).

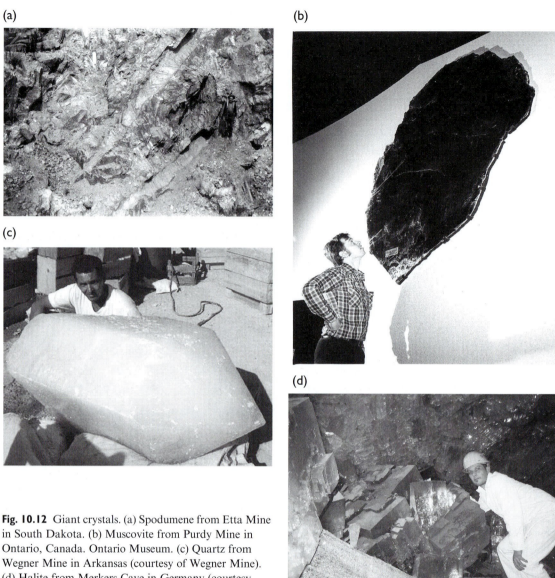

Fig. 10.12 Giant crystals. (a) Spodumene from Etta Mine in South Dakota. (b) Muscovite from Purdy Mine in Ontario, Canada. Ontario Museum. (c) Quartz from Wegner Mine in Arkansas (courtesy of Wegner Mine). (d) Halite from Merkers Cave in Germany (courtesy Frank de Wit). In each image there is a person for scale.

A few words are appropriate about the size of minerals. Most minerals in igneous and metamorphic rocks range in size from millimeters to centimeters and can be observed by eye, with a hand lens, or with an optical microscope. In sedimentary rocks and sediments mineral sizes are often much smaller, from nanometers to micrometers, and can only be distinguished with electron microscopes. Giant crystals are rare but have fascinated mineralogists (e.g., Rickwood, 1981). Large crystals occur in pegmatite dikes such as a 14 m long spodumene ($LiAlSi_2O_6$) from the Etta Mine in South Dakota (Schaller, 1916) (Figure 10.12a), or muscovite plates ($KAl_2(AlSi_3O_{10})(OH)_2$) more than 2 m in diameter from the Purdy Mine in Ontario, Canada (Harding, 1944) (Figure 10.12b). There are many large quartz crystals, including those from the Wegner Mine in Arkansas (Figure 10.12c). Halite crystals can reach sizes of over 1 m, as in Merkers Cave, Germany (Figure 10.12d). Most spectacular are recently discovered gypsum crystals in the Naica lead and silver mine in Chihuahua, Mexico, with crystals up to 15 m in length and 1 m in diameter (Plate 5a).

10.3 Various growth effects

Crystal growth leads to various crystal morphologies expressed by specific combinations of crystal forms and states of aggregation that reflect the kinetic conditions. Euhedral growth occurs if there is free space, such as in a solution or a melt. Euhedral crystals growing in igneous rocks are referred to as *phenocrysts*. An example of an orthoclase phenocryst in granite is shown in Figure 10.13 and a thin section with growth zones is illustrated in Plate 5b.

But euhedral crystals can also form in metamorphic rocks, where they grow by replacing pre-existing minerals. Andalusite, garnets, staurolite, and amphiboles are examples, and such crystals in metamorphic rocks are called *porphyroblasts*. During growth, pre-existing minerals may be completely or partially replaced by newly growing porphyroblasts. In the growth front, a film of capillary solution initiates the resorbtion and crystallization processes. Externally, porphyroblasts have the appearance of normal euhedral crystals, but sometimes they include relicts of incompletely substituted parent rock, such as in a folded schist where a newly growing andalusite crystal incorporates biotites as markers of the folds (Plate 5c). A so-called "snowball garnet" from metamorphic schist in Vermont contains a spiral trail of inclusions which display rotations during the growth history (Plate 6a).

Multicrystals consist of slightly misoriented individual grains. Good examples are quendels of quartz (Plate 6b) and roses of hematite (Plate 6c). Misorientation of these crystals can be caused if a growing crystal adsorbs microparticles or compositionally different layers. Such foreign particles do not fit a crystal's structure, but their size is comparable to the thickness of the added layer. If a new layer is deposited on a foreign particle, it is deflected and continues growing in a slightly different orientation. In experiments it was observed that the angle between lattice planes of the original crystal and a subsidiary crystal does not exceed 20–30′, but this splitting may repeat to form a fan of microcrystals with much larger deflections and apparent bending.

Crystals may contain homogeneous (solid, liquid, gas, glass-like) and heterogeneous *inclusions*. *Poikilocrystals* are particular varieties of crystals, growing in porous rocks that completely incorporate the old rock structure. Gypsum crystals growing in Sahara sand above the groundwater level contain 50% sand grains, including fine sedimentary structures (*Sahara roses*; Figure 10.14a). The rhombohedral calcite crystals of Fontainebleau (France) are also poikilocrystals, and their volume consists of 75–80% sand grains (Figure 10.14b). Figure 10.14c is a schematic cross-section through such poikilocrystals.

Inclusions are also present in igneous environments. A growing crystal may incorporate foreign particles. Sometimes the particles settle on the crystal surface and get covered with a new layer of the growing crystal, thus marking growth zones. Such a coating of mineral particles corresponds to a particular growth stage. Commonly plagioclase, quartz, and mica decorate growth zones of orthoclase megacrysts in granite (Plate 5b).

Of particular interest are fluid inclusions, because they provide information about the composition of the aqueous solution that was present during mineral formation. Generally these inclusions are small, ranging from 0.1 to 1 mm in size. At atmospheric conditions these inclusions may be liquid, liquid containing a gas bubble, liquid with a solid phase, or purely gas. Figure 10.15 illustrates an inclusion in quartz from an Alpine fissure. The liquid is water, the gas is carbon dioxide, and the solid is a cubic crystal of halite. The inclusions can be characterized by means of heating them and determining the temperature at which they homogenize, which should correspond approximately to the temperature at which the crystal formed.

Fig. 10.13 Phenocryst of orthoclase in Bergell granite, Swiss Alps. Width 15 cm.

(a)   (b)   (c)

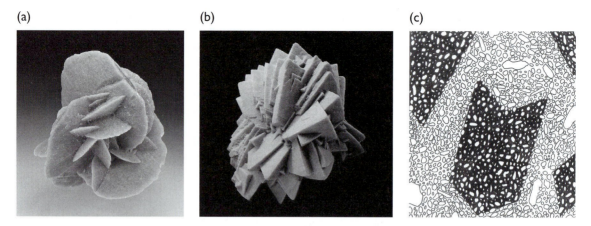

Fig. 10.14 (a) A Sahara rose is an aggregate of gypsum crystals with about 50% inclusions of sand. Morocco (courtesy D. Weinrich). Width 50 mm. (b) Calcite from Fontainebleau, France, containing up to 70% sand (courtesy Musée, Mines ParisTech, photo J. M. Le Cleac'h). Width 10 cm. (c) Schematic cross-section illustrating the structure of such poikilocrystals.

The faces of growing crystals adsorb material from the surroundings. Isomorphic substitutions depend on the chemical and physical environment that often changes during crystallization. If temperature decreases, plagioclase becomes more sodic and this process leads to what is known as *zoning*. This is easily recognized in thin sections (Plate 7a). To complicate the situation, different faces may incorporate isomorphic substitutions differently. In a macroscopic

crystal, this process leads to what is known as *sector zoning*, as is illustrated schematically for fluorite (CaF_2), where different sectors have different concentrations of rare earth elements (Figure 10.16a). The relative sizes of the sectors depend on the relative growth velocities of the faces. If octahedral faces *b* grow more slowly the corresponding sector increases in size (Figure 10.16b). The opposite is true if the cubic faces (a) grow more slowly (Figure 10.16c).

A well-known example of sector zoning is the colored quartz variety ametrine from Bolivia; sectors parallel to the negative rhombohedron $\{01\bar{1}1\}$ are iron-richer yellow citrine, and sectors parallel to the positive rhombohedron $\{10\bar{1}1\}$ are iron-poorer purple amethyst (Plate 7c). Another example is bicolored tourmaline, with the color reflecting the Mn:Fe ratio (Plate 7d).

Fig. 10.15 Fluid inclusion in quartz from an Alpine fissure. The inclusion is mainly water, with a bubble of carbon dioxide and a cube-shaped crystal of halite (from Stalder *et al.*, 1973). Width 50 μm.

(a)   (b)   (c)

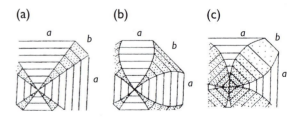

Fig. 10.16 Sketch illustrating the dependence of sector size upon relative growth velocities in fluorite (*a* denotes the cubic face and *b* the octahedral). The sectors have different concentrations of rare earth elements (from Grigor'ev, 1965). (a) Constant growth velocities. (b) Velocity of *b* decreases. (c) Velocity of *b* increases.

In tourmaline as well as ametrine the composition varies by only trace amounts but the visual effect is clear.

An observer may sometimes recognize an apparent habit characteristic of a certain mineral but, on closer inspection, discover that the actual mineral is different from what was expected, based on the morphology. In such cases, a new mineral of different structure or composition has replaced the original without changing the initial shape. This inherited shape is misleading for identification purposes and is called a *pseudomorph*. Pyrite (FeS_2) crystals are frequently replaced by limonite ($\sim FeO(OH) \cdot nH_2O$) due to oxidation at surface conditions. Pseudomorphs can preserve even small details of the initial crystal's surface, such as striations of pyrite (Plate 6d). Plate 6e illustrates an azurite crystal ($Cu_3(CO_3)_2(OH)_2$) (blue) that is partially replaced by malachite ($Cu_2(CO_3)(OH)_2$) (green). Pseudomorphs also exist in metamorphic rocks, e.g., when euhedral plagioclase is replaced by sericitic mica (Plates 7e,f). Pseudomorphs are like fossil evidence of former chemical processes and help us to infer mineral-forming processes and the geological history.

10.4 Aggregation

Sometimes crystals grow as ideal homogeneous single crystals to a macroscopic size, and the external morphology is characterized by particular crystal forms. This is rather exceptional, and more frequently intergrowths of crystals are observed. The morphology and state of aggregation depend on nucleation rate, the number of nucleation sites, and on the growth rate. All of these are complicated functions of many parameters, including temperature, chemical composition, trace amounts, and defects in the crystal structure, and in many cases relationships are not very well known (e.g., Sunagawa, 2007).

In metamorphic rocks such as marble, new crystals of calcite nucleate and grow, replacing old calcite crystallites in limestone. Growth reduces the grain boundary energies. In the end this produces a polygonal pattern with triple junctions and straight boundaries (Figure 10.17a, Plate 8a). But the boundaries in this granular aggregate are not crystal faces.

Concretions are common in porous sediments and sedimentary rocks such as sands, limestone, and shales, with quartz, calcite, pyrite, iron, aluminum, manganese oxides, and phosphorites replacing pre-existing material. The concretions range in size from fractions

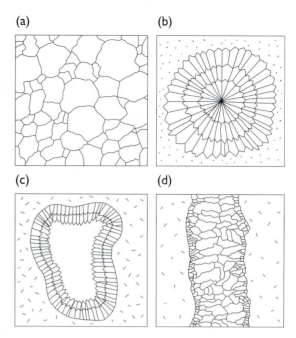

Fig. 10.17 Aggregations of crystals. (a) Section through a granular aggregate of calcite in marble with equiaxed grains and polygonal grain boundaries. (b) Radial growth of crystals in a concretion from a nucleation center. (c) Druse with growth of quartz. (d) Growth of prismatic calcite in a vein in greenschist (courtesy E. S. Denzin).

of millimeters to tens of centimeters. Smaller ones are reminiscent of fish roe (hence the name *oölite*), whereas larger ones look like peas (*pisolites*). They form in solutions, and a concentric structure develops from the nucleation sites (Figure 10.17b). Growth of concretions begins at a central nucleus. Most impressive are so-called framboidal aggregates of pyrite (Figure 10.18).

Druses form in open fractures or cavities (Figure 10.17c). They are groups of crystals growing perpendicularly or sub-perpendicularly to the substrate (Figure 10.5). In the end, druses also form spheroidal aggregates, similar to concretions. Chalcedony is a typical druse aggregate with quartz crystallites growing from the druse surface towards the center (Plate 8b). Interestingly, in chalcedony, depending on the conditions, the growth direction may switch from low angle to *c*-axis to high angle to *c*-axis, producing complex repetitions (Plate 8c). Impurities delineate bands that represent growth episodes. In a second phase, fibrous quartz precipitates in horizontal layers, through crystallization from a colloidal gel.

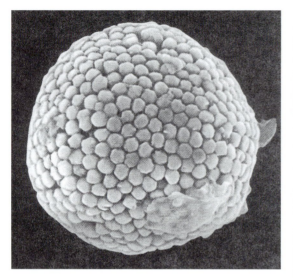

Fig. 10.18 Aggregate of framboidal pyrite in shale. SEM image. 0.01mm diameter (courtesy A. Zholnerovich, see also Zholnerovich, 1990).

Growth of *parallel-prismatic* and *fibrous aggregates* often occurs in fractures. Examples are veinlets of silky gypsum, serpentine-asbestos (chrysotile), and columnar calcite (Figure 10.17d).

Spherulitic growth occurs in druses, where growth begins from the cavity wall. Spherulites may overgrow other minerals and the linings of cavities. Kidney-shaped aggregates such as in goethite (FeOOH) (Figure 10.19), malachite, and azurite have a particularly characteristic spherulitic structure. Growth of calcite in cave stalactites and stalagmites is a special case of spherulitic growth.

Fig. 10.19 Spherulitic-botryoidal morphology of goethite (FeOOH) from Rossbach near Siegen (Germany) due to fibrous growth from a nucleation center (courtesy A. Massanek). Width 7.5 mm.

10.5 Summary

A qualitative description of crystal morphology is the crystal habit ranging from equiaxed, prismatic to platy. The slowest growing faces determine the dominant morphology. While cubic crystals are generally (but not always) equiaxed, the habit depends on conditions under which crystals grow. This has been extensively studied for snow, where the morphology depends on supersaturation and temperature. Growth velocities vary tremendously. Even for a simple system such as calcite stalactites, growth velocities range from 1 to 2000 μm/year. In a chemically complex system the composition may change during growth, e.g., during temperature changes, resulting in chemical zoning. Furthermore, different faces may preferentially adsorb different elements, leading to sector zoning, as in ametrine (amethyst-citrine varieties of quartz). Below are some terms to describe morphology:

- Phenocrysts: euhedral crystals in igneous rocks that precipitate from a melt.
- Porphyroblasts: euhedral crystals in metamorphic rocks that form by replacing pre-existing minerals.
- Skeletal and dendritic growth occurs in supersaturated conditions.
- Pseudomorphs are originally euhedral crystals that have been replaced by other minerals.

Test your knowledge

1. What is the difference between accidental intergrowths and twins?

2. Think of an example where the morphology changes with external conditions.

3. What is the difference between poikilocrystals and porphyroblasts?

4. How do druses grow? How do the concretions differ from druses?

5. Explain the origin of striations in pyrite. Discuss them on the basis of growth and symmetry.

6. What are pseudomorphs? Think of some examples.

7. Why does growth and dissolution primarily proceed from dislocations?

8. Look over the photographs in Plates 6 to 16 and identify different habits.

Further reading

Doremus, R. H., Roberts, B. W. and Turnbull, D. (eds.) (1958). *Growth and Perfection of Crystals*. John Wiley, New York.

Glikin, A. (2008). *Polymineral-Metasomatic Crystallogenesis*. Springer, New York.

Grigoriev, D. P. (1965). *Ontogeny of Minerals*. Israel Programme for Scientific Translations, Jerusalem.

Petrov, T. G., Treivus, E. B. and Kasatkin, A. P. (1969). *Growing Crystals from Solution*. Consultants Bureau, New York.

Sunagawa, I. (2007). *Crystals: Growth, Morphology and Perfection*. Cambridge University Press, Cambridge.

Part III | Physical investigation and properties of minerals

11 | X-ray diffraction

X-ray diffraction has become the most important method to identify minerals because diffraction signals are directly linked to the crystal lattice and the atomic structure. It is widely available in mineralogical laboratories and is even used by robotic spacecrafts on Mars. This chapter provides a general introduction to diffraction, how it is used for identification, and how diffraction intensities are related to the crystal structure.

11.1 Basic concepts

The notion that crystals have a lattice-based structure and that the basic building block is the unit cell was introduced in the eighteenth century. At that time the analysis of crystals was based on visual inspection, on detailed examination with a hand lens or at best a light microscope. However, visible light, with wavelengths between 400 and 700 nm, is far too coarse a probe to investigate crystal structures where information is on the scale of atoms and interatomic distances, i.e., 1–5 Å (1 Å = 0.1 nm). In this chapter we discuss experimental techniques that were developed in the early twentieth century to study crystal lattices, providing a tool for determining structural parameters and largely confirming the concepts about unit parallelepipeds, lattice structure, and symmetry suggested more than 100 years earlier. With the discovery of X-rays by C. W. Röntgen in 1895, the stage was set

for analyzing crystals at an elementary level, and this research produced unprecedented information about the solid state. Then in 1912 the famous diffraction experiment of Max von Laue established that X-rays are waves and that the suspected internal lattice structure of crystals indeed does exist (for a brief history, see Box 11.1).

Röntgen found that when electrons were accelerated in an electric field and collided with a metal anode, a very high-energy radiation was emitted. At the time, he was unable to explain the origin of this radiation (hence the name "X-rays"), and an explanation had to wait until more was known about the structure of atoms. It became clear that accelerated electrons with sufficiently high energy could displace electrons from within the inner electron shells (e.g., a K-shell electron) of an atom (Figure 11.1a). This is because the energy of the electrons within the inner

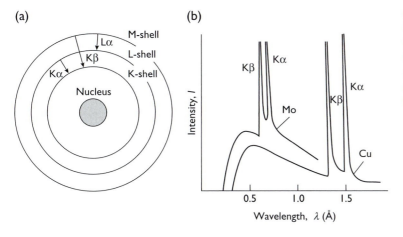

Fig. 11.1 (a) Energy transitions of inner shell electrons (not to scale). (b) Spectrum produced in an X-ray tube with a molybdenum and a copper anode. The characteristic peaks correspond to Kα and Kβ transitions.

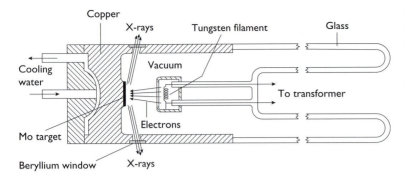

Fig. 11.2 Generation of X-rays in a modern X-ray tube. A heated filament emits electrons that are accelerated and produce X-rays when they hit an anode. The anode is cooled with water.

shells is close to, but slightly less than, the energy of the accelerated electrons. An electron from a higher shell immediately fills the electron hole, and the excess energy is released as a photon. The energy, and hence the wavelength, corresponds to the particular electronic transition of a given atom. These high-energy photons have short wavelengths λ ($\lambda = hc/eV$, where h is the Planck constant, c the speed of light, e the charge of the electron, and V the accelerating voltage) in the range 0.1–5 Å, and are called X-rays. A spectrum, obtained when a metal such as copper or molybdenum is irradiated with 50 keV electrons, is shown in Figure 11.1b. It consists of both a continuous part due to irregular energy exchanges between electrons, and a set of high-intensity peaks due to the specific energy transitions. For example, L- to K-shell transitions produce Kα X-rays, and M- to K-shell transitions produce Kβ X-rays. Note that X-rays produced from a molybdenum target have higher energies and shorter wavelengths than those produced from a copper target. This is because molybdenum has a higher atomic number. X-ray radiation, like visible

light, is part of the electromagnetic spectrum, but the shorter wavelengths of X-rays (0.1–5 Å) make them ideal for crystal structure studies.

For most applications X-rays are produced by an X-ray tube, powered by an X-ray generator. Figure 11.2 shows a schematic diagram of an X-ray tube. It consists of an evacuated glass tube in which electrons are released by heating a tungsten filament (just as in a normal light bulb). By applying a voltage, the electrons are then accelerated in a field to 40–50 keV and collide with an anode metal (molybdenum in Figure 11.2). Owing to the energy transitions in the anode, X-rays are produced and leave the tube through beryllium windows that have relatively low absorption. In an X-ray tube most of the energy of the electrons is not converted to X-rays but to heat, and thus it is necessary to cool the anode metal, usually with water. In many applications, radiation with a single wavelength is preferred, for example K. There are methods to filter the spectrum and produce nearly monochromatic X-rays.

Box 11.1 Background: Early history of X-ray crystallography

Wilhelm Conrad Röntgen discovered X-rays in Würzburg in 1895. He constructed a cathode ray tube, enclosing it in a light-tight cardboard box, and observed a peculiar phenomenon. When he sent a pulse of electrons (cathode rays) through this tube, a screen made of barium platinocyanide crystals, placed at some distance from the tube, would light up in bright fluorescence. Röntgen knew that the fluorescence was not caused by the cathode rays, for the glass tube absorbed electrons. Some other mysterious radiation of high energy had to be involved, and he named this radiation "X-rays". The rays were found to have high penetration properties, and for this reason the first practical application of X-rays was for radiography.

A turning point in crystallography came in 1912. An exceptional group of scientists were then at the University of Munich. They included: Professor Paul von Groth, a crystallographer; Professor Wilhelm Conrad Röntgen, who was now Chair of Experimental Physics; and Professor A. Sommerfeld in the

Box 11.1 (cont.)

(a) (b)

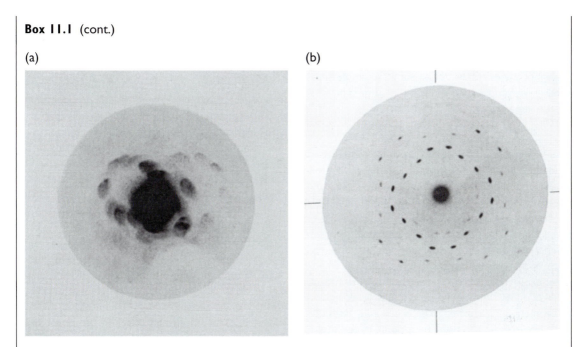

Fig. 11.3 (a) The first X-ray diffraction photograph of a copper sulfate crystal (Friedrich *et al.*, 1912). (b) A diffraction pattern of a centered sphalerite crystal with better collimation, viewed along the 4-fold symmetry axis (von Laue, 1913, reproduced by Ewald, 1962).

Department of Theoretical Physics. There were also a number of assistants and students, among them Dr. Max von Laue and Peter Debye (both Sommerfeld's assistants), and W. Friedrich and P. Knipping (graduate students of Röntgen). Crystallographers such as Groth claimed that crystals were built up of periodic arrangements of molecules, and Sommerfeld believed that X-rays were wave-like radiation. Von Laue was curious to find out what would happen if X-rays interacted with crystals. This curiosity led to the famous experiment, performed by Friedrich and Knipping on a crystal of copper sulfate, in which diffraction was observed (Figure 11.3a), i.e., intensity was not just recorded in the direction of the incoming beam but produced splotches over a wide range of angles. This proved immediately that crystals indeed had lattice character and that X-rays were waves (Friedrich *et al.*, 1912; von Laue, 1913).

Shortly after the discovery by von Laue, Friedrich and Knipping, the British physicist William Henry Bragg at the University of Leeds and his son William Lawrence Bragg, a graduate student at Cambridge University, collaborated to interpret the intensities of X-ray diffraction and were able to relate them to the crystal structure. X-ray diffraction patterns display the internal symmetry of crystals (Figure 11.3b). The Braggs explained X-ray diffraction as selective reflection on lattice planes (Bragg and Bragg, 1913) and used the diffraction data to determine the unit cell and crystal structure, first of halite and then of other minerals (Bragg, 1914). Interestingly, the Russian scientist G. Wulff, after whom the Wulff net is named, independently derived an expression that is very similar to Bragg's law (Wulff, 1913). It reminds us of the independent discovery of space-groups by A. Schoenflies and E. S. von Fedorow, and of the periodic table of elements by J. L. Meyer and D. Mendeleev. Röntgen (1901), von Laue (1914), W. H. and W. L. Bragg (1915) received Nobel Prizes in Physics for their achievements.

Ewald (1962) reviewed the early history of X-ray diffraction on the occasion of the fiftieth anniversary of von Laue's discovery. The centenary in 2014 became the UNESCO "International Year of Crystallography" in recognition of W. L. Bragg's first use of X-rays to unravel a crystal structure. It was celebrated worldwide, including postage stamps of minerals (Plate 7b) which we are using for the cover of this book.

11.2 Brief discussion of waves

Before we discuss the diffraction of waves by crystals, we need to review briefly the properties and representations of waves. Figure 11.4a shows a simple sinusoidal wave, propagating in the x-direction, which can be expressed as

$$y = A \sin x \qquad (11.1)$$

where A is the *amplitude* or maximum displacement from the line of propagation. The distance between two wave crests is the *wavelength* (λ). Figure 11.4a shows the wave in the starting position (1, solid line) and after it has propagated in the x-direction by a distance Δ (2, dashed line). This distance can be measured as a *path difference* relative to an origin, in length units (e.g., ångströms). Alternatively it can be measured as a *phase difference* (ϕ) in multiples of a wavelength, or as an angle (with 2π or $360°$ corresponding to a full wavelength).

Geometrically a propagating sinusoidal harmonic wave can also be described by a point P, rotating at constant angular velocity ω (Figure 11.4b). The projections of P (P′ on x and P″ on y) execute a simple harmonic motion. From the relative displacements $OP' = X = A \cos \phi$ and $OP'' = Y = A \sin \phi$, we can calculate the maximum displacement or amplitude with Pythagoras' theorem, $A = \sqrt{(X^2 + Y^2)}$, corresponding to the radius of the circle, and the phase $\phi = \tan^{-1}(Y / X)$. The wave is then specified by a vector f and can be conveniently represented in a Gaussian coordinate system with complex numbers, using x as the real axis and y as the imaginary axis (i).

$$f = A(\cos \phi + i \sin \phi) \qquad (11.2)$$

The length of this vector f is its amplitude A, and its direction depends on its phase ϕ. (If you are

unfamiliar with complex number algebra, don't be concerned. In this context the complex operator i is only used to keep track of X and Y separately, with $i^2 = -1$.)

Next let us consider two waves of the same amplitude A, the same wavelength λ, and propagating in the same direction, but the second wave is displaced with respect to the first (Figure 11.5). These waves may combine, as we will see in this chapter for X-rays and in Chapter 13 for light waves. The resultant wave can be obtained by addition of corresponding displacements. This has been done graphically in Figure 11.5 for three cases: one with an arbitrary phase difference $\phi = 75°$ (Figure 11.5a), one with a phase difference corresponding to a full wavelength $\phi = 2\pi = 360°$ (Figure 11.5b), and the third with a phase difference corresponding to a half wavelength $\phi = \pi = 180°$ (Figure 11.5c). The resultant wave R (dashed line) has the same wavelength as the two waves, but the amplitudes differ in the three cases, depending on the phase difference. For $\phi = 75°$, $A_{new} = 1.59 \times A$; for $\phi = 2\pi = 360°$, $A_{new} = 2 \times A$ (this is an important case which we will call *constructive interference*); and for $\phi = 2\pi = 180°$, $A_{new} = 0$ (*destructive interference*; in this case the resultant wave has no amplitude).

Instead of doing this addition graphically, we can use equation 11.2 and simply add the two wave vectors (both with the amplitude A):

$$f_1 + f_2 = A\{(\cos \phi_1 + \cos \phi_2) + i(\sin \phi_1 + \sin \phi_2)\}$$
$$= A(X + Yi)$$
$$A_{new} = (A\sqrt{(X^2 + Y^2)}$$

$$\text{and } \phi_{new} = \tan^{-1}(Y/X) \qquad (11.3)$$

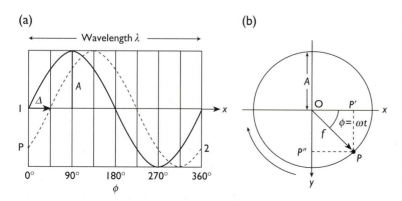

(a)

(b)

Fig. 11.4 (a) Representation of a propagating sine wave with amplitude A, wavelength λ, path difference Δ and phase difference ϕ. (b) The wave can be viewed geometrically as a harmonic oscillation with point P rotating at constant angular velocity. The wave (a) is defined by the vector f given by amplitude A and phase ϕ, or by the components X and Y in a Gaussian coordinate system.

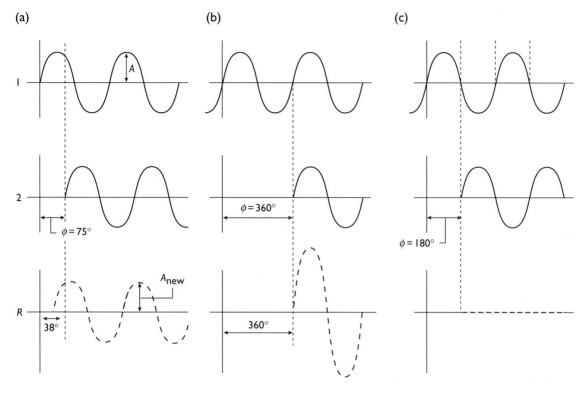

Fig. 11.5 Addition of two waves with phase differences (a) $\phi = 75°$, (b) $\phi = 360°$, and (c) $\phi = 180°$ produces a resultant wave (dashed lines).

$\phi = 75°$:

$$f_1 + f_2 = A[(\cos 0° + \cos 75°) + i(\sin 0° + \sin 75°)]$$
$$= A(1.26 + 0.97i) = A(X + Yi)$$
$$A_{new} = (A\sqrt{(X^2 + Y^2)} = A\sqrt{(1.26^2 + 0.97^2)}$$
$$= 1.59A$$

and $\phi_{new} = \tan^{-1}(Y/X) = 37.6°$ (11.4)

$\phi = 2\pi = 360°$:

$$f_1 + f_2 = A[(\cos 0° + \cos 360°) + i(\sin 0° + \sin 360°)]$$
$$= A(2 + 0i) = A(X + Yi)$$
$$A_{new} = (A\sqrt{(X^2 + Y^2)} = A\sqrt{(2^2 + 0^2)}$$
$$= 2A$$

and $\phi_{new} = \tan^{-1}(Y/X) = 0°$ (11.5)

$\phi = \pi = 180°$:

$$f_1 + f_2 = A[(\cos 0° + \cos 180°) + i(\sin 0° + \sin 180°)]$$
$$= A(0 + 0i) = A(X + Yi)$$
$$A_{new} = A\sqrt{(X^2 + Y^2)} = A\sqrt{(0^2 + 0^2)}$$
$$= 0$$

and $\phi_{new} = \tan^{-1}(Y/X) = 90°$ (11.6)

which provides the same results as those obtained graphically, but is much easier to do. We will use this method of adding waves later in this chapter. For now we reiterate the most important conclusion: if two waves have a phase difference corresponding to a multiple of a full wavelength, and such waves are added, the resultant wave has a maximum amplitude.

11.3 Laue and Bragg equations

Now we return to X-rays and crystals. In the following discussion we will assume that we have monochromatic X-rays, i.e., X-rays with a single wavelength. To explain diffraction, let us assume that a wave front of X-rays reaches a row of atoms. Each atom (or, more correctly, each electron) acts as a scattering center for a spherical wave of equal wavelength and, in direct analogy to Huygens' construction for visible light, new wave fronts form. In two dimensions these spherical waves appear as circles with wave fronts tangent to them (Figure 11.6). Each circle represents the collection of points that lie one full wavelength (measured

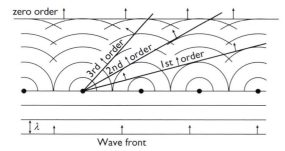

Fig. 11.6 Huygens' construction for wave fronts of different orders, produced by interaction of a wave with a row of lattice points.

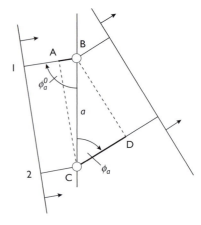

Fig. 11.7 Path difference (PD = CD − AB) produced between two wavelets, 1 and 2, scattered at two points, B and C, with a spacing a.

between wave crests) from the atoms. The most obvious tangent to these circles is parallel to the incident wave front. This is called the *zero-order* new wave front. But there are other directions into which waves are deflected. There is a *first-order* wave front that is inclined to the old wave front. There are also second- and higher-order wave fronts, and they can be drawn as tangents to wave crests from X-rays, with different arrival times (Figure 11.6).

The deflection angle, which is the change in the direction from the first-order to higher-order wave fronts, increases with wavelength and decreases as we reduce the spacing between rows of atoms. With some very simple geometry we can calculate the angles under which we observe diffraction (Figure 11.7). Consider two neighboring lattice points (B, C) with spacing a, and an incoming X-ray wave of wavelength λ that reaches the row of points at an angle ϕ_a^0 with respect to the line containing the lattice points. The two wavelets 1 and 2 will not travel the same distance from the time they formed an old wave front to the moment they are parallel again and define a new wave front.

If the path difference (PD) is an integer multiple of a full wavelength, then the waves are *in phase* after the change in direction and they reinforce each other by addition (Figure 11.5b). From Figure 11.7 we can easily derive the equation

$$PD_a = CD - AB = (\cos\phi_a - \cos\phi_a^0)\,a = n_1 \times \lambda$$
(11.7a)

This is called the *Laue diffraction condition for a one-dimensional crystal* (or *Laue equation*) and specifies the direction of the diffracted rays. The angle ϕ_a is between the row of lattice points and the diffracted ray. The integer n_1, either positive or negative, defines the order of diffraction, i.e., the number of full

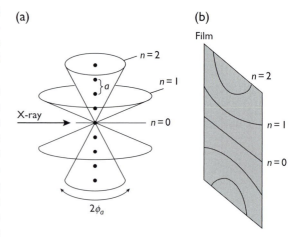

Fig. 11.8 (a) Diffraction from a one-dimensional crystal (spacing between atoms is a) is a set of concentric cones around the line of atoms. (b) The diffraction pattern of the cones can be recorded on a flat film mounted at right angles to the incident X-ray and parallel to the lattice row, producing a set of hyperbolas.

wavelengths by which waves from adjacent lattice points differ when they reach the new wave front. The maximum possible path difference between two lattice points is $2a$ ($\phi_a^0 = 0°$, $\phi_a = 180°$), and therefore n_1 can vary only within certain limits such that $n_1 < |2a/\lambda|$; that is, only a limited number of full wavelengths can fit into the maximum path difference. Diffraction from a one-dimensional crystal thus consists of a set of cones, one for each n_1, with an opening angle $2\phi_a$ (Figure 11.8a). If a plane photographic film, sensitive to X-rays, is mounted parallel to the row of lattice points of this one-dimensional crystal and perpendicular to the incident X-ray, the recorded diffraction pattern is the intersection of the cones with the film, i.e., a set of hyperbolas (Figure 11.8b).

In Figure 11.9a we illustrate an example of second-order diffraction ($n_1 = 2$) for a one-dimensional lattice in a different way. For second-order diffraction, waves scattered between adjacent lattice points have a path difference of 2λ, or a phase difference of $4\pi = 2 \times 2\pi$ (see Figure 11.7). We choose an origin (circled point) and plot the one-dimensional lattice, which is a row of points at equal distance. Then we label the phase differences (in multiples of wavelengths) for each lattice point with respect to the origin in the figure, i.e., 0, 2, 4, etc.

For a two-dimensional crystal we must add a second Laue condition to that given above (equation 11.7a). By analogy it is

$$PD_b = (\cos\phi_b - \cos\phi_b^0)\, b = n_2 \times \lambda \qquad (11.7b)$$

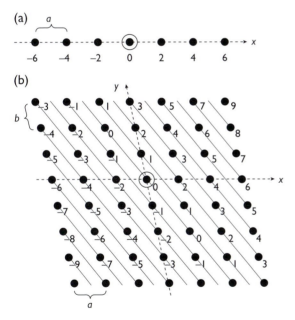

(a)

(b)

Fig. 11.9 Phase differences (in multiples of 2π) at lattice points, resulting from diffraction on a lattice when the Laue equations are satisfied. An arbitrary origin is marked with a circled point. (a) One-dimensional case, second-order diffraction: $n_1 = 2$. This means that a wave scattered at the point labeled 2 is two wavelengths behind that scattered at the origin. (b) Two-dimensional case, with $n_1 = 2$ and $n_2 = 1$. Note that all lattice points on (21) lines Miller indices have an identical phase.

where angles ϕ_b^0 and ϕ_b are between the row of lattice points with spacing b in the y-direction and incoming and diffracted ray, respectively.

In Figure 11.9b we add a second dimension to Figure 11.9a and plot a two-dimensional lattice. For diffraction we take the case where $n_1 = 2$ (second-order diffraction along x) and $n_2 = 1$ (first-order diffraction along y), and we label on these axes (dashed lines) again for each lattice point the phase differences with respect to the arbitrary origin (circled point). The two crossing directions x and y in Figure 11.9b can be extended easily to a full two-dimensional lattice by keeping in mind that, in a lattice, all points and thus all distances between lattice points in a specific direction are identical. The *total* phase difference of any lattice point with respect to the origin is $2\pi(n_1 + n_2)$, and we label the points accordingly. An interesting feature of Figure 11.9b is the presence of lines in the lattice that all have the same phase difference relative to each other. While all points of the lattice scatter in phase to satisfy the Laue equations, points on these lines have zero phase difference. If we assign Miller indices to the lattice lines with the same path difference of Figure 11.9b (reciprocal axis intercepts!), we find that $h_{Miller} = 2$, $k_{Miller} = 1$, i.e., the symbol is (21). Interestingly, these Miller indices have the same values as the Laue *diffraction order indices* n_1, n_2, and there is indeed a close relationship between the two. There is also a difference: by definition, Miller indices have no common divisor, while there is no such restriction on Laue indices. The two-dimensional model can be generalized easily to three dimensions. Thus lattice planes (hkl) exist in which all points on such planes have zero phase difference.

In Figure 11.10 we follow two waves (1 and 2) diffracting on a lattice plane at different angles (θ_i is the angle of the incident X-rays to the lattice plane, θ_d is the angle of the diffracted X-rays). Initially they are in phase, and after diffraction they establish a new wave front. As the figure illustrates, the individual paths in the three cases are different. The same path length for waves 1 and 2, and thus zero phase difference, is obtained only for the special geometry when incoming and diffracted waves are in mirror reflection geometry on the lattice plane (Figure 11.10c). This restricts the direction of incident and diffracted beam relative to the lattice planes, with all points having the same path difference in Figure 11.9b, and brings us to another interpretation of X-ray diffraction – namely, as reflection on lattice planes defined by Miller indices (*hkl*). The terms diffraction and reflection are often used interchangeably.

In Figure 11.10 we have considered a single lattice plane and investigated diffraction conditions. Real crystals, however, consist of stacks of lattice planes, and X-rays penetrate many hundreds of planes. What are the phase relations between waves reflected on two adjacent lattice planes separated by distance *d* (Figure 11.11)?

Two waves, which are initially in phase, reach a crystal. The first (1) diffracts (reflects) on the lattice plane (*hkl*) on the surface, the second (2) on the parallel plane below at a distance *d* = AC. The angle of incidence to the lattice plane is θ. The second wave has a longer path (PD = BC + CD) before the two waves establish a new wave front AD. We can easily establish a relationship (triangle ABC) sin θ = BC/AC, and correspondingly for diffraction, where the path difference has to be a multiple of the wavelength to produce constructive interference,

$$PD = 2d \sin \theta = n \times \lambda \qquad (11.8a)$$

where λ is the wavelength, θ is the angle of incidence and reflection, and *n* is an integer. The relationship is known as the *Bragg equation* (or *Bragg's law*) and was formulated in 1913 by the father and son team W. H. and W. L. Bragg. Diffraction can be viewed as reflection on lattice planes with reflection angles θ determined by the spacing of lattice planes. While von Laue's interpretation is closer to the physical nature of the diffraction process, Bragg's picture is far easier to tie in with actual experiments. If we orient a crystal such that a lattice plane (*hkl*) is in the reflection condition between incident and diffracted X-rays and measure the angle between the incident and diffracted rays we get immediate information about the spacing of corresponding lattice planes, d_{hkl}, in a crystal. The Bragg equation is often written

$$2d_{hkl} \sin \theta = \lambda \qquad (11.8b)$$

and *n*, the order of diffraction, is incorporated into the spacing *d* and the indices *hkl*. Thus, instead of talking about second-order diffraction on a lattice plane (111) with the distance *d*, say, equal to 10 Å, we talk about

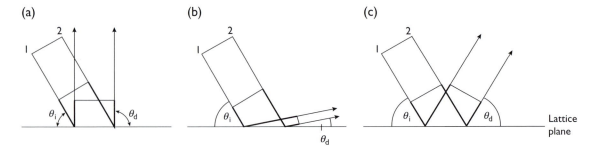

Fig. 11.10 Equal path difference requires mirror reflection on (*hk*) lines. Here, θ_i and θ_d are the angle of incidence and the angle of reflection, respectively. (a) $\theta_i < \theta_d$, path 1 < path 2. (b) $\theta_i > \theta_d$, path 1 > path 2. (c) $\theta_i = \theta_d$, path 1 = path 2.

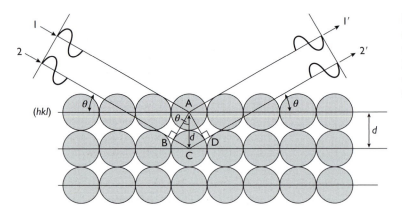

Fig. 11.11 Derivation of the Bragg equation, explaining diffraction as reflection on a stack of lattice planes with an interplanar spacing *d*. All points on the plane (*hkl*) scatter in phase.

(a)

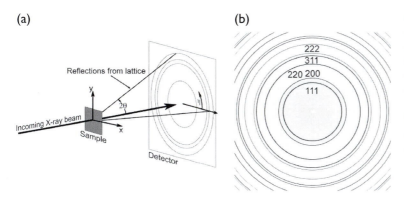

(b)

Fig. 11.12 In the powder method, diffractions from crystallites lie on cones around the primary beam with an opening angle 4θ. (a) Geometry of a synchrotron X-ray diffraction experiment on a powder with a planar detector recording Debye rings. (b) Diffraction image of cubic CeO_2. Some diffraction indices are indicated.

diffraction on (222) with d equal to 5 Å. The indices (222) correspond to the Laue indices n_1, n_2, n_3.

Bragg's law has two conditions:

1. The lattice planes (hkl) must be in a reflection orientation between the incident and diffracted X-ray waves.
2. Diffraction occurs at a specific angle that is determined by the d-spacing of the lattice planes for a given wavelength.

These conditions are seemingly very straightforward but are not easy to satisfy experimentally. If we aim a monochromatic X-ray at a crystal in some arbitrary orientation, the Bragg conditions are not satisfied and no diffraction occurs. A crystal has to be rotated to bring a particular lattice plane hkl into a reflecting position, and then the diffraction angle has to be adjusted to fit with the spacing of the lattice plane d_{hkl}. Modern computer-controlled X-ray goniometers can help to alleviate some of these problems, and they are used for special applications. Friedrich and Knipping were lucky: they got diffraction without having to rotate the crystal because they used polychromatic radiation with a wide range of wavelengths, which became known as the Laue method. Significant experimental advancement came in 1916, when P. Debye and P. Scherrer had the ingenious idea to use powders instead of single crystals.

11.4 The powder method

A powder consists of many randomly oriented small crystals or "crystallites". There will always be some crystallites with lattice planes in the right orientation to diffract (i.e., satisfying the first Bragg condition), and therefore rotation is not necessary. A powder irradiated with monochromatic X-rays of known

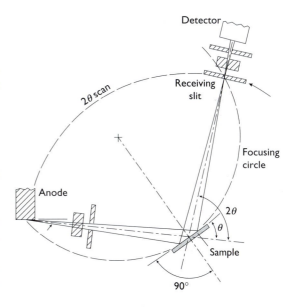

Fig. 11.13 A diffractometer scans the 2θ angle range with an electronic detector to record diffractions from a flat sample.

wavelength will produce diffracted X-rays lying on cones with an opening angle 4θ (i.e., an angle 2θ to the primary X-ray beam). We illustrate this in Figure 11.12a for monochromatic X-rays produced in a synchrotron, intersecting the diffraction cones with a two-dimensional detector and recording a set of concentric rings, each corresponding to a different d-spacing and with a different Laue index hkl. Figure 11.12b shows a diffraction image on a powder of CeO_2 and some diffraction rings are indexed.

Today the most popular powder method used in laboratories is a powder diffractometer. The powder is suspended on a flat disk, and the reflections are scanned with an electronic detector (Figure 11.13)

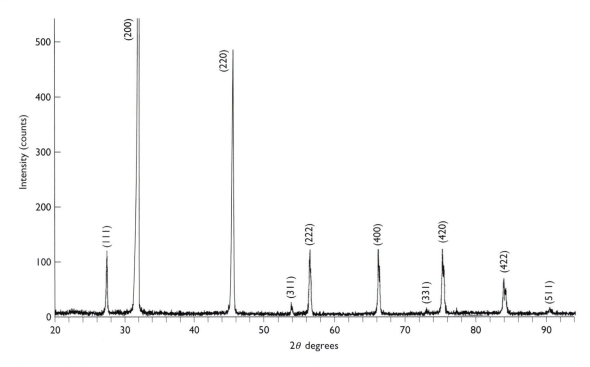

Fig. 11.14 Diffraction pattern of halite. Each peak is assigned a lattice plane on which reflection occurs. 2θ angles are indicated. Kα copper radiation is used ($\lambda = 1.5418$ Å).

which digitally records the intensity as a function of diffraction angle, as is shown for a sample of cubic halite in Figure 11.14. Each peak corresponds to diffractions from different lattice planes. The detector rotates with an angular velocity of 2θ, whereas the sample rotates at a velocity θ to maintain the reflection condition for the surface of the sample. It means that, at all diffraction angles, those lattice planes (and only those) that are parallel to the sample surface are diffracting.

From powder photographs or diffractometer scans we can obtain a list of θ angles, which can then be converted to d-spacings using Bragg's law. Spacings between lattice planes are a function of the specific lattice parameters for a given crystal and of the Miller indices hkl that define the lattice plane. We will show this relationship for an orthogonal crystal system. Figure 11.15 illustrates an orthorhombic coordinate system with axes x, y, z, and on it two parallel planes (hkl). One plane goes through the origin O. The parallel plane through A, B, C is the next adjacent one, spaced $d = $ OP, measured along the plane normal n. The lattice plane ABC is specified by axis intercepts OA $= a/h$, OB $= b/k$, and OC $= c/l$. (Remember that Miller indices are reciprocals of axis intercepts.) In the right triangle

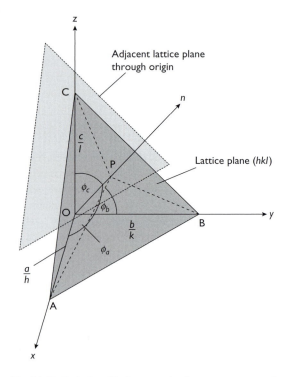

Fig. 11.15 Relationship between lattice parameters and d-spacing for orthorhombic crystals.

OAP, $\cos \phi_a = OP/OA = d \times (h/a)$ and, correspondingly, $\cos \phi_b = d \times (k/b)$ and $\cos \phi_c = d \times (l/c)$. These cosines are direction cosines specifying the direction of n in the orthogonal coordinate system x, y, z. The sum of the squares of direction cosines is 1; therefore,

$$(d^2 h^2)/a^2 + (d^2 k^2)/b^2 + (d^2 l^2)/c^2 = 1 \qquad (11.9a)$$

and

$$1/d^2 = h^2/a^2 + k^2/b^2 + l^2/c^2 \qquad (11.9b)$$

In the triclinic case, with nonorthogonal axes, the relationship is more complex. Without going through the algebra we give the result:

$$1/d^2 = \frac{[(h/a)\sin\alpha]^2 + [(k/b)\sin\beta]^2 + [(l/c)\sin\gamma]^2}{1 - \cos^2\alpha - \cos^2\beta - \cos^2\gamma + 2\cos\alpha\cos\beta\cos\gamma}$$

$$+ \frac{(2kl/bc)(\cos\beta\cos\gamma - \cos\alpha) + (2hl/ac)(\cos\alpha\cos\gamma - \cos\beta)}{1 - \cos^2\alpha - \cos^2\beta - \cos^2\gamma + 2\cos\alpha\cos\beta\cos\gamma}$$
$$(11.9c)$$

Thus, in both the orthorhombic and triclinic cases, the d-spacing is a function of lattice parameters and Miller indices of the lattice plane. The d-spacing is obtained directly from X-ray diffraction patterns (via the diffraction angle θ) and, at least for cubic crystals where $a = b = c$ and $\alpha = \beta = \gamma = 90°$, it is easy to determine the lattice parameter from powder diffraction data (see Box 11.2). For lower symmetry, other methods are used to determine lattice parameters, most commonly relying on diffraction experiments with single crystals.

11.5 Crystal identification with the powder method

As we have seen above, the d-spacings of lattice planes and therefore the 2θ *angles of a diffraction pattern* are a function of the *unit cell*. As we will see in the next section, the *intensities* of the diffraction peaks depend on the *arrangement of atoms* in the unit cell. Since the unit cell and crystal structure are diagnostic of a crystal, a powder pattern can therefore be used to identify an unknown crystalline substance. Powder patterns of all known crystalline substances have been collected, and the information on d-spacings and intensities of diffractions is published in large catalogs that contains data for over 250 000 materials, both organic and inorganic, and at different conditions. The most comprehensive database is managed by the International Centre for Diffraction Data (www.icdd.com) but

there are also open access search facilities such as the American Mineralogist Crystal Structure Database (http://rruff.geo.arizona.edu/AMS/amcsd.php). If we can match the pattern of an unknown substance, in angular locations of the diffractions and their relative intensities, with a pattern contained in the catalog, then the unknown is identified. To find the needle in this haystack, the procedure is to select first the three most intense diffraction peaks, e.g., for halite peaks select at 31.7°, 45.4°, and 56.5° 2θ in Figure 11.14, and go to a search catalog. With some luck the correct match can be established. It may be necessary to permutate the order of intensities, because the intensities depend to some extent on sample preparation and specific technique. Preferred orientation of platy minerals on diffractometer mounts often distorts the true intensity pattern. A match for halite, represented in the old ICDD format, is shown in Figure 11.16. The X-ray powder method has become the standard technique for identifying crystalline substances and is applied in every mineralogy laboratory. It is fast, safe, and accurate.

11.6 X-rays and crystal structure

So far our discussion has concentrated on the angular directions of diffracted X-rays. These directions simply depend on the lattice geometry. From the angles 2θ between incident and diffracted X-rays we obtain information about the unit cell. Ultimately we are not concerned so much about the lattice as we are about the distribution of atoms in a crystal, i.e., the crystal structure. For example, the minerals halite (NaCl), pyrite (FeS_2), and fluorite (CaF_2) all have a similar cubic unit cell ($a \approx 5.5$ Å), yet an entirely different crystal structure. Can we distinguish them, based on their diffraction patterns? The answer is "yes", and without too much additional effort we can show how diffraction intensities are related to atomic positions, using a similar approach.

Take, for example, a structure consisting of two atomic species, A and B (Figure 11.17), that can be considered as a superposition of two lattices. Each lattice obeys Bragg's law, but the two lattices generally do not scatter in phase and the intensity of the scattered wave for a reflection hkl depends on the phase shift between lattice A and lattice B. To evaluate the phase shift we return to the two-dimensional picture of Figure 11.9, except that we add a second species (open circles in Figure 11.18). The second

Box 11.2 Additional information: Determination of the lattice parameter of a cubic crystal

In the cubic system (where $a = b = c$), equation 11.9b transforms to

$$1/d^2 = (h^2 + k^2 + l^2)/a^2 \tag{11.10}$$

After squaring the Bragg equation (equation 11.8b), we can substitute for $1/d^2$ using equation 11.10:

$$\sin^2\theta = (h^2 + k^2 + l^2)\lambda^2/(4a^2) = (h^2 + k^2 + l^2)K = nK \tag{11.11}$$

Thus, the squared sines of the diffraction angles are the product of $\lambda^2/(4a^2)$, which is a constant K for a given experiment and a certain crystal, and $h^2 + k^2 + l^2$, an integer n that is the sum of three squared integers. Table 11.1 gives in column 1 a list of θ angles measured on a powder pattern for the cubic mineral halite (see Figure 11.14). From θ we then calculate $\sin^2\theta$. Each $\sin^2\theta$ is the product of a constant K and an integer n.

In the derivation of the smallest constant K, it is useful first to calculate all the differences $\varDelta$ between adjacent $\sin^2\theta$ values (third column in the table). The constant K is either the smallest $\varDelta$ or a simple fraction thereof, and it must be chosen such that all diffractions can be expressed as a product nK. If this is not possible, or if n is one of the numbers 7, 15, ... that are not sums of three squared integers, then it generally helps to use half the value of K. Slight differences in Δ values are due to experimental uncertainties.

Having established K, we can then deconvolute n into h, k, and l. For each diffraction we find indices hkl that characterize the reflecting lattice plane. It should be noted that reflections for certain lattice planes such as 003 and 122 may occur at the same angle ($n = 9$) and are not resolved in the powder technique (they are not present in the halite diffraction pattern). From K we determine the lattice parameter a:

$$a^2 = (\lambda^2/4K) = [(1.5418 \text{ Å})^2/(4 \times 0.0188)]$$
$$a = 5.62 \text{ Å} \tag{11.12}$$

This simple exercise, which any student can do with a pocket calculator, earned the Braggs the Nobel Prize in Physics for determining the size of the unit cell.

Table 11.1 Indexing of a powder pattern of halite (wavelength of X-rays: CuKα = 1.5418 Å)

$\theta_{measured}$	$\sin^2\theta$	$\varDelta \sin^2\theta$	$n = (h^2 + k^2 + l^2)$	K	h	k	l
13.68°	0.0559		3	0.0188	1	1	1
		0.0188					
15.86°	0.0747		4		2	0	0
		0.0747					
22.74°	0.1494		8		2	2	0
		0.0560					
26.95°	0.2054		11		3	1	1
		0.0188					
28.26°	0.2242		12		2	2	2
		0.0747					
33.14°	0.2989		16		4	0	0
		0.0553					
36.52°	0.3542		19		3	3	1
		0.0194					
37.68°	0.3736		20		4	2	0
		0.0745					
42.02°	0.4481		24		4	2	2
		0.0563					
45.25°	0.5044		27		5	1	1

PDF # 05-0628

d	2.82	1.99	1.63	3.26	NaCl	
I/I$_1$	100	55	15	13	Sodium Chloride	(Halite)

		[d(Å)]	I/I$_1$	hkl
Rad. CuKα$_1$	λ1.5405 Filter Ni	3.258	13	111
	I/I$_1$ Diffractometer	2.821	100	200
		1.994	55	220
Ref. Swanson and Fuyat, *NBS Circular* S39, Vol. 2, 41, 1953.		1.701	2	311
		1.628	15	222
		1.410	6	400
Sys. Cubic	S.G. Fm3m	1.294	1	331
a$_0$ 5.6402 b$_0$ c$_0$ α β γ	(225)	1.261	11	420
		1.1515	7	422
		1.0855	1	511
n: 1.542	Color Colorless	0.9969	2	440
		0.9533	1	531
		0.9401	3	600
An ACS reagent grade sample recrystallized twice from hydrochloric acid.		0.8917	4	620
		0.8601	1	533
X-ray pattern at 26 °C. *Merck Index*, 8th edn, p. 956.		0.8503	3	622
		0.8141	2	444

Fig. 11.16 ICDD index card for halite, showing values for *d*-spacings, diffraction intensities for reflections *hkl*, and additional crystallographic information. This information is now mainly available from digital databases.

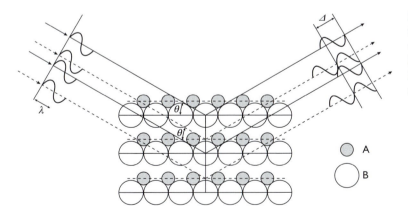

Fig. 11.17 A crystal structure with two atomic species A and B can be viewed as a superposition of two lattices. Each lattice obeys Bragg's law, but the two lattices do not scatter in phase.

lattice (open circles) is displaced by fractional coordinates x, y against the first lattice (dots). For a simple lattice, with a single atom in the corner of the unit cell, we found that the phase shifts of a lattice point relative to an origin are $2\pi(n_1 + n_2)$, where n_1 and n_2 are integers. The phase shifts for the second lattice (B, open circles) relative to the lattice in the origin (A, closed circles) are $2\pi(n_1x + n_2y)$. In general, X-ray scattering from these two lattices is not in phase. If we sum the waves, intensities will vary, depending on the relative displacements x and y. The displacements x and y correspond to atomic coordinates (see Section 7.6),

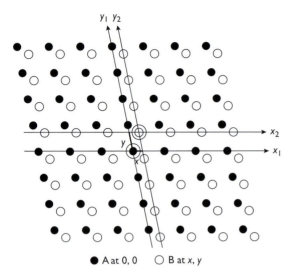

Fig. 11.18 Phase differences resulting from diffraction on a two-dimensional crystal with two atomic species indicated by closed circles (A) and open circles (B). The wave scattered on the atom B lattice (coordinates x, y) has a phase difference $2\pi(x + y)$ relative to that scattered on atom A lattice (at origin) (cf. Figure 11.9).

relative to the corners of the unit cell; n_1 and n_2 are orders of diffraction, for which we can substitute the Laue indices h and k and thus obtain for the phase shift $\phi = 2\pi(hx + ky)$.

If we consider a three-dimensional crystal structure composed of n sublattices of atoms at fractional coordinates x_i, y_i, z_i relative to the corner of the unit cell, then the phase shifts for a diffraction hkl are for each atom, by analogy with the two-dimensional case,

$$\phi_i = 2\pi(hx_i + ky_i + lz_i) \tag{11.13}$$

In order to obtain the amplitude and phase of the diffracted wave, we have to add the individual waves that are scattered from all atoms in the unit cell. In Box 11.3 we do this for the example of halite, using the method we discussed earlier (equation 11.2).

11.7 Additional atomic scattering considerations

For quantitative intensity calculations, we have to refine the model of the structure factor (Box 11.3). One complication arises from the fact that atoms have a finite size, comparable to the wavelength of X-rays.

Therefore, since X-ray scattering occurs on electrons, waves scattered on different electrons of the same atom are slightly out of phase. For example, the wavelet scattered on electron d in Figure 11.19 is $(x - y)$ out of phase relative to the center of the atom. Only wavelets in the direction of the incident beam are exactly in phase. The phase difference increases with diffraction angle 2θ. Previously we have assumed that the efficiency of atomic scattering, or the *scattering factor f*, is simply a function of the number of electrons, but it also depends on the diffraction angle 2θ. Some typical scattering factor curves are shown in Figure 11.20, which illustrates that only for $2\theta = 0°$, or for a "point atom", in which all electrons scatter in phase, does the scattering factor correspond to the atomic number. The 2θ dependence is much greater for the large O^{2-} ($r = 1.40$ Å) than for the relatively small Si^{4+} ($r = 0.42$ Å), both having 10 electrons. Owing to the scattering factor, high-angle reflections are generally weaker than low-angle reflections.

Instead of X-rays, neutrons can be used for diffraction experiments. Neutrons scatter on the nucleus and, since the nucleus is infinitely small, there is no angle dependence of the neutron scattering factor. While neutron diffraction has many advantages over X-rays, it is rarely used because it is not easily available (see Chapter 16).

For a refined description of atomic scattering, one must also consider that atoms are in constant thermal vibration and that atomic coordinates are merely average values. Another complication is that diffracted X-rays are partially polarized, and thus appropriate corrections need to be applied to account for the polarization effect. But, in principle, it is straightforward to calculate the intensity of diffractions from the atomic arrangement. It is much more difficult, and beyond the scope of this book, to achieve the inverse problem, i.e., to calculate atomic coordinates from a set of observed intensities.

11.8 Summary

The chapter introduces X-ray diffraction as the most important method for mineral identification and crystal structure determination. Every student should be able to derive the Laue equation and Bragg's law. Powder diffractometers are widely applied to investigate minerals and mineral assemblages. Powder patterns are used for phase identification. The diffraction intensities are directly linked to atomic coordinates.

Box 11.3 Additional information: Diffraction intensities and the atomic structure of halite

In Chapter 7 (see Figure 7.24b) we described the structure of halite with fractional coordinates x, y, z. These coordinates are again shown in the table below for the four Na^+ and four Cl^- that occupy the unit cell. X-rays scatter on the electrons surrounding the nucleus of an atom and the scattering strength or amplitude of the wave scattered from an atom is, to a first approximation, proportional to the number of electrons, or atomic number, i.e., 11 for Na^+ and 17 for Cl^-. This scattering amplitude is referred to as the scattering factor f_i.

The summation of individual wavelets scattered by each atom can be done graphically by adding all the waves with amplitudes f_i and phases $\phi_i = 2\pi (hx_i + ky_i + lz_i)$, or we can do it algebraically by adding the vector components, generalizing equation 11.2:

$$F_{hkl} = \sum f_i [cos2\pi(hx_i + ky_i + lz_i) + sin2\pi i(hx_i + ky_i + lz_i) = X + iY] \tag{11.14}$$

The vector F of the resultant wave is called the *structure factor*. In Table 11.2, we calculate F for (100), (200), and (111) reflections of NaCl. Having obtained components X and Y, we can then calculate amplitude $A = \sqrt{(X^2 + Y^2)}$, and phase $\phi = tan^{-1}(Y/X)$ of the diffracted wave.

We observe that in this simple structure part Y is always zero. When we add part X, we get the largest value for (200), an intermediate value for (111), and zero for (100). In (200), waves from all atoms scatter in phase (same sign), while in (111) waves from Na and Cl are out of phase. The square of the amplitudes is proportional to the observed intensities of the scattered waves. Indeed, on the powder diffraction pattern of halite shown in Figure 11.14 we find that (200) is much stronger than (111), and (100) is absent.

We noted that the amplitude for (100) is zero, and therefore we call this reflection "extinct". Some *extinctions* (i.e., lack of any diffraction intensity) are purely accidental, caused by a particular arrangement of atoms. Others are systematic and an expression of the crystal symmetry. These systematic extinctions are used to determine translational symmetries and the space-group. The (100) extinction in halite is due to the face-centered structure with a translation ½, ½, 0. For all centrosymmetrical crystal structures, in which for each atom at x, y, z there is a corresponding one at $-x$, $-y$, $-z$, the Y (imaginary, see equation 11.2) part of the structure factor is zero (as in the case of halite).

Table 11.2 Structure factor calculation for reflections (100), (200), and (111) of NaCl

i	Atomic coordinates			$X = f_i \cos 2\pi (hx_i + ky_i + lz_i)$			$Y = f_i \sin 2\pi (hx_i + ky_i + lz_i)$		
	x_i	y_i	z_i	(100)	(200)	(111)	(100)	(200)	(111)
Na(1)	0	0	0	11	11	11	0	0	0
Na(2)	½	½	0	−11	11	11	0	0	0
Na(3)	½	0	½	−11	11	11	0	0	0
Na(4)	0	½	½	11	11	11	0	0	0
Cl(1)	½	½	½	−17	17	−17	0	0	0
Cl(2)	½	0	0	−17	17	−17	0	0	0
Cl(3)	0	½	0	17	17	−17	0	0	0
Cl(4)	0	0	½	17	17	−17	0	0	0
Σ				0	112	−24	0	0	0
Amplitude:				0	112	24			
phase				0°	0°	180°			

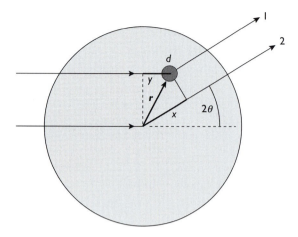

Fig. 11.19 Due to the finite size of atoms, waves scattered in different parts of the atom are not in phase. The path difference $x - y$ is a function of size and of the scattering angle 2θ.

Test your knowledge

1. How are Kβ X-rays produced?
2. Derive Bragg's law.
3. What is the d-spacing for (110) of a cubic crystal with lattice parameter $a = 5$ Å?
4. Interpret the powder pattern shown in Figure 11.14.
5. Calculate the structure factor for reflections (113) and (102) for copper (for structure, see Figure 7.24a).

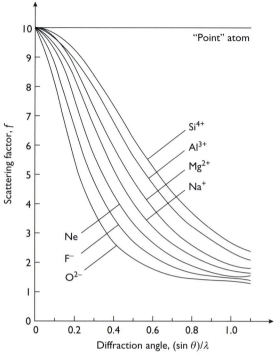

Fig. 11.20 Variation of the scattering factor with diffraction angle (represented as $\sin \theta/\lambda$) for ions of different size but all with the same number of electrons as neon (10).

Further reading

Azaroff, L. V. (1968). *Elements of X-ray Crystallography*. McGraw-Hill, New York.

Cullity, B. D. and Stock, S. R. (2001). *Elements of X-ray Diffraction*, 3rd edn. Prentice Hall, Upper Saddle River, NJ.

Glusker, J. P. and Trueblood, K. N. (2010). *Crystal Structure Analysis: A Primer*, 3rd edn. International Union of Crystallography, Oxford University Press, Oxford.

Massa, W. and Gould, R. O. (2004). *Crystal Structure Determination*. Springer.

Stout, G. H. and Jensen, L. H. (1989). *X-ray Structure Determination*, 2nd edn. Wiley, New York.

Zolotoyabko, E. (2014). *Basic Concepts of X-ray Diffraction*. Wiley-VCH, Weinheim.

12 | Physical properties

Physical properties of crystals are linked to the crystal structure and the symmetry. Many properties are anisotropic, i.e., their magnitude is different in different directions. They are described by vectors and tensors. This chapter introduces density, thermal conductivity, elastic properties, piezo- and pyroelectricity, and magnetic properties.

12.1 Vectors and tensors: general issues

In this chapter we discuss some physical properties of crystals. The physical properties of minerals are as relevant as their chemical composition, but the former have been neglected in introductory mineralogy texts. The reason for this is their complexity because, in contrast to chemical composition, many properties cannot be described by simple numbers. Physical properties are intricately linked to the structure and the symmetry of crystals. Many properties are anisotropic, i.e., they are different if the crystal is rotated (from the Greek *anisos* meaning "not the same" and *trepein* meaning "turn") and thus are directional. In this chapter we are not attempting to give a comprehensive coverage, but are trying to raise a few important issues to give you the flavor of basic concepts and to prepare for more advanced treatments. With interest in the Earth's interior, mineral physics is a rapidly growing field of mineralogy. The subjects of thermal expansion and elastic properties are essential in understanding the equation of state and stability of minerals at high pressure and temperature. Anisotropic properties such as elastic and magnetic properties are of great importance in seismology for investigating the structure of the deep Earth and for the paleomagnetic reconstruction of continental movements. But they are not only of academic interest: prospecting for mineral resources as well as for oil and gas is increasingly based on physical rather than chemical methods.

Elasticity and magnetism are some of the more difficult properties, and the brief survey can show merely where you can continue your studies, for example by reading J. F. Nye's (1957) classic book

Physical Properties of Crystals. The relationships are more transparent if some linear algebra is applied, and for this reason we introduce a few concepts concerning *tensors*.

A physical property of a material can be determined by suitable measurements and gives a relationship between two physical quantities. For example, the density relates a volume element and the corresponding mass. A more complicated property is the thermal conductivity, which relates an imposed temperature gradient to a resulting heat flux. The temperature gradient can be looked upon as a "stimulus" acting upon the material and the rate of heat flux as a "response" resulting from the interaction of the material and the stimulus. For some selections of stimulus S, a given response R is found to be unique and a linear relationship can be written, such as

$$R = PS \tag{12.1}$$

with a functional role played by the property P. In the case of density ρ we can write

$$m = \rho V \tag{12.2}$$

where m is the mass and V the volume.

Many physical properties of minerals, such as density, thermal conductivity, electrical conductivity, thermal expansion, and elasticity, can have a straightforward mathematical description. Other properties do not uniquely relate physical quantities. For example, the plastic properties of a crystal cannot be defined in terms of a unique relationship between stress and strain, but instead depend on the history of a particular crystal. We will discuss plastic deformation of crystals in Chapter 17. In this chapter we investigate some physical properties with a unique stimulus–response relationship.

If the relationship is not connected in any way to direction, such quantities are called *scalars* and are completely specified with a single number, as in equation 12.2. Most quantities, though, can be defined only with reference to directions. For example, the temperature gradient acting on a point in a crystal needs to be specified with both its magnitude and its direction. This description is commonly referred to as a *vector* and is usually represented by boldface italic type, such as ∇T for the temperature-gradient vector.

Thermal conductivity is a good example to introduce anisotropy of physical properties. Cut a slab from a quartz crystal with trigonal symmetry parallel to the *c*-axis and cover the surface with a thin layer of wax. Then apply heat with a metal pin at a point. The heat from the pin will propagate and melt the wax, creating a ridge contour of elliptical shape outlining an isotherm (Figure 12.1a). The ratio of the ellipse axes is roughly 1:2, indicating that thermal conductivity parallel to the *c*-axis is almost twice as high as the value perpendicular to it. Thermal conductivity κ relates an applied temperature gradient (a vector) to a heat flux (also a vector). If we conduct the same experiment on a slab cut perpendicular to the *c*-axis we observe a circular isotherm (Figure 12.1b). In this section thermal conductivity is the same in all directions. As we will see later, the crystal symmetry

imposes restrictions on physical properties. For example, in a cubic crystal the thermal conductivity is the same in all directions.

As an alternative to specifying a vector *v* by magnitude and direction (e.g., by an arrow), a vector can also be described with components of a rectangular coordinate system x_1, x_2, x_3, just as we have used zone indices and lattice parameters to describe a lattice direction (Figure 12.2). The components are projections of the vector on the axes. If the components of ∇T are T_1, T_2, T_3, we can write

$$\nabla T = [T_1, T_2, T_3], \text{where } T_i = dT/dx_i \qquad (12.3)$$

and the three components completely specify the vector. If the medium is isotropic and all directions are equivalent, as for example in a liquid or a glass, the stimulus and response vectors are parallel and the magnitude of the heat flux vector *q* is proportional to the temperature gradient ∇T (Figure 12.3a)

$$q = -\kappa \nabla T \qquad (12.4)$$

to account for the directional nature of heat flow. A minus sign is used to indicate that heat flows in the opposite direction to the temperature gradient. This is known as Fourier's law. The constant κ is a material property called *thermal conductivity*. In component nomenclature we obtain

$$q_1 = -\kappa T_1, \quad q_2 = -\kappa T_2, \quad q_3 = -\kappa T_3 \qquad (12.5)$$

(a)   (b)

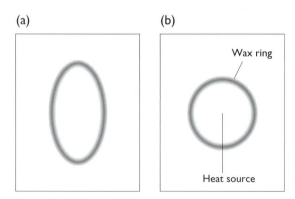

Fig. 12.1 Experiment to establish anisotropy of thermal conductivity in a crystal and its relationship to crystal symmetry. A polished surface on a slab of quartz is covered with a thin layer of wax and a pointed heat source is applied that melts the wax. (a) In a section parallel to the *c*-axis we observe an ellipse, representing an isotherm. (b) In a section perpendicular to the *c*-axis we observe a circle.

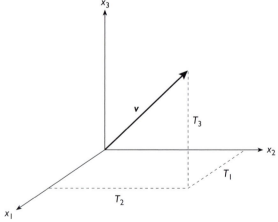

Fig. 12.2 A vector *v* (arrow) can be represented by components T_1, T_2, T_3 that are the projections of the vector on the three axes of a Cartesian coordinate system.

(a)   (b)

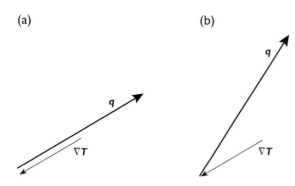

Fig. 12.3 (a) In an isotropic medium, the stimulus vector (e.g., temperature gradient ∇T) and the response vector (e.g., heat flux q) are parallel. (c) In an anisotropic medium the two vectors are in general not parallel. (The vectors have opposite signs because heat flows against a positive temperature gradient.)

in which each component of q is proportional to the corresponding component of ∇T.

For crystals, which have a lattice structure and directionality, the situation is not so simple and the vectors q and ∇T may not be parallel, because of the influence of the ordered structure on the heat flow (Figure 12.3b). Thus relations (12.5) have to be replaced by

$$q_1 = -\kappa_{11}\ T_1 - \kappa_{12}\ T_2 - \kappa_{13}\ T_3$$
$$q_2 = -\kappa_{21}\ T_1 - \kappa_{22}\ T_2 - \kappa_{23}\ T_3 \qquad (12.6)$$
$$q_3 = -\kappa_{31}\ T_1 - \kappa_{32}\ T_2 - \kappa_{33}\ T_3$$

where κ_{11}, κ_{12}, etc. (or κ_{ij}) are again constants. Each component of q is now linearly related to all three components of ∇T. Thus, in order to specify the thermal conductivity of an anisotropic crystal, we must specify the nine constants κ_{ij} (also known as coefficients), which can be written in a square array and enclosed in brackets.

$$\left[\kappa_{ij}\right] = \begin{bmatrix} \kappa_{11} & \kappa_{12} & \kappa_{13} \\ \kappa_{21} & \kappa_{22} & \kappa_{23} \\ \kappa_{31} & \kappa_{32} & \kappa_{33} \end{bmatrix} \qquad (12.7)$$

This expression denotes a *tensor of the second rank* that relates two vector quantities (Figure 12.3b). As we will see, there are tensors of higher rank, but all are linear functions of coordinates. For a second-rank tensor κ_{ij} the first suffix i gives the row (related to the heat flow) and the second suffix j the column of the coefficient (related to the temperature gradient). The summations in equations 12.6 are often abbreviated and the equations are written as

$$q_i = -\kappa_{ij} T_j \quad (i,j = 1, 2, 3) \qquad (12.8)$$

The thermal conductivity tensor $[\kappa_{ij}]$ is a physical quantity which, for a given set of arbitrary axes, is represented by nine numbers, but we will see that these are not all independent.

In general, if a property p relates two vectors $r = (r_1, r_2, r_3)$ (response) and $s = (s_1, s_2, s_3)$ (stimulus), we can write

$$r_i = p_{ij}s_j \quad (i,j = 1, 2, 3) \qquad (12.9)$$

where r_i and s_j are the components of the vectors on the three axes. The tensor $[p_{ij}]$ can be expanded to

$$\left[p_{ij}\right] = \begin{bmatrix} p_{11} & p_{12} & p_{13} \\ p_{21} & p_{22} & p_{23} \\ p_{31} & p_{32} & p_{33} \end{bmatrix} \qquad (12.10)$$

How can we visualize the second-rank property tensor p_{ij}? The representation surface (or representation quadric) is a geometrical representation of a second-rank tensor and is useful for giving us a visual image of the tensor. For symmetric second-rank tensors such as thermal conductivity (where $p_{ij} = p_{ji}$) the quadric equation of the tensor surface is

$$p_{11}x_1{}^2 + p_{22}x_2{}^2 + p_{33}x_3{}^2 + 2p_{23}x_2x_3 + 2p_{31}x_3x_1$$
$$+ 2p_{12}x_1x_2 = 1 \qquad (12.11)$$

which is the equation of an ellipsoid (tensor ellipsoid, Figure 12.4a). If we transform the coordinate system so that axes are parallel to ellipsoid axes we obtain the simple quadric (Figure 12.4b)

$$p_{11}x_1{}^2 + p_{22}x_2{}^2 + p_{33}x_3{}^2 = 1 \qquad (12.12)$$

and the corresponding tensor expression

$$[p] = \begin{bmatrix} p_{11} & 0 & 0 \\ 0 & p_{22} & 0 \\ 0 & 0 & p_{33} \end{bmatrix} \qquad (12.13)$$

with p_{11}, p_{22}, and p_{33} corresponding to the lengths of the ellipsoid axes.

12.2 Symmetry considerations

The choice of axes, i.e., the coordinate system, determines the values of the components p_{ij}, whereas the vectors r (response) and s (stimulus) are independent of the arbitrarily introduced coordinate system. F. E. Neumann (1885) found that for crystals the point-group of the property tensor must include all the symmetry operations of the point-group of the crystal, i.e., any symmetry operation that leaves the crystal

(a)

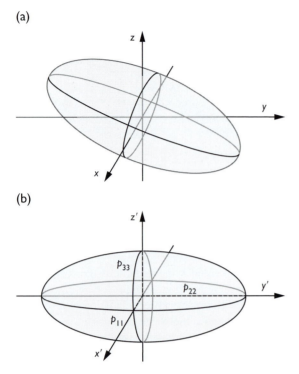

(b)

Fig. 12.4 (a) Ellipsoid representing second-rank tensor properties of a crystal. (b) The coordinate system can be transformed to bring axes parallel to ellipsoid axes.

invariant must leave the property tensor invariant. This means that for triclinic crystals there are no crystal symmetry constraints (Figure 12.5a), for cubic crystals the property quadric becomes a sphere (Figure 12.5e), and for hexagonal, trigonal, and tetragonal crystals it becomes a rotational ellipsoid (Figure 12.5d). For crystals it is desirable to express a tensor in the coordinate system that conforms with the crystal symmetry; for example, for orthorhombic crystals the axes of the property ellipsoid (x', y', z') align with the axes of the crystal (x, y, z), and for monoclinic crystals the y' property axis aligns with the y crystal axis (2-fold rotation). These crystal symmetry constraints also restrict the coefficients needed to describe the property. As mentioned earlier for symmetrical second-rank tensors $p_{ij} = p_{ji}$ (Table 12.1).

12.3 Tensors of different ranks

In the introduction to tensors, the influence of symmetry, and the representation quadric, we have dealt with second-rank tensors, such as thermal conductivity, which relate a vector stimulus to a vector response.

Unless they are restricted by crystal symmetry, second-rank tensors are specified by nine numbers (3^2), but for symmetrical second-rank tensors such as thermal conductivity this is reduced to six (Table 12.1). Vectors can also be viewed as tensors, but of the first rank, and are described by three numbers (3^1), as shown in equation 12.3. Scalars (such as density) are sometimes referred to as tensors of zero rank and are described by a single number (3^0). There are also tensors of higher order. For example, the elastic compliance S_{ijkl} is a fourth-rank tensor that relates the second-rank tensor stress (stimulus) and the second-rank tensor strain (response). It requires 81 (3^4) components for its description, but not all of these components are independent.

The rank of a tensor physical property depends on the quantity of stimulus and response that it relates. The relationship between two scalars (e.g., mass and volume) is a scalar property (density). The relationship between two vectors (e.g., thermal gradient and heat flow) is a second-rank tensor (thermal conductivity). Two second-rank tensors (e.g., stress and strain) are related by a fourth-rank tensor (elastic compliance). Whereas the representation quadric for a symmetrical second-rank tensor is an ellipsoid, those for higher-rank tensors are more complicated surfaces, some of which will be illustrated in later sections. Table 12.2 gives examples of properties that will be reviewed briefly in this chapter.

12.4 Density

Density (ρ) is a scalar material property, relating mass m and volume V

$$m = \rho V \tag{12.14}$$

and is usually measured in g/cm³. It varies widely in minerals (Table 12.3) and is therefore of important diagnostic value. It can be estimated as "heftiness" by weighing a sample in your hand. For quantitative determinations, it is necessary to determine mass and volume. Whereas mass is easy to measure with a scale, the volume of an irregularly shaped object is more difficult to assess. It can be ascertained, for example, by determining the volume displacement of water.

A related property is the specific gravity G, defined as the density divided by the density of water at 4 °C, the temperature at which water possesses its maximum density of 0.999 973 g/cm³. Thus the specific

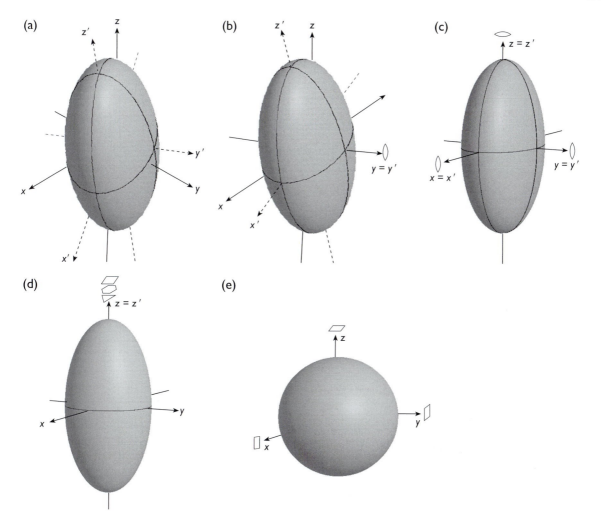

Fig. 12.5 The crystal symmetry imposes restrictions on the axes and orientation of the property ellipsoid: (a) triclinic, (b) monoclinic, (c) orthorhombic, (d) tetragonal, and (e) cubic. Crystal axes (x, y, z) and their symmetries are indicated, and principal sections of the ellipsoid with axes x'_1, x'_2, x'_3 are outlined.

gravity is numerically very close to the density. The specific gravity can be measured with a Jolly balance (Figure 12.6) by comparing the weight of a mineral in air, W_a, with that of the mineral suspended in water, W_w,

$$G = W_a/(W_a - W_w) \qquad (12.15)$$

If we know the unit cell volume (e.g., from X-ray diffraction) and the elementary occupancy of the unit cell, then we can calculate the density directly.

Take, for example, halite (NaCl) with a cubic unit cell, a lattice parameter $a = 5.639$ Å (1 cm $= 10^8$ Å), and four ions of Na^+ and Cl^- in each unit cell (Figure 7.24b). The atomic mass (i.e., the mass of one mol, or

6.023×10^{23} atoms) of Na^+ is 23 g, and that of Cl^- is 35.5 g. Thus the mass of one Na^+ is 3.82×10^{-23} g, and that of one Cl^- is 5.89×10^{-23} g. We can then calculate the density:

$$\rho = m/V$$
$$= 4 \times (3.82 + 5.89) \times 10^{-23}\text{g}/(5.639 \times 10^{-8}\text{cm})^3$$
$$= 2.17 \text{ g/cm}^3 \qquad (12.16)$$

Since the volume of most materials increases slightly with temperature, whereas the mass remains constant, the density decreases slightly with increasing temperature. Similarly, since the volume decreases with pressure, the density increases with pressure.

Table 12.1 Crystal symmetry expressed in symmetrical second-rank property tensors p_{ij} and corresponding representation quadric (see also Figure 12.5)

Crystal system	Representation quadric	Number of independent coefficients	Tensor (in xyz system)
Cubic	Sphere	1	$\begin{pmatrix} p & 0 & 0 \\ 0 & p & 0 \\ 0 & 0 & p \end{pmatrix}$
Tetragonal Hexagonal Trigonal	Rotational ellipsoid	2	$\begin{pmatrix} p_1 & 0 & 0 \\ 0 & p_1 & 0 \\ 0 & 0 & p_3 \end{pmatrix}$
Orthorhombic	General ellipsoid (axes parallel to crystal axes)	3	$\begin{pmatrix} p_1 & 0 & 0 \\ 0 & p_2 & 0 \\ 0 & 0 & p_3 \end{pmatrix}$
Monoclinic	General ellipsoid (one axis parallel to 2-fold y-axis [010])	4	$\begin{pmatrix} p_{11} & 0 & p_{13} \\ 0 & p_{22} & 0 \\ p_{13} & 0 & p_{33} \end{pmatrix}$
Triclinic	General ellipsoid (no specific relationship to crystal axes)	6	$\begin{pmatrix} p_{11} & p_{12} & p_{13} \\ p_{12} & p_{22} & p_{23} \\ p_{13} & p_{23} & p_{33} \end{pmatrix}$

Table 12.2 Tensor properties, relating a stimulus and a response

Property (rank)	Stimulus (rank)	Response (rank)
Density (0)	Mass (0)	Volume (0)
Heat capacity (0)	Temperature (0)	Mass (0)
Pyroelectricity (1)	Temperature (0)	Electric field (1)
Electrical conductivity (2)	Electric field (1)	Electric current density (1)
Permeability (2)	Magnetic field (1)	Magnetic induction (1)
Dielectric tensor (2)	Electric field (1)	Electric displacement (1)
Magnetic susceptibility (2)	Magnetic field (1)	Intensity of magnetization (1)
Thermal conductivity (2)	Temperature gradient (1)	Heat flux (1)
Thermal expansion (2)	Temperature (0)	Strain (2)
Piezoelectricity (3)	Electric field (1)	Strain (2)
Elastic compliance (4)	Stress (2)	Strain (2)
Elastic stiffness (4)	Strain(2)	Stress (2)

12.5 Thermal conductivity, thermal expansion, and specific heat

We have already introduced thermal conductivity in equation 12.4 as $q = -\kappa \nabla T$. It is relatively high for metals and minerals with substantial contributions of metallic bonding, such as graphite, where heat is transferred largely through the flow of free electrons. For ionic and covalent crystals, thermal conductivity is much lower and often strongly anisotropic. In these crystals heat is transferred through thermal vibrations. Overall, the thermal conductivity is greatest in directions of closer atomic packing. Graphite, with a layer structure, has a thermal conductivity that is four times higher within the layers with partial metallic bonding than perpendicular to them. In mica, with covalent bonding within layers and weak ionic bonding between layers, the overall conductivity is much lower than in graphite. However, the anisotropy is even more

Table 12.3 Approximate density (g/cm³) of some minerals at ambient conditions (ordered by magnitude)

Ice[a]	0.92
Sylvite	1.99
Mordenite	2.1
Halite	2.16
Graphite	2.15
Gypsum	2.33
Orthoclase	2.56
Serpentine	2.60
Albite	2.61
Quartz	2.65
Talc	2.70
Calcite	2.7
Anorthite	2.77
Muscovite	2.80
Dolomite	2.90
Enstatite	3.1
Fluorite	3.18
Garnet	3.1–4.2
Olivine	3.22–4.39
Diopside	3.3
Diamond	3.5
Corundum	4.0
Rutile	4.2–5.5
Barite	4.5
Zircon	4.68
Pyrite	5.02
Magnetite	5.18
Hematite	5.25
Iron	7.3–7.9
Galena	7.58
Cinnabar	8.18
Copper	8.95
Gold	19.3

[a] Cubic ice Ic at 0 °C and ambient pressure.

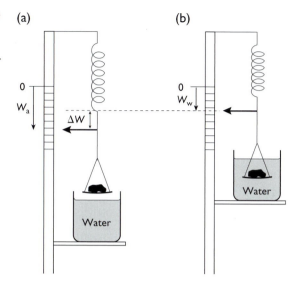

Fig. 12.6 Schematic of a Jolly balance, used to measure specific gravity by determining (a) the relative weight of a sample in air (W_a) and (b) the relative weight of the sample immersed in water (W_w).

extreme, with conductivity parallel to the (001) layer plane six times higher than that perpendicular to it. Examples of thermal conductivities are given in Table 12.4. They depend on temperature and pressure.

Thermal expansion α_{ij} is also a second-rank tensor, relating a temperature increment ΔT (a scalar) with a strain ε_{ij} (a second-rank tensor):

$$\varepsilon_{ij} = \alpha_{ij}\,\Delta T \qquad (12.17)$$

If a sphere were drawn in the crystal it would become, on change of temperature, an ellipsoid with axes proportional to $(1 + \alpha_1\,\Delta T)$, $(1 + \alpha_2\,\Delta T)$, and $(1 + \alpha_3\,\Delta T)$. Normally crystals expand in all directions with increasing temperature, as is the case for quartz (Figure 12.7a). But calcite is different and expands parallel to the c-axis, yet contracts perpendicular to it (Figure 12.7b). The abnormal behavior and high anisotropy of thermal expansion for calcite is the reason that, in calcite rocks such as marble, large local stresses are produced during thermal cycling, leading to fracturing and deterioration, reducing their suitability as a building material. Thermal expansion of single crystals is generally measured by determining lattice parameter changes with temperature by X-ray diffraction.

Values for heat capacity (or specific heat), C_p, are given in Table 12.4. The defining equation of this scalar property is

$$\Delta S = \left(C_p/T\right)\Delta T \qquad (12.18)$$

where ΔS is the entropy change, ΔT a temperature change, and T the absolute temperature (for further discussion, see Chapter 19).

12.6 Elastic properties

So far we have discussed some scalars and second-rank tensors. We now survey briefly a more complex

Table 12.4 Thermal properties of some crystals (thermal conductivity, thermal expansion, and molar heat, ordered with increasing symmetry)

Mineral	Crystal system	Temp. (K)	Thermal conductivity κ (J/(m s K))			Thermal expansion α (10^{-6} K^{-1})			Molar heat capacity, C_p (J/(mol K))
			$\|a$	$\|b$	$\|c$	$\|a$	$\|b$	$\|c$	
Gypsum	Monoclinic	310	3.16 ($\perp c$)		3.63 ($\|c$)^a	1.6	42	29	186.0
Olivine (forsterite)	Orthorhombic	300	5.84	3.38	5.06	6.6	9.90	9.8	117.9
Enstatite	Orthorhombic	300	3.27	2.72	4.31	16.4	14.5	16.8	82.1
Calcite	Trigonal	300	3.52	$= a$	4.18	−3.2	$= a$	13.3	83.5
Quartz	Trigonal	300	6.5	$= a$	11.3	14	$= a$	9	44.6
Graphite	Hexagonal	300	355	$= a$	89	−1.22	$= a$	26.7	8.536
Aluminum	Cubic	300	208	$= a$	$= a$	23	$= a$	$= a$	24.35
Copper	Cubic	273	410	$= a$	$= a$	16.7	$= a$	$= a$	24.43
Diamond	Cubic	273	138	$= a$	$= a$	0.89	$= a$	$= a$	6.109
Halite	Cubic	300	5.8	$= a$	$= a$	40	$= a$	$= a$	50.5
Garnet (pyrope)	Cubic	300	3.18	$= a$	$= a$	19.9	$= a$	$= a$	325.5

Note: $\|$, parallel to; $\perp$, perpendicular to.
^{a} Incomplete description.
Source: From Clark, 1966; Kanamori *et al.*, 1968; Horai, 1971; Beck *et al.*, 1978; Ahrens, 1995; Chai *et al.*, 1996; Grigoriev and Meilikhov, 1997.

property, elasticity, which expresses a unique relationship between two second-rank tensors: stress (σ) and elastic strain (ε). It is represented by a fourth-rank tensor.

Stress, σ_{ij}, is defined as a force dF_i in a certain direction acting on an area element dA_j:

$$\sigma_{ij} = dF_i/dA_j \qquad (12.19)$$

A more complete state of stress at a point must take account not only of one direction but of all directions, i.e., an infinite number of vectors around a point (Figure 12.8a). In three dimensions the surface of

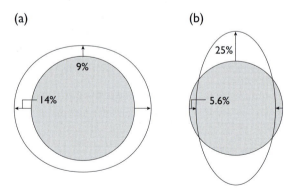

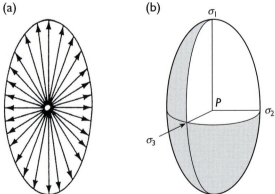

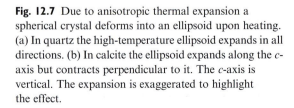

Fig. 12.7 Due to anisotropic thermal expansion a spherical crystal deforms into an ellipsoid upon heating. (a) In quartz the high-temperature ellipsoid expands in all directions. (b) In calcite the ellipsoid expands along the c-axis but contracts perpendicular to it. The c-axis is vertical. The expansion is exaggerated to highlight the effect.

Fig. 12.8 A general state of stress at a point P can be represented with a stress ellipsoid. (a) Vectors radiating from a point in a plane describe an ellipse that defines the magnitudes of stress in particular directions. (b) Three-dimensional stress ellipsoid with principal axes σ_1, σ_2, and σ_3.

these vectors around point P defines an ellipsoid, called the stress ellipsoid (Figure 12.8b).

The three orthogonal principal directions, σ_1, σ_2, and σ_3, are called principal stresses with magnitudes $\sigma_{11} > \sigma_{22} > \sigma_{33}$. The stress tensor relative to the principal axes is

$$\begin{bmatrix} \sigma_{11} & 0 & 0 \\ 0 & \sigma_{22} & 0 \\ 0 & 0 & \sigma_{33} \end{bmatrix} \qquad (12.20a)$$

or, relative to arbitrary axes,

$$\begin{bmatrix} \sigma_{11} & \sigma_{12} & \sigma_{13} \\ \sigma_{12} & \sigma_{22} & \sigma_{23} \\ \sigma_{13} & \sigma_{23} & \sigma_{33} \end{bmatrix} \qquad (12.20b)$$

If a general stress were applied to a crystal of spherical shape, the lattice would deform to an ellipsoidal shape. The resulting ellipsoid is called the *deformation*, or *strain ellipsoid* (or *stretch ellipsoid* by the mechanical community) with axes λ_1, λ_2, and λ_3 (Figure 12.9a). There may or may not be a volume change accompanying this deformation. Notice that some diameters are stretched (ellipsoid surface outside the sphere) whereas others are compressed. Figure 12.9b illustrates, for two-dimensional deformation (plane strain), shape changes during increasing deformation, keeping the volume constant.

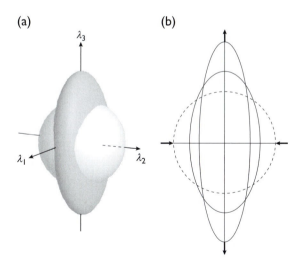

(a)

(b)

Fig. 12.9 (a) If a spherical crystal is subjected to a general stress, the crystal deforms to an ellipsoidal shape. This surface is known as the strain ellipsoid with axes $\lambda_1, \lambda_2, \lambda_3$. (b) Two-dimensional case, deforming a circle into ellipses of equal area. Extension and compression directions are indicated by arrows.

Hooke's law of elastic behavior states that deformation, or strain, imposed by a stress is proportional to that stress. This can be extended into a linear relationship between the six independent components of stress (σ_{ij}) and the six independent components of elastic strain (ε_{kl}), both symmetrical tensors (Cauchy's law). In a crystal we can express each component of stress as a linear function of all components of strain, and vice versa, to give two sets of six equations, written in matrix form as

$$\sigma_{ij} = c_{ijkl} \, \varepsilon_{kl} \qquad (12.21a)$$

or, alternatively, as

$$\varepsilon_{ij} = s_{ijkl} \, \sigma_{kl} \qquad (12.21b)$$

The fourth-rank tensor c_{ijkl} is called the *elastic stiffness* (or *elastic constant*), and the tensor s_{ijkl} is called the *elastic compliance* (or *elastic modulus*). A fourth-rank tensor has in general 81 (3^4) independent components, but this is reduced to 36 owing to the symmetry of the stress tensor σ_{ij} and the strain tensor ε_{kl}. Furthermore, for thermodynamic reasons the elastic tensor is symmetrical and this reduces the number of independent components to 21. Further reductions are due to crystal symmetry. The most generally anisotropic solid, with triclinic symmetry, has 21 independent elastic components; this reduces for crystals with cubic symmetry to 3, and for an isotropic medium (such as amorphous glass) to 2 (Table 12.5).

The fourth-rank elastic tensor is basically four-dimensional, but it is often described in a two-dimensional matrix notation suggested by W. Voigt (1928) and most handbooks list elastic constants of minerals in this notation (e.g., Nye, 1957; Simmons and Wang, 1971). Voigt represents the symmetrical

Table 12.5 Number of independent components of the elastic tensor for different crystal symmetries

Triclinic (all point-groups)	21
Monoclinic (all point-groups)	13
Orthorhombic (all point-groups)	9
Tetragonal $(4, \bar{4}, 4/m)$	7
Tetragonal $(4mm, \bar{4}m2, 422, 4/m \, 2/m \, 2/m)$	6
Trigonal $(3, \bar{3})$	7
Trigonal $(3m, 32, \bar{3}2/m)$	6
Hexagonal (all point-groups)	5
Cubic (all point-groups)	3
Isotropic (∞, ∞, m)	2

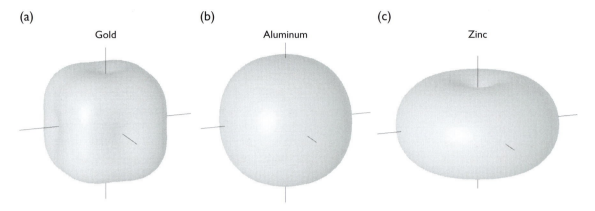

(a) Gold (b) Aluminum (c) Zinc

Fig. 12.10 Surface of the elastic stiffness for (a) gold, (b) aluminum, and (c) zinc (courtesy S. Grigull).

stress and strain tensors with a one-dimensional array of the six independent components:

$$\sigma = \{\sigma_1 = \sigma_{11}, \sigma_2 = \sigma_{22}, \sigma_3 = \sigma_{33}, \sigma_4 = \sigma_{23},$$
$$\sigma_5 = \sigma_{31}, \sigma_6 = \sigma_{12}\} \qquad (12.22a)$$

$$\varepsilon = \{\varepsilon_1 = \varepsilon_{11}, \varepsilon_2 = \varepsilon_{22}, \varepsilon_3 = \varepsilon_{33}, \varepsilon_4 = 2\varepsilon_{23},$$
$$\varepsilon_5 = 2\varepsilon_{31}, \varepsilon6 = 2\varepsilon_{12}\} \qquad (12.22b)$$

The elastic constants are then represented by a matrix of 6×6 constants C_{ij}.

As with a second-rank tensor, the directional properties of the fourth-rank elastic tensor can also be visualized as a surface. For the stiffness this surface is more complex than an ellipsoid, and even for cubic crystals it is generally anisotropic. Also here the shape of the surface has to conform with crystal symmetry. The anisotropy (ratio between largest and smallest value) can be large, even for cubic crystals, as in the case of gold (Figure 12.10a), or much smaller, as for aluminum (Figure 12.10b). Tungsten is almost isotropic. For hexagonal crystals, the elastic properties have axial symmetry, as illustrated for zinc (Figure 12.10c).

Elastic constants play a central role in the propagation of elastic waves and are of great concern to seismologists. Two types of elastic waves can be transmitted through an isotropic solid. One is called a longitudinal or *P wave* and exerts particle motions parallel to the direction of propagation (Figure 12.11a). The second is called a transverse or *S wave*, and its particle motions are perpendicular to the direction of propagation (similar to light) (Figure 12.11b). Seismologists use travel times of these waves as they pass through the Earth to explore the sedimentary crust for hydrocarbon deposits and to decipher elastic properties of the planet's deep interior.

Elastic properties are directly related to seismic velocities through what is known as Christoffel equations. In the case of an isotropic medium

$$V_P = \sqrt{(C_{11}/\rho)} \quad \text{and} \quad V_S = \sqrt{(C_{44}/\rho)} \qquad (12.23)$$

where V_P and V_S are longitudinal and transverse wave velocities, C_{11} and C_{44} stiffness coefficients, and ρ is density. For low symmetries this is more complex (e.g., Newnham, 2005; Chapman, 2010), but based on stiffness coefficients velocities can be calculated. Figure 12.12a illustrates, for olivine, that there is a difference of over 25% between the fastest (parallel to [100]) and slowest P-wave velocity (parallel to [010]). The three-dimensional velocity surface is shown in Figure 12.12b. If olivine crystals are aligned, as they often are in deformed rocks, the propagation of elastic

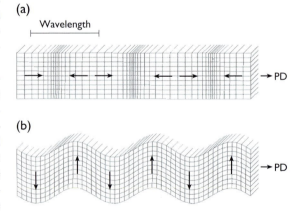

(a) Wavelength ⟶ PD

(b) ⟶ PD

Fig. 12.11 Propagation of elastic waves in a crystal. Particle motions relative to the propagation direction (PD) are shown by arrows. Wavelength is indicated. (a) Longitudinal (P) waves. (b) Transverse (S) waves.

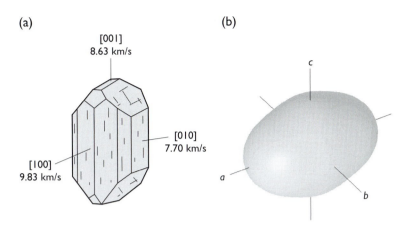

Fig. 12.12 Propagation of P waves in an olivine crystal. (a) Sketch of a crystal with P-wave velocities in the three principal directions. (b) Map of the P-wave velocity surface in the same orientation as (a) (values are squared to make variations more obvious).

waves in the aggregate is also anisotropic (e.g., Kocks *et al.*, 2000). Seismologists have established that the preferred orientation of olivine in peridotites of the upper mantle of the Earth, which was produced during convection, causes the observed anisotropy of seismic waves (e.g., Silver, 1996). For example, in the oceanic mantle underneath Hawaii, there is an azimuthal variation in P-wave velocities of over 10% (Figure 12.13).

12.7 Piezoelectricity and pyroelectricity

In crystals that lack an inversion center, there is an absolute directionality, at least for some axes. The lack of symmetry is a condition for the occurrence of such properties as optical activity (which will be discussed briefly in Chapter 14), *piezoelectricity* (from Greek *piezein*, meaning "press"), and *pyroelectricity*

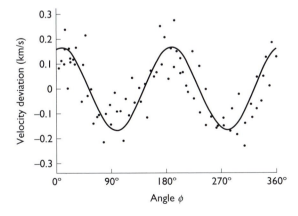

Fig. 12.13 Azimuthal variation of seismic longitudinal velocities for surface waves in the vicinity of Hawaii (data from Morris *et al.*, 1969).

(from Greek *pur*, meaning "fire"). Crystals with a center of symmetry do not display these properties.

The piezoelectric effect can be described as follows. If an electric field **E** is applied to certain noncentric crystals, the shape of the crystal changes slightly, i.e., a strain ε is produced:

$$\varepsilon_{jk} = d_{ijk} E_i \qquad (12.24)$$

E_i is a vector component and ε_{jk} a second-rank tensor; therefore, the piezoelectricity d_{ijk} is a third-rank tensor. This effect also works in reverse: By applying a stress to a crystal, we induce an electric field. The piezoelectric effect is not observed in all crystal directions.

Piezoelectricity has its basis in the crystal structure, as we will demonstrate in a simplified model for quartz, in which Jacques and Pierre Curie first observed piezoelectricity in 1880. The structure of quartz contains spiraling chains of SiO_4^{4-} tetrahedra parallel to the *c*-axis. In projection they appear as six-membered and three-membered rings of tetrahedra. The three-membered rings are more relevant in this discussion, and one such ring, with Si and O, is shown schematically in Figure 12.14a, which also displays the *a*-axes. In the undeformed crystal structure, charges of O^{2-} and Si^{4+} are balanced. However, if a slab of a quartz crystal cut parallel to the *c*-axis is stressed parallel to an *a*-axis, the charges are displaced and an electric field is induced with a surplus of negative charges on one side and of positive charges on the other (Figure 12.14b). One can apply an oscillating electric field to a quartz crystal and produce mechanical vibrations.

There is a wide range of technological applications for piezoelectricity. In transducers and pressure sensors, an applied pressure produces an electric field

(a) 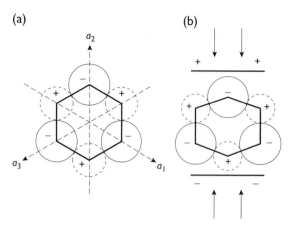 (b)

Fig. 12.14 Piezoelectricity in quartz. (a) In an undeformed quartz crystal charges of cations (Si^{4+}) and anions (O^{2-}) are balanced. (b) Compression parallel or perpendicular to an a-axis produces a shift of charges and induces an electric field.

(a) 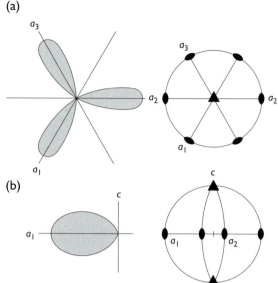 (b)

Fig. 12.15 The representation quadric of the piezoelectric tensor of quartz consists of three lobes that extend parallel to the positive a-axes.

that is then amplified and measured. Mechanical vibrations, such as those from old phonograph needles, have been recorded with quartz crystals and transformed into an electric signal. The inverse effect has applications for the precise timing of quartz watches and for tuning radio signals. If an alternating voltage is applied to an appropriately cut slice of quartz, the crystal will alternately expand and contract. The vibration frequency depends on the geometry and crystal size and is in the range $10^5 \, s^{-1}$. Quartz watches keep track of time by counting the oscillations of the alternating current whose frequency is fixed by the oscillating quartz. In radios, only signals that match the quartz oscillations are amplified, enabling fine-tuning.

The representation quadric of this third-rank piezoelectric tensor has a rather odd shape, with three lobes along the positive a-axes and a zero value in most other directions, as illustrated in Figure 12.15a,b.

Another unusual property is *pyroelectricity*, which had been observed by Theophrastus. When prismatic tourmaline crystals (point-group $3m$) are heated, opposite ends develop an opposite electric charge. A. Kundt (1883) dusted a heated tourmaline crystal with a mixture of sulfur and lead oxide powder. Owing to friction the powder particles were charged, negative for yellow sulfur and positive for red lead oxide. Positive sulfur collected on one end of the tourmaline crystal and lead on the opposite end, illustrating the effect (Figure 12.16). Pyroelectricity, a

vector property, is only possible in crystals with unique polar axes (Table 12.6).

While piezoelectricity, as well as pyroelectricity, are possible in many crystals, including minerals, the effect may be too small to be observed except in such minerals as quartz and tourmaline.

12.8 Magnetic properties

If a magnetic field strength H is applied to a crystal, it produces a magnetic moment M, i.e.,

$$M = \chi H \tag{12.25}$$

Here χ is known as the magnetic susceptibility and is a second-rank tensor. The movement of electrons produces magnetic fields in a crystal, and in this respect the most important movement is the electron spin. Each orbital may contain two electrons of opposite spin, and each spinning electron produces an electric field. However, the magnetic fields of two electrons with opposite spin in the same orbital cancel out. Therefore, in crystals with all atoms (or ions) having only paired electrons, such as Si^{4+} and O^{2-}, there is no internal magnetic field. Such crystals are called *diamagnetic*. In diamagnetic crystals only an external magnetic field may cause a weak internal field that *opposes* the external field. Therefore the susceptibility

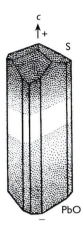

Fig. 12.16 Pyroelectricity in trigonal tourmaline. Negatively charged sulfur particles attach at one end of a heated crystal, whereas positively charged lead oxide particles attach to the opposite end.

of diamagnetic crystals is $\chi < 0$. Quartz, halite, and calcite are examples of this group.

Atoms or ions with unpaired electrons include the transition metals, whose $3d$-orbitals are only partially filled. Fe^{3+} and Mn^{2+} have the largest magnetic moments, with five unpaired $3d$-electrons. Fe^{2+} has four unpaired electrons.

	$1s$	$2s$	$2p$	$3s$	$3p$	$3d$
Fe^{2+}	↑↓	↑↓	↑↓ ↑↓ ↑↓	↑↓	↑↓ ↑↓ ↑↓	↑↓	↑ ↑ ↑ ↑
Fe^{3+}	↑↓	↑↓	↑↓ ↑↓ ↑↓	↑↓	↑↓ ↑↓ ↑↓	↑	↑ ↑ ↑ ↑
Mn^{2+}	↑↓	↑↓	↑↓ ↑↓ ↑↓	↑↓	↑↓ ↑↓ ↑↓	↑	↑ ↑ ↑ ↑

The magnetic behavior of such crystals depends on how the magnetic moments are oriented and organized within the crystal structure. A good example to illustrate this behavior is manganese oxide (MnO). It has basically the same cubic structure as halite (Figures 2.10, 7.24b), with Mn and O alternating. If we look at the magnetic moments associated with Mn, they are all aligned parallel to the direction $[1\bar{1}0]$ but adjacent dipoles along the cubic axes point in opposite directions. The magnetic structure as displayed by the dipoles no longer has cubic symmetry because dipoles are all parallel and in the (001) plane; furthermore, the unit cell is doubled along all axes owing to the alternating magnetic dipole directions. This "magnetic superstructure" cannot be measured with X-rays, since they are "blind" to magnetic spin.

Table 12.6 Symmetry groups in which pyroelectricity may be observed

Point-group	Polar direction
1	Every direction
2	[010]
m	All directions in the (010) plane
*mm*2, 3, 3*m*, 4, 4*mm*, 6, 6*mm*	[001]

However, it can be revealed by neutron diffraction, since neutrons have a magnetic moment and the interaction of magnetic dipoles produces magnetic scattering. Figure 12.17b shows two neutron diffraction patterns of MnO measured by Shull and Smart (1949), earning Shull a Nobel Prize in 1994. In the top spectrum, measured at 80 K, purely magnetic peaks 111 and 331 are indicated. At higher temperature (300 K) these peaks have disappeared because magnetic spins become randomized and the unit cell is cubic (Figure 12.17b, bottom). The transition temperature is called the Néel temperature, which for MnO is 116 K.

There are two types of magnetic behavior in crystals that have atoms with unpaired electrons:

In *paramagnetic* crystals (as in diamagnetic crystals) χ is a constant of the material and does not depend on the magnetic field. Contrary to diamagnetic crystals, paramagnetic crystals have $\chi > 0$, i.e., there is a weak attraction in an external magnetic field. For example, in fayalite (iron-olivine, Fe_2SiO_4), iron ions have a magnetic moment but moments are randomly aligned. When placed in a magnetic field, moments of Fe^{2+} will tend to align in parallel to the field, but as soon as the field is removed, thermal motion randomizes the dipoles. The magnetic susceptibility varies with crystal structure and composition and, on the whole, increases with increasing numbers of unpaired electrons. This magnetic property is made use of to separate a grain aggregate of different minerals with an electromagnet, both in the laboratory and in mining operations.

In *ferromagnetic* and *ferrimagnetic* crystals, magnetic dipoles are aligned as described above for MnO. In ferromagnetic bcc iron the dipoles are aligned in parallel, whereas in ferrimagnetic MnO the dipoles alternate in opposite directions. Ideally, in ferrimagnetic materials, magnetic moments cancel and crystals are not magnetic. However, if there is

(a)

(b)

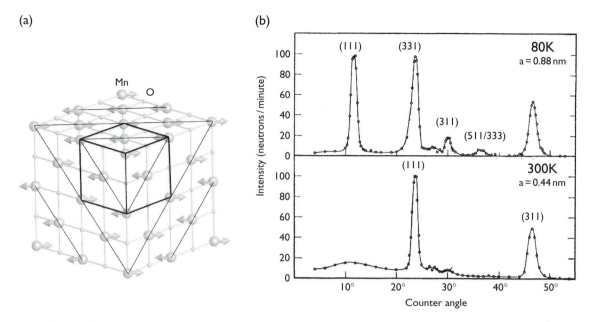

Fig. 12.17 (a) Magnetic structure of manganese oxide (MnO) with alternating dipoles, aligned along [110], which reduce the symmetry and enlarge the effective unit cell from the basic NaCl structure (cf. Figures 2.10 and 7.24). The chemical unit cell is highlighted. (b) Neutron diffraction spectra of MnO at 80 K (top) and 300 K (bottom). The ferrimagnetic structure at low temperature displays peaks such as 110 and 331, which are not compatible with the cubic structure (after Shull *et al.*, 1951; see also Roth, 1958).

some disorder and a portion of ions does not have an antiparallel partner, ferrimagnetic crystals may display weak ferromagnetic properties. A mineral that is intermediate between the two groups is magnetite $Fe^{2+}Fe_2^{3+}O_4$. Half of the Fe^{3+} (ferric iron) occupy tetrahedral interstices (8 A atoms), and the rest of the Fe^{3+} and all Fe^{2+} (ferrous iron) occupy octahedral interstices (16 B atoms) in the oxygen lattice, which is cubic close-packed. All iron ions have magnetic moments. Tetrahedral dipoles (A) point in one direction and octahedral dipoles (B) in the opposite direction. Since there are more B dipoles than A dipoles there is a net ferromagnetic behavior in the whole crystal. When a field is applied to ferromagnetic crystals the magnetic dipoles become aligned, and when the field is removed the alignment remains.

In a natural magnetic crystal, generally not all dipoles are aligned over the whole macroscopic crystal, but rather alignment is restricted to domains that may be of opposite direction and separated by boundaries. The magnetic domains can be imaged with magnetic force microscopy, a variant of atomic force microscopy (AFM, see Chapter 16). Figure 12.18b displays domains in microcrystalline magnetite that

occur as small inclusions in pyroxene (Figure 12.18a). In the case of magnetite (Figure 12.18b) the domain boundaries are determined by the microstructure. In other magnetic materials they depend simply on dipole alignment (Bloch walls) and are mobile. When a magnetic field is applied, domain walls move and change their morphology to minimize the energy. This is illustrated in Figure 12.19 for a synthetic oxide and here imaged with a polarizing light microscope. In ferromagnetic bcc iron the magnetic moments in domains may be variously oriented parallel to $\langle 100 \rangle$, $\langle 110 \rangle$, or $\langle 111 \rangle$. No particular direction is preferred, and thus the structure is overall magnetically disordered and its resultant magnetic moment is zero. However, in a strong magnetic field, those domains with dipoles parallel to the field will grow and the crystal becomes magnetic. The magnetic susceptibility of ferromagnetic crystals is extremely high. It is not simply a material property, but also depends greatly on the applied magnetic field.

As we have seen for MnO, the magnetic structure disappears above the Néel temperature. A similar behavior is observed in ferromagnetic materials. In this case the transition temperature is known as the

(a)

(b)

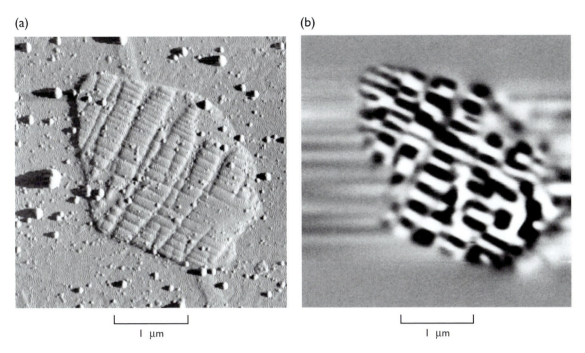

1 μm

1 μm

Fig. 12.18 (a) AFM image of magnetite inclusion in clinopyroxene from Messum (Namibia). (b) Magnetic force image of the same sample illustrating multiple domains in the 100 nm range. Black and white domains have opposite polarity (see also Feinberg *et al.*, 2005).

(a)

(b)

(c)

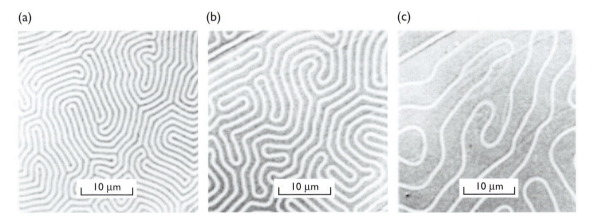

10 μm

10 μm

10 μm

Fig. 12.19 Domain structures observed in a thin plate of hexagonal $BaFe_{12}O_{19}$ for different values of the applied field normal to the plane of the plate. (a) 0 Oe, (b) 2250 Oe, and (c) 3080 Oe (1 Oe = $10^3/\pi$ A/m) (after Dillon, 1963).

Curie temperature; above it all the coupling between magnetic dipoles is lost due to intense thermal vibrations and the crystal becomes paramagnetic. In the case of metallic iron this temperature is 1043 K; for magnetite it is 858 K.

Magnetite and related oxides with the same crystal structure are called ferrites. The general composition is AFe_2O_4, with A ions including Mn, Co, Ni, Cu, and Fe^{2+}. Ferrites are characterized by a high electrical resistance (10^2–10^6 Ω cm) but a ferromagnetic behavior, which makes them suitable for cores of high-frequency coils, with many applications in the radio industry. In rocks these minerals serve another important function. The magnetic structure of a rock,

Table 12.7 Magnetic minerals and their properties

Mineral	Formula	Crystal system	Magnetic susceptibility (volume susceptibilities SI $\times 10^{-5}$)			Curie temperature (K)
			$\|a$	$\|b$	$\|c$	
Paramagnetic						
Aragonite	$CaCO_3$	Orthorhombic	−1.44	−1.42	−1.63	
Quartz	SiO_2	Trigonal	−1.51		−1.52	
Calcite	$CaCO_3$	Trigonal	−1.24		−1.38	
Halite	NaCl	Cubic	−1.36			
Diamagnetic						
Rutile	TiO_2	Tetragonal	10.5	11.2		
Ferromagnetic						
Magnetite	Fe_3O_4	Cubic	3005			858
Hematite	Fe_2O_3	Trigonal	131			958
Antiferromagnetic						
Goethite	FeOOH	Orthorhombic	27			350

Note: ‖, parallel to.

acquired by magnetite grains when an igneous rock cools below the Curie temperature, is called remnant magnetism and may be preserved for millions of years. Remnant magnetism records the orientation of a rock relative to the existing Earth's magnetic field at the time of cooling and can document rotations in the course of the geological history, for example those due to plate motions (Keffer, 1967; Banerjee, 1991). It also documents magnetic reversals, and the preserved history of such reversals has been instrumental in establishing the concept of plate tectonics. Table 12.7 gives some examples of magnetic minerals and their properties.

12.9 Summary

The chapter introduces physical properties with a unique stimulus–response relationship. These properties are generally described by tensors of different rank such as the second-rank tensor thermal conductivity that relates a temperature gradient to heat flux, or the fourth-rank stiffness tensor that relates an applied stress to a resultant strain. Tensors contain the point-group symmetry of the crystal. Anisotropy of elastic properties is significant for interpretation of seismic structures in the deep Earth. Physical properties change with temperature, which is particularly important for magnetic properties since magnetic dipoles become randomized above the Néel temperature.

Some properties to remember:

- Density (zero rank)
- Thermal conductivity (second rank)
- Elastic properties (fourth rank)
- Piezoelectricity (third rank)
- Pyroelectricity (first rank)

Test your knowledge

1. What are the basic symmetry elements of a second-rank tensor such as thermal conductivity? Draw a thermal conductivity tensor and indicate the symmetry elements on it.
2. The symmetry of a crystal has an effect on the symmetry of the second-rank tensor. Draw the thermal conductivity tensor of a tetragonal crystal and indicate the symmetry elements on it.
3. How many coefficients are needed to specify the thermal conductivity tensor of an orthorhombic crystal?
4. Cubic crystals are isotropic for second-rank tensors. Why are elastic properties anisotropic?
5. How do seismologists interpret the fact that elastic waves travel at different speeds in different directions through the upper mantle?
6. Density is a scalar property with a wide range. Sort the following minerals according to increasing density: galena, quartz, pyrite, calcite, silver, and halite.

7. Ferromagnetic properties, for example of magnetite, depend not only on crystal structure and chemical composition, but also on the microstructure. Explain the concept of magnetic domains.
8. Explain why rock magnetism is important in tectonics.

Further reading

Bhagavantam, S. (1966). *Crystal Symmetry and Physical Properties*. Academic Press, New York.

Kittel, C. (2005). *Introduction to Solid State Physics*, 8th edn. John Wiley & Sons, New York.

Nye, J. F. (1957). *Physical Properties of Crystals: Their Representation by Tensors and Matrices*. Oxford University Press, London.

O'Reilly, W. (1984). *Rock and Mineral Magnetism*. Blackie, London.

Stacey, F. D. and Banerjee, S. K. (1974). *The Physical Principles of Rock Magnetism*. Elsevier, Amsterdam.

Tarling, D. H. and Hrouda, F. (1993). *The Magnetic Anisotropy of Rocks*. Chapman & Hall, London.

13 | Optical properties

The petrographic microscope is undeniably the most important instrument to characterize minerals and rocks before more advanced techniques are applied. In this book two chapters are dedicated to microscopic analysis: Chapter 13 introduces basic principles of light and optical properties of minerals. Chapter 14 applies the microscope to identify minerals. First we discuss visible light as electromagnetic radiation, building on experience gained with X-ray diffraction. Then we introduce the optical microscope with polarizing light capabilities and the optical properties of crystals that can be described by a second-rank tensor, the optical indicatrix.

13.1 Some physical background

Optical properties are a striking expression of the anisotropic internal structure of minerals. The best way to convince yourself that this is true is to examine a thin section of a rock (a 20–30 μm thick slice, see Chapter 14) in a petrographic microscope with polarized light and compare the effects with those from a plain glass slide. The glass slide appears dull black, whereas the rock, composed of crystals, displays an intricate color pattern that changes as the thin section is rotated (see, e.g., eclogite composed of omphacite, chlorite, amphibole, garnet, and titanite in Plate 9). The change in optical properties with orientation highlights the anisotropy of crystals. Indeed, the interaction of light with crystals is directional, and this lends itself to a sophisticated analysis with a microscope that is widely used by mineralogists and petrologists. The optical properties of minerals are characteristic and serve for mineral identification. The subject of optical mineralogy is extensive and is dealt with in many excellent books. Here we provide merely a brief overview of the most important principles of crystal optics to help in understanding some of the optical features of minerals. This discussion is followed in Chapter 14 by an introduction on how to analyze and identify minerals with a petrographic microscope. Note that we will only discuss analysis of minerals in transmitted light. For opaque minerals, such as pyrite or galena, reflected light microscopy is used (see, e.g., Craig and Vaughan, 1994; Nadeau *et al.*, 2012).

Like X-rays (see Chapter 11), visible light is electromagnetic radiation resulting from the interaction of an oscillating electric field E and an oscillating magnetic field H (Figure 13.1). The sources of E are

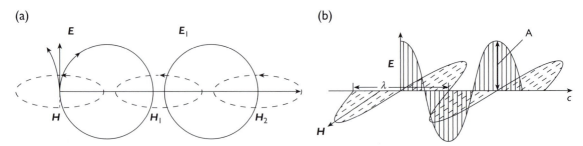

Fig. 13.1 (a) An electric field E creates a magnetic field H, and this energy transfer propagates at c, the speed of light. (b) The electromagnetic field propagates as oscillating waves with the magnetic field at right angles to the electric field. λ, wavelength; A, amplitude.

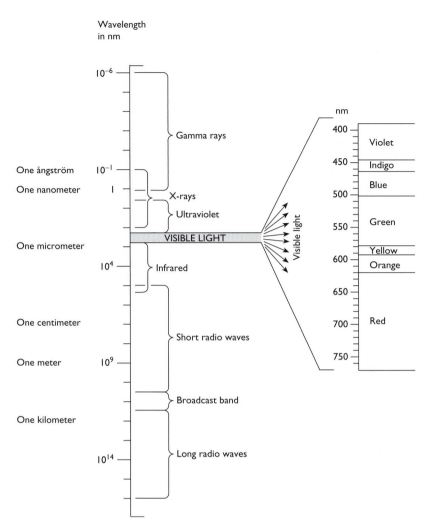

Wavelength
in nm

One ångström

One nanometer

One micrometer

One centimeter

One meter

One kilometer

10^{-6}

10^{-1}

1

10^4

10^9

10^{14}

Gamma rays

X-rays

Ultraviolet

VISIBLE LIGHT

Infrared

Short radio waves

Broadcast band

Long radio waves

Visible light

nm

400

450

500

550

600

650

700

750

Violet

Indigo

Blue

Green

Yellow

Orange

Red

Fig. 13.2 Spectrum of electromagnetic radiation as function of wavelength (nm). Visible light constitutes only a small segment (enlarged) of the entire spectrum.

electric charges, present as electrons and protons in matter. The electric field diverges from all charges and becomes the source of a magnetic field that compensates for the flowing current (Figure 13.1, left side). Magnetic and electric fields constantly create one another and spread at a velocity c, the speed at which light propagates in a vacuum (300 000 km/s). We have seen in Chapter 11 that electromagnetic waves have a sinusoidal shape ($y = A \sin \omega t$). These waves are characterized by amplitude A and wavelength λ, with the electric field E at right angles to the corresponding magnetic field H and displaced by a quarter of a wavelength (Figure 13.1, right side) (see also Figure 11.4 for a definition of wave properties).

Visible light constitutes a small segment of the large spectrum of electromagnetic radiation that ranges from short, high-energy γ-rays (less than 10^{-2} nm) to long radio waves (10^{12}–10^{17} nm = 1–10^5 km) (Figure 13.2). Only wavelengths of 400–800 nm cause a photochemical reaction in the retina of the human eye, which is registered by the brain according to intensity and wavelength. Monochromatic light consists of one single wavelength, for example the sodium D line produced by a sodium vapor lamp has a wavelength of 589.3 nm. If all wavelengths of the visible spectrum are present with equal intensity, a situation that is approached in sunlight, the human brain interprets this visible radiation as white. If some wavelengths are missing, we observe the complementary color. For example, the sky appears blue because some of the longer wavelengths (red and yellow) are preferentially scattered.

Like all electromagnetic radiation, light has the dual properties of a particle (called a photon) and of a wave. In a vacuum it propagates in a straight direction at a velocity c. The energy of a photon having mass m is related to the corresponding wavelength λ by the Planck–Einstein quantum relation

$$E = hf = hc/\lambda = mc^2 \qquad (13.1)$$

where f is the frequency, and h the Planck constant. Depending on the optical effect being considered, we will emphasize either the particle or the wave character of light in what follows.

13.2 Refractive index and optical applications

Refractive index

When we insert a thin slab of a crystal into a beam of light, the primary electric field of the light interacts with the local electric field caused by electrons and protons in the crystal. Photons propagate at the speed of light but are deflected in the crystal and thus travel on different paths. A path of a photon passing through a vacuum (Figure 13.3a) is compared with a photon path going through a crystal (Figure 13.3b). An observer of the resultant light ray will notice a time delay for those rays that passed through matter. This is not due to any change in the fundamental velocity c but rather to increased path length produced by internal scattering on electrons within the crystal. The net effect is an apparently slower propagation of light in matter compared to its speed of propagation in a vacuum. The velocity ratio of light through a vacuum (c) and through a medium (v), $n = c/v$, is

called the *refractive index* of the medium. It is a measure of the interaction between light and electrons, which increases with the number of electrons per unit volume and thus, in general, with the density. Table 13.1 gives some typical values of refractive indices. Notice that for some materials the refractive index is a single number, but for most crystals it varies with direction, as indicated by the range of values. We also find that the refractive index is a function of wavelength and of external conditions such as temperature.

Since most condensed matter has a refractive index considerably larger than 1.0 (vacuum by definition), light changes direction as it enters from air, with a refractive index close to that of vacuum, into a mineral with a much higher refractive index. The relationship known as Snell's law is applied extensively in mineral optics and is discussed in Box 13.1.

Prisms and lenses

An important application of refraction is the *prism*, with which the path of light can be changed. In a prism with two flat faces, inclined by an angle γ, Snell's law applies twice, to the entering and to the leaving light rays, causing a deflection of the light by an angle δ (Figure 13.7a). The prism angle is $\gamma = \beta_1 + \alpha_2$ and the deflection angle is $\delta = (\alpha_1 - \beta_1) + (\beta_2 - \alpha_2) = \alpha_1 + \beta_2 - \gamma$. If white light is transmitted through a glass prism, it divides light into a rainbow spectrum (Figure 13.7b), illustrating that the refractive index is not constant but depends on wavelength. This effect is known as *dispersion*.

The principle of the prism is applied in the construction of *lenses*, which can be regarded as a set of prisms with a special geometry such that parallel incident

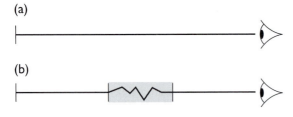

(a)

(b)

Fig. 13.3 Propagation of light in (a) a vacuum and (b) through a slab of material. The refractive index and apparent lower velocity of light in matter are due to light interacting with atoms, producing scattering and an effective longer travel distance.

Table 13.1 Refractive index n of some compounds (for $\lambda = 550$ nm)

Compound	n
Air	1.00029 at 10 °C
Water	1.33 at 20 °C
Garnet (almandine)	1.80
Quartz	1.54–1.55
Calcite	1.48–1.658
Titanite	1.606–1.644

Box 13.1 Additional information: Snell's law and refractometers

Law of refraction

Experimentally, it is observed that when light passes from a less dense into a denser medium it does not propagate in a straight line through the interface, but rather it changes direction, or is *refracted*. For example, consider the three light rays shown in Figure 13.4. At time $t = 0$ all three waves are in "phase" and define a wave front traveling at velocity v_1 through the first medium with refractive index n_1. The wave front propagates uniformly until, at $t = 1$, the first ray reaches the surface of the second medium with refractive index n_2 at C. Using the construction of Dutch mathematician Christian Huygens (1629–1695), we can visualize around each point on the interface between the two media an elementary spherical wavelet developing at a different velocity v_2. By the time wave 3 has reached the surface at B, a new wave front, BD, has developed at $t = 2$ as a tangential plane to the elementary spherical wavelets. Since AB = BC $\times \sin \alpha$, and CD = BC $\times \sin \beta$ (α being the angle between the incident wave front and the interface surface, and β the angle between the refracted wave front and the interface surface), the ratio of corresponding paths in the two media is a constant and we can write

$$AB/CD = \sin \alpha / \sin \beta = v_1/v_2 = n_2/n_1 \qquad (13.2)$$

This relationship is known as *Snell's law*, named after Dutch astronomer Willebord Snellius (1580–1626), but actually already described by Ibn Sahl from Baghdad in 984. The incident angle α is larger than the refraction angle β if light passes from a less dense into a denser medium, such as from air into glass.

A special case of Snell's law is normal incidence (Figure 13.5a). In this case no change in direction occurs, even if the refractive indices of the two media are vastly different. This fact is crucial for the observation of crystals in a microscope. Rough surfaces disperse light in all directions (Figure 13.5b). To minimize refraction and dispersion, mechanically ground thin sections of a rock, used in mineralogical analysis, are embedded in Canada balsam, a natural resin, and covered with a glass slide, both of which have refractive indices more similar to that of the crystal than to that of air (Figure 13.5c). Since glass has a flat surface there is no dispersion by refraction at the air–glass interface under normal incidence, and dispersions at the glass–Canada balsam–mineral interfaces are minimal.

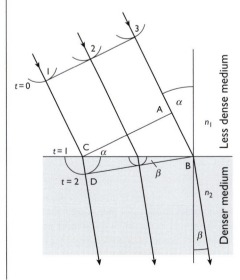

Fig. 13.4 The law of refraction. A wave front AC, traveling at velocity v_1 in a medium with refractive index n_1, reaches a plane surface interface between two media and is deflected into a new wave front BD, traveling at a slower velocity v_2 in a medium with refractive index n_2. Here α and β are the angles of incidence and refraction, respectively; t, time.

Continued

Box 13.1 (cont.)

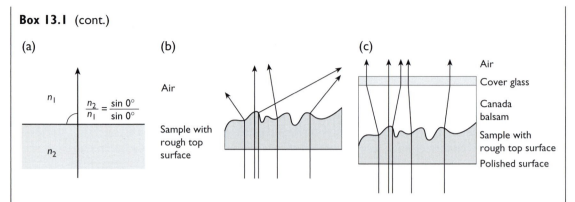

Fig. 13.5 Snell's law for (a) normal incidence with no refraction and (b) irregular scattering of light on rough surfaces. (c) The irregular scattering can be minimized by embedding minerals in a cement or oil with a similar refractive index, such as Canada balsam, and then covering the sample with a glass slide with a perfectly flat surface.

Total reflection, Abbe refractometer

If light passes from a medium with a higher refractive index (n_1) into one with a lower refractive index (n_2) there is a critical incident angle α_c where the refraction angle β becomes 90° or larger (($\sin \alpha_c/\sin 90°) = n_2/n_1$) (Figure 13.6a). In this case no light enters the second medium but is instead completely reflected. The limiting angle of total reflection, α_c, is a convenient way to measure the refractive index of liquids and polished surfaces. Such measurements are carried out in the Abbe refractometer (Figure 13.6b). It consists of a half-cylindrical piece of glass with a known refractive index N, on which is mounted a crystal or a drop of liquid with an unknown refractive index n. In this instrument the reflected light is observed with a telescope. In general, some light will be refracted and some reflected (rays 1 and 2), but if the incident angle α reaches the critical value α_c all light will be reflected (ray 3) and the observed light intensity will suddenly increase. The field of view at this angle is symmetrically divided into a brighter (rays 3–5) and a darker (rays 1–3) area. From α_c we obtain the refractive index: $n = N \sin \alpha_c$ (since $\beta = 90°$, and thus $\sin \beta = 1$, when $\alpha = \alpha_c$).

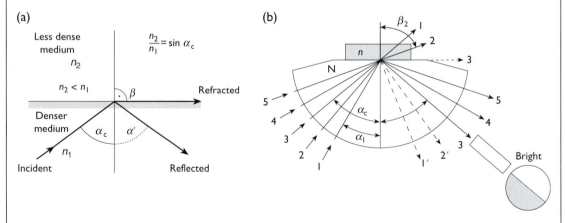

Fig. 13.6 (a) Snell's law for total reflection. If the incident angle α is at a critical value α_c, or larger, no light enters the crystal with refractive index n_2. (b) The principle of total reflection is applied in the Abbe refractometer, used to measure refractive indices of oils and crystal surfaces. A crystal or drop of liquid with refractive index n is mounted on a glass half-cylinder with refractive index N. For rays 1 and 2, some light is refracted and some is reflected. Ray 3 is at the critical angle α_c and for this, as well as all rays with larger incident angles (rays 4 and 5), all light is reflected, resulting in a bright reflection signal for rays 3–5, compared to a darker signal for rays 1–3.

(a)

$$\gamma = \beta_1 + \alpha_2$$
$$\delta = (\alpha_1 - \beta_1) + (\beta_2 - \alpha_2)$$
$$= \alpha_1 + \beta_2 - \gamma$$

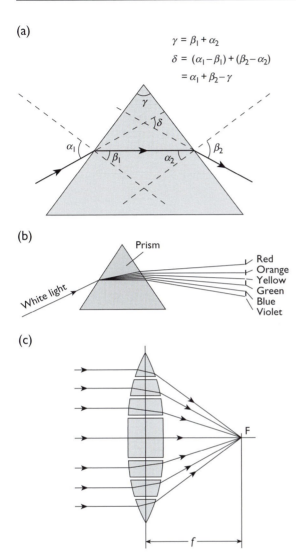

(b)

(c)

Fig. 13.7 (a) Geometry of a prism with double refraction. Here γ is the prism angle, and δ is the angle of deflection. (b) Dispersion of white light into a rainbow spectrum, because the refractive index increases with decreasing wavelength. (c) A lens can be viewed as a composite of prisms with a geometry such that parallel light rays are deflected to pass through a point F located at a distance f from the median plane of the lens.

light rays are deflected so that they all pass through a point F (Figure 13.7c). In a continuous lens (Figure 13.8a) all light that is initially parallel to the optical axis (the central normal to the median plane of the lens) is refracted to cross the optical axis at point F_1 (ray 1). Similarly, all light rays that are parallel to the optical axis after passing through the lens have

passed a point F_2 before entering the lens (ray 2). These points (F_1 and F_2) are called *focal points* and are at distances f_1 and f_2, respectively, from the median plane of the lens. The focal length f is a measure of the strength of the lens. With a short focus lens it is possible to converge light very effectively. We can use this construction to obtain the image I from an object O, whose respective heights are illustrated by the vertical arrows in Figure 13.8a. Ray 1, going through the tip of the O arrow, is parallel to the optical axis. It therefore passes through F_1 after refraction. Ray 2, which also goes through the tip of the O arrow, passes through F_2 and is thus parallel to the optical axis after refraction. The intersection of the two rays defines the tip of the image I_r. A third ray, 3, passing through the center of the lens, is not refracted. The image I_r can be recorded on a photographic film mounted at I_r or viewed on an inserted sheet of paper, documenting that it is real.

If an object is too close to the lens, i.e., within the focal point F_2, the strength of the lens is insufficient to converge light rays enough to produce an image (Figure 13.8b). The lens will, however, reduce the divergence between ray 1 and ray 2 and we can construct a *virtual* image I_v by intersection of the back extensions of the rays. This upright image cannot be recorded on a screen, but it can be transformed into a real image, for example by application of a second lens that, in combination with the first, increases the convergence. A lens often applied to record virtual images is the human eye.

This ideal geometry is only approximated in real lenses. There are two significant limitations. First, because of *spherical aberration*, the ideal lens geometry is generally satisfied only for light rays that are close to the center of the lens (Figure 13.9a). By inserting an aperture and selecting only the inner rays, overall light is reduced, but resolution of the image is increased. *Chromatic aberration* is due to the fact that different wavelengths have different refractive indices, with slightly different focal points for different colors (Figure 13.9b). Both aberrations can be minimized by constructing composite lenses in which the selection of glasses with different refractive indices compensates for these effects.

Microscopes

Microscopes, used to enlarge objects, are a combination of lenses. In a compound microscope, two lens

(a)

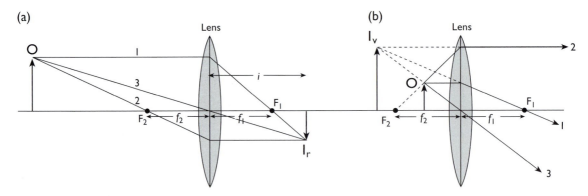

(b)

Fig. 13.8 Principles of a lens with an object O (arrow), an image I, and focal points F_1 and F_2. (a) A real image I_r is observed if the object lies outside the focal point F_2. (b) A virtual image I_v results if the object lies between the focal point F_2 and the lens.

(a)

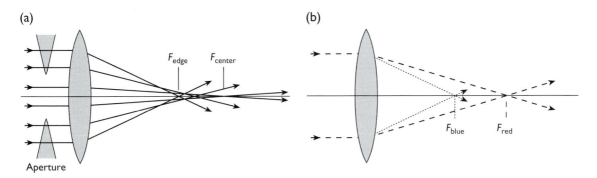

(b)

Fig. 13.9 Aberrations in lenses that limit the resolution. (a) Spherical aberration is introduced because rays passing through the edges of a lens have a focal point different from those passing through the more central part of the lens. It can be reduced by inserting an aperture that lets pass only rays close to the optical axis of the lens. (b) Chromatic aberration is due to different lens deflections because of wavelength differences. (See Figure 13.7b.)

systems are combined with the human eye (Figure 13.10a). The object ($A_0 – B_0$) lying close to, but outside, the focal point of the first (objective) lens F_1^{ob} is transformed into a reversed and enlarged real image 1 ($A_1 – B_1$). This image is inside the focal point F_2^{oc} of a second (ocular) lens, which has only sufficient power to form a virtual enlarged second image 2 ($A_2 – B_2$). The virtual image is then transformed into a real image 3 ($A_3 – B_3$), on the retina by the eye. The magnification M of the object is defined as the ratio of the apparent angular size of the object when viewed with the microscope to the angular size without the microscope. The magnification increases with increasing tube length (distance between ocular and objective lens) and with decreasing focal lengths of the lenses. Resolution and brightness are provided by the objective lens; the ocular lens contributes only magnification. Thus it is

customary to work with high objective lens magnification (2.5, 10, 25, 50×) and lower ocular magnification (2, 4, 10×) to achieve optimal conditions.

A modern petrographic microscope (Figure 13.10b, c) has more analytical features than those described above. The sample is mounted on a rotating stage. Light is emitted from a tungsten filament lamp or today also light-emitting diodes (LED), collected and condensed on the transparent object by a *condenser lens* system to achieve either a parallel incidence of light on the object or, for special applications, a convergent incidence. The *objective lens* produces an enlarged intermediate real image that is further enlarged by the ocular lens into a virtual intermediate image (dashed lines). A cross-hair with a scale is generally superposed on the intermediate real image. It is used for alignment and reference. The virtual

(a) (b) (c)

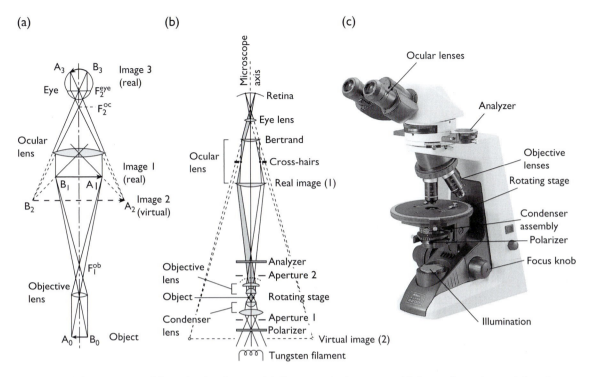

Fig. 13.10 Microscopes used for enlarging images. (a) Compound microscope with image formation and three lens systems: an objective lens, an ocular lens, and the human eye. (b) Principal components of a petrographic microscope. See text for discussion of labels. (c) A modern petrographic microscope (courtesy Nikon USA).

intermediate image is then transformed into a final real image by the eye. A first aperture is placed below the condenser lens to decrease spherical aberration (Figure 13.9a) and increase depth of focus, but this has the effect of reducing overall brightness. A second aperture in the object image plane restricts the area of the specimen viewed. The microscope axis is the central direction passing through the center of all lenses. (In modern microscopes this is not always a straight line!)

In petrographic microscopes two polarizing filters, the *polarizer* and the *analyzer*, are inserted. There is also a place to add what is known as a *compensator*. Finally an auxiliary lens, called the Bertrand lens, is used for some applications. The function of these additional features will be discussed later.

13.3 Polarization and birefringence

Polarization

In 1669 the Danish scientist Erasmus Bartholinus made a very puzzling observation. If a beam of light (and

today it is easiest to use a laser for this experiment) is aimed at a calcite cleavage face ($10\bar{1}4$) at normal incidence and penetrates a slab of some thickness, the signal splits into two components (Figure 13.11a). One component obeys Snell's law (i.e., there is no deflection because of normal incidence), but the other one does not. The experiment suggests that there are two refractive indices in crystals, causing *birefringence*. Using a polarizing filter that only allows passage of light that vibrates in a certain direction, we can determine that the two light waves leaving the crystal are polarized and vibrate perpendicularly to each other. If we observe an image, such as a label underneath a calcite crystal, we see two images (Figure 13.11b). These results indicate that the interaction of light and crystals is far more complicated than previously assumed.

In order to understand polarization, we have to consider light as a wave. In general the oscillation of the electric field has a random orientation normal to the propagation direction (illustrated by waves in three planes at different angles in Figure 13.12a). If the oscillation is confined to a single plane, we call

(a)

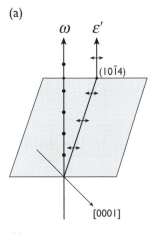

(b)

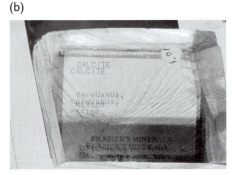

Fig. 13.11 Birefringence in calcite. (a) When an unpolarized light ray enters a crystal at normal incidence to a cleavage plane ($10\bar{1}4$), it splits into an ordinary (ω) and an extraordinary (ε) ray, which are polarized at right angles (perpendicular to the plane of the figure for ω, and parallel for ε). (b) Double refraction produces two images of a label underneath a cleavage fragment of the Iceland spar variety of calcite (from Chihuahua, Mexico).

such light *polarized* (Figure 13.12b). We will call the direction in which the light oscillates the *vibration direction*.

There are several ways to produce polarized light. In older microscopes, use is made of birefringence in calcite with the so-called Nicol prism. Today the most commonly used polarizers are Polaroid crystals, which show strong *preferential absorption*. If unpolarized light is transmitted through such a crystal, only waves vibrating in one direction are permitted to pass. This direction-dependent absorption is due to different bonding forces in different crystal directions. There are other crystals, and among them many minerals, that show direction-dependent absorption, but

(a)

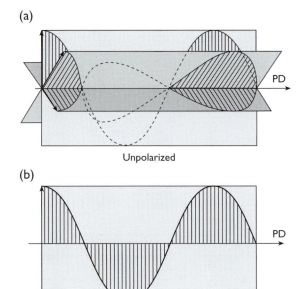

Unpolarized

(b)

Polarized

Fig. 13.12 (a) In unpolarized light the electric field oscillates in all directions perpendicular to the propagation direction PD (oscillations in three planes are shown). (b) In polarized light, vibration is restricted to a single plane.

this absorption is less complete and removes only certain wavelengths from the spectrum, resulting in color changes as the crystal is rotated in polarized light. This effect, called *pleochroism*, is subtle but easy to observe and is a very effective identification tool (see Section 13.6).

In a petrographic microscope a Polaroid filter is inserted in the light path, just above the light source, and all observations are made with polarized light. The vibration direction of the polarizer is generally chosen to be east–west.

Birefringence

On the basis of the calcite experiment it is clear that the refraction theory discussed in Box 13.1 is incomplete and needs to be re-evaluated. Consider a wave front of monochromatic light, wavelength λ, that becomes polarized after passing through a polarizing filter and enters a crystal slab at normal incidence. In Figure 13.13a, this wave is shown in a three-dimensional representation. In the crystal the polarized wave P splits into two waves, each traveling

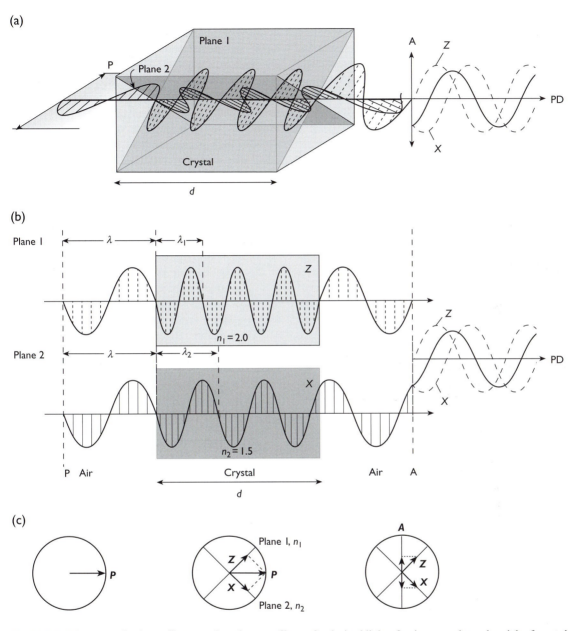

Fig. 13.13 This composite figure illustrates interference effects of polarized light after it passes through a slab of crystal of thickness d. A light wave polarized in the polarizer P, entering a crystal (shaded), splits into two components that vibrate at right angles to one another and propagate with different velocities (and wavelengths), attaining a path difference. The two waves are then brought to interference after passing through the analyzer A. PD is the propagation direction. (a) Three-dimensional representation. (b) View of the planes in which the two waves vibrate. (c) View in the propagation direction, indicating, with arrows, vibration directions X and Z.

with a different velocity v (and, correspondingly, with a different refractive index n). The waves on plane 1 and plane 2 are polarized perpendicular to one another in directions determined by the crystal structure. In Figure 13.13b the two waves are shown separately in the planes in which they are polarized, which illustrates a change in wavelength inside the crystal (λ_1 and λ_2, and correspondingly refractive indices n_1 and n_2). Figure 13.13c is a view perpendicular to the propagation direction PD. This view illustrates that the amplitude of the waves inside the crystal are obtained by projecting the original light vector $\boldsymbol{P}$ on the two vibration directions labeled X and Z. X (plane 2) is the vibration direction with the faster wave and smaller refractive index, whereas Z (plane 1) is the vibration direction with the slower wave and larger refractive index.

In order to preserve energy, an equal number of wave packets must enter and leave a given cross-section of the crystal; thus the wave frequencies f have to stay constant. Since the velocity changes (v_1 and v_2) and $f = v/\lambda$, the wavelength must change accordingly and we have two effective wavelengths (λ_1 and λ_2) in the crystal (Figure 13.13b). While the waves travel through the crystal they no longer match up. When the waves leave the crystal, their wavelengths will return to that of the original wave front, but these two still perpendicularly oscillating waves are no longer in phase, as is best seen in Figure 13.13b on the right side. The difference in the number of waves after passage through the crystal, depends on the crystal thickness d and the relative wavelengths λ_1 and λ_2 in the crystal, and is given by $d/\lambda_1 - d/\lambda_2$. After leaving the crystal, the path difference Δ in air becomes:

$$\Delta = d(\lambda/\lambda_1 - \lambda/\lambda_2) = d(c/v_1 - c/v_2) = d(n_1 - n_2)$$
(13.3)

since $n_1 = c/v_1$ and $n_2 = c/v_2$

The quantity ($n_1 - n_2$) is called *birefringence* and is, like the refractive index n, a direction-dependent material property; the path difference $\Delta = d(n_1 - n_2)$ is also called *optical retardation*. The wave with the larger refractive index propagates with a slower velocity than the wave with the smaller refractive index. In the vibration direction X, the wave has a refractive index n_2, and in the vibration direction Z a refractive index n_1 (Figure 13.13c).

The two waves are separate, vibrating in different planes, and there is no interaction between them. However, wave interference can be obtained if they are forced to vibrate in the same plane. This is done in the petrographic microscope by inserting a second polarizing filter above the crystal (see Figure 13.10b) that lets only those components of waves pass through that are parallel to the filter direction (Figure 13.13, right side). Since we use this polarizer to analyze interference effects, it is customarily called an analyzer (A).

Polarizer (P) and analyzer (A) can have parallel vibrations (*plane polarizers*), but more often their vibration directions are perpendicular to one another (*crossed polarizers*). In Figure 13.13c the case is shown for crossed polarizers.

Coherence is another condition necessary for interference. Light waves generally arrive in short strings, extending over several wavelengths. Interaction between such strings (or light pulses) produces only incoherent scattering. However, in the case discussed here, coherence is satisfied, because both component waves originated from the same wave packet and were originally in phase.

As was shown in Figure 11.5 for X-rays, there are two limiting cases for interference of waves: if two interfering waves are in phase (minima and maxima of the sinusoidal curves line up), then the resultant wave is strongly enforced (see also Figure 13.14a). If the waves are out of phase (the minima of one wave line up with the maxima of the other), addition of the two waves causes extinction (Figure 13.14b).

Let us examine the effects of crystal orientation. First we assume that polarizer (P) and analyzer (A) are parallel (Figure 13.15a–c). While examining

(a)

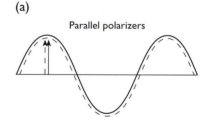

Parallel polarizers

(b)

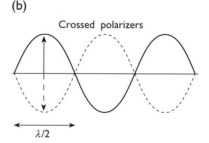

Crossed polarizers

$\lambda/2$

Fig. 13.14 Interference of two monochromatic light waves (solid and dashed sinusoidal waves) (see also Figure 11.5b,c). (a) Optimal interference conditions occur if the path difference is zero since both component waves vibrate in the same direction. (b) Extinction occurs for $\lambda/2$ path difference because the two component waves vibrate in opposite directions.

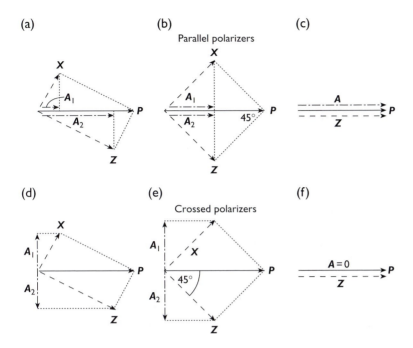

Fig. 13.15 Explanation of extinction as a crystal is rotated on the stage of a microscope. View is in the microscope axis (propagation direction), arrows are vibration vectors of light (perpendicular to the propagation direction). Parts (a)–(c) assume that the polarizer P and analyzer A are parallel. Parts (d)–(f) assume that the polarizer and analyzer are at right angles (crossed polarizers). A polarized light vector P (solid arrow) splits into two perpendicular vibrations, X and Z, in the crystal (dashed arrows) that differ depending on the rotation of the crystal (a–c and d–f). The vectors 1 and 2 resolved in the analyzer A (labeled A_1 and A_2) are shown as dot–dashed arrows. This figure is analogous to Figure 13.13c. In (a) and (d) vibration directions are in an arbitrary orientation relative to the polarizer, in (b) and (e) they are at 45°, and in (c) and (f) they are parallel.

Figure 13.15, keep in mind Figure 13.13a for the three-dimensional perspective. In the polarizer P, light vibrates in the horizontal direction (shown by a solid arrow, representing the light vector). In the crystal slab, the original light vector splits into two components, X and Z, which we construct by projecting the original light vector P on the two perpendicular vibration directions that are determined by the crystal structure (dashed vectors) (Figure 13.15a). Finally, when light passes through the horizontal analyzer A, only the vector components that are parallel to the analyzer pass (which we obtain as projections of X and Z onto the horizontal A-direction, obtaining the dot–dashed vectors, labeled A_1 and A_2). Note that both of these resultant vectors are pointing in the same direction and would add, if the two waves were in phase. As we rotate the crystal, the lengths of the projected vector components of X and Z change (Figure 13.15a–c). Special cases arise when either X or Z is parallel to the polarizer and analyzer

(Figure 13.15c). In these cases there is no contribution from one vibration (X in the case shown) and a maximum contribution from the second (Z in the figure). There is no influence of birefringence and light passes through, irrespective of crystal thickness and wavelength. For all other rotations (Figure 13.15a,b) the interference equation 13.3 applies. If the path difference produced in the crystal by birefringence is zero, or a multiple of the wavelength $\Delta = (m\lambda)$, where m is an integer, then there is positive interference.

The effect of crystal orientation on the resultant light vector becomes particularly clear if the polarizer and analyzer are at right angles (*crossed polarizers*, also sometimes referred to as "*crossed polars*", Figure 13.15d–f). The projection of the horizontal polarized light vector P is the same as before, but now we need to project vibration directions X and Z onto the vertical analyzer A (dot–dashed vectors). Note that in the case of crossed polarizers the two component vectors 1 and 2 are always opposite in sign, and vibrate in

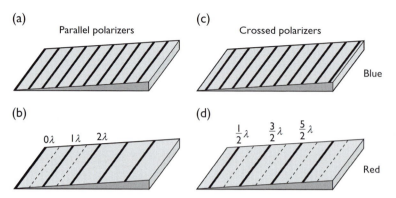

Fig. 13.16 Interference effects with monochromatic light on a crystal wedge, observed with parallel (a,b) and crossed (c,d) polarizers. There are dark (destructive interference, shaded) and light (constructive interference, black) bands that have the color of the light used, i.e., blue (a,c) and red (b,d). The bands are spaced more closely for blue light than for red light.

opposite directions. If there is no birefringence, or no crystal thickness, the two waves are out of phase (Figure 13.14b) and no light passes. For a crystal with a path difference there is interference in the analyzer plane, but the crossing of polarizer and analyzer have effectively added for crossed polarizers a path difference of half a wavelength, relative to the case of parallel polarizers (Figure 13.14a versus b). In the case of crossed polarizers interference is optimal at $\Delta = (m\lambda) + \lambda/2$, where m is an integer.

If the crystal is oriented such that a vibration direction (X or Z) is parallel to the polarizer (Figure 13.15f), the light vector is fully resolved on one vibration direction, i.e., the projection coincides with the vector (Z in this case). But there is no projection component from the other vibration direction (X), and the projection of Z on the analyzer direction is zero. This results in *extinction*, whether or not an optical retardation develops. As we rotate a crystal on the microscope stage a full turn, it goes into extinction four times, whenever a vibration direction is parallel or perpendicular to the polarizer. In the 45° position (Figure 13.15e), a maximum amount of light passes, provided there is a path difference, and this position is used for evaluating interference effects.

We now return to equation 13.3 and summarize our findings. For *plane polarized light* we have maximum constructive interference for zero path difference or an integer multiple of the wavelength. For *crossed polarizers* we have maximum constructive interference for $\lambda/2$ path difference or a multiple of wavelengths plus half a wavelength.

Next we are going to explore the effect of crystal thickness. Consider what happens if monochromatic light passes through a wedge-shaped crystal between parallel polarizers and the crystal is oriented such that the vibration directions are at 45° to polarizer and

analyzer. We observe a pattern of alternating colored and dark bands, each corresponding to a path difference of one wavelength and thus a thickness difference $d = \Delta/(n_1 - n_2)$ (Figure 13.16a,b). The first colored band (black in the figure) is at zero thickness, the second one at λ, then 2λ, etc. For short wavelengths (e.g., blue, Figure 13.16a) the bands are more closely spaced than for larger wavelengths (e.g., red, Figure 13.16b) since $\Delta = (m\lambda)$. For crossed polarizers the pattern is similar, except that it has shifted by half a wavelength and in this case the band at zero thickness is dark and the first colored band is at $\frac{1}{2}\lambda$ (Figure 13.16c,d).

For most routine analyses with a petrographic microscope, crossed polarizers are used rather than parallel polarizers and in the following discussion we will emphasize this case.

We have looked at interference effects with monochromatic light and studied the influence of crystal orientation, thickness d, birefringence ($n_1 - n_2$), and wavelength λ. For white light the situation is more complex. In this case there are waves of all wavelengths, and for each of them equation 13.3 applies. Extinction conditions for several different wavelengths (colors) on a quartz wedge with a birefringence ($n_1 - n_2$) = 0.0090 to 0.0096, depending on wavelength, are illustrated for crossed polarizers in Figure 13.17. The monochromatic experiments in Figure 13.16 represent two special cases, for a wavelength of 480 nm for blue and 710 nm for red. Note that in Figure 13.16 extinction is in gray areas (dashed lines), whereas in Figure 13.17 extinction is in dark bands. For white light, the color that is observed by the eye for each thickness is the mixture of the relative contributions of all wavelengths, which we obtain by summation of all the color patterns in Figure 13.17 (center). For instance, for a small thickness $d = 20$ μm,

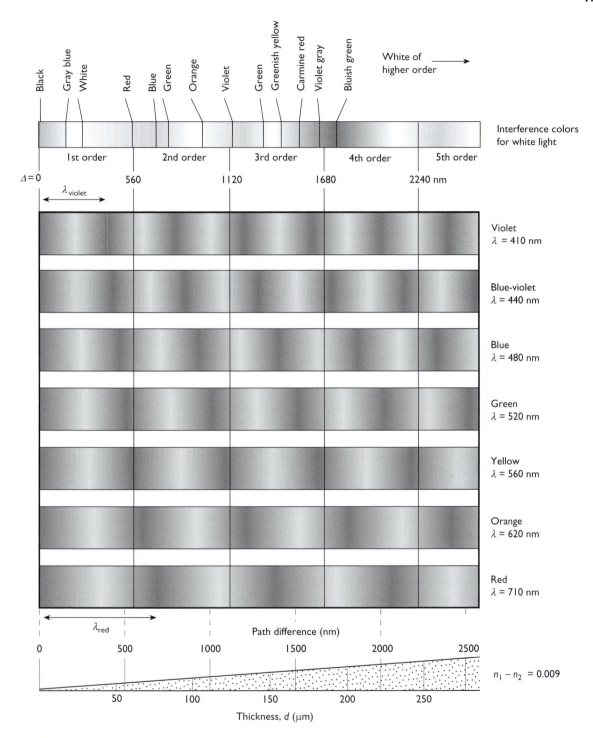

Fig. 13.17 Origin of interference colors with white light and crossed polarizers as a superposition of spectra from different wavelengths exhibited by a quartz wedge and assuming a birefringence $(n_1 - n_2) = 0.009$. Observe the shifts in extinction bands (dark) for different colors (wavelengths). For white light the sum of all color spectra applies. These so-called interference colors are indicated on top.

and a path difference 20 × 0.009 μm = 180 nm, all wavelengths contribute to the spectrum (white bands line up) and the resulting color is white. For a thickness $d = 62$ μm, and a retardation $\Delta = 62 \times 0.009$ μm = 560 nm, the wavelengths from blue to yellow are extinct, emphasizing orange and red. The resulting color is red (this color is called "first-order red"). The circumstances are similar for twice this thickness (124 μm) except that the contribution from blue is stronger, resulting in red with a purple tint ("second-order red"). For a retardation of 700 nm red and orange are extinct, but second-order blue is present, resulting in a blue interference color. For larger thicknesses with many extinctions and maximum intensities, the resultant light has many color contributions and resembles white again.

The color pattern produced by the linear effects of thickness and birefringence is shown in Plate 10a. There is a distinct *interference color* for each path difference (optical retardation) $\Delta = d(n_1 - n_2)$. Interference color charts, such as the one in Figure 13.18 and Plate 10a, are usually represented as a plot of thickness versus retardation and the birefringence is a straight line in this space. Some mineral examples are illustrated. If the thickness is known, we can use the interference color to determine the birefringence that can be applied in mineral identification with the petrographic microscope, as we will see in Chapter 14. Note that the color chart has been divided into different orders; each contains some shades of yellow, red, etc. With some practice you will be able to distinguish colors on sight. Low-order colors are very distinct, higher orders are more "washed out", with pastel shades.

The term "interference color" is somewhat misleading. Wave groups of different wavelengths in white light are not coherent and show no regular interference. The interference color is a mixture of waves with different wavelengths whose intensity has been modified by interference effects.

13.4 The optical indicatrix

In the discussion above, we established that light passing through a crystal vibrates in two different directions in most crystals and there are two refraction indices for each vibration direction. However, this behavior is different in different crystal directions. How can we visualize the variations of refraction indices with direction? To answer this question, we first perform a hypothetical experiment: we place a

light source within a crystal and turn it on for an instant to explore how far light pulses have advanced. We find that the distances the light pulses have traveled describe the surface of an ellipsoid centered on the light source (Figure 13.19a).

For those of you who have followed the discussion in Chapter 12, this pattern is due to the intrinsic physical properties of crystals. Interaction of light with a crystal is described by the relationship between two vector properties: a stimulus of electric field E, and a response as an electric displacement D:

$$D_i = \varepsilon_{ij} \, E_j \qquad (13.4)$$

where ε_{ij} is the so-called dielectric constant, which is closely related to the refractive index and is a second-rank tensor. The refractive index ellipsoid is simply the representation quadric of ε^{-1}, the inverse of the dielectric tensor. All parameters specifying the ellipsoid can be derived from ε. Three numbers are needed to describe the lengths of the axes, and three numbers are needed to describe the orientation of the ellipsoid relative to crystallographic axes.

The experiment with the light pulse cannot be realistically performed because we cannot measure such short time intervals. (In contrast, for example, to the much slower thermal conductivity, also a second-rank tensor, where we can map heat propagation as ellipsoidal isotherms as we did in Figure 12.1.) A different experiment is to focus a beam of polarized light on a crystal, determine for each direction the two vibration directions perpendicular to it, and map the refractive indices n_1 and n_2 of the vibration directions X' and Z' (Figure 13.19b).

We are switching here from the two-dimensional case to three dimensions and call the larger refractive index n_1 n'_γ and the smaller refractive index n_2 n'_α. Axes are primed to indicate that they are axes of an ellipsoid section, not necessarily principal axes of the ellipsoid. The resultant surface again describes an ellipsoid, but it is different from the "light pulse ellipsoid". This vibration direction ellipsoid is called the *optical indicatrix* and is enormously useful for evaluating anisotropic optical properties.

If we wish to study the behavior of light in a crystal, we intersect the indicatrix ellipsoid with a plane normal to the propagation direction PD of the beam of polarized light (Figure 13.19b). Most intersections of a plane with an ellipsoid are ellipses (the others are circles). The ellipse section corresponding to the three-dimensional sketch of Figure 13.19b is represented in

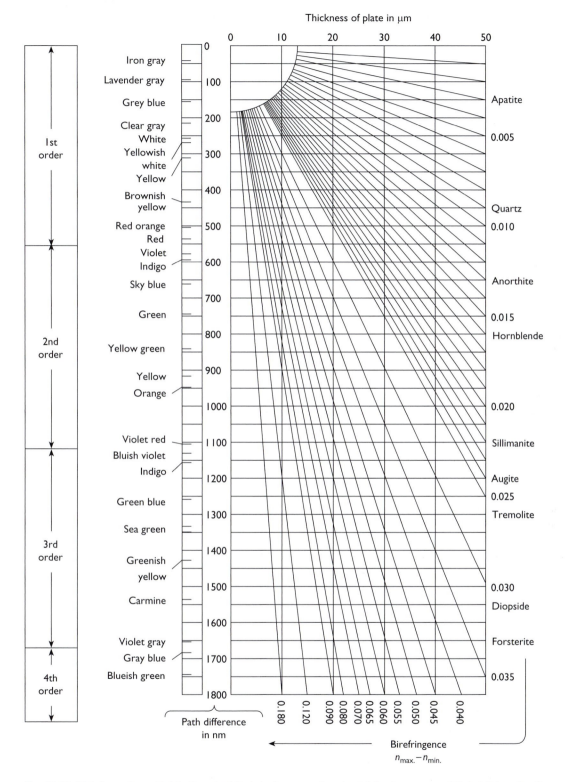

Fig. 13.18 This is a schematic black and white interference color chart (also known as a Michel–Lévy chart) as supplied with petrographic microscopes and also shown in Plate 10a. A version of the chart can be downloaded from the website for the book. The optic retardation, and thus the color, is a linear function of birefringence and thickness. If the thickness is known, as in a standard petrographic 30 μm thin section, the color correlates directly with birefringence. Some mineral examples are indicated for corresponding birefringence.

(a)

(b)

(c)

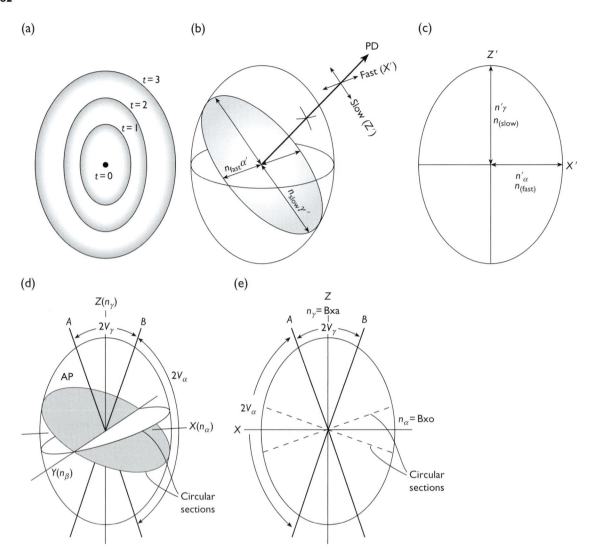

Fig. 13.19 Optical properties of crystals. (a) Light pulse ellipsoid illustrating the anisotropy of light propagation in a crystal with three surfaces displaying the light propagation at different time intervals. (b) The indicatrix ellipsoid is used to determine refractive indices and vibration directions X (n'_α) and Z (n'_γ) as a function of the propagation direction PD. An ellipse section is constructed perpendicular to PD. (c) The ellipse section has two axes representing the vibration directions; the lengths of the axes correspond to the refractive indices n'_α and n'_γ. (d) General triaxial indicatrix ellipsoid with three orthogonal axes n_α, n_β, and n_γ. Also shown are the two optic axes A and B, as well as circular sections perpendicular to these axes. The optic axial plane AP going through the optic axes and containing n_α and n_γ is indicated. (e) Axial plane section through a general indicatrix ellipsoid, with the two major axes X (n_α) and Z (n_γ), perpendicular to the intermediate axis Y (n_β). It shows the optic axes A and B, and the angle between them, which is called the axial angle or $2V$ (specifically, $2V_\alpha$ if it is measured over n_α and $2V_\gamma$ if it is measured over n_γ). Bxa and Bxo are acute and obtuse bisectrix, respectively.

Figure 13.19c. The orientations of the ellipse axes determine the vibration directions X' and Z', the lengths of the axes represent the corresponding refractive indices n'_α (fast) and n'_γ (slow), and the difference in lengths corresponds to the birefringence ($n'_\gamma - n'_\alpha$), observed in the propagation direction.

The indicatrix ellipsoid has three principal axes at right angles to one another (labeled X, Y, and Z, corresponding to refractive indices n_α, n_β, and n_γ, where $n_\alpha < n_\beta < n_\gamma$) and displays symmetry with a mirror plane perpendicular to each axis (Figure 13.19d). If the intersection of a plane with the ellipsoid contains the main axes, we call these *principal sections*, and in this case the propagation direction is along a main axis. All other sections are general sections and the ellipse has axis lengths n'_α and n'_γ, where $n_\alpha < n'_\alpha < n_\beta < n'_\gamma < n_\gamma$. An important principal section is that perpendicular to Y (n_β) (Figure 13.9e). The ellipse axes in this section correspond to n_γ and n_α and therefore the section exhibits maximum birefringence ($n_\gamma - n_\alpha$). A crystal that is in this orientation will have the highest retardation and thus display the highest possible order of interference colors. All other sections have lower-order colors.

Now consider sections that contain Y (corresponding to n_β) as one axis (Figure 13.20). At the top is a perspective drawing showing the orientation of the ellipsoid sections. Below it, the individual ellipses are shown. In the first section (1) n_γ and n_β are axes of the

ellipse. As we tilt around n_β towards n_α, the long axis gets shorter and the ellipse has n'_γ and n_β as axes (2). Tilting further we come to a point where n'_γ has the same length as n_β (3). An ellipse with two equal axes degenerates into a circle and we call this a *circular section*. In this case the two waves propagating perpendicularly to it have the same velocity and do not attain a path difference. This means that in this direction light will be extinct if observed through crossed polarizers, no matter what the rotation angle of the crystal or its thickness is. We continue to tilt around n_β but now the second axis is shorter than n_β and thus is n'_α. Finally we arrive at another principal section with n_α and n_β as axes (5).

In general, there are two such circular sections in an ellipsoid, symmetrically between n_α and n_γ, and containing n_β. They are shown with shading in Figure 13.19d and as dashed traces in Figure 13.19e. The directions perpendicular to these circular sections are called *optic axes* (labeled A and B), and the plane containing A and B is called the *axial plane* (AP).

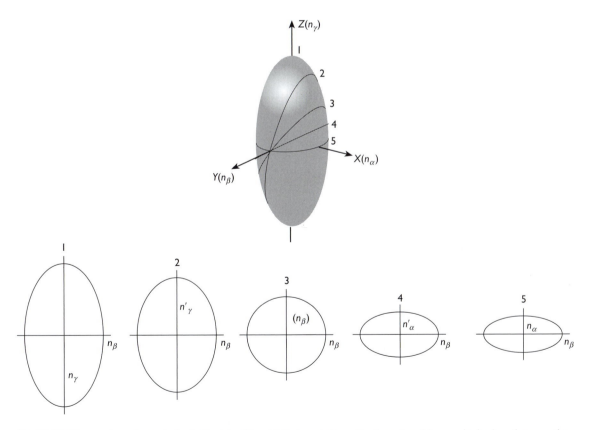

Fig. 13.20 Five sections through the indicatrix ellipsoid that contain n_β. Sections 1 and 5 are principal sections; section 3 is a circular section.

In Figure 13.19e the axial plane section containing the two optic axes is shown. In this section we find A, B, X (n_α) and Z (n_γ). It is perpendicular to n_β. The angle between the optic axes is called the *axial angle*, or $2V$; in particular, the axial angle is $2V_\alpha$ if it is measured over n_α and $2V_\gamma$ if it is measured over n_γ (Figure 13.19e).

On the basis of the geometry of an ellipsoid, the axial angle $2V_\gamma = (180° - 2V_\alpha)$ can be calculated from the values of the refractive indices along the main axes:

$$\tan V_\gamma = \sqrt{\left[\left(1/n_\alpha^2 - 1/n_\beta^2\right)/\left(1/n_\beta^2 - 1/n_\gamma^2\right)\right]} \quad (13.5)$$

Like all physical properties, the optical indicatrix has to conform to the crystal symmetry, i.e., symmetry elements of the crystal have also to be present in the indicatrix (see Figure 12.5). This condition imposes restrictions. Let us start with the highest possible symmetry, cubic. The only ellipsoid that agrees with *cubic symmetry* of three equivalent 4-fold axes is a sphere (Figure 13.21a). If the indicatrix is a sphere, the refractive index is the same in all directions and can be described with a single number, n. Cubic crystals have no direction dependence and are therefore *optically isotropic*. In cubic crystals all indicatrix sections are circular, and there is no birefringence; if

observed with crossed polarizers, cubic crystals such as garnets always appear black. (We have seen in Chapter 12 that isotropy does not necessarily extend to other properties. Indeed, elastic properties of cubic crystals are not isotropic.)

The optical properties of all other crystals are *anisotropic*. However, there are other symmetry restrictions. In *trigonal, tetragonal, or hexagonal crystals* with a single 3-, 4-, or 6-fold symmetry axis, only a rotational ellipsoid conforms to this symmetry (Figure 13.21b). In such an ellipsoid two axes are equal in length (either $n_\alpha = n_\beta$ or $n_\gamma = n_\beta$). There is a single circular section perpendicular to the unique ellipsoid axis. This axis is parallel to the crystallographic z-axis [001]. Because of the unique optic axis, such indicatrices are called *uniaxial*. The unique axis is generally referred to as n_ε (extraordinary direction), and the radius of the circular section is referred to as n_ω (ordinary direction). Each ellipse section contains the ordinary direction n_ω as one axis. There are two cases to distinguish: if the unique axis n_ε corresponds to n_γ, i.e., the long axis, the uniaxial indicatrix is called *positive* (Figure 13.22a); if the unique axis is n_α, i.e., the short axis, it is called *negative* (Figure 13.22b). In Chapter 14 we will learn methods to distinguish between these two cases. Uniaxial crystals can be

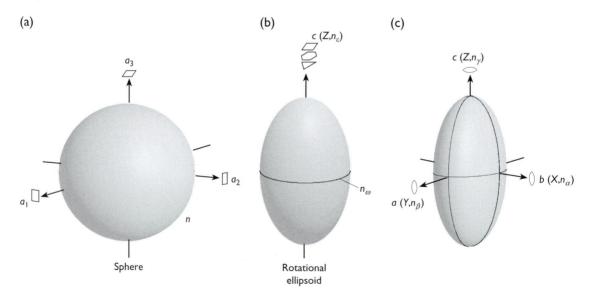

Fig. 13.21 Optical indicatrix for different symmetries. (a) Sphere in cubic crystals. (b) Rotational "uniaxial" ellipsoid in tetragonal, hexagonal, and trigonal crystals. The optic axis is parallel to c. (c) General "biaxial" ellipsoid with axes X, Y, Z in triclinic, monoclinic, and orthorhombic crystals. The orthorhombic case is shown where indicatrix axes coincide with crystal axes a, b, c. (See also Figure 12.5.)

(a) (b)

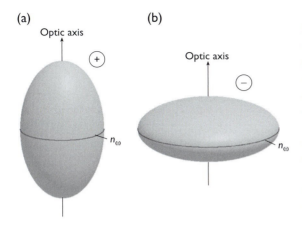

Optic axis

Optic axis

n_ω

n_ω

Fig. 13.22 (a) Indicatrix of a uniaxial positive crystal with $n_\varepsilon > n_\omega$ and an elongated "cigar shape". (b) Indicatrix of a uniaxial negative crystal with $n_\varepsilon < n_\omega$ and a "pancake shape".

viewed as special cases of biaxial crystals in which the optic axes coincide ($2V = 0°$).

Orthorhombic crystals have a triaxial ellipsoid as an indicatrix and are therefore *biaxial*, just as we saw in Figure 13.19. However, symmetry constrains the orientation of the indicatrix ellipsoid in the crystal. The main indicatrix axes, X (n_α), Y (n_β), and Z (n_γ), which are perpendicular to mirror planes in the indicatrix, have to be parallel to the crystallographic axes x, y, and z, which are perpendicular to mirror planes in the crystal (Figures 13.21c and 12.5c). There is no symmetry constraint to determine which optical direction is parallel to which crystal axis and in the figure we have chosen one possibility. Mirror planes and 2-fold rotation axes of the optical indicatrix coincide with the mirror planes and 2-fold rotation axes of orthorhombic crystals.

In *monoclinic crystals* with a 2-fold rotation axis, one of the main indicatrix axes has to be parallel to the unique crystallographic y-axis (see Figure 12.5b), i.e., it must have a principal section perpendicular to it. In *triclinic crystals* there are no symmetry constraints and the indicatrix can have any shape and any orientation relative to crystal axes (see Figure 12.5a). As we will see in Chapter 14, the orientation of the indicatrix of triclinic plagioclase depends on the chemical composition of this mineral with a solid solution, and this orientation can be used for determinative purposes. The indicatrix ellipsoid itself always has "orthorhombic symmetry", with three mirror planes at right angles.

In the discussion of optically uniaxial crystals, we have defined a positive and negative uniaxial indicatrix, depending on the overall shape of the ellipsoid (Figure 13.22). The same convention can be generalized for biaxial crystals but we cannot simply compare long and short axes n_γ and n_α. In this case we compare the long and short axes with the intermediate axis n_β. If n_β is closer to n_α than it is to n_γ then the ellipsoid is stretched and we call it positive. If n_β is closer to n_γ than it is to n_α then the ellipsoid is squashed and we call it negative. This is best defined using the axial angle, which depends on the relationship of the magnitudes of n_α, n_β, and n_γ (equation 13.5).

We return to Figure 13.19e: if $2V_\gamma$, measured over Z (n_γ), is between 0° and 90°, then we call the indicatrix biaxial positive. If $2V_\gamma$ is between 90° and 180° (i.e., $2V_\alpha$ is between 0° and 90°), then the indicatrix is biaxial negative. The principal axis, either X or Z, which bisects the acute axial angle $2V$ is called the *acute bisectrix* (Bxa), as opposed to the *obtuse bisectrix* (Bxo).

In the mineralogical literature, different symbols are used to label refractive indices, vibration directions, and indicatrix axis directions. We are using n_α, n_β, and n_γ for the main refractive indices, and X, Y, and Z for the corresponding principal indicatrix axis directions. Any *arbitrary section* of the indicatrix, except for a circular section, is still an ellipse with two axes, corresponding to the two vibration directions. The longer one, with a higher refractive index n_γ' and slower wave propagation, we call the Z' vibration direction; the shorter one, with a lower refractive index n_α' and a faster wave propagation, we call the X' vibration direction. Table 13.2 summarizes some of the relationships and conventions to describe optical properties of minerals.

13.5 Dispersion

To achieve total precision in optical analysis would require that one work with monochromatic light because the refractive index of any light ray varies slightly with wavelength. As we have seen earlier, this change of refractive index with wavelength causes dispersion of white light in a prism (see Figure 13.7b). However, for most applications dispersion effects are minor and thus white light is employed. Figure 13.23 shows dispersion curves for a borosilicate glass and for a liquid (ethyl salicylate), the latter at two

Table 13.2 Summary of conventions for describing the indicatrix

Axial angle and indicatrix geometry

Uniaxial +		Biaxial +		Neutral		Biaxial −		Uniaxial −
$0°$	$<$	$2V_\gamma$	$<$	$90°$	$<$	$2V_\gamma$	$<$	$180°$
$180°$	$>$	$2V_\alpha$	$>$	$90°$	$>$	$2V_\alpha$	$>$	$0°$

Biaxial crystals

Refractive indices	n_α	$<$	n'_α	$<$	n_β	$<$	n'_γ	$<$	n_γ
Wave speeds	Fast								Slow
Vibration directions	X		X'		Y		Z'		Z

Uniaxial crystals

n_ω (ordinary)

n_ε (extraordinary) $n_\varepsilon = n_\alpha$ uniaxial negative $n_\varepsilon = n_\gamma$ uniaxial positive

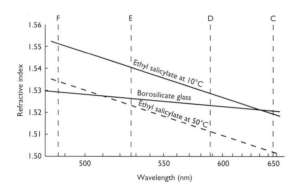

Fig. 13.23 Dispersion curves of borosilicate glass and ethyl salicylate liquid at two temperatures, illustrating the change of refractive index with wavelength. Spectral color lines C, D, E, and F are indicated. They are used in the calculation of the dispersion coefficient and the dispersive power.

temperatures. From this figure you can see that the refractive index typically decreases with wavelength, and more so for liquids than for solids. The refractive index also decreases with increasing temperature.

In biaxial crystals normally all three refractive indices show a similar dispersion behavior and the dispersion curves are parallel (Figure 13.24a). This means that the birefringence ($n_\gamma - n_\alpha$) and also the axial angle $2V$ remain fairly constant. Some minerals, such as sillimanite, display *anomalous dispersion*, and the dispersion curves for different refractive indices have different slopes (Figure 13.24b). In the case of sillimanite, with a biaxial positive indicatrix (i.e., n_β closer to n_α than to n_γ), the birefringence ($n_\gamma - n_\alpha$) decreases with increasing wavelength; also n_β moves closer to n_γ and therefore $2V$ for red (30°) is larger

than $2V$ for violet (21°), which is marked in determinative tables as ($r > v$). In the case of brookite (Figure 13.24c), for short wavelengths n_β is parallel to [100] and close to n_α, i.e., the crystal is biaxial positive; n_γ is parallel to [010], and n_α parallel to [001]. As the wavelength increases, the n_β and n_α curves cross. At this point, where $n_\beta = n_\alpha$, the indicatrix degenerates into a rotational ellipsoid ($2V_\gamma = 0°$) and, correspondingly, the mineral appears uniaxial (in spite of the orthorhombic symmetry). For larger wavelengths, the crystal becomes biaxial again, but n_β is now parallel to [001]. The axial plane has switched orientation. We will come back to this example in Chapter 14.

To describe dispersion of a material quantitatively, it is necessary to specify the index of refraction at several wavelengths. By convention, indices are usually reported for light of wavelength 486.1 nm (F), 589.3 nm (D), and 656.3 nm (C) corresponding to the Fraunhofer emission lines (Figure 13.23). If nothing is specified, 589.3 nm (n_D) is assumed, which is the light produced by a sodium vapor lamp and is in the middle of the visible spectrum. The *coefficient of dispersion* is defined as ($n_F - n_C$) and the *dispersive power* as ($n_F - n_C$)/($n_D - 1$). Table 13.3 reports some coefficient of dispersion values for different minerals. Note that diamond has an unusually high dispersion, causing color spectra when light is refracted on the surfaces of cut gems.

13.6 Pleochroism

Absorption of light may be dependent on both direction and wavelength. This is called pleochroism.

(a)

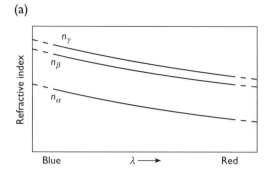

(b)

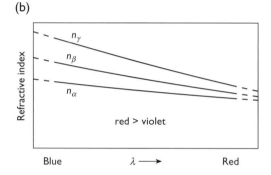

(c)

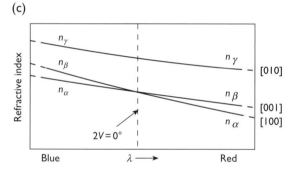

Fig. 13.24 Dispersion in biaxial crystals. (a) Normal dispersion with dispersion curves being roughly parallel. (b) Dispersion curves for sillimanite with different slopes for the three principal refractive indices. This results in a change of axial angle with wavelength, in this case red > violet. (c) Anomalous dispersion in orthorhombic brookite with two dispersion curves crossing. For the wavelength at the crossing point the mineral appears uniaxial.

In some crystals all wavelengths of light are absorbed, except for certain directions. This feature is used in the construction of polarizing filters, for example with the organic material Polaroid. In many minerals, absorption in certain directions is not complete and occurs only for certain wavelengths. For example, if a thin

Table 13.3 Coefficients of dispersion ($n_F - n_C$) for several minerals for 589.3 nm (D) (Winchell, 1929)

Sphalerite	0.08
Diamond	0.062
Sphene	0.02–0.04
Epidote	0.012–0.025
Zircon	0.022
Garnet	0.015–0.021
Calcite	0.013–0.014
Olivine	0.013
Feldspars	0.009
Quartz	0.008

slice of a crystal of the mica biotite is rotated on the microscope stage with plane polarized light (without the analyzer), colors change between different shades of brown (Plate 11a). In other biotites there may be different shades of green. Yellow to brown and black colors are observed for the often prismatic mineral stilnomelane (Plate 11b). The calcic amphibole hornblende changes color between shades of green (Plate 11c) and for those with a high ferric iron content (basaltic hornblendes) brownish colors. In sodic amphiboles such as riebeckite and glaucophane the color changes from blue to yellow and colorless (Plate 11d), which is particularly evident in a folded schist with oriented crystallites (Plate 11e). Spectacular pleochroism (yellow–pink–red) is observed in the manganese–epidote mineral piemontite, also known as piedmontite (Plate 11f). For a summary of pleochroic properties see Table 13.4 and for more examples, Appendix 3.

13.7 Summary

This chapter discusses the interaction of crystals and light. After reviewing light, lenses, and microscopes, optical properties of crystals are introduced. Refractive indices vary with direction, which becomes apparent if polarized light is used with a petrographic microscope. In crystals, light splits into components that vibrate in perpendicular directions and propagate at different speeds, causing birefringence and a wide range of interference colors that can be used for determinative purposes. The optical properties of crystals are described by the indicatrix ellipsoid having a geometry consistent with crystal symmetry, i.e., spherical for cubic crystals, rotational ellipsoid for hexagonal, trigonal, and

Table 13.4 Selected minerals with distinct pleochroism

	n_α	n_β	n_γ
Corundum	n_ε: pale yellow		n_ω: blue/purple
Humite	Yellow	Colorless	Colorless
Piedmontite	Yellow	Pink	Red
Tourmaline	n_ε: pale		n_ω: yellow, brown, or blue
Stilpnomelane	yellow	Brown, black	Brown, black
Biotite	Yellow to green	Brown	Brown
Chlorite (Mg)	Yellow	Green	Green
Hypersthene	Pink	Light yellow	Light green
Hornblende	Light green	Dark green	Brown
Glaucophane	Colorless/yellow	Light blue	Dark blue

tetragonal crystals (uniaxial indicatrix), and a general ellipsoid for lower symmetries (biaxial indicatrix).

Some important concepts:

- Birefringence ($n_\gamma - n_\alpha$)
- Interference colors
- Optic axes and axial angle $2V$
- Dispersion (e.g., red > violet)
- Pleochroism

Test your knowledge

1. Why is it advantageous to have a crystal embedded in a medium with a refractive index close to that of the crystal?
2. What is the origin of the word "birefringence"? (Describe a classic experiment.)
3. If crossed polarizers are applied and you observe a crystal of optically isotropic garnet, why is there extinction?
4. The concept of path difference and extinction is best understood with a practical example. Take a crystal of birefringence 0.002 and monochromatic light (500 nm). Using crossed polarizers, which is the first thickness for which you get extinction (not counting the trivial case of zero thickness)? See equation 13.3 and Figure 13.6.
5. Take a uniaxial positive and negative indicatrix. What are the different indicatrix sections that you may observe? Sketch them and assign labels (Table 13.2).
6. What is the relationship between $2V_\alpha$ and $2V_\gamma$? It is best to draw a sketch of the axial plane.
7. Dispersion is the dependence of the refractive index on wavelength. Why does it affect the axial angle?
8. Name some minerals with very pronounced pleochroism.

Further reading

Bloss, F. D. (1999). *Optical Crystallography*. Mineralogical Society of America Monographs, vol. 5, Washington, DC.

Gay, P. (1982). *An Introduction to Crystal Optics*. Longmans, London.

Nesse, W. D. (2012). *Introduction to Optical Mineralogy*, 4th edn. Oxford University Press, Oxford.

Phillips, R. M. (1971). *Mineral Optics: Principles and Techniques*. W.H. Freeman and Co., San Francisco.

Wahlstrom, E. E. (1979). *Optical Crystallography*, 5th edn. Wiley and Sons, New York.

Wood, E. A. (1977). *Crystals and Light: An Introduction to Optical Crystallography*, 2nd edn. Dover Publications, New York.

14 | Mineral identification with the petrographic microscope[*]

With the background of crystal optics and basics about polarized light, we are ready for a more practical description on how to use the petrographic microscope to analyze minerals and identify mineral species in thin sections of rocks. This includes determination of the refractive index, determination of birefringence, and observation of interference figures to determine the geometry of the indicatrix. In the second part of the chapter these methods are applied to the most important mineral groups.

14.1 Sample preparation

The use of a polarizing microscope to analyze the optical properties of crystals is a standard technique in mineralogy and petrography. Not only is the petrographic microscope used for identification of mineral species, but it can also help to determine structural and chemical variations in minerals – for example, in solid solutions. Every student of earth sciences should become familiar with this technique and have at least some experience with a petrographic microscope. There are two approaches to such optical studies. One approach is utilized for transparent crystals, which are analyzed with *transmitted* light. The second is used for opaque crystals. In this approach, a modified polarizing microscope is used and light is *reflected* from a highly polished surface and then analyzed using similar methods as for transmitted light (see, e.g., Nadeau *et al.*, 2012). In the following discussion we will confine ourselves to the first method, which is known as *transmitted light microscopy*. In order to follow the concepts and applications introduced in this chapter, you need to have access to a petrographic microscope and some thin sections. They are available in most geology departments.

Most minerals as they occur in rock samples are, at best, translucent. For example, if we put a chip of granite or basalt under the microscope, no light is transmitted. Thus, in order to transmit light, one of two methods must be applied. In the first method, small fragments of finely ground crystals are scattered on a glass plate, immersed in oil, and covered by a thin sheet of glass (*grain mount*) (Figure 14.1a). The second approach is to prepare a *petrographic thin section* from the rock, ideally of a thickness of 20–30 μm (Figure 14.1b). This is generally done by cutting with a diamond saw a slab 2 cm × 3 cm × 0.5 cm from the rock sample that contains the region of interest. One side of the slab is polished with a fine abrasive (silicon carbide, alumina, or diamond powders) and mounted on a glass slide with a transparent, optically isotropic cement (e.g., baked Canada balsam, called "Lakeside", which is soluble in benzene and xylene; "Crystal bond", which is soluble in acetone; or an insoluble epoxy resin). The other side of the slab is then cut and ground to the desired thickness. The ground side is covered with softer Canada balsam and a thin cover glass. The cover glass ensures normal incidence, and a flat surface prevents irregular scattering (see Figure 13.5c). By applying this method, most minerals, except opaque ore minerals such as pyrite and magnetite, become transparent and can be analyzed with the microscope.

The various lenses and types of image formation in a microscope have been discussed in Chapter 13. It is

---

[*] Optional reading. This chapter requires access to a petrographic microscope.

(a)

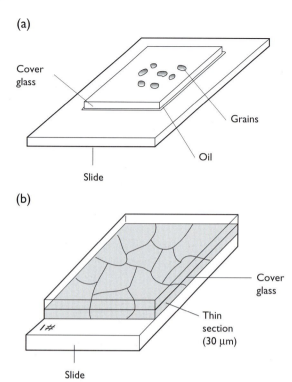

Cover glass

Grains

Oil

Slide

(b)

Cover glass

Thin section (30 μm)

Slide

Fig. 14.1 Specimens used for optical analysis of minerals. (a) Grain mount with immersion oil. (b) Petrographic thin section.

essential to have a microscope stage enabling the rotation of the sample a full 360° around the microscope axis.

14.2 Microscope alignment

Before use, a microscope needs to be aligned. There is a routine procedure that needs to be followed, and should be repeated periodically, especially if other people have used the same microscope (see Figure 13.10b,c).

- For binocular microscopes, there are two ocular lenses in the eyepiece. One ocular lens has a cross-hair and often a scale. Remove that lens, hold it towards a light source (e.g., a window), and focus the cross-hair for your eye. Insert the lens again, and then focus with that lens only on an object on the microscope stage (e.g., a dust particle on a glass slide) by raising or lowering the stage. If you now use your other eye, you will generally find that the object is no longer in focus. Focus it for that eye by

adjusting the second ocular lens (without the cross-hair). From now on both eyes should see the object in focus and, in addition, the cross-hair should be in focus.

- The center of the stage has to be exactly in the optical axis. Either tilting or translating the objective lens until the object under the cross-hair (e.g., a dust particle) does not move during stage rotation will achieve the needed centering. In general, the object will follow a circle. Move the center of the circle to the cross-hair. This alignment takes several iterations and must be performed separately for each objective lens.
- Next the condenser assembly (underneath the stage) needs to be adjusted to have the incoming light reaching the specimen parallel to the axis. To do this, close the condenser aperture until the light beam is observed, and then translate it to the center.
- Finally, the polarizer system has to be aligned. Insert the analyzer, without a sample present, and then rotate the polarizer until it is darkest.
- Usually the polarizer is oriented east–west (E–W) and the analyzer north–south (N–S), but this configuration needs to be verified. One can make use of a crystal with strong pleochroism and a distinct morphology such as a trigonal crystal of tourmaline, which is often elongated parallel to the c-axis and appears blue if the c-axis is perpendicular to the polarizer direction and yellow if it is parallel to it. One can also take a foil of Polaroid, on which the polarization direction is usually marked.

Some general comments about the use of a petrographic microscope:

- Adjust the brightness of the light to be comfortable for your eyes. Very bright light causes headaches and burns out filaments.
- Adjust apertures for particular applications. Closing the condenser aperture increases contrast and resolution, particularly at high magnification.
- Choose the magnification that is most useful for your investigation. In general, start with low magnification to get an overview, and then zero in on individual minerals with higher magnification.
- When focusing (particularly at high magnification), first bring your thin section close to the objective lens, then increase the distance until the object is in focus. This way you avoid collisions of lens and sample.

14.3 Determination of the refractive index

A petrographic microscope can be used to determine the refractive index of small crystal fragments by the *immersion method*. In this procedure, crushed grains, 0.05 to 0.2 mm in size, are mounted on a glass slide, immersed in an oil, and covered by a cover glass to ensure normal incidence (Figure 14.1a). The edges of the crystal fragments act as small prisms. If the refractive index of the crystal is higher than that of the oil, the light is dispersed inwards into the crystal (Figure 14.2a), producing a zone of increased brightness inside the contour of the crystal. By focusing on a higher level, from a plane 1 towards a plane 3 (Figure 14.2a), we notice a ring of light, called the *Becke line*, which moves concentrically into the crystal. If the refractive index of the crystal is lower than that of the oil, light is dispersed outwards, and the Becke line moves beyond the crystal into the oil (Figure 14.2b). If the refractive index of the crystal is the same as that of the oil, this line is not observed.

Therefore it is important to match the refractive index of the crystal with that of the oil. Sets of oils are available with intervals of 0.01 in refractive indices. Changing the oil around the crystal until the Becke line disappears allows for an accurate determination of the refractive index. Because of dispersion, it is sometimes advantageous to make these determinations with monochromatic light, for example by using a color filter. Oils are far more sensitive to dispersion than most crystals. In these experiments, use high magnification and close the condenser aperture. When crystal and oil are matched, the refractive index of the oil can be determined with a refractometer (either Abbe (Figure 13.6b) or Leitz–Jelley refractometers are usually available in mineralogy laboratories).

The refractive index is a diagnostic tool and is used as an identification method for minerals (Appendix 4). Often, relative values of refractive indices are also useful. For example, in a rock that contains quartz, one can determine the Becke line effect (a) between quartz in a thin section ($n_\alpha = 1.5442$, $n_\gamma = 1.5533$), which is easily recognized, and an unknown mineral, or (b) between an embedding medium (Canada balsam, $n = 1.537$) and a crystal at the edge of a thin section. If the difference in refractive indices between crystal and oil (or its surrounding) is large, the crystal is said to have a *high relief*. If it is larger than the surrounding it is called *positive relief*, if it is smaller, *negative relief*.

14.4 Use of interference colors

Determination of the true birefringence

If we insert a crystal slab (e.g., as a thin section) into the microscope between *crossed polarizers*, we can interpret optical effects with the help of the optical indicatrix. In a given thin section of a crystal, we observe a section of the indicatrix, i.e., an ellipse with axes X' and Z' and corresponding refractive indices n'_α and n'_γ. We have already described in Chapter 13 what happens if we rotate this crystal on the stage (see Figure 13.15). The original E–W vibrating light (polarizer) will split into two components parallel to

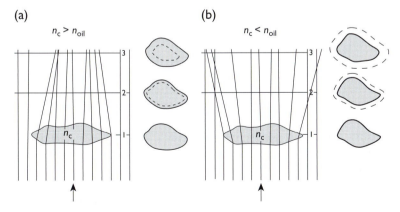

Fig. 14.2 Origin of the Becke line. (a) Edges of crystal fragments act as small prisms that deflect light towards the center if the refractive index of the crystal (n_c) is higher than that of the immersion oil (n_{oil}), and a ring of light (the Becke line, shown dashed) moves concentrically inside the crystal as the focusing takes place at a higher level. (b) If the refractive index of the crystal is lower than that of the immersion oil, a ring of light moves outside the crystal into the immersion oil with increasing defocus.

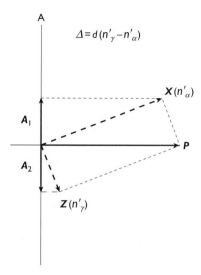

$$\Delta = d(n'_\gamma - n'_\alpha)$$

Fig. 14.3 Once more! Decomposition of a light vector **P** into two perpendicular components in the vibration directions X (n'_α) and Z (n'_γ). As light passes through the crystal a path difference $\Delta = d\left(n'_\gamma - n'_\alpha\right)$ is attained. In the analyzer the two waves are forced to vibrate in the same horizontal plane and interference occurs. P, polarizer; A, analyzer.

the vibration directions X and Z (Figure 14.3). We obtain these components as projections of the light vector **P** onto the axes X and Z. In the crystal, a path difference (retardation) $\Delta = d\left(n'_\gamma - n'_\alpha\right)$ will develop that depends on the birefringence ($n'_\gamma - n'_\alpha$) and the crystal thickness d. As we force the light through the analyzer A, only the components X and Z of the light vectors that are parallel to the analyzer direction can pass through the analyzer, and we obtain these components, A_1 and A_2, by projecting the light onto the analyzer plane (vertical).

If X or Z is parallel to either the polarizer or analyzer, we observe extinction (see Figure 13.15f). At 45° from the extinction position (see Figure 13.15e), a maximum amount of light is transmitted, and this light is used to evaluate interference colors. Thus, in general, it is necessary to rotate the crystal on the stage to bring it into a position where a maximum of light is transmitted. By comparing the color of the mineral with the color chart (Plate 10a), we can determine the retardation and, if the thickness is known (25–30 μm for a standard thin section), the value of the birefringence ($n'_\gamma - n'_\alpha$). However, what one really needs to know is the *maximum (true) birefringence* ($n_\gamma - n_\alpha$), i.e., the difference between the largest and

smallest refractive index. The true birefringence is a diagnostic material property for a mineral and can be used for its identification, but most crystals in the thin section do not display it because their indicatrix is inclined to the section.

In a thin section of a rock, most crystals are in arbitrary orientations and each grain will show different interference colors, corresponding to a different birefringence ($n'_\gamma - n'_\alpha$), rather than the maximum birefringence ($n_\gamma - n_\alpha$). For example, if the section of a biaxial crystal is close to a circular section of the indicatrix, we will observe low-order colors, even though the mineral may have a high birefringence ($n_\gamma - n_\alpha$). Among all crystals of the same mineral in a thin section, we determine ($n_\gamma - n_\alpha$) by matching the *highest-order colors* with the interference color chart. This is straightforward if a thin section contains only one mineral species (e.g., quartz in a quartzite, or calcite in a marble), but requires some practice if different mineral species are present (such as pyroxene and olivine in a basalt). This maximum birefringence can then be used for determinative purposes (e.g., Appendix 4). Plate 10c shows a 30 μm thin section image of quartzite with crystals in different orientations, observed with crossed polarizers. Colors range from black to white, and comparing with the color chart (Plate 10a) we estimate that the white crystal represents the true birefringence of quartz of about 0.009. Note that most other grains have lower-order colors.

As an exercise we could use interference colors to determine the thickness of a mineral for which the birefringence is known. Plate 10b is an image of a hexagonal quartz prism that lies on one of the prism faces. It is observed with crossed polarizers 45° from the extinction position. The image displays color fringes from the edge towards the center, corresponding to the interference color chart (Plate 10a). The increase is due to a change in thickness; in the center the color is third-order pink and green and we estimate (from the color chart) a retardation of 1700 nm. This section parallel to the c-axis is a principal indicatrix section and displays the true birefringence, which for quartz is 0.009. Extrapolating this line in Figure 13.18 to a retardation of 1700 nm we obtain a thickness of 435 μm.

Compensators (accessory plates)

It is often useful to relate the indicatrix orientation to morphological features of the crystal. For example, we may want to know for a prismatic crystal of hexagonal apatite which refractive index (n_γ or n_α) is associated with the long axis (*c*-axis). This determination could, in principle, be done with plane polarized light, by measuring refractive indices parallel and perpendicular to the long axis based on the Becke line method, but this process is cumbersome. A more efficient way is to use interference.

Petrographic microscopes are equipped with accessory *compensator* crystals, for which interference colors and orientation are known. By comparing the known optical orientation of the compensator with the unknown optical orientation of the crystal we can determine the latter.

We will introduce and explain the procedure for hexagonal prismatic crystals with the *c*-axis [0001] in the plane of the stage (similar to the quartz crystal illustrated in Plate 10b). In Figure 14.4 the prismatic elongation direction corresponds to the extraordinary indicatrix direction n_ε, the ellipsoidal indicatrix section is indicated. If n_ε is n_γ (slow direction), then the crystal is uniaxial positive (Figure 14.4a); if $n_\varepsilon = n_\alpha$ (fast direction), then it is uniaxial negative (Figure 14.4b). First we rotate a prismatic crystal on the microscope stage, using crossed polarizers, until it is extinct, which is the case when the *c*-axis is either parallel or perpendicular to the polarizer (Figure 14.4a,b). Next we rotate the stage 45° to bring the *c*-axis into the NE–SW sectors (Figure 14.4c,d) and bright colors will be observed.

The next step is to superpose the compensator crystal. Compensators are inserted at a 45° angle into the

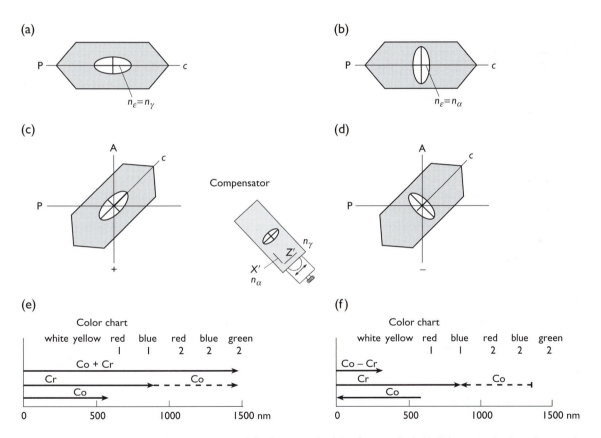

Fig. 14.4 Principle of a compensator, illustrated for hexagonal prismatic crystals. (a,b) A hexagonal prismatic crystal can be either (a) uniaxial positive if n_γ is aligned parallel to $n_\varepsilon = c$, or (b) uniaxial negative if n_α is aligned parallel to $n_\varepsilon = c$. (c,d) The crystal is rotated 45°. In this orientation the brightest interference colors are observed with crossed polarizers – for example, first-order blue (e,f; Cr for crystal). In this diagonal position a compensator crystal (Co) is superposed. In the case of (c), long ellipse axes (vibration directions) of crystal and compensator are parallel and cause addition (e; Co for compensator, Cr + Co for compensator plus crystal), producing second-order green. In (d) long ellipse axes of crystal and compensator are at right angles and cause subtraction (f), resulting in white/yellow interference colors.

light path (NW–SE). The most commonly used compensator (also known as an accessory plate) has a retardation corresponding to first-order red in the interference color chart and n'_α (fast, X') and n'_γ (slow, Z') are marked (Figure 14.4 center). (This compensator is often referred to as a gypsum plate, with a 530 nm retardation, or 1λ. The so-called mica plate has a retardation of 150 nm, or $\frac{1}{4}\lambda$. A more sophisticated compensator is a quartz wedge with variable thickness, as shown in Figure 13.16.) Superposing the compensator on our unknown crystal, we can determine the orientation of n'_α and n'_γ in the crystal, depending on whether the path difference of first-order red from the compensator is added to or subtracted from the interference colors of the crystal (Figure 14.4e,f). In the case of Figure 14.4c, the vibration ellipses of the compensator and the crystal have the same orientation. Retardations are therefore added and we observe higher-order colors (addition) (Figure 14.4e). In this case $n_\varepsilon = n_\gamma$ is parallel to the c-axis and we have verified that the crystal is uniaxial positive. In the case of Figure 14.4d, vibration ellipses of compensator and crystal are in opposite orientations. The retardation produced in the crystal is reversed as light goes through the compensator, resulting in lower-order colors (subtraction) (Figure 14.4f). Since $n_\varepsilon = n_\alpha$ is parallel to the c-axis, the crystal in Figure 14.4d is uniaxial negative. Plate 9 (second column) shows a thin section of eclogite observed with crossed polarizers at different rotations. In Plate 9 (third column) a compensator is added for the four rotations. Garnet (G) is isotropic and thus black with crossed polarizers at all rotations and becomes red with compensator. A chlorite grain (C) is gray with crossed polarizers at $0°$ and becomes yellow with compensator. This is subtraction. At $90°$ it is blue, i.e., addition.

Another example of the effect of compensators is the quartzite in Plate 10c–e. Quartz has low birefringence and thus low interference colors (black to white) in crossed polarizers (Plate 10c). With a compensator (Plate 10d) some grains become yellow (subtraction), some red (black without compensator), and one large grain is blue (addition). The colors depend on crystal orientations. After a $90°$ rotation (Plate 10e) yellow grains become blue and the blue grain becomes yellow.

Compensators thus enable us to determine the optical orientation, which is a particularly important property to know if we can relate it to crystal morphology. This will be discussed in the next section.

Extinction angle

Often minerals bear morphological markers that can be observed in thin sections. In euhedral crystals this marker may be the axis of elongation (as discussed above), or a platy surface as in the case of micas. Internal markers are cleavage traces, twin planes, or exsolution lamellae. They appear as hairline fractures. In order to have the crystal in some reference orientation, one usually chooses a crystal orientation in which either two cleavage systems or a twin and a cleavage are viewed edge-on.

Orthorhombic barite has two cleavage systems (001) and {210} (Figure 14.5a). Assume a fragment lies on the (001) cleavage and is viewed along the c-axis [001]. In this case the {210} = (210) + ($2\bar{1}0$) cleavages are viewed edge-on and appear as two sets of lines with an oblique angle of $102°$. Now, let us look at a thin section of barite, cut parallel to (010) and viewed along b = [010] (Figure 14.6a). In this case the (001) cleavage (shaded) is viewed edge-on and appears as a single, sharp line, when viewed with the microscope. The {210} cleavages are inclined to the section and the cleavage planes, over the thickness of the section, appear as diffuse bands. On focusing, for example, from M to N, the line does shift. Such oblique cleavages are usually avoided as a reference system. From Figure 14.5a and the ellipsoid in Figure 14.6a we see that Z (n_γ) is parallel to the (001) cleavage trace and perpendicular to both {210} cleavage traces, and therefore we observe extinction if a cleavage trace is parallel or perpendicular to the polarizer (E–W). This is referred to as *parallel extinction* and is common in orthorhombic minerals.

Next we look at a monoclinic amphibole crystal, such as hornblende, which has an excellent {110} cleavage (Figure 14.5b). If we observe the crystal in a thin section cut perpendicular to [001], cleavages (110) and ($1\bar{1}0$) are both edge-on, and at a characteristic $124°$ angle (Figure 14.6b). The monoclinic mirror plane (010) defines the orientation of the indicatrix ellipse in this section with Y (n_β) parallel to [010]. The axes are halfway between the cleavage traces and we observe extinction in this symmetrical position. This is called *symmetrical extinction*.

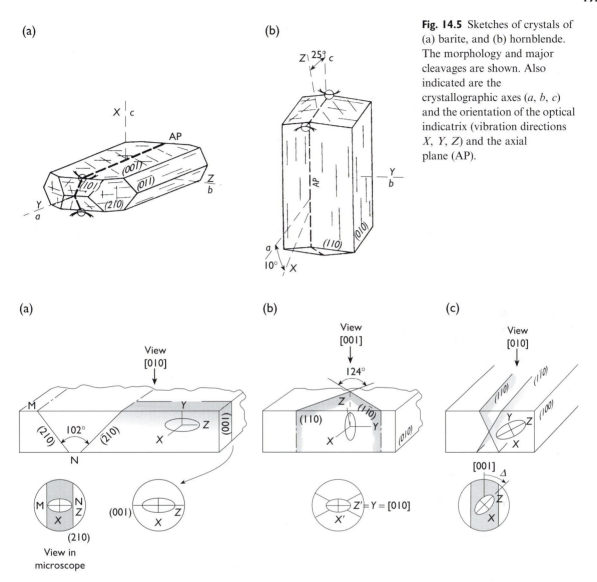

Fig. 14.5 Sketches of crystals of (a) barite, and (b) hornblende. The morphology and major cleavages are shown. Also indicated are the crystallographic axes (*a*, *b*, *c*) and the orientation of the optical indicatrix (vibration directions *X*, *Y*, *Z*) and the axial plane (AP).

Fig. 14.6 Cleavage and extinction conditions observed in thin sections. (a) Thin section of barite viewed along [010]. In this orientation (210) and ($\bar{2}$10) cleavage traces appear as parallel lines and (001) cleavage traces are perpendicular to the {210} lines. However, the (001) cleavage is viewed edge-on and lines are sharp, whereas {210} is inclined and a cleavage fracture appears as a broad band shown in the microscope views below. The long indicatrix ellipse axis *Z* is parallel to the (001) cleavage traces, resulting in parallel extinction. (b) Amphiboles are characterized by a {110} cleavage. If viewed along [001] both cleavage traces are edge-on and at a characteristic 124° angle. Ellipse axes are symmetrical between the cleavage traces, resulting in symmetrical extinction. (c) If the amphibole crystal is viewed along [010] all cleavage traces are parallel but, in this orientation, the indicatrix ellipse is inclined. The extinction angle *Δ* between *Z'* and [001] (cleavage trace) can be measured and is indicative of the chemical composition of the amphibole.

Now let us view the amphibole crystal along [010] (Figure 14.6c). In this case the {110} cleavages are not edge-on, but all the traces are parallel, extending along [001]. This section is a principal indicatrix section (*XZ*) and parallel to the axial plane. As we can see in Figure 14.5b, *Z* is inclined to [001] = *c*-axis (as defined by the parallel cleavage traces) by about 25°, producing *inclined extinction*. The angle, referred to as the *extinction angle*, is variable and we can measure it. First, we bring the cleavage trace N–S and record the rotation of the stage. Next we rotate until we observe extinction and record the rotation of the stage. The difference between the two rotation angles is the extinction angle Δ. However, since determinative tables distinguish between Z and X, we need to verify that the N–S extinction direction is Z. We do this by rotating the stage 45° clockwise (which brings the N–S direction into NE–SW) until we observe interference colors. Now we superpose the compensator. If we obtain addition, it means that we have indeed measured Z (see Figure 14.4c,d for the compensator convention). If we had observed subtraction, we would have measured X and in that case the

extinction angle to Z is the complementary angle. The angle between Z and the c-axis in a section perpendicular to [010] is diagnostic of the chemical composition of amphiboles (see Figure 14.22b).

Box 14.1 summarizes a brief step-by-step procedure on how to analyze minerals with parallel light, which is the procedure we have used so far, with light propagating through the thin section at normal incidence (perpendicular to the stage).

14.5 Observation of interference figures with convergent light

In the previous section we described how to determine the optical properties of a crystal from a thin section,

---

Box 14.1 Review: Procedures for optical analysis of minerals with parallel light

1. Prepare either a sample with grains immersed in oil of similar refractive index or a petrographic thin section that is 30 μm thick.
2. Observe the crystals in plane polarized light (polarizer below the stage only). Determine the relative refractive index with the Becke line versus oil (e.g., Canada balsam: $n = 1.537$) or a known mineral in the thin section (e.g., quartz: $n_\alpha = 1.5442$, $n_\gamma = 1.5533$).
3. Observe both color and pleochroism. Some minerals may be opaque.
4. Cross polarizers (insert the analyzer). A crystal may then appear black.
   (a) This could be due to extinction because X' (n'_α) or Z' (n'_γ) are parallel to the polarizer. If you rotate the stage, the crystal will show interference colors.
   (b) Another possibility is that you are looking at an isotropic (circular) section of the indicatrix (i.e., the microscope axis is parallel to an optic axis of the crystal). In such a case, the crystal will stay dark during rotation, but in another orientation (in another grain of the same mineral), the crystal will show interference colors.
   (c) If all crystals of the same type stay dark during rotation of the stage, then the crystal is optically isotropic and has cubic symmetry.
5. Survey the interference colors of all crystals of the same type. The highest color corresponds to the true birefringence.
6. In the event that morphological markers are present, find a suitable crystal to measure the extinction angle. Two cleavages or a cleavage and a twin ought to be perpendicular to the plane of the thin section.
   (a) Sketch the crystal.
   (b) Bring the crystal to extinction by rotating the stage. Measure the angle between the extinction direction (N–S) and the morphological marker (elongation of a needle-shaped crystal, cleavage, or twin trace). Orthorhombic, tetragonal, hexagonal, and trigonal crystals show symmetrical or straight extinction; monoclinic and triclinic crystals show generally inclined extinction.
   (c) Determine whether the direction that is N–S in step 6(b) is n'_α or n'_γ by rotating the crystal into the 45° position (NE–SW), inserting a compensator (accessory plate), and observing the change in interference colors.

using light in a single direction, i.e., the microscope axis. But the crystal and its optical properties (represented by the indicatrix) are three-dimensional. In order to determine the full shape and orientation of the indicatrix, we must either tilt the crystal on the stage or use light at an inclined angle. The former operation can be done with a universal stage, mounted onto the microscope stage. We will not discuss this rather specialized yet powerful method here but refer the interested reader to the literature (Reinhard, 1931; Slemmons, 1962; Phillips, 1971).

The second method is more widely applied. Using *convergent*, instead of *parallel*, light allows us to analyze the indicatrix in different directions. By inserting a special condenser lens underneath the specimen, one can produce a range of light that enters the sample at angles between 90° and 50° (Figure 14.7). The light is then imaged with a high-magnification objective lens close to the crystal. With this geometry the objective lens does not form an image of the object, as with parallel illumination, but of the light source. Rays entering the crystal at different angles are focused at different places in the focal plane. This so-called *interference figure* does not resolve details about the object, but rather displays interference effects at different angles. The central part of the figure corresponds to rays parallel to the microscope axis and the periphery corresponds to highly inclined rays (Figure 14.8a).

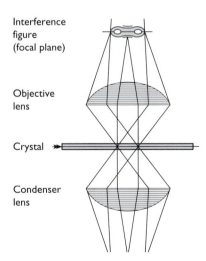

Fig. 14.7 Ray path for image formation with convergent light. The condenser lens produces incident light at different directions and the objective lens forms an interference figure in the focal plane.

The interference figure is real and can be observed without an ocular lens (simply remove it). However, generally an additional lens, the Bertrand lens, which is part of a petrographic microscope, is inserted (see Figure 13.10b).

Uniaxial interference figures

We will first discuss interference figures for *uniaxial crystals* with the optic axis aligned with the microscope axis (Figure 14.8). The interference figure can be viewed as a sort of spherical projection of the indicatrix, displaying optical properties in each direction. If the ellipse axes, viewed at a certain angle, are parallel to the polarizer or analyzer, extinction occurs; if they are not, interference colors appear corresponding to $d(n'_\gamma - n'_\alpha)$. Figure 14.8a shows simplified ray paths through the condenser lens and the indicatrix (1–4), and the projected views of the ellipses on a hemisphere for a uniaxial positive crystal.

Experimentally, this projection can be done even without an ocular lens by putting a ping-pong ball cut in half above the sample, which is covered by a sheet of Polaroid. The ping-pong ball (indicated by dots in Figure 14.8a) acts as a projection screen. In a microscope, however, the image is projected on to a plane above, and the circle above shows the ellipses (intersections with the indicatrix) on that plane, corresponding to different ray paths. A full array of ellipses is illustrated in Figure 14.8b for a uniaxial positive crystal, and in Figure 14.8c for a uniaxial negative crystal.

What happens when we pass these light rays through an analyzer? If ellipse axes are parallel or perpendicular to the analyzer, extinction occurs, which is the case for horizontal and vertical ellipses, resulting in a dark cross with N–S and E–W branches. These branches are called *isogyres*. The center, where the two branches intersect, corresponds to a circular section and is always dark – hence the name *melatope* (from the Greek *melas*, meaning "black"). When the crystal is rotated on the stage, different ellipses come into the N–S and E–W directions (Figure 14.8b,c) but the pattern remains the same (Figure 14.8d, Plate 12a,c). This is due to the axial symmetry of the indicatrix. In the sectors that are not extinct, there are concentric circles of equal interference color (*isochromes*). They are due to an increase in birefringence ($n'_\gamma - n'_\alpha$) and effective thickness d. Birefringence

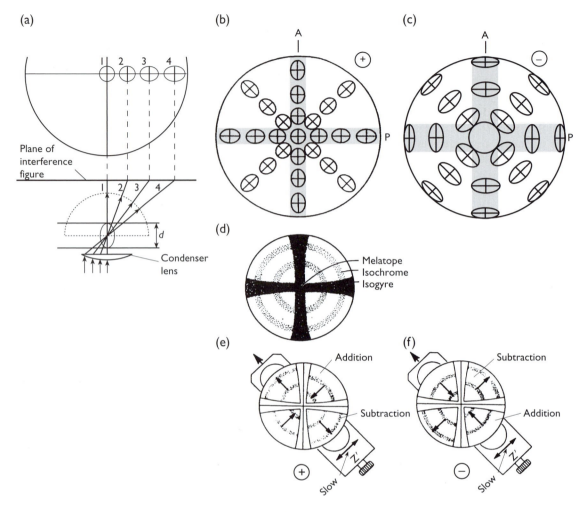

Fig. 14.8 Interference figures of uniaxial crystals. (a) Application of a special condenser lens (bottom) makes it possible to view the indicatrix in different directions by convergent light (top). Ellipse sections of the indicatrix for four different angles are shown. (b,c) Elliptical sections are projected on the plane of the microscope stage and display the geometry of the indicatrix in the case of (b) a uniaxial positive crystal and (c) a uniaxial negative crystal. A, analyzer; P, polarizer. (d) Centered uniaxial interference figure with melatope, isogyres, and isochromes (cf. Plate 12a and c). (e,f) When a 550 nm (red) compensator is inserted, isogyres will turn red and isochromes shift. (e) In the case of a uniaxial positive crystal, isochromes in the NW–SE sector move to higher-order colors and those in the NE–SW sector to lower-order colors (cf. Plate 12b). (f) The opposite is the case for uniaxial negative crystals (cf. Plate 12d). Addition and subtraction of interference colors are indicated.

is zero in the center (circular section) and increases towards the periphery (going from ray 1 to 4 in Figure 14.8a). At the same time the effective path length increases if light rays are passing through the crystal slab at an oblique angle. Ray 1 has a much shorter path than ray 4. Both effects add to an increase in interference color. We observe a color spectrum from the melatope towards the periphery

that corresponds to the interference color chart (Plate 10a). Depending on the birefringence of the mineral, the colors extend to low orders (e.g., for quartz $(n_\gamma - n_\alpha)$ = 0.009, Plate 12a) or to high orders (e.g., for calcite $(n_\gamma - n_\alpha)$ = 0.172, Plate 12c).

When a compensator plate is added, those points of the interference figure for which n_γ' in the crystal is parallel to n_γ' (Z') of the compensator (slow) will add;

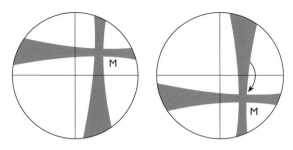

Fig. 14.9 Isogyres in an off-center uniaxial interference figure in two stage rotations. Isogyre branches stay more or less parallel to polarizer and analyzer. M, melatope.

those for which n'_α in the crystal is parallel to n'_γ of the compensator will subtract. If the crystal is uniaxial positive, there is addition in the NE–SW quadrants and subtraction in the NW–SE quadrants (Figure 14.8e, Plate 12b for quartz). The opposite is true for a uniaxial negative crystal (Figure 14.8f, Plate 12d for calcite). This method enables us to determine the optic sign very quickly. (Adding the compensators to the isogyres produces the colors of the compensator, in the case of Plate 12b,d first-order red.)

Crystals in a good orientation for centered interference figures are relatively easy to find in a thin section by searching for those with low retardation. If the optic axis is inclined, interference figures are less symmetrical and the interpretation is more difficult and requires some practice (Figure 14.9). The melatope is no longer in the cross-hair and rotates as the crystal is rotated on the stage. Also, isochromes are no longer ideal circles because thickness and birefringence no longer increase in parallel. Uniaxial interference figures all have the property in common that the isogyres remain more or less parallel during stage rotation and can usually be interpreted, even if the melatope is not within the field of view. The situation is different for interference figures of biaxial crystals.

Biaxial interference figures

The same logic used to interpret uniaxial interference figures can also be used to interpret those from *biaxial crystals*. In Figure 14.10a we look from various directions at a biaxial positive indicatrix, which is oriented with the acute bisectrix Z (n_γ) along the microscope axis (ray 1). As can be seen, the circular section (ray 3) is not in the center of the hemispherical projection. Ellipses 1 and 2 have the long axis in the N–S direction, ellipses 4 and 5 in the E–W direction. Figure 14.10b

shows all the corresponding indicatrix sections for a biaxial positive crystal, and Figure 14.10c for a biaxial negative crystal. In both cases we recognize two circular sections along the E–W line. They are again called melatopes and correspond to the optic axes A and B. But contrary to the case of uniaxial crystals the pattern does not show axial symmetry. Since ellipse axes along the N–S and E–W diameters are parallel to the polarizer and analyzer, we again observe a black cross, but contrary to the uniaxial case the vertical and horizontal isogyres are not equivalent and have different thicknesses. Figure 14.10d is an interference figure corresponding to Figure 14.10b and c (see also Plate 12e). Note, also, that the isochromes are much more complicated than in the uniaxial case, since changes in thickness and birefringence do not go in parallel. The effective sample thickness increases uniformly from ray 1 to 5, i.e., from the center to the periphery, but birefringence does not. The birefringence increases with distance from the melatope (M in Figure 14.10d).

Since the patterns in Figure 14.10b,c are not axially symmetrical, interference conditions change as we rotate the crystal. If we perform a 45° clockwise rotation we obtain Figure 14.10e,f. Again we construct isogyres by finding ellipse axes (vibration directions) that are parallel or perpendicular to the analyzer (A) (shaded in Figure 14.10e,f). The interference figure (Figure 14.10g and Plate 12f) shows that the melatope M and the isochromes have simply rotated. But the isogyres have changed from a cross into two hyperbola-shaped branches that pass through the two melatopes. At a 45° rotation, the isogyres are at a maximum distance from the bisectrix between the two melatopes (or optic axes) (Bxa) and the axial plane is clearly defined from the pattern of isochromes and the positions of the melatopes. While Figures 14.10b–f are idealized to show the full hemisphere, in an actual microscope the angular field of view is limited to an opening angle of at most 63° (with a 0.85 numerical aperture objective lens and $n = 1.50$). For larger angles no isogyres are visible at the 45° position.

In oriented biaxial interference figures (Figure 14.10d,g and Plate 12e,f), with the acute bisectrix (Bxa) (cf. Figure 13.19e) parallel to the microscope axis, we wish to determine whether the mineral is biaxial positive (Figure 14.10b,e) or biaxial negative (Figure 14.10c,f). This can be done, as in the case of uniaxial minerals, by using compensators. We need to determine whether the direction normal to the axial

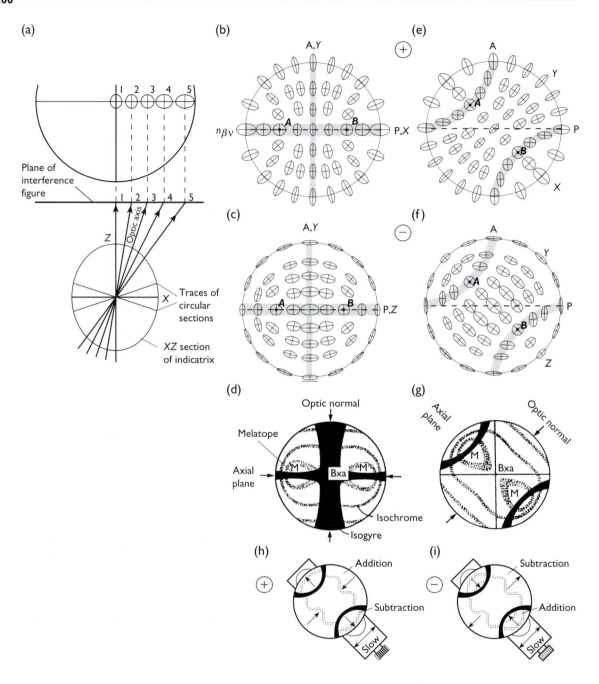

Fig. 14.10 Interference figures of biaxial crystals. (a) Projection of the indicatrix by using convergent light. Note the change in shape of ellipse sections between the center (ray 1) and the periphery (ray 5) with a circular section in between (ray 3). (b,c) Elliptical sections are projected onto the plane of the microscope stage and display the geometry of the indicatrix for biaxial positive (b) and biaxial negative (c) crystals. (d) Centered biaxial interference figure with the acute bisectrix (Bxa), isogyres (black), melatopes (M), and isochromes corresponding to (b) and (c) (cf. Plate 12e). (e,f) Same as (b,c) but rotated 45°clockwise. (g) Interference figure corresponding to the rotated orientation (e,f) (cf. Plate 12f). (h,i) Shifts in isochromes when a compensator is inserted for biaxial positive (h) and biaxial negative (i) crystals (cf. Plate 12g). A, analyzer; P, polarizer; *A, B* optic axes; *X, Y, Z* indicatrix axes. (b,c,e,f) are projections of the whole sphere, not just the region that is observed with the microscope (d,g).

plane (optic normal) is n_α or n_γ. We observe first the black cross (Figure 14.10d), and then open the isogyres by rotating the stage, keeping track of the rotation of the melatopes (Figure 14.10g). At 45° we insert the compensator crystal and check in which quadrant addition or subtraction of interference colors occurs (Figure 14.10h,i and Plate 12g). The interpretation is similar to the uniaxial case (see Figure 14.8e,f). In the case of a positive biaxial indicatrix, there is addition in the NE and SW sectors (where large ellipse axes of crystal and compensator line up), whereas for a negative biaxial indicatrix, there is subtraction.

Melatopes indicate the position of the optic axes. The line connecting the two melatopes defines the axial plane, the distance between the two melatopes is indicative of the axial angle $2V$. If $2V$ is small, melatopes are close together and close to the Bxa. If $2V$ is near 65° (in the case of muscovite, $2V$ is around 40°, Plate 12f), melatopes are near the periphery, and for larger angles they are outside the field of view. Such observations can be used to estimate the axial angle.

Centered Bxa interference figures of biaxial crystals such as those shown in Figure 14.10d,g and Plate 12e–g are easy to interpret, at least if the axial angle is not too large. However, it is often difficult to find crystals in that orientation. They correspond neither to crystals with a very low retardation (such as circular sections in the uniaxial case), nor to crystals with a very high retardation, but rather are of intermediate retardation.

However, also for biaxial crystals the circular section is easy to find because of the low retardation (and interference color), and it can also be interpreted. If we view a crystal along one of the optic axes (one melatope), isogyres are not very symmetrical, but at least one branch of the isogyres always passes through the center of the figure and one melatope is in the center. When the interference figure is rotated such that the axial plane is at 45° (Figure 14.11), isogyres are more or less curved, and the opening angle is a function of $2V$. If the isogyre appears more or less as a straight diagonal line, $2V$ is 90°. If it appears as a cross, $2V$ is 0° (uniaxial). For intermediate angles $2V$ can be estimated, using the calibration on Figure 14.11. In this orientation the second branch of isogyres is visible only for low axial angles (i.e., 15° in Figure 14.11). The position of the axial plane is obvious, and the optic sign of the crystal is therefore quickly determined with a compensator. This is the preferred section for routine analysis of biaxial crystals.

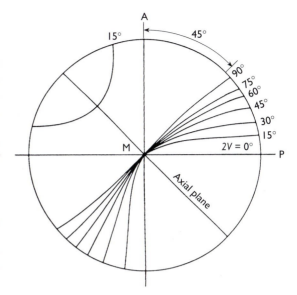

Fig. 14.11 Superposed optic axis interference figures for different axial angles, with one melatope (M) in the center, in the 45° position. The axial angle can be determined from the curvature of the isogyres.

Dispersion

Dispersion is expressed in interference figures, since the orientation and shape of the indicatrix depends to some extent on wavelength (see Section 13.5). This is particularly the case for anomalous dispersion, where the birefringence changes with direction. A good example is the monoclinic K-feldspar sanidine. In high-temperature sanidine the axial plane is (010), i.e., the mirror plane of this monoclinic mineral, and mirror symmetry has to apply to all colors. Indeed, the interference figure (Plate 12h) shows mirror symmetry, but the isogyres show a blue fringe on the inside and a red fringe on the outside. Therefore $2V$ is larger for violet than for red, which is abbreviated as $r < v$. Because of the mirror symmetry, this is called inclined dispersion. In low-temperature sanidine the axial plane is perpendicular to (010) and the axial plane pivots on [010]. In this case the mirror symmetry no longer applies to colors, but the 2-fold rotation does. This produces color fringes related by a rotation about the acute bisectrix (Plate 12i).

In Figure 13.24c and the discussion in Chapter 13, we looked at the anomalous dispersion of orthorhombic brookite (TiO_2), where the axial plane for blue light is (001) (Plate 12j), for green light it appears uniaxial (Plate 12k), and for red light it is (001)

(Plate 12l), which is best seen in interference figures with monochromatic light.

Optical activity

Finally a few words should be said about *optical activity*. If polarized light passes through a crystal that lacks a center of symmetry, the plane of polarization is rotated to either the left or the right, depending upon whether the crystal is left or right handed. The amount of rotation depends on thickness and wavelength. For quartz it is 22.1°/mm when viewed along the *c*-axis with yellow light (589 nm). For a 30 μm thin section this is meaningless, but for a thicker slab viewed with crossed polarizers the mineral is not extinct, as one would expect for a uniaxial crystal. This rotation leads to interference figures in which the isogyres do not extend to the center and are replaced by a bright circular region (Figure 14.12a). In monochromatic light this bright region can be brought to extinction when the analyzer is rotated from the crossed position. These abnormal interference figures can be used to confirm the hand of rotation. Monochromatic interference figures with a λ/4 compensator superposed show a spiral (so-called Airy's spiral, Figure 14.12b). The sense of the spiral is the same as that of the crystal (left handed). Optical activity is particularly pronounced in organic crystals such as sucrose. Most biologically grown organic crystals are left handed, while synthetic crystals vary in handedness.

The procedure for analyzing crystals with interference figures is summarized in Box 14.2.

14.6 Characteristics of important rock-forming minerals

Having learnt about the optical properties of minerals and how to use the petrographic microscope, we now survey the optical properties of the most important rock-forming minerals. Even though we have not yet discussed these minerals systematically, it is useful to know how to recognize quartz, feldspars, micas, amphiboles, and carbonates in thin section early on. For mineral names, their structure, and composition, refer to later chapters in Part V, which contain tables with optical properties of the most important minerals, and to the tables in Appendices 3 and 4. Box 14.3 summarizes the general procedure of thin-section analysis and gives examples.

(a)
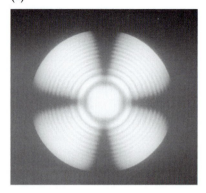

(b)
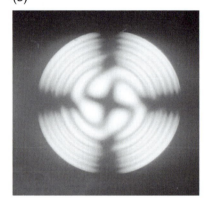

Fig. 14.12 Interference figures illustrating optical activity of quartz. (a) Quartz slab of 4 mm thickness viewed along the *c*-axis. Note that isogyres do not extend to the center. (b) Quartz slab with λ/4 compensator plate superposed produces an Airy's spiral. The rotation sense of the spiral corresponds to the handedness of the crystal (left-handed). (Courtesy O. Medenbach.)

Distinguishing quartz from feldspar

Feldspar and quartz are the commonest crustal minerals and occur together in rocks of many different kinds such as a granitic gneiss in Plate 13a,b. They are colorless and transparent and look similar in plane polarized light (Plate 13a), except that feldspars may display cleavage traces. Even though both have a similar birefringence (gray to white in standard thin sections), optical identification of the two minerals is relatively simple with cross-polarized light (Plate 13b). The most important distinguishing features (in order of importance) are the following:

- *Undulatory extinction*: Quartz commonly shows irregular or patchy extinction, dividing each grain

---

Box 14.2 Review: Determination of the optical character by means of interference figures

1. Find a crystal in the thin section with an optic axis parallel to the microscope axis. Such a crystal is easily selected because those orientations have no retardation and, with crossed polarizers, stay extinct during rotation of the stage. After identifying a crystal, go to high magnification. Make sure that the objective lens is centered and that the area of interest does not move during stage rotation.

2. Obtain an interference figure by inserting the condenser lens, which must be close to the thin section. Also insert the Bertrand lens (or remove the ocular lens).

3. *Uniaxial crystals*: The interference figure of a uniaxial crystal has axially symmetrical isochromes. Their color pattern corresponds to the birefringence and can be compared with the color chart. With a thin section of standard thickness, for quartz the highest-order color is yellow; for calcite, with a high birefringence, isochromes show many orders of color. The cross of isogyres is oriented N–S and E–W and does not change during rotation of the stage. The optical character (+ or −) can be determined by inserting compensator plates at any rotation of the stage.

   If the optic axis is inclined to the microscope axis, the melatope (and the cross) are not centered and move during stage rotation. However, the isogyre branches stay as straight lines and displace in a parallel fashion; such patterns can therefore be interpreted.

4. *Biaxial crystals*

   (a) In biaxial crystals isochromes are not circular, but instead display more complicated curves, centered around the optic axes. Isogyres are curved and change their curvature during rotation. In optic axis figures, they divide the field into three, but only two of those fields are generally visible. If there is subtraction in the SE sector with an acute angle of the isogyre arc, the crystal is biaxial positive.

   (b) It is usually possible to distinguish between uniaxial and biaxial crystals, even in inclined sections (in biaxial crystals isogyres change their curvature during stage rotation), but it may be very difficult to determine the optical character unless the section is nearly normal to an optic axis or Bxa. Interference figures with the Bxo or n_β close to the microscope axis show broad isogyres that change rapidly with stage rotation. They are ambiguous and are best avoided.

   (c) In biaxial minerals, we can determine, or at least estimate, the optical angle $2V$. This can be done in a section normal to an optic axis using the curvature of the hyperbola. Alternatively, we can measure the distance between the axes in a section normal to the Bxa. With a high magnification objective lens (e.g., numerical aperture 0.85, as marked on the lens), the total field of view is 63°. For larger $2V$, it can be estimated from the speed with which the hyperbolas move outward during a rotation of the stage. (High precision determination of axial angles requires a universal stage.)

---

into domains (Plate 13b, center). This effect is not seen in feldspar.

- *Twinning*: Quartz is rarely twinned in an optically obvious fashion, whereas feldspars are most commonly twinned on a variety of twin laws. Particularly in plagioclase, grains are divided into lamellar stripes with different extinction (Plate 13b).

- *Cleavage*: Quartz lacks cleavage. Feldspars have two good cleavages – (001) and (010) – more or less at 90°. Cleavage may be difficult to observe in the high-temperature alkali feldspar sanidine.

- *Relief*: Refractive indices in quartz and the commoner feldspars are close to that of Canada balsam (1.54). Relative to balsam, all alkali feldspars have negative relief. Most plagioclase feldspars have positive relief, increasing with anorthite (Ca) content (Figure 14.13). Quartz ($n_\varepsilon = 1.553$, $n_\omega = 1.544$) always shows slight positive relief against balsam.

- *Optical symmetry and sign*: Quartz is *uniaxial positive*. Feldspars, being monoclinic or triclinic, are *biaxial*. K-feldspars are optically negative, with

---

Box 14.3 Review: General procedure for thin section analysis

1. *Observe the thin section with plane polarized light at low magnification*
   - Opaque minerals (pyrite, magnetite, hematite, etc.)
   - Colored minerals (amphiboles, biotite, chlorite, tourmaline)
   - Pleochroism (examples same as above)
   - High–low relief (low: quartz, feldspars, apatite; high: calcite, titanite, garnet, zircon)
2. *Observe the thin section with crossed polarizers at low magnification (interference colors)*
   - Very high birefringence (calcite, titanite)
   - High birefringence (olivine, pyroxene, muscovite)
   - Medium birefringence (biotite, amphiboles)
   - Low birefringence (quartz, feldspars)
   - Very low birefringence (zeolites, apatite)
   - Isotropic (garnets, spinel)
3. *Detailed analysis with plane polarized light at high magnification*
   - Becke line (close aperture!)
   - Cleavage (feldspars, amphiboles, pyroxenes)
4. *Detailed analysis with crossed polarizers at high magnification*
   - Interference figures (condenser close to sample, find optic axis figure)
   - Interpretation with compensator
5. *Some mineral characteristics*
   - Quartz: undulatory extinction
   - Microcline: cross-hatched twinning
   - Plagioclase: lamellar twins (determine calcium content (anorthite, An) from extinction angle)
   - Pyroxenes/amphiboles: cleavage, extinction angle
   - Calcite/dolomite: very high birefringence

---

either large $2V$ (orthoclase and microcline) or small $2V$ (sanidine) angles. Plagioclases have large $2V$ angles, and the optic sign depends on composition.

- *Alteration*: Quartz does not alter to other minerals and is typically clear and colorless in thin section. Feldspar is subject to alteration and is commonly cloudy or "dirty-looking" in comparison with adjacent quartz. Such alteration is often to sericite mica and is usually concentrated in patches or along cleavages (e.g., Plate 7e,f). When alteration-free, feldspars are generally glass clear, like quartz.

Alkali feldspars

At high temperature *alkali feldspars* form a solid solution series between albite ($NaAlSi_3O_8$) and orthoclase ($KAlSi_3O_8$). In plutonic rocks (those cooled slowly within the Earth), intermediate feldspars in this series exsolve into two-phase crystals called *perthites*. In many of these rocks, the individual exsolved parts

can be seen with the microscope as wavy lamellar features (Plate 13c).

Twinning (see Chapter 9) in monoclinic alkali feldspars (such as orthoclase and sanidine) is confined to simple doublets (a common law is the *Carlsbad* law, with twin axis [001] and composition plane (010)) (Plate 13d). The triclinic feldspars (microcline) commonly show lamellar twinning on the *Albite* law (twin axis perpendicular to (010), composition plane (010)) and the *Pericline* law (twin axis [010], composition plane ($h0l$)). (Note that "Albite" and "Pericline" are capitalized if the names refer to twin laws, rather than to the mineral.) In microcline, fine lamellae of Albite and Pericline twins intersect to give a characteristic "cross-hatched" appearance (Plate 13e). Listed below are some diagnostic optical properties:

- In *orthoclase* (*low sanidine*), only simple doublet twinning is observed. It is optically negative with large $2V_\alpha$. The axial plane is parallel to [010] = Z (Figure 14.14a, Plate 13d).

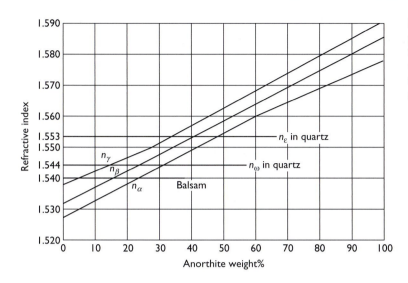

Fig. 14.13 Refractive indices for plagioclase (n_α, n_β, and n_γ) as a function of anorthite content. The values for quartz (n_ω and n_ε) and Canada balsam are also shown for reference.

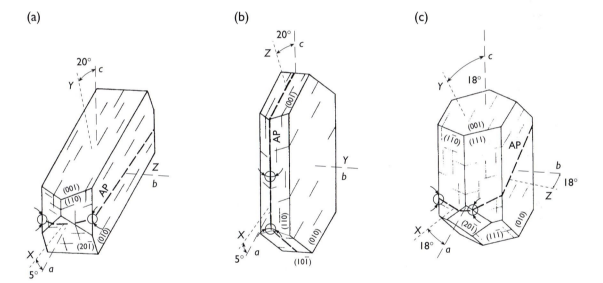

Fig. 14.14 Morphology and orientation of the optical indicatrix in alkali feldspars: (a) orthoclase, (b) sanidine, and (c) microcline (vibration directions X, Y, Z, the axial plane AP, crystal axes a, b, c).

- In *low albite* (plutonic, metamorphic) lamellar twinning in one or two sets is common, but there are exceptions. Low albite is optically positive, with very high $2V_\gamma$ (80°).
- High albite (volcanic) is similar to low albite but optically negative with moderate $2V_\alpha$ (about 45°).
- *Sanidine* from volcanic rocks ("*high sanidine*") may display simple doublet twinning and is usually clear and glassy. The optic axial plane is at a right angle to

the twin composition plane (contrary to orthoclase) and perpendicular to [010] = Y (Figure 14.14b). Sanidine is optically negative with a small $2V_\alpha$ (about 18°). It can be confused with quartz, but quartz is uniaxial positive.
- *Microcline* has characteristic cross-hatched twinning (Plate 13e) and is optically negative, with large $2V_\alpha$. The axial plane is similar to orthoclase (Figure 14.14c) but orthoclase does not display cross-hatched twinning.

Plagioclase feldspars

Plagioclase is generally twinned with diagnostic lamellae, most commonly according to the Albite law with (010) as composition plane and twin plane and the Albite–Carlsbad law with [001] as twin axis and also (010) as composition plane. An example is shown in Plate 13f. Thus the twin planes are parallel to the (010) cleavage, yet the (001) cleavage deflects between twins. Particularly in metamorphic plagioclase, also Pericline twins are observed and some may be mechanically induced.

The plagioclase feldspars form a solid solution series between albite ($NaAlSi_3O_8$) and anorthite ($CaAl_2Si_2O_8$). The Ca:(Ca + Na) ratio, commonly referred to as *anorthite content* or *An-content*, is an important diagnostic feature in determining the rock type or the metamorphic grade. Therefore, optical properties are often used as a rapid method of estimating the composition. Optically, two main types of plagioclase feldspar are distinguished: high-temperature, disordered plagioclases are typical of phenocrysts in extrusive volcanic rocks, and low-temperature, ordered plagioclases are typical of slowly cooled feldspars in plutonic and metamorphic rocks. In triclinic plagioclase there are no symmetry constraints on the orientation of the optical indicatrix. Indeed, as Figure 14.15 illustrates, the optical directions, including the axial plane (dashed line), rotate greatly between albite and anorthite, while the crystallographic axes barely change. Better than the sketches in Figure 14.15, the stereographic projection in Figure 14.16 documents the large rotations of indicatrix axes X, Y, and Z, and optic axes A and B, relative to the crystal structure, between albite (Ab) and anorthite (An). As a reference coordinate system the c-axis [001] of this triclinic mineral is placed in the center and the pole to (010) in the E-point. Some other lattice planes, corresponding to twin laws, are indicated. There are two curves: the dashed lines are for plutonic and metamorphic plagioclase, with an ordered structure, and the dotted lines for volcanic plagioclases with a disordered structure.

On the basis of the indicatrix orientation, various determinative curves have been constructed and are used to determine the chemical composition of plagioclase. The commonest and simplest methods are as follows.

- *Refractive indices*: To confirm the presence of plagioclase more sodic than medium oligoclase (An 20), the values of *refractive indices* should be

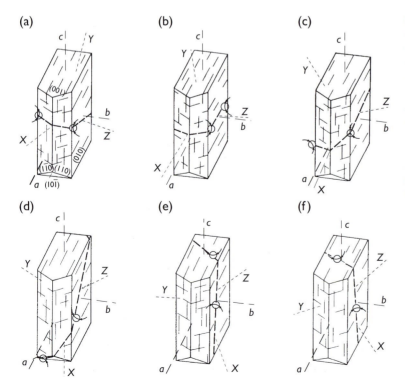

Fig. 14.15 Morphology and orientation of the optical indicatrix in plagioclase. Note that the crystal morphology and lattice are very similar for each of the solid solution members, but the orientation of the indicatrix changes greatly between albite and anorthite, as is best visible in these sketches by the orientation of the axial plane (dashed line). (a) Albite, (b) oligoclase, (c) andesine, (d) labradorite, (e) bytownite, and (f) anorthite.

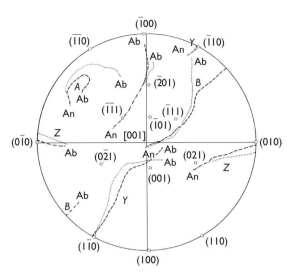

Fig. 14.16 Stereographic projection of triclinic plagioclase with [001] in the center and the pole to (010) in the E-point. Shown is the rotation of the optical indicatrix (vibration directions *X*, *Y*, *Z*) and the optic axes (*A* and *B*) as a function of plagioclase composition from albite (Ab) to anorthite (An). Dashed curves are for plutonic and metamorphic plagioclase (ordered structure) and dotted curves for volcanic plagioclase (disordered structure). Note that some optical directions change their orientation by over 90°. Such measurements are best accomplished with a petrographic microscope, equipped with a universal stage. Some crystallographic directions relevant to plagioclase twin laws are indicated as reference (data from Burri *et al.*, 1967).

compared with those of quartz and balsam, using Becke lines (cf. Figure 14.13).

- *Extinction angles on sections perpendicular to [100] (Michel–Lévy technique)*: In this section *both* (001) cleavages and (010) twins (Albite law) are viewed edge-on and appear as sharp lines. They do not sway sideways, as focus is slightly raised and then lowered. When (010) is parallel to the N–S vibration direction of the analyzer, the twin lamellae are uniformly illuminated (Figure 14.17a). (Similar looking Pericline twins do not show this feature and cannot be used for this method.) When the thin section is rotated clockwise by an angle Δ, one set of twins becomes extinct (the other set becomes extinct if the section is rotated by the same angle counterclockwise). Extinction angles are measured between n'_α and the trace of (010) for Albite twins. It is necessary to identify whether the extinct N–S direction is n'_α and not n'_γ. This can be done by rotating the

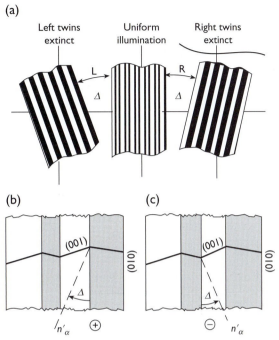

Fig. 14.17 Michel–Lévy technique for determining the extinction angle. (a) Lamellar Albite twins show uniform illumination if they are viewed edge-on and the trace is parallel to the analyzer. By rotating the thin section clockwise or counterclockwise by an angle Δ, one set of the twins is brought to extinction. (b,c) Definition of the extinction angle Δ between the trace of the (010) twin and n'_α in sections perpendicular to [100] (i.e., traces of (010) twins and (001) cleavage fractures are viewed edge-on), and convention for positive (b) and negative (c) extinction angles. View is looking down [100].

extinct direction 45° clockwise (it will show interference colors) and inserting the compensator. If you observe subtraction, it is n'_α. Figure 14.17b explains how to identify a positive extinction angle (n'_α in the acute angle between (001) and (010) traces, as is the case for calcic plagioclase). Figure 14.17c illustrates the case for a negative angle (n'_α in the obtuse angle between cleavages, as is the case for sodic plagioclase). The extinction angle varies by more than 60° between albite and anorthite (Figure 14.18) and the determinative curves are different for volcanic and plutonic plagioclase. The extinction angle is a quick method for determining the composition of plagioclase.

Plagioclase in igneous rocks often displays compositional zoning that reflects the cooling history.

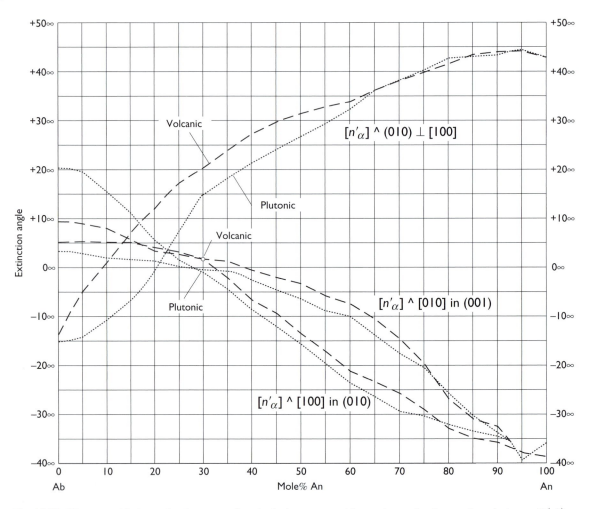

Fig. 14.18 Diagram with determinative curves for plagioclase composition, using extinction angles relative to $X(n'_\alpha)$. One set of curves displays the extinction angle in crystals that have [001] perpendicular to the section and view the (001) cleavage and (010) twin plane edge-on (for the definition of the extinction angle, see Figure 14.17b). A second set of curves shows the variations of extinction angle in (001) and (010) cleavage flakes relative to [100] (trace of second cleavage or twin plane (Albite twin law)). Plutonic plagioclase is represented by dotted lines, and volcanic plagioclase by dashed lines. $n'_\alpha \wedge (010)$ indicates angle between n'_α and (010) (data from Burri *et al.*, 1967).

This is expressed in variable extinction, often concentric, in phenocrysts (Plate 7a).

- *Extinction angles on cleavage fragments (grain mounts)*: Plagioclases have (010) and (001) cleavages at almost 90°. In this method, cleavage flakes of crushed plagioclase are mounted in suitable immersion oil. Fragments lying on the (001) cleavage, which is usually better developed and will show uniform birefringence, and the composition planes of Albite twins will be parallel to the trace of the second cleavage (010). The extinction angle of the fast n'_α direction with respect to this trace is

measured and compared with the determinative curve (Figure 14.18). Fragments lying on the (010) cleavage are recognized by the lack of lamellar Albite twinning. They are not suitable for this method.

- *Axial angles*: Determination of the axial angle $2V_\gamma$ (Figure 14.19) is less diagnostic than the extinction angle because of the sinusoidal variation with An content and the large difference between volcanic and plutonic varieties for sodium-rich plagioclase. For example, a $2V$ of 90° could be interpreted as a plutonic oligoclase (An 18), andesine (An 32) or

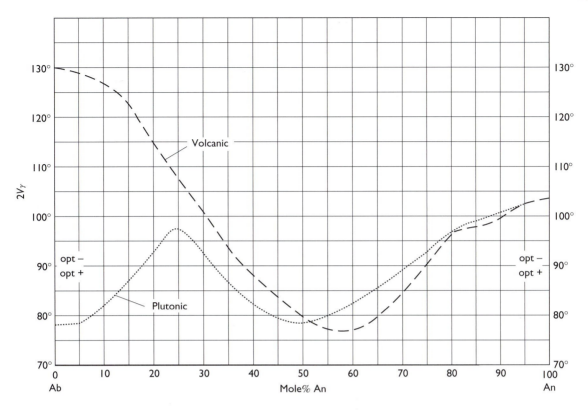

Fig. 14.19 Diagram showing the variation of the optic axial angle $2V_\gamma$ with anorthite content. Plutonic plagioclase is represented by dotted lines, and volcanic plagioclase by dashed lines (data from Burri *et al.*, 1967).

bytownite (An 71), or as a volcanic andesine (An 38) or bytownite (An 75).

Olivine

The olivines are a solid solution between forsterite (Mg_2SiO_4) and fayalite (Fe_2SiO_4). They are common constituents of basalts and ultramafic rocks such as peridotite. Olivine is the main component of the upper mantle of the Earth. All olivines are biaxial with a large axial angle. Olivine is orthorhombic and the axial plane is (001) (Figure 14.20a). Twins on (010) are common. Diagnostic features are:

- Usually clear, compared to slightly shaded in pyroxenes with which olivine often coexists in basalts (Plate 14a).
- High birefringence, higher than pyroxenes: 0.035–0.052 (second-order colors, Plate 14b).
- Large $2V_\gamma$: 82° (forsterite) to 134° (fayalite). See the determinative diagram in Figure 14.20b.
- Cleavage: imperfect or absent, compared to pyroxenes (Plate 14a).

- Commonly altered hydrothermally to fine-grained products such as iddingsite (yellow/orange) in volcanic rocks (Plate 14c), and serpentine, talc, or chlorite in metamorphic rocks (Plate 14d).

Pyroxenes and amphiboles

Pyroxenes and amphiboles are so-called chain silicates (Chapter 30) and span a wide range of compositions, with solid solutions mainly between magnesium, iron, calcium, and sodium. The calcium- and sodium-free pyroxenes are usually orthorhombic (*orthopyroxenes*). The remaining pyroxenes, generally calcium- or sodium-bearing, are all monoclinic (*clinopyroxenes*). Most of the amphiboles are monoclinic, except orthorhombic *anthophyllite*. The commonest pyroxene is called *aluminous diopside* (previous name *augite*) and the most common amphibole is *hornblende*. Both have highly variable chemical compositions and optical properties.

Most pyroxenes and amphiboles are prismatic and elongated along $c = [001]$ (Figure 14.21). All show very good prismatic cleavages ({210} in orthopyroxenes

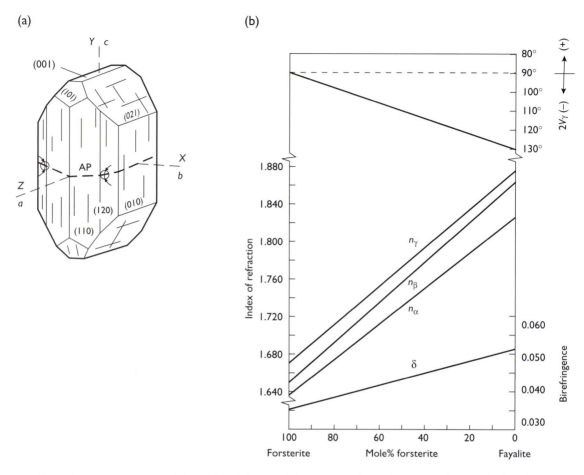

Fig. 14.20 Optical properties of olivine. (a) Relationship between crystallographic axes and optical indicatrix. (b) Variation of $2V_\gamma$, refractive indices and birefringence δ between forsterite (Mg_2SiO_4) and fayalite (Fe_2SiO_4).

(Figure 14.21a,b), {110} in clinopyroxenes (Figure 14.21c–e), and amphiboles (Figure 14.21f)). The cleavages are inclined, almost at right angles (92°–93°) in pyroxenes and at a 124° angle in amphiboles. As we have discussed earlier (Figure 14.6c), on prismatic sections (containing the c-axis [001]) the cleavage traces are parallel to each other. Large clinopyroxene crystals in Plate 14e,f show chemical sector zoning that is expressed in the interference colors. Observe the ~90° cleavage in the crystal on the top, best visible with plane polarized light (Plate 14e).

In orthopyroxenes, the optic axial plane is parallel to (100) (Figure 14.21a,b). In most clinopyroxenes as well as amphiboles, it is parallel to (010) (Figure 14.21d–f). Exceptions are calcium-poor pigeonite and clinoenstatite, where the axial plane is perpendicular to (010) (Figure 14.21c). In orthopyroxenes all prismatic sections (with c = [001] in the plane of

the section) have parallel extinction since $Z = [001]$ (Figure 14.21a,b). In clinopyroxenes extinction is parallel in (100) sections and inclined in (010) sections (Figure 14.21c–e). The extinction angle between $Z(n_\gamma)$ and [001] is used for determinative purposes. The ranges for extinction angles in clinopyroxenes in sections cut parallel to (010) are shown in Figure 14.22a. In this most useful section, cleavages are inclined about 45° to the section, in two directions, which can be ascertained by focusing (Figure 14.6c), and (100) twins are viewed edge-on (see below).

Orthopyroxenes in igneous rocks often undergo chemical exsolution during cooling which is expressed in lamellar structures (Plate 15a,b).

The common amphiboles contain calcium (tremolite–actinolite series) and some are, in addition, aluminous (hornblendes). Alkali-bearing varieties occur in both metamorphic rocks and igneous rocks. With the

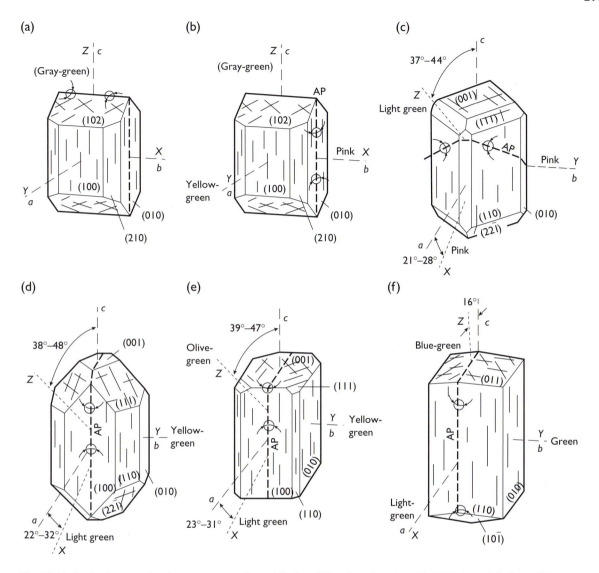

Fig. 14.21 Optical orientation in pyroxenes and amphiboles. (Vibration directions X, Y, Z, the axial plane (AP), crystal axes a, b, c, color of pleochroism). (a) Enstatite, (b) hypersthene, (c) pigeonite, (d) diopside, (e) augite, and (f) hornblende.

exception of some of the alkali varieties (e.g., riebeckite in which the axial plane is perpendicular to (010)), amphiboles have (010) as the axial plane (see Figure 14.21f). On section (010), the extinction angle between Z and c = [001] (cleavage trace) ranges, in all but alkali amphiboles, between 0° and 30° (Figure 14.22b).

Most amphiboles, except tremolite and anthophyllite, show distinct pleochroism (see Table 13.4). Green colors are typical for calcic amphiboles (Plate 11c) and blue colors for sodic amphiboles (Plate 11d,e). The colors become stronger with increasing iron

content. Tremolite may occur in metamorphic talc schists (Plate 15c,d). Prismatic crystals cross-cut the matrix. If they are viewed down the long axis [001] the typical amphibole 120° cleavage is observed.

Micas

Micas, as well as the related chlorites (Chapter 29), are easy to identify in thin sections through the following obvious optical properties:

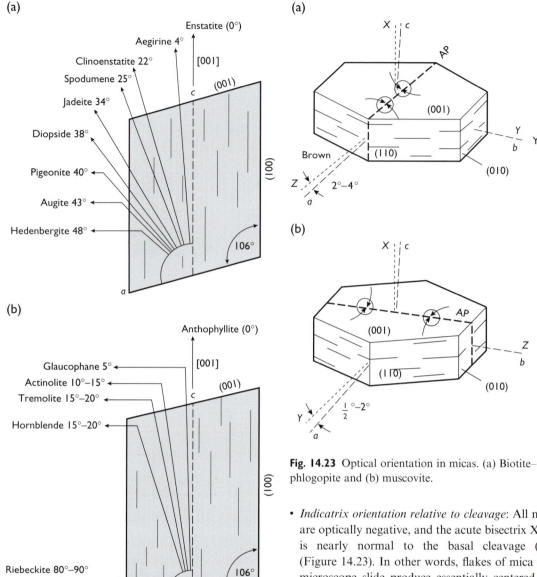

(a)

Enstatite (0°)
Aegirine 4°
Clinoenstatite 22°
Spodumene 25°
Jadeite 34°
Diopside 38°
Pigeonite 40°
Augite 43°
Hedenbergite 48°

[001]
c
(001)
(100)
106°
a

(b)

Anthophyllite (0°)
Glaucophane 5°
Actinolite 10°–15°
Tremolite 15°–20°
Hornblende 15°–20°
Riebeckite 80°–90°

[001]
c
(001)
(100)
106°
a

Fig. 14.22 Extinction angles $Z(n_\gamma)$ to [001] in a (010) section for (a) common pyroxenes and (b) common amphiboles.

- *Cleavage*: An excellent single (001) cleavage is observed in all sheet silicates.
- *Pleochroism*: All iron-bearing sheet silicates, including micas as well as chlorites, have distinct pleochroism, with colors ranging between green and brown (Plate 11a).

(a)

X ∥ c
AP
(001)
Brown
(110)
Z
2°–4°
a
Y
b
Yellow
(010)

(b)

X ∥ c
AP
(001)
(1̄10)
Z
b
Y
$\frac{1}{2}$°–2°
a
(010)

Fig. 14.23 Optical orientation in micas. (a) Biotite–phlogopite and (b) muscovite.

- *Indicatrix orientation relative to cleavage*: All micas are optically negative, and the acute bisectrix X (n_α) is nearly normal to the basal cleavage (001) (Figure 14.23). In other words, flakes of mica on a microscope slide produce essentially centered Bxa interference figures and this is a good material on which to practice observing interference figures (Plate 12e–g). The $2V_\alpha$ angle varies with composition, and it is largest in aluminous micas such as muscovite (30°–45°) and small in iron and magnesium micas such as biotite. When mica in thin section is viewed on-edge, as it usually is, one observes parallel extinction because Y is in the cleavage plane and is parallel to the cleavage trace.
- *Birefringence*: In most micas, birefringence is 0.03–0.04. Maximum birefringence is seen in sections cut at right angles to the cleavage, and these produce interference colors up to third order in standard thin

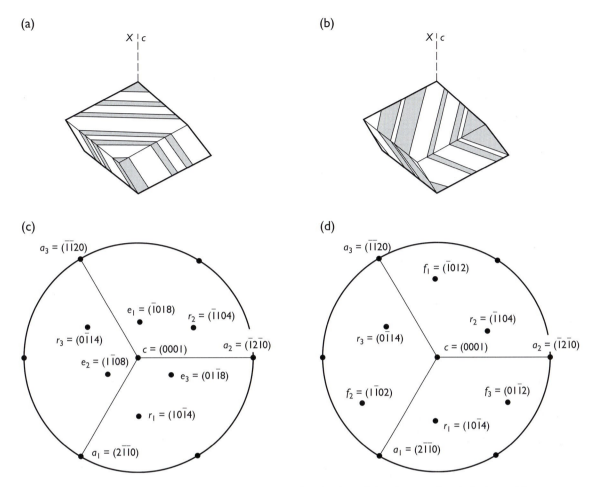

Fig. 14.24 Twinning in (a,c) calcite and (b,d) dolomite. (a,b) Sketches of the cleavage rhombohedron with a lamellar twin system indicated. (c,d) Stereographic projections illustrating the disposition of poles to cleavage planes $r = \{10\bar{1}4\}$ and twin planes $e = \{01\bar{1}8\}$ and $f = \{01\bar{1}2\}$.

sections. Basal sections (sections cut parallel to the cleavage plane) have low birefringence and produce Bxa interference figures with relatively small $2V_\alpha$. Chlorites are distinguished from biotites (Plate 16a, b) by much lower birefringence and abnormal interference colors, blueish or brown, due to anomalous dispersion (Plate 15e,f).

• *Appearance at extinction*: In thin section, micas cut across the cleavage and showing high birefringence have a mottled or twinkling appearance at and very near the extinction position. This is a distinctive characteristic of mica in thin section (Plate 16a,b).

Calcite and dolomite

Calcite and dolomite (Chapter 24), as well as the much rarer orthorhombic aragonite, are, like all carbonate

minerals, recognized by a very high birefringence (0.15–0.2) (see Plates 8a, 12c,d). Aragonite is biaxial, $2V_\alpha = 18°$. Rhombohedral carbonates – among them, calcite and dolomite – are uniaxial negative and would be difficult to distinguish were it not for the characteristic twinning that is present in most crystals from carbonate-bearing rocks. The most common twinning, $\{01\bar{1}8\} = e$ in calcite (Figure 14.24a,c, Plate 2d) is at a shallow angle to the basal plane, and $\{01\bar{1}2\} = f$ in dolomite (Figure 14.24b,d) at a much larger angle to the basal plane. These twins are generally mechanically induced. The angles between twin traces, cleavage traces, and in particular the angles between the c-axis and a twin trace, are best seen in stereographic projections (Figure 14.24c,d) and are used to distinguish between the two carbonate minerals (Table 14.1).

Table 14.1 Characteristic angles for distinguishing calcite and dolomite

	Calcite	Dolomite
Angle between twin traces	44°	80°
Angle between c-axis and twin trace	55°	20°

Plate 16c shows a marble with large grains and polygonal grain boundaries. The whitish colors are very high orders on the right side of the color chart (Plate 10a) and not to be mistaken for first-order white observed for quartz and feldspar. More distinct colors are observed for crystals oriented with the optic axis close to the thin section normal. Plate 16d is a limestone with a fine-grained calcite matrix and larger

grains of dolomite growing from fractures. This process will be discussed in Section 24.5.

Some common orthosilicate minerals

In conclusion of this survey of optical properties of rock-forming minerals, a few words should be said about orthosilicate minerals, which are fairly common in metamorphic rocks and will be discussed extensively in Chapter 28.

Garnets are orthosilicates and have cubic symmetry. Therefore they are optically isotropic, with no birefringence. With plane polarizers they are transparent but with crossed polarizers they are black, at all stage rotations.

Epidote is common in low-grade metamorphic rocks and often occurs as prismatic crystals such as in calcsilicates (Figure 14.25a, Plate 16e,f). Epidote

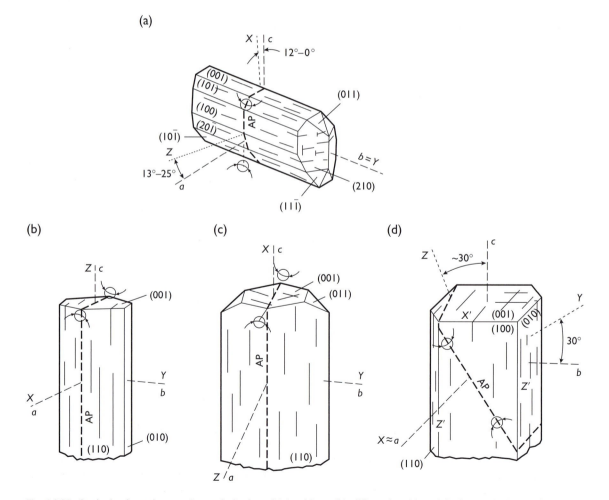

Fig. 14.25 Optical orientations and morphologies of (a) epidote, (b) sillimanite, (c) andalusite, and (d) kyanite.

has anomalous interference colors with very bright patterns that do not agree with the color chart (Plate 10a). The bright crystals in chlorite schist are also epidote (Plate 15f). Epidote is a case of high anomalous colors compared to chlorite in the same section with low anomalous colors (brown).

Of great interest in metamorphic petrology are the three aluminosilicate minerals sillimanite, andalusite, and kyanite, all of composition Al_2SiO_5. Their occurrence in aluminous schists depends on temperature–pressure conditions (see Section 28.6). They generally form prismatic crystals (Figure 14.25b–d). Sillimanite and andalusite are orthorhombic and have parallel extinction (Figure 14.25b,c). Sillimanite is often fibrous and has a perfect (010) cleavage. It is biaxial positive, with a small $2V_\gamma$ ($20°$–$30°$). Andalusite is biaxial negative, with a large $2V_\alpha$ ($80°$–$85°$); it displays weak pleochroism. It often occurs as porphyroblasts (Plate 5c). Plate 17a,b shows large prismatic crystals of sillimanite (S). The birefringence is somewhat higher than quartz and feldspar. Biotite (B) is reacting to form small sillimanite needles called fibrolite (F). Triclinic kyanite is easily distinguished from the orthorhombic aluminosilicates by inclined extinction (Figure 14.25d). The thin section in Plate 17c,d is a special case where the three minerals coexist, which defines temperature and pressure. There is a large andalusite porphyroblast (A). Two crystals of kyanite (K1 and K2) are marked. They are in different orientations and have a different relief, which is typical of kyanite (hence the old name "disthene"). On the bottom right is a cluster of sillimanite needles (S). Such a thin section is like a snapshot of chemical reactions that occurred millions of years ago.

Finally, Plate 17e,f shows a thin section of a biotite–muscovite schist with a large crystal of staurolite $(Fe^{2+},Mg)_2(Al,Fe^{3+})_9O_6(SiO_4)_4OOH$, another orthosilicate. It is of brown color. In thin sections it displays yellow to brown pleochroism.

With all this background about optical properties of minerals, let us return briefly to the eclogite in Plate 9. In plane polarized light (left column) we can already distinguish five minerals. One, with weak greenish pleochroism and a platy habit is chlorite (C). It is quite abundant. Then there is a slightly brownish phase with higher relief, not very clear morphology. This is the pyroxene omphacite (P), cleavage traces are almost at $90°$. Then there is a similar one, slightly greenish-blue but with $120°$ cleavage (A). This is a sodic amphibole. Also there is a large grain of similar relief as

omphacite, slightly colored. This is garnet (G). Finally, we observe some crystallites with very high relief. They are titanite $CaTi(SiO_4)O$ (T), also an orthosilicate.

By using crossed polarizers (second column) we see immediately that garnet is black at all rotations and this is diagnostic. Titanite has extremely high colors, well beyond the color chart (gray-white), and thus very high birefringence like very few other minerals. We also observe a lot of brownish colors. This is chlorite with anomalous interference colors (compare with Plate 15f). Then there are crystals with blue and red colors corresponding to patches with intermediate relief and green-blueish colors with plane polarized light. This is omphacite which originally dominated this eclogite but was partially transformed to amphibole and chlorite. By inserting the compensator (right column) we notice that there is addition when a pyroxene platelet is perpendicular to the compensator (SW–NE) and subtraction when it is parallel, indicating that n_α of omphacite is parallel to the platelet.

For more information on optical properties of common minerals, consult the tables in Chapters 22–31 and Appendices 3 and 4.

14.7 Summary

The chapter provides an introduction on how to use the petrographic microscope to determine the main optical properties of minerals such as refractive index (with Becke line), interference colors, birefringence, extinction angle, and geometry of indicatrix (uniaxial and biaxial interference figures). A section is then dedicated to help identify the main rock-forming minerals based on optical properties.

Test your knowledge

1. Extinction depends on the orientation of vibration directions relative to the polarizer and analyzer. Review what happens if a crystal is rotated through a full $360°$.
2. Interference figures are at first very difficult to understand, since they are a combination of interference phenomena and extinction. Take the uniaxial positive interference figure in Figure 14.8b and label all the axes of the ellipses (n_α, n'_α, n'_γ, n_γ).
3. Now do the same for a biaxial positive crystal in Figure 14.10b (n_α, n'_α, n_β, n'_γ, n_γ).
4. How does one recognize quartz in a thin section?

5. How can you distinguish between microcline and plagioclase?

6. How could you tell olivine from plagioclase in a thin section of basalt?

7. Pyroxenes and amphiboles are both chain silicates. List some diagnostic features that help to identify them and to distinguish between the two.

8. Calcite and quartz are both uniaxial. Give a list of diagnostic differences in optical properties between the two minerals.

Further reading

Barker, A. J. (2014). *A Key for Identification of Rock-forming Minerals in Thin Section.* CRC Press, Boca Raton, FL.

Deer, W. A., Howie, R. A. and Zussman, J. (2013). *Introduction to the Rock-forming Minerals*, 3rd edn. Mineralogical Society, Twickenham.

Nesse, W. D. (2012). *Introduction to Optical Mineralogy*, 4th edn. Oxford University Press, Oxford.

Perkins, D. and Henke, K. R. (2003). *Minerals in Thin Section*, 2nd edn. Prentice Hall, Upper Saddle River, NJ.

Tröger, W. E. (1982). *Optische Bestimmung der gesteinsbildenden Minerale,* ed. H. U. Bambauer, F. Taborsky and H. D. Trochim, 2 vols, 5th edn. Schweizerbart, Stuttgart.

Vernon, R. H. (2004). *A Practical Guide to Rock Microstructure.* Cambridge University Press, Cambridge.

15 | Color

Color has been mentioned as a key property for identifying minerals in hand specimens (Chapter 5). Here we look in more detail at the causes of color such as color centers (in purple fluorite), crystal field transitions (in red almandine garnet), and charge transfer transitions (in rose quartz). Fluorescence and phosphorescence help identify minerals such as scheelite. Periodic submicroscopic features such as exsolution lamellae in plagioclase and stacking of spheres in opal produce striking color effects.

15.1 Overview

The color of a mineral is our perception of the wavelengths of light that are either reflected or transmitted through the material and that reach our eye. Color is one of the most striking features of minerals and is most readily observed (e.g., Loeffler and Burns, 1976). There are many reasons why a mineral displays a particular apparent color, all related to the interaction of light with the crystal. Light may be transmitted, absorbed, scattered, refracted, or reflected by a crystal (Nassau, 1980). As we will see, however, color is generally not a bulk property determined by the general structure, as for example is the refractive index, but rather depends on the trace elements present, or on mineral defects. For example, a mineral with the general composition Al_2O_3 may be white (as corundum, Plate 1e), red (as ruby, Plate 1f), or blue (as sapphire, Plate 1g), with only very minor differences in composition. The same is true for quartz, basically SiO_2, which can be colorless-transparent (Plate 2a), brown (as smoky quartz, Plate 2b), black (as morion), purple (as amethyst, Plate 2c), yellow (as citrine, Plate 7c, 20c), pink (as rose quartz, Plate 20d), or green (as chrysoprase). A summary of different causes of color in minerals is given in Table 15.1.

15.2 Absorption

If white light is transmitted through a crystal without absorption, the crystal appears clear and colorless. If some wavelengths are preferentially absorbed, the combination of the remaining spectrum is perceived

as color. For example, in the corundum variety ruby, the colors violet, green, and yellow are preferentially absorbed, leaving a spectrum composed largely of blue and red that gives rise to the typical dark-red ruby color.

We have already discussed the interaction of electromagnetic radiation and atoms in Chapters 11 and 13. When highly energized electrons hit an atom, they displace inner electrons (e.g., from the K- to the L-shell), and when an electron returns to the ground state, the gained energy is emitted as X-rays. These inner electron transitions require high energies and are associated with very short wavelengths, but basically similar transitions occur at much longer wavelengths. Transitions in the energy levels of outer electrons can be in the visible range. For example, the energy of a light photon may be absorbed and used to displace an electron to a higher energy level, leaving a lower energy level vacancy (Figure 15.1a). Absorption is generally energy dependent and is particularly high for energies corresponding to electron transitions. When the photon returns to the ground state, radiation corresponding to the energy difference ΔE is emitted (Figure 15.1b). Absorption in the visible range is controlled largely by electron transitions between different energy levels, such as crystal field transitions, molecular orbital transitions, and transitions caused by defects called color centers. These transitions are discussed in turn in the remainder of this section. In the infrared range, molecular vibrations of water (H_2O) and carbon dioxide (CO_2) generally cause absorption. At yet longer wavelengths, absorption is caused by

Table 15.1 Causes of colors in important minerals

Mineral name	Gem names[a]	Color	Origin of color[b]
Fluorite		Purple	Color centers
Halite		Blue, yellow	Color centers
Topaz	Topaz	Blue, yellow	Color centers
Corundum	Ruby	Red	Cr^{3+} (CF)
	Sapphire	Blue	$Fe^{2+} \rightleftharpoons Ti^{4+}$ (CT)
Garnet	Spessartine	Yellow-orange	Mn^{2+} (CF)
	Uvarovite	Green	Cr^{3+} (CF)
	Almandine	Dark red	Fe^{2+} (CF)
Beryl	Emerald	Deep green	Cr^{3+} (CF)
	Aquamarine	Blue-green	$Fe^{2+} \rightleftharpoons Fe^{3+}$ (CT)
	Morganite	Pink	Mn^{2+} (CF)
	Heliodore	Yellow	$O^{2-} \rightleftharpoons Fe^{3+}$ (CT)
Cordierite	Iolite	Blue	$Fe^{2+} \rightleftharpoons Fe^{3+}$ (CT)
Kyanite	Kyanite	Blue	$Fe^{2+} \rightleftharpoons Ti^{4+}$ (CT)
Tourmaline	Rubellite	Pink	Mn^{3+} (CF)
	Verdelite	Green	$Mn^{2+} \rightleftharpoons Ti^{4+}$ (CT)
Quartz	Amethyst	Violet	$Fe^{3+}-O^-$ (color centers)
	Citrine	Yellow	Fe^{3+} (color centers)
	Rose quartz	Pink	$Fe^{2+} \rightleftharpoons Ti^{4+}$ (CT, inclusions)
	Smoky quartz	Brown	Al (color centers)
Olivine	Peridote	Green	Fe^{2+} (CF)
Turquoise	Turquoise	Blue	Cu^{2+} (CF)

[a] Separate gem names for colored minerals are indicated.
[b] CF, crystal field transition; CT, charge transfer (molecular orbital) transition.

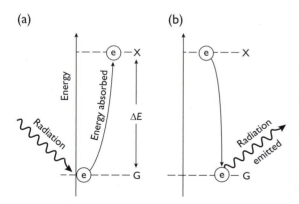

Fig. 15.1 (a) Radiation that displaces electrons (e) from a ground state (G) into a higher energy level (X) causes preferential absorption. (b) When the electron returns to the ground state, radiation is emitted.

lattice vibrations. We will discuss some of these effects in more detail in Chapter 16, but introduce here some concepts that are pertinent to color in minerals. A typical absorption spectrum for beryl is shown in

Figure 15.2. Note the two large absorption peaks that occur in the visible range.

Crystal field transitions

Crystal field transitions are electronic transitions between partially filled 3d-orbitals of transition elements (titanium, vanadium, chromium, manganese, iron, cobalt, nickel, copper) or partially filled 4d-orbitals in lanthanides and actinides. These elements are particularly active in color development because the outer orbitals of these elements contain unpaired electrons. Iron, for example, is a common component and is responsible for the color of many minerals. Since the presence of these elements causes color, they are known as *chromophore elements*.

Let us look at an iron atom with six d-electrons distributed over five orbitals. Three d-orbitals (d_{xy}, d_{xz}, and d_{yz}) have the shape of butterflies and are diagonal between two axes (Figure 15.3a). Two d-orbitals ($d_{x^2-y^2}$ and d_{z^2}) are aligned with the axes

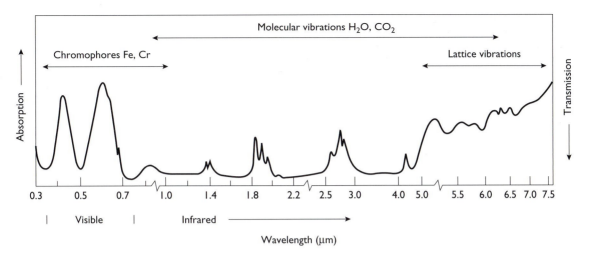

Fig. 15.2 Absorption spectrum of beryl.

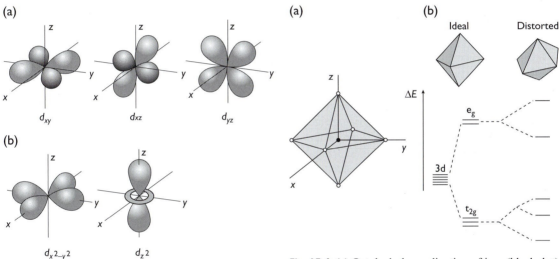

Fig. 15.3 The five d-orbitals have different geometries (see also Figure 2.3c).

Fig. 15.4 (a) Octahedral coordination of iron (black dot) by oxygens (open circles). (b) Repulsion of d-electrons by oxygen ions causes splitting of energy levels. The diagonal orbitals t_{2g} have lower energies than those aligned along the axes, e_g. If the octahedron is distorted further splitting occurs.

(Figure 15.3b). In an isolated atom, the energy levels of all states are identical. In a crystal structure, iron is surrounded by anions and they induce an electrical field about the cation; this is known as a "crystal field". In the case of a transition metal ion with partially filled orbitals there is a nonuniform interaction between d-orbitals and neighboring anions. Take, for example, Fe in octahedral coordination and surrounded by six oxygen atoms (Figure 15.4a). Oxygen atoms surrounding the iron ion cause electrons in diagonal orbitals (d_{xy}, d_{yz}, and d_{xz}), referred to as t_{2g} orbitals, to have a lower energy because they are

further apart from the O^-. Electrons in orbitals aligned with the axes $\left(d_{x^2-y^2} \text{ and } d_{z^2}\right)$, referred to as e_g orbitals, have a higher energy because they are closer to oxygen and negatively charged oxygen ions repel the electrons. The crystal field of an octahedral atom with d-electrons splits into two different energy levels in the case of undistorted octahedra (Figure 15.4b). The energy difference ΔE between the levels corresponds to wavelengths of photons in the visible light range. Thus, for example, a light photon that has

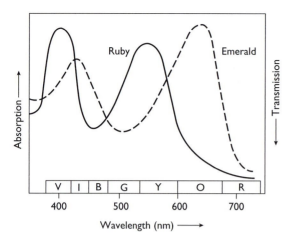

Fig. 15.5 Absorption spectra for ruby and emerald with absorption peaks in the visible range. V, violet; I, indigo; B, blue; G, green; Y, yellow; O, orange; R, red.

the energy ΔE may displace a d_{xy} electron into a d_{z^2} level and in the process be absorbed.

In general, there are several polyhedra that can exist in a mineral structure; also, coordination polyhedra are often distorted, and this distortion causes additional splitting of $3d$-orbitals, resulting in many energy levels and thus a complicated absorption behavior, although the splitting discussed above is prevalent.

Consider the examples of ruby (corundum, Plate 1f) and emerald (beryl, Plate 18b). Pure corundum has the composition Al_2O_3 and pure beryl the composition $Be_3Al_2Si_6O_{18}$. Neither of these pure varieties contains transition elements, and therefore they are colorless or white. However, in ruby and emerald, traces of Cr^{3+} substitute for octahedral Al^{3+}, causing crystal field splitting with two absorption peaks in the spectrum (Figure 15.5 and Figure 15.2). The corundum structure is ionic, and the crystal field splitting is strong. The splitting causes absorption with one peak in violet and a second one in yellow. Subtracting yellow and violet from the white spectrum leaves red, orange, and blue, producing the dark red color typical of ruby. In the beryl structure, as in all silicates, there is a covalent component to bonding and similar Cr^{3+} splitting is weaker than in corundum. Thus, absorption bands are displaced towards lower energies and higher wavelengths, with peaks in blue and orange, resulting in the complementary green color in emerald.

The crystal field interactions are influenced by several factors, including the specific transition element that is present, its oxidation state, the coordination and exact geometry of the site, and the type of bonding. The amount of the transition element also influences the strength of the color. For example, in ruby only 1–2% chromium creates a dark-red color, and emeralds typically contain only up to 0.5% chromium. In other minerals, crystal field transitions that produce color are due to the major elements present, such as iron (Fe^{2+}) in the garnet almandine ($Fe_3Al_2(SiO_4)_3$, dark-red color) and olivine (($Mg,Fe)SiO_4$, green color). The blue color of turquoise is attributed to crystal field transitions due to copper (Cu^{2+}). In tourmaline traces of iron (Fe^{3+}) and chromium (Cr^{3+}) produce a green color and traces of manganese (Mn^{3+}) a pink color. In tourmaline crystals from Brazil and Madagascar a variation of trace elements during growth and different concentrations on different faces can lead to spectacular color zoning (Plate 7d).

Molecular orbital transitions

Molecular orbital transitions (or charge transfer transitions) occur if valence electrons transfer back and forth between adjacent cations that have variable charges, sharing orbitals. Assume that differently charged cations occupy adjacent sites A and B. An electron of the ion with the lower charge on site A absorbs radiation and gains energy, allowing it to transfer to the site occupied by the ion of higher charge. When the electron falls back to its lower energy positions, the cations revert to their original charge. Common transitions are $Fe^{2+} \rightleftharpoons Fe^{3+}$ and $Ti^{4+} \rightleftharpoons Fe^{2+}$, and the energies associated with this "hopping" correspond to those of visible light photons. In both examples the energy difference matches the energy of red light that is absorbed, and the resulting color is blue. Mineral examples for $Fe^{2+} \rightleftharpoons Fe^{3+}$ transitions are beryl ($Be_3Al_2Si_6O_{18}$) as aquamarine (Plate 18a) and cordierite (($Mg)_2Al_4Si_5O_{18} \cdot nH_2O$), and for $Ti^{4+} \rightleftharpoons Fe^{2+}$ transitions examples are kyanite (Al_2SiO_5) and corundum (Al_2O_3 (as sapphire)). In all these cases the chromophore elements occur only as traces.

Color centers

In some minerals, color is caused by structural defects, most commonly vacancies or interstitial impurities that constitute *color centers*. In fluorite (CaF_2), for example, a fluorine atom may be missing because it was knocked out by high-energy radiation, or because

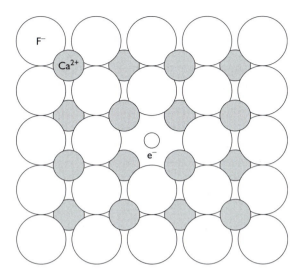

Fig. 15.6 A color center in fluorite is produced by an F^- vacancy filled by a free electron e^-.

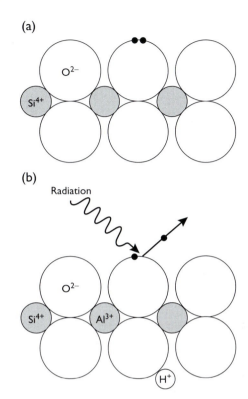

Fig. 15.7 Color centers in quartz. (a) Structure of ideal quartz. The black dots represent electrons on the oxygen anion. (b) A Si^{4+} is replaced by Al^{3+} and H^+. High-energy radiation removes an electron to balance the charge difference around O^{2-}.

of excess Ca^{2+} during growth. An electron then substitutes for F^- to maintain charge balance (Figure 15.6). This free electron is controlled by surrounding ions and can exist in different energy levels. Movement of electrons between these states can cause absorption colors and also *fluorescence*, which is emission of visible light of different wavelength.

Color centers also produce colored quartz, such as smoky quartz, amethyst, and citrine. In smoky quartz, some Si^{4+} are replaced by Al^{3+}, usually coupled with some interstitial H^+ to maintain neutrality (Figure 15.7). Radiation can expel an electron from an oxygen ion adjacent to Al^{3+}, and the resulting unpaired electron in oxygen can have different energy levels, similar to the free electron in fluorite. In amethyst and citrine, traces of iron produce color centers. Both colors are apparent in the sector-zoned variety ametrine (Plate 7c). Intense radiation (such as γ-rays) with an energy of at least ΔE_1 is necessary to dislodge electrons (Figure 15.8) to an excited state X, but less energetic radiation (such as sunlight) of an energy ΔE_2 suffices to overcome an activation barrier and return the electron to its stable ground state G. Colors in these minerals fade if they are exposed to sunlight.

15.3 Fluorescence and phosphorescence

When fluorite (and many other minerals) are irradiated with ultraviolet radiation, they re-emit light in the

visible range. This behavior, named after the observations on fluorite (CaF_2), is called *fluorescence*, and is generally applied to the process of emission of electromagnetic radiation produced by energy transitions, not just in the visible range. It is widely used for mineral identification (e.g., Schneider, 2007). Fluorescence is a special case of *luminescence* where energy transitions have been caused by incident electromagnetic radiation. There are other causes of luminescence, such as chemical reactions, radioactive decay, or an electric current.

Fluorescence and *phosphorescence* differ only in the amount of time it takes for electrons to return to their ground states. With fluorescence, vacant lower energy positions are filled within small fractions of a second. Phosphorescent materials, however, continue emitting light significantly after the exciting radiation has been turned off, sometimes for hours.

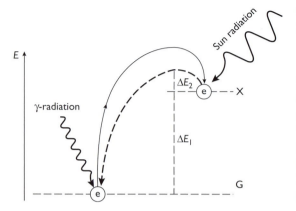

Fig. 15.8 Color centers are produced by high-energy radiation. It takes lower energy radiation to overcome the threshold to return the structure to the ground state.

Since the process of displacing electrons into higher energy configurations absorbs electromagnetic radiation, then, as the electrons return to the ground state, they emit radiation. The emitted radiation is always of lower energy than the radiation used to displace the electron, and of a specific wavelength, corresponding to the difference between the excited state and the ground state. This process is most transparent for X-rays and we have discussed it briefly in Chapter 11. Many transitions contribute to fluorescence in the visible range, and the spectra are not as sharp as those in the X-ray range. In addition, similarly to color, visible fluorescence depends critically on trace elements and defects but is nevertheless a diagnostic property used in mineral identification and even for prospecting in the field. Common activator elements are chromium, manganese, uranium, and tungsten, while other elements such as iron, cobalt, and nickel suppress fluorescence. Plate 18d shows a group of white fluorite crystals in ordinary light. When irradiated with short wavelength (ultraviolet) light, violet radiation is emitted (Plate 18e).

Another example of a mineral with strong fluorescence is scheelite ($CaWO_4$), an important tungsten ore that is difficult to distinguish from carbonates in hand specimens (Plate 18f) but is immediately recognized by bright white fluorescence when irradiated with ultraviolet light (Plate 18g). Diamond, when irradiated with high-energy X-rays, produces green-yellow fluorescence, with the intensity depending on defects in the structure (cf. Plate 23b). A list of mineral examples showing strong ultraviolet fluorescence or phosphorescence is given in Table 15.2.

15.4 Dispersion

Diamond is generally colorless but, as we noted in Chapter 14, this mineral has an unusually high dispersion of its refractive index with wavelength (0.062, see Table 14.3). White light is dispersed into a rainbow spectrum, and on a properly cut crystal (a "brilliant") the process of refraction is repeated many times (see Figure 34.3), resulting in a sparkling, brilliant color pattern (Plate 23a).

15.5 Luster

Scattering and reflection of light by crystals is perceived as *luster*, a qualitative term to describe the interaction of crystal surfaces and light. There are two main types of luster – metallic and nonmetallic – and there is some range between these two types. Minerals with metallic luster reflect light like metals and are generally opaque, even in thin sections. In crystals with metallic bonding, energy gaps between ground states and excited states of electrons are very small and variable. There are a large number of possible excited states with energies in the range of the visible spectrum. This means that light photons of most wavelengths are immediately absorbed at the surface of the crystal and then re-emitted as visible light, resulting in almost complete reflection.

In ionic and covalent crystals, these band gaps are not available and light enters the crystal. Minerals with nonmetallic luster are, in general, light colored and transmit light, at least to some extent. The different qualities of nonmetallic luster depend on the refractive index, with a high index associated with brilliant luster as in diamond. Nonmetallic luster can be divided into varieties such as *vitreous* (glassy luster), *pearly* (displaying iridescence effects parallel to cleavage planes), *greasy* (oily-like luster caused by microscopically rough surfaces resulting from irregular fracture), or *adamantine* (brilliant luster due to a high dispersion and refractive index).

15.6 Microstructure

Submicroscopic microstructural features, especially if they are periodic, often add very characteristic color effects. Several examples are well known for their

Table 15.2 Minerals with pronounced fluorescence and phosphorescence (*P*). Assumed activator elements for specific colors are indicated.

Mineral	Shortwave ultraviolet	Longwave ultraviolet
Halite	Red (Mn)	Red (w) (Mn)
Fluorite (P)	White/yellow (org), blue (Eu)	White/yellow (org), blue (Eu)
Calcite (P)	White, red (Mn), yellow, green (U), blue (Eu)	White, red (Mn), yellow, green (U), blue (Eu)
Aragonite (P)	White, yellow, green (U)	White, yellow, green (U)
Barite	White, red (w), yellow	White, red (w), yellow
Apatite	Orange-yellow (Mn), blue (Eu)	Orange-yellow (Mn)
Autunite	Green (U)	Green (U)
Scheelite	White, yellow, blue (W)	Brown, yellow
Corundum	Red (w) (Cr)	Red (Cr)
Anthophyllite	Red (Mn)	Red (Mn)
Benitoite	Blue	
Tremolite	Orange-yellow (Mn)	Orange (Mn)
Wollastonite	White, orange (Mn), yellow, blue	White, orange (Mn), yellow, blue
Willemite (P)	Green (Mn), yellow (Cu)	Green (Mn)
Microcline	Red, blue (Eu)	Blue (Eu)
Albite	Blue	White, blue
Sodalite	Orange (S)	Orange (S)
Sphalerite	Orange (Mn)	Red, orange (Mn), blue (Cu, Ag)
Uvarovite	Red (Cr)	
Witherite	White, yellow, blue	White, yellow, blue
Selenite	White, yellow, blue	White, yellow, blue

Note: w, weak fluorescence; org, organic traces.

striking visual appeal. Dispersed inclusions of hematite in the silica variety jasper, for example, add a sparkling golden-brown effect, called *tiger-eye*. Fibrous rutile inclusions in corundum (Figure 15.9a) produce a radiating pattern known as *star sapphire* (see Plate 32a). Dark carbon inclusions in some emeralds from Columbia produce a hexagonal star-shape pattern called *trapiche* (see Plate 32b). Color effects in some feldspars are due to a separation into lamellae of different composition during cooling of these minerals, which form a homogeneous solid solution at high temperature. The so-called *schiller effect* is observed in the alkali feldspar *moonstone* (with potassium-rich and sodium-rich lamellae), in the sodic plagioclase *peristerite*, and the more calcic plagioclase *labradorite* (both with sodium-rich and calcium-rich exsolution lamellae) (see Plate 21f). The lamellar structure is best seen with a transmission electron microscope (Figure 15.9b). The processes of compositional zoning of lamellae during cooling (referred to as *exsolution*) will be discussed in more detail in Chapter 20.

Amorphous opal often consists of minute spherules of silica gel arranged in a close-packed array which can be observed with a scanning electron microscope (Figure 15.10). The spherules are similar in size to the wavelength of visible light, and interaction of light with the array of spherules produces diffraction effects that result in a play of color that is green and blue for smaller spherules and red for larger spherules (Plate 20g).

15.7 Summary

Color in most minerals can be attributed to absorption of light photons and corresponding displacements of electrons. Crystal field transitions are electronic transitions between partially filled 3*d*-orbitals of transition elements. Molecular orbital transitions occur if valance electrons transfer back and forth between adjacent cations. Color centers are caused by structural defects, most commonly vacancies and interstitials. Some minerals emit visible light if irradiated by

(a)

(b)

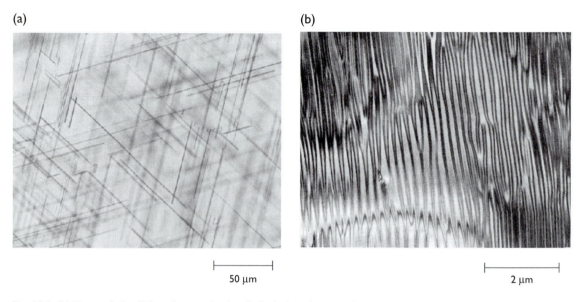

50 μm

2 μm

Fig. 15.9 (a) Transmission light micrograph of rutile inclusions in corundum, producing the star-shaped pattern in star sapphire (cf. Plate 32a) (courtesy T. Mitchell). (b) TEM image of the exsolution microstructure in labradorite from Labrador, displaying alternating sodium-rich and sodium-poor lamellae formed during exsolution (cf. Plate 21e) (courtesy H.-U. Nissen).

Fig. 15.10 Scanning electron microscope (SEM) image of the microstructure of opal, showing a regular stacking of spherical particles. The diameter of the spheres is approximately 3000 Å (courtesy R. Wessicken).

higher energy ultraviolet radiation, including fluorite and scheelite. This is called fluorescence. Periodic microstructural features in the range of visible light can produce color effects such as exsolution lamellae in feldspar (e.g., moonstone) or an array of spheres in opal.

Test your knowledge

1. Review the concept of crystal field splitting for a transition metal in octahedral coordination.
2. In ruby and in emerald, color is due to chromium. Why do they have different colors?
3. How are color centers produced in smoky quartz?
4. Why is fluorescence generally not a safe method for mineral identification?
5. In labradorite, what is the reason for the wide color range, between red and blue?

Further reading

Nassau, K. (2001). *The Physics and Chemistry of Color*, 2nd edn. Wiley, New York.

Robbins, M. (1994). *Fluorescence: Gems and Minerals under Ultraviolet Light*. Geoscience Press, Phoenix, AZ.

16 | Advanced analytical methods*

X-ray diffraction and optical microscopy are essential backgrounds for mineralogy. This chapter briefly introduces techniques for high-resolution imaging, advanced diffraction, and spectroscopic analysis to at least highlight some of the possibilities that are available, generally in more specialized facilities. Results from such studies have already been presented, such as transmission electron microscope images to show dislocations in deformed crystals, neutron diffraction spectra to document magnetic structures, and absorption spectra to reveal causes of color. For more details the reader is referred to an extensive literature.

16.1 Overview

Mineralogy relies on a quantitative characterization of minerals and many different techniques are available. We have discussed X-ray diffraction (see Chapter 11) and analyses with the petrographic microscope (see Chapters 13 and 14). In this chapter we will describe briefly some other methods. Reading it you will not become an expert, but at least you will have an idea of how to pursue more in-depth studies. Some introductory books that give you an introduction into advanced analytical methods are given in "Further reading" at the end of this chapter. You may want to skip the chapter for now and return to it after knowing more about mineral systems (Part V), in order to better appreciate the discussion of examples.

The petrographic microscope furnishes some immediate answers to mineralogical inquiries but there are obviously limitations. One of them is the spatial resolution, which is limited by the wavelength of visible light. For high-resolution imaging, electron microscopes provide unique opportunities. Two types of electron microscopes are applied. The *scanning electron microscope* (SEM) studies crystal surfaces at high magnification. It can be used to determine morphology and compositional variations on a very fine scale (<1 μm) (e.g., Reed, 2010). The *transmission electron microscope* (TEM) is used to investigate microstructures and defects, such as dislocations, antiphase boundaries, exsolution lamellae, and microtwins. Over 50 years, since its first use for mineralogical studies, the TEM has revolutionized mineralogy by documenting that many minerals are heterogeneous on a very fine scale. It has also led to the discovery of many new minerals that exist only as submicroscopic particles. Taking advantage of its high resolution it is possible to use the TEM for imaging individual atoms. There are other methods for imaging at the microscopic scale. Atomic force microscopy and X-ray tomography provide three-dimensional images with resolutions reaching the nanometer scale.

Mineralogists must be familiar with X-ray powder diffraction, a very important and reliable laboratory technique for mineral identification. X-ray powder diffractometers are available in many university geology departments, in most mining companies, and in materials science laboratories. While the method is used mainly for identification purposes, it can also be used to determine precise lattice parameters that may be indicative of composition and the history of formation of a mineral. *Crystal structure determination* is a more specialized application of X-ray diffraction. Originally it was based largely on diffraction data from single crystals. More recently, however, the quantitative analysis of powder diffraction spectra

---

* Optional reading for those who plan mineralogical laboratory work.

has become equally important (including the Rietveld method and pair distribution function). At one time, crystal structure analysis was a significant component of mineralogical studies. Today it is pursued mainly in chemistry, where new crystalline compounds are produced every day and methods have been streamlined and automated.

While X-ray powder diffractometers are widely used in research laboratories, large synchrotron facilities provide new opportunities. Taking advantage of the high intensity and collimation of synchrotron X-rays, combined with a large range of wavelengths, opened possibilities to study phase transformations of materials *in situ* at high temperatures and high pressures, even conditions corresponding to those at the center of the Earth. For special applications, neutron, rather than X-ray diffraction, has distinct advantages. Neutrons are barely absorbed by most elements and penetrate not only large samples but also pressure vessels and heating devices, making *in situ* experiments much easier. We have mentioned in Chapter 12 the significance of neutron diffraction for investigating magnetic properties.

Clearly the *chemical composition* of a mineral is of utmost importance, and many techniques are available for such studies. Traditionally, minerals were dissolved in acids such as hydrochloric or hydrofluoric, and various reagents were added to the solution for gravimetric, calorimetric, or titrational analyses of different elements. Today wet-chemical analyses are often replaced by physical methods. One example is X-ray fluorescence, which is applied widely in quantitative analyses of solid minerals and rocks and is very sensitive for trace elements. Perhaps the most important instrument for modern mineralogical research is the electron microprobe. It is a specialized SEM and has become an indispensable tool for determining local compositions of a crystal, or variations within a crystal, on areas less than 5 μm in diameter. Newer field emission electron microprobes allow analytical precision of 1–10 parts per million at a spatial resolution of 200–700 nm.

There is a whole range of so-called *spectroscopic techniques* that use energy transitions of electrons or vibrations of atoms or molecules to derive information on the composition and structural state. With *infrared spectroscopy* the hydrogen speciation in minerals can be determined (H^+, OH^- or H_2O). *Raman spectroscopy* is used to investigate the nature of chemical bonding. The isotopic composition of crystals can

be determined with *mass spectrometers*, and also now with *ion-probes* on the nanometer scale. *Nuclear magnetic resonance* (NMR) is particularly sensitive to ^{29}Si and ^{27}Al isotopes and has been used to investigate Al–Si order in silicates, which is difficult to determine with X-ray diffraction. *Mössbauer spectroscopy* is applied to investigate the oxidation state of iron and much work has been done to determine the Fe/Mg distribution and ordering in pyroxenes.

In the following sections we discuss some principles and capabilities of the most important experimental methods, dividing the discussion into high-resolution imaging, diffraction, and spectroscopy. Admittedly, this is a very brief overview, illustrating for which purposes the various techniques can be used in mineralogy.

16.2 High-resolution imaging

Transmission electron microscopy

The resolution obtained with visible light with a wavelength range of 400–700 nm is at best ~1000 nm = 1 μm. This value is very large compared with the size of the unit cell, or even more with the size of interatomic distances, and for details we need to use radiation with a shorter wavelength. Short wavelength X-rays and γ-rays, unfortunately, cannot be used for imaging because the refractive index of matter for those wavelengths is close to 1.0 and this means that no lenses can be designed.

This situation leads us to consider accelerated electrons. The energy E of an accelerated electron is $E = Ve = \frac{1}{2}mv^2$, where V is the electric potential, e the charge of the electron, m the mass, and v the velocity. In diffraction, electron microscopy, and particularly in spectroscopy, different energy units are used to characterize radiation. Box 16.1 gives some correspondences and summarizes relationships. The wavelength λ of an accelerated electron is given by the de Broglie equation:

$$\lambda = h/mv = h/\sqrt{(2mVe)}$$
$$= \sqrt{(1.5/V(\text{in volts}))}\text{nm} \qquad (16.1)$$

where V is the voltage, h the Planck constant, and c the speed of light. (This is not strictly correct because the mass varies with velocity and a relativistic correction needs to be applied.) Wavelengths are very short (0.0037 nm for 100 kV and 0.00087 nm for 1000 kV). In contrast to X-rays of similar wavelength, accelerated

---

Box 16.1 Additional information: Energy conversions

$E = hf = hc/\lambda = V e$

where E is the energy, f is the frequency, λ is the wavelength, V is the voltage and

$h = 6.6237 \times 10^{-34}$ J s (the Planck constant)
$c = 2.998 \times 10^8$ m/s (speed of light)
$e = 4.803 \times 10^{-10}$ esu (charge of electron)

Joule	Cal	$f\,(\text{s}^{-1})$	λ (m)	eV
1	0.239	1.510×10^{33}	1.986×10^{-25}	0.624×10^{-19}
4.184	1	6.318×10^{33}	4.745×10^{-26}	2.612×10^{19}
6.624×10^{-34}	1.583×10^{-34}	1	2.998×10^8	4.133×10^{-15}
1.986×10^{-25}	4.746×10^{-26}	2.998×10^8	1	1.239×10^{-6}
1.602×10^{-19}	3.829×10^{-29}	2.419×10^{14}	1.986×10^{-25}	1

---

electrons have a charge, and their path can be changed by an electric field. Therefore, one can construct electromagnetic lenses that work in a manner analogous to that of optical lenses. The limit of resolution of an electron microscope is not determined by the wavelength but by the spherical aberration of the lens, and for the highest resolution microscopes it lies in the range of 0.1 nm, similar to the size of atoms. Electrons are easily absorbed by matter, requiring that electron microscopes operate in high vacuum and that only very thin samples be transmitted. Special techniques have been developed to prepare thin slices, less than 1 μm thick.

A typical ray path for a TEM is shown in Figure 16.1a. The fundamental optical principles of image formation by the objective lens are the same as those in a light microscope (see Figure 13.10a). By means of a condenser lens, a beam of electrons is focused on the specimen. After the beam has passed through an objective lens an enlarged first intermediate image is formed. This is further enlarged by an intermediate lens to form a second intermediate image, and again by a projector lens to a final highly enlarged image that can be viewed on a fluorescing phosphor screen, recorded on a photographic film, or captured digitally.

Contrast in the image arises from interaction of the accelerated electrons with the crystal. As you can imagine, the contrast is particularly high around defects in the crystal where the local charge balance is disturbed and electrons are deflected. Therefore, the

TEM is the ideal instrument to image structural defects such as dislocations (see, e.g., Figure 9.5), twin boundaries, and stacking faults (see Figure 9.6). In Figure 16.2a we show twin boundaries (TB) and antiphase boundaries (APB) in perovskite, all of which appear as bright lines. APBs display fringes.

However, since we are dealing with waves interacting with a crystal lattice, image formation is really more complex and can be viewed as a 2-fold diffraction process. A first diffraction pattern of the object is formed in the back focal plane of the objective lens. The waves continue their travel after the diffraction pattern to form a magnified intermediate image of the object. If the strength of the intermediate lens is reduced, a second diffraction pattern (Figure 16.1b), instead of a second intermediate image (Figure 16.1a), is produced in the back focal plane of the intermediate lens. This second intermediate diffraction pattern is then projected onto the imaging device by the projector lens. Figure 16.2b shows the single-crystal diffraction pattern of perovskite, corresponding to the image in Figure 16.2a. Each spot corresponds to diffraction on a lattice plane hkl and can be labeled accordingly. From the distances and angles between diffraction spots we can obtain lattice parameters. The crystal diffraction pattern displays the symmetry of the crystal viewed in that direction.

While contrast from defects is generally expressed as dark lines, and in some cases as bright lines on a gray background, diffraction on the lattice also

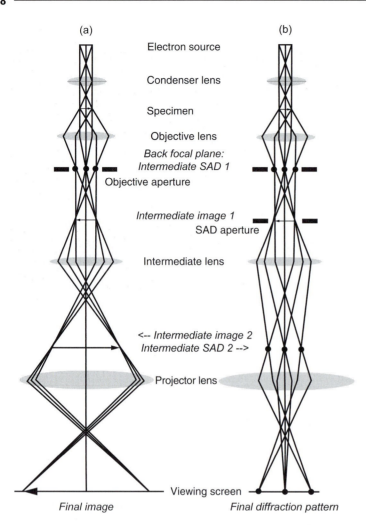

(a)   (b)

Electron source

Condenser lens

Specimen

Objective lens

Back focal plane:
Intermediate SAD 1
Objective aperture

Intermediate image 1
SAD aperture

Intermediate lens

<-- Intermediate image 2
Intermediate SAD 2 -->

Projector lens

Viewing screen

Final image   *Final diffraction pattern*

Fig. 16.1 Ray path in a transmission electron microscope. (a) Imaging mode. (b) Diffraction mode.

produces contrast that is present even in a perfect crystal. So-called *Bragg fringes* (labeled BF in Figure 16.2a) are associated with diffraction on a set of lattice planes and depend on orientation and thickness. They usually are more or less parallel to an edge. When a crystal is tilted relative to the beam, Bragg fringes move, whereas contrast from defects may change the appearance but will stay in a particular location.

As is illustrated in Figure 16.1, apertures of various sizes can be inserted in the back focal plane of the objective lens, as well as the plane of the first intermediate image. They serve different purposes. First they reduce the angular spread of the electron beam and thus spherical aberration. With the objective aperture (or diffraction aperture) we can reduce the number of *hkl* rays that contribute to the image. Usually just one *hkl* or only the primary electron beam is selected. With the selected area diffraction

(SAD) aperture we can select the part of the image from which we want to record a diffraction pattern. The use of these apertures is very important for the characterization of microstructures. We can obtain a structural identification of regions in the image and also investigate the effect of an *hkl*-diffracted wave on the image contrast to identify dislocation types and boundaries.

We demonstrate the use of the objective aperture for the case of the carbonate mineral dolomite (CaMg$(CO_3)_2$), which has small domains of a calcium-rich phase with an ordered superstructure. The strong spots on the diffraction pattern inserted in Figure 16.3 are due to the basic dolomite structure. The weak and elongated reflections halfway inbetween are due to the superstructure with a different ordering scheme. An image can be formed with the primary electron beam as illustrated in Figure 16.4a. This is called a

(a)

(b)

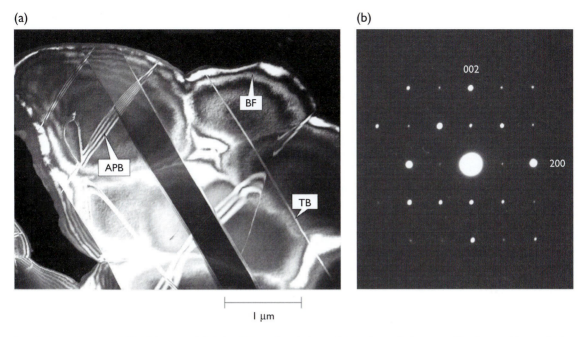

I μm

Fig. 16.2 (a) Darkfield TEM image with twin (TB) and antiphase (APBs) boundaries, as well as Bragg fringes (BF) in perovskite. (b) Corresponding diffraction pattern: the primary beam is in the center and some of the diffracted beams are assigned lattice plane indices (from Hu *et al.*, 1992).

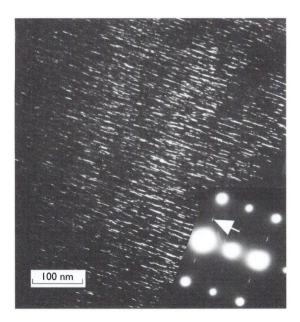

Fig. 16.3 Darkfield image of ordered domains in iron-rich, low-temperature dolomite. The inserted diffraction pattern shows strong spots due to the basic dolomite structure and weak elongated spots due to a superstructure. The image is taken with a weak ordering reflection in the diffraction pattern (indicated by arrow) and highlights the domains that have superstructure (bright areas) (from Wenk *et al.*, 1991).

brightfield image. Alternatively, an aperture can be placed around a particular diffraction spot *hkl*, excluding the primary beam (Figure 16.4b). Such a *darkfield* image contains primarily information from structural features contributing to the diffraction spot. For example, Figure 16.3 is a darkfield image of dolomite with the aperture placed on a weak diffraction spot (arrow) representing an ordering structure in dolomite. The darkfield image shows that the ordered superstructure is present only in narrow domains (bright areas).

In modern TEMs, lattice planes can be easily resolved, and, under favorable conditions, even individual atoms may be imaged. But at these high resolutions, contrast is by no means easy to interpret, since it depends greatly on focus and thickness. Usually experimental images are compared with simulated images, on the basis of diffraction theory. Figure 16.5a shows a brightfield high-resolution image of a staurolite crystal, obtained with many reflections and viewed along [001]. Figure 16.5b compares a corresponding simulated image for the same microscope conditions and a thickness of 100 Å, with good agreement. Obviously one does not image single atoms, but rather columns of atoms in the direction of the beam. The darkest spots correspond to columns of iron atoms.

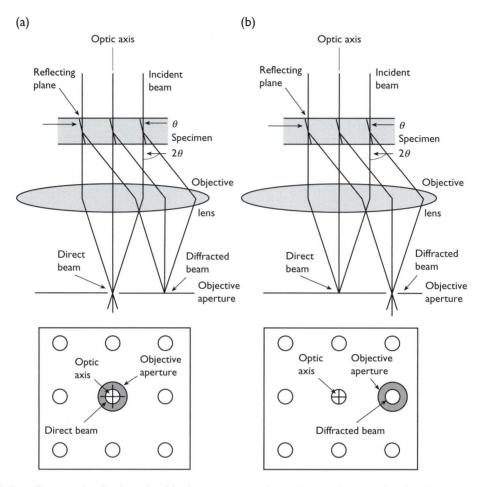

Fig. 16.4 Ray diagrams showing how the objective aperture can be used to obtain (a) a brightfield image if it is placed over the primary electron beam, and (b) a darkfield image if it is placed over a diffracted beam. (In modern microscopes darkfield images are generally not obtained by translating the aperture, but by tilting the primary beam, which is equivalent.) Below the ray diagrams is a view of the back focal plane perpendicular to the optic axis.

By viewing a crystal in different directions, it is possible to reconstruct the three-dimensional crystal structure from two-dimensional projections. Two sections of the three-dimensional electron potential distribution are shown in Figure 16.6a (xy section at $z = 0$) and Figure 16.6b (xy section at $z = 0.25$). To the right are corresponding sections of the crystal structure (Figure 16.6c,d). As is apparent, dark features in the experimental distribution correspond to atoms; not only cations (Fe^{2+}, Al^{3+}, Si^{4+}), but even O^{2-} are resolved. This direct method of structure determination is called electron crystallography. It is anticipated that in the future three-dimensional structure determinations will be performed on small domains of crystals, only 10 unit cells wide, an increase in resolution that is

important for gaining structural information on small and heterogeneous minerals such as clays.

Scanning electron microscopy

The scanning electron microscope (SEM) has only a single lens system which condenses the electron beam to a size of <1 μm on the sample. This beam is oscillated over the sample through the function of two cathode ray tubes. The signals are recorded electronically as a function of the position of the beam on the sample (Figure 16.7). Accelerating voltages range between 5 and 50 kV, which is considerably lower than voltages applied with a TEM. The lower voltage implies less sample penetration, and thus SEM images

(a)   (b)

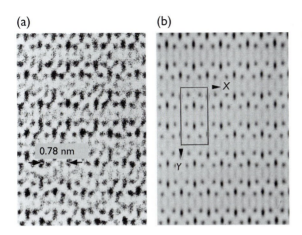

0.78 nm

Fig. 16.5 (a) High-resolution TEM image of a staurolite crystal, viewed along the *c*-axis. (b) Image simulation for a crystal thickness of 100 Å. The unit cell of staurolite is outlined in the simulation.

can provide only information about the few micrometers immediately below the surface. The images are composed of signals recorded at different spots, with no lens system beyond the sample. Because of this there is a large depth of focus, and the SEM is widely used for investigations of surface morphology.

There are various types of signals produced by the electron beam. *Secondary electrons* (SE) originate from a very thin surface layer of the specimen. Edges

and small particles have the highest SE yield, and the image displays an exaggerated topography with many details, as in the polished section of plagioclase from a volcanic rock, displaying the porosity around the large crystal, shown in Plate 19a-SE. Some electrons are *backscattered* (BE) after interacting with the sample surface. The intensity of the backscattered signal depends on the material (crystals with high atomic numbers scatter more efficiently), the orientation of the crystals (a diffraction contrast), and the surface topography, which is largely a shadow effect in the ray path. Plate 19a-BE is a BE image illustrating oscillatory zoning of plagioclase with regions rich in calcium (high atomic number) appearing brighter.

There are other signals that are sometimes explored. Incident electrons produce electronic transitions in the sample. If the transitions involve inner electrons, characteristic X-rays are emitted that give information on chemical composition and the concentric zoning structure for calcium (Plate 19a-Ca) that correlates with the BE image (Plate 19a-BE). If transitions involve bonding electrons associated with color centers, visible light may be emitted (cathodoluminescence, CL), and indeed the CL image best depicts growth zones (Plate 19a-CL). In addition, backscattered electrons produce a diffraction pattern that can be used for identification and determination of crystal orientation (electron backscatter diffraction, EBSD).

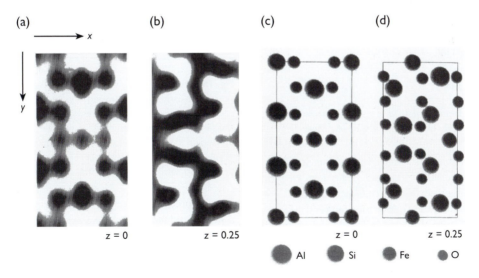

(a)   (b)   (c)   (d)

x

y

z = 0   z = 0.25   z = 0   z = 0.25

Al   Si   Fe   O

Fig. 16.6 Three-dimensional reconstruction of the structure of staurolite from five high-resolution images (after Downing *et al.*, 1990). Two *xz* sections through the experimental electron potential distribution are shown: (a) *z* = 0 and (b) *z* = 0.25. This is compared with sections through the crystal structure (c,d). The dark areas are regions of high electron density and thus correspond to atom positions. Fe, Si, Al as well as O are resolved.

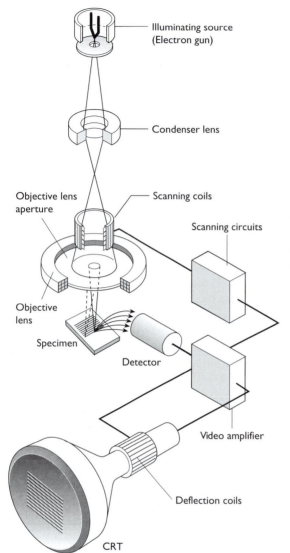

Fig. 16.7 Schematic view of a scanning electron microscope. CRT, cathode ray tube (courtesy JEOL).

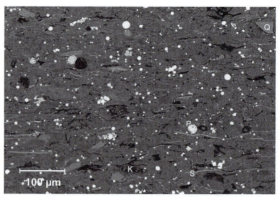

Fig. 16.8 SEM-BE image of Kimmeridge shale on which sheet silicates (S), quartz fragments (Q), pyrite aggregations (P), and kerogen clusters (K) can be recognized (see Vasin *et al.*, 2013).

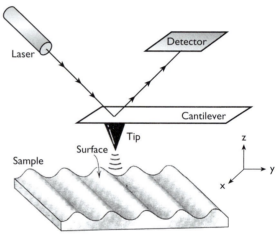

Fig. 16.9 Schematic view of an atomic force microscope. A tip, consisting of a diamond crystal, is attached to a cantilever and makes quasi-contact with the surface of the sample. The position of the cantilever is detected with an optical system. The sample is translated in *x*, *y*, and *z* with a piezoelectric system to keep the force between tip atoms and sample surface atoms constant.

Figure 16.8 is a BE image of shale on which sheet silicates (S), quartz fragments (Q), pyrite aggregations (P) and kerogen clusters (K) can be recognized.

The SEM has many applications in mineralogy and is the ideal instrument to document sample morphology and compositional variations. Compared with the TEM, larger surfaces can be surveyed.

Atomic force microscopy

An entirely different way to investigate the surface structure of a crystal, down to atomic resolution, is atomic force microscopy (AFM). A tip, consisting of a sharp fragment of diamond or silicon nitride crystal and mounted on a flexible cantilever, is brought into close contact (<20 Å) with the sample (Figure 16.9). If atoms in the sample and on the tip are in very close contact, then electrostatic repulsion by van der Waals forces occurs. This repulsion balances any applied force and the cantilever bends, rather than forcing the atoms in the crystal tip closer to the sample atoms.

Fig. 16.10 High-resolution atomic force microscope image of atoms in the surface of muscovite. The dots represent potassium atoms that are spaced at about 5.2 Å (courtesy W. Schmahl).

The deflection of the cantilever is determined with a laser beam that is reflected on the back side of the cantilever, recording the light signal with a split photodiode detector. The sample is scanned under the tip with piezoelectric translators in the x, y, and z directions. The force at the tip is held constant by lowering or raising the position of the tip, and the corresponding variations in z as a function of x and y produce an image that expresses the surface topography. The maximum resolving power of an AFM is 10 Å in the plane of the sample and 1 Å perpendicular to the surface. With this resolution it is possible to image the arrangement of individual atoms, as in the surface of a muscovite crystal (Figure 16.10). AFM images of van der Waals forces and magnetic forces have been shown in Figure 12.18 as an example for recording magnetic domains in magnetite.

X-ray tomography

The three-dimensional (3D) microstructure of minerals and mineral aggregates can be investigated with X-ray tomography. This technique relies on preferential absorption and is nondestructive. A focused monochromatic X-ray beam passes through a sample and an image is created (Figure 16.11). Either the X-ray signal is recorded directly or it is converted by a scintillator screen to visible light and then enlarged by an optical lens. Images are taken in different orientations by rotating the sample and from two-dimensional images the 3D structure is reconstructed. There are laboratory X-ray tomography instruments with a resolution of <1 μm and synchrotron X-ray beamlines that are dedicated to tomography and have high brilliance and tunable wavelength. Plate 19b is a 3D image of pyrite crystallites in shale, including an octahedral crystal. A special technique is nanotomography, with a resolution of <10 nm, and energy-sensitive detectors that can record chemical components, such as focusing on radiation near absorption edges (e.g., Withers, 2007). Plate 19c is an image of strands of iron (red) in a silicate perovskite aggregate produced experimentally by melting at high pressure and temperature, relevant for the lowermost mantle of the Earth.

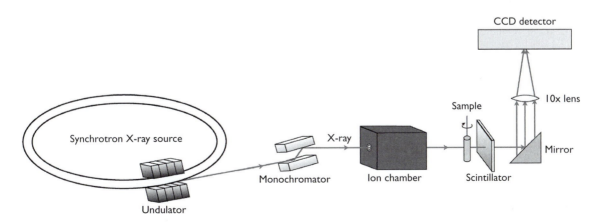

Fig. 16.11 Geometry of an X-ray tomography experiment with a focused X-ray beam, a sample that is rotated on a stage, and an image recorder.

(a)

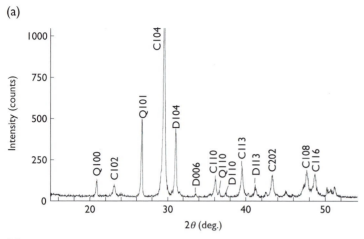

(b)

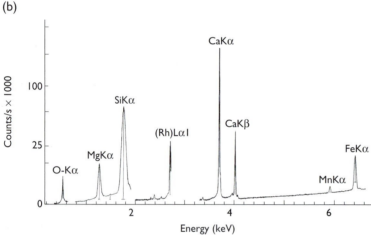

Fig. 16.12 Impure limestone, composed of 70% calcite (C), 20% dolomite (D), and 10% quartz (Q). (a) X-ray powder diffraction pattern used for phase identification. Some diffraction peaks are indexed. $CuK\alpha$ radiation. (b) X-ray fluorescence spectrum used for chemical analysis.

16.3 Diffraction

Synchrotron X-rays

X-ray diffraction has been discussed extensively in Chapter 11. In most experiments single crystals or powders are irradiated with monochromatic X-rays of wavelengths between 0.7 and 2 Å, depending on the anode material of the X-ray tube. In a diffraction pattern, the location of the peak (i.e., the Bragg angle θ in the case of a powder) is related to the lattice parameters, and the intensity of a diffraction peak depends on the positions of atoms in the unit cell. Powder diffraction is the standard method of mineral identification, which is a straightforward procedure for single-phase materials, but can also be used to identify mixtures of several minerals, including their volume proportions, for example by applying the Rietveld method (e.g., Rietveld, 1969) to analyze diffraction patterns such as that of an impure limestone with calcite, dolomite, and quartz (Figure 16.12a).

While X-ray powder diffractometers are widely available, another type of X-ray source, a synchrotron, is limited to large facilities. However, for special applications it may be necessary to use the very intense and highly focused X-ray beams of synchrotrons. Synchrotron X-rays are produced by a continuous release of energy when electrons are accelerated in a storage ring to speeds close to that of light, and their trajectory is deflected by a magnetic field. A synchrotron beam can have an intensity more than 10 orders of magnitude greater than a conventional X-ray tube and microbeams can be produced that probe areas less than a micrometer in diameter. Unlike X-rays produced by conventional X-ray tubes, synchrotron X-rays have a broad and continuous energy (or wavelength) range.

Monochromatic X-rays, used for diffraction experiments, can be produced by diffracting continuous X-rays on the lattice plane of a crystal with a specific d-spacing. In such monochromators, Bragg's law $2d_{hkl} \sin \theta = \lambda$ (equation 11.8b) is applied by fixing d and θ to obtain a certain λ value. Different wavelengths are used for different applications. Short wavelengths (high energy, e.g., 0.1 Å = 124 keV) provide high sample penetration without much absorption. For other applications, such as the investigation of bonding characteristics and surface structures, a continuous spectrum is often used (see the discussion of spectrometric methods).

Synchrotron X-rays have become particularly important in the investigation of phase transformations at high pressure. A small sample is squeezed between a pair of diamonds to pressures that can exceed 500 GPa (Figure 16.13a). Additionally, with a laser beam, the sample can be heated to temperatures exceeding 5000 K. With such diamond anvil cells, conditions can be reproduced that are equivalent to any place in the Earth, including the inner core. The X-ray beam penetrates the diamond and produces a diffraction pattern of the sample that can be recorded *in situ* (Figure 16.13b), from which the phases can be identified. With diamond anvil cells it has been established that a major component in the Earth's lower mantle is bridgmanite, a magnesium silicate with a perovskite structure, and that iron in the inner core exists in a hexagonal close-packed form.

Naturally, access to synchrotrons is limited and relies on proposals that have to be submitted far in advance, yet thousands of researchers and many students take advantage of the opportunities. There are many synchrotron facilities around the world (see, e.g., www.lightsources.org/regions).

Neutron diffraction

Through electromagnetic interaction, X-rays are scattered by electrons surrounding the nucleus of an atom. As a result, heavy atoms scatter X-rays more efficiently than do light atoms. Correspondingly, absorption is high for heavy atoms and low for light atoms. Therefore, X-rays are not very suitable for investigating the atomic positions of light atoms such as hydrogen in the crystal structure, or for differentiating between atoms of similar atomic number (and similar number of electrons) – such as Si and Al, which are important components of many silicate minerals.

(a)

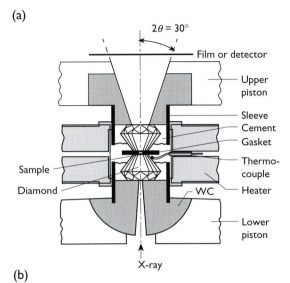

(b)

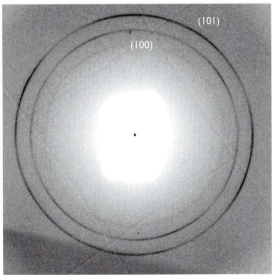

Fig. 16.13 (a) Diamond anvil assembly used for investigations at very high pressure. X-rays penetrate the diamond and sample without much absorption. The sample can be heated with an external heater (shown) or heated internally with a laser. WC is tungsten carbide (after Manghnani and Syono, 1987). (b) Diffraction image of hexagonal close-packed iron-9%Si at 403 GPa pressure and a temperature of 5910 K, corresponding to conditions of the Earth's inner core, measured at beamline BL10XU of SPring-8 in Japan (courtesy K. Hirose and S. Tateno; see also Tateno *et al.*, 2015).

Neutrons also have a wave character, with a wavelength range similar to that used for X-ray diffraction (see, e.g., Bacon, 1975). Neutrons have

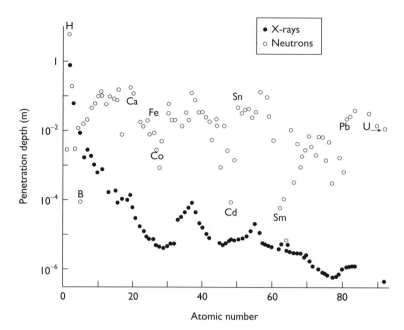

Fig. 16.14 X-ray and neutron penetration depths. The penetration depth corresponds to the thickness when the intensity has been reduced to 40%. Wavelength is 0.14 nm.

no charge, and their interaction with electrons is minimal. They interact with the nucleus via short-range nuclear forces, rather than electrical forces. Since nuclei are very small as compared with the distance between atoms, a beam of neutrons that travels through a crystal shows relatively little absorption for most elements, as compared with a beam of X-rays (Figure 16.14). Boron and cadmium are exceptions, and those elements are used for shielding. This reduction in absorption has the related disadvantage that scattering is much reduced. Compounding this problem is the low intensity of neutron beams, requiring long exposure times. For these reasons, neutron diffraction is used only to determine material properties that cannot be obtained by other means.

Elastic neutron scattering is used to analyze the equilibrium atomic structure, in much the same way as with X-rays. Since neutron diffraction depends on nuclear forces, different isotopes scatter differently. An extreme case is that of hydrogen and deuterium, where the scattering factors have opposite signs. The scattering power of some atoms and their isotopes, represented as spheres with different radii, is shown in Figure 16.15. This figure highlights again the monotonous increase in scattering power with atomic number for X-rays (top) but the very irregular behavior for neutrons. Isotopes with negative scattering amplitudes are white circles.

Since hydrogen has a very low X-ray scattering power, neutron diffraction is used to determine the position of hydrogen (usually substituting deuterium for hydrogen) in the structure of both organic and inorganic crystals. Among minerals, zeolites have been investigated extensively by neutron diffraction. The characterization of aluminum and silicon ordering (elements with very similar X-ray scattering factors) has been another application, for example in feldspars and zeolites. Neutrons, unlike X-ray photons, have a magnetic moment, and this property can be used to determine magnetic structures, such as the alignment of magnetic dipoles in manganese oxide (see Section 12.8). An additional benefit is that, because of the low absorption, environmental cells can easily be placed in the neutron beam, allowing phase transformations to be studied *in situ* at high and low temperature and pressure.

Diffraction is only one application of neutron scattering. Recording *inelastic scattering* provides information about bonding. Neutron tomography relies on absorption patterns. The resolution is much worse than X-ray tomography but very large samples can be investigated destruction-free, e.g., to obtain three-dimensional images of fossils in sedimentary rocks.

Similarly to synchrotron X-rays, neutrons are produced at large facilities. One method of producing neutrons is in a nuclear reactor by means of fission of atoms (largely uranium) in the reactor fuel

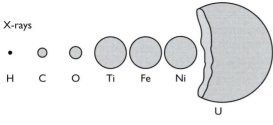

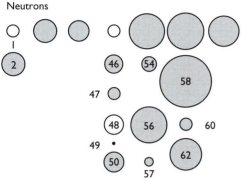

Fig. 16.15 Scattering amplitudes for some atoms and isotopes for X-rays (top) and neutrons (bottom). For neutrons, the top row represents natural abundance and below it some specific isotopes. White circles have a negative scattering factor.

(Figure 16.16a). Another method is with an accelerator. Accelerated high-energy protons collide with a heavy-metal target, such as tungsten, and during the collisions spallation neutrons are released (Figure 16.16b). The protons arrive in bursts, and spallation neutrons are generated in pulses of 20 to 50 per second. This pulsed character makes it possible to record the timespan between the generation of a neutron and its arrival in the detector. The velocity v is related to the wavelength λ according to the de Broglie relation

$$\lambda = h/mv \qquad (16.2)$$

The mass of the neutron m is 1.67×10^{-27} kg, h is the Planck constant (see Box 16.1). For a wavelength of 1 Å the velocity is about 4000 m/s. The time is proportional to the distance the neutron traveled and inversely proportional to its velocity, and therefore inversely proportional to the wavelength. Unlike reactor neutrons, where a monochromatic beam is usually selected, for spallation neutrons all wavelengths are used and the time-of-flight (TOF) spectrum is recorded. A single detector at a given Bragg angle θ records a whole diffraction spectrum simultaneously, with λ rather than θ as a variable. This makes TOF neutrons very efficient.

Access to neutrons is limited. Nevertheless, because of their unique advantages, mineralogists often travel to those facilities to perform experiments. Presently there are two facilities in the USA for neutron scattering experiments: the Spallation Neutron Source (SNS) at Oak Ridge National Laboratories and the National Institute for Standards (NIST) reactor in Gaitherburg, MD. In Europe, neutron diffraction is more popular, with large user facilities in the UK (ISIS), France (ILL, Grenoble, and LLB, Saclay), Germany (FRM, Munich), Russia (JINR, Dubna), and the European

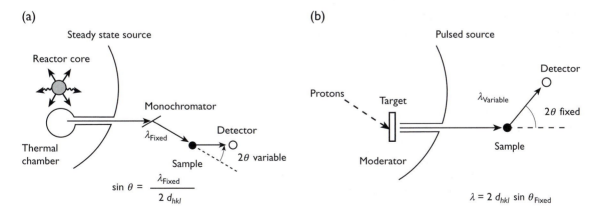

Fig. 16.16 (a) Neutrons produced in a reactor, and after passing through a monochromator producing monochromatic radiation, which is diffracted on the sample and recorded by a detector that scans the diffraction pattern. (b) Spallation neutrons are produced when pulsed high-energy neutrons hit a tungsten target. The time-of-flight of the polychromatic neutrons is recorded by a stationary detector.

spallation source under construction in Lund, Sweden. There are also large facilities in Japan (e.g., J-PARC) and Australia (ANSTO) (see, e.g., www .neutron.anl.gov/facilities.htm).

16.4 Spectroscopic methods

Spectroscopic techniques analyze energy differences between a ground state and excited states (Figure 16.17). We have already discussed the general principle in chapters on X-ray diffraction (see Chapter 11) and color (see Chapter 15). To summarize, an incident beam of radiation is either absorbed by matter or can cause emission or scattering of radiation from the material. Absorption and emission of energy arise if the incident radiation causes changes in the energy level (electronic, vibrational, or nuclear). The energy E is related to frequency f, or wavelength λ, according to the equation introduced earlier ($E = hf = hc/\lambda$; equation 16.1), where h is the Planck constant and c is the velocity of light in vacuum. Depending on the application, either energies (in kJ/mol or eV), frequencies (in Hz, i.e., in s^{-1}), wavelengths (in Å or nm), or wave numbers (the inverse of the wavelength, in cm^{-1}) are used (see Box 16.1). The energy differences between ground and excited states cover a wide range of the frequency/wavelength spectrum from radio waves (10^6 Hz/10^2 m) to γ-rays (10^{20} Hz/10^{-12} m). Energy transitions due to vibrations of molecular groups correspond to infrared radiation, while transitions of inner shell electrons correspond to X-rays. With spectroscopic methods the local structural environments of atoms in crystals can be investigated.

X-ray fluorescence

X-ray fluorescence (XRF) makes use of fluorescence of X-rays due to characteristic energy transitions. A high-energy X-ray tube (e.g., Rh or W radiation) produces polychromatic X-rays that are used to excite electrons in the sample and produce secondary X-rays, characteristic of the elements present in the sample (Figure 16.18). The X-ray spectrum is generally recorded with an energy-dispersive detector. In order to record X-rays from elements with low atomic numbers that are easily absorbed (such as Na–Si), measurements are often performed in a vacuum. Contrary to an electron microscope, large samples in the form of powders can be analyzed, similar to those used in a powder diffractometer. XRF has become the most important method for quantitative rock analysis. XRF is very sensitive for trace element analysis, in the parts per million (ppm) range.

Figure 16.12b shows an XRF scan of the same impure limestone sample for which an X-ray diffraction pattern is displayed in Figure 16.12a. Peaks are identified according to element and electron transition (remember that Kα means a transition from the L- to the K-shell, Lα a transition from M to L, etc.; see Figure 16.19). There are peaks for O, Mg, Si, Ca,

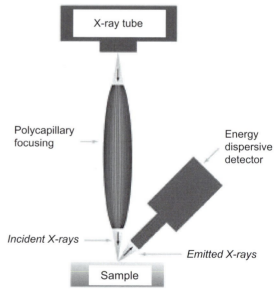

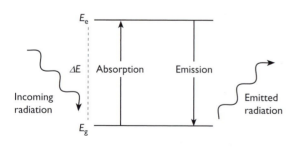

Fig. 16.17 Energy of incoming radiation is preferentially absorbed by creating an excited state with higher energy E_e. As the system returns to the ground state with energy E_g, the gained energy is emitted as radiation. The energies of absorption and emission are characteristic of structural features.

Fig. 16.18 Schematic of an X-ray fluorescence spectrometer. X-rays irradiate a sample, producing characteristic fluorescent X-rays that are analyzed for wavelength.

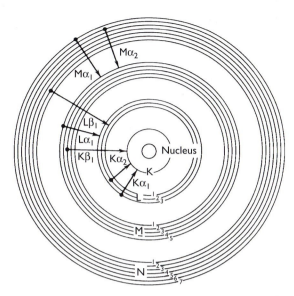

Fig. 16.19 Energy transitions producing X-ray fluorescence that can be used for chemical analysis.

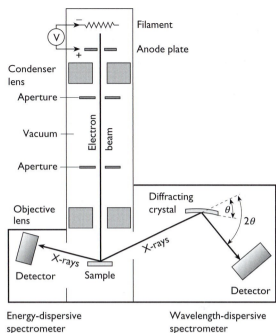

Fig. 16.20 Schematic of an electron microprobe, with an electron beam focused on the sample and producing characteristic X-rays that are recorded either with an energy-dispersive spectrometer or a wavelength-dispersive spectrometer.

which are the major elements in this sample, and small peaks for Mn and Fe, which are traces. The Rh peak originates from the X-ray tube. The XRF spectra give information about the chemical elements that are present, whereas X-ray diffraction (XRD) patterns identify the minerals in the sample. XRF spectra are quantified by comparing measured spectra with a standard.

Electron microprobe

An SEM that is optimized for quantitative chemical analysis, rather than imaging, is the *electron microprobe* (Figure 16.20). In the electron microprobe, electrons are accelerated to 10–20 keV. An electron beam about 10 μm in diameter is focused on the sample by means of a condenser–objective lens system. The sample consists of a flat polished surface, and emitted X-rays are analyzed. There are two types of detector. Energy-dispersive detectors use a semiconductor crystal that converts X-ray photons to electric pulses that vary with photon energy. The pulses are then processed and identified according to energy. This technique is fast but somewhat limited in resolution and sensitivity. A second detector type relies on Bragg's law. The X-rays of variable wavelength diffract on an analyzer crystal with known *d*-spacing (e.g., (0002) of graphite at *d* = 3.36 Å). A detector scans the θ angle range, but this time the wavelength is unknown and the *d*-spacing

is fixed. The electron microprobe provides relative numbers of atomic concentrations. In order to get absolute data, spectra of unknown samples need to be compared with known standards measured under identical conditions. Quantitative analyses with the electron microprobe are generally limited to elements with atomic number of sodium or higher, although some instruments can even analyze oxygen. Major elements, present in concentrations of 0.05 weight% or higher, are measured easily. Small trace amounts are more difficult to quantify. It is also difficult to distinguish between ions in different oxidation states, such as Fe^{2+} and Fe^{3+}. An electron microprobe can also produce maps such as the one shown for calcium recorded with an SEM (Plate 19a-Ca).

Mass spectrometry

Chemical analyses can also be done with a mass spectrometer. In this procedure, the sample is vaporized and the vapor is introduced into an ionization chamber (Figure 16.21), where the atoms are ionized. The ions are then accelerated, and the ion beam is injected

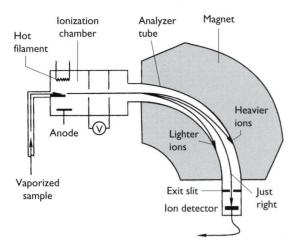

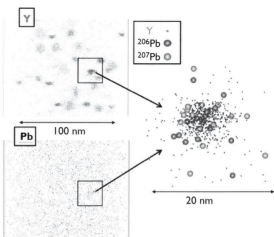

Fig. 16.21 Mass spectrometer in which ions of different weight are deflected by a magnet.

Fig. 16.22 Atom probe ion tomography with SHRIMP of zircon, illustrating clusters of Y and Pb at the nm scale (after Valley *et al.*, 2014).

into a magnetic field. The magnetic field deflects the path of the ions to follow the curvature of an analyzer tube. The amount of deflection depends on the mass and on the charge of the ions, lighter ions being deflected more than heavier ions. Those within a certain weight range will strike a detector at the far end of the instrument. Each ion that arrives in the detector produces a pulse in current and the strength of the electric current is related to the amount of ion present.

While originally mass spectrometers used large samples, more recently ion microprobes have been developed that focus an ion beam (typically cesium or oxygen) on the target material to produce a plasma that is then introduced into a mass spectrometer. Such ion microprobes (called SHRIMP for Sensitive High Resolution Ion MicroProbe) can provide isotope analyses at the micrometer scale (e.g., Holden *et al.*, 2009). Valley *et al.* (2014) were able to map nanoclusters (Figure 16.22) of lead inside zircons, and were able to determine ages of 4.4 gigayears on volumes of <500 nm^3.

Since mass spectrometers are sensitive to mass (rather than electron transitions), they can be used to measure relative amounts of different isotopes of elements in minerals and are applied mainly for radiometric dating and determination of stable isotope distributions.

Infrared and Raman spectroscopy

Infrared (IR) and Raman spectroscopy both involve the use of light to investigate the vibrational behavior of crystals (e.g., the stretching of bonds, and the

rotations of molecules such as water). Vibrational spectra occur typically in the approximate energy range 0–80 kJ/mol (0–4000 cm^{-1}). They give information on structural properties such as symmetry, bond lengths and angles, and coordination polyhedra. These spectra are also used widely in identification of molecules and of coordination polyhedra in crystals with strong covalent bonds, such as H_2O, CO_3, SO_4, and SiO_4.

In *infrared absorption spectroscopy*, broadband IR radiation is passed through a sample and the intensity of the transmitted light is measured as a function of wavelength (Figure 16.23a). Absorption occurs at specific frequencies (or wave numbers, w, measured in cm^{-1}), corresponding to the energy differences ΔE between the vibrational energy levels.

In a *Raman experiment* an incident monochromatic laser beam with a fairly high energy in the visible range (20 000 cm^{-1} wave number) is passed through the sample and the scattered light is analyzed by a spectrometer (Figure 16.23b). The incident photon with energy E excites the investigated system (molecule or crystal) to a short-lived ($<10^{-14}$ s) "virtual state" V_0 or V_1, which decays with the release of a photon (Figure 16.24, top). Most of the incident light is scattered with no energy change, i.e., producing a strong signal at the energy E of the incident light (Rayleigh scattering) (Figure 16.24, bottom). But some of the incident photons may gain or lose a small amount of energy ΔE from the vibrational modes in

(a)

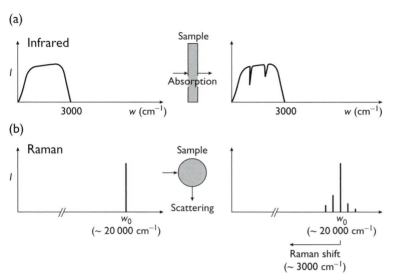

(b)

Fig. 16.23 Schematic representation of (a) an IR experiment with incident light (I) being absorbed by the sample and a detector measuring the change in intensity with frequency (in wavenumbers, w) and (b) a Raman experiment with incident laser radiation of wavenumber w_0 entering the sample and the scattered light then analyzed for energy (wavenumber w).

the crystal. Two vibrational modes in Figure 16.24, top, $n = 0$ and $n = 1$, are raised to energies V_0 and V_1, respectively, by the incident photons, releasing photons with energies $E + \Delta E$ and $E - \Delta E$. This gives rise to peaks at $E + \Delta E$ (anti-Stokes lines) and $E - \Delta E$ (Stokes lines) in the spectrum (Figure 16.24, bottom). (There are also higher-level vibrational modes, producing additional peaks.) Since the population of

molecules is larger at $n = 0$ than at $n = 1$, Stokes lines are always stronger than anti-Stokes lines, and it is customary to measure Stokes lines in Raman spectroscopy. Only the absolute energy shift with respect to the Rayleigh line is given, and, as in IR absorption spectroscopy, these shifts are usually expressed as wave numbers (cm^{-1}).

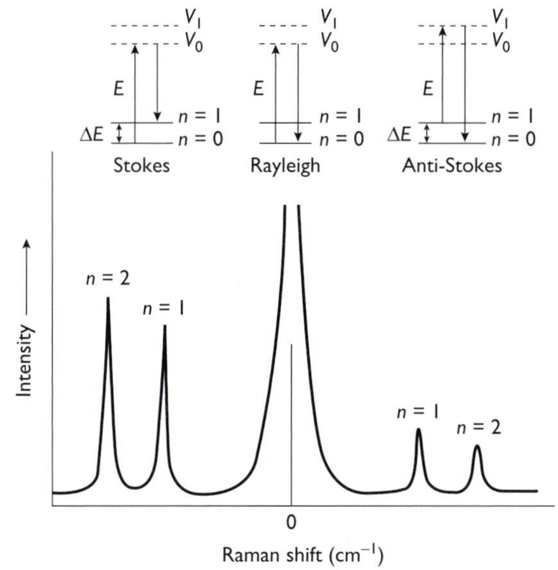

Though there are similarities between the two techniques, there are also fundamental differences between them. IR absorption occurs if an electric field can excite a vibration (or, in other words, if the vibration induces an electric polarization). Raman scattering, on the other hand, takes place due to polarizability variations during vibrations. This difference is best understood with examples.

A simple case is the carbon dioxide (CO_2) molecule. (Carbon dioxide is not a mineral on the Earth but comprises large volumes on the surface of Mars.) Figure 16.25 displays the molecule in the ground state (a) and in two vibrational states (b and c). Atomic displacements as in (b) produce a shift in the charge

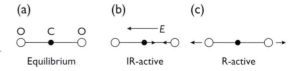

Equilibrium   IR-active   R-active

Fig. 16.25 Vibrational modes for the CO_2 molecule. (a) Equilibrium molecule. (b) Asymmetrical displacements produce an IR-active mode. (c) Symmetrical displacements of oxygen atoms produce polarizability and a Raman (R) signal.

Fig. 16.24 Origin of a Raman spectrum due to energy transitions, with a Rayleigh peak at the wave number of the incident radiation and Stokes and anti-Stokes peaks with negative and positive Raman shifts, respectively.

center and are therefore sensitive to an electric field E. On the other hand, there is no variation in polarizability. Thus this vibration is IR active but Raman inactive. In (c), charges are symmetrically balanced but the polarizability of the molecule changes. This vibration is IR inactive but Raman active. Box 16.2 analyzes some features of infrared and Raman spectra of calcite.

There are numerous other applications of IR and Raman spectroscopy in mineralogy. For example, Raman spectroscopy has been applied extensively to identify gases inside fluid inclusions (such as CO_2, H_2S, CO, CH_4, and especially N_2 and O_2 which have no IR signature). The SiO_2 stretching frequency depends on the polymerization state of the SiO_4^{4-} tetrahedra and this frequency can be used to study the structure of silicate glasses and melts. IR spectroscopy has been applied to distinguish between OH in the structure and randomly oriented H_2O molecules – for example, as fluid inclusions or bubbles. Wet synthetic quartz displays a sharp absorption band at 1920 nm owing to molecular water as inclusions (Figure 16.28). Multiple weak absorption peaks at 2250 nm are caused by OH groups substituting for oxygen. Raman spectroscopy has been used in the destruction-free analysis of gemstones, particularly in determining whether gems are naturally grown or synthetic. It is also applied to identify polymorphs of minerals, for example quartz versus coesite or graphite versus diamond, and this was critically important for recognizing ultra-high-pressure metamorphism.

Raman scattering peaks are much sharper than IR absorption peaks (e.g., Figure 16.26) and the low-frequency vibrations (below $c.$ 400 cm^{-1}) are specific to the structures. On the other hand, IR spectra are easier to interpret quantitatively and are preferred in determinations of, for example, the amount and speciation of water in minerals. In a loose sense, IR spectroscopy is used to determine how much of a vibrational molecule (e.g., H_2O or OH) is present in a structure, whereas Raman scattering gives information about the detailed structure causing the vibrations.

X-ray absorption spectroscopy

We have already mentioned that energy transitions of inner shell electrons result in emission of radiation in the X-ray wavelength range and that this phenomenon is used for chemical analysis. Overall, X-ray absorption by matter increases with increasing wavelength (decreasing energy), but there are some sharp peaks in the absorption spectrum at energies corresponding to displacements of electrons. The ejection of an electron from the ground state is not a simple process if an atom is surrounded by neighboring atoms and depends on the arrangements of those neighbor atoms. Therefore, absorption edges are not sharp but display a fine structure with regular modulations (Figure 16.29). While the location of the absorption edge is characteristic of the element, details in the pre-edge depend on the oxidation state and details in the smoothly undulating decreasing intensity provide information about the local structural environment of the absorbing atom (XANES, X-ray absorption near edge structure). Further away from the edge, modulations arise from interference effects of scattered electron waves with neighboring atoms. Such an extended X-ray absorption fine structure (EXAFS) spectrum, generally measured with synchrotron X-rays with tunable energy, can be interpreted by inverting the pattern, thus providing a radial distribution function. This curve gives the probability of finding a second atom as a function of distance from the absorbing atom.

For example, manganese oxides have characteristic octahedral coordination with Mn–O distances of 1.9 Å and correspondingly a large peak in the distribution function (Figure 16.30). There are additional peaks at 2.8 Å and 3.4 Å corresponding to Mn–Mn distances (indicated by arrows). The 2.8 Å distance corresponds to the distance of cations in edge-sharing octahedra (e.g., as illustrated for the NaCl structure in Figure 2.10c). The 3.4 Å distance corresponds to the distance of cations in octahedra linked over corners. This will later become more evident, when we discuss crystal structures of the oxide minerals. In the structure of rutile (see Figure 27.7b) each cation has two 2.8 Å cations and four 3.4 Å cations as closest neighbors. In manganese oxides there is a great deal of variety, with layers, tunnels, and frameworks of octahedra (see Figure 27.8). In each mineral the relative numbers of edge-sharing and corner-sharing octahedra are different, and correspondingly so are the relative intensities in the radial distribution function derived from EXAFS which can thus be used to identify the minerals, even if they are only present in thin crusts such as dendrites.

Figure 16.31 shows an EXAFS spectrum of the clay mineral nontronite with a peak at 1.4 Å corresponding to an octahedral Fe–O distance, whereas the peak at 2.7 Å corresponds to an Fe–Fe distance. It would be

Box 16.2 Additional information: Infrared and Raman spectroscopy of calcite

A more complicated case illustrating the difference between infrared (IR) absorption spectroscopy and Raman scattering spectroscopy is provided by the CO_3^{2-} groups in crystals of carbonates. In crystals with molecular ions (e.g., carbonates, borates, sulfates, and others) the molecular groups have strong covalent bonding and are only weakly disturbed by neighboring cations. Distortions of the CO_3^{2-} group gives rise to high-frequency vibrations within the molecule (so-called *internal vibrations*). These vibrations have energies that are very characteristic of the molecule and are easy to identify. There are also low-frequency vibrations owing to the interaction of the molecule with other ions. These *external modes* consist not of distortions of the molecular group but rather of librations (rotations) or translations of the whole group with respect to the other ions of the crystal. The external modes are structure specific but are more difficult to attribute.

Figure 16.26 compares an IR absorption spectrum (a) and a Raman scattering spectrum (b) of calcite. The IR has some low-frequency absorption bands and Raman some scattering peaks that are characteristic of calcite. There are also peaks at high frequencies due to the internal modes of CO_3^{2-}. The free triangular CO_3^{2-} group has four internal modes of vibration and thus four different vibrational frequencies (Figure 16.27). In a calcite crystal there are two CO_3^{2-} groups in the unit cell, and vibrations are coupled with two possibilities for each mode, one in which the pairs vibrate the same way (left) and one where the vibrations are opposite (right). A stretching of oxygen atoms (Figure 16.27a) does not polarize the molecule, and there is no infrared absorption for the 1087 cm^{-1} frequency. However, if the two CO_3^{2-} groups stretch in tandem, polarizability produces a strong Raman peak at 1088 cm^{-1} (peaks are slightly shifted, compared to the free CO_3^{2-} frequency, due to the influence of the environment in a crystal). Mode 2 (Figure 16.27b) shifts oxygen charges relative to carbon, and, if two groups are vibrating in the same direction, this produces a shift in overall charges and absorption of electromagnetic waves. This mode is sensitive to IR, though not to Raman. Indeed, there is a signal near 879 cm^{-1} in the IR spectrum only. The two remaining modes (Figure 16.27c,d) produce both an IR and a Raman signal, depending on whether the displacements are opposite (IR) or equal (Raman). There are peaks for these modes, both in the IR and in the Raman spectra near 1432 cm^{-1} and 714 cm^{-1}.

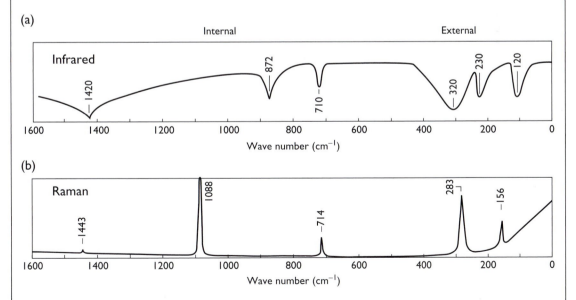

Fig. 16.26 (a) IR absorption and (b) Raman spectrum of calcite. The vibrational modes of CO_3 groups are indicated (from Bischoff *et al.*, 1985).

Continued

Box 16.2 (cont.)

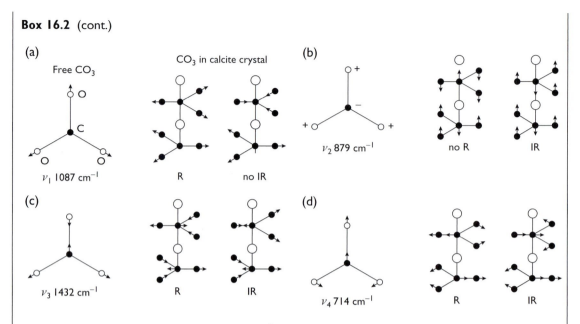

Fig. 16.27 Four different vibrational modes for CO_3^{2-} groups, both for free CO_3^{2-} and for CO_3^{2-} in calcite, with interaction between groups (after Nakamoto, 1997, and Wilkinson, 1973).

very difficult to obtain such structural information from normal X-ray diffraction experiments of the poorly crystalline clay mineral with limited long-range order.

Since the absorption edge is element specific, EXAFS spectra are successfully used to determine the radial distribution function and structural

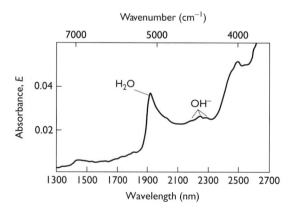

Fig. 16.28 Near-IR spectra of a typical synthetic quartz. The absorbance is measured parallel to c. Strong absorption at 1920 nm is due to molecular H_2O. Multiple peaks at 2250 nm are due to OH groups. Bands at 1410 nm and 2500 nm are due to both H_2O and OH (from Aines *et al.*, 1984).

environment for each element, not only for crystals but also for glasses, melts, and liquids.

Nuclear magnetic resonance

Atomic nuclei have a positive electric charge and can be thought of as spinning around an axis. Such spinning of a charged particle produces a magnetic field. The magnetic dipole that forms is described with the nuclear spin quantum number m_s. For nuclei with even mass numbers, the value of m_s is 0 and there is no magnetic moment. Nuclei with odd mass numbers have $m_s = n/2$, where n is an integer $\left(\text{e.g., } {}^{17}O : \frac{5}{2}, \right.$ ${}^{27}Al : \frac{5}{2}, {}^{29}Si : \frac{1}{2} \left. \right)$, and there is a magnetic moment.

In the absence of an external magnetic field, all spin states of the nucleus have the same energy. An applied magnetic field interacts with the magnetic moment of the nucleus and splits the energy levels into groups. If $m_s = \frac{1}{2}$ there are two energy levels, $-\frac{1}{2}$ and $+\frac{1}{2}$, and the nucleus behaves as a magnetic dipole. By placing a sample in a large static magnetic field and applying radio-frequency radiation, a split of the energy level is produced that increases with increasing magnetic field (Figure 16.32). Magnetic resonance occurs if the

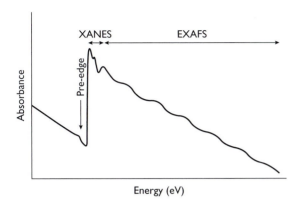

Fig. 16.29 Detailed view of the K absorption edge, as investigated with synchrotron X-rays. The spectrum can be divided into a pre-edge, the main absorption peak (XANES: X-ray absorption near edge structure) and the smoothly decreasing intensity (EXAFS: extended X-ray absorption fine structure).

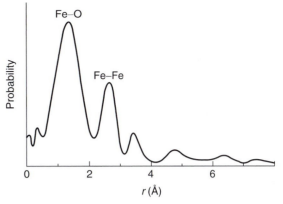

Fig. 16.31 EXAFS spectrum of nontronite, showing a pair distribution function with a peak at 1.4 Å corresponding to an octahedral Fe–O distance and a peak at 2.7 Å corresponding to Fe–Fe. It illustrates the environment around the iron atom (after Putnis, 1992).

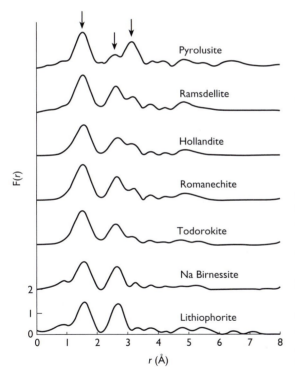

Fig. 16.30 Radial distribution function, obtained from an inversion of EXAFS spectra, for different manganese oxide minerals. The peak at 1.9 Å (arrow) corresponds to an octahedral Mn–O distance, peaks at 2.8 Å and 3.4 Å (arrows) correspond to Mn–Mn distances for edge- and corner-sharing octahedra, respectively (from McKeown and Post, 2001).

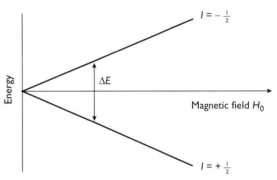

Fig. 16.32 Splitting of the $I = +\frac{1}{2}$ and $I = -\frac{1}{2}$ energy levels when a magnetic field $\boldsymbol{H}$ is applied to a ^{27}Al nucleus in a crystal.

applied radio frequency is equal to the energy difference between the spin levels. The exact resonance frequency of an isotope depends on the local chemical and crystallographic environment because electrons in the near environment shield the nucleus to varying degrees from the applied magnetic field. The resonance spectrum therefore contains information about the environment of atoms, and it can be used, for example, to determine the occupancy of an element in different structural sites.

Nuclear magnetic resonance (NMR) examines the properties of a specific isotope. In geological systems, ^{29}Si and ^{27}Al have been of particular interest. Investigations of feldspars, zeolites, and silicate glasses have provided information on Al–Si order and the

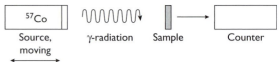

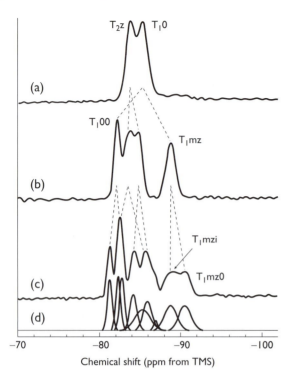

Fig. 16.34 Schematic of a Mössbauer spectrometer. A γ-ray source is vibrating relative to the sample, and a detector measures the absorbed γ-rays as a function of the vibration velocity.

Fig. 16.33 ^{29}Si NMR spectra for feldspars of various symmetry. (a) Synthetic monoclinic $SrAl_2Si_2O_8$ feldspar (space-group I2/c) with two nonequivalent Si positions, (b) anorthite $CaAl_2Si_2O_8$ at 400 °C (symmetry I $\bar{1}$) with four nonequivalent positions, (c) anorthite at 25 °C (symmetry P$\bar{1}$) with eight nonequivalent positions and corresponding splitting of peaks. In (d) the spectrum (c) has been deconvoluted into individual peaks. TMS is tetramethyl silane standard (from Phillips, 2000).

distribution of Al over tetrahedrally and octahedrally coordinated sites. Plagioclase can have different symmetries, depending on composition and thermal history, resulting in different NMR spectra (Figure 16.33). The synthetic $SrAl_2Si_2O_8$ feldspar is monoclinic, with two nonequivalent Si positions and thus two peaks in the ^{29}Si NMR spectrum (Figure 16.33a). Disordered anorthite ($CaAl_2Si_2O_8$) at 400 °C is triclinic (space-group $I\,\bar{1}$) with four nonequivalent Si positions (Figure 16.33b), and fully ordered anorthite at 25 °C is also triclinic but with a superstructure (space-group $P\,\bar{1}$) and eight nonequivalent Si positions and accordingly a more complex NMR spectrum (Figure 16.33c,d).

Mössbauer spectroscopy

NMR spectroscopy is concerned with small differences in nuclear spin energy levels in a magnetic field, corresponding to low-frequency radio waves. In contrast, Mössbauer spectroscopy analyzes the core nuclear energy levels of specific excited atomic nuclei, and these energy differences are very large, corresponding to γ-rays. When γ-radiation is absorbed or emitted by an atomic nucleus, the momentum of the system must be preserved. The recoil of the nucleus emitting γ-rays thus changes the energy (frequency) of the emitted radiation. Under special circumstances and only for certain nuclei, recoil-free absorption and emission may take place, with sharp resonance of the energy. This so-called Mössbauer effect can be used to probe nuclear energy levels sensitive to the local atomic environment of the nuclei and the magnetic and electric fields of the crystal. It is a short-range probe, and is sensitive to (at most) the first two coordination shells, but has an extremely high energy resolution that enables the detection of small changes in the atomic environment. The Mössbauer effect is present only in a few isotopes. For mineralogical systems the ^{57}Fe isotope is by far of most interest.

A Mössbauer spectrometer is relatively simple (Figure 16.34). It consists of a radioactive source that emits γ-rays. ^{57}Co decays to the excited state of ^{57}Fe ($m_s = \frac{5}{2}$), which then relaxes over a $\frac{3}{2}$ state to the ground state $\frac{1}{2}$. Because of selection rules, only the $\frac{3}{2} \rightarrow \frac{1}{2}$ transition has a Mössbauer effect, and it emits 14.4 keV γ-rays that are used to probe ^{57}Fe in the sample. The ^{57}Co source is attached to a vibration mechanism that produces a Doppler shift to the emitted γ-ray energy, which enables one to vary the energy continuously over a small energy range. The modulated γ-rays pass through the sample, where the component with the appropriate energy is absorbed. If the emitted energy coincides with a transition energy level in the crystal, resonant absorption occurs. A detector records the intensity of the signal as a function of the source velocity (vibration frequency).

If the nucleus in the source and the sample has the same environment and the source is stationary with respect to the sample, i.e., the nucleus is unperturbed,

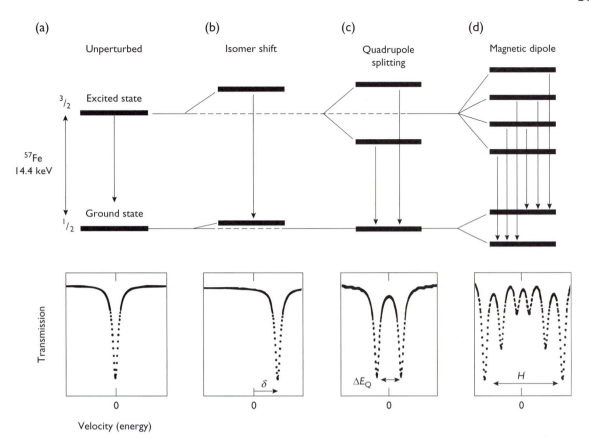

Fig. 16.35 Schematic illustration of interactions for ^{57}Fe nuclei, showing energy level diagrams and Mössbauer spectra for (a) unperturbed nucleus, (b) electric monopole interactions, (c) electric quadrupole interactions, and (d) magnetic dipole interactions causing shifts and splitting of peaks (after McCammon, 2000). ***H***, magnetic field.

resonant absorption will occur at the 14.4 keV energy (Figure 16.35a). If the environment is different, a shift occurs (isomer or chemical shift) that depends on the energy differences between source and absorber nuclei, such as difference in valence state, spin state, and coordination of the absorber atom (Figure 16.35b). This shift can be used to investigate the oxidation state and the coordination of ^{57}Fe in the mineral sample. In the spectrum one observes a single line that is shifted by δ from the reference point. We have already discussed splitting of energy levels of atoms with d-electrons in the context of colors (see Figure 15.4). This splitting depends on the valence and spin state of the absorber atom, as well as on coordination and distortion of the crystallographic site (Figure 16.35c). In the absorption spectrum a single line splits into several lines, according to the energy levels (quadrupole splitting). Finally, magnetic splitting arises through a dipole interaction between the nuclear

magnetic dipole moment and a magnetic field at the nucleus (Figure 16.35d).

The most important application of Mössbauer spectroscopy for minerals has been the determination of oxidation states of iron. Such analysis cannot be done with X-ray diffraction. Since the isomer shift increases with coordination number, even trace amounts of iron can be identified on individual sites. This is illustrated for garnet in Figure 16.36a. The spectrum has a large quadrupole doublet (at –0.7 and 3.0 mm/s) corresponding to Fe^{2+} (ferrous iron) in the dodecahedral site (92% relative area), and a smaller one (at 0.3 mm/s) corresponding to Fe^{3+} (ferric iron) in the octahedral site (8% relative area). The area is directly proportional to the amounts present.

Mössbauer spectroscopy has been used extensively to study Fe^{2+}/Mg distribution in pyroxenes (chain silicates) and other minerals as a function of temperature. In pyroxenes, iron and magnesium can occupy

(a)

(b)

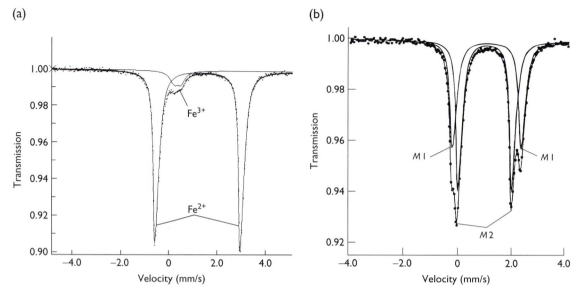

Fig. 16.36 Mössbauer spectra to determine oxidation state of iron. (a) Garnet from an eclogite (pyrope 66%–almandine 28%–grossular 6%) at room temperature. The spectrum consists of two quadrupole doublets corresponding to Fe^{2+} in the dodecahedral site (large peaks) and Fe^{3+} in the octahedral site (small peak) (courtesy C. McCammon). (b) Synthetic orthopyroxene (enstatite 20%–ferrosilite 80%) at room temperature. The spectrum has two quadrupole doublets corresponding to the $M1$ and $M2$ sites. The relative area of each doublet corresponds to the Fe^{2+} content, i.e., 55% Fe^{2+} is on the $M2$ site and 45% Fe^{2+} on the $M1$ site (after McCammon, 1995). Dots are the measured spectrum, lines define the deconvolution.

two octahedral sites with different environments, $M1$ and $M2$ (see Chapter 30). A Mössbauer spectrum for orthopyroxene at room temperature displays two quadrupole doublets, one corresponding to Fe^{2+} in the $M1$ site and the other corresponding to Fe^{2+} in the $M2$ site (Figure 16.36b). The relative area of each doublet is roughly equal to the proportion of iron on each site, hence this spectrum shows that approximately 45% of iron is in the $M1$ site and 55% in the $M2$ site.

16.5 Summary

This chapter provides a very brief introduction of some advanced techniques that are used to characterize minerals to determine their chemical composition, crystal structure, bonding characteristics, and microstructure. It will not make you an expert and you need to consult the extensive literature, as well as associate with scientists who are specialists. Below is a list of some techniques and their uses:

- Light microscopy for general surveys.
- X-ray diffraction for further identification, lattice parameters, and crystal structure.

- High-resolution imaging of microstructures with TEM, SEM, AFM, and X-ray tomography.
- Synchrotron X-rays for diffraction, spectroscopy, and tomography. An important application is with diamond anvil cells for ultra-high-pressure experiments.
- Neutron scattering for H in structures, Al–Si ordering, magnetic structure.
- Chemical analysis by XRF, microprobe, and SEM.
- Isotope analysis with mass spectrometers. A recent advance is SHRIMP.
- Spectroscopic methods to investigate structural environments:
  - IR/Raman are used for molecular vibration studies,
  - EXAFS to investigate coordination in materials with poor long-range order,
  - NMR to determine Al–Si distribution in minerals and glass,
  - Mössbauer to determine the iron oxidation state or distribution over atomic sites.

Test your knowledge

1. Why is thin section analysis the most important and efficient method of identification?

2. Which mineral group relies almost exclusively on identification by X-ray diffraction?

3. Diffraction angles and diffraction intensities are both indicative of specific structural parameters. Explain which parameters (review Chapter 11 if necessary).

4. What are some advantages of neutron diffraction? What are its disadvantages?

5. What are the most important techniques to determine the chemical composition of minerals and rocks?

6. Explain the difference between a TEM and an SEM and give some applications for each.

7. A wide range of spectroscopic techniques are used to characterize structural features of minerals. List some techniques according to the energy differences that they analyze.

8. Which spectroscopic techniques rely on absorption? Which ones rely on scattering?

9. Name techniques to investigate the oxidation state of iron.

Further reading

High-resolution imaging

Transmission electron microscopy (TEM)

Buseck, P. R. (ed.) (1992). Minerals and reactions at the atomic scale: transmission electron microscopy. *Rev. Mineral.*, **27**, 1–36.

Fultz, B. and Howe, J. M. (2012). *Transmission Electron Microscopy and Diffractometry of Materials*, 4th edn. Springer.

Goodhew, P. J., Humphreys, J. and Beanland, R. (2000). *Electron Microscopy and Analysis*, 3rd edn. CRC Press, Boca Raton, FL.

McLaren, A. C. (2005). *Transmission Electron Microscopy of Minerals and Rocks*, 2nd edn. Cambridge University Press, New York.

Reimer, L. and Kohl, H. (2008). *Transmission Electron Microscopy*, 5th edn. Springer.

Wenk, H.-R. (ed.) (1976). *Electron Microscopy in Mineralogy*. Springer-Verlag, Berlin.

Scanning electron microscopy (SEM)

Goldstein, J. I., Newbury, D. E., Eichlin, P., Joy, D. C., Lyman, C. E., Echlin, P., Lifshin, E., Sawyer, L. and Michael, J. (2007). *Scanning Electron Microscopy and X-ray Microanalysis*, 3rd edn. Springer-Verlag.

Reed, S. J. B. (2010). *Electron Microprobe Analysis and Scanning Electron Microscopy in Geology*, 2nd edn. Cambridge University Press, Cambridge.

Atomic force microscopy

Haugstad, G. (2012). *Atomic Force Microscopy.* Wiley, New York.

X-ray tomography

Alshibli, K. A. and Reed, A. H. (eds.) (2010). *Applications of X-ray Microtomography to Geosciences.* Wiley, New York.

Stock, S. R. (2008). *Micro Computed Tomography: Methodology and Applications.* CRC Press, Boca Raton, FL.

Diffraction

X-ray diffraction and crystal structure determination

Cullity, B. D. and Stock, S. R. (2001). *Elements of X-ray Diffraction*, 3rd edn. Prentice Hall, Upper Saddle River, NJ.

Egami, T. and Billinge, S. J. L. (2012). *Underneath the Bragg Peaks: Structural Analysis of Complex Materials*, 2nd edn. Pergamon, Oxford.

Glusker, J. P. and Trueblood, K. N. (2010). *Crystal Structure Analysis: A Primer*, 3rd edn. IUCr Texts, Oxford University Press, Oxford.

Mittemeijer, E. J. and Welzel, U. (eds.) (2013). *Modern Diffraction Methods.* Wiley-VCH, Weinheim.

Stout, G. H. and Jensen, L. H. (1989). *X-ray Structure Determination.* Wiley, New York.

Suryanarayana, C. and Norton, M. G. (2014). *X-ray Diffraction: A Practical Approach.* Springer, New York (reprint of 1998 edition published by Plenum Press).

Young, R. A. (1993). *The Rietveld Method.* Oxford University Press, Oxford.

Zolotoyabko, E. (2014). *Basic Concepts of X-ray Diffraction.* Wiley-VCH, Weinheim.

Synchrotron X-rays

Hemley, R. J. (ed.) (1998). *Ultra-high Pressure Mineralogy: Physics and Chemistry of the Earth's Deep Interior.* Reviews in Mineralogy, vol. 37. Mineralogical Society of America, Washington, DC.

Manghnani, M. M. and Syono, Y. (1987). *High Pressure Research in Mineral Physics.* Geophysical Monograph 39, American Geophysical Union.

Willmott, P. (2011). *An Introduction to Synchrotron Radiation: Techniques and Applications.* Wiley, New York.

Neutron diffraction

Bacon, G. E. (1975). *Neutron Diffraction.* Oxford University Press, Oxford.

Wenk, H.-R. (ed.) (2006). *Neutron Scattering in Earth Sciences.* Reviews in Mineralogy and Geochemistry, vol. 63. Mineralogical Society of America, Washington, DC.

Spectroscopic techniques

X-ray fluorescence and X-ray spectroscopy

Hawthorne, F. C. (ed.) (1988). *Spectroscopic Mineralogy and Geology.* Reviews in Mineralogy and Geochemistry, vol. 18. Mineralogical Society of America, Washington, DC.

Hren, J. J., Goldstein, J. I. and Joy, D. C. (eds.) (1979). *Introduction to Analytical Electron Microscopy.* Plenum Press, New York.

Mass spectrometry

Becker, J. S. (2007). *Inorganic Mass Spectrometry: Principles and Applications.* Wiley, New York.

Infrared and Raman

Ferraro, J. R., Nakamoto, K. and Brown, C. W. (2002). *Introductory Raman Spectroscopy*, 2nd edn. Academic Press, New York.

Larkin, P. J. (2011). *Infrared and Raman Spectroscopy: Principles and Spectral Interpretation.* Elsevier.

Sherwood, P. M. A. (2011). *Vibrational Spectroscopy of Solids.* Cambridge University Press, Cambridge.

Vandenabeele, P. (2013). *Practical Raman Spectroscopy: An Introduction.* Wiley, New York.

Wilson, E. B., Decius, J. C. and Cross, P. C. (1980). *Molecular Vibrations: The Theory of Infrared and Raman Vibrational Spectra.* Dover Books, New York.

Nuclear magnetic resonance (NMR)

Levitt, M. (2008). *Spin Dynamics: Basics of Nuclear Magnetic Resonance*, 2nd edn. Wiley, New York.

Sanders, J. K. M. and Hunter, B. K. (1987). *Modern NMR Spectroscopy.* Oxford University Press, Oxford.

Mössbauer

Cranshaw, T. E., Dale, B. W., Longworth, G. O. and Johnson, C. E. (1986). *Mössbauer Spectroscopy and Its Applications.* Cambridge University Press, Cambridge.

17 | Mechanical properties and deformation

Chapter 12 introduced stiffness and compliance as properties relating an applied stress and resulting strain. However, this applied only to elastic strain resulting in a reversible distortion of the lattice. Here we address briefly the issue of plastic deformation of minerals, which is significant for structural geology, tectonics, and geodynamics. Topics are dislocations, intracrystalline slip, and mechanical twinning.

17.1 Stress–strain

In Chapter 12 we explored the intrinsic physical properties of minerals that uniquely relate physical quantities and are described by a single number or a set of coefficients. Density, thermal conductivity, and elastic properties are examples. These properties depend on composition and structure; they also may vary with temperature and pressure. The *mechanical* properties of minerals are more complicated because they are greatly modified during a crystal's history. In this chapter we will explore how crystals deform plastically.

We start our discussion with a few basic definitions. If we apply a compressive force F to the surface of a crystal, we impose a deformation apparent in a change in length (Δl) and correspondingly by a change in area (Figure 17.1a). Force per surface area is called *stress* σ and the resulting deformation ($\Delta l/l$) is called *strain* ε (see, e.g., Means, 1976). We have introduced stress and strain in Chapter 12 as second-rank tensors. Both are highly directional.

In Figure 17.1b we show the evolution of observed strain as a function of applied stress for a material. The stress–strain curve generally has three segments. In the first segment (regime 1), strain increases almost linearly with stress. If the stress is removed, the strain returns to its original value. In this reversible *elastic range*, deformation occurs by compression of bonds and can be described fully with the elastic tensor. No bonds are broken in this regime. In some directions the crystal compresses more easily than in others based on structural characteristics. The strain usually does not exceed 1%.

(a)

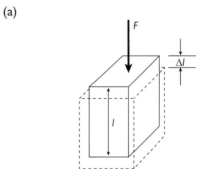

(b)

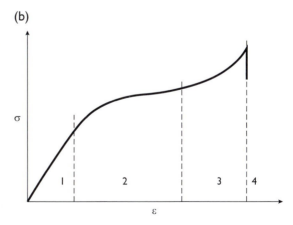

Fig. 17.1 Deformation of a crystal. (a) Application of a compressive stress causes shortening of length Δl. (b) Typical stress (σ)–strain (ε) curve with three regions, prior to failure stress. Region 1 is elastic deformation, region 2 is plastic deformation in easy slip, region 3 is caused by microstructural work-hardening, and failure occurs at 4.

If strain is increased further, in regime 2, the stress–strain curve starts to bend and the same increment in stress produces a larger increase in strain. In this region dislocations become active and move freely through the crystal, causing permanent changes in the microstructure and shape of the crystal. A release of stress in this regime does not return the crystal to its original shape. This irreversible deformation is described as *plastic* or *ductile*. Slip deformation is discussed in more detail in the next section, and in it we will make use of our earlier discussion of linear defects in Chapter 9. Recall that dislocations are linear defects, and that, on the basis of the local lattice distortion, they can be classified as either edge or screw dislocations. In regime 3, stress increases more rapidly again. This is because dislocations start multiplying and interfering with each other. This regime is called *work-hardening*. Ultimately, at strain level 4, the material has reached its ultimate strength and fails by fracturing.

17.2 Deformation by slip

Two observations led to the discovery of dislocations by G. I. Taylor in 1934. First, it had long been known that when a large stress is applied to a crystal, the crystal deforms on crystallographically defined slip planes (*hkl*) and displacements occur in crystallographically defined slip directions [*uvw*] (Figure 17.2). The slip direction is a lattice line in the slip plane. Second, it was observed that slip occurs at relatively low stresses, much lower than those required to break all bonds across a slip plane. With the propagation of dislocations, such universal instantaneous bond breakage is not necessary.

The latter hypothesis by Taylor (which is now universally accepted) is illustrated in Figure 17.3, where a

lattice is deformed to a new shape by applying a shear stress that propagates an edge dislocation across the slip plane. At each instant only one bond is broken, but at the end of the process (right side of Figure 17.3), the top half of the crystal has been displaced over the bottom half by one lattice unit. Such deformation involves less energy input, and therefore a lesser stress, than that required to produce an instantaneous offset. Analogies can be seen in the way that a large carpet can be moved easily by propagating a ruck (pucker or fold) across the carpet surface; or in the movement of a caterpillar (Figure 17.4), whereby a hump, beneath which the caterpillar's pedicles are not in contact with the ground, propagates along the length of the caterpillar, similar to the broken bonds in the dislocation core.

When an axial force *F* (in compression or tension) is applied to a crystal of cross-sectional area *A*, only that component of the force that is resolved on the slip plane ($A \cos \psi$) and in the slip direction ($\cos \lambda$) causes shear deformation (Figure 17.5a), leading to a relationship for the resolved shear stress τ of

$$\tau = (F/A) \cos \psi \cos \lambda \qquad (17.1)$$

where ψ is the angle between the force and the slip plane normal, and λ is the angle between the force and the slip direction. This law, named after E. Schmid (1924), indicates that the optimal orientations for the slip plane and slip direction are at 45° to the axial force, i.e., cos (45°) cos (45°) = 0.5. If the force is

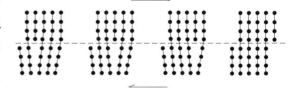

Fig. 17.3 Edge dislocation propagating left to right through a crystal and causing a permanent deformation. Note that only one bond at a time is broken in the dislocation core.

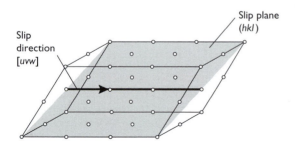

Fig. 17.2 Slip plane (*hkl*) and slip direction [*uvw*] defining a slip system in a triclinic crystal.

Fig. 17.4 Analogy of the movement of a caterpillar to dislocation slip.

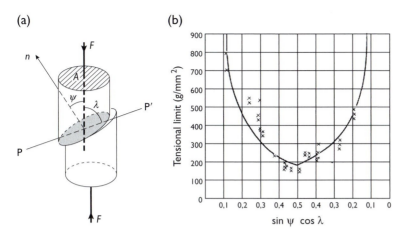

Fig. 17.5 Schmid's law. (a) Applying an axial stress causes shear deformation on the slip plane (*hkl*) (shaded) with a displacement in the slip direction [*uvw*] (P–P′). For explanation of symbols, see the text. (b) Schmid's law for a crystal of zinc deformed in different directions in tension. Solid line is Schmid's law (equation 17.1). (Redrawn from Rosbaud and Schmid, 1925.)

perpendicular or parallel to the slip plane, one of the cosines is zero, the resolved shear stress becomes zero, and the crystal cannot be deformed. Figure 17.5b illustrates the Schmid law for a crystal of zinc, the material on which it was first discovered. Deformation occurs only if the resolved shear stress reaches a critical value τ_c. The critical resolved shear stress τ_c is a material constant for a given crystal and slip system.

Crystals may have a single slip system or may have several slip systems. There are symmetrically equivalent slip systems (owing to crystal symmetry), and independent slip systems. Slip systems are described by general notation {*hkl*} ⟨*uvw*⟩, which indicates that the slip plane is a member of the form {*hkl*} and the slip direction is symmetrically equivalent to direction [*uvw*]. For example, {111} ⟨1$\bar{1}$0⟩ slip in fcc metals is not just on the (111) plane, but also on (11$\bar{1}$), (1$\bar{1}$1), ($\bar{1}$11), and ($\bar{1}$ $\bar{1}$ 1). (The planes ($\bar{1}$ $\bar{1}$ 1), ($\bar{1}$1$\bar{1}$), (11$\bar{1}$), and (11$\bar{1}$) are related by a center of symmetry, and are therefore equivalent.) In addition, on each octahedral plane the slip can be in three equivalent directions (e.g., for (111): [1$\bar{1}$0] [0$\bar{1}$1], and [$\bar{1}$01]), adding up to a total of 12 systems, not counting those related by centrosymmetry. (Review Chapters 7 and 8 for nomenclature.) While these 12 slip systems are symmetrically equivalent, they are not equivalent with respect to the applied stress and that system for which the resolved shear stress is largest will operate preferentially. If systems are not related by symmetry, for example basal slip (0001) ⟨$\bar{1}$2$\bar{1}$0⟩ and prismatic slip {10$\bar{1}$0} ⟨$\bar{1}$2$\bar{1}$0⟩ in hcp metals, the different slip systems have different critical shear stresses that vary with temperature and other factors. In zinc, for example, basal slip has a lower critical shear stress than prismatic slip at low temperature, whereas at high temperature both systems operate with similar ease. Some important slip systems in crystals are listed in Table 17.1.

Note that both slip planes and slip directions have generally simple indices, corresponding to lattice planes with large *d*-spacings and lattice directions with closely spaced lattice points. Therefore, during movements of dislocations, small incremental displacements

Table 17.1 Important slip systems in crystals

Crystal	Slip plane	Slip direction	Equivalent systems
fcc metals	{111}	⟨1$\bar{1}$0⟩	12
bcc metals	{110}	⟨1$\bar{1}$1⟩	12
hcp metals	(0001)	⟨$\bar{1}$2$\bar{1}$0⟩	3
	{10$\bar{1}$0}	⟨$\bar{1}$2$\bar{1}$0⟩	3
	{10$\bar{1}$1}	⟨$\bar{1}$2$\bar{1}$0⟩	6
Halite	{110}	⟨$\bar{1}$10⟩	3
Galena	{100}	⟨011⟩	6
	{110}	⟨$\bar{1}$10⟩	3
Pyrite	{100}	⟨001⟩	6
Calcite	{10$\bar{1}$4} = r	⟨2$\bar{2}$0$\bar{1}$⟩	3
	{01$\bar{1}$2} = f	⟨2$\bar{2}$0$\bar{1}$⟩ ⟨$\bar{2}$02$\bar{1}$⟩	6
Dolomite	(0001) = c	⟨$\bar{1}$2$\bar{1}$0⟩	3
	{01$\bar{1}$2} = f	⟨2$\bar{2}$0$\bar{1}$⟩ ⟨$\bar{2}$02$\bar{1}$⟩	6
Quartz			
LT	(0001) = c	⟨$\bar{1}$2$\bar{1}$0⟩	3
HT	{10$\bar{1}$0} = m	⟨$\bar{1}$2$\bar{1}$0⟩	3
HT	{10$\bar{1}$1} = r	⟨$\bar{1}$ $\bar{1}$23⟩	12
Olivine			
LT	(100)	[001]	1
HT	(010)	[100]	1

Note: LT, low temperature; HT, high temperature.

occur that are energetically favorable. In the case of fcc metals, the slip plane is the close-packed {111} plane and the slip direction is the close-packed $\langle 1\overline{1}0 \rangle$ direction, as we have already discussed in Chapter 2 (Figure 2.7).

17.3 Dislocation microstructures

We have seen that dislocations are necessary for ductile deformation of crystals. Dislocations are present in most crystals, and even during ideal crystal growth conditions may reach densities of about 10^6 dislocations/cm^2 (e.g., Hull and Bacon, 2011). During deformation the number of dislocations generally multiplies, and one of the mechanisms of such deformation multiplication is called a *Frank–Read source*. Assume that a dislocation propagates on a slip plane and runs into two obstacles (A and B in Figure 17.6). These obstacles may be either dislocations that run across the slip plane or included particles. Under a stress, the segment AB will increasingly bow out, eventually sweeping right around the pinning obstacles and pinching off to create both a loop and a new version of the original segment (labeled A'B' in Figure 17.6, right side). The process can repeat, creating sets of loops, as for example in aluminum (Figure 17.7a).

Dislocation microstructures within a crystal become increasingly complicated during deformation, and there is interaction between the dislocations, mostly in a way that inhibits the free movement of the dislocations as the density of dislocations increases. As dislocations multiply and interact by forming tangles and clusters (Figure 17.7b), a greater stress is necessary to deform the crystal, causing *work-hardening*. Correspondingly, the stress–strain curve becomes steeper (regime 3, in Figure 17.1b). When the stress reaches a critical limit, the crystal will fail by rupture.

At low temperatures, the movement of dislocations is restricted to the particular slip planes, and

dislocations are often concentrated in slip or deformation bands (Figure 17.7c). If more than one slip system is present, "cross-slip" of dislocations from one slip system to another one may occur and can lead to dense tangles of dislocations and concentrations of strain energy.

Strain energy can be reduced if dislocations are able to move out of their slip planes. This can be achieved at higher temperatures by a mechanism that involves diffusion of vacancies (e.g., Poirier, 1985). Take, for example, a crystal with an edge dislocation and a lattice vacancy (Figure 17.8a). If the vacancy diffuses into the core of the dislocation (Figure 17.8b,c), the dislocation has essentially *climbed* from the original slip plane to the next higher lattice plane. Assume that a crystal contains two dislocations of opposite sign in slip planes of the same type, but at different levels. These dislocations may move on top of each other by means of slip (Figure 17.9a) and then, through climb, come close together and ultimately join, thereby eliminating each other and reducing the total dislocation density and the strain energy. For dislocations of equal sign, it is energetically preferable for the dislocations to align on top of each other (Figure 17.9b), an organization that can again be achieved by a combination of slip and climb. This mechanism organizes dislocations and creates from an irregular distribution (Figure 17.9c) regular networks known as *subgrain boundaries* (Figure 17.9d). The dislocations cause a misorientation θ of the two domains across a subgrain, ranging from 1° to 10°. This process, in which the dislocation energy is reduced by climb, is called *recovery*. A typical microstructure with subgrain boundaries in quartz is shown in Figure 17.7d.

Microstructures such as those shown in Figure 17.7 are typical of certain deformation conditions. While slip bands and tangles are typical of low-temperature deformation, the significance of regular networks and loops is that they are associated with thermal activation and diffusion. These features are often observed in naturally deformed rocks and are indicative of the metamorphic environment during deformation. The TEM has been used widely to investigate dislocations in minerals (e.g., McLaren, 1991).

17.4 Mechanical twinning

Apart from slip, caused by movements of dislocations, there is a second way in which some crystals deform. H. W. Dove (1860) observed that when a stress is

Fig. 17.6 Schematic illustration of a Frank–Read source that multiplies dislocations around obstacles (A and B) by creating loops.

(a)

(c)

(b)

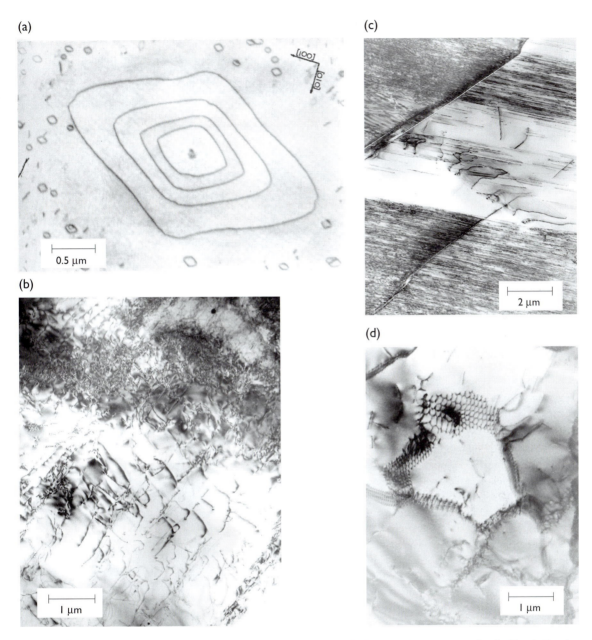

(d)

Fig. 17.7 TEM images of dislocation microstructures in crystals. (a) Dislocation loops as a result of a Frank–Read source in aluminum (from Westmacott *et al.*, 1962). (b) Tangles of dislocations with work-hardening at low temperature in greenschist facies calcite from the Central Alps (from Barber and Wenk, 1979). (c) Predominantly screw dislocations propagating on a slip plane in an experimentally deformed crystal of dolomite (from Barber *et al.*, 1981). (d) Network of dislocations in quartz from amphibolite facies mylonite in the Bergell Alps. This is a low-energy configuration due to climb at high temperature.

applied with a knife edge to a wedge of a calcite crystal, part of the crystal flips into a new orientation (Figure 17.10a). The old orientation and the new orientation have the lattice plane e = (01$\bar{1}$8) in

common, and the new orientation is related to the old one by a mirror reflection. Geometrically the two domains are therefore in a twin relationship (Figure 17.10b). Contrary to slip, which can be macroscopically

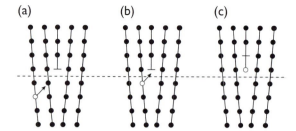

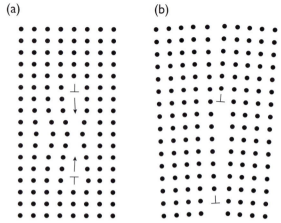

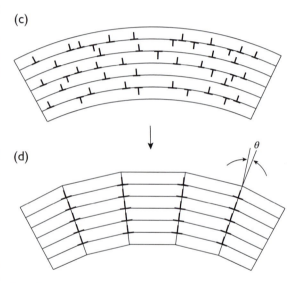

Fig. 17.8 Dislocation climb. A vacancy in (a) moves into the dislocation core, thereby causing the dislocation to climb to the next higher lattice plane (b,c).

viewed as a continuous process with an arbitrary deformation (Figure 17.11a), the deformation by twins is specific to the twin law and, once twinning has occurred, deformation ceases (Figure 17.11b). Mechanical twinning is very common in deformed calcite and, since Dove's discovery, it has been observed in many crystals, some of which are listed in Table 17.2 (see also Klassen-Neklyudova, 1964).

17.5 Polycrystal plasticity

When minerals deform in the Earth they are not isolated crystals subjected to a well-defined stress, but rather exist in an aggregate and linked to neighbors across grain boundaries. A typical case is the metamorphic marble composed of calcite in Plate 16c. The lamellar structures in many grains are deformation twins. Since individual grains have different orientations, also these lamellae are in different directions. We cannot see dislocations with an optical microscope, but imagine similar structures as in Figure 17.7b. In each crystal, slip systems and mechanical twinning have been active, resulting in shape changes of the rock aggregate. This is not the place to explore such complex deformation patterns, a branch of science called polycrystal plasticity (e.g., Kocks *et al.*, 2000) but keep in mind that also in macroscopic rocks deformation is controlled by what happened in individual mineral grains.

17.6 Summary

Plastic deformation of crystals occurs largely by movement of dislocations on slip systems and by mechanical twinning. Slip systems as well as mechanical twins are defined by rational planes (*hkl*) and directions [*uvw*]. In axial compression the optimal

Fig. 17.9 (a) Owing to climb, two dislocations of opposite sign can annihilate each other. (b) Dislocations of equal sign align on top of each other because this is a low-energy configuration. (c) In a deformed crystal, dislocations of opposite signs are arranged irregularly on slip planes. (d) During recovery, pairs of opposite sign annihilate each other, and remaining dislocations of equal sign arrange in subgrain boundaries with misorientation θ between the two parts. ⊥ indicates the dislocation core.

orientation is when slip plane and slip direction are at 45° to the compression axis (Schmid's law). At low temperature, dislocations interact and form tangles, causing work-hardening. At higher temperatures, diffusion of vacancies causes climb and this reduces dislocation densities.

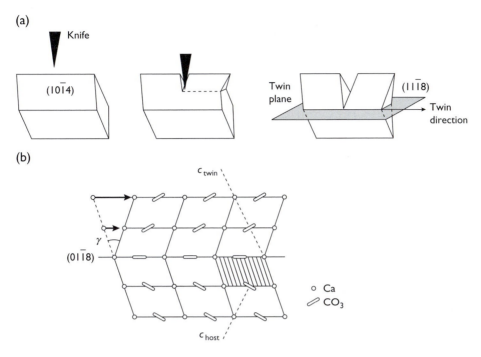

Fig. 17.10 (a) Mechanical twinning of calcite can be produced by applying a stress with the edge of a knife blade to a cleavage fragment. (b) After applying a shear γ (arrows), the structures are in a twin relationship; $01\bar{1}8$ is the twin plane.

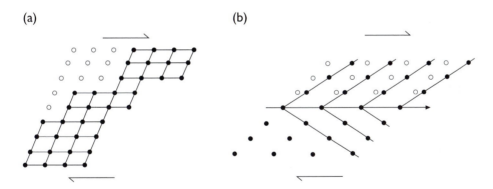

Fig. 17.11 (a) In slip, the amount of deformation is arbitrary and can be large. (b) During mechanical twinning, deformation is fixed.

Test your knowledge

1. Why are dislocations required to deform a crystal?
2. Assume that a crystal of ice has a single slip system (0001) $\langle\bar{1}2\bar{1}0\rangle$ and is deformed in compression. What is the optimal orientation of slip plane and slip direction relative to the compression direction?
3. There is a basic difference between slip (glide) and climb. Describe the difference geometrically and review the conditions under which the two mechanisms are active.
4. Mechanical twinning, though not as common as slip, is nevertheless an important mechanism in many minerals. Review mechanical twinning in quartz (Chapter 9) and relate it to the α–β phase transformation.

Table 17.2 Mechanical twinning in some crystals

	Twin plane	Twin direction
fcc metals	$\{111\}$	$\langle\bar{2}11\rangle$
bcc metals	$\{\bar{1}\,\bar{1}\,2\}$	$\langle111\rangle$
Zinc and hcp metals	$\{2\bar{1}\,\bar{1}2\}$	$\langle2\bar{1}\,\bar{1}\,3\rangle$
	$\{10\bar{1}2\}$	$\langle\bar{1}011\rangle$
Calcite	$\{01\bar{1}8\} = e$	$\langle0\bar{4}41\rangle$
Dolomite	$\{01\bar{1}2\} = f$	$\langle0\bar{1}11\rangle$
Quartz	$\{10\bar{1}0\}$	$[0001]$
Corundum	$\{01\bar{1}2\}$	$\langle0\bar{1}11\rangle$

Further reading

Barber, D. J., Wenk, H.-R., Hirth, G. and Kohlstedt, D. (2009). Dislocations in minerals. Chapter 96 of *Dislocations in Solids*, Vol. 16, J. Hirth and L. Kubin Eds. Elsevier, Amsterdam, pp. 171–232.

Hirth, J. P. and Lothe, J. (1982). *Theory of Dislocations*, 2nd edn. Wiley, New York.

Hull, D. and Bacon, D. J. (2011). *Introduction to Dislocations*, 5th edn. Butterworth-Heinemann, Oxford.

Karato, S.-I. and Wenk, H.-R. (eds.) (2002). *Plastic Deformation of Minerals and Rocks*. Reviews in Mineralogy and Geochemistry, vol. 51. Mineralogical Society of America.

Kocks, U. F., Tomé, C. and Wenk, H.-R. (2000). *Texture and Anisotropy: Preferred Orientations in Polycrystals and Their Effect on Materials Properties*. Cambridge University Press, Cambridge.

Means, W. D. (1976). *Stress and Strain: Basic Concepts of Continuum Mechanics for Geologists*. Springer-Verlag, Berlin.

Poirier, J. P. (1985). *Creep of Crystals: High-temperature Deformation Processes in Metals, Ceramics and Minerals*. Cambridge University Press, Cambridge.

Reed-Hill, R. E., Hirth, J. P. and Rogers, H. C. (1965). *Deformation Twinning*. Metallurgical Society Conferences, vol. 25. Gordon and Breach, New York.

Part IV | Mineral-forming processes

18 | Mineral genesis

Mineral genesis describes some of the conditions under which minerals form in different environments. A primary link is to petrology: minerals precipitate from solutions in sediments or crystallize from magmatic melts in igneous rocks. The original minerals may transform to new mineral assemblages during metamorphism at high temperature and pressure. Under special conditions rare minerals may accumulate and form ore deposits. The morphology, composition, and properties of crystals is often linked to the conditions under which they grow.

18.1 Overview

The term "genesis" (from the Greek *genesis*, meaning "a productive cause") is synonymous with *origin*, and in this section we will describe the life cycle of minerals. Genesis refers to both primary crystallization and the subsequent history of minerals, which may include structural transitions, changes in texture (e.g., grain coarsening), exsolution processes, and chemical reactions (e.g., oxidation). Both "syngenetic" and "epigenetic" aspects depend on the geological environment and are governed by physical and chemical laws.

Genetic mineralogy relates directly to petrology, economic geology, physics (especially thermodynamics), and chemistry. Ultimately a mineral and its properties and composition cannot be understood in isolation from its environment. Whereas some processes are relatively simple and have been studied in much detail (e.g., the crystallization of clinopyroxene phenocrysts from a mafic magma), others are still very puzzling and lack a quantitative physical explanation (e.g., the formation of dolomite in sedimentary rocks, or the nucleation and growth patterns of minerals in metamorphic rocks).

A *mineral deposit* is a geological body that forms under specific conditions and contains characteristic minerals or mineral assemblages. Within the deposit, a mineral may be scattered throughout different rocks and ores, or occur independently as segregations, lenses, strata, and veins of various shapes. There are deposits that include numerous rock units and show a fascinating diversity of minerals (e.g., skarns or layered mafic intrusions), and others (e.g., rock-salt formations) that consist of a single rock type with very little variation in mineralogy. Mineral deposits are commercially exploited as the natural source of metals and other useful materials.

18.2 Mineral-forming environments

In a physical and chemical sense, mineral nucleation and crystal growth may occur in any of several different systems. These systems include aqueous solution, gas, a mixture of gas and liquid, colloidal solution, magma (mainly the subject of igneous petrology), and solid or solid with solution as films along grain boundaries (the subject of metamorphic petrology).

Aqueous solutions

Aqueous solutions occur due to processes operating in the interior of the Earth (endogenic) or on the Earth's surface (exogenic). The former type of solution is classified as hydrothermal, and the latter as surface (vadose) solutions or brines.

There are several sources of water for the *hydrothermal solutions*. First, crystallizing magma chambers release volatile components that migrate to the surrounding country rocks and form liquid mineralized aqueous solutions. Second, the reactions of dehydration and decarbonation in the deeper parts of the crust produce water and carbon dioxide. Many types of sedimentary rock, such as clays and siliceous limestones, undergo dehydration and decarbonation processes during diagenesis and regional metamorphism. Finally, controversial processes of mantle degassing may be a source of water. It has been proposed that some hydrocarbons could escape from the Earth's mantle, and that their oxidation produces water and carbon dioxide when they rise to the crustal levels. Such processes may be described by the reaction

$$CH_4 + 2O_2 \rightarrow 2H_2O + CO_2$$

Such reactions usually produce energy that heats the fluids as well as the country rocks.

Not all water sources for hydrothermal solutions lie beneath the Earth's surface. Another source is surface (meteoric) water. Investigations have shown that surface water can migrate to depths of over 500 m, heating up and dissolving mineral components and exchanging atoms and ions with adjacent rocks. Another source is seawater circulated through fissures in the oceanic crust and responsible for much of the hydrothermal activity at mid-oceanic ridges, seamounts, and some back-arc basins.

As the solutions move along their pathways, some components are removed from adjacent rocks while others are added indirectly from magmatic sources. In hydrothermal solutions, components are transferred predominantly as aqueous complexes (Table 18.1). If the solution is basic (high pH) quartz is easily dissolved, transported, and precipitated as veins where pressure and temperature are lower. The composition of the hydrothermal solution is often derived from the rock through which the solution flows and, in the case of magmatic hydrothermal solutions, from the type of magma that gave rise to the fluids initially. Quartz is the most important hydrothermal mineral. Calcite veins form where metamorphic solutions acquired a large partial pressure of CO_2 from mineral reactions, from dissolution of carbonaceous rocks, or from mantle sources. There are many mineral deposits that crystallized from hydrothermal solutions. Examples are accumulations of various sulfide ores containing pyrite (FeS_2), chalcopyrite ($CuFeS_2$), sphalerite (ZnS), and galena (PbS).

The *surface aqueous solutions* are of several different types. There are ground, karst, and soil waters that precipitate carbonates – for example, calcite and aragonite in stalactites and stalagmites in karst caves. Also, there are lacustrine, oceanic, and lagoon waters that can produce beds of evaporite minerals, such as halite, gypsum, and some varieties of limestone.

Waters in different environments have a distinct isotopic signature, and hydrogen and oxygen isotopes incorporated into minerals can be used to trace the origin of a particular water sample. Figure 18.1 illustrates deviations δ of the ratios of heavier and lighter isotopes of water ($^{18}O/^{16}O$ and $^2D/^1H$) in various environments from that of present-day seawater (also referred to as "standard mean ocean water" or SMOW), i.e., $\delta^{18}O$ (in ‰) $= [(^{18}O/^{16}O_{sample})/(^{18}O/^{16}O_{SMOW}) - 1] \times 1000$ and δD (in ‰) $= [(^2D/^1H_{sample})/(^2D/^1H_{SMOW}) - 1] \times 1000$. A negative value indicates that isotopes are lighter than seawater and a positive value indicates that they are heavier. When H_2O evaporates from seawater, the vapor is enriched with light isotopes that require less energy to vaporize than do the heavier isotopes. Also, during condensation and rainfall, heavier isotopes are concentrated in the rain, and, with increasing distance from the ocean, the δ values become smaller and smaller. This is indicated by the "Meteoric water" line in Figure 18.1. Metamorphic and magmatic waters are considerably heavier than meteoric water. Waters

Table 18.1 Examples of complex ions that transport metals in hydrothermal solutions

Chemical element	Form of transfer (complexes)	Conditions
Copper	$CuCl_2^-$	Slightly acid and neutral solutions, at relatively high temperatures
	$Cu(HS)_2^-$	Neutral to alkaline solutions, at relatively low temperatures
Molybdenum	$NaHMoO_4$	Highly alkaline solutions, at temperatures below 450 °C
	$KHMoO_4$	Neutral solutions, at temperatures below 450 °C
Gold	$AuCl_2^-$	At temperatures of about 350–450 °C
	$Au(HS)_2^-$	At temperatures below 350 °C
Silver	$AgCl_2^-$	At temperatures above 200–250 °C
	$Ag(HS)_2^-$	At temperatures below 250 °C

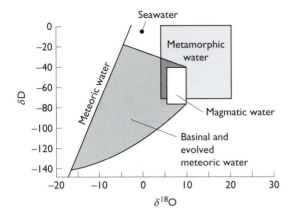

Fig. 18.1 Oxygen and hydrogen isotopic composition of water in different geological environments. The deviations δ are the ratios of heavier and lighter isotopes of oxygen ($^{18}O/^{16}O$) and hydrogen ($^{2}D/^{1}H$) relative to seawater (see text) (data from Hoefs, 1987).

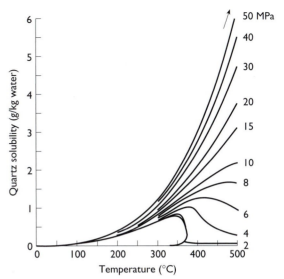

Fig. 18.2 Solubility of quartz (SiO_2) in water as a function of temperature for different pressures (based on Fournier, 1985).

in sedimentary basins are intermediate. This signature is inherited by minerals that contain water, for example sheet silicates, and one can determine whether a kaolinite was formed by surface weathering or by alteration through magmatic solutions. The isotope ratios in minerals also provide information about temperatures during crystallization.

Gas

Gas is a relatively rare crystallization environment, but some minerals are known to precipitate under such conditions. For example, hematite (Fe_2O_3), sal-ammoniac (NH_4Cl), realgar (AsS), and native sulfur may crystallize from volcanic gases (Plate 19d) and precipitate as skeletal crystallites (Plate 19e). Also ice crystals frequently grow from vapor, producing dendritic snowflake patterns (see Figure 10.10).

Fluid

Most geological processes are confined to the pressure–temperature range within which water occurs as a liquid phase or supercritical fluid, the latter being neither gas nor liquid. In many cases, fluid mixtures of CO_2 and H_2O are active in mineral-forming processes, especially during the formation of skarns and metamorphism of limestone under high pressure and temperature conditions. Similar fluids cause alteration of dunites and peridotites, producing amphiboles, serpentine, or talc, in association with calcite, magnesite, or

dolomite. Water can dissolve substantial amounts of minerals, particularly at higher pressure and temperature. This is illustrated for quartz in Figure 18.2. Zircon dissolves preferentially in alkaline solutions, whereas calcite and apatite dissolve in acid solutions.

Colloidal solutions

Colloidal solutions are a typical mineral-forming environment in ocean floor silts rich in clay minerals, aluminum, iron, and manganese hydroxides. Minerals can also crystallize from colloidal systems in hot aqueous solutions at surface conditions. An example is silica gel formation in thermal springs in areas of recent volcanic activity and subsequent precipitation of amorphous opal from the gels, as in Yellowstone National Park, USA (Figure 18.3). We have mentioned crystallization of silica minerals in chalcedony (Section 10.4, Plates 8b,c).

Magma

Magma is not a simple melt, such as melts of pure substances like water (from ice) or melted sugar (from crystalline sugar), where the liquid composition corresponds completely to that of the crystals. It is a mixture, and even the compositions of granites and many other plutonic rocks do not accurately

Fig. 18.3 Precipitation of silica from colloidal solutions in Fountain Paint Pot, Yellowstone National Park, Wyoming, USA.

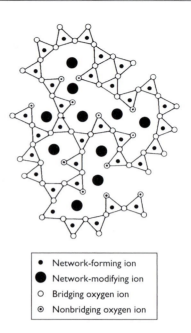

- ● Network-forming ion
- ⬤ Network-modifying ion
- ○ Bridging oxygen ion
- ◉ Nonbridging oxygen ion

Fig. 18.4 Idealized scheme of the structure of a silicate melt with network-forming silicate tetrahedra and network-modifying large cations or ion groups (see also Lee, 1964).

correspond to those of their parental magmas, for reasons that will be explained later. A magma has properties of a liquid as well as of a solution. Anion groups, in the form of coordination polyhedra, are "dissolved" in magmas in much the same way as complex ions are in aqueous solutions. The anion groups are locally organized into clusters, but without long-range atomic order, as in a crystal structure. Silicate melts contain primarily $(Si,Al)O_4$ tetrahedra, sometimes isolated, but more often linked into irregular groups (Figure 18.4). There are ion groups such as SiO_7^{6-}, $n(SiO_3)^{2n-}$, $Si_6O_{18}^{12-}$, $n(Si_4O_{11})^{6n-}$, MgO_6^{10-}, CaO_6^{10-}, AlO_4^{5-}, and SiO_4^{4-}. They represent nuclei and building blocks for subsequent crystallization of silicates. The relative proportion of tetrahedra connected into individual building blocks increases with increasing silica content. Silicic magmas are significantly more polymerized than their mafic counterparts. Magmas also contain large cations such as K^+ and Na^+.

Solid systems

Solid systems can be either amorphous or crystalline. An example of the former is volcanic glass, which is not very stable and tends to devitrify with time. In crystalline systems three types of processes can be distinguished. First there are polymorphic transitions (e.g., diamond to graphite, high quartz to low quartz, aragonite to calcite), which do not involve a change in chemical composition of the mineral. Amorphous phases also recrystallize to minerals of the same bulk composition (e.g., opal transforms to quartz). A second process involves more complicated reactions, with transformations of a precursor mineral, or minerals, into new phases of different composition. Examples are pseudomorphs, such as limonite replacing pyrite (Plate 6d). A third type comprises replacement processes, such as the growth of so-called porphyroblasts in metamorphic rocks, like garnet growing in a gneiss and replacing most of the pre-existing minerals and incorporating a few as inclusions (Plate 6a). These transformations are generally associated with a thin molecular film of water along grain boundaries which transports elements to crystallization sites. In the example of garnet porphyroblasts, growth can be associated with shear deformation with inclusion patterns documenting rotations (Figure 18.5 and Plate 6a).

18.3 Types of mineral crystallization

Why do minerals form? The main reason for crystallization of new minerals from pre-existing phases is that they are more stable than the melts, or solutions, or pre-existing minerals. Crystallization occurs during undercooling of liquids, gases, and supersaturation of solutions associated with changes in pressure,

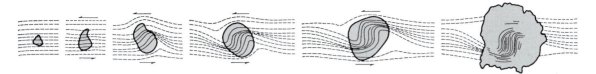

Fig. 18.5 Growth of a garnet porphyroblast (enclosed by dark line and shaded) in schist (after Spry, 1969). The growth stages are shown sequentially, moving from left to right and top to bottom. Initially, a nucleus grows in a schist with foliation. The garnet incorporates some minerals such as biotite, and as the schist is deformed the garnet rotates progressively and continues to grow. The incorporated schistosity can be used to infer the deformation process and the sense of shear. After deformation has ceased, the garnet continues to grow, replacing the schist (compare with Plates 5c and 6a).

temperature, or the concentration of chemical components. We will discuss each of these conditions in more detail in Chapter 19.

We can classify crystallization based on the volume that the new crystal occupies and distinguish between *free space crystallization*, *metasomatism*, and *recrystallization*. Some crystals grow freely in a gas, a melt, or a solution. Examples are sulfur growing in volcanic gas (Plate 19d,e), porphyritic feldspars growing in a magma (Figure 10.6), and amethyst growing in a hydrothermal solution. These crystals usually display euhedral habits.

Metasomatism is a powerful geological process that leads to the formation of compositionally diverse ores and rocks. Metasomatism is defined as a process of simultaneous capillary dissolution and crystallization, by which a new mineral completely or partially replaces an initial mineral, often changing the chemical composition (e.g., Lindgren, 1933). Metasomatic substitutions proceed not only through intragranular film solutions, but also through lattice diffusion. Examples of metasomatic growth are pseudomorphs and poikiloblasts, growing in a solid granular pegmatite. An example of a metasomatic rock is *greisen*, a quartz–mica aggregate that forms when granite is subjected to hydrothermal solutions:

$$3K(AlSi_3O_8) + 2H^+ \rightarrow KAl_2(AlSi_3O_{10})(OH)_2 + 6SiO_2 + 2K^+$$
Microcline   Muscovite   Quartz

Simultaneously with this reaction, greisens are often supplied with tin, which crystallizes in the form of the oxide mineral cassiterite (SnO_2), thus creating an important ore for tin.

Recrystallization implies that new crystals replace those formed earlier. This process may be accompanied by an increase or a decrease in grain size, and it may or may not involve compositional changes. As a rule, recrystallized minerals have fewer chemical impurities. Recrystallization proceeds in solid state and is driven either by the chemical free energy or by deformation defects in the crystal structure. Diagenesis and metamorphism involve recrystallization and polymorphic transformations, the former at very low temperature and the latter at higher temperature. Marble, as shown in Plate 8a and Figure 10.17a, is a recrystallized limestone.

18.4 Types of mineral deposits

Minerals form at virtually every step in the geological rock cycle, by both internal and external processes (Figure 18.6). Feldspars crystallize in slowly cooling plutonic magma or in the groundmass of a rapidly quenched volcanic rock. When exposed on the surface, the feldspars are no longer stable and alter during weathering to clay minerals. In evaporite lakes, halite and gypsum precipitate from saturated aqueous solutions. Solutions rich in carbonate and silica form calcite and quartz, which cement sand grains to sandstone. During recycling of sedimentary rocks, clay minerals transform to mica at higher pressure and temperature. At greater depth and in contact with a magma, limestone undergoes metamorphic reactions that produce such minerals as olivine and tremolite.

Depending on the environment, minerals may form continuously or discontinuously, and a single mineral may dominate or several minerals may be present. The boundaries of a mineral deposit mark occurrence limits of a characteristic mineral (or, in the case of economic mineral deposits, a desirable concentration level of some useful component). Several classification principles have been suggested for mineral deposits based on their genesis, mineral type, morphology of mineral bodies, mineralization scale, economic factors, and whether the processes involve a number of stages or a single stage. The most useful classifications seem to be those that apply the genetic principles.

We apply a simplified genetic scheme for the major mineral deposits, as given in Table 18.2. A first division is into deposits formed by endogenic (i.e., internal) and exogenic (i.e., external) processes. Then we distinguish between magmatic and metamorphic deposits. Magmatic deposits include all those that are related directly to the cooling and crystallization of magmas (igneous deposits), or to activity of postmagmatic fluids and gases (hydrothermal deposits). Exogenic geological processes form sedimentary mineral deposits and deposits in weathering zones. There are also intermediate types such as the metal-bearing benthic muds of the Red Sea. While the mud itself is an exogenic geological product, the metals were clearly introduced by endogenic processes, i.e., heating of seawater.

Table 18.2 Main genetic types and groups of mineral deposits

Types	Genetic groups (chapter reference)
Endogenic	
Magmatic	Igneous *(Chapters 21, 28, 30)*
	Pegmatite *(Chapter 21)*
	Skarn *(Chapter 28)*
	Hydrothermal *(Chapter 26)*
Metamorphic	Metamorphic *(Chapters 28, 30)*
Exogenic	
Supergene	Vadose *(Chapter 22)*
	Weathering and oxidation zones *(Chapters 25, 27)*
Sedimentary	Mechanical *(Chapter 24)*
	Chemical *(Chapters 23, 24)*
	Biogenic *(Chapters 24, 25)*
Endogenic– exogenic	Hydrothermal–sedimentary *(Chapter 24)*

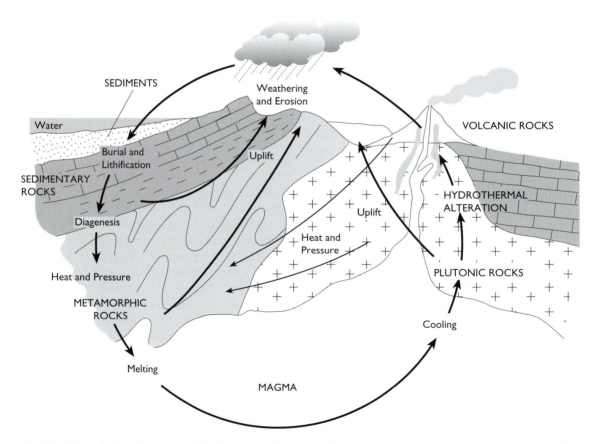

Fig. 18.6 Geological rock cycle as originally proposed by James Hutton over 200 years ago, showing the relationship of internal and external processes and the main mineral- and rock-forming environments.

Some specific genetic processes will be discussed in conjunction with the relevant mineral or group of minerals in the chapters to follow. However, it should be remembered that no genetic process is exclusive to a particular mineral group – there are many exceptions and overlaps.

18.5 Multistage processes, generations, and parageneses

The formation of any mineral deposit is a long-term multistage process. Minerals crystallize in a certain order that reflects changes in pressure, temperature, and chemical composition of the crystallization environment. Depending on the local geological conditions, and particularly on global processes related to plate tectonics, mantle plumes, and mantle convection, these changes may happen slowly, abruptly, or rhythmically. A mineral can crystallize over a single time interval, or its growth may be interrupted by periods of recess, dissolution, and precipitation of other minerals. Assemblages of crystals and grains of the same mineral that formed at different times are called *generations*, as illustrated for celestite in Figure 18.7. A first generation grows on limestone. This is then covered by a layer of sulfur on which a second generation nucleates. The grains and crystals

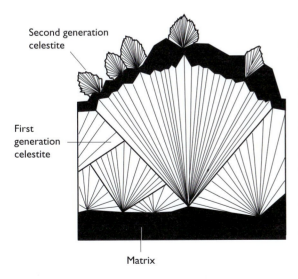

Fig. 18.7 Two generations of celestite (based on Yushkin, 1968). Crystals of the first generation grew on the substrate in a cavity in limestone. They were later covered by a coarse-grained aggregate of sulfur (black) and finally by a second generation of celestite.

of the same mineral belonging to different generations may differ in morphology, size, composition, and properties. Also, they will most likely have different relationships with coexisting minerals. An assemblage of minerals that crystallizes nearly contemporaneously, and in similar physical and chemical conditions, is called a *paragenesis*, a term introduced by August Breithaupt in 1849. The investigation of mineral parageneses allows petrologists to deduce physical and chemical conditions of their formation. This branch of mineralogy was developed largely by T. F. W. Barth (1962), P. Eskola (1946), D. S. Korzhinskii (1970), J. B. Thompson (1959), and F. J. Turner (1981).

18.6 Typomorphism of minerals

Some minerals typically occur in specific rock types. F. Becke (1903) described them as *typomorphic minerals*. For example, staurolite occurs in metamorphic rocks. When minerals crystallize, their constitution, crystal morphology, and properties are related to the environment in which they nucleate and grow, and to the physical and chemical conditions of crystallization. The ways these factors are linked were described in the classical monograph of A. Fersman (1939).

For example, systematic changes in the morphology of quartz crystals from fissures in the Swiss Alps are related to temperature and pressure conditions, as well as the composition of fluids from which these crystals grew (Figure 18.8). The conditions were inferred mainly from a detailed study of fluid inclusions in quartz. Typical quartz morphologies are illustrated in Figure 18.9. A doubly terminated type with prism $\{10\bar{1}1\}$ and steep rhombohedra $\{30\bar{3}1\}$ and $\{03\bar{3}1\}$ (Figure 18.9a) is typical of the northern low-temperature (200 °C) diagenetic zone. Here, the fluid was rich in heavy hydrocarbons released from sedimentary rocks during their diagenesis. At slightly higher temperature the hydrocarbons transformed to methane (CH_4) and in this gas-rich hydrothermal solution edge-growth with skeletal crystals and frequent scepter overgrowths occurred (Figure 18.9b, Plate 3f). Such forms resulted from rapid crystallization when gas was released from the solution during pressure drops. Further south, the classic symmetric prismatic-rhombohedral quartz, with dominant $\{10\bar{1}1\}$ and subordinate $\{0\bar{1}11\}$, is observed (Figure 18.9c). This morphology corresponds to a

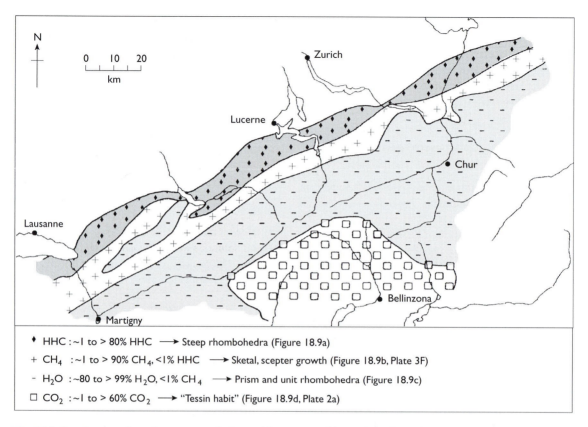

Fig. 18.8 Regular zonation of quartz morphology with metamorphic grade in the Swiss Alps. The zonation also correlates with the composition of fluid inclusions. Percentages are mol%; HHC, heavy hydrocarbons. (After Mullis, 1991; see also Mullis *et al.*, 1994.)

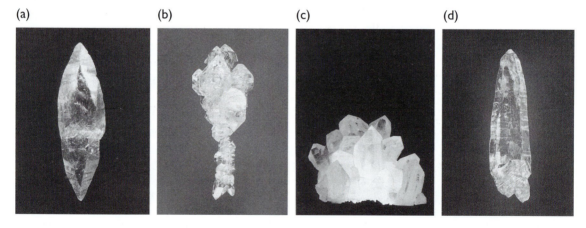

Fig. 18.9 Morphological types of quartz from Alpine fissures. (a) Bipyramidal steep rhombohedral habit. (b) Skeletal and scepter growth. (c) Prismatic quartz with unit rhombohedra $\{10\bar{1}1\}$. (d) Tessin habit with steep rhombohedral faces and prism (from Mullis, 1991).

fairly pure solution of water, temperature of 350 °C and pressure of 400 MPa. At higher temperatures (450 °C) and low fluid pressures (100 MPa), in the rapidly uplifting and eroding southern part, where CO_2 was the major gas phase, quartz attained a stepped morphology known as the "Tessin habit", incorporating the prism and steep rhombohedra $\{30\bar{3}1\}$ $\{03\bar{3}1\}$ (Figure 18.9d, Plate 2a). It has been

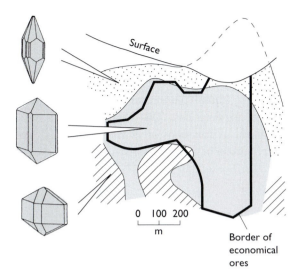

Fig. 18.10 Habit changes in cassiterite in the hydrothermal tin deposits of the Komsomolsk group in eastern Siberia, illustrated in a cross-section. Cassiterite changes from a barrel-shaped equiaxed habit, over a prismatic-columnar habit to an acicular habit. The border of the economical ores is outlined in black line and this zone coincides with columnar cassiterite habit (based on Evzikova, 1984).

shown that individual crystals grew continuously over 5 million years (e.g., from 20 to 15 million years ago).

Morphological features of a mineral can frequently be used as a prospecting criterion and this has been extensively employed in Russia. For example, in cassiterite (SnO_2) crystals from a hydrothermal tin deposit in eastern Siberia, the crystal habit changes from equiaxed and barrel-shaped (at depth), to columnar/prismatic in the central part, to acicular near the surface (Figure 18.10). During the movement of ore-forming solutions from depth to the surface the crystals precipitate at various temperatures, and this is expressed in the habit. In the central part of the vein system crystals display a columnar habit and precipitation is most extensive. Such a regular relationship between morphology and occurrence can be of practical importance in answering such questions as:

- How deeply is a vein eroded? (If a vein contains many equiaxed crystals, it is deeply eroded.)
- How far does a vein extend? (The more slowly cassiterite changes its habit, the further the vein extends.)

- How close is one to the most productive part of the vein system? (With many crystals of intermediate morphology we are close to the core of the ore system.)

18.7 Summary

The environment under which minerals form depends on physical and chemical conditions. While minerals in igneous rocks crystallize in a liquid magma, for many mineral-forming environments aqueous solutions play an important role. This can be on the surface (vadose) or in solutions passing through rocks on fractures and grain boundaries at depth. Hydrothermal solutions are particularly significant for ore deposits. The growth morphology is often linked to physical–chemical conditions as illustrated for quartz and cassiterite.

Test your knowledge

1. Name some sources of water in hydrothermal solutions.
2. Describe a typical paragenesis of minerals under high-grade metamorphic conditions and also under low-grade metamorphic conditions.
3. Volatile phases are mainly H_2O and CO_2. Describe some conditions where they participate in the formation of minerals and give some mineral examples that contain these phases.
4. Give some examples of mineral formation at various stages of the rock cycle.
5. Review each major type of mineral deposit and describe its formation as a result of geological processes.
6. What does the term "typomorphism" of minerals imply?

Further reading

Barth, T. F. W. (1962). *Theoretical Petrology*, 2nd edn. Wiley, New York.
Korzhinskii, D. S. (1970). *Theory of Metasomatic Zoning.* Oxford University Press, Oxford.
Turner, F. J. (1981). *Metamorphic Petrology: Mineralogical, Field, and Tectonic Aspects*, 2nd edn. McGraw-Hill, New York.

19 | Considerations of thermodynamics

Part III dealt with physical properties of minerals. Equally important are chemical aspects and their influence on mineral stability and phase relations. This chapter reviews basic aspects of thermodynamics to describe equilibria as functions of entropy, enthalpy, and internal energy and to understand chemical reactions in mineral systems.

19.1 Background

Minerals form by chemical reactions over a wide range of conditions, with temperature, pressure, and chemical potentials of all components being the most important variables. The principles of thermodynamics, developed in chemistry to quantify chemical transformations, are directly applicable to these reactions. The formal derivation of thermodynamic relationships will not be covered here and it is assumed that the reader has some background in elementary chemistry. Many of the quantitative derivations are not necessary to follow the rest of this book. Yet, at the end of this chapter, a student should be familiar with phase diagrams, and how they are related to the chemical properties of minerals.

We introduce some basic concepts and illustrate them with mineral examples. There are three main laws of thermodynamics that were formulated in the nineteenth century. The *first law*, based on the recognition by Robert Mayer in 1840 that heat (ΔQ) is equivalent to mechanical work (ΔW), states that a change in the total internal energy of a system (ΔE) is equivalent to the heat transferred into the system minus the work performed by the system; that is, $\Delta E = \Delta Q - \Delta W$. The total value of internal energy (E) cannot be readily quantified, and in most cases we need to know only how E changes during a process or reaction.

The *second law of thermodynamics*, proposed by Rudolf Clausius in 1850, can be formulated in several different ways. Unlike in an ideal (reversible) process, the heat absorbed by a system undergoing an irreversible process is not equal to the work performed on the system. Part of the energy is always lost in the process

and cannot be retrieved. "Irreversibility" is reflected through changes in the value of entropy S, which is a measure of the degree of disorder in the system. Thus, for all irreversible processes, the *second law of thermodynamics* can be formulated as $\Delta S > Q/T$, and for reversible processes, it can be stated as $\Delta S = Q/T$, where T stands for absolute temperature. For example, two bodies at different temperatures will exchange heat, such that heat flows from the hotter to the colder body.

Both internal energy and entropy characterize the state of a system, and they are independent of how that state has been reached. If C is the heat capacity, a scalar property that specifies the heat maintained by a substance, then for any given compound whose entropy is known at a temperature T_1, an absolute value of S at T_2 can be found by integrating its heat capacity over the T_1–T_2 interval. It has been proved experimentally that all pure and perfectly ordered crystalline substances (i.e., all substances excluding solid solutions, glasses, and crystals with defects) have the same entropy at absolute zero temperature. This statement, known as the *third law of thermodynamics*, provides a useful reference frame. These fundamental laws of thermodynamics have profound implications for geological processes of all magnitudes.

It should be emphasized, however, that *kinetics* (i.e., the rate at which reactions take place) is also important for mineral reactions. Many minerals do not crystallize in their stability field and are not in chemical equilibrium. On cooling, reaction rates slow down, often to immeasurable rates at low temperature, and minerals that are stable at high temperature are preserved at room temperature. For example, cristobalite and tridymite, high-temperature SiO_2 phases, precipitate in seawater; sanidine, the disordered K-feldspar in volcanic rocks, forms in sediments; and aragonite, the high-pressure polymorph of $CaCO_3$ is the main constituent of seashells. At higher temperatures, as during the crystallization of many metamorphic and igneous rocks, a closer approximation to equilibrium exists. However, most minerals collected in the field or analyzed in the laboratory are presently not in chemical equilibrium, and this is used by petrologists to establish the conditions present during rock formation.

Minerals are products of chemical reactions or polymorphic transformations. These always involve an assemblage (*system*) of *phases* and chemical *components* (such as elements, oxides, etc.). There are open and closed systems, depending on whether matter can or cannot enter or leave the system. Thermodynamic principles allow us to analyze real or imaginary phases and chemical transformations within the system. A *phase* is a homogeneous and physically distinct part of the thermodynamic system that may be isolated mechanically from the system. It may be a solid, a liquid, or a gas. Figure 19.1 shows a pressure versus temperature phase diagram for H_2O. Ice (hexagonal), vapor, and water are three coexisting

phases at $T = 298.16$ K (0.01 °C) and $P = 0.61$ Pa, which is the invariant point, or triple point, for water. There are several high-pressure polymorphs of ice.

Each mineral in a solidifying magma is a separate phase; the melt itself is a separate phase as long as it is homogeneous. A homogeneous gas mixture that has separated from the melt is another phase.

Thermodynamic *components* are chemical constituents that can be used to describe completely the chemical compositions of the phases of a system. In principle we could use individual elements as chemical components, but it is simpler to use compounds. For example, the system containing the three phases ice, vapor, and water has one component, namely H_2O. The system gypsum plus anhydrite (as in some sedimentary rocks) can be regarded as having two components, $CaSO_4$ and H_2O. These components may be connected in the following manner by a chemical reaction:

$$CaSO_4 + 2H_2O \rightleftharpoons CaSO_4 \cdot 2H_2O$$
Anhydrite $\qquad\qquad$ Gypsum

In the system calcite + gypsum (which is found in corroded marbles of old statues and buildings) the following reaction can proceed:

$$CaCO_3 + SO_3 + 2H_2O \rightleftharpoons CaSO_4 \cdot 2H_2O + CO_2$$
Calcite $\qquad\qquad\qquad$ Gypsum

CaO, CO_2, SO_3, and H_2O can be chosen as thermodynamic components of this system. The number of components ($c = 4$) is the number of phases ($p = 5$) minus the number of chemical reactions between the phases ($n = 1$), i.e.,

$$c = p - n \tag{19.1}$$

The stable state of a mineral is determined by the energy minimum rule. The maximum number of minerals that could coexist in equilibrium is determined by the phase rule. Both will be discussed in the following sections.

The liquid–vapor boundary terminates at a critical temperature and pressure, called the critical point. Beyond the critical point there is a single state, neither gas nor fluid, a so-called supercritical fluid with exceptional properties. We will return to it in Section 26.3.

19.2 Energy minimum in a system

A mineral is stable when it coexists in equilibrium with other minerals and chemical compounds. The symbol $\rightleftharpoons$ indicates the state of equilibrium. For example,

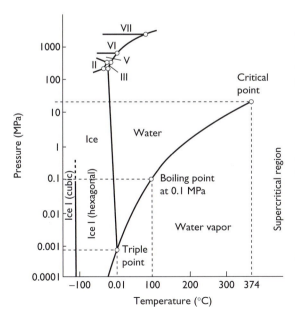

Fig. 19.1 P–T phase diagram of H_2O illustrating stability fields of ice, water, and vapor. All three phases coexist at the triple point. The different polymorphs of ice are indicated by roman numerals.

C $\rightleftharpoons$ C
Diamond   Graphite

$CH_4 + O_2 \rightleftharpoons$ C $+ 2H_2O$
                  Diamond

In the first example, diamond coexists in equilibrium with graphite. In the second, it coexists with a mixture of methane, oxygen, and water. These two very different thermodynamic systems reach equilibrium at some energy minimum that is different in the two cases. The mineral's stability is not determined by the minimum of the mineral's own energy, but by that of the system as a whole.

Under different conditions the same thermodynamic system will spontaneously reach its stable (equilibrium) state at a different energy minimum. The values of these minima may be defined by using thermodynamic potentials or internal energies (Table 19.1). These chemical potentials – such as entropy, enthalpy, and Helmholtz or Gibbs potentials – are material constants, similar to the physical properties discussed in Chapter 12. They depend on chemical composition, bonding, and crystal structure, as well as on temperature and pressure. Values for these potentials are listed for standard conditions in handbooks (e.g., Wagman *et al.*, 1982; Robie and Hemingway, 1995; Gottshalk, 1997). Standard conditions usually refer to a temperature of 25 °C (298.15 K) and a pressure of 1 bar (0.1 MPa). In the following discussion we will sometimes use absolute values. Often relative values are important, for example energy differences between reactants and reaction products. If potentials change, for example with temperature and pressure, differential expressions are useful. In

Box 19.1 notations, constants, and conversion factors used in thermodynamic calculations are summarized.

19.3 The simplest thermodynamic calculations and diagrams

Dolomite–quartz–diopside reaction, enthalpy, and entropy

The variation of *enthalpy H* in a system corresponds to the heat exchange of an isobaric process or reaction. Enthalpy can be considered as thermal or kinetic energy due to atomic vibrations. If the temperature is 25 °C and the pressure 0.1 MPa, then

$$\Delta H^0 = \sum n_i \, \Delta H_j^0(\text{products}) - \sum n_i \, \Delta H_j^0(\text{reactants})$$
(19.2)

where ΔH_j^0 are the standard enthalpy values for individual substances, and n_i are the molar fractions of the components in a balanced chemical reaction. Enthalpy is measured in J/mol (or in kcal/mol) units, and, because calculations are in multiples of moles, the results are in J (or in kcal) units. Let us calculate ΔH^0 for a process at 25 °C and 0.1 MPa for the reaction between dolomite and quartz to form diopside, which is a typical reaction for metamorphism of limestones at high temperature and pressure:

$CaMg(CO_3)_2 + 2SiO_2 \rightleftharpoons CaMg(Si_2O_6) + 2CO_2$
Dolomite   Quartz   Diopside   Gas

The standard enthalpy values for these substances are listed in Table 19.2 and, applying equation 19.2, we obtain

Table 19.1 Definition of some thermodynamic potentials

	General relationships (equilibrium state)	Equilibrium restrictions	Spontaneous process		Equilibrium states
			Possible	Impossible	
Entropy, S	$dS = dQ / T$	$dE = 0; dV = 0$	$dS > 0$	$dS < 0$	$dS = 0$
Internal energy, E	$dE = T\,dS - P\,dV$	$dS = 0; dV = 0$	$dE < 0$	$dE > 0$	$dE = 0$
Enthalpy, H	$dH = dE + P\,dV$	$dS = 0; dP = 0$	$dH < 0$	$dH > 0$	$dH = 0$
Helmholtz potential, F	$F = E - TS$; $dF = -S\,dT - P\,dV$	$dV = 0; dT = 0$	$dF < 0$	$dF > 0$	$dF = 0$
Gibbs potential, G	$G = E + PV - TS$; $G = H - TS$; $dG = -S\,dT + V\,dP$	$dT = 0; dP = 0$	$dG < 0$	$dG > 0$	$dG = 0$

Q, heat; P, pressure; V, volume; T, temperature (K).

Box 19.1 Additional information: Notation, constants, and conversion factors used in thermodynamic calculations

Symbol	Explanation
T	absolute temperature in kelvin, 25 °C = 298.15 K
K	kelvin, the unit of absolute temperature
Mol	mole, the amount of a substance corresponding to a gram formula weight
P	pressure in pascals: the standard atmosphere is equal to $1.013\ 25 \times 10^5$ Pa; 1 kg/cm^2 is equal to $0.980\ 655 \times 10^5$ Pa
p_{CO_2}	partial pressure of CO_2
p_{H_2O}	partial pressure of H_2O
V	volume (in cm^3)
V^0_{298}	molar volume of 1 mole of a substance at 1 bar pressure and 298.15 K (cm^3/mol = J/(MPa mol))
0	superscript here and in the following expressions indicates standard state
ΔV^0_{298}	change of total molar volume of a substance as a result of a process (in cm^3)
$C^0_{P,298}$	heat capacity at constant pressure P 1 bar and temperature 25 °C (in J/(mol degree), or in cal/(mol degree))
$\Delta C^0_{P,298}$	change of a total heat capacity of a substance as a result of a process (in J/degree, or in cal/degree)
S^0_f	entropy of formation of a substance from the elements in their reference state (in J/(mol degree), or in cal/(mol degree))
ΔS^0	change of entropy as a result of a process (in J/degree, or in cal/degree)
S^0_T	entropy at temperature T (in J/mol, or in cal/mol)
$\Delta_f H^0$	enthalpy of formation of a substance from the elements in their reference states (in J/mol, or in kcal/mol)
ΔH^0	change of enthalpy as a result of a process (in J, or in kcal)
ΔH^0_T	change of enthalpy as a result of a process, at temperature T (in J, or in kcal)
$\Delta_f G^0$	Gibbs free energy of formation of a substance from the elements in their reference states (in J/(mol degree), or in kcal/(mol degree))
ΔG^0	change of Gibbs free energy as a result of a process (in J, or in kcal)
ΔG^0_T	change of Gibbs free energy as a result of a process, at temperature T (in J, or in kcal)
$G^0_{T,P}$	change of Gibbs free energy as a result of a process, at temperature T and pressure P (in J, or in kcal)
k	equilibrium constant
k_{red}	equilibrium constant, reduced to standard conditions
a	activity
f	fugacity
E^0	electromotive force (emf) for an electrochemical cell at standard conditions (in V)
Eh	emf between electrode in any state and H_2 electrode in standard state (in V)
pH	logarithm of the activity of hydrogen ions in a solution
R	gas constant, 8.3145 J/(degree mol), or 0.848 kg/(degree mol) (or 1.9872 cal/(degree mol))
F	Faraday constant, 96.485 J/(V mol) (or 23.062 kcal/(V mol))
$\ln x$	$2.302585 \log x$
$R \ln x$	$4.57567 \log x$ (in cal/(degree mol))
1 bar	10^5 Pa = 0.1 MPa (1 kbar = 100 MPa), 1 MPa = 1 J/cm^3
1 J	2.390×10^{-4} kcal (1 MPa cm^3/mol)
1 cal	4.18 MPa cm^3/mol = 4.184 J

Table 19.2 Enthalpies and heat capacities for diopside, CO₂, dolomite, and quartz

	Enthalpy, $\Delta_f H^0$ (kJ/mol)	Heat capacity, C^0_{P298} (J/(mol degree))
Diopside	−3210.68	166.36
CO₂	−393.51	37.11
Dolomite	−2325.97	157.78
Quartz	−910.69	46.64

$$\begin{aligned}
\Delta H^0 &= [(-3210.68 \text{ kJ/mol}) \times 1 \text{ mol} \\
&\quad + (-393.51 \text{ kJ/mol}) \times 2 \text{ mol}] \\
&\quad - [(-2325.97 \text{ kJ/mol}) \times 1 \text{ mol} \\
&\quad + (-910.69 \text{ kJ/mol}) \times 2 \text{ mol}] \\
&= (-3997.70 \text{ kJ}) - (-4147.35 \text{ kJ}) \\
&= +149.65 \text{ kJ} \quad (19.3)
\end{aligned}$$

This reaction (left to right) is *endothermic*, which means that heat is absorbed.

When heat is added to a substance (dQ), the temperature increases (dT). The scalar property that specifies the heat maintained is called the *heat capacity C* and is described with the expression

$$dQ = C \, dT \quad (19.4)$$

Heat capacities are measured in J/(mol degree), and they are listed in Table 19.2. We can calculate the change in heat capacity of the system

$$\begin{aligned}
\Delta C^0_{P\,298} &= 1 \times 157.78 + 2 \times 46.64 - 1 \times 166.36 \\
&\quad - 2 \times 37.11 \text{ k J/degree} \\
&= -6.49 \text{ k J/degree}
\end{aligned}$$

Heat capacities vary slightly with temperature. It makes a difference if heat is added at constant pressure or constant volume, and usually the heat capacity at constant pressure C_P is specified. If we add heat Q to a closed system during a reversible change in its state at constant pressure, we increase the enthalpy correspondingly. Therefore we can write

$$dH = C_P \, dT \quad (19.5)$$

The most common form of *mechanical work W* of a thermodynamic system is to expand against the constant pressure P of the surroundings. Mathematically, this is written as $\Delta W = P \, \Delta V$, where ΔV is the volume change. The first law of thermodynamics can then be expressed as

$$dE = dQ - P \, dV \quad (19.6)$$

i.e., the change of internal energy E is the difference between thermal energy Q and the expansion against pressure.

The *entropy S* is involved in all thermodynamic potentials. It is measured in J/(degree mol) (or in cal/(degree mol)). Entropy is not easily defined, but it can be viewed as a measure of internal disorder. It increases, for example, if a substance transforms from a highly ordered crystalline state to a liquid or a gaseous state. But it can also vary within the solid state. For example, going from fully ordered microcline to sanidine with a disordered Si–Al distribution, S changes from 995.83 to 1100.94 J/(degree mol). And transforming to a feldspar glass with no long-range order in a regular lattice increases S to 1206.75 J/(degree mol).

Among the Al_2SiO_5 polymorphs, kyanite has the lowest entropy (92.17 J/(degree mol)). It is higher for andalusite (93.22 J/(degree mol)), and sillimanite (96.19 J/(degree mol)), but the differences are much smaller than for K-feldspar, since all three of these polymorphs are ordered crystal structures.

The change of entropy in a system (ΔS^0) is used in many thermodynamic calculations. The system entropy is equal to the weighted sum of entropies of formation (S^0_f) for the reaction products minus that for the reactants

$$\Delta S^0 = \sum n_i S^0_f \text{ (products)} - \sum n_i S^0_f \text{ (reactants)} \quad (19.7)$$

where n_i are the molar fractions of the components.

The second law of thermodynamics stipulates that in any reversible process the change in entropy of the system (dS) is equal to the heat received by the system (dQ) divided by the absolute temperature T:

$$dS = dQ/T \quad (19.8a)$$

In an irreversible process, the change in entropy of the system is larger than the heat received by the system divided by the absolute temperature T:

$$dS > dQ/T \quad (19.8b)$$

If we substitute the expression for dQ from equation 19.4 into equation 19.8a we obtain

$$dS = C(dT/T) \quad (19.9)$$

At an arbitrary temperature and pressure, enthalpy $\Delta H_{T,P}$ values may be calculated from the following equation:

$$dH = T dS \text{ (change in heat)} + V dP \text{ (mechanical work)} \quad (19.10)$$

For a constant pressure (e.g., $P = 0.1$ MPa) and temperature T a simplified equation may be used (cf. equation 19.5):

$$\Delta H_T = \Delta H^0 + \Delta C_{P,298}^0 (T - 298) \qquad (19.11)$$

where ΔH^0 and $\Delta C_{P,298}^0$ are, respectively, enthalpy and heat capacity at 298 K.

Let us now return to the dolomite–quartz reaction forming diopside and CO_2 but at a temperature of 500 °C (773 K). Using equation 19.11 and values of +149.65 for ΔH^0 and −6.49 for $\Delta C_{P,298}^0$ from above, we obtain

$$\begin{aligned}
\Delta H_{773} &= +149.65 \text{ kJ} + [(-6.49 \text{J/degree}) \\
&\quad \times (773 - 298) \text{ degree}] \\
&= +149.65 \text{ kJ} - 3.08 \text{ kJ} \\
&= +146.57 \text{ kJ} \qquad\qquad (19.12)
\end{aligned}$$

Again, the value is positive and the reaction is endothermic.

Table 19.1 gives equilibrium restrictions for enthalpy: $dS = 0$ and $dP = 0$. In the example above, P is constant and therefore $dP = 0$, whereas the entropy of the system greatly increases because a gas phase forms. Thus $dS \neq 0$, and the equilibrium restriction for using S is not satisfied. At this point in our discussion, we can draw no conclusions as to whether this reaction is possible or not, at either 25 or 500 °C. We will explore this question in the next section.

Calcite–aragonite transformation: Gibbs free energy

In order to account for changes in both entropy and enthalpy, J. Willard Gibbs defined a new function, which we now call the *Gibbs free energy* (or Gibbs potential) G. This function is independent of variations in pressure and temperature and can be stated as

$$G = H - TS \qquad (19.13)$$

where H is enthalpy and S is entropy.

The Gibbs free energy depends only on the state of a system, not on how this state has been attained. The change in free energy associated with the formation of a compound from its constituent elements under standard conditions (25 °C and 0.1 MPa) is termed the standard Gibbs free energy of formation (G_f^0) and is measured in J/mol (or in kcal/mol). Gibbs potentials for elements are by definition zero. The change of Gibbs free energy of a chemical reaction in the standard state (25 °C and 0.1 MPa) is

$$\Delta G^0 = \Delta H^0 - T\,\Delta S^0 \qquad (19.14)$$

From Table 19.1, a spontaneously occurring reaction requires that $dG < 0$; hence changes in entropy and enthalpy work together to define whether or not the process may occur spontaneously. Reactions that result in more order ($dS < 0$) will occur spontaneously if the heat expelled during the process is greater than the increase in G due to the decrease in entropy. And, vice versa, reactions absorbing heat ($dH > 0$) will occur if the decrease in the Gibbs potential due to the increase in entropy outweighs the increase in G due to the heat absorption.

Indeed, in comparison with H (Table 19.1), the conditions necessary for the application of the Gibbs potential (namely pressure and temperature invariability) are most easily understood and most readily measured. Therefore evaluations of the Gibbs potential are frequently used to analyze the formation processes of various substances. A change of G in a system during some process (ΔG^0 of the process) at 25 °C (≈ 298 K) and 0.1 MPa pressure (standard conditions) is equal to

$$\Delta G^0 = \sum n_i \Delta G_f^0 \text{ (products)} - \sum n_i \Delta G_f^0 \text{ (reactants)} \qquad (19.15)$$

The ΔG_f^0 values (free energy to form compounds from elements in the reference state) for different substances are listed in reference books mentioned earlier. Here we apply the concept of Gibbs free energy to the polymorphic reaction

$$\begin{array}{ccc}
CaCO_3 & \longrightarrow & CaCO_3 \\
\text{Calcite} & & \text{Aragonite}
\end{array}$$

Values for S_f^0, ΔG_f^0, and molar volume V for aragonite and calcite are as follows:

	S_f^0 (J/(degree mol))	ΔG_f^0 (kJ/mol)	V_{298}^0 (cm³/mol)
Aragonite	88.62	−1128.33	34.15
Calcite	92.68	−1129.30	36.93

Using equation 19.15, ΔG^0 for the phase transition of one mole of calcite to aragonite at standard conditions (25 °C and 0.1 Pa) is

$$\begin{aligned}
\Delta G^0 &= (-1128.33 \text{ kJ/mol}) \times 1 \text{ mol} \\
&\quad \text{Aragonite (final composition)} \\
&\quad - (-1129.30 \text{ kJ/mol}) \times 1 \text{ mol} = +0.970 \text{ kJ} \\
&\quad \text{Calcite (initial composition)} \qquad (19.16)
\end{aligned}$$

Since the ΔG^0 value is larger than zero, the reaction must proceed in the direction opposite to that indicated by the arrow in the calcite–aragonite reaction. In other words, calcite is stable at standard conditions, whereas aragonite is not. At equilibrium ΔG^0, the free energy of the reaction, is zero. This can be used to determine the pressure at which calcite and aragonite are in equilibrium at 25 °C.

We first consider the difference in molar volume V^0 at 298 K between aragonite and calcite, which is

$$\Delta V_{298}^0 = 1 \text{ mol} \times 34.15 \text{ cm}^3/\text{mol}$$
$$\text{Aragonite (final composition)}$$
$$- 1 \text{ mol} \times 36.93 \text{ cm}^3/\text{mol} = -2.78 \text{ cm}^3$$
$$\text{Calcite (initial composition)} \qquad (19.17)$$

Since aragonite has the smaller molar volume, it is favored at high pressure. Using this value, we can then calculate the pressure P at which aragonite and calcite are in equilibrium at 25 °C, i.e., the pressure at which the free energy of the reaction $\Delta G_{P,T}^0 = 0$. To obtain the Gibbs free energy ΔG_P^0 at pressure P we have to add to the Gibbs free energy at standard conditions ΔG_T^0 obtained in equation 19.17, the mechanical work due to volume change:

$$\Delta G_{P,T}^0 = \Delta G_T^0 + P \,\Delta V_{298}^0 \qquad (19.18)$$

Because at equilibrium $\Delta G_{P,T}^0 = 0$, we can write: $0 = +0.970 \text{ kJ} + P \, (-2.78 \text{ cm}^3)$ and correspondingly, $P = 0.970 \text{ kJ}/2.78 \text{ cm}^3 = 349 \text{ J/cm}^3$. Using the conversion factor in Box 19.1 (1 MPa = 1 J/cm^3) we obtain the equilibrium pressure of 349 MPa for the transformation from calcite to aragonite at 25 °C. This gives us one point on the line that separates the stability fields of calcite and aragonite in the temperature–pressure phase diagram (Figure 19.2a).

For differential temperature and pressure changes during a reaction we can use the following expression of the first law:

$$\Delta G_{T,P}^0 = \quad \Delta V dP - \Delta S^0 dT \qquad (19.19)$$

At equilibrium, $\Delta G_{T,P}^0 = 0$ and, correspondingly (using equation 19.4),

$$dP/dT = \Delta S^0/\Delta V = \Delta H_T^0/(T \,\Delta V) \qquad (19.20)$$

This relationship, known as the Clausius–Clapeyron equation, defines the slope of the equilibrium line on the pressure–temperature phase diagram at the position of the point calculated above. Because only one variable (P or T) can be assigned arbitrarily along this line, it is said to be univariant.

From the thermodynamic data for calcite and aragonite we can calculate that $\Delta S^0 = 4.06$ J/degree, and we have obtained that $\Delta V^0{}_{298} = 2.78$ cm^3 (equation 19.17). Now, assuming that ΔS and ΔV are unaffected by temperature and pressure, we can write

$$dP/dT = (4.06 \text{ J/degree})/2.78 \text{ cm}^3$$
$$= 1.46 \text{ MPa/degree} \qquad (19.21)$$

With this we can finish constructing the P–T phase diagram shown in Figure 19.2a. Note that in this

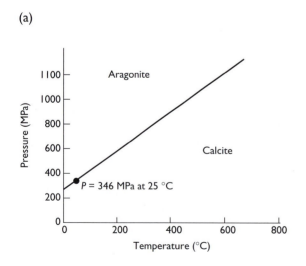

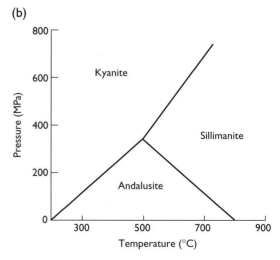

Fig. 19.2 P–T phase diagrams. (a) System CaCO$_3$, illustrating stability fields of aragonite and calcite. (b) System Al$_2$SiO$_5$, showing stability fields of polymorphs kyanite, andalusite, and sillimanite.

diagram the line along which the two minerals coexist is more or less straight because in solids the volume difference between two polymorphs is roughly constant and does not change much with pressure and temperature.

Figure 19.2b gives a phase diagram for the aluminosilicate polymorphs andalusite, kyanite, and sillimanite (Al_2SiO_5), which are important components of metamorphic rocks and whose presence can be used to estimate P–T formation conditions. Also in this diagram phase boundaries are straight because we are dealing with solids, but the Clausius–Clapeyron slopes between the phases are very different. The boundary kyanite/sillimanite has a positive slope whereas the boundary andalusite/sillimanite has a negative slope. The Clausius–Clapeyron slope is significant because it determines the change of stability of a mineral with pressure.

Stability of malachite and azurite: mass action law and partial gas pressure phase diagrams

The thermodynamic potentials and phase diagrams, such as those shown in Figure 19.2, refer only to the ultimate results of a process; they do not provide any information on how easy (or difficult) it is for a system to reach equilibrium. In other words, a system may be far from equilibrium, but it is prevented from attaining equilibrium by the sluggishness of the process. There are countless examples of such sluggish processes in the geological environment, including devitrification of volcanic glasses and the metastable existence of high-pressure and high-temperature minerals (e.g., diamond and coesite).

To a first approximation, the rate of a reaction such as $aA + bB \rightleftharpoons cC + dD$ depends on the concentrations a, b, c, d of the components A, B, C, D that are involved. At equilibrium, the rates of the forward (left side) and backward (right side) reactions are the same, and this condition is stated in the general mass action law that introduces the equilibrium constant k:

$$\frac{a_C^c\, a_D^d}{a_A^a\, a_B^b} = k \qquad (19.22)$$

where a_x are activities of the components. Activities are used if the components are dissolved in a solution, or if a mixture of gases is present. The pressure of air, for example, is the sum of the partial pressures of its individual components such as nitrogen, oxygen, and water vapor. The sum of all partial pressures is the total pressure, and the sum of activities adds to 1.0. The activities for solids and liquids involved in the reaction are 1.0.

For ideal gases ($VP = RT$ for one mole of gas, where R is the gas constant; see Box 19.1), there is a simple correspondence between the Gibbs free energy G and the chemical reaction equilibrium constant k:

$$\Delta G_{T,P}^0 - \Delta G^0 = RT \ln k$$
$$= 2.303\ RT \log k \qquad (19.23a)$$

where $\Delta G_{T,P}^0$ is the free energy change of the reaction at any state and ΔG^0 is the free energy change in the standard state (logarithms to base 10 are generally preferred for such calculations). At standard conditions (T = 298 K and $\Delta G_{T,P}^0 = 0$), then, we have (for ΔG^0 in kJ)

$$\log k^0 = -0.1750\,\Delta G^0 \qquad (19.23b)$$

In the oxidation zone of copper ores, some malachite and azurite may form. Can we predict which of them is more stable? Garrels and Christ (1990) considered the transformation of one to the other with the reaction

$$3Cu_2(CO_3)(OH)_2 + CO_2 \rightleftharpoons 2Cu_3(CO_3)_2(OH)_2 + H_2O$$

| Malachite | Gas | Azurite | Liquid |

This reaction is shown in the phase diagram of partial pressures $p_{O_2} - p_{CO_2}$ (Figure 19.3a).

The equilibrium constant of this reaction, for a partial CO_2 pressure p_{CO_2}, is

$$k = \frac{a_{azurite}^2\, a_{water}^1}{a_{malachite}^3\, p_{CO_2}} \qquad (19.24)$$

and, because activities of all pure solid and liquid substances are equal to 1,

$$k = 1/p_{CO_2} \qquad (19.25)$$

With free energies of formation (G_f^0) for azurite = -1429.56 kJ/mol, malachite = -900.41 kJ/mol, $H_2O = -237.23$ kJ/mol, and $CO_2 = -394.37$ kJ/mol, we can determine the free energy ΔG^0 for the malachite–azurite reaction, -0.75 kJ/mol, and from it the $\log k^0$ value (equation 19.23b), 0.13. The partial pressure of CO_2 at which the two minerals coexist in equilibrium is thus (equation 19.25)

$$\log p_{CO_2} = -0.13 \qquad (19.26)$$

Since p_{CO_2} at near-surface conditions is about $10^{-3.5}$ MPa ($\log p_{CO_2} = -3.5$), malachite is the more stable mineral (see the phase diagram in Figure 19.3a).

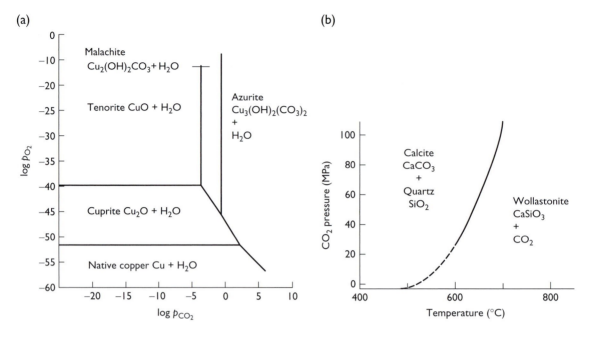

Fig. 19.3 (a) $p_{O_2} - p_{CO_2}$ phase diagram, illustrating the stability of copper minerals in the presence of water at 25 °C and 0.1 MPa (pressure is in bars, 1 bar = 0.1 MPa). (b) $P–T$ phase diagram for the reaction $CaCO_3 + SiO_2 \rightleftharpoons CaSiO_3 + CO_2$.

Garrels and Christ (1990) cited an interesting example for this mineralogical equilibrium: on many old paintings the sky has a greenish color. Paints made from blue azurite have transformed over time to the more stable green malachite.

Similar approaches can be applied to any mineral assemblage. Let us consider quartz-bearing limestones undergoing metamorphism. At high temperature the silicate mineral wollastonite ($CaSiO_3$) forms as a result of the following reaction:

$$CaCO_3 + SiO_2 \rightleftharpoons CaSiO_3 + CO_2$$
Calcite   Quartz   Wollastonite

The equilibrium constant k of this reaction is numerically equal to the partial pressure of CO_2:

$$k = p_{CO_2} \text{ or } \log k = \log p_{CO_2} \tag{19.27}$$

From equation 19.23b it follows that

$$\log p_{CO_2} = -\Delta G^0_{T,P}/(2.303RT) \tag{19.28}$$

Using this equation we can calculate the $P–T$ phase diagram shown in Figure 19.3b. Because the value of ΔG depends linearly on temperature and logarithmically on pressure, the equilibrium line separating the stability fields is not a straight line as it was between

calcite and aragonite, but rather is curved. This shape is typical of reactions involving a gas phase (e.g., CO_2).

19.4 Electrolytes and Eh–pH phase diagrams

Many minerals form in aqueous solutions – for example, in sedimentary mineral-forming environments and in hydrothermal processes. Let us review the chemistry of electrolytes. We can write the following equilibrium equation for the dissociation of water:

$$H_2O \rightleftharpoons H^+ + OH^-$$

Solutions where $a_{H^+} > a_{OH^-}$ are called acid, and those where $a_{H^+} < a_{OH^-}$ are termed alkaline. Solutions where $a_{H^+} = a_{OH^-}$, such as pure water (above), are neutral. For very dilute solutions, the activity of H_2O is a constant (1.0) and we obtain for the equilibrium constant

$$k = a_{H^+} \times a_{OH^-} \tag{19.29}$$

At 25 °C, k has a value of 1.0×10^{-14}. Since a_{H^+} is generally much smaller than 1.0, pH = $-\log a_{H^+}$ has been introduced as a measure to describe the concentration of hydrogen ions in a solution, with pH ranging between 0 and 14. Solutions with a low pH value are acidic, while those with a high pH are alkaline. The pH for pure water at 25 °C is 7.0 $(a_{H^+} = a_{OH^-} = \sqrt{10^{-14}})$. Most natural waters have a pH between 4 and 9.

Many reactions taking place near the Earth's surface in aqueous solutions can be described in terms of the energetics of the exchange of electrons and the activity of free H^+. When an atom or ion gains an electron, its valence is decreased and the element is said to be reduced. If it loses an electron, it is oxidized.

The ease with which the loss or gain of an electron takes place depends on the energy required to dislodge an electron from an outer shell and is measured as electrical work. This process is best understood in terms of a galvanic cell consisting of two half-cells, each with a metal electrode and an electrolyte (Figure 19.4). For example, in one half-cell we may have a zinc electrode with Zn^{2+} in solution in sulfuric acid, whereas in the other cell we have a copper electrode with Cu^{2+} in solution. Metal ions are prevented from mixing by a porous partition. If the two electrodes are connected, reactions occur. At the zinc electrode, metallic Zn gives up two electrons and dissolves as Zn^{2+}. The electrons flow from the zinc anode to the copper electrode, where Cu^{2+} in solution pick up two electrons and deposit as metallic Cu on the cathode. Sulfate ions migrate through the porous partition to maintain electrical neutrality of the solutions. The overall reaction is

$$Zn + Cu^{2+} \rightleftharpoons Cu + Zn^{2+}$$

We can divide this into two half-reactions, each taking place in one of the half-cells:

$$Zn \rightleftharpoons Zn^{2+} + 2e^- \text{ (oxidation)}$$

and

$$2e^- + Cu^{2+} \rightleftharpoons Cu \text{ (reduction)}$$

The combined reaction is referred to as a redox (reduction–oxidation) reaction. In the half-reactions described above, Zn is liberating electrons while Cu is collecting electrons. There is a flow of electrons from the zinc to the copper electrode, and a voltage V is recorded between the two half-cells. The electrical potential generated by the half-reactions is called the electromotive force (emf or E). As electrons flow from the Zn anode to the Cu cathode, the activity of Zn^{2+} increases and that of Cu^{2+} decreases until equilibrium is reached and the flow of electrons stops. At equilibrium the electrochemical cell obeys the mass action law (equation 19.22) and

$$k = a_{Zn^{2+}}/a_{Cu^{2+}} \qquad (19.30)$$

The equilibrium constant (k) and thus the emf (E) are a function of the ionic activities and of temperature. Standard potentials (E^0) are at 25 °C, 1 molal concentration, and a pressure of 1 bar (= 0.1 MPa) in the case of gases. The standard potential difference for the Zn–Cu cell is 1.1 V. In order to have a uniform scale of half-reaction potentials, one must define an arbitrary origin, for which chemists have chosen the half-reaction

$$\frac{1}{2}H_2 \rightleftharpoons H^+ + e^-$$

(known as the hydrogen half-reaction) as having zero potential. Standard redox potentials E^0 of half-reactions are then expressed against the hydrogen half-cell. In the case of the $Zn \rightleftharpoons Zn^{2+} + 2e^-$ half-reaction, E^0 is 0.76 V and that for $2e^- + Cu^{2+} \rightleftharpoons Cu$ is -0.34 V, with the difference between the two being 1.1 V. Table 19.3 lists some potentials for redox reactions observed in minerals.

The free energy difference ΔG^0 of a redox reaction is, according to the first law of thermodynamics,

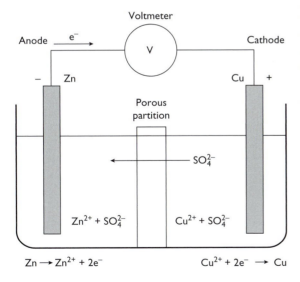

Anode

e^-

$-$ Zn

Voltmeter

V

Cathode

Cu $+$

Porous partition

SO_4^{2-}

$Zn^{2+} + SO_4^{2-}$

$Cu^{2+} + SO_4^{2-}$

$Zn \rightarrow Zn^{2+} + 2e^-$

$Cu^{2+} + 2e^- \rightarrow Cu$

Fig. 19.4 Galvanic cell with two electrodes, a zinc anode and a copper cathode. The two half-cells are divided by a porous partition and a voltmeter registers the potential.

Table 19.3 Standard potentials E^0 (in volts) and ΔG^0 (in kJ/mol) at 25 °C for some reactions in acidic solutions

	E^0	ΔG^0		E^0	ΔG^0
$K^+ + e^- \rightarrow K$	-2.93	-282.4	$Cu^{2+} + e^- \rightarrow Cu^+$	0.16	-15.44
$Ca^{2+} + 2e^- \rightarrow Ca$	-2.87	-552.7	$Cu^{2+} + 2e^- \rightarrow Cu$	0.34	64.85
$Na^+ + e^- \rightarrow Na$	-2.71	-261.9	$Cu^+ + e^- \rightarrow Cu$	0.54	50.21
$Zn^{2+} + 2e^- \rightarrow Zn$	-0.76	-147.3	$Cl_2 + 2e^- \rightarrow 2Cl^-$	1.36	-130.96
$Fe^{2+} + 2e^- \rightarrow Fe$	-0.41	-84.9	$Au^{3+} + 3e^- \rightarrow Au$	1.50	433.46
$2H^+ + 2e^- \rightarrow H_2$	0.0	0.0	$Au^+ + e^- \rightarrow Au$	1.68	163.18

strictly related to the electrical work and thus to the difference in redox potential E^0 by

$$\Delta G^0 = nFE^0 \tag{19.31}$$

where F is the Faraday constant (96.489 J/(V mol)), and n is the number of electrons transferred in the reaction. Note that equation 19.31 refers to oxidation reactions, such as the zinc oxidation reaction with electrons in the right-hand side. For reducing reactions (oxidized form $+ ne^- \rightarrow$ reduced form), equation 19.31 must be rewritten as $\Delta G^0 = -nFE^0$. In the case of the Cu–Zn cell, $\Delta G^0 = -212.54$ kJ and

$$
\begin{aligned}
E^0 &= \Delta G^0/nF \\
&= -212.54\,\text{kJ}/(2\ \text{mol} \times 96.485\ \text{kJ/V mol}) \\
&= -212.54\,\text{kJ}/(192.98\ \text{kJ/V}) = -1.10\ \text{V} \quad (19.32)
\end{aligned}
$$

Combining equations 19.23a and 19.31 we obtain

$$\Delta G^0_{T,P} = E^0/nF + RT \ln k \tag{19.33}$$

where $\Delta G^0_{T,P}$ is the free energy change of the reaction at any state that we can convert to emf for an electrode in any state relative to the H_2 electrode in standard state and call Eh:

$$
\begin{aligned}
\text{Eh} &= \Delta G^0_{T,P}/nF = E^0 + (RT/nF)\ln k \\
&= E^0 + (2.303\ RT/nF)\log k \quad (19.34a)
\end{aligned}
$$

Eh is the redox potential in the solution at any nonstandard condition (pressure, temperature, or concentration of compounds in the solution), R is the gas constant, and F is the Faraday constant. Now, if $T = 273.15$ K, we can simplify equation 19.34a by entering numerical values for R, F (Box 19.1), and T and n, the number of electrons, to give

$$\text{Eh} = \left[E^0 - (0.059/n)\right]\log k \tag{19.34b}$$

Let us first explore the stability of water by investigating the equilibrium

$$2H_2O^{(l)} \rightleftharpoons 2H_2^{(g)} + O_2^{(g)}$$

where superscript "l" denotes liquid and "g" gas. As molecular oxygen is one of the strongest oxidizing agents, and by far the most common found in nature, we can define the upper limit of Eh through the half-reaction (assuming $p_{O_2} = 0.1$ MPa $= 1$ bar):

$$2H_2O^{(l)} \rightleftharpoons 4H^{+(aq)} + 4e^- + O_2^{(g)}$$

where superscript "aq" denotes in solution. Solving this (equations 19.30 and 19.34a), we obtain (p_{O_2} and the activity of pure water are unity and $n = 4$)

$$
\begin{aligned}
\text{Eh} &= E^0 + (0.059/4)\left[\log\left(p_{O_2} a_{H^+}^4\right)/a_{H_2O}^4\right] \\
&= E^0 + (0.059/4)\,4\log a_{H^+} = E^0 - 0.059\text{pH}
\end{aligned}
$$
$$\tag{19.35}$$

This equation defines a straight line in Eh–pH space (dot-dashed in Figure 19.5 (top)), identifying fields where water is stable (below the line), and where O_2 forms (above the line). The slope of the line is -0.059 V per pH unit. The intercept E^0 is obtained from the free energy difference (equation 19.31), using values given in Table 19.4 for the dissociation reaction of water above (G for elemental gas is zero).

$$
\begin{aligned}
\Delta G^0 &= (4\ \text{mol} \times 0\ \text{kJ/mol} + 4\text{mol} \times 0\ \text{kJ/mol} \\
&\quad + 1\ \text{mol} \times 0\ \text{kJ/mol}) \\
&\quad - (2\ \text{mol} \times -237.23\ \text{kJ/mol}) \\
&= +474.47\ \text{kJ} \quad (19.36)
\end{aligned}
$$

and correspondingly $E^0 = \Delta G^0/4F = 474.47$ V/(4 $\times$ 96.485) = 1.23 V.

If the hydrogen partial pressure is 0.1 MPa $= 1$ bar (and the oxygen pressure negligible), we can write a half-reaction that defines the lower stability of water (versus formation of hydrogen gas):

Table 19.4 Gibbs free energies of formation of some ions and compounds (in kJ/mol)

	ΔG^0
Fe^{2+}	−84.94
Fe^{3+}	−10.53
Fe_2O_3	−740.99
Fe_3O_4	−1014.2
H_2O	−237.23

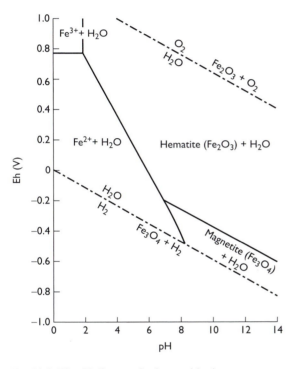

Fig. 19.5 Eh–pH diagram for iron oxides in water at 25 °C and 0.1 MPa, showing stability fields of hematite, magnetite, Fe^{3+}, and Fe^{2+}. The boundary for Fe^{3+} and Fe^{2+} is for an activity of 10^{-6}.

$$H_2^{(g)} \rightleftharpoons 2H^{+(aq)} + 2e^-$$

Solving this in a similar way we obtain (all free energies are zero) Eh $= 0 - 0.059$ pH. This line is shown dot-dashed in Figure 19.5 (bottom). Between the two dot-dashed lines is the stable field of water.

Next we would like to explore under which conditions magnetite and hematite are in equilibrium with water. For the oxidation of magnetite to hematite, we consider the reactions

$$2Fe_3O_4 + \tfrac{1}{2}O_2^{(g)} \rightleftharpoons 3Fe_2O_3$$

$$H_2O^{(1)} \rightleftharpoons 2H^{+(aq)} + \tfrac{1}{2}O_2^{(g)} + 2e^-$$

$$2Fe_3O_4 + H_2O^{(1)} \rightleftharpoons 3Fe_2O_3 + 2H^{+(aq)} + 2e^-$$
$$\text{Magnetite} \qquad\qquad\qquad \text{Hematite}$$

The addition of the first two reactions serves to eliminate O_2 gas and substitute for it H^+ and electrons as variables. Following the same procedure as before (equations 19.34 and 19.35), and using free energies from Table 19.4, we obtain for the magnetite–hematite reaction:

$$\begin{aligned}
\Delta G =\ & (3 \text{ mol} \times -740.99 \text{ kJ/mol}) \text{ (hematite)} \\
& - (2 \text{ mol} \times -1014.2 \text{ kJ/mol}) \text{ (magnetite)} \\
& - (1 \text{ mol} \times -237.23 \text{ kJ/mol}) \text{ (water)} \\
=\ & 42.66 \text{ kJ}
\end{aligned}$$

Now, in accordance with equations 19.31 and 19.34b, with $n = 2$ (2 electrons), activities for solids and water equal to 1, $\log(a_{H+}) = -$ pH, and $\log x^2 = 2 \log x$:

$$\begin{aligned}
\text{Eh} =\ & [42.66/(2 \times 96.485)] \\
& + (0.059/2) \times \log[(a_{Hem}^3\, a_{H+}^2)/(a_{Mgt}^2\, a_{Water})] \\
=\ & [0.221] + (0.059/2) \times [\log(1) - 2\text{ pH}] \\
=\ & 0.221 - 0.059 \text{ pH} \qquad\qquad (19.37)
\end{aligned}$$

This line is also shown in Figure 19.5. As we would expect, magnetite is more stable at lower pH and Eh than hematite.

Magnetite and hematite are slightly soluble in water. For magnetite we can write the reaction

$$3Fe^{2+(aq)} + 4H_2O^{(1)} \rightleftharpoons Fe_3O_4 + 8H^{+(aq)} + 2e^-$$

In this case, and following the procedure in equation 19.37 but activity for Fe^{2+} variable,

$$\begin{aligned}
\Delta G =\ & (1 \text{ mol} \times -1014.2 \text{ kJ/mol}) \text{ (magnetite)} \\
& - (3 \text{ mol} \times -84.94 \text{ kJ/mol}) \text{ (Fe}^{2+}) \\
& - (4 \text{ mol} \times -237.23 \text{ kJ/mol}) \text{ (H}_2\text{O)} \\
=\ & 189.54 \text{ kJ} \\[4pt]
\text{Eh} =\ & [189.54/(2 \times 96.485)] \\
& + (0.059/2) \times \log[(a_{Mgt}^3\, a_{H+}^8)/(a_{Fe^{2+(aq)}}^3\, a_{Water}^4)] \\
=\ & [0.982] + (0.059/2) \times [\log(1) - 8\text{ pH} \\
& - 3 \log a_{Fe^{2+(aq)}}] \\
=\ & 0.982 - 0.236 \text{ pH} - 0.089 \log a_{Fe^{2+(aq)}} \quad (19.38)
\end{aligned}$$

This is again a straight line but of a different (steeper) slope, and the position of the line (but not its slope) depends on the activity of Fe^{2+}. In Figure 19.5 the line for the reaction is drawn for an activity of 10^{-6} mol ions/liter, and a similar line is added for the equilibrium of hematite with Fe^{2+}.

Such Eh–pH diagrams are widely used in investigations of mineral equilibria, where aqueous solutions are present and oxidation–reduction reactions occur. Note that the condition Eh = E^0 is not a statement of equilibrium, but rather implies that the activities of all products and all reactants in a solution are unitary. The true equilibrium is achieved only when Eh = 0, ΔG_r = 0, and pressure and temperature remain constant. Eh–pH diagrams help us to understand the equilibrium of minerals and ions in seawater, in weathering oxidation processes, and in hydrothermal transformations.

19.5 Phase rule

The phase rule relates the number of minerals (phases) to the number of components in a system, and the number of possible reactions. It can be proven that in any thermodynamic system the number of phases p, the number of components c, and the number of degrees of freedom f are related in the following manner:

$$p = c + m - f \tag{19.39}$$

where the number m denotes the external parameters that have an effect on the state of the system. The number of degrees of freedom f corresponds to a number of thermodynamic parameters that are allowed to change without affecting the state of the system and its phase composition. In mineralogical systems such parameters are usually represented by external pressure, temperature, and chemical potentials.

The more complex the chemical composition of a geological system, and the more external factors affecting it, the greater the number of minerals expected to occur in this system. For those geological systems where the formation of minerals is only temperature and pressure dependent ($m = 2$):

$$p = c + 2 - f \tag{19.40}$$

The cooling of a magmatic melt is an example. Such a system ideally has an invariable bulk chemical composition. The external factors are largely temperature and pressure changes that cause crystallization

processes in the melt. Both pressure and temperature may be arbitrary in value, i.e., $f = 2$. Therefore,

$$p = c \tag{19.41}$$

This equality, introduced by V. Goldschmidt as the mineralogical phase rule, states that in geological systems, where temperature and pressure are the only external factors that may vary arbitrarily, the maximum number of minerals that may coexist in equilibrium is equal to the number of chemical components.

In a P–T phase diagram (e.g., Figure 19.2a,b) the stability fields of minerals have two degrees of freedom, since both pressure and temperature change independently. The lines separating stability fields correspond to one degree of freedom because only one parameter (either temperature or pressure) can vary independently. Junctions of lines (triple points) have no degrees of freedom. The equilibrium conditions described above are correspondingly called divariant, univariant, and invariant, respectively.

For example, in Figure 19.2b the triple point corresponds to the invariant conditions of kyanite, sillimanite, and andalusite coexistence. The lines indicate the univariant coexistence conditions of mineral pairs (kyanite $\rightleftharpoons$ andalusite; kyanite $\rightleftharpoons$ sillimanite; sillimanite $\rightleftharpoons$ andalusite). The fields between the lines correspond to the bivariant stability conditions of one mineral. There are three phases (three minerals) and one component (Al_2SiO_5) in the andalusite–kyanite–sillimanite system. At the invariant point, f is zero. Thus, the highest possible number of coexisting phases p is 3 ($p = 1 + 2 - 0 = 3$). Similarly, anywhere on the line of a univariant equilibrium $f = 1$. Thus p is equal to 2 ($p = 1 + 2 - 1 = 2$). Finally, within any field between the lines, $f = 2$. Therefore, p is equal to 1 ($p = 1 + 2 - 2 = 1$).

One of the major drawbacks of equation 19.40 is that many geological systems behave as open systems that are capable of exchanging volatile components with their surroundings, and hence c becomes a variable. Dmitriy Korzhinskii (1959) suggested that the phase rule had to be modified to account for the number of components that migrate in or out of the system, i.e., that behave as the mobile components M:

$$p = c + 2 - f - M \tag{19.42}$$

In this modified form, the mineralogical phase rule states that, in geological systems, the maximum number of minerals that may coexist in equilibrium is equal to the number of inert components.

19.6 Summary

The chapter reviews thermodynamic principles and explains how the three laws of thermodynamics can be used to understand phase relations. Based on intrinsic chemical properties such as enthalpy, entropy, and Gibbs free energy, phase diagrams can be constructed such as temperature–pressure. This is illustrated for the reactions dolomite + quartz $\rightleftharpoons$ diopside + CO_2, calcite $\rightleftharpoons$ aragonite, andalusite $\rightleftharpoons$ kyanite $\rightleftharpoons$ sillimanite and malachite $\rightleftharpoons$ azurite. More complex electrolyte systems are significant for hydrothermal systems as well as mineral dissolution and weathering. Finally the phase rule relates the number of mineral phases to chemical components and degrees of freedom (such as temperature and pressure).

Test your knowledge

1. Review with a mineral example the two basic laws of thermodynamics.
2. What are thermodynamic phases and components in geological systems?
3. Which potential is the most useful in geological contexts? Illustrate it with a mineral system.
4. What is the difference between E^0 and Eh?
5. Describe the phase rule with an example.

Further reading

Douce, A. P. (2011). *Thermodynamics of the Earth and Planets.* Cambridge University Press, Cambridge.

Faure, G. (1998). *Principles and Applications of Geochemistry*, 2nd edn. Prentice Hall, Englewood Cliffs, NJ.

Nordstrom, D. K. and Munoz, J. L. (2006). *Geochemical Thermodynamics*, 2nd edn. The Blackburn Press, Caldwell, NJ.

Wood, B. J. and Fraser, D. G. (1976). *Elementary Thermodynamics for Geologists.* Oxford University Press, Oxford.

20 | Phase diagrams

With the background in thermodynamics we are now ready to approach some phase diagrams to appreciate conditions under which minerals form. Most important are crystallization from a melt, phase transitions (relevant to the deep Earth), and solid solutions, including exsolution.

20.1 Introduction

Mineral (phase) equilibrium diagrams show the limits of stable existence of minerals at different conditions. They are plotted either on the basis of thermodynamic calculations and the phase rule, as illustrated in the previous chapter, or as a graphic representation of experimental results.

Phase diagrams, as a rule, are plotted as a function of two variables such as:

- temperature T versus total pressure P (Figure 19.2);
- temperature T versus partial pressure p (Figure 19.3);
- oxidation–reduction potential Eh versus pH (Figure 19.5);
- temperature T versus composition of system X.

Sometimes three or four variables are used, especially for complex compositions. Be aware that most phase diagrams of geological systems assume idealized situations.

The principles of interpretation of phase diagrams are straightforward. They allow us to determine, for example, which phase is in equilibrium at a certain temperature and pressure. Temperature–composition phase diagrams, however, which describe crystallization of a magma and subsolidus exsolution processes in minerals, deserve further elaboration.

20.2 Diagrams for crystallization from a melt

The binary system diopside ($CaMgSi_2O_6$)–anorthite ($CaAl_2Si_2O_8$) (Figure 20.1) is a classic example and has direct application to our understanding of crystallization processes in basaltic magmas. There are two components ($CaMgSi_2O_6$ and $CaAl_2Si_2O_8$) and one free parameter (temperature, $m = 1$) in this system. The solid lines correspond to the univariant equilibrium between two phases ($p = c + 1 - f = 2 + 1 - 1 = 2$ for equation 19.39), which are a mineral (anorthite or diopside) and a melt.

At point E, where the two univariant lines meet, the number of degrees of freedom is zero, while the number of coexisting phases is three ($p = c + 1 - f = 2 + 1 - 0 = 3$): diopside, anorthite, and melt. E is called the eutectic point and corresponds to the lowest temperature at which the melt and solid phases can coexist. The two univariant lines marking the onset of crystallization constitute the *liquidus* curve, whereas the horizontal line that meets the liquidus at E is called the *solidus*. The solidus line marks the temperature limit below which only solid phases are stably present.

For a composition X (in weight%) and a temperature above T_1 the melt is homogeneous and consists of 25% of the $CaMgSi_2O_6$ component and 75% of the $CaAl_2Si_2O_8$ component. When the temperature decreases and becomes less than T_1 (1490 °C at point I_1), anorthite begins to crystallize. With further cooling, the composition of the melt shifts to the left, since anorthite has been extracted as crystals. The relative amounts of crystals and remaining melt can be derived with the help of the lever rule shown with the diagram on the right-hand side of Figure 20.1. (It is called the lever rule because it balances crystals and liquid according to their relative amounts.) For example, at temperature T_2 the distance $I_2 - A$ is divided into $I_2 - X_2$ (crystals) and $X_2 - A$ (melt).

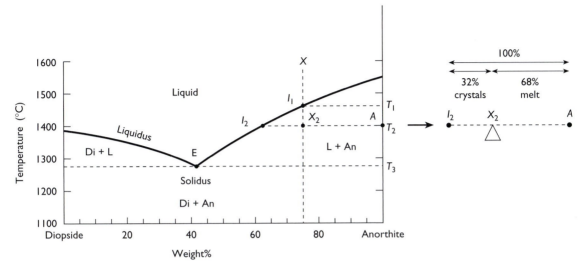

Fig. 20.1 Eutectic melting in the system diopside ($CaMgSi_2O_6$)–anorthite ($CaAl_2Si_2O_8$). Di, diopside; An, anorthite; L, liquid. For other symbols, see the text (data from Bowen, 1915).

At temperature T_2 (1400 °C) the crystals of anorthite make up 32% while the melt occupies 68%. The composition of the melt corresponds to that at point I_2 (40% $CaMgSi_2O_6$ and 60% $CaAl_2Si_2O_8$). As the temperature decreases, more anorthite crystallizes, the volume of the melt decreases, and its composition changes towards $CaMgSi_2O_6$. At temperature T_3 (1270 °C) the anorthite crystals make up 52%, and the melt 48%, again applying the lever rule. The composition of the remaining melt is 42% $CaAl_2Si_2O_8$ and 58% $CaMgSi_2O_6$.

At the eutectic temperature T_3 diopside may coexist with anorthite and melt in equilibrium, but the slightest loss of heat causes the simultaneous crystallization of the two minerals. The temperature remains constant until both minerals are fully crystallized. Only at this point does the system continue to cool.

The resultant rock has a microstructure with euhedral anorthite crystals (which crystallized freely in the melt as *phenocrysts*) within a mass of intergrown anorthite and diopside that crystallized at the same time (the *groundmass*). The ratio of total diopside and anorthite in the aggregate is the same as the initial ratio between the corresponding components in the melt X, i.e., 75% anorthite and 25% diopside. However, within the groundmass the proportion is different and corresponds to the composition of the melt at the eutectic E-point (58% diopside and 42% anorthite). Melting of this gabbroic rock will follow the same route in reverse.

Eutectic crystallization can result in a groundmass microstructure if many nucleation sites exist. If nucleation is limited, unusual intergrowths of the two minerals form that have a worm-like or cuneiform appearance.

Many mineral pairs are known to crystallize eutectically and some are listed in Table 20.1. We will discuss some more complex melting diagrams in Chapter 30.

20.3 Pressure–temperature phase diagrams and implications for the Earth mantle

Figure 20.2 shows a P–T phase diagram of Mg_2SiO_4, based on information from Anderson (1967) and Fei *et al.* (2004), which applies to the Earth's mantle and is very relevant for seismology because olivine (Mg, Fe)$_2$SiO$_4$ is the major component of the *upper mantle* and converts at around 15 GPa to wadsleyite of the same composition $Si(Mg,Fe)_2O_4$ but with a spinel

Table 20.1 Eutectic points for some mineral pairs (percentages of phases in groundmass given in parentheses)

Orthoclase (75.5)–quartz (24.5)	990 °C
Albite (96.5)–diopside (3.5)	1085 °C
Anorthite (42)–diopside (58)	1270 °C
Diopside (88)–forsterite (12)	1387 °C
Spinel (29)–forsterite (71)	1725 °C

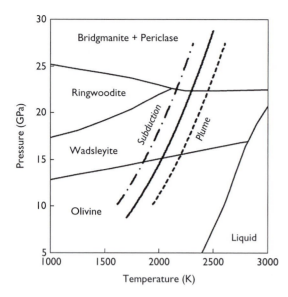

Fig. 20.2 *P–T* phase diagram of Mg$_2$SiO$_4$ (data from Fei *et al.*, 2005). This diagram is relevant to the Earth's mantle. Solid line is the average Earth geotherm, dot-dashed line is a colder subducting slab, displaced 100 K, and dashed line is a hot uprising plume.

crystal structure. At higher pressure and lower temperature wadsleyite converts to ringwoodite, also of the same composition but a different structure. The zone with these two minerals is called the *transition zone* and occurs between 410 and 650 km depth. Above 25 GPa ringwoodite and wadsleyite react to form bridgmanite (Mg,Fe)SiO$_3$ with a perovskite structure and ferropericlase (Mg,Fe)O. These two minerals dominate the lower mantle which is, volume-wise, the largest part of the Earth. We will return to this in Chapter 38.

During these phase transitions minerals acquire a higher density with pressure by having more compact crystal structures. In (Mg,Fe)SiO$_3$ perovskite, for example, silicon is in octahedral coordination, contrary to olivine and most other silicates. The density changes occur spontaneously during the phase transformations and cause velocity changes of seismic waves which are visible in seismic tomography maps (e.g., Schubert *et al.*, 2001; Fowler, 2005). It should be mentioned that the phase diagram in Figure 20.2 is for pure Mg$_2$SiO$_4$. In reality the minerals contain 10–30% iron, which causes shifts in phase boundaries (with increasing iron content boundaries move to lower pressure).

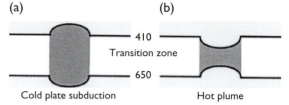

Fig. 20.3 (a) Due to the opposite Clapeyron slopes for transition zone minerals, the range of stability is enlarged for a cold subducting slab relative to the average mantle, with a bulge. (b) The transition zone is reduced for a hot uprising plume. Numbers indicate depth in km.

In the phase diagram (Figure 20.2) we note that the line of the olivine–wadsleyite transformation is inclined to the left (in terms of thermodynamics it has a positive Clapeyron slope). This means that heat is released (exothermic reaction). On the other hand, the ringwoodite–bridgmanite (spinel–perovskite) reaction has a negative Clapeyron slope, i.e., heat is consumed (endothermic reaction) and such reactions are less spontaneous.

In the phase diagram the average Earth geotherm (Dziewonski and Anderson, 1981) is drawn as a solid line. To the left of it is a dot-dashed line displaced by 100 K towards colder temperatures as one would expect for a cool subducting slab. The phase transitions are displaced and, due to the opposite Clapeyron slopes, the transition zone bulges relative to the average mantle, thickening about 10 km (Figure 20.3a). In contrast, in an upwelling hot plume (dashed line in the phase diagram) the transition zone is reduced (Figure 20.3b). This gives different seismic signatures for zones with subduction and upwelling which have indeed been observed.

20.4 Melting behavior of solid solutions

Solid solutions have a peculiar melting behavior. For example, in the olivine system (Fe$_2$SiO$_4$–Mg$_2$SiO$_4$), which is an example of a disordered and homogeneous solid solution, fayalite (Fe$_2$SiO$_4$) has a melting point that is over 685 °C lower than that of forsterite (Mg$_2$SiO$_4$) (Figure 20.4).

When a melt of composition 50% Fe$_2$SiO$_4$–50% Mg$_2$SiO$_4$ (*X*) cools to temperature T_1 (on the upper curve, which corresponds to the liquidus), crystallization begins. The composition of the olivine crystal is given by the lower curve (the solidus) at the point x_1.

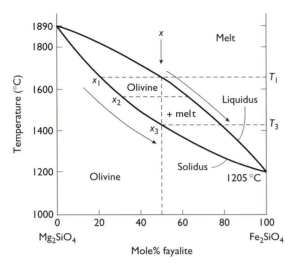

Fig. 20.4 Melting behavior in the olivine solid solution forsterite (Mg_2SiO_4)–fayalite (Fe_2SiO_4) with liquidus and solidus curves.

Since the crystal is enriched in magnesium (80% Mg_2SiO_4, 20% Fe_2SiO_4), the remaining melt becomes enriched in iron. Upon further cooling, the composition of the melt changes along the liquidus and that

of the crystal along the solidus (from point x_1 to point x_3). In this phase diagram there is melt above the liquidus, melt and crystal at conditions between the liquidus and solidus, and only crystal below the solidus. Using some thermodynamic arguments we can analytically derive this rather complicated diagram; an example of such a derivation is illustrated for another important system, plagioclase feldspar, in Box 20.1.

The composition of an olivine grain that interacts with the cooling melt changes gradually towards Fe_2SiO_4. If our system behaved as ideal, and the crystals continuously equilibrated with the melt to yield more iron-rich compositions, the crystallization process would finally cease at temperature T_3, where the composition of the crystals is identical with that of their parental liquid. However, in natural systems some of the early precipitated crystals are removed from the system by gravitational settling, or grow so rapidly that they have no chance to equilibrate. In such cases, lesser amounts of the low-melting point component are extracted from the melt to equilibrate the early crystals, and the liquid becomes even more enriched in that component. As a result, even when temperature T_3 is reached, there is still some melt

---

Box 20.1 Additional information: Derivation of the melting diagram for plagioclase (albite ($NaAlSi_3O_8$)–anorthite ($CaAl_2Si_2O_8$)

We write two chemical equilibria between crystal c and melt m, one for species A ($NaAlSi_3O_8$) and one for species B ($CaAl_2Si_2O_8$):

$$X_c A_c \rightleftharpoons X_m A_m \qquad (1 - X_c)B_c \rightleftharpoons (1 - X_m)B_m \qquad (20.1)$$

where X and $(1 - X)$ are mole fractions. Corresponding equilibrium constants are given by

$$k_A = X_m/X_c, \qquad k_B = (1 - X_m)/(1 - X_c) \qquad (20.2)$$

The change of an equilibrium constant with temperature T is, at constant pressure P, according to the Gibbs–Helmholtz equation,

$$(d \ln k/dT)_p = \Delta H/RT^2 \qquad (20.3)$$

We integrate over temperature and obtain for the equilibrium constants:

$$k_A = \exp(\Delta H_A/R)[(1/T_A) - (1/T)] \qquad (20.4a)$$
$$k_B = \exp(\Delta H_B/R)[(1/T_B) - (1/T)] \qquad (20.4b)$$

where ΔH_A and ΔH_B are molar heats of fusion (enthalpies) of the pure phases A and B, and T_A and T_B are their melting points. We next substitute k values from equation 20.4 into equation 20.2:

$$X_m = \exp(\Delta H_A/R)[(1/T_A) - (1/T)]X_c \qquad (20.5a)$$
$$1 - X_m = \exp(\Delta H_B/R)[(1/T_B) - (1/T)](1 - X_c) \qquad (20.5b)$$

Continued

Box 20.1 (cont.)

For plagioclase feldspar we know the melting points and the heat of fusion:

Albite : $T = 1370$ K $\Delta H = 53.22$ kJ/mol
Anorthite : $T = 1823$ K $\Delta H = 121.34$ kJ/mol

Now we solve equations 20.5a and 20.5b for X_m and X_c. For example we obtain

T (K)	X_c	X_m
1500	0.6221	0.9326
1600	0.4126	0.8075
1700	0.2294	0.5682

and construct the T–X phase diagram (Figure 20.5), where X is the mole fraction of albite. The experimental results of Bowen (1913) are more or less identical.

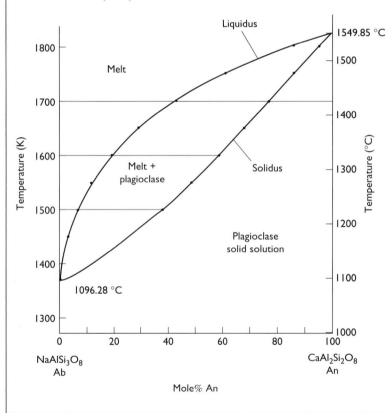

Fig. 20.5 Liquidus and solidus melting curves for plagioclase feldspar albite (NaAlSi$_3$O$_8$)– anorthite (CaAl$_2$Si$_2$O$_8$). Dots are calculated values in 50 degree temperature intervals; lines are experimentally determined curves.

available, and crystallization continues further towards fayalite. Crystallization of olivine in the geological environment most typically follows a nonequilibrium path producing crystals with a magnesium-rich core and an iron-rich rim. Many ferromagnesian silicates (micas, amphiboles, pyroxenes) show a similar behavior

(e.g. Plate 14f). In plagioclases, anorthite melts at a higher temperature than albite (see Figure 20.5). Correspondingly, plagioclase crystals often display a zoning pattern with calcium- and aluminum-enriched compositions in the core, and zones with progressively increasing sodium content towards the rim (see Plate 7a).

20.5 Exsolution

At high temperatures, alkali feldspars form a continuous solid solution with a homogeneous crystal. As we have discussed briefly in Chapter 9 (concerning crystal defects) and will further emphasize in Chapter 21 (when we turn our attention to feldspars), the attraction of unlike atoms in a solid solution leads to an ordering upon cooling (see Figure 9.11). Vice versa, the attraction of like atoms causes separation of a homogeneous crystal into local domains of differing composition, and we call this exsolution. Ordering and exsolution occur after a host crystal has solidified, and are therefore called a subsolidus transformation.

The subsolidus behavior of a solid solution crystal can be understood by considering variations of the

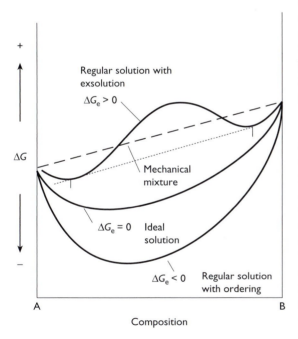

Fig. 20.6 Gibbs free energy of a solid solution A–B as a function of composition at a given temperature. For a mechanical mixture the composition dependence is a straight line (dashed). In an ideal solution ΔG results from the entropy of mixing. For a regular solution there is an excess energy ΔG_e that depends on the interaction between atoms. It can be positive (with a tendency for exsolution), or negative (with a tendency for ordering). The dotted line is the tangent to the regular solution with exsolution, defining the two compositions that are in equilibrium.

Gibbs free energy ΔG with composition and evaluating which system has the lowest free energy. If a crystal were a simple "mechanical mixture" of two components A and B, we could estimate the free energy of an intermediate composition at a given temperature T by a linear interpolation of the end members (dashed line in Figure 20.6). This is not realistic because, even if there is no interaction between substituting atoms, the different possibilities of arranging A and B on structural sites introduces a configurational entropy of mixing that reduces the free energy of intermediate compositions (the "ideal solution" curve in Figure 20.6). The solid solution Mg–Fe in olivine comes close to such ideal behavior because Mg and Fe have the same charge, similar size, and similar electronic configurations. In other systems there is interaction between substituting atoms. If unlike atoms A and B attract each other, the free energy is further reduced (the excess free energy ΔG_e is negative) and ordering occurs, with an intermediate structure that has a lower free energy than the end members. If there is repulsion between unlike atoms, the excess free energy ΔG_e is positive and the total free energy is augmented, resulting in a bulge in the free energy curve. Let us look at this second case in more detail.

Figure 20.7a shows free energy curves for a solid solution A–B at three temperatures. Notice that with increasing temperature ($T''' < T'' < T'$) the free energy curves more closely approximate that of the ideal solution. The composition of the two phases X and Y that are in equilibrium are determined by constructing the tangent to the two minima of the free energy curve. For temperature T'' the two compositions are X'' and Y''; for T''' they are X''' and Y'''. For temperature T' there is only a single minimum in the free energy curve and therefore a single composition corresponding to that of the mixture. Figure 20.7b is a phase diagram plotting the composition of equilibrium phases as a function of temperature. The curve labeled "equilibrium solvus" indicates the composition of stable phases at some temperature. Above the solvus the crystal is homogeneous; below it, the crystal consists of two phases. If we know the composition of the two phases, A_x and B_x, in Figure 20.7b, we can estimate the temperature T_x at which exsolution occurred.

There are different mechanisms for exsolution that may take place when a homogeneous solid solution is cooled below the solvus. Each process requires diffusion, at least on a local scale, and exsolution is

(a)

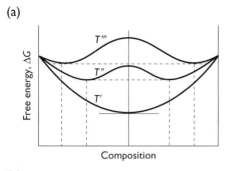

(b)

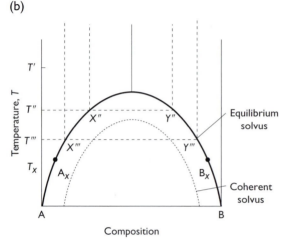

Fig. 20.7 (a) Gibbs free energy for a solid solution with a tendency for exsolution, for three temperatures, T', T'', and T'''. (b) T – composition phase diagram, illustrating the solvus curve below which exsolution occurs. For definition of symbols, see the text.

therefore sluggish. Homogeneous crystals can be preserved if they are quenched rapidly, as in some volcanic rocks. Some mechanisms occur by nucleation of the new, pure phases. Since nuclei form only when

the system is cooled below the equilibrium solvus, the actual "coherent" solvus curve (dotted line in Figure 20.7b) is depressed and exsolution occurs at a lower temperature. Undercooling is not required, if nuclei form on defects, such as dislocation clusters. Other exsolution mechanisms occur homogeneously throughout the crystal. The different exsolution mechanisms produce characteristic microstructures. All exsolution microstructures are generally lamellar along the plane, with minimum misfit between the two structures, and the size of the lamellae is larger for slowly cooled crystals (Figure 20.8a) than it is for those that are rapidly quenched (Figure 20.8b).

Exsolution phenomena have been observed in many igneous and metamorphic minerals. Some examples are given in Table 20.2. Crystal structures of one or all of the exsolved phases may differ significantly from that of the parental homogeneous compound, for example decomposition of $(Fe,Ti,Mg)_3O_4$ (titanomagnetite) into magnetite and ilmenite.

20.6 Summary

Phase diagrams are used to visualize conditions under which minerals occur. They are either based on thermodynamic calculations or on experiments. Important conditional parameters are temperature and pressure. They are plotted relative to composition. In later chapters we will explore two- and three-dimensional phase diagrams. Here we introduce simple two-dimensional diagrams: cooling of a melt composed of anorthite and diopside with a eutectic point; temperature–pressure phase transitions in the Earth's mantle and temperature effects with solid solutions for olivine and plagioclase that can lead to exsolution during cooling.

(a)

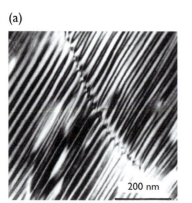

(b)

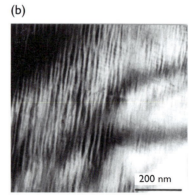

Fig. 20.8 TEM images illustrating exsolution in lunar pigeonite. (a) Coarse exsolution lamellae parallel to (001). (b) Wavy fine exsolution structure due to rapid cooling (from Champness and Lorimer, 1971).

Table 20.2 Examples of exsolution in minerals

Albite	K-feldspar	(Na–K)
Nepheline	Kalsilite	(Na–K)
Albite	Anorthite	(Na–Ca)
Augite	Pigeonite	(Ca–Mg/Fe)
Augite	Hypersthene	(Ca–Mg/Fe)
Hematite	Ilmenite	(Fe–Ti)
Ilmenorutile	Columbite	(Nb, Ti, Fe, Mn, Sn, Sc)
Bornite	Chalcocite	(Fe–Cu)
Bornite	Chalcopyrite	(Fe–Cu)
Chalcopyrite	Cubanite	(Fe–Cu)
Pyrrhotite	Pentlandite	(Fe, Ni, Co)
Kamacite	Taenite	(Fe–Ni)
Calcite	Strontianite	(Ca–Sr)

Test your knowledge

1. Follow the cooling path as olivine of intermediate composition crystallizes from a melt.
2. Calculate the composition of melt and a plagioclase crystal in equilibrium with the melt for a temperature of 1500 K. Use the equations in Box 20.1.
3. Alkali feldspars are a typical example of a system where exsolution occurs upon cooling. What happens as an albite with 10% orthoclase content crosses the solvus?
4. Describe some microstructures observed in minerals that have undergone exsolution.

Further reading

Christian, J. W. (2002). *The Theory of Phase Transformations in Metals and Alloys (Parts I+II)*. Pergamon, New York.

Ehlers, E. G. (1972). *The Interpretation of Geological Phase Diagrams*. W. H. Freeman, San Francisco.

Gasparik, T. (2013). *Phase Diagrams for Geoscientists: An Atlas of the Earth's Interior*, 2nd edn. Springer.

Hillert, M. (2007). *Phase Equilibria, Phase Diagrams and Phase Transformations: Their Thermodynamic Basis*, 2nd edn. Cambridge University Press, Cambridge.

Morse, S. A. (1980). *Basalts and Phase Diagrams: An Introduction to the Quantitative Use of Phase Diagrams in Igneous Petrology*. Springer-Verlag.

Part V | A systematic look at mineral groups

21 | Important information about silica minerals and feldspars. Their occurrence in granites and pegmatites

In Part V we will discuss the properties and occurrences of important minerals. There will first be a discussion of their general structures, then some details about the minerals in that group, and finally a section on typical geological occurrences or significant applications. We decided to start our discussion of specific minerals not with the standard classification introduced in Table 5.1. Instead we start with the most common minerals in the continental crust, quartz and feldspar. We must understand these first in the context of rock-forming processes and mineral parageneses. Without quartz and feldspar it is difficult to discuss igneous, metamorphic, and sedimentary mineral-forming processes. There are complexities, with many silicon dioxide polymorphs and a wide range of feldspars with solid solutions. The mineral descriptions will be followed with a section on granites and pegmatites in which quartz and feldspars dominate.

21.1 Silica minerals

The silica minerals, with an overall composition SiO_2, include many polymorphs (Table 21.1). *Quartz* is the most common member, occurring both in a trigonal low-temperature form (α-quartz) and a hexagonal high-temperature form (β-quartz). Other important silica polymorphs are α- and β-tridymite, α- and β-cristobalite, coesite, and stishovite. Opal is a solid silica gel containing a large amount of water. The stability fields of some silica polymorphs are shown in Figure 21.1.

In all these polymorphs, except for stishovite, silicon is surrounded (coordinated) by four oxygen atoms, forming a tetrahedral SiO_4^{4-} group (Figure 21.2a). The corners of the tetrahedron represent oxygen atoms, and a silicon atom is in the center. Often the atoms are omitted in the representation and only the corners of the tetrahedron are shown. The edges represent O–O bonds (Figure 21.2b). A tetrahedron that is viewed along an apex looks like a trigonal pyramid (Figure 21.2c). The bonds between silicon and oxygen are complex ionic–covalent. The

tetrahedra build up an infinite three-dimensional framework by sharing each oxygen corner with another tetrahedron. There are many ways such linkages can be made, and the manner in which linkage occurs determines the resultant crystal structures of the various polymorphs. Their symmetry may be cubic, hexagonal, tetragonal, trigonal, orthorhombic, or monoclinic. Although the basic building unit, the tetrahedron, is very simple, the ways of combining the tetrahedra in a crystal structure are intricate and diverse.

Tridymite is the simplest silica polymorph structure in which SiO_4^{4-} tetrahedra link to hexagonal rings and the rings link further to an infinite hexagonal net (Figure 21.3a). Apices of tetrahedra point alternately upwards and downwards and the A layers connect with B layers that are the mirror image of A layers. In Figure 21.3b the B layer is shown on the right side and illustrates that a downward-pointing tetrahedron in the B layer is exactly above an upward-pointing tetrahedron in the A layer. This results in a set of 6-fold "tunnels" bordered by the silicate rings and

Table 21.1 Silica minerals and feldspars, with some diagnostic properties; important minerals are given in italics

Mineral Formula	System Space-group	Morphology *Cleavage*	H	D	Color	n	Δ	2V (dispersion)
Silica minerals								
Quartz (α) SiO_2	Trigonal $P3_121$	Pris./Pyr.	7	2.65	Clear, (violet, yellow, brown)	1.544–1.553	0.009	(+)
Tridymite (α) SiO_2	Orthorhombic $C222_1$	Platy (0001) *(010)*	7	2.26	White	1.47–1.48	0.004	+35
Cristobalite (α) SiO_2	Tetragonal $P4_32_12$	*{100}, {111}*	6–7	2.32	Clear	1.48–1.49	0.003	(−)
Coesite SiO_2	Monoclinic $C2/c$	Tab. $(01\bar{1}0)$	7–8	2.92	Clear, white	1.59–1.60	0.005	+64 (r < v)
Stishovite SiO_2	Tetragonal $P4_2/mnm$		7–8	4.29	Clear	1.80–1.83	0.027	(+)
Opal $SiO_2 \cdot nH_2O$	Amorphous		5.5–6	2.1	White, colored	1.3–1.45		
Feldspars								
Microcline $KAlSi_3O_8$	Triclinic $C\bar{1}(c=7.6$ Å$)$	Platy (010) *(001), (010)*	6–6.5	2.56	White, red, green	1.518–1.526	0.006	−60–84 (r > v)
Orthoclase $KAlSi_3O_8$	Monoclinic $C2/m(c=7.6$ Å$)$	Platy (010) *(001), (010)*	6–6.5	2.55	White, red, green	1.518–1.530	0.006	−60–80 (r > v) AP n(010)
Sanidine $KAlSi_3O_8$	Monoclinic $C2/m(c=7.6$ Å$)$	Platy (010) *(001), (010)*	6	2.58	Clear, (yellow, gray)	1.518–1.532	0.006	−0–20 AP p(010)
Albite low $NaAlSi_3O_8$	Triclinic $C\bar{1}(c=7.6$ Å$)$	Platy (010) *(001), (010)*	6	2.61	Clear, white	1.529–1.539	0.010	+77
Albite high $NaAlSi_3O_8$	Triclinic $C\bar{1}(c=7.6$ Å$)$	Platy (010) *(001), (010)*	6	2.61	Clear, white	1.527–1.534	0.007	−50
Anorthite $CaAl_2Si_2O_8$	Triclinic $P\bar{1}(c=15.2$ Å$)$	Platy (010) *(001), (010)*	6–6.5	2.77	White, (green)	1.575–1.590	0.013	−77
Celsian $BaAl_2Si_2O_8$	Triclinic $P\bar{1}(c=15.2$ Å$)$	Platy (010) *(001), (010)*	6.5	3.4	White	1.585–1.595	0.010	−65–(+)95 (r > v)

Notes: H, hardness; D, density (g/cm³); n, range of refractive indices; Δ, birefringence; 2V, axial angle for biaxial minerals. For uniaxial minerals (+) is positive and (−) is negative. Acute 2V. If 2V is negative the mineral is biaxial negative and 2V is $2V_\alpha$; if it is positive, the mineral is biaxial positive and 2V is $2V_\gamma$. Dispersion r < v means that acute 2V is larger for violet than for red. AP, axial plane; n normal, p parallel. *Space-group*: c, lattice parameter. *Morphology*: Pris., prismatic; Pyr., pyramidal; Tab., tabular. *Colors*: Secondary colors are given in parentheses.

extending parallel to the *c*-axis of this hexagonal structure. Every third layer is exactly on top of the first layer. Such a stacking sequence of layers may be designated as AB–AB–AB. This (slightly idealized) β-tridymite is a high-temperature polymorph and occurs in volcanic rocks. However, this structure crystallizes easily and is also favored (kinetically, not thermodynamically) at low temperature, when silica, for example, precipitates from seawater as skeletons of diatoms. The six-membered ring is one of the most stable polymerized forms of silica in water and we will encounter it again in the structure of sheet silicates such as clays.

Cristobalite (also occurring in volcanic rocks and siliceous sediments) is a modification of the tridymite structure. In this case all layers are the same (i.e., adjacent layers are not related by mirror symmetry), but displaced to obtain connectivity between layers. The scheme is illustrated in Figure 21.3c,

with three layers A, B, and C. The fourth layer is again exactly above the first layer and defines a stacking order ABC–ABC rather than AB–AB as in hexagonal tridymite. This is analogous to hexagonal and cubic close-packing in metals; indeed the ideal structure of cristobalite is also cubic (Figure 21.4a).

Coesite is a rare high-pressure polymorph of SiO_2 that occurs at meteorite impact sites and in ultrahigh-pressure metamorphic rocks such as Dora Maira in Italy and Dabie Shan in China. In the monoclinic structure of coesite, rings of four tetrahedra are linked to chains, and the lower chains are linked by a second higher set of chains, connecting the free tetrahedral apices (Figure 21.4b). Silicon in coesite is in tetrahedral coordination as in cristobalite and tridymite.

At higher pressure coesite transforms to *stishovite* (Figure 21.1). The structure is fundamentally different: each silicon atom is surrounded by six oxygen atoms in the shape of an octahedron, and these octahedra are linked by sharing edges and corners similar to the titanium oxide mineral rutile (see Chapter 27). The ionic radius ratio for $Si^{4+}:O^{2-}$ conforms to tetrahedral coordination (Chapter 2). However, at very high pressures, oxygen atoms are no longer ideally spherical and then octahedral coordination allows for denser packing. Stishovite has a much higher density (4.4 g/cm^3) than the low-pressure silica minerals (about 2.6 g/cm^3), and even coesite (2.9 g/cm^3). Stishovite has only been found in meteor craters where siliceous rocks have been transformed by shock pressures during a meteorite impact. It may also occur in the Earth's lower mantle, transforming from siliceous subducting sediments. At such ultrahigh pressures conditions, there may also be an orthorhombic SiO_2 phase with $CaCl_2$-type structure (e.g., Nomura *et al.*, 2010).

β-quartz is hexagonal. As was briefly discussed in Chapter 3, it exists only above 573 °C. It is a rare mineral. Figure 21.5a shows a projection of

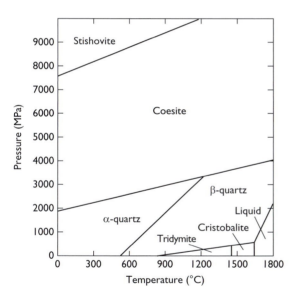

Fig. 21.1 Pressure–temperature phase diagram of SiO_2 with stability fields of polymorphs.

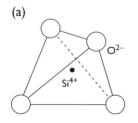 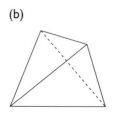 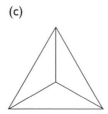

Fig. 21.2 SiO_4^{4-} tetrahedron. The O atoms are at the apices and the Si atom is in the center. (a) Representation with ions shown. (b) Representation without ions. (c) View from above an apex.

(a)

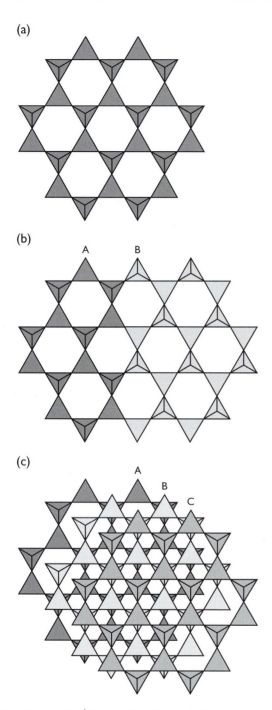

(b)

A   B

(c)

A

B

C

Fig. 21.3 (a) SiO_4^{4-} tetrahedra linked to a hexagonal net with apices pointing alternately up and down. (b) Linkage of A nets and mirrored B nets (right side) in the hexagonal structure of ideal tridymite. (c) Stacking of three layers, A, B, and C by tetrahedral linkage in the structure of ideal cristobalite.

high-temperature hexagonal β-quartz along the c-axis. As in tridymite and cristobalite, the structure of quartz consists of an infinite three-dimensional framework of linked SiO_4^{4-} tetrahedra. Each oxygen corner of a tetrahedron is shared with another one and we can only count it as one-half per tetrahedron, resulting in an overall oxygen content of $4 \times \frac{1}{2} = 2$ and one silicon per tetrahedron, i.e., a molecular composition of SiO_2. Upon cooling, tetrahedra become tilted, resulting in a less symmetrical trigonal structure (Figure 21.5b) (α-quartz). This transformation involves only slight displacements of atoms, with no breakage of bonds, and is therefore instantaneous and reversible.

The mineral *α-quartz* comprises about 12 volume% of the Earth's crust and, after feldspars, is the second most abundant mineral. The framework structure of α-quartz is more complex than that of *β-quartz* or either tridymite or cristobalite. Careful inspection of the structure reveals two 3-fold spirals of tetrahedra, resembling a screw (indicated by arrows). Tetrahedra are brought to coincidence by a counterclockwise $120°$ rotation and a one-third translation (z-coordinates of Si^{4+} in the center of tetrahedra in the lower right side are labeled and equivalent ones are marked with the same shading). Depending on the handedness of the screw, two varieties of quartz can be distinguished: left-handed α-quartz with a counterclockwise rotation (Figure 21.5b) and right-handed α-quartz with a clockwise rotation for the spirals (Figure 21.5c).

Both the symmetry and the handedness are expressed in the morphology. Ideally, crystals of hexagonal β-quartz show the six well-developed faces of a hexagonal prism $m = \{10\bar{1}0\}$ and are capped above and below by a bipyramid $r = \{10\bar{1}1\}$ (Figure 21.6a). (Conventionally, letters are often assigned to label crystal forms.) In rare cases the prism faces are missing or are very small, and this gives β-quartz the distant resemblance of diamond (the so-called "Herkimer diamond", Figure 21.6b).

In trigonal α-quartz six faces at every apex of the crystal consist of two rhombohedra. One rhombohedron $r = \{10\bar{1}1\}$ is generally larger than the other $z = \{01\bar{1}1\}$, which is rotated $60°$ relative to $\{10\bar{1}1\}$ (Figure 21.6c,d). The handedness is expressed in the disposition of minor forms such as $x = \{5\bar{1}\bar{6}1\}$ (Figure 21.6c versus d). A characteristic feature of α-quartz is a horizontal striation on the prism faces (see Figure 10.4b, Plate 2a).

(a)

(b)

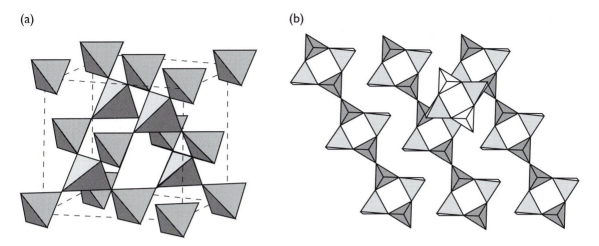

Fig. 21.4 (a) Tetrahedra in the cubic face-centered arrangement of cristobalite. (b) Structure of coesite with emphasis on chains.

The striation originates from minute growth steps during combined prismatic and rhombohedral growth. We have seen earlier that the morphology of quartz is also indicative of formation conditions (see Figures 18.8 and 18.9).

Twin intergrowths are very common in α-quartz (see discussion in Chapter 9, Box 9.1). Brazil twins are an intergrowth of left- and right-handed quartz, both related by a center of symmetry (see Figure 9.10c). Dauphiné twins in trigonal α-quartz are related by a 2-fold rotation about the *c*-axis (see Figure 9.10b). The twin laws are named after the localities where specimens were first found.

During the phase transformation from hexagonal to trigonal quartz, displacement of oxygen atoms may occur in one of two possible directions (as indicated by differently sized arrows in Figure 21.7a). The two possible configurations are related by a 60° (180°) rotation about the *c*-axis, i.e., they have a twin relationship. Different parts of a crystal may choose one or the other trigonal arrangement, resulting in Dauphiné twins. Such *transformation twins* on a fine scale have been observed by electron microscopy at temperatures in the vicinity of the α–β phase transformation (Figure 21.8).

Yet another origin for Dauphiné twins in quartz is mechanical. If we project the crystal structure along the *c*-axis (Figure 21.7b), we notice that, in α-quartz, alternating tetrahedral apices (oxygen atoms) along the spiral of tetrahedra are closer and further removed from the rotation axis, respectively. If a shear stress is applied, this order can be reversed

by a slight displacement, without breaking any bonds (top and bottom part in Figure 21.7c). Such configuration pairs are called *deformation* (or *mechanical*) *twins* because they are caused by shearing stress.

The structural features of quartz are expressed in its properties. Strong, partially covalent bonds and the three-dimensional framework are responsible for the great hardness of quartz (Mohs scale 7) and its lack of cleavage and conchoidal fracture. The open framework is the reason for the low density (2.65 g/cm³) of quartz. The good transparency of quartz crystals, the low refractive index (about 1.5), and the vitreous luster can also be attributed to the bonding. Since the structure of α-quartz lacks mirror planes and a center of symmetry, α-quartz displays piezoelectricity (Chapter 12) and optical activity (Chapter 13).

21.2 Feldspars

Feldspars (Table 21.1), minerals with the general formula $XAl(Si,Al)Si_2O_8$ with X = K, Na, Ca, deserve a special place among minerals because of their great abundance in the Earth's crust (about 43 vol.%). If we ignore feldspars, we cannot understand rocks. They occur in almost every metamorphic and igneous rock and show subtle variations in crystal structure that allow petrologists to gain information about the chemical and physical conditions prevailing during rock formation. Feldspars have fascinated generations of mineralogists, and in the 1950s there

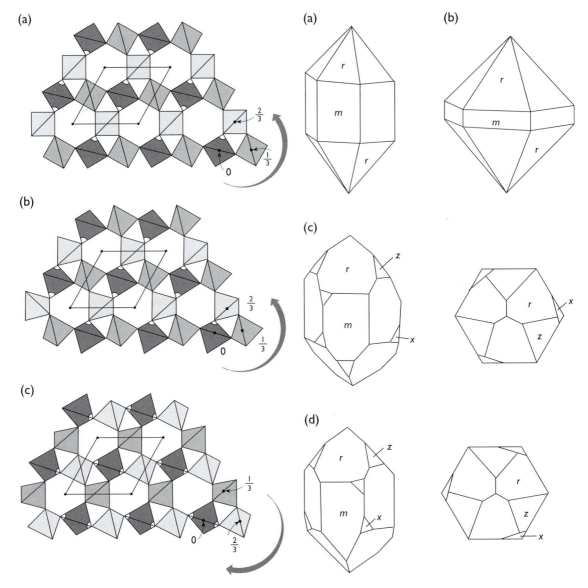

Fig. 21.5 Projection of the structure of quartz along the *c*-axis, with spirals of SiO$_4^{4-}$ tetrahedra (indicated by arrows). Tetrahedra are at different levels and do not connect to a three-membered ring. Fractions in the lower right-hand corner label the *z*-axis atomic coordinates of Si atoms and tetrahedra at different levels are shaded differently. The hexagonal unit cell is indicated. (a) Hexagonal left-handed β-quartz, (b) trigonal left-handed α-quartz, and (c) trigonal right-handed α-quartz.

Fig. 21.6 Morphology of quartz. (a) Hexagonal prism $m = \{10\bar{1}0\}$ and hexagonal pyramid $r = \{10\bar{1}1\}$. (b) Hexagonal bipyramid (dominating in "Herkimer diamond"). (c,d) Prism and two rhombohedra $r = \{10\bar{1}1\}$ and $z = \{01\bar{1}1\}$ and a trapezohedron $x = \{51\bar{6}1\}$ in trigonal α-quartz. (c) Left-handed and (d) right-handed forms, based on the disposition of the trapezohedron.

were revealing pioneering studies into the heterogeneous crystal structures of feldspars (see Box 21.1).

The structure of feldspars is related to that of silica minerals: the SiO$_4^{4-}$ tetrahedron is the main building element and tetrahedra are linked to a three-dimensional framework, though the framework is less symmetrical than in quartz. One difference between feldspars and silica minerals is that in the former some

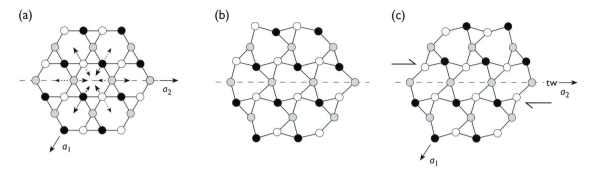

Fig. 21.7 Schematic view of the structure of (a) hexagonal β-quartz and (b) trigonal α-quartz. Projection along the *c*-axis. Only silicon atoms are shown and shades indicate different levels along *c*. The triangles correspond to the spirals in Figure 21.5 marked with arrows. (c) Dauphiné twin (top–bottom) of α-quartz produced by shear; tw is the twin plane. Arrows in (a) indicate schematically the displacement of silicon atoms that occurs during the β–α phase transformation.

0.5 μm

Fig. 21.8 TEM image illustrating triangular Dauphiné twins in the vicinity of the α–β phase transition (courtesy G. Van Tendeloo).

of the Si^{4+} are replaced by Al^{3+}. In feldspars all aluminum is in tetrahedral coordination, and the feldspars are therefore chemically referred to as *alumosilicates* as opposed to *aluminosilicates* in which at least some Al^{3+} is in octahedral coordination. The SiO_2 framework in silica minerals is charge balanced. To balance the charges in feldspars, caused by the Si–Al substitution, additional cations need to be added and these are placed in the cavities within the framework. Large alkali and alkaline earth atoms are introduced, leading to considerable variation in chemical composition. Most feldspars in rocks are solid solutions with variable amounts of potassium, sodium, calcium, and

some barium. Their names and compositional ranges are summarized in Figure 21.9a in a triangular diagram (instructions on how to use and read ternary diagrams have been given in Section 4.5). The pure end members are anorthite ($CaAl_2Si_2O_8$; Ca-feldspar, abbreviated An); albite ($NaAlSi_3O_8$; Na-feldspar or Ab); and the K-feldspars ($KAlSi_3O_8$) sanidine, orthoclase, and microcline (usually referred to as Or). The composition of feldspars is often expressed in percentages (given as subscripts) of the three end members anorthite, albite, and orthoclase, for example $An_{10}Ab_{85}Or_5$. The series between albite and orthoclase is known as the *alkali feldspars*. The series between albite and anorthite is known as the *plagioclase feldspars* or simply *plagioclase*. Additional varietal names are assigned to intermediate plagioclases on the basis of their composition: oligoclase (An_{10}–An_{30}), andesine (An_{30}–An_{50}), labradorite (An_{50}–An_{70}), and bytownite (An_{70}–An_{90}).

This range in composition for plagioclase can be understood if we consider some basic principles of crystal chemistry: Na^+ and Ca^{2+} have similar radii and substitution is easy, accounting for the wide occurrence of intermediate compositions. The substitution is special because it replaces atoms of similar size but different charge. To balance the overall charge, a simultaneous replacement of Si^{4+} by Al^{3+} has to occur, resulting in a chemical composition of $NaAlSi_3O_8$ for albite and $CaAl_2Si_2O_8$ for anorthite.

In the substitution of alkali ions, the ionic radius of Na^+ is considerably smaller than that of K^+, and substitution of the two ions is only possible at higher

(a)

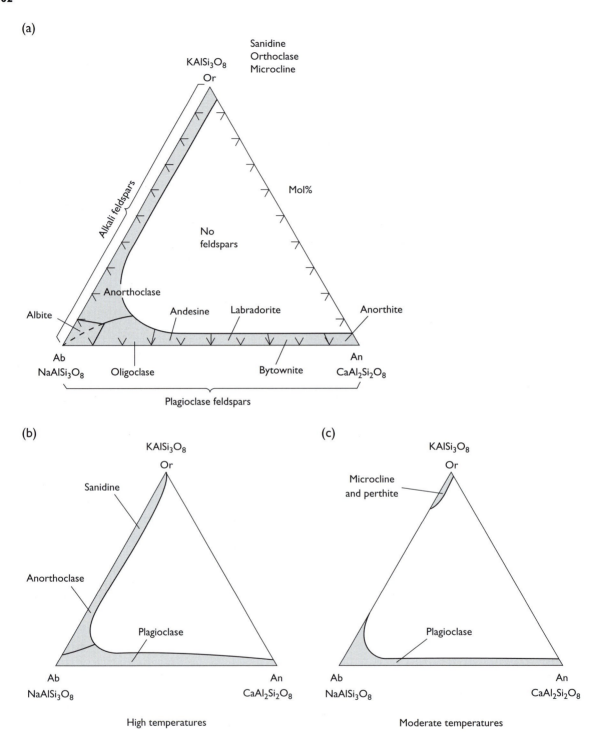

Fig. 21.9 Ternary representation of feldspar compositions. (a) Mineral names. (b) Range of feldspar compositions in typical volcanic rocks (high temperature). (c) Range of feldspar compositions in metamorphic rocks (low temperature).

Box 21.1 Background: Feldspars, a historical note

It is an understatement to claim that feldspar structures are complicated. In fact, many detailed structural features of this most common mineral, including the crystal structure of intermediate plagioclase and the plagioclase phase diagram, still elude a satisfactory interpretation. The basic structure of feldspar was determined from X-ray diffraction data in 1933 by W. H. Taylor at the University of Cambridge, UK. Subsequently Cambridge became a focus for feldspar investigations, with permanent and visiting researchers, such as Helen Megaw (who dared a first interpretation of Al–Si ordering in these silicates), S. G. Fleet and Paul H. Ribbe (who were the first to document chemical heterogeneities by electron microscopy), and Joe V. Smith (who later published a most comprehensive book on feldspars). Different groups, approaching the feldspar problem from the experimental side, worked in the USA at the University of Chicago (Julian R. Goldsmith, Fritz Laves, and, later, Joe V. Smith) and the Geophysical Laboratory at Washington, DC (Tom F. W. Barth). Particularly between the universities of Cambridge and Zurich (where Laves later resided) a not always friendly competition developed, as we can see in the published discussion between Laves and Megaw in one of the first feldspar conferences in Spain, in 1961. Laves and Goldsmith begin their 1961 paper entitled "Polymorphism, order, disorder, diffusion and confusion in the feldspars" by pointing out "there is a good deal of discussion on the problems of the feldspars and a certain amount of confusion is apparent. With the risk of adding to the confusion by further contributing to the discussion, we should like to stress anew some old ideas ..." It illustrates the enormous complexities in these minerals, produced by competing processes of ordering with attraction of unlike atoms (Si and Al) and exsolution with attraction of like atoms (K and Na). If you have time you may find it fascinating (and sometimes amusing) to read some of the early papers on feldspar structures and then compare them with the present state of knowledge, summarized, for example, in the book by Smith and Brown (1988), or the proceedings of a more recent feldspar conference (e.g., Parsons, 1994).

temperatures ($>660\,°C$). At lower temperatures, phase separation by exsolution occurs, which we discuss below. No feldspars exist that are intermediate in composition between K-feldspar and Ca-feldspar. Figure 21.9b shows the range of feldspar compositions in typical volcanic rocks, and Figure 21.9c displays the much more restricted range in metamorphic rocks.

The structure of feldspars is a framework of tetrahedra with Si and Al as cations. In open cages there are the large cations. We will first look at alkali feldspars. In Figure 21.10a the structure is projected on the plane (001). The distribution of Al and Si over the tetrahedral structural sites (called T sites) is disordered at high temperatures and ordered at low temperatures. In the disordered structure, where all T-sites have the same probability to be occupied by Si or Al, you notice a vertical mirror plane. Often the structure is idealized by omitting oxygens and showing only Al and Si atoms, connecting them by lines (an oxygen atom is somewhere near the center of the line; Figure 21.10b).

In the high-temperature polymorph sanidine ($KAlSi_3O_8$), above about 530 °C, all four tetrahedral

sites are occupied randomly by Al (1/4) and Si (3/4). Such an arrangement has a mirror plane parallel to (010) and therefore monoclinic symmetry (space-group $C\,2/m$) (Figure 21.11a). At lower temperatures some tetrahedral sites show a preference for Si, others for Al. The ordering pattern is controlled by the fact that adjacent Al tetrahedra exert strong repulsions, and thus such an arrangement is very unstable (the Al–O–Al avoidance rule; Loewenstein, 1954). At first there is a tendency for Al to occupy T_1 sites and for Si to occupy T_2 sites. This is the case in orthoclase and it preserves the monoclinic symmetry (mirror plane in Figure 21.11a). Finally, when ordering is complete, Al occupies all $T_1(m)$, and this eliminates the mirror plane (Figure 21.11b) and the fully ordered polymorph *microcline* is triclinic (space-group $C\bar{1}$). (A C-centered triclinic unit cell is used for microcline to maintain the relationship with the monoclinic C-centered unit cell of sanidine.) There are four different T sites, labeled $T_1(0)$, $T_1(m)$, $T_2(0)$ and $T_2(m)$.

Often two regions in the crystal behave differently during ordering. Al atoms can choose the T_1 site on

(a)

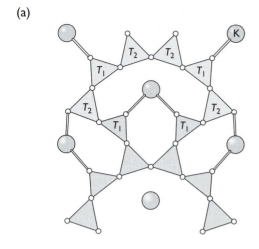

(b)

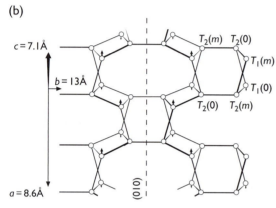

Fig. 21.10 (a) Structure of sanidine using Si–Al tetrahedra and projected on to the $a-b$ plane (001). Oxygen atoms are shown by small circles, potassium by large circles. (b) Simplified representation of Si–O–Si bonds only, represented by lines. Circles are tetrahedral atoms (silicon or aluminum) and tetrahedral (T) sites are labeled. Axes a and b are indicated as well as the (010) mirror plane. The c-axis is oblique to the plane.

either the left or the right side of the monoclinic mirror plane (010). The two regions are related by a mirror reflection and appear as twins (Figure 21.12a). Such (010) twins are called "Albite twins". Al could also choose different T_1 sites in regions related by a 2-fold rotation axis ($y = [010]$) and such rotational twins are called "Pericline twins". During ordering of monoclinic K-feldspar twins are often combined, resulting in a cross-hatched pattern that is common in triclinic microcline (Figure 21.12b, Plate 13e).

One application of the Al/Si ordering state in alkali feldspars to broader geological questions is in the

determination of the cooling history of igneous rocks. A good example is provided by Tertiary intrusions into Precambrian gneisses in Colorado, USA. Before the intrusion, all alkali feldspar in the gneisses was triclinic microcline. In the vicinity of the contact, however, the microcline became disordered, and we now see a boundary between triclinic microcline and monoclinic orthoclase that represents a 500 °C isotherm (Figure 21.13). Notice that in some places the isotherm is close to the contact with the surface, suggesting that this contact is nearly vertical, whereas in other locations it is far away from the surface contact, indicating that the intrusion is at a shallow level.

Next we explore the structure of plagioclase and here we look at chains parallel to [001] (Figure 21.14). Ordered albite (NaAlSi$_3$) has the same structure as ordered microcline (KAlSi$_2$O$_8$). Tetrahedral motifs along the z-axis repeat every 7 Å (Figure 21.14a). In anorthite (CaAl$_2$Si$_2$O$_8$), with two Al and two Si atoms, the Al–O–Al avoidance rule dictates a different ordering pattern. Al and Si tetrahedra must alternate to avoid Al–O–Al bonds, causing a doubling of c (Figure 21.14b). This 14 Å anorthite with an ordered Al–Si distribution is described with a body-centered spacegroup ($I\bar{1}$). Since the unit cell has changed in size, the structure of anorthite is called a "superstructure". At intermediate plagioclase compositions there is a complicated mixture between the albite and anorthite ordering patterns, resulting in long-range ordered superstructures with periodicities up to 70 Å. In pure anorthite Ca atoms attain an ordered distribution by slight displacements below 230 °C, resulting in a primitive 14 Å unit cell ($P\bar{1}$).

While Al^{3+}/Si^{4+} ordering in sanidine gives rise to twinning in microcline with domains that are in a different orientation, ordering in anorthite can cause a domain structure in which different regions within the macrocrystal are translated with respect to each other, depending upon which T site in the structure an aluminum atom prefers (Figure 21.15a). Boundaries between domains that are related by translation are called "antiphase domain boundaries" or APBs (see Chapter 9). Like the lamellar twins in microcline, APBs in anorthite are indicative of an ordering phase transition. APBs are undetectable under visible light but can be imaged with the transmission electron microscope (Figure 21.15b) because the structure is distorted along these boundaries.

Whereas substitution of Al^{3+} for Si^{4+} leads to ordering with symmetry changes and superstructures,

(a)

(b)

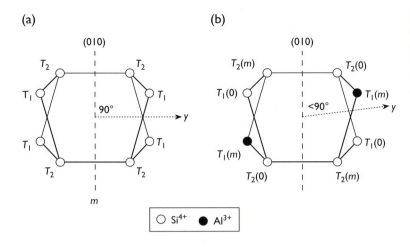

Fig. 21.11 Details of the Al/Si distribution in (a) sanidine (disordered) and (b) microcline (ordered). Only a portion of the structure is shown. The unit is centrosymmetric; *m*, mirror plane.

(a)

(b)

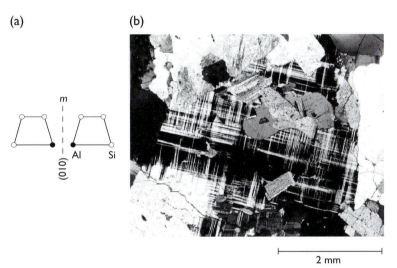

2 mm

Fig. 21.12 (a) Albite (010) twinning as a result of Al/Si ordering in microcline; *m*, mirror plane. (b) Photomicrograph of cross-hatched Albite and Pericline twinning in microcline (from granite near Prescott, Arizona, USA; crossed polarizers).

substitutions among the large cations K^+, Na^+, and Ca^{2+} that occupy irregularly coordinated sites in the framework tend to cause chemical phase separation into pure end members at low temperature, i.e., exsolution processes, which we have discussed in Chapter 20.

A temperature–composition phase diagram for alkali feldspars is shown in Figure 21.16. It displays the polymorphic phase transformations due to Al/Si ordering in pure K-feldspar. Similar transformations occur in albite: in low albite, Al/Si is ordered; in high albite it is progressively disordered, but, owing to distortion of the framework, which is not stabilized by the large K^+ as in orthoclase, the disordered structure of high albite is still triclinically distorted. Only at much higher temperatures does a displacive transformation expand the triclinic structure to a monoclinic disordered form (so-called "monalbite"), which is isostructural with sanidine.

At high temperatures (above 700 °C) there is a continuous solid solution between alkali feldspars of different compositions and the structure of sanidine. But if such a homogeneous crystal, say of composition $Ab_{60}Or_{40}$, is cooled below the solvus curve, it is no longer stable and decomposes into regions that are sodium-rich (albite) and regions that are potassium-rich (sanidine, orthoclase, microcline), resulting in characteristic intergrowths called *perthites*. The geometry of these intergrowths depends on the bulk composition and on the cooling rate. Some perthite structures show very coarse lamellae that can be distinguished in hand specimens, as in some pegmatites, though this is rare. Others are much finer and may be seen only with a petrographic microscope (microperthites, Plate 13c) or an electron microscope (cryptoperthites, including moonstones). In electron microscope images, morphologies vary,

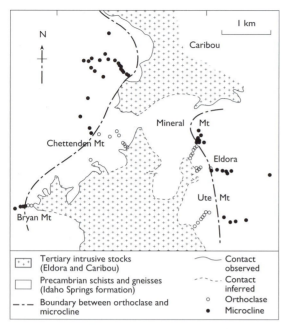

Fig. 21.13 Changes in the ordering pattern of K-feldspar in Precambrian gneisses around a Tertiary intrusion in Colorado, USA (based on Steiger and Hart, 1967).

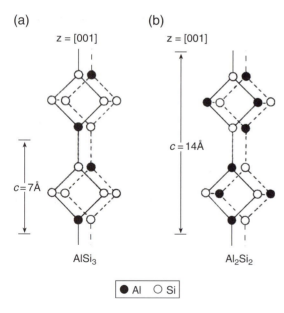

Fig. 21.14 Al–Si distribution in (a) ordered albite and (b) anorthite. Chains of tetrahedra extend along [001].

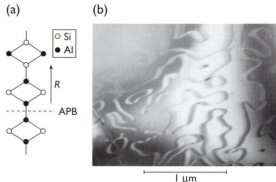

Fig. 21.15 Ordering in domains giving rise to antiphase boundaries (APBs) in anorthite where the two sides are offset by a translation R. (a) Schematic structure. (b) TEM image of "b" APBs in lunar anorthite (from Müller *et al.*, 1972).

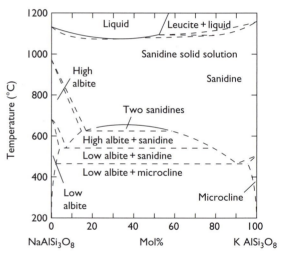

Fig. 21.16 Simplified temperature–composition phase diagram for alkali feldspars at atmospheric pressure (based on Smith and Brown, 1988).

exsolved phases and the host. Preferred orientations of lamellae are $(\overline{6}01)$ and $(\overline{3}01)$, which minimize the strain energy. Sanidine is usually homogeneous or contains cryptoperthites. Another alkali feldspar *anorthoclase*, also generally homogeneous, is a triclinic high-temperature feldspar with a composition intermediate between sanidine and albite ($Or_{90}Ab_{10}$–$Or_{50}Ab_{50}$).

In plagioclase (solid solution albite–anorthite), phase transformations are very sluggish because of the requirement of charge neutrality with coupled substitution of sodium by calcium and silicon by aluminum. For this reason it is very difficult to establish

ranging from parallel, slightly wavy bands (Figure 21.17a) to coarser zigzag domains (Figure 21.17b). In both cases there is fine twinning of the triclinic sodium-rich phase. In general there is a tendency to minimize mismatch across interfaces between the lamellar

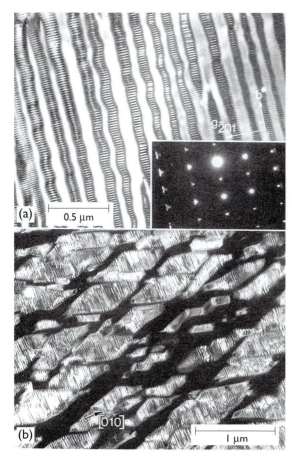

Fig. 21.17 Lamellar exsolution structures in (a) cryptopertite (moonstone) from Sri Lanka with a very fine sinusoidal compositional modulation and (b) coarser exsolution in microperthite from Mogok, Myanmar, with a zigzag morphology. In both cases the triclinic sodium-rich phase is twinned. TEM images (from Champness and Lorimer, 1976).

phase diagrams experimentally and they are still based largely on empirical observations of feldspar assemblages in natural rocks of known origin. The proposed phase diagram for plagioclase (Figure 21.18) is much less well founded than that for alkali feldspars (Figure 21.16). Nonetheless, it does illustrate that subsolidus relations are extremely complex in this solid solution series. There are at least three miscibility gaps, in oligoclase (peristerite gap), in labradorite (Bøggild gap), and in bytownite (Huttenlocher gap). Exsolution lamellae are very small (<1 μm) and can be imaged only by electron microscopy. However, it is these lamellae that are responsible for the diffraction of light, causing a remarkable play of colors, exhibited by some plagioclases (e.g., peristerite and labradorite; Plate 21e). In metamorphic rocks, intergrowths of pure albite and anorthite have been observed (Plate 21f) suggesting that these are the stable phases that are in equilibrium at lower temperatures.

Low albite ($C\bar{1}$), body-centered anorthite ($I\bar{1}$) and primitive anorthite ($P\bar{1}$) all have an ordered Al–Si distribution. In $P\bar{1}$ anorthite Ca atoms show positional order.

21.3 Brief description of silica minerals (see also Table 21.1)

Quartz is found in euhedral single crystals in druses and cavities. More often it occurs in granular aggregates, as rock-forming anhedral, semitransparent grains with vitreous or greasy luster such as in granite, sandstones, and ore deposit veins. It is a major component of many felsic (silica-rich) igneous, metamorphic, and sedimentary rocks. We will discuss its occurrence in granite and pegmatite below. The

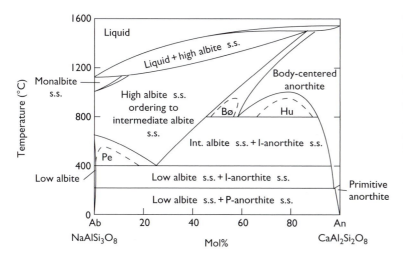

Fig. 21.18 Simplified temperature–composition phase diagram for plagioclase (based on Smith and Brown, 1988). Pe, Bø, and Hu refer to the Peristerite, Bøggild, and Huttenlocher exsolution gaps, s.s. to solid solution, P to primitive anorthite structure, and I to body-centered anorthite structure. Monalbite is monoclinic.

(a)　　　　　　　　　　　　　　　　　　　　　(b)

Fig. 21.19 (a) Quartz from Arkansas (courtesy O. Medenbach). Width 105 mm. (b) SEM image of opal from Hungary with close-packing of spheres. On the surface there is some transformation of amorphous opal to crystalline cristobalite. The diameter of the spheres is approximately 6000 Å (courtesy R. Wessicken). Width 30 μm.

quartz variety *chalcedony* consists of fibrous and cryptocrystalline aggregates, often concentric and banded (see Plate 8b,c). *Agate* is a variety of chalcedony. In thin sections of rocks, quartz is generally recognized by undulatory extinction (Plate 13b).

Quartz occurs in a wide variety of colors. Ideal pure quartz is colorless and transparent, and has an appearance similar to that of ice (see Figure 21.19a and Plate 2a). The Greeks originally named it *krystallos* (meaning "ice"), and thought it represented an ice variety that forms at high pressure and therefore does not melt. Quartz is an old German mining name. Fe^{3+}-bearing varieties are either purple-violet (*amethyst*, Plates 2c and 7b,c) or yellow (*citrine*, Plates 7c and 20c). Brown and black varieties (*smoky quartz* and *morion*, Plates 2b and 20a,b) are due to traces of aluminum, substituting for silicon. Green *chrysoprase* forms in altered ultramafic rocks and contains inclusions of oxidized Ni-minerals, while *aventurine* contains inclusions of goethite. The reason for the pink color of some varieties of *rose quartz* (Plate 20d) has been revealed only recently as being due to submicroscopic inclusions of a fibrous nanocrystalline material related to the rare mineral dumortierite $(Al_7O_3(BO_3)(SiO_4)_3)$ (Goreva *et al.*, 2001). As we have discussed in Chapter 18, the morphology of quartz changes with conditions of formation (see Figures 18.9 and 18.10).

There are various uses for quartz. It is industrially significant in the production of piezocrystals, though most of them are presently synthetic. The colored varieties are used as gemstones. Quartz is used for the production of glass, and pure material has gained significance in the manufacturing of fibers for optical applications. The Miñas Gerais province of Brazil, which contains extensive pegmatite ores, is a major source of pure quartz.

The high-pressure silica polymorphs **coesite** and **stishovite** are rare in crustal rocks and confined to unusual ultra-high-pressure metamorphic conditions, such as coesite in the Dora Maira massif in the Western Alps and Dabie Shan in China, the transformation of quartz during meteorite impact at the Vredefort site in South Africa or the Barringer (Meteor) Crater in Arizona, USA, and in meteorites. Due to their small size they are difficult to recognize in hand specimens. In thin sections coesite has low birefringence and is biaxial (to distinguish it from quartz), stishovite has a much higher refractive index and birefringence than quartz. Both minerals may occur in mantle subduction zones and stishovite, as well as a high-pressure SiO_2 phase with an orthorhombic $CaCl_2$ type structure, may be significant in the lower mantle.

The high-temperature polymorphs **cristobalite** and **tridymite** occur in siliceous volcanic rocks, including obsidian. Cristobalite may occur as aggregations in druses (Plate 20e) and tridymite forms hexagonal platelet-like crystals (Plate 20f). Both have very low birefringence and lower refractive indices than quartz. The two minerals also occur quite commonly in siliceous deep ocean sediments, transforming metastably from silica gel and forming microcrystalline cristobalite-tridymite (CT) aggregates, converting later, during burial, to stable quartz in cherts. The metastable CT mixtures are recognized by X-ray diffraction. The name cristobalite derives from the Cerro San Cristobal in Mexico.

Opal ($SiO_2.nH_2O$) is an unusual silica mineral. It is amorphous, in the sense that it does not have a long-range ordered crystal structure. In addition to SiO_2 it contains large quantities of water. This solid hydrogel varies in composition. The gel exists as minute spheres that are 1500–8000 Å in size and often have a regular packing (see Figures 15.10 and 21.19b). This close-packed arrangement of spheres acts as a diffraction grating for visible light and produces a characteristic play of colors (opalescence, Plate 20g). Opals that have brown, green, yellow (Plate 20h), and black colors contain impurities of natural pigments such as green garnierite, brown iron hydroxides, black manganese oxides, etc. Opal is usually found as compact translucent, vitreous masses, veinlets, and colloform aggregates. The name opal derives from the Sanskrit word *upala*, meaning "precious stone".

Opal crystallizes from geyser water during surficial weathering of feldspars and other silicates; it precipitates in the coastal zones of marine basins owing to coagulation of silica gels. Even though opal is the first silica phase to precipitate in seawater, it is highly unstable and transforms during diagenesis first to cristobalite–tridymite and later to quartz. Opal constitutes the solid tissue of diatoms, flagellatae, some radiolaria, siliceous sponges, some gastropods, and other invertebrates (see Chapter 25), and it often replaces the remains of plants in soils. Opal rocks such as *opoka* (silica clay) and *diatomite* are applied as filters and used in ceramics manufacturing, as a metal-polishing abrasive, and as refractory materials in the chemical, food, and petroleum industries. Gem opals are used as decorative stones. Famous occurrences of such precious opals are in Central Australia (e.g., Coober Pedy, Lightning Ridge). When present in aggregate, opal is an undesirable component of concrete because of a reaction with cement that causes deterioration ("silica poisoning") (see Chapter 35).

21.4 Brief description of feldspars (see also Table 21.1)

As with silica minerals, the framework structure of feldspars is characterized by the strong covalent–ionic

bonds within a cellular framework. Bonding causes a high hardness (Mohs scale 6) and the open framework low densities of 2.5–2.8 g/cm^3 for most of these minerals. The bond type and composition of feldspars determine their vitreous luster and transparence or translucence. Feldspars are mostly white because transition elements generally do not enter their structures. We may identify feldspar minerals on the basis of their color, luster, cleavage, and standard hardness; however, often it is difficult to distinguish the different kinds of feldspar, even alkali feldspar and plagioclase in hand specimens. It is most efficient to use thin sections and a petrographic microscope (see Table 21.1 and Chapter 14), or X-ray diffraction in combination with electron microprobe analyses.

Alkali feldspars

In this subgroup, which includes *microcline*, *orthoclase*, and *sanidine*, Na^+ commonly substitutes for K^+ and a general chemical formula may be written as $(K,Na)AlSi_3O_8$. As we have seen, Na^+ substitutes for K^+ in significant proportions only at high temperatures because of the significant difference between their ionic radii, and homogeneous alkali feldspars of intermediate composition are preserved only by rapid cooling in volcanic rocks. Volcanic sanidines are usually richest in sodium, whereas metamorphic microclines contain the least amount of sodium.

The crystals of alkali feldspar polymorphs are similar in habit. Orthoclase crystals from granites are often very simple and close to equant owing to a combination of (010) and (001) pinacoids and a monoclinic prism {110} (Figures 21.20a and 21.21; see also Figure 10.13). The minerals have (010) and (001) cleavages, with an interfacial angle that is almost 90°. Adularia is a variety of K-feldspar, with a habit dominated by the {110} prism (Figure 21.20b, Plate 21a) which occurs in hydrothermal veins and fissures.

There are many types of twins in alkali feldspars, in addition to the Albite and Pericline transformation twins discussed above (see also Chapter 9, Figures 9.9b and 21.12b, and Plate 13e). A common growth twin in orthoclase and sanidine is the Carlsbad twin, with one crystal rotated 180° about the [001] axis (Figures 9.9f and 21.20c).

While most alkali feldspars are white, gray, or pink (owing to submicroscopic hematite inclusions, Plate 21c), some beautiful emerald-green or

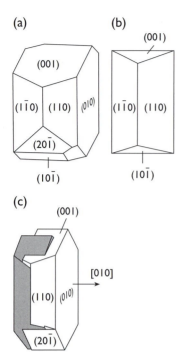

Fig. 21.20 K-feldspar morphology. (a) Typical habit of orthoclase. (b) Habit of adularia. (c) Carlsbad twin in orthoclase.

Fig. 21.21 Euhedral crystals of orthoclase from Wieza, Poland (courtesy A. Massanek). Width 95 mm.

bluish-green varieties of microcline, called *amazonite*, exist (Plate 21b). The origin of this color is still uncertain. It is possibly due to an isomorphous substitution $K^+Si^{4+} \rightleftharpoons Pb^{2+}Fe^{3+}$ or $K^+Al^{3+} \rightleftharpoons Pb^+Fe^{3+}$ which causes distortion in the structure and introduces color centers.

Well-developed crystal faces and cleavage surfaces in alkali feldspars have a vitreous luster, while fracture surfaces have a greasy luster. Sometimes orthoclase has *perthite* exsolution on a submicroscopic scale, resulting in diffraction of visible light and a beautiful glimmering luster with pearly optical effects. Such perthites are called *moonstone*. A similar play of colors is typical for some anorthoclases, so-called *larvikites*.

Alkali feldspars are rock-forming minerals of plutonic rocks such as granites, granite pegmatites, and syenites (alkali feldspar-rich plutonic rocks); they also occur as phenocrysts in felsic volcanic rocks such as rhyolites. These minerals also crystallize in hydrothermal conditions predominantly within ore-bearing veins or aureoles of metasomatic alterations of country rocks. Alkali feldspars are common in metamorphic rocks as a result of profound transformation of sediments or other initial rocks. Potassium feldspars also form during diagenesis in limestones, sandstones, and shales ("authigenic sanidine"). The minerals undergo different types of alteration when they react with water: muscovitization or sericitization (sericite is a fine-grained variety of muscovite that may deviate from the ideal composition of muscovite, Plate 7d,e) at high temperatures, and kaolinitization due to weathering at low temperatures. Muscovitization may be characterized schematically by the reaction:

$3KAlSi_3O_8 + 2H_2O \rightleftharpoons$
  K-feldspar
      $KAl_2(AlSi_3O_{10})(OH)_2 + 6SiO_2 + 2KOH$
      Muscovite   Quartz   In solution

The weathering processes can be described by the reaction:

$2KAlSi_3O_8 + 3H_2O \rightleftharpoons$
  K-feldspar
      $Al_2(Si_2O_5)(OH)_4 + 4SiO_2 + 2KOH$
      Kaolinite   Quartz   In solution

Alkali feldspars are used as raw materials in the ceramics industry. For such purposes granite pegmatites are preferred; but since most pegmatite deposits are quite small and rare, granites rich in orthoclase and microcline are also employed.

Plagioclase feldspars (or plagioclase)

The composition of plagioclase feldspars is usually explained in terms of the coupled substitution $Na^+Si^{4+} \rightleftharpoons Ca^{2+}Al^{3+}$. The composition of plagioclase is generally expressed as percentage of anorthite (An_0 = albite = $NaAlSi_3O_8$, An_{100} = anorthite = $CaAl_2Si_2O_8$).

The crystal morphology of plagioclase is similar to that of alkali feldspars. Multiple (polysynthetic) twins are very typical but not on such a fine scale as the transformation twins in microcline. According to the Albite law, twinning with a composition plane (010) is most common (see Figure 9.9b and Plate 13f).

The plagioclases are white, green, grayish-lilac, or dark gray (due to inclusions). Plagioclase with compositions corresponding to the exsolution gaps may display iridescent colors of blue, red, yellow, or green ("schiller color"). Best known are *peristerites* (of oligoclase composition) and *labradorites* (Plate 21e). The color of peristerite and labradorite has an origin similar to that of opal, but, instead of spheres, fine exsolution lamellae act as diffraction gratings for light. Labradorite with iridescent colors is the only plagioclase that is used as a decorative stone.

The plagioclases have perfect cleavages parallel to (010) and (001), with an angle between the cleavage surfaces of about 86°. The cleavage surfaces (usually one of them, sometimes both) exhibit striation caused by the polysynthetic twinning. Optical properties are described in Section 14.6.

Plagioclases are major rock-forming minerals in felsic (i.e., silica-rich), intermediate, and mafic, as well as alkali igneous rocks (granites, granite pegmatites, diorites, gabbro, and their volcanic analogs). In these rocks all varieties of plagioclases are found (sodium- and silicon-rich varieties in felsic rocks; calcium- and aluminum-rich varieties in mafic rocks; pure albite in alkali rocks). Very calcic plagioclase (anorthite) is a common component of lunar basalts. The largest plagioclase crystals (oligoclase measuring up to several meters on a side) occur in granite pegmatites. Albite is characteristic for hydrothermal deposits. Sugar-like masses of albite replace earlier-formed minerals of granites, syenites, and pegmatites, signifying the process of metasomatic "albitization". Albite also occurs as plate-like crystals in druses, in acicular accretions in cavities, and in open fractures (Plate 21d). Albite is often accompanied by rare-metal mineralization, in particular by zirconium, niobium, tantalum, beryllium, and rare earth elements. Plagioclases are the main components of metamorphic rocks such as gneisses and amphibolites. Sandstones and graywackes contain terrigenous material consisting of plagioclase. Calcium-rich plagioclases are ubiquitous in stony and stony-iron meteorites as well as lunar rocks. During meteoritic impact they frequently undergo amorphization.

21.5 The origin of granite

When discussing quartz and feldspars the rock *granite* comes immediately to mind. It is composed largely of plagioclase, alkali feldspar, and quartz. Because of the high silica content, it is known as a siliceous (or felsic) rock, contrary to basalt, for example, which is called a mafic rock (deficient in silica and rich in Mg and Fe). Because granitic rocks are leucocratic (light-colored, from Greek *leucos*, meaning "white") and contain only small amounts of mafic minerals (such as amphibole or mica), the relative amounts of quartz (Q), plagioclase (P), and alkali feldspar (A) are used to classify granitic rocks; their composition is represented in a Q–P–A ternary diagram (Figure 21.22). Depending on the Q–P–A composition, different names are assigned. There are at least five major compositional varieties of granitic rocks. So-called *normal granites* consist of quartz (20–60 vol.%), alkali feldspars (30–75%) and plagioclase (20–50%), and colored minerals – biotite or hornblende (7–10% of the total). In *diorite* and *tonalite*, alkali feldspars are minor; hornblende is a common colored constituent. *Alkali feldspar granites* and *syenites* contain only small amounts or no plagioclase. *Granodiorite*, intermediate between normal granite and tonalite, is the most common granitic rock.

If we take a melt of the average composition of the Earth's mantle and crust and cool it, magnesium- and iron-rich minerals such as olivine and pyroxene will precipitate first, followed by calcic plagioclase. The residual melt becomes enriched in silica and alkalis and will finally approach a granitic composition. It was therefore speculated that granites were the product of such compositional differentiation from a magma of more mafic composition. However, geologists made the unexpected observation that granitic rocks in the Earth's continental crust are very common in plutons but rare as volcanic rocks, whereas the more mafic basalts are common as volcanic rocks and rare as plutonic rocks. This heterogeneous occurrence of the two rock types makes their origin through differentiation of a uniform magma unlikely. Most granitic plutons are presently believed to represent remelted crustal material.

Some phase diagrams summarize results of experimental work and give an approach to understanding processes of granite crystallization. Contrary to binary phase diagrams, which we discussed in Chapter 20, granite constitutes a ternary system, with

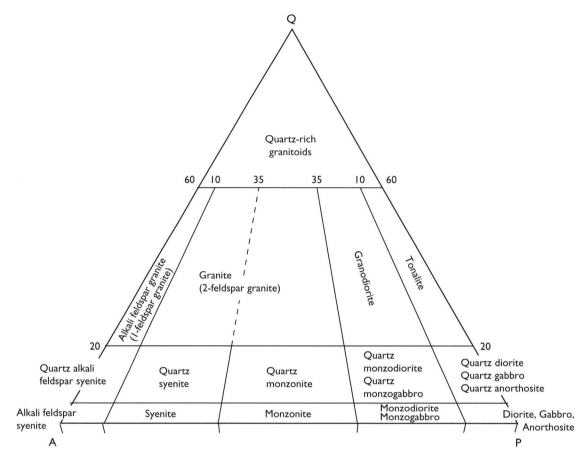

Fig. 21.22 Q (quartz)–A (alkali feldspar)–P (plagioclase) triangular diagram to represent the composition of granitic rocks and corresponding rock names (based on Streckeisen, 1976). Values are in vol.%.

quartz, albite, and K-feldspar as major components. (All plagioclase is represented as albite to simplify the relationships.) One of the first phase diagrams of this system was published and discussed by N. Bowen and O. F. Tuttle in 1950 to explain crystallization in a system composed of SiO_2, Al_2O_3, K_2O, and Na_2O. Figure 21.23a is a simplified version of this diagram in the absence of water and at atmospheric pressure, with the albite–orthoclase–quartz triangle as the base and temperature as the ordinate (incongruent melting of orthoclase to leucite plus liquid is omitted). The top surface is the liquidus temperature, i.e., the equilibrium temperature for coexisting melt and crystals. When a melt cools below this rather complicated surface, the first crystals will appear. The topography displays three temperature hills at each of the corners of the triangle. The highest hill is at SiO_2 and pure silica has a melting point of 1713 °C, as compared to albite at 1118 °C and

sanidine at 1200 °C (at atmospheric pressure and in the absence of water).

The binary $NaAlSi_3O_8$–$KAlSi_3O_8$ (front) side of the ternary diagram (Figure 21.23a, comparable to Figure 21.16) explains the order of crystallization of alkali feldspars in granites. This graph has a temperature minimum for the melt at 1062 °C. Consider crystallization of a liquid with composition X. Falling temperature cools the liquid towards the liquidus line. At the intersection of this line (point 1, temperature T_1) sanidine enriched in sodium begins to crystallize. It has a composition $1'$. Further cooling causes growth of sanidine crystals and their reaction with the liquid. With decreasing temperature the melt compositions shift along the liquidus curve towards point 2, whereas crystal compositions shift along a solidus curve towards point $2'$ (temperature T_2), which marks the end of crystallization of the liquid. As long as the crystals of feldspar remained in equilibrium with

(a)

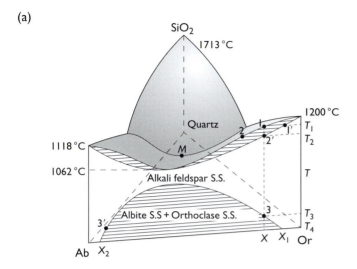

(b)

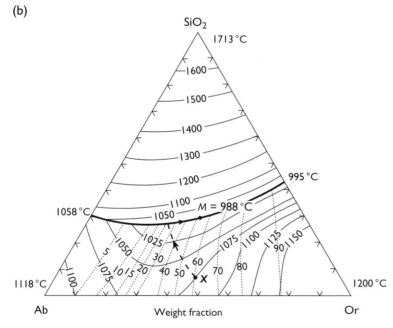

Fig. 21.23 Liquidus in the ternary system quartz–plagioclase (albite)–K-feldspar (orthoclase) dry and at atmospheric pressure (data from Barron, 1972). (a) Three-dimensional representation with temperature as the vertical axis. S.S. is solid solution. (b) Projection on the compositional triangle, indicating isotherms of the liquidus (solid lines, in °C), alkali feldspar compositions at the solidus (dotted lines, in % Or), and crystallization path of a composition X (dashed). (The diagram is simplified and does not include incongruent melting of orthoclase to leucite plus liquid.) M is the eutectic point of the system, with the minimum melting temperature.

the magma throughout the entire process, the composition of feldspar corresponds to a homogeneous composition X, equivalent to the composition of the initial melt.

Further cooling of homogeneous albite–sanidine solution crystals causes no phase changes until temperature T_3 (point 3) is reached. At that point, after intersecting the solvus, feldspar crystals exsolve into two phases. They are sanidine containing some sodium (composition 3), and albite with some potassium (composition 3'). They often form lamellar perthite structures (see, e.g., Plate 13c). On further cooling, compositions of these two phases migrate

(in equilibrium conditions) down each side of the solvus. For example, at temperature T_4, compositions of coexisting feldspars are X_1 and X_2.

The binary diagram in Figure 21.23a is schematic. A more detailed phase diagram for alkali feldspars was introduced in Figure 21.16. In addition to the liquidus, solidus, and solvus curves, it contains boundaries between disordered sanidine and high albite, and ordered low albite and microcline.

Now let us return to the ternary diagram. It has a gently dipping surface of alkali feldspar, with a valley dipping towards the back. This surface intersects the much steeper surface of the quartz peak, where quartz (or one of the high-temperature silica polymorphs) crystallizes first. The line separating the fields is a depression and is called the *cotectic line* by analogy with the eutectic point, introduced in Chapter 19. Along the cotectic line, alkali feldspars and quartz crystallize simultaneously. Point *M* is the temperature minimum along the cotectic line. Liquids of any composition in Figure 21.23a will finish crystallization on the cotectic line.

Crystallization in the ternary system is better explained in Figure 21.23b, a projection on the compositional triangle with solid lines representing contours of the liquidus temperature. The composition of the alkali feldspar that is in equilibrium with the melt is illustrated with dotted lines. Take, for example, a melt composition *X* in Figure 21.23b. Crystallization begins with sanidine of composition Or_{60}. Orthoclase is removed from the melt and correspondingly the melt becomes enriched in Ab and SiO_2, moving down the steepest gradient towards the cotectic line. During this crystallization, the alkali feldspar becomes more albite-rich and, when the melt reaches the cotectic line, the equilibrium composition of the feldspar is $Or_{25}Ab_{75}$. At this point quartz starts to crystallize and the melt shifts in composition towards *M*, which is the eutectic point of this system and the lowest

temperature that a dry (volatile-free) granitic melt can have (988 °C). Below this temperature, quartz and alkali feldspar precipitate simultaneously.

Microstructures that are consistent with such simultaneous crystallization are occasionally observed in what is known as "graphic texture", a characteristic intergrowth of quartz and feldspar that is found in some granitic rocks, especially pegmatites. The structure displays spindle-shaped, curved, or wedge-like grains of quartz within large crystals of microcline or orthoclase (Plate 21c). They resemble cuneiform symbols, hence the name "graphic".

We have used the ternary phase diagram to explain crystallization during cooling of a melt. It can also be used to explain the melting of solid rock. If melting occurs in overall equilibrium, the first melt of any composition in this ternary system will have a composition at *M*.

All systems shown in Figures 21.16 and 21.23 are for dry conditions and atmospheric pressure. The presence of water significantly changes what is shown in these diagrams, mainly by lowering the liquidus and thus the melting point. For a dry melt, the minimum melting temperature in the system $NaAlSi_3O_8$–$KAlSi_3O_8$ is 1062 °C (Figure 21.24a). At a water

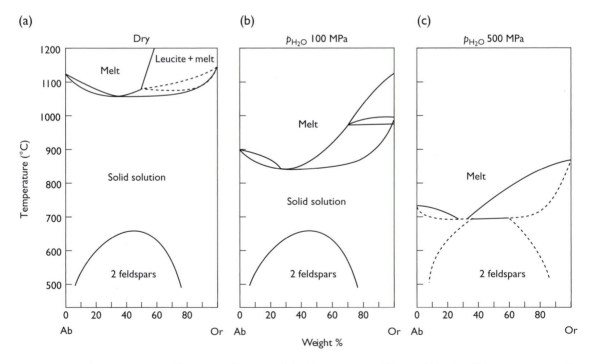

Fig. 21.24 Temperature–composition phase diagrams of the binary system albite–sanidine for different pressures of water vapor: (a) dry, (b) 100 MPa, (c) 500 MPa. (Data from Bowen and Tuttle, 1950, and Yoder *et al.*, 1956.)

Plate 1a Sulfur from Caltanisetta in Sicily, Italy (courtesy O. Medenbach). Width 120 mm.

Plate 1b Cinnabar on calcite from Erzberg in Steiern, Austria (courtesy O. Medenbach). Width 14 mm.

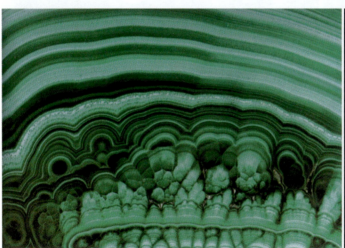

Plate 1c Botryoidal malachite from Tsumeb, Namibia (courtesy J. Arnoth and W. Gabriel). Width 50 mm.

Plate 1d Turquoise, Arizona, USA (courtesy D. Weinrich). Width 50 mm.

Plate 1e Colorless transparent corundum, Mogok, Myanmar (courtesy D. Weinrich). Height 12 mm.

Plate 1f Ruby from Jegdalek, Afghanistan (courtesy A. Massanek). Width 35 mm.

Plate 1g Sapphire from Ratnapura, Sri Lanka (courtesy D. Weinrich). Height 27 mm.

Plate 2a Clear quartz with red rutile needles from Val Corno TI, Switzerland (courtesy T. Schüpbach). Width 14 mm.

Plate 2b Smoky quartz from Central Alps, Switzerland (courtesy G. Ivanyuk). Width 120 mm.

Plate 2c Amethyst geode from Artigas, Uruguay (courtesy D. Weinrich). Width 9 cm.

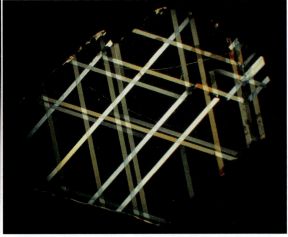

Plate 2d Twinning in calcite: view of a thin slab along the *c*-axis with cross-polarized light. Width 6 mm.

(e)

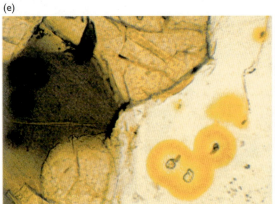

(f)

Plate 2e,f Cordierite with halos around monazite inclusions (clear crystal) and biotite with halos around zircon inclusions (brown crystal) in granulite from Namaqualand, South Africa. (e) Plane polarizers; (f) same as (e) but with crossed polarizers (courtesy O. Medenbach). Width 0.7 mm.

Plate 3a Halite from salt mines of Bex, VD, Switzerland (courtesy J. Arnoth and W. Gabriel). Width 70 mm.

Plate 3b Magnetite from Itabira, Brazil (courtesy J. Arnoth and W. Gabriel). Edge 20 mm.

Plate 3c Pyrite cube with smaller twins from Abasaguas, La Rioja, Spain (courtesy D. Weinrich). Cube is 20 mm on edge. Notice striations.

Plate 3d Pyrite dodecahedron, Rio Marina, Elba Island, Italy (courtesy D. Weinrich). Width 30 mm.

Plate 3e Epitaxial rutile prisms growing on a plate of hematite. Binntal, Switzerland (courtesy W. Gabriel). Width 15 mm.

Plate 3f Skeletal growth of quartz with edges dominating. Engstligenalp, Switzerland (courtesy T. Schüpbach). Height 7 cm.

Plate 4a Dendritic copper from Tsumeb, Namibia (courtesy O. Medenbach). Width 34 mm.

Plate 4b Dendritic growth of manganese oxides (MnO_2) on bedding planes of Solnhofen limestone (Germany). Growth initiates from fractures (courtesy E. H. Wenk). Width 55 mm.

(c)   (d)

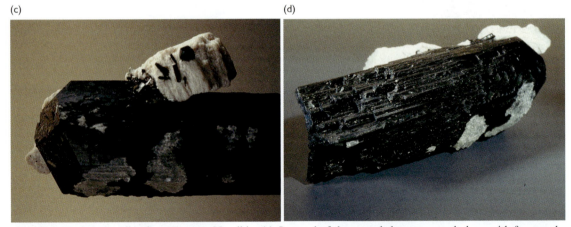

Plate 4c,d Black tourmaline from Erongo, Namibia. (c) One end of the crystal shows a morphology with faces and trigonal symmetry. (d) The opposite end shows extensive dissolution with no faces, but ridges that display the trigonal symmetry (courtesy J. Arnoth and W. Gabriel). Length 45 mm.

Plate 5a Gypsum from cave in Naica mine, Chihuahua, Mexico (courtesy A. VanDriessen).

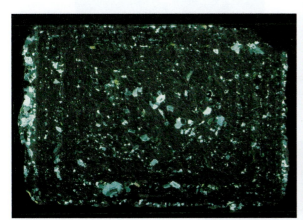

Plate 5b Phenocryst of orthoclase incorporating other minerals during growth. Plagioclase, quartz, biotite, and hornblende inclusions decorate concentric growth zones in megacryst from Cathedral Peak granite, Sierra Nevada, California, USA. Photographed with crossed polarized light, the orthoclase is almost in extinction. Width 30 mm.

Plate 5c Porphyroblast of andalusite in pelitic schist with folding. Val Forno GR, Switzerland. Crossed polarizers. Width 10 mm.

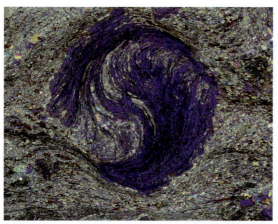

Plate 6a Garnet crystal, growing as a porphyroblast in metamorphic schist from Vermont. The garnet rotated as it overgrew mineral grains in a deforming matrix and it displays a spiral trail of inclusions (photomicrograph with crossed polarized light and compensator, courtesy J. Christensen; see also Christensen *et al.*, 1989). Width 30 mm.

Plate 6b Multicrystal smoky quartz from Central Alps with slightly misoriented crystals (quendel growth) (courtesy T. Schüpbach). Height 50 mm.

Plate 6c Hematite rose with adularia feldspars. Gotthard Pass, TI, Switzerland (courtesy T. Schüpbach). Diameter 14 mm.

Plate 6d Pseudomorphs of limonite after pyrite, preserving striations (courtesy P. Gennaro). Width 35 mm.

Plate 6e Malachite (green) pseudomorphically replacing an azurite crystal (blue); from Tsumeb, Namibia (courtesy O. Medenbach). Length 4 cm.

Plate 7a Zoned and twinned plagioclase phenocryst with groundmass in andesite from Marysville Buttes, California, USA. Width 12 mm. Crossed polarizers.

Plate 7b Quartz on a Swiss stamp celebrating the UNESCO International Year of Crystallography 2014. This image is also on the book cover (courtesy T. Schüpbach).

Plate 7c Sector zoning in the quartz variety ametrine from the Anahi mine in Bolivia, with purple sectors of amethyst (Fe-poor) and yellow citrine (Fe-rich). Displayed is a section cut through the tip of a crystal (courtesy M. Weibel). Width 40 mm.

Plate 7d Growth sectors with different colors in tourmaline from Madagascar. The colors correspond to different trace amounts of Fe and Cr (courtesy J. Arnoth). Width 22 mm.

(d)

(e)

Plate 7e,f Old granite subjected to later metamorphism, with plagioclase laths (clear) replaced partially by sericite (brown). Also hornblende is replaced by chlorite and epidote (bright colors with crossed polarizers). (e) Plane polarized light, (f) crossed polarizers. Austroalpine nappes in Engadine, Switzerland. Width 5 mm.

Plate 8a Marble fabric with triple junctions. New York Mountains, California, USA (crossed polarizers). Width 5 mm.

Plate 8b Chalcedony with various growth structures. First, fibrous microcrystalline quartz grows in shells perpendicularly to the cavity walls, presumably precipitating from a vapor phase. The next stage is marked by horizontal layers of fibrous microcrystalline quartz, probably from a colloidal solution. In the last stage, coarse crystalline quartz precipitates from aqueous hydrothermal solutions. Sample from Brazil (courtesy J. Arnoth and W. Gabriel). Width 85 mm.

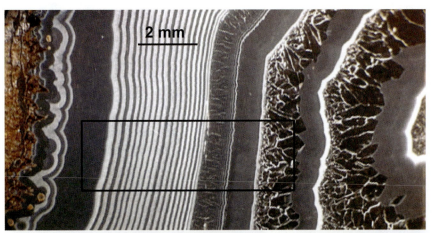

2 mm

Plate 8c Chalcedony from Brazil with growth zones; thin section images. Top illustrates a complicated pattern with coarse and fine quartz crystals. Bottom is a detail with crossed polarizers and compensator documenting zones with c-axes more or less parallel to the growth direction (blue) and zones with a-axes parallel to the growth direction (yellow) (courtesy J. Grimsich).

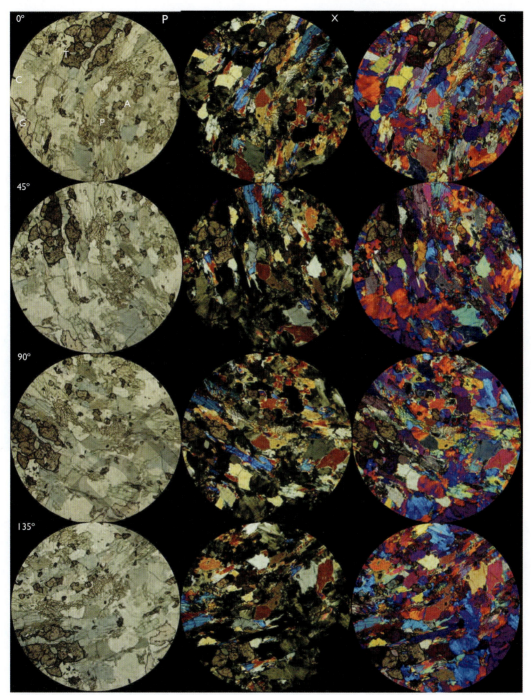

Plate 9 Eclogite from Panoche Pass, California, USA. Garnet G, titanite T, omphacite (pyroxene) O, sodic amphibole A, and chlorite C. First column plane polarized light (P). Second column crossed polarizers (X). Third column crossed polarizers with compensator ("gypsum plate") inserted (G). Micrographs taken at four rotations. Diameter 3 mm.

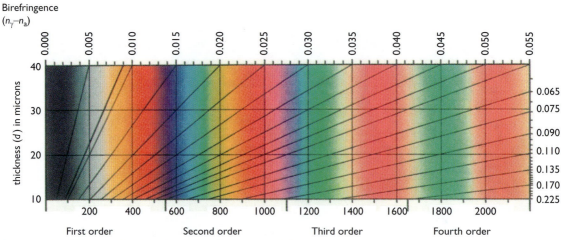

Birefringence
$(n_\gamma - n_a)$

Plate 10a Interference color chart (path difference versus thickness).

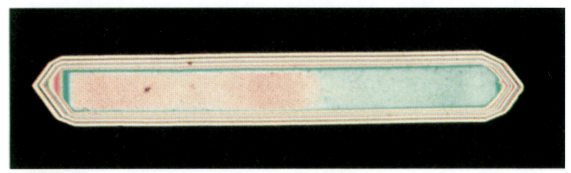

Plate 10b Interference colors observed on a quartz prism in the microscope. The *c*-axis of the crystal was at 45° (NW–SE). Crossed polarizers.

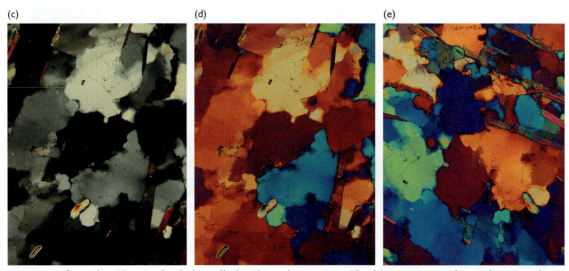

Plate 10c–e Quartzite: (c) crossed polarizers display the grain structure; (d) with compensator inserted some grains display addition (blue) and some show subtraction (yellow). The grain in the center is red, corresponding to the black grain on left. (e) Sample rotated 90° inverts the colors, except for the central grain which remains red. Width 1 mm.

Plate 11 (a) Pleochroism in biotite from micaschist (Val Bregaglia, Switzerland) with colors ranging from dark brown to light brown. (b) In stilpnomelane colors range from yellow to brown and black (California Coast Ranges, USA). (c) In hornblende of amphibolite colors range from dark green to light green (Val Bregaglia). (d) In riebeckite rosettes colors range from yellow to dark blue (Engadine, Switzerland). Width a–d 4 mm. (e) In a folded blueschist from northern California, glaucophane is either blue or yellowish, depending on crystal orientations. Width 10 mm. (f) Most spectacular pleochroism is observed in piemontite, also known as piedmontite with colors ranging from purple to red and yellow. Width 1 mm.

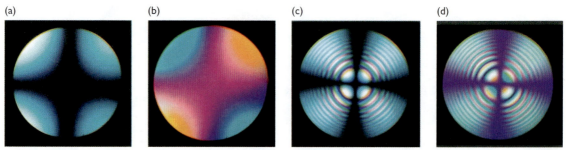

Plate 12 (a)–(d). Centered interference figures on uniaxial crystals with standard thickness (30 μm).

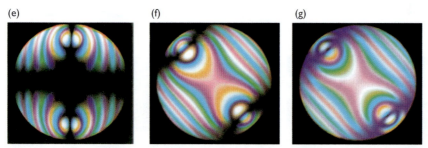

Plate 12 (e)–(g). Centered interference figures on biaxial crystals with standard thickness.

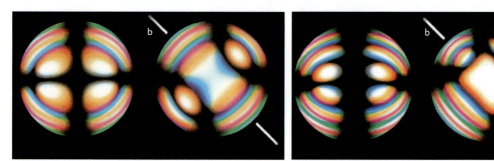

Plate 12h Inclined dispersion in high sanidine $r < v$, axial plane (010); b is the axis [101].

Plate 12i Horizontal dispersion in low sanidine $r > v$, axial plane perpendicular to (010).

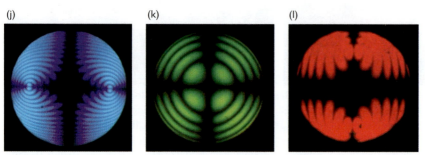

Plate 12j–l Dispersion in brookite, viewed along the a-axis. (j) Blue (axial plane (100)); (k) green (unaxial); and (l) red light (axial plane (001)).

(a)   (b)

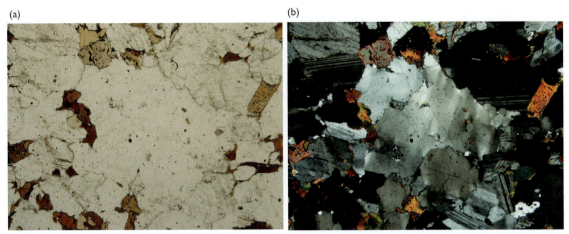

Plate 13a,b Granodiorite from Palm Canyon, California, USA. In plane polarized light (a) plagioclase and quartz are difficult to distinguish. Biotite is brown. With crossed polarizers (b) plagioclase shows lamellar twinning and quartz is recognized by undulatory extinction. Width 4 mm.

Plate 13c Large twinned orthoclase crystal with perthic exsolution and a hornblende inclusion. Granite from Santa Rita Mountains, Arizona, USA. Crossed polarizers. Width 4 mm.

Plate 13d Twinned orthoclase (center) and quartz with undulatory domains. Bergell granite, Switzerland. Crossed polarizers. Width 4 mm.

Plate 13e Granite with microcline (cross-hatched twinning) and quartz (e.g., dark crystal at bottom left). Cripple Creek, Colorado, USA. Crossed polarizers. Width 10 mm.

Plate 13f Doubly twinned plagioclase in olivine gabbro from Skaergaard, Greenland. Crossed polarizers. Width 4 mm.

(a)

(b)

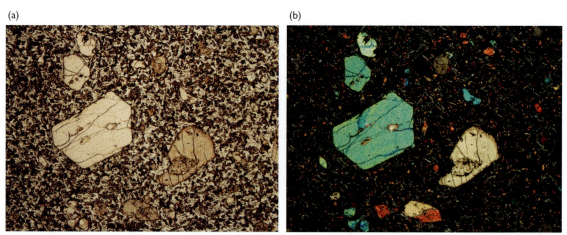

Plate 14a,b Basalt with olivine phenocrysts (large colored crystals) and plagioclase and pyroxene in groundmass from Oelberg, Siebengebirge, Germany. Plane and crossed polarizers. Width 4 mm.

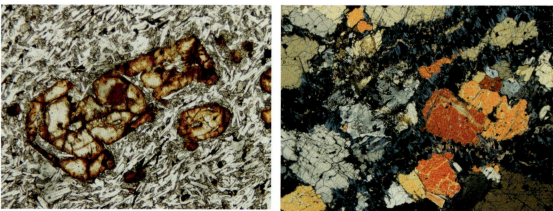

Plate 14c Iddingsite (red/yellow) replacing olivine in basalt. East of Santa Rosa, California, USA. Plane polarizers. Width 2 mm.

Plate 14d Olivinite becoming serpentinized (blueish). Val Bregaglia, Switzerland. Crossed polarizers. Width 4 mm.

(e)

(f)

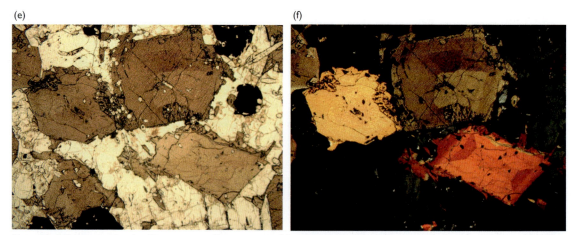

Plate 14e,f Sector zoning in clinopyroxene (brown crystal *bottom right*). The large crystal *above* it displays the 89° {110} cleavage traces. Laths are plagioclase. Walhola alkaline rocks, Otago, New Zealand. Plane and crossed polarizers. Width 2 mm.

(a)

(b)

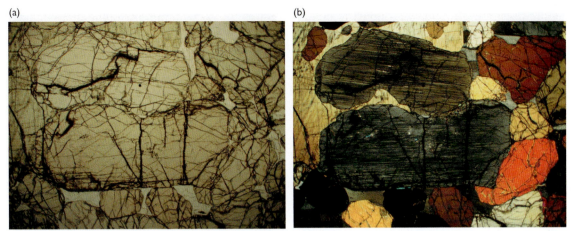

Plate 15a,b Exsolution lamellae in orthopyroxene from pyroxenite in the ultramafic zone of the Stillwater Complex, Montana, USA. Plane and crossed polarizers. Width 8 mm.

(c)

(d)

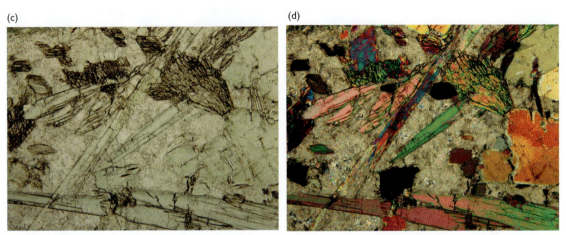

Plate 15c,d Prismatic tremolite in calcsilicate rock. Amphibole cleavage is visible in crystals cut normal to [001]. Plane and crossed polarizers. Width 4 mm.

(e)

(f)

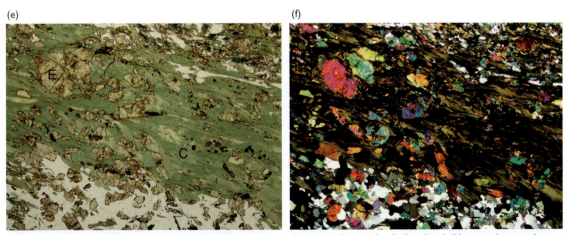

Plate 15e,f Chlorite schist with epidote (E) and chlorite (C) from Val Bregaglia, Switzerland. Plane and crossed polarizers. Width 4mm.

(a)

(b)

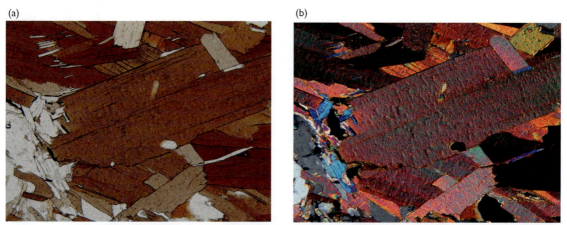

Plate 16 (a) Biotite crystals in plane polarized light display pleochroism. (b) With crossed polarizers a mottled texture is observed that is unique for biotite and distinguishes it from, e.g., stilpnomelane. Plane and crossed polarizers. Val Bregaglia, Switzerland. Width 1 mm.

Plate 16c Calcite in marble from Palm Canyon, southern California, USA. Note the high interference colors and the presence of lamellar twins due to deformation. Crossed polarizers. Width 10 mm.

Plate 16d Diagenetic replacement of limestone (dark, fine grained) by euhedral crystals of dolomite, mainly along stylolites and fractures. Sample is from the Lost Burro Formation, southeastern California, USA (see also Plate 24b and Wenk and Zenger, 1983). Crossed polarizers. Width 4 mm.

(e)

(f)

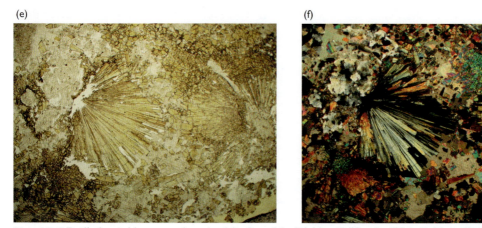

Plate 16e,f Radiating epidote crystals and calcite from Mt. Diablo, California, USA. Epidote displays abnormal interference colors, for example brilliant green colors on top right. Plane and crossed polarizers. Width 10 mm.

(a)

(b)

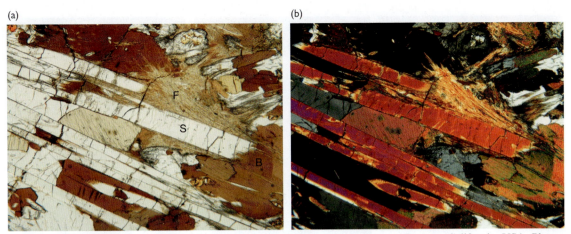

Plate 17a,b Sillimanite (S), some fibrolitic (F), and biotite (B) in gneiss from Deep Canyon, California, USA. Plane and crossed polarizers. Width 4 mm.

(c)

(d)

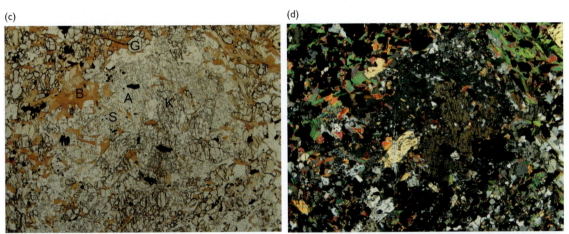

Plate 17c,d Pelitic schist with all three aluminosilicates coexisting, from Cataeggio, Alps, Italy. A large porphyroblast of andalusite (A) includes crystals of kyanite (K) and a needle of sillimanite (S). B is biotite and G is garnet. Plane and crossed polarizers. Width 4 mm.

(e)

(f)

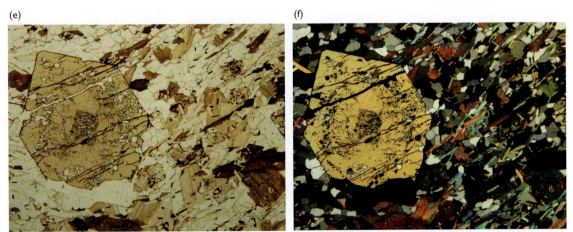

Plate 17e,f Euhedral staurolite phenocryst (yellow) growing in biotite schist from Ontario, Canada. Plane and crossed polarizers. Width 4 mm.

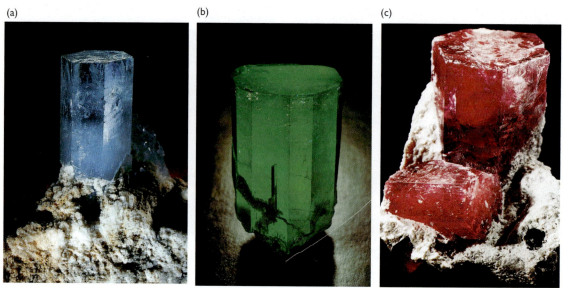

Plate 18a–c Beryl with different colors. (a) Aquamarine from Pech, Afghanistan, 40 mm high (courtesy O. Medenbach). (b) Emerald from Coscuez Mine, Colombia, 1759 carats (courtesy E. and H. Van Pelt and J. R. Sauer). (c) Morganite from Utah, USA, 35 mm high (courtesy O. Medenbach).

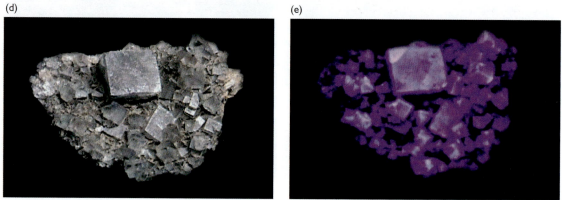

Plate 18d,e Fluorescence of fluorite from Durham, England (courtesy M. J. Thompson). (d) Daylight; (e) longwave ultraviolet. Width 60 mm.

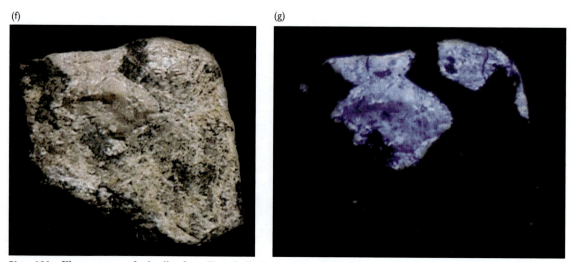

Plate 18f,g Fluorescence of scheelite from Trumball, Connecticut, USA (courtesy M. J. Thompson). (f) Daylight; (g) shortwave ultraviolet. Width 60 mm.

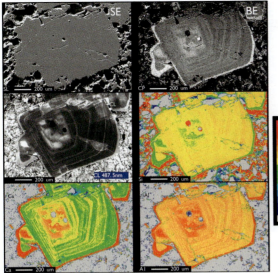

Plate 19a SEM images of zoned plagioclase in a volcanic rock. SE: Secondary electron image with topographic contrast, BE: backscattered electron image displaying compositional contrast, CL: cathodoluminescence (CL) image with growth zones, Si: silicon X-ray map, Ca: calcium X-ray map, Al: aluminum X-ray map. (From Takakura *et al.*, 2001, courtesy JEOL.)

Plate 19b X-ray tomography of Qusaiba shale with octahedral pyrite crystallites and spheroidal pyrite aggregates (see also Kanitpanyacharoen *et al.*, 2011). Width 200 μm.

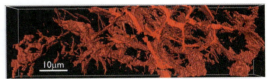

Plate 19c X-ray nanotomography with network of iron channels (red) in MgSiO$_3$ perovskite aggregate formed by melting at high pressure and temperature (courtesy W. Mao, see also Shi *et al.*, 2013).

Plate 19d Sulfur fumarole from Mt. Papandayan in Java, Indonesia (courtesy G. Kourounis).

Plate 19e Detail of sulfur dendrites (courtesy G. Kourounis). Height 10 mm.

Plate 20a Smoky quartz and octahedral fluorite from Voralp UR, Switzerland (courtesy T. Schüpbach). Width 20 mm.

Plate 20b Smoky quartz from Zinggenstock UR, Switzerland (courtesy T. Schüpbach). Height 7 cm.

Plate 20c Citrine from Alto de Cruzes, Colombia (courtesy D. Weinrich). Height 3 cm.

Plate 20d Rose quartz with striated dark tourmaline from Sapuchaia, Minas Gerais, Brazil (courtesy E. and H. Van Pelt, J. Sauer Collection). Width 15 cm.

Plate 20e Cristobalite in obsidian from Coso Hot Springs, Inyo Co. California, USA (courtesy D. Weinrich). Width 8 cm.

Plate 20f Hexagonal platy tridymite crystals from Vechec, Slovakia (courtesy D. Weinrich). Width 15 mm.

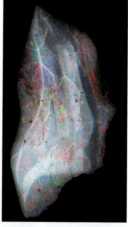

Plate 20g Opal from Coober Peedy, Australia (courtesy G. Ivanyuk). Width 30 mm.

Plate 20h Yellow opal from Borzsony Mountains, Hungary (courtesy D. Weinrich). Width 5 cm.

Plate 21a Adularia from Lukmanier TI, Switzerland. Width 10 cm.

Plate 21b Amazonite from Colorado (iRocks.com photo; Arkenstone specimen). Width 2 cm.

Plate 21c Graphic texture with cuneiform intergrowth of quartz and orthoclase. Sample from Kola Peninsula, Russia. Width 15 cm.

Plate 21d Albite with quartz from Lincoln Co., New Mexico, USA (courtesy D. Weinrich). Width 6 cm.

Plate 21e Labradorite from Labrador, Canada, with "schiller" due to lamellar exsolution. Width 50 mm.

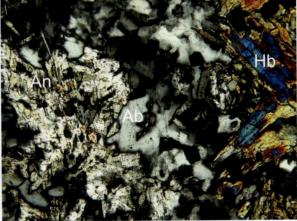

Plate 21f Thin section of amphibolite with intergrowth of albite (Ab) and anorthite (An) from the NE Bergell Alps. Hb is hornblende. Crossed polarizers (see Wenk, 1979). Width 1.5 mm.

Plate 22a (left) Silver from Freiberg, Germany (iRocks.com photo; Arkenstone specimen). Height 17 cm.

Plate 22b (right) Crystalline gold from Eagle's Nest Mine, Placer County, California, USA (courtesy D. Weinrich). Height 5 cm.

Plate 22c Gold crystals on quartz from Zalanta, Romania (courtesy O. Medenbach). Width 11 mm.

Plate 22d Gold nugget, Klondike, Alaska, USA (courtesy D. Weinrich, 1066115). Width 11 mm.

Plate 22e Mercury, Socrates Mine, California, USA (courtesy D. Weinrich). Width 50 mm.

Plate 22f Widmanstätten pattern, an intergrowth of kamacite and taenite, observed in meteoritic iron. Gibeon meteorite, Namibia (courtesy O. Medenbach). Width 80 mm.

Plate 23a Octahedral diamond in kimberlite from Russia. Brilliant-cut carat stone in insert (courtesy E. and H. Van Pelt, W. Larson Collection).

Plate 23b Synthetic diamond imaged when exposed to a high-intensity synchrotron X-ray beam, producing yellow optical luminescence and corresponding reflections (courtesy A. Freund, see also Hoszowska *et al.*, 2001). Diameter 10 mm.

Plate 23c Halite with blue zoning (K+S Winterschall, Hessen, Germany). Width 20 cm.

Plate 23d Fluorite with growth steps, Xianghualing Mine, Hunan, China (courtesy D. Weinrich 1041907a). Width 14 cm.

Plate 23e Prismatic–scalenohedral calcite from Cumbria, UK (courtesy O. Medenbach). Width 48 mm.

Plate 23f Tabular calcite with basal–prismatic morphology from Guangdong, China (courtesy Museum of Geology, Ministry of Geology, Beijing, China). Width 70 mm.

Plate 24a Rhombohedral dolomite from Eugui, Spain. Width 12 cm.

Plate 24b Limestone (dark) being replaced by dolomite (bright) along veins and fractures. Lost Burro Formation, Nevada, USA. Compare with thin section in Plate 16d.

Plate 24c Twinned aragonite from Miglanilla, Spain (courtesy O. Medenbach). Width 34 mm.

Plate 24d Botryoidal azurite (blue) and malachite (green) from Bisbee, Arizona, USA (courtesy J. Scovil, Rice Northwest Museum). Width 12 cm.

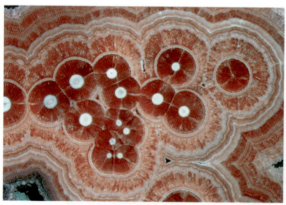

Plate 24e Rhodochrosite from Minas Capillitas, Andalgo, Argentina (courtesy A. Massanek). Width 14 cm.

Plate 24f Rhombohedral rhodochrosite from Sweet Home Mine, Mount Bross, Colorado, USA (courtesy D. Weinrich). Width 8 cm.

Plate 25a Euhedral crystals of monazite in an alpine vein from Vamlera, Central Alps, Italy (courtesy F. Bedogne and R. Maurizio). Width 12 mm.

Plate 25b Apatite from Cerro de Mercado, Durango, Mexico (courtesy O. Medenbach). Width 20 mm.

Plate 25c Torbernite from Vogtland, Germany (courtesy O. Medenbach). Width 18 mm.

Plate 25d. Pyrite pyritohedron on quartz from Peru (courtesy O. Medenbach). Width 30 mm.

Plate 25e Marcasite from Sparta, Illinois, USA (courtesy J. Arnoth and W. Gabriel). Diameter 9 cm.

Plate 25f Sphalerite on dolomite from Binntal VS, Switzerland, 111 twin (courtesy O. Medenbach). Width 32 mm.

Plate 26a Tetrahedrite from Wilroth, Westerwald, Germany (courtesy O. Medenbach). Width 29 mm.

Plate 26b Cube-octahedral galena from Harz, Germany (courtesy O. Medenbach). Diameter of crystal 38 mm.

Plate 26c Chalcopyrite and dolomite, Dreislar, Winterberg, Germany (courtesy D. Weinrich). Width of aggregate 5 mm.

Plate 26d Chalcopyrite with typical oxidation colors on siderite from Wilroth, Westerwald, Germany (courtesy O. Medenbach). Width 36 mm.

Plate 26e Realgar with calcite from Shimen, Hunan Province, China (courtesy A. Massanek). Width 80 mm.

Plate 26f Orpiment (yellow) with realgar (red) in dolomite from Lengenbach VS, Switzerland (courtesy W. Gabriel). Width 5 cm.

Plate 27a Prismatic stibnite from Herja, Romania (courtesy O. Medenbach). Width 5 cm.

Plate 27b Platy molybdenite with quartz from Kingsgate, Queensland, Australia (courtesy O. Medenbach). Width 25 mm.

Plate 27c Cross-sectional piece of a sulfide chimney recovered from an active black smoker recovered from the Juan de Fuca Ridge in the Pacific Ocean. Walls show concentric zones of mineralization with Zn, Fe sulfides (Z) (sphalerite, wurtzite) and Cu, Fe sulfides (C) (chalcopyrite) as well as pyrite and marcasite in the interior, and sulfates (S) (barite, anhydrite) and oxides (O) (amorphous silica and goethite) near the outer walls. The inner conduit with clays is also indicated (I) (courtesy D. Kelley). Width 60 cm.

Plate 27d Perovskite in apatite from Afrikanda, Kola Peninsula, Russia (courtesy G. Ivanyuk). Width 50 mm.

Plate 27e Twinned crystals of loparite from Lovozero, Kola Peninsula, Russia (courtesy G. Ivanyuk). Width 50 mm.

Plate 27f Fibrous-spheroidal goethite from Devon, UK (courtesy O. Medenbach). Width 35 mm.

Plate 28a Rutile on goethite pseudomorphs from Binntal VS, Switzerland (courtesy, T. Schüpbach). Height of crystal 9 mm.

Plate 28b Acicular rutile on quartz from Alpe Cavradi GR, Switzerland (courtesy O. Medenbach). Width 20 mm.

Plate 28c Framework of prismatic rutile from Simplon Pass VS, Switzerland (courtesy W. and D. Leicht). Width 2 cm.

Plate 28d Anatase (large crystal 9mm long) on quartz from Val Russein GR, Switzerland (courtesy, T. Schüpbach).

Plate 28e Octahedral cuprite with calcite from Tsumeb, Namibia (courtesy O. Medenbach). Width 12 mm.

Plate 28f Sapphire from Sri Lanka (courtesy E. and H. Van Pelt). Width 3.5 cm.

Plate 29a Olivine from Mogok, Myanmar (courtesy J. Scovil, W. Larson Collection). Height 43 mm.

Plate 29b Topaz, from Ghun Dao mine in Pakistan, in calcite (courtesy E. and H. Van Pelt, W. Larson Collection). Height 50 mm.

Plate 29c Garnet (spessartine) with trapezohedral morphology in quartz from Val Codera, Italian Alps (courtesy F. Bedogne and R. Maurizio). Width 20 mm.

Plate 29d Zircon from Lovozero, Kola Peninsula, Russia (courtesy O. Medenbach). Width 30 mm.

Plate 29e Epidote on quartz with amianth needles. Loetschental VS, Switzerland (courtesy T. Schüpbach). Width 4 cm.

Plate 29f Kyanite in muscovite schist from Pizzo Forno TI, Switzerland (courtesy T. Schüpbach). Width 12 cm.

Plate 30a Titanite, Furka, Switzerland (courtesy, T. Schüpbach). Width 6 mm.

Plate 30b Prismatic beryl (aquamarine) with muscovite from Dusso Gilgit, Pakistan (courtesy E. and H. Van Pelt, Smithsonian Collection). Height 50 mm.

Plate 30c Black tourmaline (ilvaite) with hematite rose, Gotthard TI, Switzerland (courtesy T. Schüpbach). Width 2 cm.

Plate 30d Bicolored tourmaline from Himalaya mine, Pala, California, USA (courtesy E. and H. Van Pelt, W. Larson Collection). Height 100 mm.

Plate 30e Pyrophyllite, Graves Mountain, Georgia, USA (courtesy D. Weinrich). Width 8 cm.

Plate 30f Muscovite with beryl from Hunza Valley, Pakistan (courtesy D. Weinrich). Width 18 cm.

Plate 31a Chrysotile asbestos, resembling a textile fabric, from Uschione, Italian Alps (courtesy F. Bedogne and R. Maurizio). Width 30 cm.

Plate 31b Clinochlore (green) with diopside from Taeschalp VS, Switzerland (courtesy T. Schüpbach). Diameter of chlorite 5 mm.

Plate 31c Platy clinochlore from Saas Fe VS, Switzerland (courtesy T. Schüpbach). Diameter of platelets up to 5 mm.

Plate 31d Diopside (up to 4 mm in length) with garnet (red) and chlorite (flakes), Zermatt VS, Switzerland (courtesy T. Schüpbach).

Plate 31e Leucite from Redina, Italy (courtesy O. Medenbach). Width 60 mm.

Plate 31f Mellite, Usti, Czech Republic (iRocks.com photo; Arkenstone specimen). Width 20 mm.

Plate 32a Star sapphire in cabochon cut; 32 carat (courtesy E. and H. Van Pelt, W. Larson Collection).

Plate 32b Trapiche emerald with pyrite from Muzo, Columbia (courtesy E. and H. Van Pelt). Width 3 cm.

Plate 32c Trapiche emerald from Muzo with cabochon cut, 7.9 carat (courtesy H. Hänni).

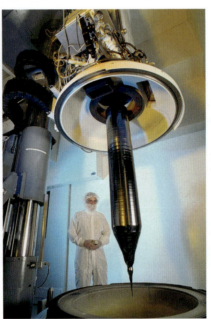

Plate 32d (left) Boule of synthetically grown ruby (courtesy D. Belakovskiy).

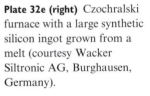

Plate 32e (right) Czochralski furnace with a large synthetic silicon ingot grown from a melt (courtesy Wacker Siltronic AG, Burghausen, Germany).

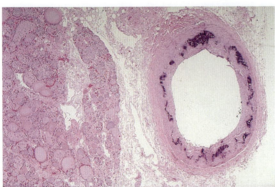

Plate 32f Micrograph displaying a section through a human coronary artery. The dark region is calcified tissue that is common in arteriosclerosis (courtesy E. C. Klatt). Width 300 μm.

Plate 32g 2 mm diamond from kimberlite in Zimbabwe with inclusions of garnet and diopside. The inclusion assemblages suggest that those diamonds equilibrated at 1100–1300 °C and 5–6 GPa, i.e., at a depth of 180–200 km (courtesy M. G. Kopylova; see also Kopylova *et al.*, 1997).

pressure of 100 MPa the minimum melting temperature is reduced to 850 °C; at 500 MPa it is reduced to 700 °C (Figure 21.24b,c). There are other changes. For example, at $p_{H_2O} = 500$ MPa, there is no longer a field for a homogeneous alkali feldspar solid solution at high temperature (Figure 21.24c) and, upon cooling, albite and sanidine will crystallize simultaneously.

Keep in mind that all the interpretations given above are for equilibrium crystallization processes. The diversity of real conditions during crystallization of a granitic melt gives rise to many compositional, textural, and structural varieties of granitic rocks. Kinetic factors, such as sluggish nucleation, and material diffusion may prevent crystallization at equilibrium conditions.

21.6 Pegmatites

Usually the term *pegmatite* is applied to veins, dikes, and lenses of coarse-grained quartzo-feldspathic rocks. The dikes are sometimes 10 or more meters in thickness and hundreds of meters in length. Pegmatites are composed of the same minerals as the rocks to which they are related by origin, and often occur in spatial proximity. Granite pegmatites are the most abundant variety. They consist of feldspars (microcline, orthoclase, albite-rich plagioclase), quartz, micas (muscovite and biotite), and accessory minerals. Syenite pegmatites are much less abundant. Granite pegmatites are commonly composed of very large crystals, much larger than those in the host rock. We will discuss only granite pegmatites below.

Feldspars are the dominant minerals in pegmatites, making up 50–70% of the volume, and quartz comprising 20–40% of the volume. Among the common minor minerals are muscovite and biotite. For industrial applications it is significant that pegmatite minerals often concentrate boron, phosphorus, uranium, rare earth elements, lithium, beryllium, cesium, and tantalum. Pegmatites are the main source for some of these elements.

The grain size and composition of pegmatites often differ within a vein. The largest mineral grains are typical for the central blocky parts of the veins. It is in such pegmatites that the largest crystals occur. For example, feldspar crystals exceeding 10 m in size have been found in Norway, and huge plates of mica (muscovite) weighing 85 tonnes in India and Canada (Figure 10.12b), beryl crystals weighing up to 300 tonnes in Brazil, and spodumene crystals exceeding

16 m in length and 90 tonnes in weight in South Dakota, USA (Figure 10.12a).

Presumably, pegmatites formed at a depth of about 6–8 km. Their genesis is a subject of controversy. One school of thought (see, e.g., London, 2008) proposes that they are the crystallization products of residual melts intruding into fractures in country rocks. Such residual melts are relatively rich in volatiles (H_2O, HF, HCl, B_2O_3, etc.). Crystallization starts at 900–800 °C, and the major minerals (feldspars and quartz) form in the 800–600 °C temperature range at a late igneous stage. Later minerals crystallize from subcritical aqueous solutions.

A very different origin has been proposed by others (see, e.g., Cameron *et al.*, 1949; Jahns and Burnham, 1969). Their interpretation is based on the frequent chaotic structure of the pegmatite veins in which recrystallization phenomena are widespread, with resorbtion and chemical substitution of earlier-crystallized minerals. In this scenario, pegmatites result from profound transformation of felsic rocks under the influence of hydrothermal solutions, rather than by direct magmatic crystallization.

On the basis of mineral assemblages, several types of granite pegmatites may be distinguished. The most important types are mica-bearing pegmatites, topaz–beryl pegmatites, and albite–spodumene pegmatites. The mica-bearing pegmatites are exploited for feldspars, which are raw materials for the ceramic industry, and for muscovite. Piezoquartz and gemstones are obtained from topaz–beryl pegmatites. Albite–spodumene varieties are mined for lithium, cesium, tantalum, and other rare-element ores.

21.7 Summary

The chapter first introduces the crystal structures of silica minerals and feldspars. Both are characterized by SiO_4^{4-} tetrahedra that are three-dimensionally linked to form infinite frameworks. In the case of silica, the chemical composition is very simple: SiO_2, but there are six different polymorphs with very different structures which form at various temperature–pressure conditions. Feldspar, the most common mineral in the Earth's crust, has a wide range of compositions ranging between $KAlSi_3O_8$–$NaAlSi_3O_8$–$CaAl_2Si_2O_8$. The structures are similar, with an open framework that can accommodate large K, Na, and Ca ions. At high temperatures they form continuous solid solutions but during cooling there is often exsolution, expressed in microstructures. The chapter discusses the origin of granite and pegmatite, rock types in which feldspars and quartz are major components. The ternary Q(uartz)–A(lkalifeldspar)–P(lagioclase) diagram is used for classification of granites. Water pressure has a strong influence on melting in granitic systems. Pegmatites are coarse-grained dykes that host many gem minerals.

Important minerals to remember

Silica minerals

Name	System	Formation conditions
α-Quartz	Trigonal	Low temperature
β-Quartz	Hexagonal	High temperature
Cristobalite	Cubic	High temperature
Tridymite	Hexagonal	High temperature
Coesite	Orthorhombic	High pressure
Stishovite	Tetragonal	Very high pressure

Feldspars

Name	Formula	System
Microcline	$KAlSi_3O_8$	Triclinic
Orthoclase	$KAlSi_3O_8$	Monoclinic
Sanidine	$KAlSi_3O_8$	Monoclinic
Albite	$NaAlSi_3O_8$	Triclinic
Anorthite	$CaAl_2Si_2O_8$	Triclinic
Plagioclase solid solution	$(Na,Ca)(Si,Al)AlSi_2O_8$	Triclinic

Test your knowledge

1. Discuss the silicon tetrahedron: size, charge, radius ratio, character of bonding, and representation.
2. Do cristobalite and tridymite form at low temperature, high temperature, or both? Why?
3. Explain three mechanisms by which Dauphiné twinning in quartz may develop.

4. Review the names of major feldspars and their chemical formulas (refer to the ternary diagram, Figure 21.9a).
5. Explain Al/Si order in K-feldspar and anorthite, and explain how defects form during the ordering phase transformation.
6. What are the important twin laws observed in feldspars (review section on twinning in Chapter 9)?
7. What is the influence of water pressure on the alkali feldspar phase diagram?
8. Review the ternary system $KAlSi_3O_8$–$NaAlSi_3O_8$–SiO_2.
9. Which elements are concentrated in granite pegmatites?

Further reading

Silica minerals and feldspars

Deer, W. A. and Howie, R. A. (2006). *Rock-forming Minerals, Volume 4B, Framework Silicates: Silica Minerals, Feldspathoids and Zeolites*, 2nd edn. The Geological Society of London.

Deer, W. A., Howie, R. A. and Zussman, J. (2001). *Rock-forming Minerals, Volume 4A, Framework Silicates: Feldspars*, 2nd edn. The Geological Society of London.

Heaney, P., Prewitt, C. T. and Gibbs, G. V. (eds.) (1994). *Silica: Physical Behavior, Geochemistry and Materials Applications*. Reviews in Mineralogy, vol. 29. Mineralogical Society of America, Washington, DC.

Ribbe, P. (ed.) (1983). *Feldspar Mineralogy*. Reviews in Mineralogy, vol. 2, 2nd edn. Mineralogical Society of America, Washington, DC.

Smith, J. V. and Brown, W. L. (1988). *Feldspar Minerals, Volume 1*. Springer-Verlag, Berlin.

Occurrence in granites

McBirney, A. R. (2007). *Igneous Petrology*, 3rd edn. Jones and Bartlett, Sudbury, MA.

Tuttle, O. F. and Bowen, N. L. (1958). *Origin of Granite in the Light of Experimental Studies in the System $NaAlSi_3O_8$–$KAlSi_3O_8$–H_2O*. GSA Memoir, 74. Geological Society of America, Washington, DC.

22 | Simple compounds. Unusual mineral occurrences

Most of the minerals in this section are very rare. In total, they make up less than 0.0002 weight% of the Earth's crust. However, not only are their structures interesting and well known in materials science, they are also of great economic interest. The minerals gold, platinum, osmium, iridium, and silver are the principal sources for the elements of the same name. A significant portion of the sulfur used in the chemical industry is mined in the form of native sulfur. Minerals such as diamond and graphite are used because of their unique properties (hardness, electrical conductivity, and refraction).

22.1 Background about metals and intermetallics

After the discussion of some of the most common but also the most complex minerals in Chapter 21, we now examine some chemically and structurally very simple compounds that are rare and form only under unusual conditions. In this chapter we discuss minerals of native elements such as graphite (C), diamond (C), copper (Cu), gold (Au), and silver (Ag) and solid solutions of these elements (for instance, Au–Ag and Au–Cu) (Table 22.1). There are also intermetallic compounds with ordered crystal structures that differ from the end members. Often, these intermetallic compounds form during cooling as a result of ordering of high-temperature solid solutions of the same compositions. Some intermetallic minerals are awaruite ($FeNi_3$) and auricupride (Cu_3Au), both occurring in serpentinites.

For many years it was generally assumed that only a few elements could exist in nature in the native form because of their chemical inertness. Just a few decades ago no mineralogist would have imagined that elements such as aluminum, cadmium, and silicon could form in nature. Yet recent investigations with electron microscopes and electron microprobes have led to the discovery of minerals whose existence was thought to be impossible in rocks and ores. Certainly they occur under unique conditions and are found in infinitesimally small

quantities and tiny grains, which generally cannot be seen, even with an optical microscope.

22.2 Crystal structures and relationships to morphology and physical properties

The crystal structures of native metals are relatively simple and can be described by a close-packing arrangement (see Figures 2.6 and 2.7). The structures of native *gold*, *copper*, *silver*, and *aluminum* are (at room temperature and atmospheric pressure) cubic close-packed, with a stacking sequence ABC–ABC; *zinc*, *ruthenium*, and *osmium* are hexagonal close-packed with a stacking AB–AB. The less metallic the character of the bonding, the further the structure deviates from ideal close-packing.

In contrast with these metals, native *sulfur* has a molecular structure. Each molecule consists of a ring of eight atoms with covalent bonding and a complete charge balance. In the most common orthorhombic polymorph, the rings (shaded in Figure 22.1) are stacked on top of each other to form "pillars" that are differently oriented in space. Bonding between the rings and the pillars is of the van der Waals type.

The structures of two natural carbon polymorphs (*diamond* and *graphite*), along with a phase diagram, are shown in Figure 22.2 (see also Figure 2.12). The diamond structure has strong covalent bonds in all

Table 22.1 Minerals of native elements, with some diagnostic properties; important minerals are given in italics (elements in parentheses indicate partial substitutions)

Mineral Formula	System	Morphology Cleavage	H	D	Color/luster Streak	n	Δ	$2V$
fcc metals								
Aluminum Al	Cubic	Micr.	4	2.7	White/metallic Gray	opaque		
Copper Cu	Cubic	{111} {100}	3	8.7	Red/metallic Red/metallic	opaque		
Gold Au (Ag, Pd, Cu)	Cubic	{111} {100}	2.5	19.2	Yellow/metallic Yellow	opaque		
Taenite (γ-Fe) Fe, Ni	Cubic	Platy	5–5.5	8.1	Silver, white-gray/metallic Gray	opaque		
Lead Pb	Cubic	Oct., Cub.	1.5	11.3	White-gray/metallic Gray	opaque		
Osmiridium Os (Ir)	Cubic	Micr.	6–7	21	Gray/metallic Gray	opaque		
Palladium Pd	Cubic	Micr.	4, 5	12	Gray/metallic Gray	opaque		
Platinum Pt (Fe, Pd, Rh)	Cubic	{111} {100}	4	21.5	Gray/Metallic Gray	opaque		
Rhodium Rh	Cubic	Micr.		12.4	Gray/metallic Gray	opaque		
Silver Ag (Au)	Cubic	{111} {100}	3	10.5	White, gray/ metallic *White*	opaque		
bcc metals								
Kamacite (α-Fe) Fe (Ni)	Cubic	{001}	4	7.7	Gray/metallic Gray	opaque		
hcp metals								
Zinc Zn	Hexag.	Micr. (0001)	2	7	White/metallic White-gray	opaque		
Other metals								
Mercury Hg	Rhomb.	(Liquid at 25 °C)		13.6		opaque		
Bismuth Bi	Trig.	Micr. *(0001)*	2	9.7	Pink/metallic	opaque		
Arsenic As	Trig.	Eq. *(0001)*	3.5	5.7	Gray/metallic Black	opaque		
Nonmetals								
Graphite C	Hexag.	Platy (0001) *(0001)*	1	2.2	Black/metallic Gray	1.93–2.07		
Diamond C	Cubic	{111}{*111*}	10	3.52	Clear, yellow, blue	2.411–2.447		
α-Sulfur S	Ortho.	Pris., Platy (001) *{101}{110}*	1.5–2	2.0	Yellow White	1.96–2.25	0.288	+70

Notes: H, hardness; D, density (g/cm^3); n, range of refractive indices; Δ, birefringence; $2V$, axial angle for sulfur.
System: Hexag., hexagonal; Ortho., orthorhombic; Rhomb., rhombohedral; Trig., trigonal; fcc, face-centred cubic; bcc, body-centered cubic; hcp, hexagonal close-packed.
Morphology: Cub., cubic; Eq., equiaxed; Micr., microscopic; Oct., octahedron; Pris., prismatic.

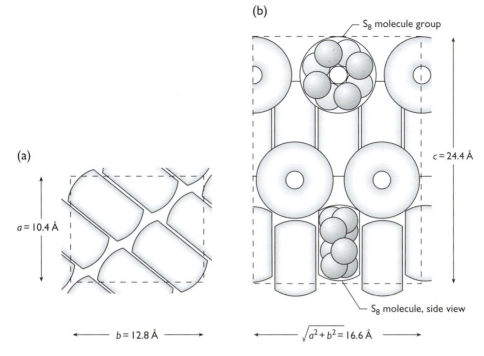

Fig. 22.1 Structure of orthorhombic sulfur with covalently bonded S_8 rings linked by van der Waals forces. (a) View along the *c*-axis, with rings stacked roughly along [110]. (Only the bottom part of the unit cell is shown in this projection.) (b) Projection of the unit cell on (110), illustrating again the stacks of rings, some parallel to (110) and others parallel to ($1\overline{1}0$). Dashed lines show the unit cell.

three dimensions. Four neighboring atoms in the form of a tetrahedron surround every atom. The structure of graphite consists of hexagonal carbon sheets. Within a layer, carbon atoms are connected by strong covalent bonds, whereas weak van der Waals bonds operate between layers (dashed lines in Figure 22.2). If we turn the structures of these two minerals to make a 3-fold axis of diamond parallel to a 6-fold axis of graphite we can see an important difference with the same six-membered rings being flat in graphite and goffered in diamond (Figure 22.3).

A relatively newly identified and rare form of carbon is *fullerenes*; in spherical form its molecule consists of many carbon atoms, each bonded to three others to form five- and six-membered rings to cover spheres with an unusual 5-fold symmetry (Figure 22.4). Fullerenes exist in trace amounts in some meteorites. Carbon nanotubes are a type of fullerenes which have an extremely large length-to-diameter ratio and have extraordinary properties. Graphene is pure carbon in the form of a very thin, nearly transparent sheet, one atom thick.

Cubic close-packed crystals such as gold often display the forms of a cube, octahedron (Figure 22.5 and Plate 22c), dodecahedron, or a combination of these, but in nature the morphology is much more complex. Metals often occur as dendrites (silver in Figure 22.6, copper in Plate 4a), wire-shaped morphology (silver in Plate 22a), or more irregular aggregates (gold in Plate 22b).

The native nonmetallic minerals have a more diverse morphology. Sulfur crystallizes in orthorhombic symmetry as dipyramid-like polyhedra combined with the prismatic and pinacoidal faces (Plate 1a) or as dendrites (Plate 19e). The crystals of graphite are, as a rule, thin platelets parallel to (0001). Diamond occurs as octahedral (Plate 23a,b), dodecahedral, and (more rarely) cube-shaped crystals.

Native gold, silver, and copper have typical metallic luster, high density, high electrical and thermal conductivity, are malleable, and lack cleavage. In contrast, sulfur is brittle, has a low density (2.05–2.08 g/cm^3) and low hardness (Mohs scale about 3), and easily ignites and melts owing to its molecular structure.

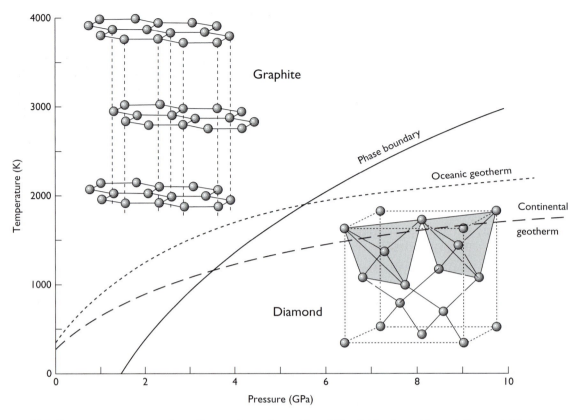

Fig. 22.2 Temperature–pressure phase diagram for carbon with stability fields of graphite and diamond. Their structures are shown in the corresponding stability fields. The dashed curve is an average geothermal gradient for continental crust and the dotted curve that for oceanic crust.

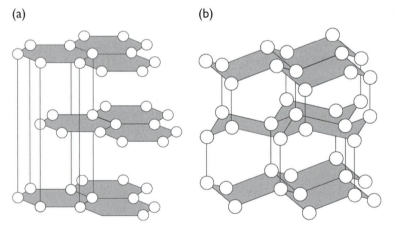

Fig. 22.3 Comparison of the structures of diamond with graphite by turning one of the 3-fold axes of diamond parallel to the 6-fold axis of graphite. Both display six-membered rings of carbons, but in the case of graphite they are on a flat plane (a), and in the case of diamond they are goffered (b).

Graphite has a perfect cleavage along the plane of the hexagonal sheets (0001) and also high electrical conductivity parallel to this plane because of the quasi-metallic nature of the bonding, with some free electrons existing between carbon layers. Individual carbon layers are easily rubbed off, contributing to the low hardness of graphite and its use in writing tools. Finally, diamond possesses an extremely high

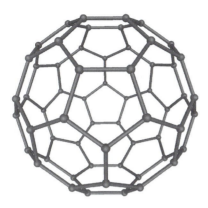

Fig. 22.4 Structure of the carbon polymorph spherical fullerene, a molecule of 60 carbon atoms arranged by linking of five- and six-membered rings.

Fig. 22.6 Skeletal dendrites of silver from Wolkenstein, Erzgebirge, Saxony, Germany (courtesy A. Massanek). Width 20 mm.

Fig. 22.5 Octahedral crystals of gold from Mariposa County, California, USA (courtesy J. Arnoth). Width 15 mm.

hardness (10) because of the unusual compactness of its structure and the covalent character of the bonds.

22.3 Brief description of important minerals of the native elements

Native **copper** has only minor impurities of silver, gold, and iron. It occurs as nodules, as single irregular grains, as wire-like and plate-like crystals in fractures, and as dendrites (Plate 4a). The mineral is red in color and often covered with black, green, and blue films of copper oxides. In order to identify copper, it is useful to scratch it or determine its streak color. Copper has metallic luster in fresh pieces, can be scratched with a knife (hardness 2.5–3), and has a high density (8.4–8.9 g/cm^3).

Copper occurs in assemblages with other copper minerals such as cuprite, tenorite, malachite, and chrysocolla in oxidation zones of sulfide ores. It serves as a copper ore, but as a native element it is economically not important and most copper is extracted from minerals such as chalcopyrite.

Gold contains variable amounts of silver, palladium, rhodium, copper, tellurium, and bismuth. If gold contains more than 10% silver it is called *electrum*. The mineral forms fine platelets, irregular grains, and inclusions in quartz and sulfides (pyrite, arsenopyrite, tennantite, and tetrahedrite). Regular gold single crystals are rare. Exceptional octahedral crystals have been found in Romania (Plate 22c) and in California, USA (see Figure 22.5). In sediments (placers), gold is found as nuggets of various shapes and sizes ranging from fractions of a gram to tens of kilograms in weight (Plate 22d). Gold is soft (Mohs scale 2–3) but has an extremely high density of about 16–18 g/cm^3 (pure gold is 19.3 g/cm^3).

Gold has long been an international standard, and many currencies are still backed by it. Gold is used mainly in the manufacture of jewelry and there are applications in the electronics industry. In recent years its value has undergone large fluctuations owing to changes in supply and demand, caused in particular by political factors, and gold production has varied accordingly. The main industrial deposits are medium-temperature (rarely low-temperature) hydrothermal veins, often associated with tectonic events. Other large gold deposits are secondary, either owing to extensive chemical weathering in a tropical climate (e.g., Carajas in Brazil) or river placers (e.g., Western Australia).

Silver is a rare mineral that forms thin platelets, sheets, skeletal dendrites (Figure 22.6), and wire-like crystals (Plate 22a) in fractures of ore bodies. This mineral has a silver-white color, metallic luster, and hook-shaped (hackly) fracture surfaces. Silver occurs in medium-temperature hydrothermal deposits and oxidation zones. When exposed to air it oxidizes easily and the surface blackens. The alteration product is greasy and rubs off.

In nature, **iron** occurs as two cubic polymorphs, native iron or *kamacite* (bcc) and *taenite* (austenite is the corresponding metallurgical name, fcc). Kamacite is essentially a meteoritic equivalent of native iron, enriched in nickel (up to 5–8%), while taenite also occurs in meteorites but has up to 70% nickel. Both polymorphs are extremely rare minerals. Impregnations and larger concentrations of iron are found in some terrestrial and lunar basalts.

In iron and stony-iron meteorites, the iron polymorphs are characteristically intergrown with a hatched pattern (known as the Widmanstätten pattern; see Plate 22f). Probably these aggregates form by separation of an initially homogeneous nickel–iron solid solution into two phases, one relatively poor in nickel and another nickel-rich. While these minerals are rare in the Earth's crust, bcc and fcc iron are the major phases of commercial steel. The solid inner core of the Earth is likely to consist of hexagonal close-packed iron. Minerals in meteorites and the conditions of their formation will be discussed in more detail in Chapter 38.

For a long time **platinum** was considered to be a single mineral of variable composition, and only recently could a distinct submicroscopic structural and chemical heterogeneity be documented. The color of platinum grains and nuggets is overall gray, ranging from silver-white to greenish-black shades, and the density fluctuates from 15 to 19 g/cm^3, depending on composition.

Platinum is generally a complex solid solution of platinum, iridium, rubidium, osmium, palladium, iron, and nickel. This mineral is extremely rare and occurs in mantle-derived ultramafic rocks as minute inclusions. Examples of such deposits are the Bushveld Complex in South Africa and the Stillwater Complex in Montana, USA. In the Earth's crust platinum

undergoes phase transformations and interacts with fluids, resulting in a whole series of minerals such as intermetallic tetraferroplatinum (tetragonal PtFe) and isoferroplatinum (cubic Pt$_3$Fe), native osmium and iridium, sperrylite (PtAs$_2$), cooperite (PtS), etc. Weathering of ultramafic rocks and subsequent erosion produce placers from which these minerals are mined.

Mercury is liquid at ambient conditions. Droplets of mercury form during thermal decomposition of cinnabar (HgS) and have been observed in such deposits, most notably the Almaden mine in Spain and the Socrates mine in northern California, USA (Plate 22e).

Sulfur usually has orthorhombic symmetry. When sulfur precipitates from volcanic vapors it crystallizes as euhedral crystals with a combination of dipyramidal, prismatic, and pinacoidal faces (Plate 1a) or as dendrites (Plate 19e); in sedimentary rocks it appears as amorphous masses, accretions, veinlets, and druses. Sulfur is translucent, yellow or greenish-yellow in color, and has a greasy luster in aggregates and on fracture surfaces. It is brittle, soft (hardness 1–2), and light (density 2.05–2.08 g/cm^3), and can be ignited with a match.

The main economic deposits of sulfur are products of volcanic sublimation, but sulfur also forms in sediments owing to bacteria-initiated decay of hydrogen sulfide. Sulfur is used mainly for manufacturing of sulfuric acid and as a fungicide in agriculture.

Graphite occurs in two polytypes, hexagonal and trigonal. It is greasy to touch, and typically forms dark-grayish and black-colored masses. More rarely it forms euhedral plate-like hexagonal crystals (with a submetallic luster), as in some marbles and schists. Graphite has a very low density and hardness (1) and leaves a black streak on paper.

Commercially valuable deposits of graphite are found in nepheline syenites and metamorphic rocks. Graphite is used as a technical lubricant, as electrodes in electrical equipment, in metallurgy (high carbon steels), and as a neutron absorber in nuclear reactors.

Diamond is a carbon polymorph that is stable at high pressure (see Figure 22.2). At atmospheric pressure and with oxygen present it burns to CO$_2$ at 850 °C. Diamonds crystallize mainly as octahedra with faces that are frequently covered with numerous growth and dissolution steps. Since these steps often cannot be seen with the naked eye, the faces seem to be curved or spherical; twins in diamond are common. Some large and famous diamond crystals are listed in Table 34.3.

Diamond has the highest hardness of all known minerals (10). It has a good cleavage on {111} and a higher density than graphite (3.50–3.53 versus 2.2 g/cm^3, respectively). Diamonds display a diversity of colors, although colorless transparent crystals are most common. Colored diamonds can be blue (owing to replacement of carbon by boron), yellow (when nitrogen substitutes for carbon), red, orange, brown, green, and black (because of graphite inclusions). When exposed to X-rays it may display yellow luminescence (Plate 23b). Diamond has a high refractive index (2.42), and a striking dispersion (0.044) which causes a play of colors, especially in cuts called "brilliants" (Plate 23a).

Primary diamond deposits are in kimberlites and lamproites (see below). Diamond also occurs in placers of various composition and age, sometimes accompanied by platinum and gold.

Diamond is the most precious gemstone. It is also widely used as an ultrahard material for manufacturing of drilling heads, cutting tools, perforators, and abrasive instruments. About 75–80% of mined diamonds are used for industrial applications. The best-known diamond deposits are in South Africa, Botswana, Australia, Canada, India, Siberia (Russia), and Brazil. The distribution and economic importance of diamonds will be discussed in more detail in Chapter 34.

22.4 Unusual conditions of formation

Elements such as platinum, osmium, iridium, and, to some extent, gold are chemically inert and do not react under normal circumstances. However, their formation often requires a low oxygen fugacity in the mineral-forming environment. This condition is sometimes satisfied in the Earth's mantle. Examples are platinum in ultramafic rocks, and diamonds in kimberlite pipes and eclogites.

Native metals of the platinum group also occur in olivine peridotite complexes in a cratonic environment (e.g., Bushveld in South Africa, Sudbury in Canada, Stillwater in Montana (USA), and Monchegorsk and Talnakh in Russia), or in ophiolites in orogenic belts (e.g., the Ural Mountains). In primary ore deposits many of these minerals are solid solutions of complicated composition. During their subsequent geological history these primary products of magmatic crystallization were brought to crustal levels and recrystallized under the influence of fluids and hydrothermal solutions. Complex solid solutions were replaced by simpler ones such as platinum–iron and iridium–osmium.

Subsequently, native metals platinum, palladium, and intermetallic compounds Pt_3Fe, $PtFe$, Pt_2NiFe, and Pt_2FeCu formed, followed by arsenides and sulfides (e.g., $PtAs_2$, PtS, $CuFeS_2$, FeS, $FeNi_4S_8$). Parallel to these transformations primary ferromagnesian silicate minerals were replaced by amphibole, serpentine, talc, and chlorite, and the native elements are often the sole reminder of the early mantle origin.

Diamonds form within the mantle at depths of around 650 km. Those diamonds that have been found in the uppermost crust were first uplifted into magma chambers at 200–150 km before finally being carried to the surface. Classic diamond deposits are vertical pipes (diatremes) of kimberlite, which is an ultramafic igneous rock (Figure 22.7). The deepest diatremes can be traced in quarries and mines for more than 1 km. They consist of volcanic vents filled with a breccia containing a mixture of fragments and xenoliths of country rocks,

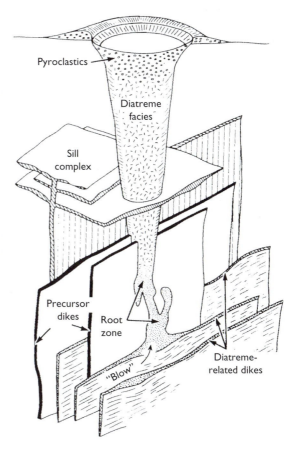

Fig. 22.7 Structure of a kimberlite body with early precursor dikes and sills and a pyroclastic diatreme originating at a root zone (after Mitchell, 1986).

as well as fragments of rocks carried from depths of 130 km and more, from the upper mantle. All rock fragments are cemented by volcanoclastic material and alkaline ultramafic tuffs. The total diamond content of a diamond-bearing pipe of commercial value rarely exceeds a tenth of a gram per tonne of rock and only a few diatremes contain diamond crystals.

The age of kimberlite pipes ranges from Archean to Cenozoic, and yet the diamonds within them are all 2–3 billion years old, judging from the few available geochronological data. The formation of pipes is related to the rapid intrusion of alkaline-ultramafic magmas along narrow conduits, which can occur only at high pressure.

Diamonds also form in eclogites and some schists at great crustal depth and conceivably high tectonic stresses. Particularly in deep continental areas with low geothermal gradients (see Figure 22.2) diamond is the stable carbon polymorph, but becomes metastable when these unusual metamorphic rocks rise to the surface. For kinetic reasons they survive as long as the environment is not oxidizing. Such metamorphic diamonds have been found in China (Dabie Shan), the Urals, and Germany (Erzgebirge).

Diamonds also crystallize from carbon during meteorite impacts, both in meteorites and in the rocks underlying impact craters. In the Barringer ("Meteor") crater in Arizona, USA, for example, diamond occurs as intergrowths within yet another hexagonal form of native carbon, lonsdaleite. Judging from the occurrence of associated stishovite, the high-pressure silica polymorph, shock pressures during these impact events have exceeded 8 GPa.

22.5 Summary

In the chemical classification of minerals, those with elemental compositions are an important group, though there is a wide range of bonding types and crystal structures. Metal elements have generally cubic close-packed structures and high densities. They will be discussed in more detail in Chapter 33. Metallic iron is not stable at crustal conditions but constitutes the core of the Earth (both liquid and solid). It also reaches the Earth from meteorites where two iron–nickel phases (kamacite and taenite) form intricate intergrowths (Widmanstätten

pattern) (see Chapter 37). Sulfur is the most common of the elemental minerals. It has a molecular structure and is mainly linked to volcanic vapors. Carbon has mainly three polymorphs of entirely different structures: diamond with covalent bonding forms at very high pressure and is brought to the surface in kimberlite pipes. Graphite consists of hexagonal carbon sheets that are linked by weak van der Waals bonds. Fullerenes (not a mineral) form large spherical structures with five and six-membered carbon rings.

Important native elements to remember

Element	Symbol	System
Gold	Au	Cubic (fcc)
Copper	Cu	Cubic (fcc)
Silver	Ag	Cubic (fcc)
Platinum	Pt	Cubic (fcc)
Iron	Fe	Cubic (fcc and bcc)
Sulfur	S	Orthorhombic
Diamond	C	Cubic
Graphite	C	Hexagonal

Test your knowledge

1. What are the structural differences between the metal solid solutions and intermetallics? (Review also Chapter 6.)
2. Describe the bonding that is found in graphite and sulfur. Compare it with bonding in copper and diamond.
3. What can you say about the formation conditions of native copper, gold, sulfur, graphite, and diamond?

Further reading

Dawson, B. (1980). *Kimberlites and their Xenoliths.* Springer-Verlag, Berlin.
Field, J. E. (ed.) (1992). *The Properties of Natural and Synthetic Diamond.* Academic Press, London.
Mitchell, R. H. (1995). *Kimberlites, Orangeites, and Related Rocks.* Plenum Press, New York.
Orlov, Y. L. (1977). *The Mineralogy of the Diamond.* John Wiley, New York.

23 | Halides. Evaporite deposits

About 120 minerals are halide compounds, characterized by the presence of halogen ions (Cl^-, Br^-, F^-, and I^-). Most important are fluorides (e.g., CaF_2: fluorite) and chlorides (e.g., NaCl: halite and KCl: sylvite). Halides are characterized by dominating ionic bonding, with relatively simple structures, such as the cubic structure of halite which has been discussed in detail. Most halides precipitate in supersaturated saline brines and the dependence of reactions on composition and concentration has become a classic topic in aqueous chemistry. Potassium salts are mainly used as fertilizers; sodium salts, apart from their value as food additives, are used to melt ice on roads and to extract chlorine for chemical applications.

23.1 Common compositional and structural features of halides

Chlorine and fluorine are chemically very active elements and ionize easily by incorporating an electron. The Cl^- and F^- anions are fairly large and bond readily with metallic cations. The most widespread halogen compounds that occur in nature as minerals are the fluorides and chlorides of alkali and alkaline earth elements (sodium, potassium, calcium, magnesium, and strontium).

Some halide minerals (e.g., bischofite ($MgCl_2 \cdot 6H_2O$) and carnallite ($KMgCl_3 \cdot 6H_2O$)) may contain molecular water in their crystal structure. This situation is particularly typical for the magnesium and aluminum fluorides, where water molecules compensate for the relatively small sizes of Mg^{2+} and Al^{3+} as compared with Cl^- and F^-.

The crystal structures of the halides are very diverse, but simple structures with ionic bonding dominate, as in halite (NaCl), fluorite (CaF_2), and cesium chloride (CsCl, not a mineral) (Chapter 2). We have already described the halite structure as a combination of two cubic face-centered lattices, one with Na^+ and an origin at coordinates 000, and one with Cl^- at an origin $0\frac{1}{2}0$ (see Figures 2.10b and 7.24b). Each sodium ion is surrounded by six chloride ions, and vice versa. The NaCl structure can be represented as a system of octahedra that share edges and corners

(Figure 23.1). The structure can also be described as cubic close-packing of large Cl^-, with Na^+ located in octahedral interstices.

The crystal structure of fluorite is most easily remembered as a primitive cubic lattice of fluorine with calcium in alternate body centers (Figure 23.2a), resembling a Rubik's cube that lacks half of the octants. Since the ionic radius of calcium is relatively large as compared with that of fluorine, the coordination number of calcium is 8, and the coordination polyhedron is a cube. In the fluorite structure the cubic coordination polyhedra share edges. Alternatively, the structure can be viewed

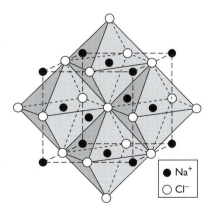

Fig. 23.1 Structure of halite (NaCl) represented as a framework of edge-sharing coordination octahedra (cf. Figure 2.10).

as an fcc array of calcium with fluorine in each one-eighth cube of the array. This representation highlights that each fluorine atom is surrounded tetrahedrally by four calcium atoms (Figure 23.2b).

Another typical halide structure is that of CsCl, which is a primitive cubic lattice of chloride ions

(a)   (b)

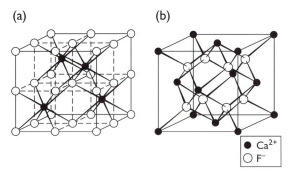

Fig. 23.2 Structure of fluorite (CaF_2). (a) Structure viewed as a primitive lattice of fluorine, with calcium in alternate one-eighth cubes. (b) Structure viewed as face-centered cubic array of calcium with fluorine in one-eighth cubes. Here each fluorine is surrounded by four calcium atoms (conventional unit cell).

(coordinates 000) and cesium in the body center $\frac{1}{2}\frac{1}{2}\frac{1}{2}$ (Figure 23.3). As in fluorite, the cation is coordinated by eight anions and the coordination polyhedron is a cube. The compound CsCl does not occur in nature, but the mineral salammoniac (NH_4Cl), which does form in saline lakes, has the same basic structure. In the latter mineral, instead of a single cation, the ammonium group NH_4^+ is surrounded by four Cl^-.

There are more complicated halide structures. For example, in bischofite the coordination octahedra (Mg $(H_2O)_6)^{2+}$ are linked over Cl^- (Figure 23.4a), and atacamite is built up of an octahedral framework with

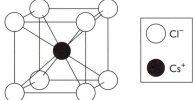

Fig. 23.3 Structure of CsCl, isostructural with salammoniac (NH_4Cl).

(a)   (b)

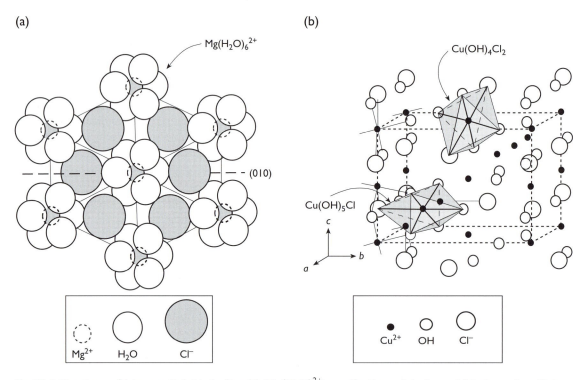

Fig. 23.4 Structures of (a) monoclinic bischofite with $Mg(H_2O)_6^{2+}$ coordination polyhedra and (b) atacamite with two different types of distorted coordination polyhedra that are linked (one of each type is shaded). The orthorhombic unit cell is shown by dashed lines.

two different types of alternating coordination poly-hedra, $Cu(OH)_4Cl_2$ and $Cu(OH)_5Cl$ (Figure 23.4b).

The ionic character of the bonds and the polyhedral structures determine the crystal morphology of a majority of the halide compounds. The symmetry of the halides is often cubic or pseudocubic and the crystal habit is isometric, with forms such as {100} (cube) (halite in Figure 1.6b and Plate 3a, and fluorite in Plates 18d 23d), {111} (octahedron) (fluorite in Plate 20a), and {110} (dodecahedron) dominating. Also, cleavages are highly symmetrical planes (e.g., {100} in halite and {111} in fluorite). We have already discussed that the excellent cleavage of halite inspired Hauy to develop the concepts of unit cell and lattice in the eighteenth century (see Figure 7.3a).

Owing to the ionic bonding and lack of transition elements, many halide minerals are colorless, trans-parent, with refractive indices less than 1.5, and have a vitreous luster. Halides with traces of iron, manganese, and Cu^{2+} are blue, green, yellow, or red colored. Colorless halite becomes blue when irradiated with X-rays or γ-rays. Fluorite emits visible light when irradiated with ultraviolet radiation (see also Chapter 15 and Plate 18d,e).

23.2 Brief description of halide minerals

Minerals are described here in the same order as listed in Table 23.1.

Fluorite (CaF_2) may contain some rare earth elements (cesium, yttrium, etc.) that take part in the two following complex isomorphic substitutions: $Ca^{2+} \rightleftharpoons Y^{3+} (or\ Ce^{3+})^+ F^-$ (fluoride occupies intersti-tial sites) and $2Ca^{2+} \rightleftharpoons Ce^{3+} + Na^+$. Fluorite crystals are of cubic symmetry with octahedral (Plate 20a), cube-octahedral, or cubic habit (Plate 23d, Figure 23.5). The mineral often forms intergrowths, druses, radiating or granular aggregates, thin coatings, or occurs as scattered isolated grains. Fluorite may be transparent or opaque. Its color varies from white to inky-blue, green, or violet, depending on the concentration of rare earth elements that cause color centers in the structure. The mineral fluoresces in ultraviolet light (Plate 18d,e). Fluorite has a vitreous luster on crystal faces and a greasy luster on fracture planes. The mineral has a hardness 4 on the Mohs scale and perfect octahedral cleavage. The fracture surfaces of large crystals often display cleavage cracks intercrossing at 60°. Small grains and impregnations of fluorite are common in high-temperature hydrothermal veins. Large quantities of fluorite (massive granular and radiaxial aggregates) that are a source of the mineral for the chemical industry occur in medium- and low-temperature hydrothermal deposits. Large, pure crystals are found in cavities and druses of granite pegmatites. Such crystals are used as a raw material for optical applications.

Halite (NaCl), also called *rock salt* or *table salt*, is cubic in symmetry and crystals are cubic or cube-octahedral in shape (Figure 1.6b and Plate 3a). Halite generally forms granular massive and layered aggregates in sedimentary rocks. The color of the mineral is white (clear) or, sometimes, spotty inky-blue (due to color centers related to $Na^+ \rightleftharpoons Na^0$ isomorphism, Plate 23c). Halite has a vitreous or greasy luster, hardness 2, and a cubic cleavage. It is generally an evaporite mineral.

The importance of halite as a food additive and preservative is well known. However, the main indus-trial application for halite is as road salt. The chemical industry uses halite to produce soda, hydrochloric acid (HCl), metallic sodium, sodium hydroxide (NaOH), and chlorine. Halite is also used in metallurgy.

Sylvite (KCl) has the same structure as halite, is also cubic in symmetry, and its crystals are cube shaped. Sylvite occurs in some evaporite deposits as granular masses of dirty-brown and red color and forms under similar conditions as halite. This mineral is the last precipitate of evaporite lakes. Unlike halite, sylvite has a bitter-salty taste. Sylvite is used to pro-duce potassium fertilizers and potassium chemicals for medical, photographic, and cosmetic applications.

Carnallite ($KMgCl_3 \cdot 6H_2O$), a hydrous chloride mineral, occurs as white or pink aggregates. The min-eral is very hygroscopic and has a specific bitter-salty taste. Carnallite is a constituent of saline sedimentary rocks. Like sylvite, the mineral is used for potassium and magnesium extraction and as a fertilizer.

23.3 Origin of halide minerals

The conditions of halide crystallization and their sta-bility are affected by two factors: (1) the abundance of halogen atoms in the Earth's crust, and (2) the chem-ical properties of the compounds. The abundance of the halogens in the continental crust parallels the electron affinities and energies of formation of NaX-like binary compounds (Table 23.2). The ionic radii of the halogens increase in the reverse order.

Table 23.1 Halide minerals with some diagnostic properties; important minerals are given in italics

Mineral Formula	System	Morphology Cleavage	H	D	Color Streak	n Pleochroism	Δ	2V (Dispersion)
Fluorite CaF$_2$	Cubic	{100}; {111} *{111}*	4	3.18	Clear, violet, yellow, green	1.434		
Cryolite Na$_3$AlF$_6$	Ps. Cubic (Monocl.)	Eq. (001), *{110}*	2.5	3.0	White, pink, brown	1.33–1.34	0.001	+43 (r < v)
Halite NaCl	Cubic	{100} *{100}*	2	2.1	Clear, yellow, red, blue	1.56		
Sylvite KCl	Cubic	{100} *{100}*	2	1.9	White, yellow, red	1.50		
Salammoniac NH$_4$Cl	Cubic	{110}, {211} *{111}*	1–2	1.53	Clear, yellow, brown	1.66		
Chlorargyrite AgCl	Cubic	{100}	1.5	5.5	Clear, violet, brown, gray	2.10		
Carnallite KMgCl$_3$·6H$_2$O	Ortho.	Gran. Fibr.	2.5	1.6	Clear, white, yellow, red	1.47–1.49	0.027	+66 (r < v)
Bischofite MgCl$_2$· 6H$_2$O	Monocl.	Fibr. [001] *{110}*	1–2	1.60	Clear, red	1.49–1.53	0.034	+80 (r > v)
Atacamite CuCl(OH)$_3$	Ortho.	Pris. [001] *(010)*	3–3.5	3.75	Green *Green*	1.83–1.88 *yellow-green*	0.05	−75

Notes: H, hardness; D, density (g/cm³); *n*, range of refractive indices; Δ birefringence; 2V, axial angle for biaxial minerals. Acute 2V is given in the table. If 2V is negative the mineral is biaxial negative and 2V is 2V_α; if it is positive, the mineral is biaxial positive and 2V is 2V_γ. Dispersion r < v means that acute 2V is larger for violet than for red.

System: Monocl., monoclinic; Ortho., orthorhombic; Ps. pseudo.

Morphology: Eq., equiaxed; Fibr., fibrous; Gran., granular; Pris., prismatic.

Table 23.2 Relationship of abundance, electron affinity, energy of formation, and ionic radius for halogens

Property	Halogen			
	F	Cl	Br	I
Abundance in continental crust (ppm, weight)	625	130	2.5	0.5
Electron affinity (kcal)	95	86	84	76
Energy of NaX formation (kcal/g mol)	136	98	91	77
Ionic radius (Å)	1.19	1.81	1.96	2.20

Fig. 23.5 Cubic twinned crystals of fluorite from Freiberg, Saxony, Germany (courtesy A. Massanek). Width 20 mm.

Melting points decrease in the same order, whereas volatility and solubility increase. Correspondingly, fluoride minerals form typically in high-temperature endogenic processes. Fluorite occurs in granites, pegmatites, hydrothermal veins, and skarns at igneous contacts. Chloride minerals form in both endogenic and exogenic conditions. Since chlorine concentrations in surface waters (such as in lakes, oceans, and thermal springs) are very high, evaporites are the most typical deposits of chloride minerals. Bromide and iodide minerals are exotic because the ionic radii of Br^- and I^- are large and their electron affinities and binding energies are low, reducing their stability. Most Br^- and I^- is dissolved in surface water.

Evaporites in marine basins

Most halides, except fluorides, are present in sedimentary rocks called *evaporites* because they are the result of the evaporation of water. Evaporites are chemical precipitates that crystallize in supersaturated solution and generally concentrate at the bottom of a basin. An evaporation process is most effective in an arid and hot climate and in closed or partially closed sedimentary basins, where solutions become highly concentrated in ions. An example has been the Mediterranean basin (Figure 23.6), which dried up in the Late Tertiary because of a dry climate and little freshwater input from rivers. Salt was resupplied from the Atlantic Ocean through the Strait of Gibraltar and occasional flooding. The evaporites in marine basins include precipitates of various salts (halite, anhydrite, potassium, and magnesian salts, some carbonates, primary borate aggregations, soda, sodium sulfates, saltpeter) that comprise well-stratified beds of different thickness, composition, and structure, depending on whether they are of marine or continental origin.

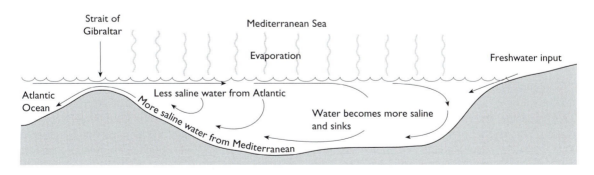

Fig. 23.6 Circulation of water in the Mediterranean basin. The influx of freshwater from rivers does not balance evaporation and there is presently an influx of seawater from the Atlantic Ocean through the Strait of Gibraltar. In the Triassic period, this influx was interrupted and consequently the Mediterranean Sea dried out, with extensive formation of evaporites.

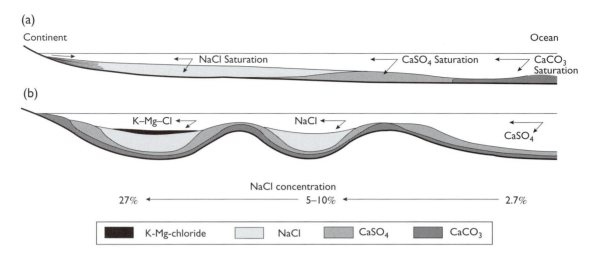

(a)

Continent

Ocean

NaCl Saturation

CaSO₄ Saturation

CaCO₃ Saturation

(b)

K–Mg–Cl

NaCl

CaSO₄

NaCl concentration

27% ← 5–10% ← 2.7%

K-Mg-chloride | NaCl | CaSO₄ | CaCO₃

Fig. 23.7 Sedimentation of evaporite minerals on the continental shelf, with a regular sequence – calcite, gypsum, halite, sylvite/carnallite – with increasing NaCl concentration. (a) Shallow shelf; (b) deeper shelf with basins.

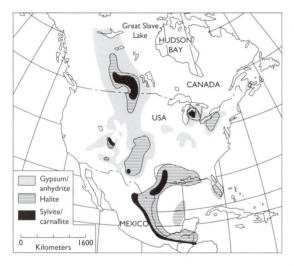

Fig. 23.8 Large Permian evaporite deposits in North America (data from Kesler, 1994).

In marine basins the water composition is close to that of seawater. Salinity of the recent ocean is about 35 g/l and the dried residue contains (in weight%): NaCl 77.8; $MgCl_2$ 10.9; $MgSO_4$ 4.7; $CaSO_4$ 3.6; K_2SO_4 2.5, with minor amounts of carbonates ($CaCO_3$ 0.3%), bromides ($MgBr_2$ 0.2%), borates, and iodates. On a continental shelf, seawater can become increasingly saturated, with consequent precipitation of first calcite, then gypsum, halite, and finally sylvite and carnallite (Figure 23.7). If one evaporated the whole

ocean, the seafloor would be covered with a 60 m thick layer of salt. The composition of seawater has changed over geological history and varies somewhat from ocean to ocean. However, the simple evaporation of the ocean could not have created the enormous saline beds with thicknesses of tens to hundreds of meters that appear in several salt deposits around the world. Today it is generally recognized that there are also tectonic prerequisites for a large evaporite deposit: the basin must subside over geological time, as is the case in rift valleys and continental graben structures. In the Paleozoic era, the ocean flooded much of the continents, and reefs isolated shallow seas. Examples are the Permian Zechstein of Central Europe, North American basins such as Williston (Canada), Michigan (USA), and Gulf Coast (Figure 23.8), and the Late Tertiary Mediterranean mentioned above. Rift valley basins include the Triassic salt deposits of the Rhine graben, the Danakil depression in Ethiopia, and the Cambrian salt stocks in the Persian Gulf.

The order of crystallization in an evaporating lagoon is controlled by the rules of aqueous solution physical chemistry. The geological implications of evaporite systems were first pointed out by chemist J. H. van't Hoff (1912) using the example of Stassfurt (Germany), which is part of the large Permian Zechstein deposit in peripheral basins of the North Sea in Central Europe (Box 23.1). Figure 23.9 shows the general crystallization sequence. When an amount of seawater is dried to 60–70% of its original volume,

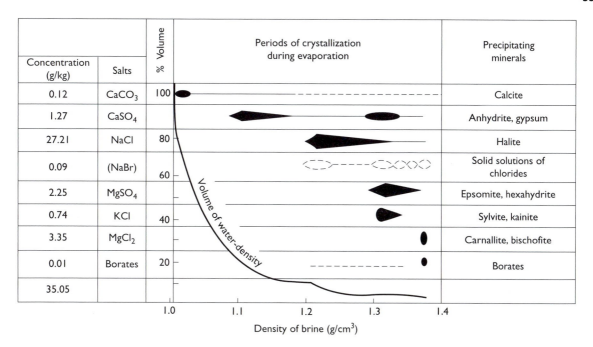

Fig. 23.9 Crystallization sequence of evaporite minerals from seawater based on Valyashko (1962). Also shown is the volume of water and the density of the residual brine.

Box 23.1 Analytical focus: Crystallization from aqueous solutions during evaporation

Van't Hoff (1905, 1909) derived the crystallization sequence in a system with five components $NaCl$–KCl–$MgCl_2$–Na_2SO_4–H_2O, which has since then become classic in solution chemistry. It is assumed that the solution is saturated with respect to NaCl, i.e., the solution contains halite crystals. There are many other possible phases in addition to halite (Table 23.3).

Table 23.3 Compositions of the main evaporite minerals in the system NaCl–KCl–MgCl$_2$–Na$_2$SO$_4$–H$_2$O that exist in a hydrous solution saturated with NaCl

Name	Formula	Letters on Figure 23.10
Aphthitalite	$Na_2K_4(SO_4)_2$	aph
Bischofite	$MgCl_2 \cdot 6H_2O$	bi
Blödite	$Na_2Mg(SO_4)_2 \cdot 4H_2O$	bl
Carnallite	$KMgCl_3 \cdot 6H_2O$	c
Epsomite	$MgSO_4 \cdot 7H_2O$	e
Hexahydrite	$MgSO_4 \cdot 6H_2O$	hx
Kainite	$KMg(SO_4)Cl \cdot 3H_2O$	ka
Kieserite	$MgSO_4 \cdot H_2O$	ks
Leonite	$K_2NaMg_2(SO_4)_4 \cdot 8H_2O$	l
Picromerite	$KMg(SO_4)_2 \cdot 6H_2O$	p
Sylvite	KCl	sy
Thenardite	Na_2SO_4	t

Box 23.1 (cont.)

(a)

(b)

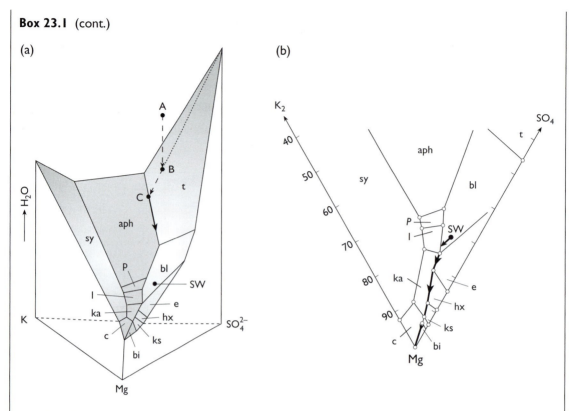

Fig. 23.10 Phase diagram for the system NaCl–KCl–MgCl$_2$–Na$_2$SO$_4$–H$_2$O, saturated with NaCl at 25 °C. (a) Three-dimensional diagram with K:Mg:SO$_4^{2-}$ as base, and water content at saturation as fourth axis. Stability fields of phases are indicated. (b) Enlarged portion of the ternary system, illustrating changes during evaporation of seawater. For abbreviations of phase compositions, see Table 23.3. SW, seawater.

We represent the stability field in a ternary diagram with the water content at saturation as a fourth axis and K, Mg, and SO$_4^{2-}$ as end members (Figure 23.10a). Take a composition A rich in SO$_4^{2-}$. During evaporation, the water content decreases without changing the composition in K, Mg, and SO$_4^{2-}$. When the surface is reached (point B), thenardite (t) will start to crystallize and, on further evaporation, the composition of the solution will change, descending the thenardite field, away from the thenardite composition, until the boundary with aphthitalite (aph) is reached (C). At that point both thenardite and aphthitalite crystallize and the composition of the solution changes along the borderline between the two stability fields (arrow).

Seawater (SW) has a composition that is much richer in Mg than point A (K:Mg:SO$_4^{2-}$ = 6:61:33). As you can see, it is close to a region with many phases. In order to better follow the processes during the evaporation of seawater, we enlarge the lower corner of the ternary diagram and display it as a projection on the triangle K–Mg–SO$_4^{2-}$ (Figure 23.10b).

With evaporation of seawater (SW) the first mineral to precipitate is blödite (bl). Then the composition path of the solution follows the valley between epsomite (e) and kainite (ka), hexahydrite (hx) and kainite, kieserite (ks) and kainite, and finally bischofite (bi) and carnallite (indicated by the arrows in Figure 23.10b). Similar trends have been observed in the Stassfurt evaporite sequence (Figure 23.11). The system that we have discussed does not include any calcium, nor does it have any carbonate or borate anion groups; therefore minerals such as calcite, gypsum, and anhydrite are not included and in reality processes are far more complex. In the phase diagram the evaporation path never passes through the field of sylvite. In nature, the initial composition in a specific evaporite basin can differ from that of ideal seawater and temperatures may vary. Secondary processes of recrystallization may also change the primary evaporation products.

calcite forms. When the brine becomes more concentrated, sulfates, gypsum, and anhydrite precipitate, followed by halite. Subsequently, along with the precipitation of magnesium and potassium sulfates, sylvite crystallizes. Later, carnallite and finally bischofite crystallize. This scheme is idealized because in reality the thickness of beds varies widely, depending on local changes in sea level and climate.

Continental salt lakes

Continental evaporite deposits are typical of deserts. The amount of water present may be quite variable, and the composition of continental lakes varies much more than that of seawater because it is controlled by chemical weathering of the surrounding surface and penetration of the local rocks by groundwater. There are soda, sulfate, boron, and nitrate lakes, with an abundance of rare minerals. The brines of some continental salt lakes are enriched in lithium, bromine, or iodine and the lakes are mined for these elements.

The Mojave Desert in California, USA, is an excellent example of an evaporite formation that has evolved over a period of more than 20 million years (Smith, 1979). Particularly interesting is a group of evaporite lakes east of the Sierra Nevada that document the climatic history of the last 100 000 years. At present the average rainfall is 8 cm/year in this region, whereas the evaporation rate exceeds 150 cm/year. Volcanic rocks provided a source of rare elements, which became concentrated in the evaporating brines. As one descends from the source near the Sierra Nevada Mountains through the chain of evaporite lakes Owens, Searles, Panamint, and finally Death Valley, the composition of the deposit changes systematically. The Searles Lake deposit is particularly rich in potassium, boron, sodium, chloride, lithium, and sulfate, and is actively mined for these elements and compounds. Evaporites are distributed over a depth of over 100 m. Temperature and pressure gradients, as well as variations in brine concentrations, produce many different evaporite minerals, all indicative of the local chemical and physical conditions under which they formed. (Some examples are sulfohalite ($Na_6ClF(SO_4)_2$), thenardite (Na_2SO_4), borax ($Na_2B_4O_5(OH)_4 \cdot 8H_2O$), burkeite ($Na_2CO_3(SO_4)_2$), gaylussite ($Na_2Ca(CO_3)_2 \cdot 5H_2O$), glaserite ($K_2SO_4$), hanksite ($Na_3K(SO_4)_9(CO_3)_2Cl$), trona ($Na_3H(CO_3)_2 \cdot 2H_2O$), and pirssonite ($Na_2Ca(CO_3)_2$)). Note that, while the major cations are Na^+, K^+, and Ca^{2+}, these minerals have chloride, sulfate, carbonate, and borate components. Related evaporites of the Clayton Valley in Nevada, USA, are mined for lithium. Also in the Mojave Desert is the older boron mineralization of the Kramer deposit, one of the world's largest boron deposits. (Some boron minerals are borax ($Na_2(B_4O_5(OH)_4) \cdot 8H_2O$), kernite ($Na_2(B_4O_6(OH)_2) \cdot 3H_2O$), tincalconite ($Na_2(B_4O_5(OH)_4) \cdot 3H_2O$), ulexite ($NaCa(B_5O_6(OH)_6)$ $5H_2O$), and colemanite ($Ca(B_3O_4(OH)_3) \cdot H_2O$).) The boron concentration is attributed to hot springs associated with volcanic activity.

Secondary salt deposits

Conditions that can provide a regular crystallization order as proposed by Van't Hoff are rarely realized in nature. Often one observes a repetition and alternation of the same layers, an absence of others, changes of bedding periodicity, and wedges of clay and other rocks found in an evaporite sequence. All these factors prove that the initial marine basin dried under changing geological conditions. Periodic freshwater flooding or transgressive episodes with fresh brine influx produced waters that dissolved the halide salts from their primary deposits and redeposited them as secondary deposits in the upper parts of a sedimentary sequence. Furthermore, salt minerals are very ductile and, owing to their low density, buoyant in a sedimentary sequence. Therefore, deeply buried salt beds may tectonically intrude upwards, as documented by the structures of anticlinal *salt domes*, which constitute some of the major salt deposits (Figure 23.11). By contrast, gypsum-bearing beds and other sulfates are generally left in place and are transformed by the increasing pressure and temperature during burial into less hydrated or even entirely anhydrous minerals. This dehydration process can be illustrated with the following reactions:

$$CaSO_4 \cdot 2H_2O \rightarrow CaSO_4 + 2H_2O$$
$$\text{Gypsum} \qquad \text{Anhydrite} \quad \text{Water}$$

$$2Na_2Mg(SO_4)_2 \cdot 4H_2O \rightarrow 2Na_2Mg(SO_4)_2 \cdot 2.5H_2O + 3H_2O$$
$$\text{Blödite} \qquad \text{Loeweite} \qquad \text{Water}$$

$$2Na_2SO_4 + Na_2Mg(SO_4)_2 \cdot 4H_2O \rightarrow Na_6Mg(SO_4)_4 + 4H_2O$$
$$\text{Thenardite Blödite} \qquad \text{Vanthoffite} \quad \text{Water}$$

Such processes release large amounts of water that dissolve halide salts from the primary beds and penetrate the country rocks. They illustrate how evaporite sediments continually change their mineralogical and chemical composition and structure.

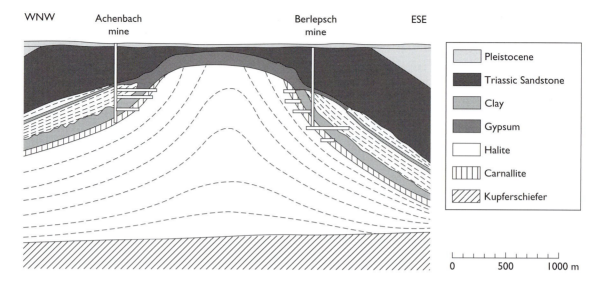

Fig. 23.11 Cross-section through the diapiric salt dome at Stassfurt, Germany, which is mined mainly for carnallite. The salt intrudes Triassic sandstone. Mineral deposits are indicated.

23.4 Commercial deposits

Large commercial deposits of halite, sylvite, carnallite, kainite, bischofite, and mirabilite ($Na_2SO_4 \cdot 10H_2O$) exist in many countries. The largest deposits of the potassium salts (sylvite, carnallite) are situated in the Williston Basin (Canada), in Michigan and along the coast of the Gulf of Mexico (USA) (Figure 23.8), in the Urals (Russia), in the Permian Zechstein Basin of Central Germany (Stassfurt), and along the Rhine graben in France. Famous halite deposits are in Poland (with the historic salt mines of Velichka), Ethiopia (Danakil depression along the Red Sea), Central Germany (e.g., around Halle), India, and Russia (Solikamsk and Berezovsk). Potassium salts are mainly used as fertilizer, whereas sodium chloride is used as road salt to melt ice.

23.5 Summary

The main halide minerals have very simple cubic strucures: NaCl combines two face-centered cubic lattices of cations and anions, in fluorite cations have an fcc lattice and fluorine occupies tetrahedral interstices. In CsCl (not a mineral) Cl forms a simple primitive lattice and Cs occupies the centers of the cubes. There are more complex chlorides with additional water as in bischofite. Most halides (with the exception of fluorite) occur as evaporite minerals in marine basins and continental lakes. They form large deposits that are mined for road salt and chemical applications (halite) and fertilizer (sylvite/carnallite). Crystallization from aqueous solutions during evaporation is explained.

Important halide minerals to remember

Name	Formula	System
Fluorite	CaF_2	Cubic
Halite	NaCl	Cubic
Sylvite	KCl	Cubic
Carnallite	$KMgCl_3 \cdot 6H_2O$	Orthorhombic

Test your knowledge

1. How does the melting point change with cation radius (e.g., NaCl, KCl, AgCl)? (Review also Chapter 2.)
2. Why are most halides colorless and transparent?
3. Compare the origin of fluorite and halite and discuss the reasons for differences.
4. Describe the crystallization sequence in an evaporite basin.
5. What is the composition of present-day seawater? If you could evaporate all oceans, how thick would the salt layer be?

6. Under which conditions do very large evaporite deposits form?
7. What are some industrial applications of evaporite minerals?

Further reading

Braitsch, O. (1971). *Salt Deposits: Their Origin and Composition*. Springer-Verlag, Berlin.

Chang, L. L. Y., Howie, R. A. and Zussman, J. (1995). *Rock-forming Minerals, Volume 5B, Non-silicates: Sulphates, Carbonates, Phosphates, Halides*, 2nd edn. The Geological Society of London.

MacKenzie, F. T. (2005). *Sediments, Diagenesis, and Sedimentary Rocks: Treatise on Geochemistry,* 2nd edn. Elsevier Science.

24 | Carbonates and other minerals with triangular anion groups. Sedimentary origins

Carbonates are the primary representative of compounds with an $(XO_3)^{n-}$ radical. About 170 different carbonate minerals are known. Most of them are simple salts of carbonic acid (H_2CO_3), such as calcite ($CaCO_3$) and dolomite ($CaMg(CO_3)_2$). Others contain additional anions, for example, malachite ($Cu_2(CO_3)(OH)_2$), and a few are mixed chemical compounds. We have mentioned two of these, burkeite ($Na_2CO_3(SO_4)_2$) and hanksite ($Na_2K(SO_4)_9(CO_3)_2Cl$), in Chapter 23. The $(XO_3)^{n-}$ group is also present in nitrates and borates. In some borates, the BO_3^{3-} groups are isolated as in calcite, but in most the planar triangular groups are linked to form chains and sheets and even form tetrahedra. The most common carbonates, calcite and dolomite, make up nearly 2.5% of the volume of the Earth's crust, most of it in sedimentary and metasedimentary rocks, which will be discussed in some detail. The main industrial application of calcite from limestones and marbles is for cement production (see Chapter 35). Carbonatites are economically significant for rare earth and phosphor deposits.

24.1 Characteristic features of composition and crystal chemistry of carbonates and borates

By their chemical nature, the carbonate minerals are the most stable salts of carbonic acid. This acid is relatively weak and prefers to bond with elements of low ionization potential (sodium, potassium, calcium, strontium) that are not too small in size (such as lithium). For the sodium ion, with a single charge, water molecules are inserted into the structure to shield the effect of Na+ on the CO_3^{2-} complex. Molecular H_2O acts as a buffer between Na^+ and CO_3^{2-}. By contrast, cations of high ionization potential such as Bi^{2+}, Cu^{2+}, and rare earth elements form only if OH^- groups, F^-, or O^{2-} are present to weaken the CO_3^{2-} complex. An example is bastnäsite ($CeCO_3F$). The carbon:oxygen radius ratio is very small (<0.18 Å$/1.40$ Å ~ 0.13) and produces a triangular coordination (see Chapter 2), which is the characteristic structural motif of the CO_3^{2-} complex (Figure 24.1a).

Many carbonates crystallize in the rhombohedral system, but orthorhombic and monoclinic carbonates also exist. At least at low temperature, triangular CO_3^{2-} groups do not have rotational freedom. In calcite and dolomite they occupy planes parallel to (0001) in which the CO_3^{2-} triangles all point in the same direction (Figure 24.1b). This feature causes the crystal structures of the carbonates and their physical properties to be highly anisotropic, with large differences between properties parallel and perpendicular to the c-axis.

The structure of calcite is related to that of NaCl. First we take the halite structure and align it along a body diagonal [111] (Figure 24.2a). We then replace each chloride ion by a CO_3^{2-} triangle, with triangles aligned perpendicular to the body diagonal and pointing in opposite directions in alternate planes. Finally, we compress the structure along the body diagonal to produce rhombohedral rather than cubic symmetry (Figure 24.2b). By analogy with halite, we

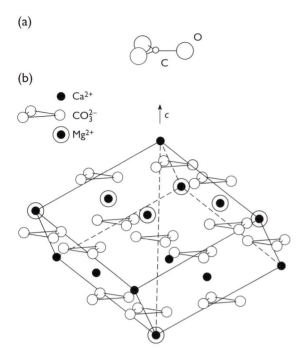

Fig. 24.1 (a) The triangular CO_3^{2-} group is a basic building block of all carbonate minerals. (b) In rhombohedral carbonates, layers of CO_3^{2-} groups (pointing in opposite direction) alternate with layers of cations. In dolomite, shown here, layers of Ca^{2+} alternate with layers of Mg^{2+}. The cleavage rhombohedron $\{10\bar{1}4\}$ is outlined. (Note that this is not the structural unit cell.)

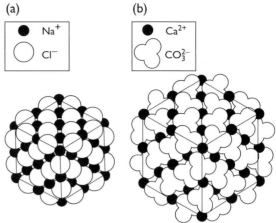

Fig. 24.2 The structure of calcite can be viewed as a distorted NaCl structure. (a) The cubic NaCl structure, viewed along the body diagonal [111]. (b) The distorted NaCl structure, with chloride ions replaced by CO_3^{2-} groups and compression along the body diagonal. View is along the [0001] axis.

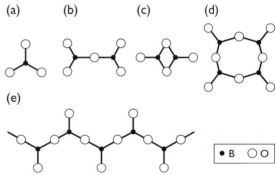

Fig. 24.3 Polymerization of BO_3 groups in borates.

can therefore view the calcite structure as cubic close-packing of CO_3^{2-} groups, with Ca^{2+} in "octahedral" interstices. The analogy goes so far that even the excellent {100} cleavage in halite is preserved; it is now a rhombohedral $\{10\bar{1}4\}$ cleavage, illustrated by the cell in Figure 24.1b, which is not the conventional unit cell. The second polymorph of $CaCO_3$, aragonite, with orthorhombic symmetry, can be viewed as hexagonal close-packing of CO_3^{2-} groups with Ca^{2+} in interstices, although this analogy is not perfect and the structure is considerably distorted. Figure 24.1b shows the structure of dolomite $(CaMg(CO_3)_2)$, where layers of Ca^{2+} alternate with layers of Mg^{2+}. In calcite all cations are Ca^{2+}.

We have encountered calcite and aragonite in earlier chapters. In Chapter 3, for example, we introduced carbonates as examples of minerals exhibiting isomorphism and polymorphism. For small cations the calcite structure is preferred, but for large cations

the aragonite structure is more stable. However, for $CaCO_3$ both structures are possible. At higher temperature and low pressure calcite is stable, and at high pressure and low temperature aragonite is stable (for the phase diagram, see Figure 19.2a).

A few carbonate minerals have chain- or sheet-like structures. For instance, the evaporite minerals nahcolite $(NaHCO_3)$ and trona $(Na_3(CO_3)H(CO_3)\cdot2H_2O)$ have a chain structure in which hydrogen bonds induce the polymerization of the CO_3^{2-} groups as $(HCO_3)_2^-$ or $H^+(CO_3)_2^{2-}$. Sheet-like structures have been observed in uranium, bismuth, and lead carbonates.

Polymerization of triangular (and tetrahedral) groups is most prevalent in borates (Figure 24.3). The triangular BO_3^{3-} groups may be isolated as CO_3^{2-} groups

Table 24.1 Common carbonate, nitrate, and borate minerals with some diagnostic properties; most important minerals are given in italics

Mineral Formula	System	Morphology Cleavage	H	D	Color Streak	n Pleochroism	Δ	2V (Dispersion)
Carbonates								
Calcite group								
Calcite CaCO₃	Rhomb.	{10Ī4} etc. {10Ī4}	3	2.72	Clear, white, yellow, etc.	1.486–1.658	0.172	(−)
Magnesite MgCO₃	Rhomb.	{10Ī4} etc. {10Ī4}	4–4.5	3.0	White, gray	1.60–1.70	0.191	(−)
Siderite FeCO₃	Rhomb.	{10Ī4} etc. {10Ī4}	4–4.5	3.89	Yellow, brown	1.75–1.87	0.240	(−)
Rhodochrosite MnCO₃	Rhomb.	{10Ī4} etc. {10Ī4}	4	3.5	Pink, red	1.70–1.81	0.218	(−)
Smithsonite ZnCO₃	Rhomb.	{10Ī4} etc. {10Ī4}	5	4.4	Clear, gray, green	1.73–1.85	0.228	(−)
Dolomite group								
Dolomite CaMg(CO₃)₂	Rhomb.	{10Ī4} etc. {10Ī4}	3.5–4	2.9	Clear, white, gray	1.59–1.68	0.179	
Ankerite CaFe(CO₃)₂	Rhomb.	{10Ī4} etc. {10Ī4}	3.5–4	3.0	White, brown	1.52–1.75	0.19	(−)
Kutnahorite CaMn(CO₃)₂	Rhomb.	{10Ī4} etc. {10Ī4}	3.5–4	3.1	White, pink	1.54–1.73	0.19	(−)
Aragonite group								
Aragonite CaCO₃	Ortho.	Pris. [001] (010) poor	3.5–4	2.95	White, yellow, brown	1.53–1.69	0.156	−18
Witherite BaCO₃	Ortho.	Eq., Fibr. [001] (010)	3.5	4.28	White, gray, yellow	1.53–1.68	0.148	−16
Strontianite SrCO₃	Ortho.	Pris. [001] {110}	3.5	3.7	Clear, white, yellow	1.52–1.67	0.150	−8
Cerussite PbCO₃	Ortho.	Eq., Pris. [100] {110}	3–3.5	6.5	White, gray, black	1.80–2.08	0.274	−8
Soda carbonates								
Nahcolite NaHCO₃	Monocl.	Pris. [001]	2.5	2.2	White	1.38–1.58	0.20	−75

Mineral / Formula	System	Morphology	H	D	Color / pleochr.	n	Δ	2V
Natron $Na_2CO_3 \cdot 10H_2O$	Monocl.	Platy (010), (100)	1–1.5	1.44	Clear, white, gray	1.41–1.44	0.035	−71 (r < v)
Thermonatrite $Na_2CO_3 \cdot H_2O$	Ortho.	Microcryst.	1–1.5	2.2	Clear, white, gray	1.42–1.52	0.10	−48
Trona $Na_3CO_3HCO_3 \cdot 2H_2O$	Monocl.	Tab. (001), (100) {110}	2.5–3	2.17	White, gray, yellow	1.41–1.54	0.128	−76
Other carbonate minerals								
Azurite $Cu_3(CO_3)_2(OH)_2$	Monocl.	Tab. (001), (011)	3.5–4	3.8	Blue / *Blue*	1.73–1.84	0.108	+68 (r > v)
Bastnäsite $(Ce,La,Y)CO_3F$	Hex.	Tab. (0001)	4–4.5	5.1	Yellow, brown	1.72–1.82	0.10	(+)
Malachite $Cu_2CO_3(OH)_2$	Monocl.	Botr., Fibr. [001], (201)	4	4	Green / *Green*	1.66–1.91	0.254	−43 (r < v)
Nitrates								
Niter (saltpeter) KNO_3	Ortho.	Fibr. [001] {011}	2	2.0	Clear, white, gray	1.34–1.51	0.171	−7
Nitratite $NaNO_3$	Trig.	Eq. {10$\bar{1}$1} {10$\bar{1}$1}	1.5–2	2.27	White, yellow	1.47–1.59	0.248	(−)
Borates								
Borax $Na_2B_4O_5(OH)_4 \cdot 8H_2O$	Monocl.	Pris. (100), {110}	2–2.5	1.8	Clear, white	1.45–1.47	0.025	−40
Ludwigite $Mg_2Fe(BO_3)O_2$	Ortho.	Fibr.	5–6	3.6	Green, black / *Black*	1.85–2.02 *Green–green–brown*	0.016	+0–5 (r ≫ v)
Kernite $Na_2B_4O_6(OH)_2 \cdot 3H_2O$	Monocl.	Eq., Pris. [001] (001), (100)	2.5	1.92	Clear, white	1.45–1.49	0.034	−80 (r > v)
Kotoite $Mg_3(BO_3)_2$	Ortho.	Gran. {110}	6.5	3.1	Clear, white	1.65–1.67	0.02	+22
Tincalconite $Na_2B_4O_5(OH)_4 \cdot 3H_2O$	Trig.	Powder	1	1.88	White	1.46–1.47	0.01	(+)

Notes: H, hardness; D, density (g/cm^3); n, range of refractive indices; pleochr., pleochroism; Δ, birefringence; 2V, axial angle for biaxial minerals. For uniaxial minerals (+) is positive and (−) is negative. Acute 2V is given in the table. If 2V is negative the mineral is biaxial negative and 2V is 2V$_α$; if it is positive, the mineral is biaxial positive and 2V is 2V$_γ$. Dispersion r < v means that acute 2V is larger for violet than for red.

System: Monocl., monoclinic; Ortho., orthorhombic; Rhomb., rhombohedral; Trig., trigonal; Hex., hexagonal.

Morphology: Botr., botryoidal; Eq., equiaxed; Fibr., fibrous; Gran., granular; Pris., prismatic; Tab., tabular; Microcryst., microcrystals.

(Figure 24.3a), or combined to form $B_2O_5^{4-}$ (Figure 24.3b) or $B_2O_4^{2-}$ (Figure 24.3c) pairs, $B_4O_8^{4-}$ rings (Figure 24.3d), or $B_2O_5^{4-}$ chains (Figure 24.3e). Charges are balanced by additional cations. Larger cations are present in spaces between polymerized sheets, chains, and rings, and in frameworks. In this respect the borates have some similarities with silicates.

A list of the important carbonates, nitrates, and simple borates is given in Table 24.1.

24.2 Morphology and properties of carbonates

The majority of carbonates, such as calcite, magnesite, siderite, and dolomite, have rhombohedral symmetry. A morphology with the cleavage rhombohedron $\{10\bar{1}4\}$ as the growth form is common, but other forms are also observed. Some typical habits for calcite are shown in Figure 24.4. They may be classified as rhombohedral habits in which a rhombohedral form dominates (Figure 24.4a–c) (a rhombohedron is a form of the type $\{10\bar{1}l\}$ with rhombuses as faces), scalenohedral habits (Figure 24.4d–f) (a scalenohedron is a form of the type $\{21\bar{3}1\}$ with triangles as faces), and prismatic types (Figure 24.4g–i). Plate 23e shows a crystal with scalenohedral–prismatic habit; in Plate 23f crystals are platy and the basal face dominates. Calcite crystals in veins sometimes grow to enormous size, such as those with rhombohedral morphology in an Alpine vein (Figure 24.5). Dolomite frequently displays morphology with the cleavage rhomb $r = \{10\bar{1}4\}$ (Plate 24a). Orthorhombic carbonates may display pseudohexagonal morphology largely because of twinning, such as in aragonite (Plate 24c) or cerussite.

The carbonates with more complicated compositions and more complex structures have a less symmetrical morphology. Prismatic nahcolite, tabular trona, prismatic (Plate 6e) and tabular azurite, and fibrous malachite occurring as botryoidal aggregates (Plate 1c) are examples.

Many carbonate minerals are colorless or white, but those with manganese are mainly pink (rhodochrosite, Plate 24e,f), and with copper are mainly green (malachite) or blue (azurite) (Plates 1c, 6e, and 24d). Iron-containing carbonates are often yellow (ankerite and siderite) and brown because the iron oxidizes in inclusions of iron oxides and hydroxides.

A distinct property of carbonates is their high birefringence, which expresses the pronounced anisotropy of their crystal structure. The difference between the maximum refractive index of calcite (1.658) and the minimum index (1.486) is 0.172. Birefringence was first discovered with transparent calcite crystals called Iceland spar, where a "double" image is produced if light passes through the crystal (see Figure 13.11b). When carbonates are examined with a polarizing microscope, the high birefringence with high-order interference colors is the most diagnostic feature (Plates 8a and 16c,d).

An easy way to distinguish some carbonates is to observe their reaction with hydrochloric acid and other solutions. When a drop of diluted HCl is applied, calcite reacts very intensely and "fizzing" is observed ($2HCl + CaCO_3 \rightarrow H_2O + CO_2 + CaCl_2$). Dolomite, on the other hand, reacts only with concentrated acid, while magnesite reacts only when heated. Staining thin sections with an alizarin sulfonate solution produces colors characteristic of carbonate minerals and particularly helps to distinguish between calcite and dolomite (for a review of staining techniques, see Friedman, 1959). More detailed characterizations rely on chemical and X-ray diffraction analyses.

24.3 Brief description of important carbonate minerals

Calcite ($CaCO_3$) often contains minor amounts of magnesium, iron, and manganese, and its more complete chemical formula would be $(Ca,Mg,Fe,Mn)CO_3$. The limits of isomorphic miscibility between the carbonate end members $CaCO_3$ and $MgCO_3$ are shown in the temperature–composition phase diagram, Figure 24.6. At room temperature ionic substitution is very limited. Calcite is found in druses and in single crystals of diverse habit (rhombohedral, prismatic, platy, and more complex; see Figures 24.4 and 24.5). It also forms solid granular masses and veinlets. The crystals are usually transparent or translucent and white in color. Calcite has cleavage along the three directions of a rhombohedron $r = \{10\bar{1}4\}$, vitreous or pearly luster (on the cleavage surfaces), and, with a hardness of 3 on the Mohs scale, it can be scratched with a knife.

The bulk of calcite is found in limestone and has a chemical or biological origin. It is the dominant carbonate mineral in more recent sedimentary rocks. Calcite that precipitated in warm seawater is high in magnesium and may transform into aggregates of

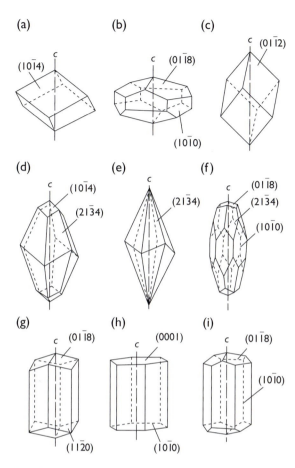

Fig. 24.4 Typical habits of calcite; forms are indicated. (a–c) Rhombohedra dominating, (d–f) scalenohedra dominating, (g–i) prisms dominating.

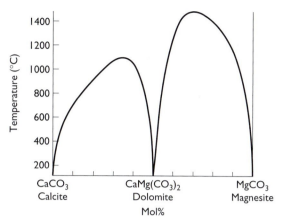

Fig. 24.6 Temperature–composition phase diagram of the system $CaCO_3$–$MgCO_3$. Above the curve the phase is homogeneous (data from Goldsmith and Heard, 1961).

Fig. 24.5 Large rhombohedral calcite crystals in an Alpine vein from Gonzen SG, Switzerland. The edge of individual crystals measures 40–70 cm.

pure calcite and dolomite during diagenesis (a low-temperature form of metamorphism involving groundwater, which is discussed in more detail later in this chapter). Regional and contact metamorphism transforms limestone to marble. Calcite also occurs in skarns and in medium- and low-temperature hydro-thermal deposits. Calcite is the main component of the rare igneous rock carbonatite.

Limestone and marble are used for the production of lime and particularly of cement (combined with other components, see Chapter 35). In the past, lime-stone and marble were favorite building materials for Ancient Greek temples, medieval cathedrals, and bar-oque palaces.

Rhodochrosite ($MnCO_3$) can be identified by its pale-pink color (Plate 24e,f), but since its color may also be white, gray, or greenish-gray (just as with calcite and dolomite), positive identification can be difficult. The mineral crystallizes in some hydrother-mal deposits and in sedimentary manganese deposits.

Magnesite ($MgCO_3$) is an end member of the magnesite–siderite isomorphic series. It forms spotty colored grayish-white solid masses of granular texture in dolostones that have recrystallized and were affected by hydrothermal solutions. Magnesite also occurs as white porcelain-like veins that form during weathering of serpentinites. The mineral is used as a magnesium ore and as a refractory material in the ceramics industry.

Siderite ($FeCO_3$) is found in brown rhombohedral crystals with a highly vitreous luster and in granular aggregates within hydrothermal medium-temperature veins. Often saddle-morphology is observed, with curved surfaces. Most siderite is formed chemically in sedimentary rocks as late stages of hydrothermal alteration. The mineral is an iron ore. It is easy to identify siderite by its rusty yellow-brown color and high density of about 3.9 g/cm^3; however, nonoxidized crystals of siderite are white in color.

Smithsonite ($ZnCO_3$) is an oxidation product of sphalerite ores and occurs as colloidal, botryoidal, earthy masses and rarely as larger white, greenish, or brown crystals.

Dolomite ($CaMg(CO_3)_2$) has a structure that is related to that of calcite, but layers of calcium and magnesium alternate along the c-axis (see Figure 24.1b). There is some solid solution with calcite but only at high temperature (Figure 24.6). When dolomite coexists with calcite in metamorphic rocks, the composition of the two minerals can be used to determine the temperature of crystallization (Figure 24.6). Another solid solution is Mg–Fe, with the iron end member being the mineral ankerite $CaFe(CO_3)_2$, which has the same structure as dolomite. Dolomite crystals are often rhombohedral (Plate 24a), and sometimes curved (saddle dolomites) in habit. The color is white, brownish-gray, or rusty brown and the hardness is 4. The luster and cleavage of dolomite and calcite are very much alike, but the two can be distinguished with the hydrochloric acid test, with the alizarine sulfonide test, or by X-ray diffraction.

Dolomite occurs in low- and medium-temperature hydrothermal deposits and in sedimentary rocks. Only in some rare cases, when dolomite crystallizes from highly saline (more than 15% salt) water of lagoons and lakes, is it a primary mineral of sedimentary rocks. More often dolomite is the result of secondary diagenetic processes transforming magnesian calcite into dolomite.

Cerussite ($PbCO_3$) is a secondary product of lead ore oxidation, forming fine-grained solid and dense gray aggregates. The crystals are semitransparent, grayish-white, or even black, and have adamantine luster on their faces and greasy luster on fracture surfaces. The fracture surfaces are uneven or shell-like. The mineral is a lead ore, although not an important one.

Aragonite ($CaCO_3$) crystallizes in the orthorhombic system. Single crystals are tabular or prismatic in habit, but pseudohexagonal twinned intergrowths

are more common. These intergrowths usually have indented angles and sutures along the composition planes (Plate 24c). Aragonite also forms oolites and colloidal masses. The color is typically white, yellowish-white, brown, or gray, the luster is vitreous or greasy, and the hardness is 3.5–4.

Aragonite crystallizes as precipitates of mineral springs, in the oxidation zone of sulfide ores and weathering crusts, in marine chemical sediments, and in karst caves. Major quantities of the mineral are found in coral skeletons and mollusk shells (see Chapter 25). As we have mentioned above, aragonite is the high-pressure polymorph of $CaCO_3$ and has been found in some high-pressure–low-temperature metamorphic rocks in subduction zones. In most occurrences it is metastable and crystallizes for kinetic reasons.

Malachite ($Cu_2(CO_3)(OH)_2$) is a mineral of the oxidation zone of chalcopyrite and other copper sulfide deposits. The mineral forms botryoidal (Plate 1c, 24d), kidney-shaped aggregates, sometimes fibrous and thin coatings in fractures and caverns. Malachite is frequently found as pseudomorphs after native copper, cuprite, and azurite (Plate 6e), and is generally associated with copper ores. Its characteristic color is bright green (Plate 1c). Lighter and darker green malachite layers may alternate with blue bands of chrysocolla, a semi-amorphous siliceous mineral of approximate composition $(Cu,Al)_2(H_2Si_2O_5)(OH)_4 \cdot nH_2O$. Malachite is applied as a paint and is valued as a rather rare, decorative semiprecious stone.

Azurite ($Cu_3(CO_3)_2(OH)_2$) occurs as granular, crystalline, and colloform aggregates of dark-blue and sky-blue color and is often accompanied by malachite (Plates 6e, 24d). The mineral is a component of many copper ores and is used in the production of blue paint.

Bastnäsite ($(Ce,La,Y)CO_3F$) is a yellow to reddish brown hexagonal mineral. It occurs in alteration zones of alkaline rocks and in carbonatites, such as Mountain Pass, southeast California, USA, where it is mined for rare earths.

24.4 Formation conditions of carbonates

Carbonates occur in many types of rock and in very diverse mineral deposits. Their major occurrence, however, is in *sedimentary rocks* (limestone and dolostone). We will discuss the sedimentary mineral formation in some detail in the next section. The most abundant carbonate mineral, calcite, is often associated with organisms (Chapter 25).

Carbonatites (e.g., Bell, 1989; Wall and Zaitsev, 2004) are relatively rare carbonate rocks of igneous origin. In the upper mantle, at deep levels where ultramafic and alkaline magmas nucleate, carbon exists as a native element, as carbonates, and as hydrocarbons. In silicate melts that are rich in carbon, a carbonate liquid may separate. The carbonate liquid is immiscible with the silicate portion of the melt. Crystallization of such a melt, generally at crustal levels, produces carbonatite dikes and stocks that consist largely of calcite, dolomite, or other carbonate minerals. Carbonatites form as plugs in some volcanoes in the Recent East African rift systems. Very rarely alkaline fluorine–chlorine–carbonic lavas pour out on to the surface (e.g., at Oldoinyo Lengai in northern Tanzania), and such lavas crystallize to form rocks consisting of halite, fluorite, nyerereite ($Na_2Ca(CO_3)_2$), and gregoryite (Na_2CO_3). In ancient, deep-seated massifs that are presently exposed, primary carbonatites and surrounding silicate rocks are often altered owing to the interaction with aqueous-carbonic alkaline fluids. Carbonatites are often rich in rare earth elements (e.g., bastnäsite), niobium–tantalum oxides (e.g., pyrochlore), and zirconium oxides (e.g., baddeleyite).

In *hydrothermal deposits* carbonates are common and form under both medium- and low-temperature conditions in a neutral–alkaline environment. The most widespread carbonates of gold-bearing and polymetallic veins are calcite, dolomite, and siderite; rarer components of hydrothermal veins are rhodochrosite, kutnahorite, witherite, and strontianite.

In general the formation of the carbonates as a medium-temperature hydrothermal alteration of ultramafic rocks is common and is caused by solutions enriched in CO_2. Two reactions illustrate the formation of magnesite and calcite:

$$4(Mg, Fe)_2SiO_4 + 2H_2O + CO_2 + O_2 \rightleftharpoons$$
Olivine
$$Mg_3Si_2O_5(OH)_4 + MgCO_3 + 2Fe_2O_3 + 2SiO_2$$
Serpentine　　　　　Magnesite　Hematite　　Quartz

$$5CaMgSi_2O_6 + H_2O + 3CO_2 \rightleftharpoons$$
Diopside
$$Ca_2Mg_5Si_8O_{22}(OH)_2 + 3CaCO_3 + 2SiO_2$$
Tremolite　　　　　　Calcite　　Quartz

At *surface conditions*, calcite and aragonite form in caves as stalactites and stalagmites, crystallizing from supersaturated solutions (Figure 24.7). They also precipitate as travertine and tufa from hot springs that penetrate limestone. Splendid examples are at Pamukale (Turkey) (Figure 24.8), Garm-Chashma (Pamirs, Tajikistan), and at Mammoth Hot Springs

Fig. 24.7 Calcite forming as stalactites (top) and stalagmites (bottom) in a cave at Carlsbad Caverns, New Mexico, USA. Draperies in stalactites (right side) form when water runs down an edge. The so-called "popcorn" structure at the bottom is due to calcite growing from evaporation of a thin meniscus of water (courtesy M. Queen).

Table 24.2 Abundance, by vol.%, of rocks deposited on the Earth's surface

	Clay	Clastic	Carbonate	Evaporite	Volcanic
Platforms	46	22	24	2.8	4.5
Geoclinal belts	38	18	21	0.3	21

Source: From Ronov and Yaroshevsky, 1969.

Fig. 24.8 Calcite tufa deposit associated with hot springs at Pamukale, Turkey.

in Yellowstone National Park (Wyoming, USA). Secondary carbonates may form as *weathering crusts* in ore deposits and can be used as indicators of primary ore minerals. For example, smithsonite is a product of sphalerite alteration, cerussite forms owing to galena oxidation, and malachite and azurite are secondary minerals of copper sulfides.

In *metamorphic rocks* carbonates are products of recrystallization and metasomatic alteration of sedimentary rocks. Calcite and dolomite are the major constituents of marbles and often coexist in chemical equilibrium (see Figure 24.6).

If limestones consist only of calcite, increasing metamorphic grade (i.e., increasing temperature and pressure) will transform these rocks into pure marbles (as in Carrara, Italy). But if primary sedimentary rocks are not pure and contain quartz in addition to calcite, new stable silicate minerals will form at different metamorphic conditions. We have already discussed the reaction

$$CaCO_3 + SiO_2 \rightleftharpoons CaSiO_3 + CO_2$$
Calcite　　Quartz　　Wollastonite

in Chapter 19, and we will revisit the metamorphism of carbonate rocks in Chapter 30.

24.5 Carbonates in sedimentary rocks: chemical and biological origins

Carbonate rocks make up approximately one-quarter by volume of all sediments (Table 24.2) and carbonate minerals are, after silicates, the most important sedimentary component. The major carbonate rock is limestone; dolostone is less common, and carbonate evaporites are relatively rare. Carbonates cannot exist in seawater below a critical depth, because the high CO_2 pressure causes dissolution (Figure 24.9). The CO_2 content in seawater has changed somewhat in accordance with the CO_2 partial pressure in the atmosphere over the course of geological history. Presently the critical depth of carbonate dissolution is about 4.2 km in the Pacific Ocean and about 4.7 km in the Atlantic Ocean. Carbonate sediments are typical of continental shelves and of coral reefs. The map in Figure 24.10 documents their principal present-day distribution, which can be seen to be mainly in shallow oceans of the temperate and tropical zones.

Interestingly, worldwide the ratio of dolomite to calcite decreased from the Precambrian era to the Quaternary period (Figure 24.11). Today dolomite forms only in minute amounts in a few very special

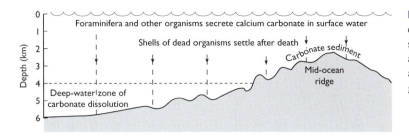

Fig. 24.9 Diagram outlining carbonate formation in oceans at shallow sea levels, for example along mid-oceanic ridges and reefs, and carbonate dissolution at depths greater than 4 km.

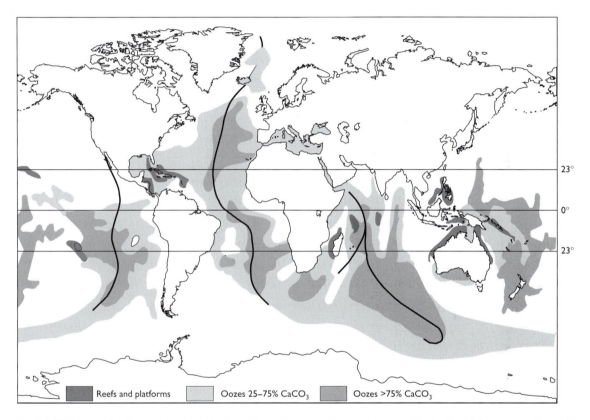

Fig. 24.10 Present-day formation of calcite in reefs, platforms, and oceanic oozes. Notice the high concentrations of calcite in clays along mid-oceanic ridges (dark lines). (Compiled from data of Rogers, 1957; Davies and Gorsline, 1976; and others.)

environments. The reason for this change is still unclear and is known as "the dolomite problem" of sedimentary petrology.

Some researchers attribute the change to variations in seawater composition, assuming that the earliest chemically precipitating carbonates were dolomite. Subsequently, chemical dolomites were accompanied by biogenic (algal) dolomites and limestone in the Precambrian era. After the beginning of the Cambrian period, dolomites yielded their dominance to limestones. The dolomites continued

to form only in arid continental areas, whereas limestones formed in both the oceans and continental basins, and in arid and humid climates. During the Mesozoic and Cenozoic, biogenic limestone formed not only within a shallow coastal zone, but also in deep-water zones owing to the large biomass of plankton organisms building their skeletons of calcium carbonate.

The problem with this evolutionary theory is that present-day seawater also has a composition favoring precipitation of dolomite, as is illustrated in a phase

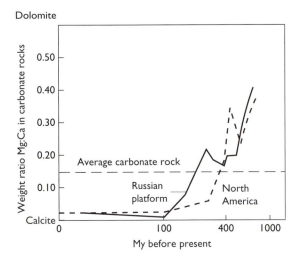

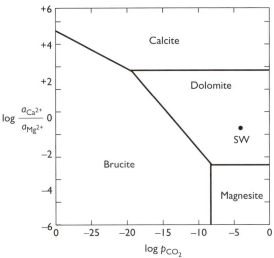

Fig. 24.11 Prevalence of dolomite and calcite in carbonate rocks over the course of geological history for North America and the Russian platform. Composition of average carbonate rock (dashed line) is shown for reference.

Fig. 24.12 Phase diagram with partial pressure of CO_2 and activity ratio of Mg^{2+}:Ca^{2+} in an aqueous solution, displaying stability fields of calcite, dolomite, magnesite, and hydromagnesite. Also indicated is the composition of present-day seawater (SW) (data from Garrels *et al.*, 1960).

diagram of calcium and magnesium activities versus the CO_2 pressure (Figure 24.12), and yet calcite forms dominantly. The formation of calcite in the stability field of dolomite is due to the kinetic difficulty of crystallizing the ordered dolomite structure. Therefore, many carbonate petrologists believe that most dolomites are secondary, replacing primary limestone during subsequent diagenesis. An example from a Devonian carbonate belt is the Lost Burro Formation in southeastern California, USA, where large euhedral dolomite crystals replace calcite limestone (Plates 16d and 24b).

Continental carbonate sediments include clastic carbonate rocks, caliche, soil (particularly loess), shallow-water sediments of lakes and lagoons situated in arid areas, and evaporites. The term *caliche* is applied to surface crusts enriched in calcite that cover soils. These crusts are formed in arid climates by lime-rich solutions that percolate to the surface through capillaries and cracks and then evaporate.

In some lakes and marine lagoons located in arid regions, the precipitation of carbonates proceeds owing to the evaporation of water. The Dead Sea is one of the most saline basins of the world. The coastal zone of the Dead Sea is covered with a firm gypsum crust, and gypsum continuously precipitates from its surface waters. However, below a depth of 3–6 m all gypsum is replaced by calcite, perhaps because of

sulfate-reducing bacteria (Neev and Emery, 1967). The primary precipitating carbonate in the Dead Sea is aragonite. It forms periodically when the temperature of the water reaches its annual maximum value. The resulting sediments are composed of alternating calcite and aragonite layers.

The three carbonates calcite, dolomite, and aragonite occur in silts of Lake Balkhash (Kazakhstan). Figure 24.13 shows a southwest to northeast cross-section with mineral composition (top) and corresponding magnesium content, calcium/magnesium activities, and pH of water (bottom). Aragonite forms when the pH is higher than 8.9, and the Mg^{2+} content is greater than 90 mg/l. Dolomite precipitates when the Ca:Mg ratio reaches a very low value (< 0.1), the Mg^{2+} content is about 250 mg/l, and the pH value reaches 9.0. In the Balkhash silts there is a regular sequence from southwest to northeast: calcite; calcite + aragonite; calcite + aragonite + dolomite.

Other examples of dolomite formation from highly concentrated brines in evaporitic environments are the recent deposits in the Coorong lagoon (southern Australia), in the sandy plains that adjoin the southern coast of the Persian Gulf, and on banks near the Bahamas and Florida. Dolomite in all these localities is accompanied by anhydrite and gypsum.

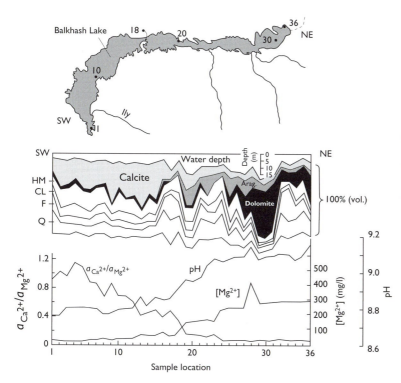

Fig. 24.13 Systematic mineralogical changes in silts of a southwest to northeast cross-section of Lake Balkhash, Kazakhstan (data from Verzilin and Utsalu, 1990). Shown are the mineral composition, activities of Ca^{2+} and Mg^{2+}, Mg^{2+}-concentration $[Mg^{2+}]$, and pH. Arag., aragonite; HM, hydromica; Cl, chlorite; F, feldspar; Q, quartz. Sample locations numbered on the map (top) are shown on the ordinate.

Another example of a carbonate evaporite where dolomite is precipitating is Deep Springs Lake in southeastern California, USA. Deep Springs Lake has no drainage and covers an area of about $13~km^2$ with a water depth of only 30 cm. The dense brine has a pH of 9.5 to 10.0, and dolomite prevails in the composition of bottom silts. The silts, salt crusts, and thin coatings of the lake consist of sodium–calcium carbonates (gaylussite, trona, pirssonite, nahcolite, thermonatrite), halite, sylvite, and various sulfates.

Carbonates in oceans may originate from terrigenous (clastic) material that is transported from the coast or submarine slopes, or they may be the remnants of skeletons of organisms. Carbonates produced through biogenic processes are the most important.

The biogenic carbonates are (in order of abundance) aragonite, calcite, and magnesian calcite. The $MgCO_3$ component is usually less than 1 mol% in aragonite but can reach up to 25 mol% in biogenic calcite (Table 24.3). Typically, the carbonate skeletons of deep-water organisms contain less than 1–2 mol% $MgCO_3$, and corresponding silts are low in magnesium. Shallow-water carbonate sediments may have varying quantities of $MgCO_3$, averaging 5 mol%. Magnesian calcite is not stable and converts over

geological time, during diagenesis or metamorphism, to calcite and dolomite.

Authigenic minerals form in place, rather than being transported. This may happen in bottom silts and primary sediments due to the interaction of primary chemical, biogenic, and terrigenous minerals with porous solutions during the transformation of the sediments into sedimentary rocks. The crystallization sequence during this process, called *diagenesis*, is regular (Figure 24.14). At the very first stages of diagenesis the polymorphic transformation of aragonite to calcite takes place. Only very rarely is aragonite preserved in ancient limestone, usually when clays or organic matter isolate it from aqueous solutions.

We have already discussed the transformation of calcite to dolomite. Dolomite forms because of either the dissociation of initial Mg-rich calcite into calcite and dolomite

$$10(Ca_{0.8}Mg_{0.2})CO_3 \rightleftharpoons 2CaMg(CO_3)_2 + 6CaCO_3$$

or the metasomatic replacement of calcite by dolomite under the influence of secondary porous solutions rich in magnesium.

Carbonates can dissolve in the early diagenetic stages. For instance, calcite skeletons in recent alluvium are easily dissolved in humid regions because fresh river waters ordinarily are acidic (pH > 7). Similarly, a high CO_2 content in cold seawater causes calcium carbonate to dissociate:

Table 24.3 Most abundant organisms with magnesian calcite skeletons

	$MgCO_3$ (mol%)
Red carbonate algae	10–25
Echinodermata	10–15
Bryozoa (corals)	10–11
Benthic foraminifera	1–15
Crustacea	1–5

$$H_2O + CO_2 \rightleftharpoons H_2CO_3$$

$$CaCO_3 + H_2CO_3 \rightleftharpoons Ca^{2+} + 2HCO_3^-$$

Diagenesis applies not only to carbonate minerals. Silica minerals transform from amorphous opal to cryptocrystalline chalcedony and ultimately to quartz. Clay minerals such as kaolinite and montmorillonite change to mica (hydromica and sericite, later muscovite) and chlorite. There is a regular progression in zeolites from phillipsite to clinoptilolite, analcime, and laumontite and, in a final stage of diagenesis, plagioclase feldspar and epidote appear. These and other diagenetic transformations are illustrated in Figure 24.14. Dissolution and reprecipitation are only one side of the diagenetic transformation of sedimentary carbonate rocks. The activity of bacteria,

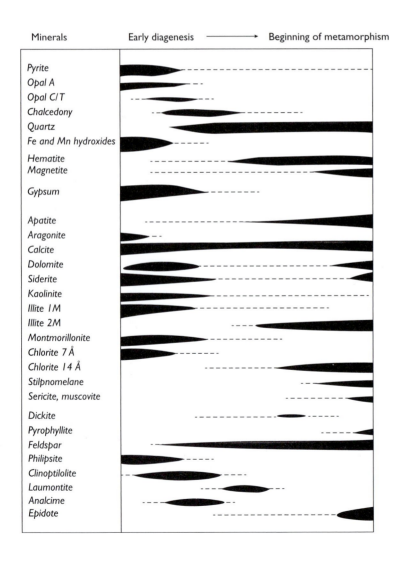

Fig. 24.14 Authigenic minerals typical of various stages from early diagenesis to the beginning of metamorphism (data from Logvinenko and Orlova, 1987).

decomposition of organic material, and variations in CO_2 and O_2 concentrations add further complexities. Diagenesis is followed by other transformations during which the mineral composition of a rock and its textural–structural features develop further.

24.6 Summary

Carbonates are a large mineral group. They are all characterized by structures with triangular CO_3^{2-} groups. In the simplest case, calcite, the structure can be visualized as an fcc lattice of Ca^{2+} ions, with CO_3^{2-} groups in octahedral interstices with triangles aligned perpendicular to a body diagonal. The cube is then distorted by shortening along this direction. Borates and nitrates also have triangular anion groups but they are often linked to form rings and chains. Calcite and dolomite form in water, but mostly not during evaporation (stalactites in caves and tufas around hot springs are exceptions) but by biological origins, e.g., as skeletons of algae, foraminifera, corals, and crustacea, accumulating as limestones. Limestones convert to marble when subjected to metamorphic recrystallization. More rare is the igneous occurrence of carbonates as carbonatites, where they are mined as rare earth ores. Limestones are of huge instrustrial significance as the main component of cement production.

Important carbonate minerals to remember

Name	Formula	System
Calcite	$CaCO_3$	Rhombohedral
Aragonite	$CaCO_3$	Orthorhombic
Siderite	$FeCO_3$	Rhombohedral
Dolomite	$CaMg(CO_3)_2$	Rhombohedral
Malachite	$Cu_2CO_3(OH)_2$	Monoclinic
Bastnäsite	$(Ce,La,Y)CO_3F$	Hexagonal

Test your knowledge

1. Compare the similarities and differences between the structures of halite and calcite.

2. What other minerals besides carbonates have XO_3^- groups as structural elements?

3. Why does dolomite pose an enigma to sedimentologists?

4. Why do carbonates not crystallize in the ocean at depths more than 4 km?

5. Discuss the major geological conditions under which carbonates form.

6. Name some organisms with calcite or aragonite skeletons.

7. What are the major uses of calcite?

8. Write a reaction involving carbonates in progressive metamorphism.

9. Write an equation for the dissolution of calcite, for example in a karst environment.

Further reading

Carbonate and borate minerals

Chang, L. L. Y., Howie, R. A. and Zussman, J. (1995). *Rock-forming Minerals, Volume 5B, Non-silicates: Sulphates, Carbonates, Phosphates, Halides*, 2nd edn. The Geological Society of London.

Lippmann, F. (1973). *Carbonate Minerals*. Springer-Verlag, Berlin.

Morse, J. W. and MacKenzie, F. T. (1990). *Geochemistry of Sedimentary Carbonates*. Elsevier Science.

Reeder, R. J. (ed.) (1983). *Carbonates: Mineralogy and Chemistry*. Reviews in Mineralogy, vol. 11. Mineralogical Society of America, Washington, DC.

Sedimentary occurrence

Boggs, S. (2013). *Petrology of Sedimentary Rocks*, 2nd edn. Cambridge University Press, Cambridge.

Tucker, M. E. (2001). *Sedimentary Petrology: An Introduction to the Origin of Sedimentary Rocks*, 3rd edn. Blackwell Science, Oxford.

Tucker, M. E. and Wright, V. P. (1990). *Carbonate Sedimentology*. Blackwell Science, Oxford.

Garrett, D. E. (1998). *Borates: Handbook of Deposits, Processing, Properties, and Use*. Academic Press, San Diego, CA.

25 | Phosphates, sulfates, and related minerals. Apatite as a biogenic mineral

In Chapter 24 we discussed the triangular plane CO_3^{2-} and BO_3^{2-} groups as fundamental building blocks of carbonates and borates. In phosphates and sulfates the fundamental building block is the PO_4^{3-} or the SO_4^{2-} tetrahedron, respectively. Related to phosphates and sulfates are arsenates, vanadates, and tungstates, with AsO_4^{3-}, VO_4^{3-}, and WO_4^{2-} tetrahedra, respectively. Important minerals are apatite ($Ca_5(PO_4)_3(F,OH,Cl)$), barite ($BaSO_4$), anhydrite ($CaSO_4$), gypsum ($CaSO_4 \cdot 2H_2O$), and scheelite ($CaWO_4$). Apatite occurs in bones and teeth and we will use this chapter to discuss some biogenic mineral-forming processes. The minerals in these groups are of considerable economic interest. Apatite is a major source of phosphorus used mainly as fertilizer. Gypsum is used as a building material, and scheelite is the major tungsten ore.

25.1 Phosphates, arsenates, and vanadates

Phosphates and related minerals are numerous and some important examples are listed in Table 25.1. The crystal structures are rather complicated in detail, and we will not elaborate on them. However, it is noteworthy that several phosphate structures are identical with silicate structures, and these we will study in more depth in Chapters 28–31. For example, berlinite ($AlPO_4$) is isostructural with quartz (SiO_2), triphyline ($LiFePO_4$) with olivine (Mg_2SiO_4), and xenotime (YPO_4) with zircon ($ZrSiO_4$). Mostly though, phosphate and sulfate coordination polyhedra are isolated, whereas in silicates, the tetrahedra are generally polymerized to form sheets, chains, and frameworks.

The vanadate vanadinite ($Pb(VO_4)_3Cl$) is isostructural with apatite and there are limited isomorphic substitutions of phosphorus, vanadium, and arsenic. Vanadates may also have other coordination polyhedra such as VO_5^{5-}, VO_6^{7-}, and $V_2O_8^{6-}$. Many of these minerals contain additional OH^-, F^-, and Cl^-, and molecular water. The main cations are Ca^{2+}, Al^{3+}, Fe^{2+}, Cu^{2+}, Co^{3+}, and Ni^{3+}. The "uranium micas" such as carnotite ($K_2(UO_2)_2(VO_4)_2 \cdot 3H_2O$) contain UO_2^{2+}, with uranium in 6-fold valency.

25.2 Brief description of important phosphate minerals

Monazite ($CePO_4$) usually contains other rare earth elements besides cerium. In addition, some thorium (with a rather complicated isomorphic scheme: $Ce^{3+} + Ce^{2+} \rightleftharpoons Ca^{2+} + Th^{4+}$, or $Ce^{3+} + P^{5+} \rightleftharpoons Th^{4+} + Si^{4+}$) and uranium may be present in the mineral. The maximum contents of ThO_2 and UO_2 are 32 weight% and 7 weight%, respectively.

Monazite is found in thick tabular and isometric isolated crystals. Monazite crystals from granite pegmatites (Plate 25a) morphologically resemble garnet, but monazite has a cleavage and a lower hardness (Mohs scale 5.5). Monazite grains found in gneisses are minute and are recognized only with a petrographic microscope. Monazite inclusions in cordierite produce pleochroic halos due to damage from γ-radiation (Plate 2e, f). This feature is diagnostic of both cordierite and monazite (zircon that produces pleochroic halos in biotite does not produce them in cordierite). Local concentrations of monazite in river and coastal sands are mined as sources of thorium and cerium.

Apatite ($Ca_5(PO_4)_3(F,OH,Cl)$) sometimes contains strontium (up to 15 weight% SrO), cerium (up to 12

Table 25.1 Phosphates and related minerals with some diagnostic properties; important minerals are given in italics

Mineral Formula	System	Morphology Cleavage	H	D	Color Streak	n Pleochroism	Δ	2V (Dispersion)
Monazite group								
Monazite $Ce(PO_4)$	Monocl.	Tab. (100) *(001)*	5–5.5	5.1	Yellow, brown *Yellow, brown*	1.80–1.84	0.045	+6–19
Xenotime $Y(PO_4)$	Tetrag.	Pris. [001] *{100}*	4–5	4.8	Brown, red-brown *Pale brown*	1.72–1.82	0.095	(+)
Apatite group								
Fluor-, hydroxy- and chlor-apatite $Ca_5(PO_4)_3(F, OH, Cl)$	Hexag.	Pris. [0001] (0001), {10$\bar{1}$0} *(001)*	5	3.2	Clear, green, yellow, violet	1.63–1.65	0.001	(−)
Pyromorphite $Pb_5(PO_4)_3Cl$	Hexag.	Pris. [0001]	3.5–4	6.8	White, green *White, yellow*	2.05–2.06	0.011	(−)
Vanadinite $Pb_5(VO_4)_3Cl$	Hexag.	Pris. [0001]	3	7.0	Yellow, brown, orange *White, yellow*	2.35–2.42	0.066	(−)
Vivianite group								
Annabergite $Ni_3(AsO_4)_2 \cdot 8H_2O$	Monocl.	Fibr. [001] *(010)*	2.5–3	3.1	Yellow-green *Green*	1.62–1.69	0.065	− 84 (r > v)
Erythrite $Co_3(AsO_4)_2 \cdot 8H_2O$	Monocl.	Fibr. [001] *(010)*	2.5	2.95	Pink *Pink*	1.63–1.70	0.072	90 (r > v)
Vivianite $Fe_3(PO_4)_2 \cdot 8H_2O$	Monocl.	Pris. [001] *(010)*	2.5	2.68	Clear-white	1.58–1.63	0.047	+80–90 (r < v)
Uranium micas								
Autunite $Ca(UO_2)_2(PO_4)_2 \cdot 8{-}12H_2O$	Tetrag.	Platy (001) *(001)*	2–2.5	3.1	Yellow-green, yellow *Clear-yellow-yellow*	1.55–1.58	0.024	− 10–30 (r ≫ v)
Carnotite $K_2UO_2(VO_4)_2 \cdot 3H_2O$	Monocl.	Platy (001) *(001)*	3–4	4.5	Yellow, green *Yellow*	1.75–1.95	0.200	− 46 (r < v)
Torbernite $Cu(UO_2)_2(PO_4)_2 \cdot 8{-}12H_2O$	Tetrag.	Platy (001) *(001)*	2.5	3.2	Green	1.58–1.59	0.01	(−)
Other phosphate minerals								
Turquoise $CuAl_6(PO_4)_4(OH)_8 \cdot 4H_2O$	Tricl.	Micr. *(001)*	5–6	2.7	Blue, blue-green	1.61–1.65	0.04	+40 (r ≪ v)

Notes: H, hardness; D, density (g/cm³); n, range of refractive indices; Δ, birefringence; $2V$, axial angle for biaxial minerals. For uniaxial minerals (+) is positive and (−) is negative. Acute $2V$ is negative the mineral is biaxial negative and $2V$ is $2V_\alpha$; if it is positive, the mineral is biaxial positive and $2V$ is $2V_\gamma$. Dispersion r < v means that acute $2V$ is larger for violet than for red.

System: Hexag., hexagonal; Monocl., monoclinic; Tricl., triclinic; Tetrag., tetragonal.

Morphology: Fibr., fibrous; Micr., microscopic; Pris., prismatic; Tab., tabular.

weight% Ce_2O_3), and other elements. Hexagonal apatite often forms well-developed colorless, green, or blue crystals of prismatic habit. Pinacoidal faces usually terminate the crystals (Plate 25b). The mineral's luster varies from vitreous to greasy. Hardness is 5 (softer than a knife blade). Apatite is uniaxial negative and has a low birefringence and refractive index. In thin sections, apatite has a characteristic mottled texture.

Apatite occurs in a wide range of rocks and deposits as isolated crystals and grains, usually small in size (1–2 mm). The largest and most perfect apatite crystals are found in granite pegmatites and marbles. The main use of apatite is as a fertilizer in the agricultural industry. The industrially important concentrations of apatite are in nepheline-bearing alkaline rocks. In alkaline rocks the impregnations, compact veins, lenses, and segregations of sugar-like apatite compose extensive deposits in which the apatite content may reach 80 weight% over tens of kilometers, such as in the Khibini massif of the Kola Peninsula, Russia.

In sedimentary rocks the mineral forms layers called phosphorites, which often contain nodules of fine acicular–radial structure. Other components of phosphorites are shell and bone fragments and other apatite-bearing organic remains. Apatite is a main constituent of the solid tissue of vertebrates and of brachyopod skeletons. (We will discuss the biomineralogical origins of apatite later in this chapter.)

The term **uranium micas** is applied to a series of minerals such as torbernite ($Cu_2(UO_2)_2(PO_4)_2 \cdot 10H_2O$) (Plate 25c), autunite ($Ca(UO_2)_2(PO_4)_2 \cdot 10H_2O$), carnotite ($K_2(UO_2)_2(VO_4)_2 \cdot 3H_2O$), and others. These minerals occur as earthy, powdery aggregates and more rarely as well-formed platy crystals, resembling mica, in the weathering crusts of uranium deposits. Brilliant colors (yellow, yellowish-green, green), mica-like cleavage, pearly luster on the cleavage planes, and high radioactivity are characteristic features. Carnotite is a major ore of vanadium and, in the USA, the principal ore of uranium.

Turquoise ($CuAl_6(PO_4)_4(OH)_8 \cdot 4H_2O$) (Plate 1d) is usually cryptocrystalline and forms blue to bluish-green aggregates and is easily recognized by its color and relatively high hardness (6). Turquoise is a secondary mineral that occurs in veins within altered volcanic rocks. It is used as a gemstone.

Erithrite ($Co_3(AsO_4)_2 \cdot 8H_2O$) and **annabergite** ($Ni_3(AsO_4)_2 \cdot 8H_2O$) are rare secondary minerals that usually form as thin crusts that are alteration products. While they have no economic significance, the minerals are used by prospectors and are indicative of cobalt and nickel ores. Erithrite and annabergite are easily recognized by their striking pink and pale-green colors, respectively.

25.3 Sulfates and tungstates

Sulfate minerals are the natural salts of sulfuric acid H_2SO_4. They include acid, basic, and intermediate salts and crystallohydrates (see Table 25.2). All these minerals have isolated complex anion groups SO_4^{2-} in the form of tetrahedra in their structures. In gypsum, double columns of stacked SO_4^{2-} tetrahedra are linked to columns of Ca^{2+}, and columns of water molecules. Columns extend along the c-axis (a c projection is shown in Figure 25.1). This structural arrangement produces an excellent (010) cleavage. Sulfates are not widespread in nature; the most abundant minerals are gypsum, anhydrite, and barite. There are about 200 sulfate minerals. In exogenic environments, sulfates such as thenardite, epsomite, anhydrite, and gypsum form evaporites. Some sulfates (among them barite, gypsum, and anhydrite) also occur in hydrothermal ore deposits. Tungstates, with WO_4^{2-} tetrahedra instead of SO_4^{2-}, are closely related to sulfates.

25.4 Brief description of important sulfate and tungstate minerals

The two isostructural minerals **barite** ($BaSO_4$) and **celestite** ($SrSO_4$) form a limited solid solution. Crystals of these minerals are rhombic–prismatic or tabular in habit, often transparent and white, yellow, brown (when inclusions of limonite are present), or blue in color, with vitreous luster. Stepped fracture surfaces mark perfect cleavages on three planes (parallel to prism and pinacoid faces). Barite often occurs as crested aggregates (Figure 25.2), frequently associated with fluorite. Barite is recognized by its high density of about 4.5 g/cm^3.

Celestite and barite are often found in druses and granular aggregates. Barite forms in medium- and low-temperature hydrothermal deposits, while celestite develops as secretions in sedimentary rocks.

Barite is used to manufacture paints and in the chemical, rubber, and paper industries. The high density makes barite useful as an additive to drilling mud in deep drilling projects. It is also used to cap oil and

Table 25.2 Sulfates, tungstates, and related minerals with diagnostic properties; important minerals are given in italics

Mineral Formula	System	Morphology Cleavage	H	D	Color Streak	n Pleochroism	Δ	2V (Dispersion)
Barite group								
Barite $BaSO_4$	Ortho.	Tab. (001) (001), {210}	3–3.5	4.48	White, gray, yellow, red	1.64–1.65	0.012	+36 (r < v)
Celestite $SrSO_4$	Ortho.	Pris., Tab. (001), {210}	3–3.5	4.0	Clear, white, blue	1.62–1.63	0.009	+51 (r < v)
Anhydrite group								
Anglesite $PbSO_4$	Ortho.	Eq. (010), (001)	2.5–3	6.3	Clear, white, blue	1.88–1.89	0.017	+60–75 (r ≪ v)
Anhydrite $CaSO_4$	Ortho.	Eq., pris. [010] (010), (001)	3–3.5	2.9	Clear, white	1.57–1.61	0.044	+42 (r < v)
Gypsum								
Gypsum $CaSO_4 \cdot 2H_2O$	Monocl.	Tab. (010) {011} (010), (100)	1.5–2	2.3	Clear, white	1.52–1.53	0.009	+58 (r > v)
Various sulfates								
Alunite $KAl_3(SO_4)_2(OH)_6$	Ps. Cubic (Rhomb.)	{10$\overline{1}$1} (0001)	3.5–4	2.7	White, clear, yellow	1.57–1.59	0.020	(+)
Epsomite $MgSO_4 \cdot 7H_2O$	Ortho.	Fibr.–pris. [001] (010), {011} (010)	2–2.5	1.68	Clear, white	1.43–1.46	0.028	−51 (r < v)
Jarosite $KFe_3(SO_4)_2(OH)_6$	Trig.	Tab. (0001)	2.5–3.5	3.1	Orange-brown *Yellow*	1.72–1.82 *Clear-yellow*	0.10	(−)
Mirabilite $Na_2SO_4 \cdot 10H_2O$	Monocl.	Fibr. [010] (100)	1.5–2	1.49	Clear	1.39–1.40	0.004	−80 (r > v)

Table 25.2 (*cont.*)

Mineral Formula	System	Morphology Cleavage	H	D	Color Streak	n Pleochroism	Δ	2V (Dispersion)
Polyhalite $K_2Ca_2Mg(SO_4)_4 \cdot 2H_2O$	Tricl.	Tab. (010) {101}	3	2.77	Red, white, yellow	1.55–1.57	0.020	–62 (r < v)
Thenardite Na_2SO_4	Ortho.	Tab. (010) (010)	2–3	2.67	Clear, white, brown	1.47–1.48	0.015	+83
Tungstates and molybdates								
Scheelite $CaWO_4$	Tetrag.	Eq. {101}	4.5	6.0	Gray, white, yellow	1.92–1.93	0.016	(+)
Wolframite (Fe, Mn)WO_4	Monocl.	Platy (101) (010)	5–5.5	6.7–7.3	Brown, black *Yellow-brown,* *black*	2.26–2.42	0.16	+76
Wulfenite $PbMoO_4$	Tetrag.	Platy (001) (001)	3	6.8	Yellow-orange	2.28–2.41	0.122	(–)

Notes: H, hardness; D, density (g/cm³); *n*, range of refractive indices; birefringence; 2V, axial angle for biaxial minerals. For uniaxial minerals (+) is positive and (–) is negative. Acute 2V is given in the table. If 2V is negative the mineral is biaxial negative and 2V is $2V_\alpha$; if it is positive, the mineral is biaxial positive and 2V is $2V_\gamma$. Dispersion r < v means that acute 2V is larger for violet than for red.

System: Monocl., monoclinic; Ortho., orthorhombic; Ps., pseudo; Rhomb., rhombohedral; Tricl., triclinic; Trig., trigonal; Tetrag., tetragonal.

Morphology: Eq., equiaxed; Fibr., fibrous; Pris., prismatic; Tab., tabular.

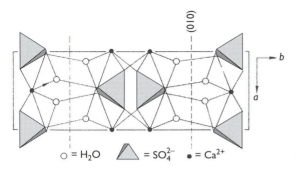

Fig. 25.1 (001) projection of the structure of gypsum with isolated SO_4^{2-} tetrahedra. The (010) cleavage plane is indicated by dashed lines, unit cell by solid lines.

Fig. 25.2 Aggregation of barite crystals with crested morphology from Dreisla, Germany (courtesy O. Medenbach). Width 90 mm.

gas wells. Celestite is used for sugar manufacturing, in pyrotechnology, and in pharmaceuticals.

Anhydrite ($CaSO_4$) is a typical mineral of evaporite deposits. In association with gypsum it forms massive marble-like layers of various thicknesses within a sedimentary series. Unlike calcite, anhydrite does not react with hydrochloric acid. This orthorhombic mineral is also found as grains in medium-temperature hydrothermal sulfide ore deposits. Crystals are ordinarily white, gray, or lilac in color and display excellent cleavage on (010), (100), and (001) planes. Anhydrite also occurs in some metamorphic rocks. The mineral is a raw material for cement production.

Anglesite ($PbSO_4$) is generally a product of galena oxidation. It is often found as massive granular and colloform aggregates, druses, and crusts. Crystals of anglesite are prismatic–tabular in habit and colorless or white and gray in color. Anglesite may be difficult to distinguish from isostructural barite since it also

has a high density (6.3 g/cm^3). Contrary to barite it is often associated with galena.

Gypsum ($CaSO_4 \cdot 2H_2O$) forms flattened and prismatic crystals with monoclinic symmetry. Distinctive swallow-tail twins are common (Figure 25.3a). The habit ranges from tabular to prismatic (Figure 25.3b). Crystals are colorless and transparent with vitreous or pearly luster. The mineral has a perfect (010) cleavage. The aggregates of gypsum are white, clear, or translucent. Its hardness is 1.5–2 in single crystals and reaches 3 in granular masses.

Gypsum forms as a low-temperature hydrothermal mineral, crystallizing from meteoric water circulating in sandstones and clays, but the major occurrence of the mineral is in evaporite deposits. As a solution becomes supersaturated, gypsum, anhydrite, and halite precipitate out consecutively and form alternating layers of these minerals (see Chapter 24). Sedimentary gypsum is found in fine-grained massive rocks. These rocks often contain gypsum as veins, geodes, and nests of euhedral crystals or as parallel fibrous aggregates. Gypsum also grows on the surface of clay or sand. Such crystals incorporate clay particles and sand grains, and exhibit a flower-like morphology (Sahara roses), which we have discussed in Chapter 10 (see Figure 10.14a). Gypsum growing in cavities can form huge euhedral crystals (Plate 5a). Gypsum is recognized by the shape of its crystals and twins, its perfect cleavage, and its low hardness. It is used mainly in the production of cement and plaster for the construction industry.

Alunite ($KAl_3(SO_4)_2(OH)_6$) forms by hydrothermal alteration of felsic volcanic rocks. Its formation is a result of the interaction between volcanic rocks and sulfuric hydrothermal solutions. Feldspars, for example, transform to alunite according to the following reaction:

$$3KAlSi_3O_8 + 2SO_4^{2-} + 10H^+ \rightleftharpoons$$
K-feldspar
$$KAl_3(SO_4)_2(OH)_2 + 9SiO_2 + 2K^+ + 4H_2O$$
Alunite

Alunite is found in solid, massive, chalcedony-like layered aggregates of patchy color and in loose masses that replace volcanic rocks. These aggregates have a hardness ranging from 3 to 7. The mineral is an aluminum ore and a source of potassium.

Wolframite is a general name for minerals that belong to the isomorphic series ferberite $FeWO_4$–hübnerite $MnWO_4$. The symmetry is monoclinic. Wolframites

(a)

(b)

Fig. 25.3 (a) Platy crystal of gypsum with swallow-tail twin from eastern Mojave Desert, California, USA (photograph P. Gennaro). Width 100 mm. (b) Prismatic crystals of gypsum from Eisleben, Saxony, Germany (courtesy O. Medenbach). Width 110 mm.

Fig. 25.4 Tetragonal crystals of scheelite from the Erzgebirge at Zinnwald (now Cinovec in the Czech Republic) (courtesy A. Massanek). Width 50 mm.

directions, forms isometric, not elongate or tabular crystals, and is much less dense than wolframite.

Scheelite ($CaWO_4$) forms mainly in quartz veins (together with wolframite, cassiterite, and sulfides) and in skarns. It occurs as tetragonal bipyramidal crystals of white, yellowish-white color with resinous luster (Figure 25.4). Scheelite's hardness is 4.5, and it displays {111} cleavage. It is easily mistaken for quartz and calcite. Bright, blue and white fluorescence is the best diagnostic property (Plate 18f,g). Scheelite is the major tungsten ore.

25.5 Biogenic processes

Above, we described properties and origins of the most important minerals from phosphates, arsenates, vanadates, sulfates, and tungstates. Among them apatite is unique because it composes bones and teeth. In this context we will discuss some biogenic mineral-forming processes, including other biogenic minerals.

Biogenic mineral deposits form in surface environments as transformations of primary organic aggregates or as a result of biochemical processes. Biogenic minerals are not minerals in the strict conventional sense because life is involved. However, organisms produce many of the same substances that form inorganically in rocks and therefore a discussion of these organic processes is appropriate. Biogenic minerals originate from living organisms or with their assistance; all crystallize within living organisms as a result of cell activity and are surrounded by organic material (see also Chapter 36).

Volume-wise the most important biogenic minerals are *calcite* and *aragonite* composing mollusk shells, coral skeletons, and foraminifera (Table 25.3). The vast

form black, dark-brown, or reddish-brown tabular and prismatic roughly striated crystals. Their hardness is about 5. Wolframites are easily cleaved on a (010) plane. Cleavage planes show an adamantine or a strong semimetallic ("old mirror") luster. The streak is brown or pale brown, similar to that of sphalerite. The density of wolframites is nearly 7 g/cm^3. These minerals form in quartz veins and are sometimes found in placers. They are mined as a tungsten ore.

Diagnostic properties of wolframite include crystal morphology, cleavage character, "mirror" luster, and high density. Having similar streak, color, and perfect cleavage, wolframite can be mistaken for sphalerite, but the cubic sphalerite has a {111} cleavage in several

Table 25.3 Mineralogical composition of solid plant and animal tissue (see also Table 36.2)

Composition	Plant or animal examples
Silica (opal, chalcedony, quartz)	Radiolaria, siliceous sponges, diatomaceous algae
Carbonates	
Calcite	Archeocyatha, foraminifera, stromatoporoids, carbonate sponges, echinoderms, brachiopods, belemnites, ostracods, coccolithophora, cyanophycerae, purple algae, some mollusk shells, egg shells of birds and reptiles
Calcite crystals	Eyes of trilobites and brittle stars
Aragonite	Corals, shells of mollusks and cephalopods
Aragonite transforming into calcite	Corals, bryozoa, gastropods, pelecypods
Phosphates	
Apatite	Bones, teeth, scales of vertebrates, brachiopods
Struvite	Kidney and gall stones
Sulfates	
Barite, gypsum	Ear stones of animals
Oxalates	
Whewellite, weddellite	Kidney and gall stones
Phosphate-bearing carbonates	Brachiopods
Magnetite	In brain tissue of birds and insects (carrier pigeons, bees, etc.), bacteria (*Magnetospirillum magnetotacticum*). Magnetite is used for navigation and orientation
Fe-hydroxides	Shells of diatoms, pediculates of Protozoa

majority of limestones are composed of biominerals. Mother-of-pearl (nacre) in sea shells is composed of layers of aragonite (Figure 25.5) and calcite, separated by the protein conchialine. A very unusual application of minerals by organisms is the use of calcite single crystals as lenses in the eyes of the long-extinct trilobites. Recently it has been discovered that calcite crystals are distributed over the body of the eyeless invertebrate brittle star (*Ophiocoma wendtii*), a relative of the starfish. The crystals act as powerful microlenses that collect light and focus it on nerve bundles. The many crystallites thus form a compound eye. Calcite also forms egg shells.

The second important biomineral is *silica*. The algae-like radiolaria produce intricate skeletons of spheroidal shape (Figure 25.6a), which remind us of some crystal structures such as fullerene (Figure 22.4). They have been abundant in oceans since the Cambrian. Originally the skeletons are amorphous silica, similar to opal. Dead skeletons then accumulate at the seafloor, later transform to tridymite/cristobalite, and then to quartz. Other organisms with silica cell walls are diatoms, another, more recent group of algae (Figure 25.6b). Without radiolaria and diatoms

there would be no siliceous sediments, such as cherts, which later transform into quartzites in metamorphic environments.

Iron-hydroxides such as *goethite* also form shells of diatoms and pediculates of protozoa, contributing the red color of radiolarian chert. There is an interesting interaction between some bacteria and goethite at the nanocrystalline level. The association affects the charge and the stability of both components. Natural organic matter with bacterial activity is more resistant to degradation when adsorbed to iron hydroxides and microbial exopolymer production actually deposits iron oxides. This interaction at the nanometer scale has been studied both in natural environments and in the laboratory (e.g., Chan *et al.*, 2009).

Returning to phosphates, the bones and teeth of vertebrates consist of fine fibers or platy crystals of *hydroxyapatite* closely related to the mineral apatite with an idealized formula $Ca_5(PO_4)_3(OH)$. In reality the composition is more complex, with various substitutions, and we can only give a bulk formula such as $(Ca_5(PO_4)_2(OH,F,Cl,CO_3))$. Carbonate-hydroxyapatite crystallites in bone are suspended in organic collagen, a

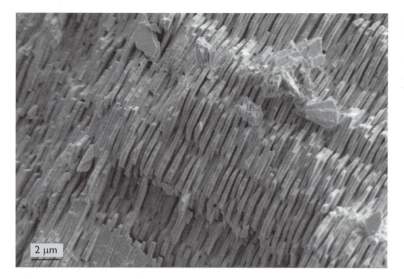

Fig. 25.5 Scanning electron microscope image illustrating stacking of platelet-shaped crystals of aragonite in the nacre (mother-of-pearl) shell of abalone (courtesy T. Teague).

(a)   (b)

Fig. 25.6 SEM images of skeletons of (a) an actinommid radiolarian from the Miocene, diameter is 55 μm (courtesy K. Finger, UCMP), (b) the diatom *Paralia sulcate*, length is ~50 μm (courtesy G. Hallegraeff, IMAS, University of Tasmania).

protein. The apatite crystals, which often do not exceed 10 nm in length, form up to 70 weight% of the dried bone, protein making up the remaining 30%. Microstructures have been investigated in great detail in tooth enamel (e.g., Robinson *et al.*, 1995). Figure 25.7 is an SEM image of strongly aligned apatite crystallites in mouse tooth enamel, with bundles of crystallites interwoven to produce a strong structure (e.g., Moradian-Oldak, 2012). With the TEM, individual crystals 50–200 nm in length can be resolved, documenting that crystallite size increases with age (Figure 25.8). Very recently it

has been discovered that teeth of limpets, a composite of goethite nanofibers (Fe) and proteins, have almost the tensile strength of the strongest artificial fibers (Barber *et al.*, 2015).

Sulfates *barite* and *gypsum* form ear stones of mammals, the phosphate *struvite* ($NH_4MgPO_4\ 6H_2O$) is present in kidney and gall stones (Chapter 36).

Even a mineral such as *magnetite*, which is generally considered to be representative of high-temperature rock-forming conditions, has been found as small grains in the tissues of salmon, bees, butterflies,

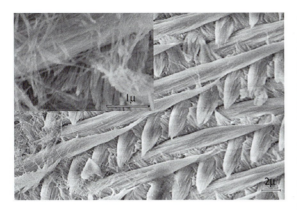

Fig. 25.7 SEM image of etched surface of mouse incisor tooth enamel showing the interwoven arrangement of bundles of carbonated apatite crystallites (courtesy of J. Moradian-Oldak).

tortoises, pigs, and birds. Carrier pigeons rely on magnetite magnetism to orient themselves during cloudy weather when they cannot use the Sun or stars. Magnetite serves as an amazing biomineralogical backup navigation system. Magnetite crystals also occur in primitive bacteria such as *Magnetospirillum magnetotacticum* as minute crystals often aligned in chains (Figure 25.9). Each crystal is a single magnetic domain (see Chapter 12) and thus constitutes a dipole, enabling the bacteria to align in magnetic fields (Stolz, 1992). They occur near the oxic/anoxic transition zone. Above it iron is ferric and insoluble, below it is mostly ferrous sulfide. However, within the transition zone there is soluble ferrous iron, and bacteria may find this optimal zone more efficiently by using the inclination of the Earth's magnetic field to point them in a downward direction. Once the cells die, magnetite inclusions are deposited in sediments and become stable carriers of remnant magnetization.

In all, about 80 different minerals occur within fossil and recent animals and plants.

Other biogenic processes involve bacteria. Large deposits of native sulfur, manganese oxides, and hydroxides, and iron ores have been attributed to bacterial activity. Bacteria are also involved in weathering processes with sulfide oxidation and transformation of kaolinite into bauxites. New methods of electron microscopy reveal the importance of bacteria in the formation of sedimentary rocks and ores as far back as the Precambrian era. Fossilized cyanobacteria have been identified in Precambrian rocks composed of jasper (Kursk group, Russia), and in Cambrian

(a)

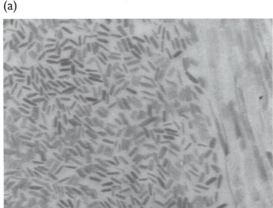

(b)

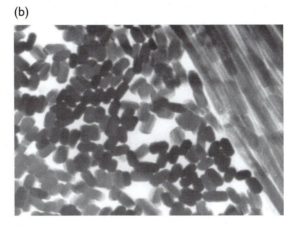

Fig. 25.8 TEM images of needle-shaped hydroxyapatite crystals in enamel of rat incisors shown in two layers. On the left side crystals are viewed edge-on. On the right side they are viewed longitudinally. (a) Small crystals in young animal, (b) larger crystals in older animal. Width 2 μm. (Courtesy C. Robinson, Dental Institute, University of Leeds.)

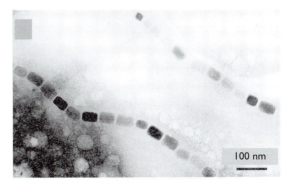

Fig. 25.9 TEM image of chains of single domain magnetite crystals in *Magnetospirillum magnetotacticum* bacteria (from Devouard *et al.*, 1998).

limestones (UK). Cyanobacteria and purple bacteria are known in Cambrian phosphorites (Mongolia).

Biomineralogy is a new but rapidly evolving field. We will discuss some additional aspects of this topic in Chapter 36. For more information on phosphates and sulfates, and biomineralogical processes see Further reading.

25.6 Summary

Phosphates and sulfates form structures with PO_4^{3-} and SO_4^{2-} tetrahedra and rather complex structures. Arsenates (with AsO_4^{3-}), tungstates (with WO_4^{2-}), and molybdates (with MnO_4^{2-}) are related. There is a wide variety of minerals and compositions. Some occur at high temperature in alkaline igneous and metamorphic rocks (e.g., apatite, monazite, xenotime, scheelite, and wolframite), others are mainly evaporite minerals (gypsum and anhydrite). Apatite also forms vertebrate bones and teeth and is used as an example to explain biogenic mineral processes.

Important phosphate and sulfate minerals to remember

Name	Formula	System
Phosphates		
Apatite	$Ca_5(PO_4)_3(F,OH,Cl)$	Hexagonal
Monazite	$Ce(PO_4)$	Monoclinic
Xenotime	$Y(PO_4)$	Tetragonal
Turquoise	$CaAl_6(PO_4)_4(OH)_8 \cdot 4H_2O$	Triclinic
Sulfates		
Barite	$BaSO_4$	Orthorhombic
Anhydrite	$CaSO_4$	Orthorhombic
Gypsum	$Ca(SO_4) \cdot 2H_2O$	Monoclinic

Test your knowledge

1. Compare the main building unit of carbonates and borates with that of phosphates and sulfates.
2. Review the processes of biomineralogy and give examples of minerals in vertebrates, bees, radiolaria, and foraminifera.
3. What are the major uses of gypsum and of apatite?

Further reading

Phosphates and sulfates

Alpers, C. N., Jambor, J. L. and Nordstrom, D. K. (eds.) (2000). *Sulfate Minerals: Crystallography, Geochemistry, and Environmental Significance.* Reviews in Mineralogy, vol. 40. Mineralogical Society of America, Washington, DC.

Chang, L. L. Y., Howie, R. A. and Zussman, J. (1996). *Rock-forming Minerals, Volume 5B, Non-silicates: Sulphates, Carbonates, Phosphates, Halides.* The Geological Society of London.

Nriagu, J. O. and Moore, P. B. (eds.) (1984, 2011). *Phosphate Minerals.* Springer-Verlag, Berlin.

Biominerals

Banfield, J. F. and Nealson, K. H. (eds.) (1997). *Geomicrobiology: Interactions between Microbes and Minerals.* Reviews in Mineralogy, vol. 35. Mineralogical Society of America, Washington, DC.

Driessens, F. C. M. and Verbeek, R. M. H. (eds.) (1990). *Biominerals.* CRC Press, Boca Raton, FL.

Lowenstam, H. A. and Weiner, S. (1989). *On Biomineralization.* Oxford University Press, Oxford.

26 | Sulfides and related minerals. Hydrothermal processes

About 500 minerals belong to the sulfides and related arsenides. Among the most important sulfides are pyrite (FeS_2), chalcopyrite ($CuFeS_2$), and sphalerite (ZnS). Sulfides are of great industrial importance and are the major ores for copper, zinc, lead, mercury, bismuth, cobalt, nickel, and other nonferrous metals. Although pyrite is not used as an ore for iron, it is used to produce sulfuric acid and it is also an important gold ore, containing small fragments of native gold as inclusions. Many sulfides form by hydrothermal processes and some examples will be discussed.

26.1 Crystal chemistry

Sulfides are generally subdivided into three chemical classes: (a) simple sulfides that are salts of HS (e.g., sphalerite, ZnS); (b) salts of thioacids, which are oxygen-free acids with sulfur playing the role of oxygen (e.g., pyrargyrite, Ag_3SbS_3, is the silver salt of the sulfoacid H_3SbS_3); and (c) polysulfuric compounds (persulfides) that can be considered as salts of the polysulfuric acid H_2S_2, which contains the bivalent S_2^{2-} group (pyrite FeS_2 is an example). The closest analogs to sulfides are arsenides and their complex compounds (arsenide–sulfides) such as realgar (AsS) and arsenopyrite (FeAsS).

The structural properties of sulfides are determined by bonding between a metal and sulfur, which is highly ionizing. The S^{2-} ionic radius is large (1.84 Å) compared with that of metals, and most sulfide structures do not correspond to simple close-packing of anions with cations in interstices (sphalerite (ZnS), galena (PbS), cinnabar (HgS), pyrrhotite (FeS), and chalcopyrite ($CuFeS_2$) are exceptions), though close-packing is never ideal. The large sulfur ions easily become polarized, and complex anion groups (S_2^{2-}) may form due to covalent pairing. In general, sulfides display a great diversity of crystal structures with complex bonding (ionic–metallic–covalent) in which metallic bonding always plays a considerable role. Among the numerous structures are polyhedral types, types with isolated molecular groups (S_2), and types in which sulfur is linked to bands and sheets (Table 26.1).

A typical example of a *polyhedral sulfide* is galena, with the same structure as halite (NaCl), in which both cations and anions have a coordination number of 6. The structure can be viewed as close-packing of S^{2-} with Pb^{2+} in octahedral interstices. Polyhedral structures are also found in sphalerite, wurtzite, pyrrhotite, nickeline, and chalcopyrite. In the conventional unit cell of sphalerite small Zn^{2+} are arranged on an fcc lattice and large S^{2-} occupy alternate eighth cubes (Figure 26.1a). However, the origin of the unit cell can be shifted to put S^{2-} into the corners (Figure 26.1b), and in this case it becomes obvious that Zn^{2+} are in tetrahedral coordination (Figure 26.1c). Chalcopyrite is structurally closely related to sphalerite but Cu^{2+} and Fe^{2+} atoms are ordered in layers over the positions of the fcc lattice, basically stacking a CuS cube over an FeS cube and this destroys the cubic symmetry and produces a tetragonal structure. Polyhedral sulfides generally have a cubic (or pseudocubic) morphology, with cubes, octahedra, or tetrahedra as dominant forms.

Group sulfides have S_2^{2-} groups, rather than isolated S^{2-} atoms. An example is pyrite. This structure also has a resemblance to NaCl, if we consider that Cl^- is

Table 26.1 Sulfide minerals with some diagnostic properties; important minerals are given in italics

Mineral Formula	System	Morphology Cleavage	H	D	Color/luster Streak	n	Δ	2V (Dispersion)
Polyhedral sulfides								
Bornite Cu_5FeS_4	Tetrag. (Ps. Cubic)	Cub., Tet.	3	5.1	Bronze, blue (oxidized)/metallic; *Gray-blue*	opaque		
Chalcocite Cu_2S	Ortho.	Tab. (001) {110}	2.5–3	5.7	Gray-blue/metallic; *Gray-black*	opaque		
Chalcopyrite $CuFeS_2$	Tetrag.	{111}	3.5–4	4.2	Brass-yellow/metallic; *Green-black*	opaque		
Cinnabar HgS	Trig. (Ps. Cubic)	Eq. Cub. {$10\overline{1}0$}	2–2.5	8.1	Red; *Scarlet*	2.91–3.27	0.359	(+)
Galena PbS	Cubic	Cub. {100}	2.5	7.4	Gray/metallic; *Gray-black*	opaque		
Nickeline (Niccolite) NiS	Hexag.	Xls. rare	5.5	7.5	Green-red/metallic; *Brown-black*	opaque		
Pentlandite $Fe4Ni4CoS_8$	Cubic	Xls. Rare {111}	3–4	4.8	Bronze-brown/metallic; *Light bronze*	opaque		
Pyrrhotite $Fe_{(1-x)}S$	Hexag. (Monocl.)	Tab. (0001) (0001)	4	4.6	Bronze/metallic; *Gray-black*	opaque		
Sphalerite α-ZnS	Cubic	Tet. {110}	3–4	4.1	Yellow, brown, black/metallic; *Brown,yellow*	2.37		
Tetrahedrite $Cu_{12}(SbS_3)_4S$	Cubic	Cub.	3–4	4.9	Gray, silver/metallic; *Red-brown, black*	opaque		
Wurtzite β-ZnS	Hexag.	Pris., Tab. (0001), {$10\overline{1}0$}	3.5–1	4.1	Brawn, orange-yellow/(metallic); *Yellow, brown*	2.36–2.38	0.022	(+)

Group sulfides

	System	Morphology	H	D	Color	n	Δ	2V
Arsenopyrite FeAsS	Monocl.	Pris. or Eq. {101}	5.5–6	6.0	Gray-silver/metallic *Black*	opaque		
Cobaltite CoAsS	Ortho. (Ps. Cubic)	Eq. (100)	5.5	6.2	White, pink/metallic *Gray-black*	opaque		
Marcasite FeS_2	Ortho.	Tab. (001) {101}	6–6.5	4.8	Yellow/metallic *Green-black*	opaque		
Pyrite FeS_2	Cubic	{100} {210}	6–6.5	5.1	Yellow/metallic *Green-black*	opaque		
Realgar As_4S_4	Monocl.	Pris. [001] (010)	1.5–2	3.5	Red, orange *Orange*	2.46–2.6	0.015	–40 r ≫ v

Band sulfides

	System	Morphology	H	D	Color	n	Δ	2V
Stibnite Sb_2S_3	Ortho.	Pris. [001] (010)	2–2.5	4.6	Lead-gray, black/metallic *Lead-gray*	opaque		

Sheet sulfides

	System	Morphology	H	D	Color	n	Δ	2V
Molybdenite MoS_2	Hexag.	Platy (0001) (0001)		4.7	Silver/metallic *Gray-green*	opaque		
Orpiment As_2S_3	Monocl.	Platy (010) (010)	1.5–2	3.49	Yellow *Yellow*	2.4–3.0	0.6	–76

Notes: H, hardness; D, density (g/cm^3); *n*, range of refractive indices; Δ, birefringence; 2*V*, axial angle for biaxial minerals. For uniaxial minerals (+) is positive and (–) is negative. Acute 2*V* is given in the table. If 2*V* is negative the mineral is biaxial negative and 2*V* is 2V_α; if it is positive, the mineral is biaxial positive and 2*V* is 2V_γ. Dispersion r < v means that acute 2*V* is larger for violet than for red; anom., anomalous dispersion or birefringence. AP, axial plane.

System: Hexag., hexagonal; Monocl., monoclinic; Ortho., orthorhombic; Ps., pseudo; Trig., trigonal; Tetrag., tetragonal.

Morphology: Cub., cubic; Eq., equiaxed; Pris., prismatic; Tab., tabular; Tet., tetrahedron.

Color: Metallic luster is also included; parentheses indicate submetallic or occasionally metallic.

(a)

(b)

(c)

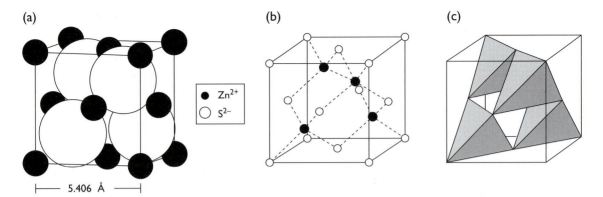

├─── 5.406 Å ───┤

Fig. 26.1 Sphalerite (ZnS) as an example of a polyhedral sulfide with tetrahedral coordination. (a) Ions are shown according to relative size. In the conventional unit cell, Zn^{2+} are distributed over an fcc lattice. (b) The origin is shifted so that the S^{2-} are in the corners, thus better displaying the tetrahedral coordination of Zn^{2+}. Ions are represented as small spheres. (c) The polyhedral representation of the sphalerite structure with four corner-sharing tetrahedra.

replaced by S_2^{2-} groups directed along body-diagonals of the cube (Figure 26.2). Fe^{2+} occupies an fcc lattice, with each ion coordinated by six sulfur atoms in a tilted octahedral coordination. The oblique sulfur groups reduce the high symmetry in NaCl and PbS. Even though pyrite crystallizes frequently as cubes, a surface striation indicates that a 4-fold rotation axis does not exist (cubic point-group $2/m\bar{3}$; see Figure 10.4a and Plate 3c).

While in pyrite the group is S_2^{2-} (Figure 26.3b), other groups are found, such as AsS^{3-} (e.g., cobaltite, Figure 26.3c) and AS_4^{4-} (e.g., skutterudite, Figure 26.3d). In these complexes, atoms share outer electrons to attain an inert gas configuration, acquiring additional electrons from metal cations. Realgar (As_4S_4) is a special case of a purely covalent group sulfide. In realgar, atoms are linked by covalent bonds to form two rings, and this molecular structure is charge-balanced without additional metal ions (Figure 26.3a). Van der Waals bonds connect the groups, which is the reason for the low hardness (Mohs scale 1.5–2) and low melting point (310 °C) of realgar.

Band-like structures are typical for orthorhombic stibnite (Sb_2S_3) (Figure 26.4) and bismuthinite (Bi_2S_3). The "corrugated" bands parallel to the crystallographic axis [100] are infinite and are held together by Sb–S bonds between the bands. The length of bonds that link bands is 3.1–3.6 Å as compared to 2.5–2.8 Å for Sb–S bonds within bands. The chain character is expressed in the prismatic and acicular morphology along the Sb_2S_3 ribbons and in a perfect (010) cleavage that is parallel to the bands (see Plate 27a).

Molybdenite (MoS_2) is an example of a *sheet structure* with a Mo layer sandwiched between two layers of S atoms (Figure 26.5). Within the Mo layer there is good metallic bonding, with high conductivity. The S–Mo–S sheets are linked by weak van der Waals bonds, resulting in an excellent cleavage and low hardness of this mineral, similar to graphite (see Plate 27b).

Many sulfides are reminiscent of metals with a pronounced metallic luster and variously shaded metallic colors (mostly gray-black and yellow, and more rarely red and blue), resembling iron, aluminum,

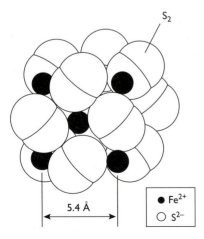

S_2

├─ 5.4 Å ─┤

● Fe^{2+}
○ S^{2-}

Fig. 26.2 Pyrite is an example of a sulfide with S_2 groups in its structure. Iron ions in pyrite are arranged as in an fcc lattice. Each Fe is coordinated octahedrally by six S_2 groups.

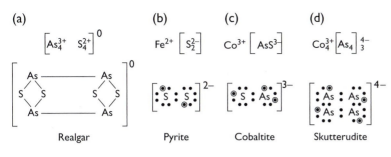

Fig. 26.3 Schematic representation of some group sulfides. Formula is on top and electron distribution at bottom. Circles indicate shared electrons. (a) Realgar, (b) pyrite, (c) cobaltite, and (d) skutterudite.

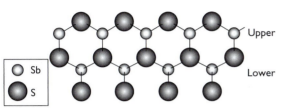

Fig. 26.4 Structure of stibnite, an example of a band-like sulfide.

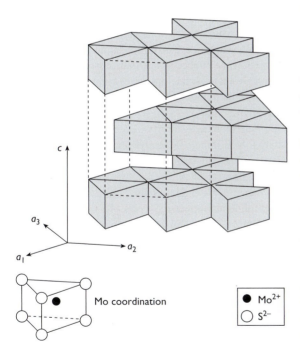

Fig. 26.5 Structure of molybdenite, showing sulfur linked in a sheet.

or brass. Sulfides are electrical conductors due to the prevailing metallic bonding, which is also responsible for the luster. In fact, sulfides are potentially both ionic and electrical conductors. Correspondingly, adjacent grains of two different sulfides that occur in the same ore form a galvanic micro-cell with an electric potential when in contact with water. Electrochemical reactions take place, with ions partially migrating into a solution and creating a dispersion of metals around sulfide ore deposits that can be used for prospecting. We will discuss this at the end of this chapter.

26.2 Brief description of important sulfide minerals

Minerals in this section are listed in alphabetical order.

Arsenopyrite (Fe(AsS)) contains minor substitutions of nickel, tin, and cobalt for iron. Its symmetry is orthorhombic. Elongated or wedge-shaped crystals with a rough striation are typical. The mineral is steel-gray or slightly bronze in color, has a striking metallic luster, a hardness of 6, and sometimes displays a cleavage. Arsenopyrite produces a "garlic" smell when hammered. It occurs in high- and medium-temperature deposits, and in the latter is often accompanied by native gold. It is used for arsenic extraction.

Bornite (Cu_5FeS_4) occurs as irregularly shaped pockets and veinlets in copper sandstones and as thin secondary coatings on chalcopyrite. In fresh pieces the mineral has a characteristic red to pink color, but it oxidizes rapidly and is often covered with brown, blue, or green films of secondary minerals such as covellite. The mineral forms in low-temperature hydrothermal deposits, in copper sandstones, and in the zone of secondary sulfide enrichment within oxidized copper sulfide ores. It is used as a copper ore.

Chalcocite (Cu_2S) occurs as solid masses, nodules, and veinlets within copper sandstone. It has a grayish-black color, submetallic luster, and a rough fracture surface. The mineral is usually covered with a thin coating of brightly colored malachite or azurite. It is recognized by its color, by the forms of its occurrence, and by the colored coatings. A shiny scratch distinguishes it from minerals of the tetrahedrite group. Chalcocite is a low-temperature hydrothermal mineral and also occurs in zones of ore oxidation with secondary sulfide enrichment.

Chalcopyrite ($CuFeS_2$) commonly occurs in granular aggregates, veinlets, and nodules. Euhedral pseudocubic crystals are rare. The mineral has a greenish-yellow ("brass yellow") color with a specific tint that is difficult to describe and needs to be experienced. This tint distinguishes chalcopyrite (Plate 26c) from the similar pyrite (Plate 25d). Chalcopyrite has uneven fracture surfaces, metallic luster, and a greenish-black streak. It is easily oxidized and is covered with iridescent blue, green, and red films of bornite and covellite (Plate 26d). Chalcopyrite grains are often partially or completely replaced by iron hydroxides, malachite, and azurite on the surface and along fractures. Chalcopyrite is an important copper ore. It is common and is found in industrially valuable concentrations in magmatic sulfide ores, medium-temperature hydrothermal deposits, copper sandstones, and skarns.

Cinnabar (HgS) is found as crimson-red and carmine-red grains in porous sandstones, graywackes, and marbles. It may also occur in stibnite- and fluorite-bearing quartz veins. The low-temperature polymorph has hexagonal symmetry, high-temperature metacinnabar is cubic. Large cinnabar grains display rhombohedral morphology, often with interpenetration twins and pseudocubic cleavage (see Figure 9.8b, Plate 1b). The luster of individual grains is adamantine, whereas that of aggregates is greasy or matte. Cinnabar is recognized by its color, red streak, high density (8.1 g/cm^3), and the association with calcite, fluorite, stibnite, and quartz. Cinnabar forms in low-temperature hydrothermal deposits and is the main ore for mercury.

Cobaltite (Co(AsS)) has crystals similar to those of pyrite and also forms granular coatings in ores, but, unlike pyrite, it is pinkish-gray in color and displays a cleavage. Cobaltite occurs mainly in skarns and other high-temperature hydrothermal rocks. In oxidation zones, cobaltite is gradually replaced by pink erythrite ($Co_3(AsO_4)_2·8H_2O$) powder. It is a cobalt ore.

Galena (PbS) has a cubic structure and often a cubic or cube-octahedral morphology and a perfect cubic cleavage. There are limited isomorphic substitutions of silver, tin, thallium, selenium, and tellurium. If you have ever seen the almost perfect galena cubes or octahedra in small druses, you will never forget them (see Figure 8.1e and Plate 26b). Galena often occurs with quartz, calcite, sphalerite, and chalcopyrite. Its color is lead-gray, and its luster varies from metallic to faded metallic. It has a low hardness and a high density (7.4 g/cm^3). Galena forms in hydrothermal deposits, skarns, and stratiform deposits. The mineral is a major lead ore but silver, bismuth, and thallium are also extracted from galena.

Marcasite (FeS_2) is orthorhombic and forms spear-shaped crystals with rhombic bipyramids that are frequently twinned. It occurs as radiating aggregates and concretions (Plate 25e) and is often intergrown with pyrite. The mineral has a pale-yellow color, slightly darker than that of pyrite, and has a metallic luster. Its hardness is 6. Marcasite is found in sedimentary rocks and more rarely as a late-stage phase in some hydrothermal deposits. Like pyrite, it is used to manufacture sulfuric acid.

Molybdenite (MoS_2) crystallizes in the hexagonal system as platy crystals with gray color of slightly bluish tint, bright metallic or greasy luster, and perfect (0001) cleavage (Plate 27b). The mineral is easily identified by its low hardness (1) and greasy touch. Graphite, which it resembles, has a darker color and a more intense streak. The molybdenite streak becomes greenish when rubbed. Rose-shaped, scaly, or massive aggregates of molybdenite are common. Molybdenite occurs in different high-temperature deposits. Commercially important concentrations are associated with high-temperature hydrothermal deposits and skarns. It is mined for molybdenum sulfide (a lubricant) and molybdenum.

Nickeline (NiS) occurs in hydrothermal uranium-bearing deposits as fine-grained kidney-shaped aggregates of light copper-red color. In oxidation zones nickeline is often replaced by powdery annabergite masses ($Ni_3(AsO_4)_2·8H_2O$), pale green or green in color.

Orpiment (As_2S_3) occurs as elongated platy crystals and sunflower-like radiating aggregates, or in the form of earthy masses. The mineral has a bright golden-yellow color (Plate 26f) and a perfect tabular cleavage, with a pearly luster on cleavage surfaces. Orpiment forms in low-temperature hydrothermal deposits together with realgar, quartz, and calcite and is, like realgar, an arsenic ore.

Pentlandite approaches the formula $Fe_4Ni_4(Co,Ni, Fe,Ag)S_8$. The identification of this mineral is often difficult. It occurs in fine-grained chalcopyrite–pyrrhotite aggregates in ultramafic and mafic rocks. It resembles chalcopyrite and pyrrhotite in color but the strong metallic luster and perfect octahedral cleavage

(a)   (b)

Fig. 26.6 Morphology of sulfide minerals. (a) Tabular pyrrhotite from Trebca, Kosovo (courtesy O. Medenbach). Width 60 mm. (b) Tetrahedral crystal of sphalerite from Dzeskazgan, Kazakhstan (courtesy L. Arnoth).

distinguish it. Pentlandite almost never occurs with pyrite. The mineral forms in magmatic copper–nickel sulfide ores and is the principal ore of nickel.

Pyrite (FeS_2) may contain some cobalt, nickel, arsenic, copper, and antimony. The mineral often occurs as cubes {100} (see Plate 3c) and dodecahedra {210} with pentagons as faces (so-called *pyritohedra*, Plate 3d) or combinations of the two forms (Plate 25d). The faces display characteristic striations that are due to a microscopic alternation of {100} and {210} growth. Pyrite has a straw yellow or golden-yellow color (hence the name "fool's gold"), strong metallic luster, and a high hardness (6.5, higher than glass). These properties aid in its identification and distinguish it from the similar chalcopyrite. When hammered, pyrite produces a sulfurous smell. During oxidation, iron hydroxide minerals called "limonite" replace pyrite. Pseudomorphs after pyrite, which maintain details of the pyrite morphology, are common (Plate 6d). Pyrite is the most abundant sulfide mineral and is present in a majority of rocks and deposits. It is used mainly in the production of sulfuric acid. It is an undesirable component of iron ores. Sometimes pyrite is used as a gold ore when it contains fine dispersions of gold.

Pyrrhotite has a variable composition approaching FeS, but generally is found with some iron deficiency due to structural defects. Chemical formulas range from $Fe_{0.83}S$ to FeS. Pyrrhotite occurs in solid fine-grained masses with a dull metal-like luster and uneven fracture surfaces. Crystals are rare and have a tabular pseudohexagonal habit (Figure 26.6a). Pyrrhotite is easily weathered and becomes covered with

a brown film that hides its natural color. Some pyrrhotites have weak magnetic properties. A diagnostic property of pyrrhotite is the bronze color on fresh surfaces. Beginners often mistake pyrrhotite for pyrite, but the latter generally occurs as good crystals and has a more golden-yellowish color. Pyrrhotite forms at a wide range of conditions and is therefore very widespread in nature, although easily overlooked. The mineral has no practical application; on the contrary, it is an undesirable component for many metallurgical technologies.

Realgar (As_4S_4) forms bright reddish-orange prismatic crystals with an adamantine luster (Plate 26e,f). It also occurs as granular masses and veinlets in calcite-bearing rocks. The mineral is brittle and has uneven and rough fracture surfaces. It occurs in low-temperature hydrothermal deposits and is accompanied by yellow orpiment (As_2S_3) within quartz and calcite veins and as veinlets in shales. The association with yellow orpiment, to which it alters under the influence of light, and the orange color and streak distinguish realgar from cinnabar. Realgar is mined for arsenic.

Sphalerite ((Zn,Fe)S) may contain minor amounts of manganese, cadmium, mercury, and tin. The mineral is cubic and forms tetrahedral crystals, often euhedral (Figure 26.6b, Plate 25f). The equiaxed grains have a submetallic luster on cleavage surfaces. Since the cleavage is dodecahedral {110} with six nonparallel planes, reflections occur in many directions as the sample is rotated. Many properties depend on the isomorphic substitutions: iron-free varieties are transparent, colorless, greenish, or honey-yellow and have a striking adamantine luster; samples that are low in iron but contain manganese are reddish-brown in color; and iron-rich sphalerites are brownish-black and black with a semimetallic luster and weaker cleavage. Powders and streaks of all varieties are pale to dark brown. The mineral is easily recognized on the basis of color, luster, and cleavage, as well as its association with galena, quartz, and calcite. Sphalerite generally occurs in hydrothermal deposits, skarns, and stratiform ores. It is mined for zinc and cadmium.

Stibnite (Sb_2S_3) crystallizes in the orthorhombic system as columnar or needle-shaped crystals with a striation parallel to the elongation (Plate 27a). The color is dark gray, but tarnished crystal faces are bright indigo-blue. The luster is metallic. Stibnite cleaves parallel to the elongation, and cleavage planes often display a transverse striation. A characteristic property of stibnite is that it burns when struck against a hard surface. (Stibnite powder is a component of the tips of modern matches.) The mineral forms in hydrothermal deposits in association with cinnabar, quartz, and fluorite. Stibnite is an ore of antimony.

Tetrahedrite ($Cu_{12}(SbS_3)_4S$) and **tennantite** ($Cu_{12}(AsS_3)_4S$) form a continuous solid solution. Silver, zinc, and iron may substitute for copper, and bismuth and tellurium for arsenic and antimony. The most abundant minerals are copper–arsenic and copper–antimony compounds, which are similar in color and morphology. Crystals are rare (Plate 26a), and irregular aggregates are more common. The minerals have a gray-black color and a dull metallic luster. Their identification in hand specimens is often difficult because the minerals lack any outstanding properties. The faded luster, irregular grain shape, and association with pyrite, sphalerite, and quartz are characteristic. They resemble chalcocite but are more brittle, and a scratch with a sharp needle has a dull rather than shiny appearance. The tetrahedrite minerals are constituents of medium-temperature (sometimes low-temperature) hydrothermal deposits and, in particular, of gold deposits.

26.3 Sulfide genesis and hydrothermal deposits

Only rarely do sulfides form by primary magmatic processes. In nature, sulfide minerals crystallize predominantly from aqueous solutions at temperatures below 600 °C.

Magmatic ore-forming processes

Let us first look briefly at magmatic metal ore-forming processes. When magma crystallizes during cooling, different minerals form at different temperatures. Some crystallization processes in simple melts have been discussed in Chapters 19–21. Early crystallizing minerals are often heavier than the melt and sink to the bottom of the magma chamber, resulting in *crystal fractionation* and the formation of cumulate layers (Figure 26.7, bottom). Layers of this type are particularly important in mafic and ultramafic intrusions, such as the Bushveld complex in South Africa, with layers of chromite and magnetite containing platinum and vanadium minerals (see also Chapter 33).

Some magmas dissociate in the course of their crystallization history into two *immiscible liquids* of different composition. One magma is a typical silicate magma, while the other is rich in metal sulfides or oxides (Figure 26.7, top). The immiscible sulfide melt is heavier than the silicate melt and sinks to the bottom of the chamber with droplets and pockets coalescing to form large deposits of nickel-rich sulfides. Crystallization of the dominant silicate melt produces dunites, peridotites, and gabbros. Crystallization of the sulfide melt produces segregations, dispersed in the already

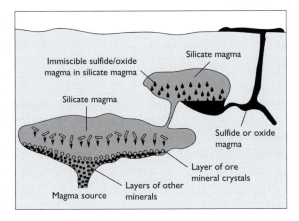

Fig. 26.7 Idealistic sketch of magmatic ore-forming processes. At the bottom of the figure is a silicate magma chamber, where a component has crystallized and subsequently sunk to the bottom. In regions such as in the Bushveld complex (South Africa), the result can be accumulating layers of ore minerals such as chromite. At the top of the figure is a magma chamber with two immiscible liquids. The sulfide liquid forms droplets in the mafic silicate magma and accumulates at the bottom. An example is the nickel–copper mineralization at Sudbury, Canada (after Kesler, 1994).

existing silicate rock, with such minerals as pentlandite, chalcopyrite, and pyrrhotite. Tectonic movements may cause the sulfide melt to intrude along fractures. An unusual example of such a sulfide deposit is the nickel–copper mineralization at Sudbury (Canada), where a large meteorite impact probably caused formation and emplacement of a mafic magma.

Hydrothermal ore-forming processes

Hydrothermal sulfide deposits are more common than magmatic deposits and form because of precipitation of minerals from largely aqueous or carbonaceous–aqueous supercritical fluids and liquid solutions. The depths of their occurrence range from 5 km to surface conditions. The solutions can be magmatic water released during crystallization of a silicate melt, metamorphic water that becomes free during mineral reactions of progressive metamorphism, sedimentary water that is expelled during compaction, atmospheric water penetrating rocks (meteoric water), or seawater. The different types of hydrothermal solutions can be distinguished on the basis of their isotopic composition (see Figure 18.1). In order for the solutions to become mobile, there must be rock porosity or fracturing. Hydrothermal solutions are often saline and very reactive, causing alteration of the country rock adjacent to the fracture.

Take, as a typical example, a fracture in granite (Figure 26.8, from an old textbook in ore mineralogy). As a result of chemical interactions between the feldspars of granites and the ore-transporting aqueous solutions that pass through them, the granite becomes altered and the solution changes composition. Some of the reactions that take place during the alteration of various feldspars are

$3NaAlSi_3O_8 + 2H^+ + K^+ \rightarrow$
Albite
$\qquad KAl_2Si_3AlO_{10}(OH)_2 + 3Na^+ + 6SiO_2$
$\qquad$ Muscovite $\qquad\qquad$ Quartz

$3KAlSi_3O_8 + 2H^+ \rightarrow$
Microcline
$\qquad KAl_2Si_3AlO_{10}(OH)_2 + K^+ + 6SiO_2$
$\qquad$ Muscovite $\qquad\qquad$ Quartz

$CaAl_2Si_2O_8 + 4F^- + 4H^+ \rightarrow$
Anorthite
$\qquad Al_2SiO_4F_2 + SiO_2 + CaF_2 + 2H_2O$
$\qquad$ Topaz $\qquad$ Quartz Fluorite

Muscovite is often in a fine-grained form called *sericite*, and associated with minerals such as topaz and fluorite. The altered granitic rocks are known as *greisen* (from the German *Gries*, meaning "grit"). The term was first applied by miners in Saxony in the seventeenth century to granular, altered quartz–muscovite rocks that contain cassiterite (SnO_2) and scheelite. Even then, it was noticed that the altered rock had formed by a metasomatic transformation of granite. Today it is clear that greisens accompany most high-temperature hydrothermal veins that cross granites, and they constitute considerable volumes in shallow granite intrusions. In the case of the vein from the Erzgebirge near Freiberg, Germany, shown in Figure 26.8, a layer of lepidolite (Li-mica) separates the quartz vein containing the ore minerals from the host rock. The greisens are significant sources of tin (cassiterite), tungsten (scheelite and wolframite), and, sometimes, gemstones (beryl, topaz, tourmaline).

While the host rock is altered and solutions circulate through open fractures, new minerals precipitate from the fracture walls of an earlier generation of minerals. This has been illustrated in the classic drawings of W. Maucher (1914), which have been reproduced in many textbooks. Figure 26.9 shows mineralization in a sericitic gneiss of the Erzgebirge. Crystallization of fine prismatic quartz on the fracture surface of the greisen host rock is followed by sphalerite (black), with some arsenopyrite. A younger generation of quartz with more massive crystals follows. The last minerals that

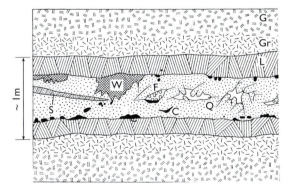

Fig. 26.8 Schematic illustration of a hydrothermal vein in granite (G) from the Erzgebirge, Germany, with a zone of altered sericite gneiss (greisen, Gr), lepidolite mineralization (a Li-mica; L), and a quartz vein (Q, dotted) containing cassiterite (C, black), scheelite (S), wolframite, and fluorite (after Beck, 1909).

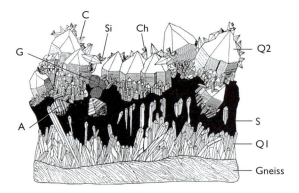

Fig. 26.9 Druse in sericitic gneiss from Freiberg in Saxony, Germany. Fine prismatic quartz (Q1) is overgrown by sphalerite (S), striated orthorhombic crystals of arsenopyrite (A), large stubby quartz (Q2), and galena (G). On top of the younger quartz are small crystals of calcite (C), siderite (Si), and chalcopyrite (Ch). (Drawing from W. Maucher, 1914.)

precipitate are scalenohedral calcite, saddle-shaped siderite, and chalcopyrite.

The crystallization sequence of hydrothermal solutions is very complicated, and depends on temperature, pressure, composition, pH, redox potential Eh, and concentration. All these factors can change easily, both locally and with time. They are influenced by the velocity of the hydrothermal solutions moving through the host, mixing of different solutions, temperature gradients, and reactions with the underlying rocks. It is, therefore, impossible to draw general conclusions. Nevertheless, hydrothermal alterations have been classified according to their relationship with igneous processes and by the temperature conditions under which they take place.

Hydrothermal ore deposits are divided into plutonic, volcanic, and so-called *telethermal* deposits, which are far removed from igneous activity. Depending on temperature and pressure, hydrothermal ores have characteristic chemical and mineralogical compositions and morphology (Table 26.2). We have introduced the temperature–pressure phase diagram of water in Figure 19.1, with emphasis on the structure of ice. Figure 26.10 expands this diagram for the liquid region, showing different isochors v_0, which are volumes (in cm^3) occupied by a gram of water (Smith, 1963). For example, the line 1.0 corresponds to "usual" water, while the line 0.9 corresponds to a more dense substance, and the line 3.17 corresponds to a less

condensed substance. In this diagram we can identify gaseous, liquid, and supercritical phases of water.

Let us first review some features of *plutonic hydrothermal deposits*. In Chapter 21 we explored the formation of granitic rocks and pegmatites. With increasing crystallization (and decreasing temperature) a granitic magma becomes enriched in water and elements that do not substitute in the structures of major rock-forming silicates. This includes many heavy-metal ions. Pegmatites and aplites form at the transition from late magmatic to hydrothermal crystallization. They have characteristics of dikes formed by igneous intrusions, as well as those of veins precipitating from aqueous solutions.

At high temperatures (400–600 °C), water is in a supercritical state (see Figure 19.1). Crystallization produces an increase in vapor pressure, causing shattering of rocks and formation of a highly permeable fracture zone, for example adjacent to a granitic intrusion. Hydrothermal veins range in thickness from 0.1 to 4 m and can often be traced over distances of more than 750 m. Veins often form complicated networks with various crystallization sequences. The siliceous supercritical water penetrates surrounding rocks and quartz, cassiterite (SnO_2), and scheelite ($CaWO_4$) precipitate. Such so-called catathermal deposits have long been mined for tin in the Erzgebirge (at Zinnwald, now Cinovec in the Czech Republic), and in Thailand, Malaysia, and Bolivia.

Skarns are special types of plutonic hydrothermal deposits that occur along contacts between granites and limestones or marbles. They form a polymineralic deposit of variable thickness (from 1–2 cm to hundreds of meters). The skarns are typically composed of calc-silicate minerals such as garnets, diopside, vesuvianite, epidote, diopside, and wollastonite. Skarns are metasomatic rocks that form at a depth of 3–7 km under the influence of hot supercritical solutions that are expelled from the magma. These solutions contain calcium and magnesium originating from the marbles, as well as silicon, aluminum, and sodium originating from the granites. Skarns are often important as a source of tungsten, tin, lead, zinc, copper, and beryllium ores. *Skarn* is an old Swedish miners' term meaning "waste rock" because miners disposed of the skarn rock, extracting only the metal ore.

Large hydrothermal systems are associated with shallow granitic intrusions, generally underlying stratovolcanoes (Figure 26.11). The granitic intrusion is

Table 26.2 Examples of ore-bearing hydrothermal deposits

Type of associated magmatism	Temperature conditions	Type of deposit	Nonmetallic minerals	Elements extracted	Occurrence	Example
Plutonic	Catathermal 300–500 °C	Porphyry copper	Quartz, muscovite	Cu, Mo, Au	Veins	Bingham, UT, USA, Butte, MT, USA
		Greisen	Quartz, muscovite	Sn, Bi, Ta, Nb	Veins	Harz, Erzgebirge, Germany and Czech Republic, Thailand, Bolivia
		Skarns (contact metamorphism)	Quartz, calcite	Bi, Zn, Pb W	Veins	Hidalgo, Mexico, Peru Hunan, China, Vostok, Russia
Volcanic	Mesothermal 150–300 °C	Massive sulfides (black smokers)	Calcite, barite, anhydrite	Cu	Veins	Cyprus, Kuroko, Japan
	Epithermal 50–200 °C	Antimony–mercury–arsenic	Quartz, calcite	Hg, As, Sb	Veins	Monte Amiata, Italy
Telethermal	Mesothermal 200–300 °C	Greenstones	Pyrite, quartz	Au	Veins	Mother Lode, CA, USA
	Epithermal 50–200 °C	Mississippi Valley		Pb, Zn, Ag	Strata-bound beds, etc.	Joplin, MO, USA
		Kupferschiefer (black shales)		Cu	Strata-bound beds	Germany, Michigan, USA, Zambia

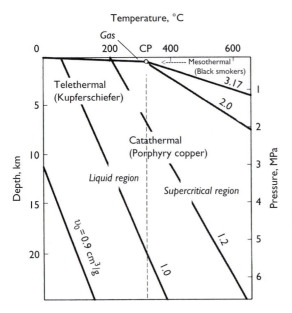

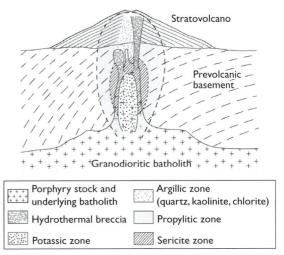

Fig. 26.11 Cross-section through a stratovolcano with a porphyry stock, and zones of alteration, associated with mineralization (after Sillitoe, 1973).

Fig. 26.10 Pressure temperature phase diagram for water with regions of hydrothermal, gas (pneumatolytic), and thermofluid (supercritical) systems for pure water. Isochore lines v_0 are explained in the text. CP, critical point (after Bulakh, 1977).

surrounded by an extensive zone of alteration and a system of veins that contain the metal ores. At depth there is a potassic alteration zone containing K-feldspar, quartz, and biotite, followed by a sericitic (quartz, sericite, pyrite) and argillic alteration (quartz, kaolinite, chlorite) at shallower levels. An alteration zone with chlorite, epidote, and carbonates extends all the way into the stratovolcano. These so-called "porphyry coppers" contain some of the world's largest copper and molybdenum deposits, often with gold as a byproduct. Porphyry copper deposits are generally along young convergent margins such as in North America (Butte, MT; Bingham, UT; Cananea, Mexico), Chile, New Guinea, and the Philippines (see also Figure 33.3a–c). There is a regular metal zonation in plutonic hydrothermal systems. For example, in the mineralization of Cornwall (southeast UK), a zone of tin is followed by one of copper, then lead–zinc, and finally iron, with increasing distance from the granite contact (Figure 26.12).

Volcanogenic massive sulfides are lenses of iron, copper, zinc, and lead sulfide minerals deposited as sediments where hydrothermal systems vented onto the seafloor as hot springs. The surface deposits are underlain by a system of feeder veins through which the solution reached the surface (Figure 26.13). The deposits are underlain by shallow volcanic rocks that provided the heat and to some extent the hydrothermal water. There is a systematic evolution of mineral deposition with increasing temperature. First "black ore" rich in sphalerite, galena, pyrite, and barite precipitates as relatively cool (200 °C) hydrothermal solutions mixed with cold seawater. Subsequently, hotter (300–350 °C) solutions replace the earlier deposited minerals with chalcopyrite in the lower part of the deposit ("yellow ore"). Still hotter, copper-undersaturated solutions dissolve some chalcopyrite to form pyrite-rich bases. During this whole process silica minerals and hematite precipitate at the peripheral parts of the hydrothermal system to form ferruginous chert. Ancient massive sulfide deposits have been mined for millennia in Cyprus (which derives its name from the Greek word for copper (*kypros*), and large deposits are also found in Kuroko (northern Japan).

Such mineral-forming processes are observed today along oceanic ridges. Pillars and cones, up to 50–100 m high, are composed of calcite, barite, anhydrite, and pyrrhotite. Black smoke emerges from the centers of the pillars (Figure 26.14), and hence they have been given the name *black smokers*. The black smoke contains microscopic grains of pyrrhotite and amorphous

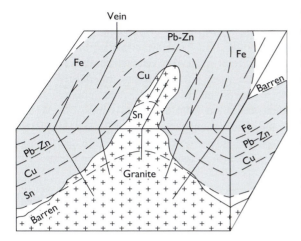

Fig. 26.12 Zoned mineralization around a granitic intrusion in Cornwall, UK, with zones of tin followed by copper, lead–zinc, and iron (after Hosking, 1951).

silica, mixed with zinc and copper sulfides. If the smoke is white, it contains particles of amorphous silica, anhydrite, and barite. The associated steam has a pH of about 4, a temperature of 300–400 °C, and is enriched in various metals such as copper, zinc and lead, and methane. Chimneys are usually covered with nodes of bacterial mats, worms (*Riftia pachyptila*, up to 1.5–2 cm in length), and large mollusks (*Calyptogena magnifica*, up to 25 cm in length). In addition, other worms such as *Alvinella pompejana* live on or near the pillars and cones. These and other organisms have adapted to exist in such unusual environments at high temperatures and in chemically active solutions, and, in fact, sulfides form in the skeletons of some of them. A cross-section of a

black smoker chimney from the Juan de Fuca Ridge in the northeast Pacific Ocean displays a core with bacterial mats and clays, an interior zone with pyrite, sphalerite, marcasite, and wurtzite, and exterior walls with barite, anhydrite, and amorphous silica (Plate 27c). Near black smokers, the sediments become saturated with metals and transform to sulfides during diagenesis. Very active modern processes of sulfide precipitation have been observed at the bottom of the Red Sea.

Hydrothermal-sedimentary deposits are generally telethermal, i.e., far removed from igneous activity. The deposits form simultaneously with sediment deposition and diagenesis and are generally of low temperature (<200 °C, epithermal). The composition of the pore water is changed by metal-rich solutions that penetrate into the sediments. As the sediments are buried they become compacted and recrystallize. Chemical reactions then cause precipitation of sulfide minerals. Depending on the composition of the hydrothermal solutions, stratiform deposits of copper, polymetallic lead–zinc–iron ores, and iron–manganese deposits may form. The hydrothermal solutions can have various origins, related to volcanic activity, plutonic intrusions, or metamorphic processes.

Stratiform deposits are the most common sources for copper, lead, and zinc. They occur in shales, sandstones, and carbonate rocks. Sulfides comprise the intergranular space of sedimentary and volcanoclastic rocks, forming layered lenses and beds of sulfides within those rocks. A characteristic feature of stratiform ores is the relatively simple mineral assemblage with pyrite, chalcopyrite, bornite, chalcocite, galena, and sphalerite, impregnating the host sedimentary

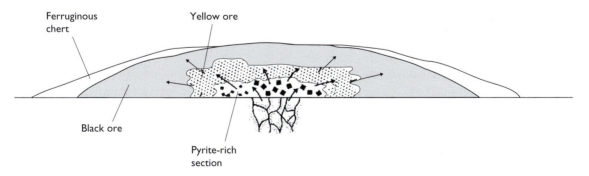

Fig. 26.13 Volcanogenic massive sulfide deposit with a fracture system through which hydrothermal solutions percolate and precipitate, with increasing temperature, black ore (sphalerite, galena, pyrite, and barite), yellow ore (chalcopyrite), and finally pyrite. In the peripheral part of the deposit, silica minerals and hematite precipitate to form ferruginous chert composed of cryptocrystalline silica (after Evans, 1993).

Fig. 26.14 Active black smoker located on the Endeavour Segment of the Juan de Fuca Ridge in the Pacific Ocean at a depth of 2270 m. Pillar structure emanating black smoke, the chimney is 8 m tall. See also Plate 27c (courtesy Deborah Kelley, see also Kelley *et al.*, 2001).

rocks as fine-grained dispersions and veinlets. Strati-form ore deposits show great areal extent, and even those with low concentrations constitute large reserves. An example is the Mississippi Valley type lead–zinc deposit, where basinal brines precipitate dissolved elements in their flow path.

Large sedimentary copper deposits are the Kup-ferschiefer shale that underlies an area of 20 000 km², extending from northern UK to Poland; the White Pine shale in Michigan (USA); and the African Copper Belt extending from the southern region of the Democratic Republic of the Congo to northern Zambia, where copper mineralization is up to 100 m thick. In the so-called *Keweenaw-type* deposits (named after the Keweenaw Peninsula of Michigan) native copper and silver fill vesicles and pores in submarine basalts and conglomerates.

There are some mesothermal (>250 °C) sediment-ary deposits that formed at deeper crustal levels. The greenstone-hosted hydrothermal system of the Mother Lode in the Sierra Nevada foothills of Cali-fornia, USA, originated at depths of up to 10 km. Waters were released at a late stage of the intrusion

of the Sierra Nevada batholith, while subduction and thrusting stacked plates of the upper crust on top of one another. Gold is concentrated in iron-rich wall rocks, rather than the veins themselves, and was deposited when gold–sulfur ions in solution reacted to form iron sulfide minerals, predominantly pyrite, removing sulfur from the complex ions. The Mother Lode system is an extensive, over 200 km long, frac-ture system that is now filled mainly with quartz veins. We will return to some of these issues in Chapter 33.

26.4 Weathering and oxidation of sulfides

Most sulfides are easily oxidized in water- and oxygen-rich environments in surface conditions. Three agents are involved: oxygen, electrochemical processes, and bacteria.

Oxidation occurs in that part of an ore deposit that is above a certain oxygen threshold. For example, oxidation of chalcopyrite to covellite can be described with the following reaction:

$$CuFeS_2 + Cu^{2+} + SO_4^{2-} \rightleftharpoons 2CuS + Fe^{2+} + SO_4^{2-}$$
Chalcopyrite   Covellite   In solution

Oxidation reactions involved in the formation of goethite are:

$$CuFeS_2 + 4O_2 \rightleftharpoons Cu^{2+} + Fe^{2+} + 2SO_4^{2-}$$
Chalcopyrite   In solution

and

$$6Fe^{2+} + 6SO_4^{2-} + H_2O + 1\frac{1}{2}O_2 \rightleftharpoons$$
In solution
$$2FeOOH + 4Fe^{3+} + 6SO_4^{2-}$$
Goethite   In solution

and:

$$4Fe^{3+} + 2H_2O + 3O_2 \rightleftharpoons 4FeOOH$$
In solution   Goethite

Electrochemical processes take place if sulfide minerals are in contact in a damp environment. A pair of grains, such as pyrite and sphalerite (Figure 26.15), acts like a galvanic micro-cell. With the two minerals in contact with an acidic aqueous solution, we can write two dissociation reactions:

$$ZnS \rightleftharpoons Zn^{2+} + S + 2e^-$$
$$FeS_2 + 2e^- \rightleftharpoons Fe + 2S^-$$

The standard redox potentials (see Chapter 19) for the sphalerite reaction is $E_0^{(a)} = +0.12$ V and for the pyrite

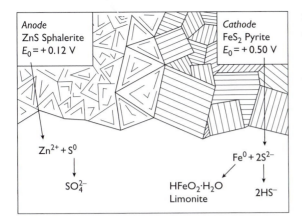

Fig. 26.15 Electrochemical processes during oxidation. Sulfide minerals in contact can act as a galvanic micro-cell, causing decomposition along grain boundaries. Here the anode is sphalerite (ZnS) and the cathode is pyrite (FeS$_2$).

reaction $E_0^{(b)}$ = +0.50 V. ZnS is the electron donor, and thus the anode, while pyrite is the cathode. There is an electrical current flowing between the two minerals, and they are decomposing along grain boundaries, the initial electromotive force being 0.38 V.

Biochemical processes can contribute to oxidation (see also discussion in Chapter 25). The concentration of bacteria in surface solutions can reach over 100 cells per milliliter of solution or per gram of ore. A living bacterial cell is an active electron acceptor because biochemical processes inside it operate more efficiently when these electrons are accepted. A redox potential exists between an inner part and the surface of a bacterial cell, which thus can play the role of an active oxidizing agent. The electromotive force in galvanic mineral micro-cells reaches a maximum (0.60–0.65 V) only when bacteria take part in the oxidation process. Bacteria of the genus *Thiobacillus* are especially active and oxidize many sulfide minerals.

These combined electrochemical and biochemical processes lead to the formation of oxidized ores above primary sulfide deposits, and the oxidized zones can range in thickness from a fraction of a meter to hundreds of meters, depending on local factors such as climate, relief, groundwater level, fracturing, and composition of the original ore. In Tsumeb (Namibia) such oxidation zones are exceptionally large and extend to a depth of several hundred meters. Oxidation zones have been used as prospecting tools. The ferrous ochers (yellow), copper oxides, and carbonates (green and blue) are easily recognized on the surface. Oxidation zones often concentrate metals and can be important secondary deposits. Industrially valuable concentrations of copper, zinc, and lead occur in such secondary deposits. The oxidized cover of iron sulfide ores may contain important concentrations of gold, such as in the Serra dos Carajás (Brazil).

26.5 Summary

Sulfide minerals have very diverse crystal structures. Some have polyhedral structures, such as galena with the same structure as halite. Also, in sphalerite, S^{2-} occupies an fcc lattice, but Zn are in tetrahedral, rather than octahedral interstices. Geometrically the pyrite structure is related to NaCl but the "anions" are sulfide groups S_2^{2-}. In the sulfide stibnite, chain units form and correspondingly the mineral is often prismatic or fibrous. In molybdenite, sulfur atoms are bonded to form sheets. Sulfide minerals are important as metal ores and form during magmatic (e.g., the nickel–copper mineralization in Sudbury) as well as hydrothermal processes (e.g., porphyry–copper deposits at Bingham or volcanogenic systems vented onto the seafloor, as in Cyprus).

Important sulfide minerals to remember

Name	Formula	System
Chalcocite	Cu_2S	Orthorhombic
Chalcopyrite	$CuFeS_2$	Tetragonal
Cinnabar	HgS	Trigonal
Galena	PbS	Cubic
Sphalerite	ZnS	Cubic
Wurtzite	ZnS	Hexagonal
Marcasite	FeS_2	Orthorhombic
Pyrite	FeS_2	Cubic
Realgar	As_4S_4	Monoclinic
Stibnite	Sb_2S_3	Orthorhombic
Molybdenite	MoS_2	Hexagonal

Test your knowledge

1. Discuss the structural divisions of sulfides.
2. Name some examples of close relationships between physical properties and crystal structure.

3. What are some processes of oxidation at surface conditions?
4. Describe the principles of hydrothermal ore formation.
5. Explain some important hydrothermal ore deposits.
6. Prepare a table of important elements (Cu, Co, Ni, Zn, Hg, As, Mo) and list sulfide minerals used for their extraction.

Further reading

Vaughan, D. J. (ed.) (2006). *Sulfide Mineralogy and Geochemistry*. Reviews in Mineralogy, vol. 61. Mineralogical Society of America, Washington, DC.

Vaughan, D. J. and Craig, J. R. (1978). *Mineral Chemistry of Metal Sulfides*. Cambridge University Press, New York.

27 | Oxides and hydroxides. Review of ionic crystals

Over 300 minerals are oxides and hydoxides. Silicon oxides have already been discussed in Chapter 21. While quartz composes nearly 12 vol.% of the Earth's crust, another oxide is just as important, crystalline H_2O (ice). It forms polar ice shields (crucial for the Earth's climate) and glaciers. It also builds up the polar caps on Mars and may form a large part of some satellites of Saturn, Jupiter, and Uranus. In contrast to quartz and ice, iron oxides and hydroxides contribute only about 0.2% to the crust of the Earth, but the minerals magnetite ($Fe^{3+}Fe_2^{2+}O_4$), hematite ($Fe_2^{3+}O_3$), and goethite ($Fe^{2+}OOH$) are major constituents of iron ores, and without iron the present state of our civilization is unimaginable. In many oxides ionic bonding dominates and as an introduction this chapter will review ionic structures.

27.1 Overview

Immediately after X-ray diffraction was discovered, the new technique was applied to determine the crystal structure of oxides and hydroxides, and this research served as the basis for the development of crystal chemistry. By 1915 the structure of spinel ($MgAl_2O_4$) had been determined, followed in 1916 by those of rutile (TiO_2), anatase (TiO_2), and corundum, in 1919 by that of brucite ($Mg(OH)_2$), and in 1920 by that of periclase. These ionic structures became models for such principles as close-packing of anions and helped to establish the basic concepts of ionic structures formulated in the Pauling rules and discussed in Chapter 2.

Later the more complex structures of goethite (FeOOH), gibbsite ($Al(OH)_3$), boehmite (AlOOH), diaspore (AlOOH), and manganese oxides such as cryptomelane $\left(KMn_7^{4+}Mn^{3+}O_{16}\right)$ and hollandite $\left(Ba\,Mn_6^{4+}\,Mn_7^{3+}O_{16}\right)$ were resolved. A great diversity of oxide and hydroxide structures was found to exist, and this diversity can be attributed to the broad variation in chemical compositions and to the presence of chemical bonds of various types.

In this chapter we will first revisit some principles of ionic structures, then look in more detail at structures of

some oxide minerals, and finish with brief descriptions of important oxides. Oxides are important ores for iron, aluminum, chromium, and titanium. They are also important ceramic materials and are synthesized for many applications. Alumina (Al_2O_3), zirconia (ZrO_2), magnesia (MgO), and spinel are "structural ceramics". Because of their outstanding high-temperature strength and low density, oxides have recently been applied in the manufacture of car engines and turbine blades. Other "functional ceramics" are used because of unique electronic properties. Perovskites are known for their ferroelectric and high-temperature superconducting properties and are components of actuators, sensors, and other electronic devices.

27.2 Ionic crystal structures

As we have gone through the mineral groups in previous chapters we have encountered many of the basic ionic structure types. Before advancing to the structurally more complex silicates it is appropriate to look back and review the principles of bonding and coordination and to summarize the most important ionic structures. Oxides are a good mineral group with which to undertake such a review: bonding in oxides is largely ionic and for practically all types of ionic structures there are oxide examples, although we will add a few sulfides because there are better-known sulfide minerals for some types of ionic structures. The structures we will discuss here are a minimal set that every student should remember after having taken a class in mineralogy.

Before starting this discussion it is useful to clarify the difference between *mineral* and *structure*. A structure refers to the atomic arrangement of atoms in the unit cell. Structures are often named after the mineral in which they have first been discovered, particularly so for simple structures. The first structure determination was done by W. L. Bragg on halite (NaCl) (see Chapter 11). Thus, this cubic structure with alternating cations and anions is called the "*halite structure*" (Figure 7.24b). It also applies to galena (PbS), periclase (MgO), and others. Similarly, in this chapter we will talk about *spinels* and *perovskites*. The spinel mineral is $MgAl_2O_4$ but there are many other minerals and compounds with the same structure yet different composition. Compounds such as chromite ($FeCr_2O_4$) are often referred to as "spinels", though this usage is mineralogically not quite correct. The same applies to perovskites. The relatively rare perovskite mineral is $CaTiO_3$, but "perovskite structures" are numerous.

According to the principles derived by V. M. Goldschmidt and L. Pauling (see Chapter 2), several rules must be followed in ionic structures. (1) Anions are relatively closely packed; in many cases in outright cubic or hexagonal close-packed arrangements. (2) Even if not ideally close-packed, anions form regular coordination polyhedra about cations (tetrahedra, octahedra, cubes, and dodecahedra), i.e., they surround cations in a close-packed arrangement and the coordination polyhedra are linked. (3) Cations are generally smaller and occupy interstices in the anion sublattice. Small cations (e.g., Si^{4+}) occupy tetrahedral interstices (coordination number 4), medium-sized ones (Al^{3+}, Mg^{2+}, Fe^{2+}) occupy octahedral interstices (coordination number 6), and large ones (Ca^{2+}) are in cubic (coordination number 8) or cuboctahedral (coordination number 12) interstices.

Let us first consider simple ionic compounds of the type A–X, where A stands for the cation and X for the anion. We can write a simple matrix for the types of close-packing (cubic or hexagonal) and occupied interstices (octahedral or tetrahedral) to arrive at the four most important crystal structures, each with

some mineral examples (Figure 27.1, Table 27.1). The importance of these four simple ionic structures cannot be overemphasized: of all known ionic A–X compounds, over 50% crystallize in the halite structure (Figure 27.1a), 10% in the nickeline structure (Figure 27.1b), and 10% in the sphalerite structure (Figure 27.1c). The wurtzite structure (Figure 27.1d) is more rare.

The four structure types highlighted in Table 27.1 cannot accommodate very large cations. For this purpose the CsCl structure (not a mineral) is better suited: the anions are in a simple cubic lattice with the cations in the center of the cubes (Figure 27.2), and the coordination number is 8. Twenty percent of A–X compounds are in the CsCl structure.

The concept of close-packing applies also to many ionic structures of a more complicated type $A_nB_mX_p$. For example, in corundum (Al_2O_3) and isostructural hematite (Fe_2O_3) and ilmenite ($FeTiO_3$), the oxygen ions display hexagonal close-packing and cations are in two-thirds of the octahedral interstices (Figure 27.3).

In cubic spinel ($MgAl_2O_4$), the oxygen close-packing is cubic and the close-packed plane is (111). Interstices are occupied by Mg^{2+} and Al^{3+}. Interestingly, and contrary to the conventional rule of radius ratio, the usually larger magnesium cations occupy smaller tetrahedral sites (here, the Mg^{2+} radius is about 0.49 Å as compared with 0.72–0.89 Å in most minerals), whereas aluminum is accommodated in octahedral sites (with an ionic radius of about 0.53 Å) (Figure 27.4a). Therefore, close oxygen packing in spinel is not ideal, and tetrahedral and octahedral polyhedra are deformed somewhat in this structure. Figure 27.4b shows the cubic unit cell with eight tetrahedral ions distributed on corners, face centers, and centers of alternate eighth cubes. Sixteen octahedral ions are at more irregular positions. There are a large number of mineral oxides with spinel structure, some of which are listed in Table 27.2. Note that this table indicates whether a particular cation is in an octahedral or tetrahedral interstice. The Si–Mg spinel ringwoodite is a high-pressure phase that does not occur in crustal rocks but is thought to be an important phase in the transition zone of the Earth's mantle and has been found in meteorites (see discussion in Chapter 20 and Figure 20.2).

An interesting derivative of a close-packed structure is perovskite ($CaTiO_3$). Oxygen within the perovskite structure is in a cubic close-packed arrangement

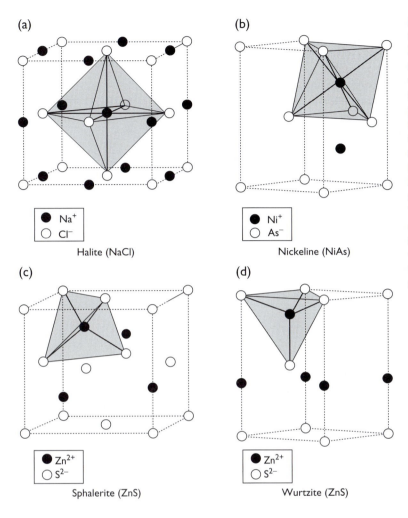

(a)

Na⁺
Cl⁻
Halite (NaCl)

(b)

Ni⁺
As⁻
Nickeline (NiAs)

(c)

Zn²⁺
S²⁻
Sphalerite (ZnS)

(d)

Zn²⁺
S²⁻
Wurtzite (ZnS)

Fig. 27.1 Important ionic structures of the A–X type. (a) Halite (NaCl), anions in cubic close-packing with cations in octahedral interstices. (b) Nickeline (NiAs), anions in hexagonal close-packing with cations in octahedral interstices. (c) Sphalerite (ZnS), anions in cubic close-packing with cations in tetrahedral interstices. (d) Wurtzite (ZnS), anions in hexagonal close-packing with cations in tetrahedral interstices. The unit cell of the fcc and hcp anion sublattices is shown with dashed lines and can be compared with Figure 2.10a,b (note that in the case of nickeline this is not the standard unit cell).

Table 27.1 Structure types with close-packing of anions and cations in interstices, with examples. Structure type is given in bold face.

Occupied interstices	Close-packing	
	Cubic	Hexagonal
Octahedral (CN = 6)	**Halite** (NaCl)	**Nickeline** (NiAs)
	Periclase (MgO)	Pyrrhotite (FeS)
	Wüstite (FeO)	Jaipurite (CoS)
	Galena (PbS)	Breithauptite (NiSb)
Tetrahedral (CN = 4)	**Sphalerite** (ZnS)	**Wurtzite** (ZnS)
	Ice (H₂O)	Greenockite (CdS)
	Metacinnabar (HgS)	ZnO (synthetic)

Note: CN, coordination number.

(i.e., fcc), except that in every second layer one oxygen ion is missing, leaving a large cavity (in the center of Figure 27.5a). Ti^{4+} occupies octahedral interstices, and the large Ca^{2+} cation occupies the cavity left vacant by the missing oxygen ion. This site is coordinated by 12 oxygen ions (Figure 27.5b). The perovskite structure is unique among the close-packed oxides in its ability to accommodate very large cations such as the rare earth elements. At the high pressures in the Earth's lower mantle, the silicate mineral olivine transforms first to ringwoodite with spinel structure (in the transition zone) and at higher pressure to the silicate perovskite structure (bridgmanite) and periclase:

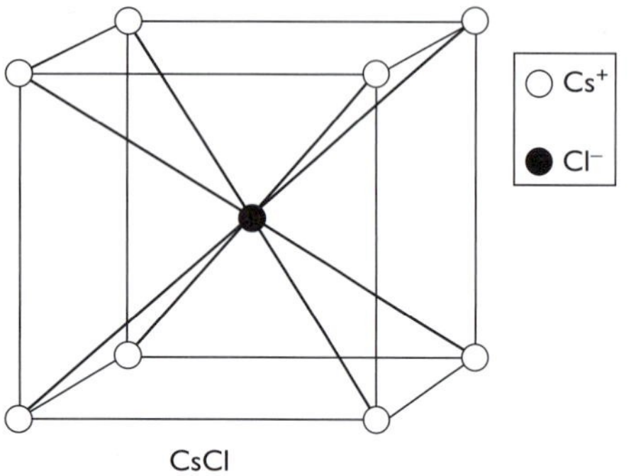

○ Cs⁺

● Cl⁻

CsCl

Fig. 27.2 Structure of CsCl with coordination number 8. The coordination polyhedron and the unit cell coincide.

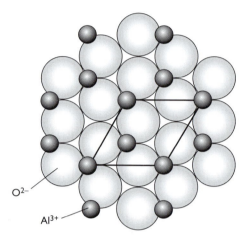

O^{2-}

Al^{3+}

Fig. 27.3 (0001) projection of the structure of corundum (Al_2O_3) with hexagonal close-packing of oxygen and aluminum ions in two-thirds of the octahedral interstices. Only one oxygen ion layer is shown. Hexagonal unit cell is outlined.

(a)

(b)

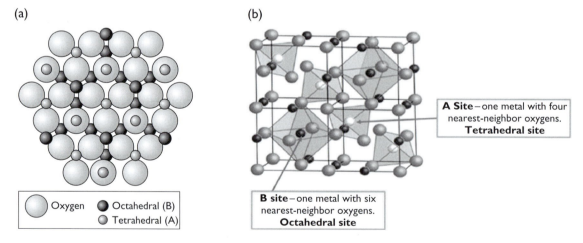

○ Oxygen ● Octahedral (B) ○ Tetrahedral (A)

A Site – one metal with four nearest-neighbor oxygens. **Tetrahedral site**

B site – one metal with six nearest-neighbor oxygens. **Octahedral site**

Fig. 27.4 Structure of spinel of general formula AB_2O_4. Oxygen atoms form a cubic close-packed arrangement: A cations occupy tetrahedral interstices, and B cations occupy octahedral interstices. (a) Close-packed oxygen layer parallel to (111) with occupancy of cations above and below. (b) Cubic unit cell with location of tetrahedral and octahedral cations (some tetrahedra are outlined) and oxygen atoms (see also Figure 12.17a for magnetic ordering in magnetite).

Table 27.2 Some compounds with spinel structure and the distribution of cations in tetrahedral and octahedral interstices

Name	Formula	Tetrahedral (8 sites)	Octahedral (16 sites)
Magnetite	$Fe^{2+}Fe_2^{3+}O_4$	Fe^{3+}	Fe^{2+}, Fe^{3+}
Spinel	$MgAl_2O_4$	Mg^{2+}	Al^{3+}
Magnesioferrite	$MgFe_2^{3+}O_4$	Fe^{3+}	Fe^{3+}, Mg^{2+}
Ulvite	$Ti^{4+}Fe_2^{2+}O_4$	Fe^{2+}	Fe^{2+}, Ti^{4+}
Chromite	$Fe^{2+}Cr_2O_4$	Fe^{2+}	Cr^{3+}
Gahnite	$ZnAl_2O_4$	Zn^{2+}	Al^{3+}
Jacobsite	$MnFe_2^{3+}O_4$	Mn^{2+}	Fe^{3+}
Ringwoodite	Mg_2SiO_4	Si^{4+}	Mg^{2+}

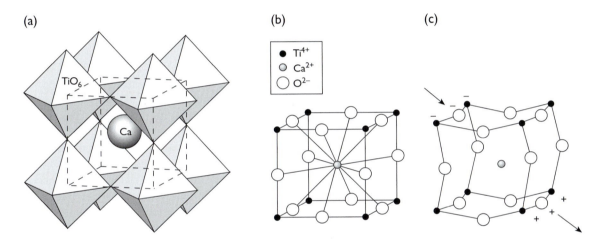

Fig. 27.5 (a) Polyhedral representation of the structure of perovskite, showing the large 12-fold coordinated Ca^{2+} in the center of the cubic unit cell and the smaller 8-fold coordinated Ti^{4+} in the corners. (b) Representation of the perovskite structure with small spheres. (c) The perovskite structure distorts when stress is applied, creating an electric field (piezoelectricity).

This will be discussed in more detail in Chapter 38 (e.g., Figure 38.4).

$$Mg_2SiO_4 \rightarrow SiMg_2O_4 \rightarrow MgSiO_3 + MgO$$
Olivine   "Spinel"   "Perovskite" Periclase

If the 12-fold coordinated cation is smaller than oxygen, the structure becomes distorted by the tilting of octahedra, particularly at lower temperatures, giving rise to many phase transformations. The modified structure may be tetragonal, orthorhombic, or monoclinic, in some cases without a center of symmetry (Figure 27.5c). Noncentric perovskites may display piezoelectricity or ferroelectricity, and synthetic perovskites are applied as electronic sensors. The structure of the newly discovered high-temperature superconductors (HTC) is closely related to that of perovskites. In these superconductors Cu substitutes for Ti in layers that display an unusual superconductivity at higher temperature. Some examples of natural and synthetic perovskites are given in Table 27.3.

Some other oxide structures are easier to visualize if the cation packing is considered. For example, the structures of such unlike oxides as molecular CO_2 (Figure 27.6a) and ionic ZrO_2 (baddelyite, Figure 27.6b) can be viewed as cubic close-packing of carbon and zirconium, respectively. Even the more open structure of cristobalite (SiO_2), which has been discussed in Chapter 21 as consisting of rings of tetrahedra, can be viewed as silicon in an fcc arrangement with additional silicon atoms in alternating eighth cubes (Figure 27.6c). In fact, cation packing has been suggested as an alternative system for classifying ionic structures (O'Keefe and Hyde, 1985).

27.3 More complex oxide structures

The structures discussed above have an equal or similar number of cations and anions. We relied mainly on the concept of close-packing of large anions with small cations in interstices. Also in these structures, each cation is surrounded by anions with the geometry of coordination polyhedra, but these are very densely packed and share faces. It would be difficult to build a model of the halite structure with polyhedra because it would consist of two sets of edge-sharing octahedra. Other oxides, particularly those with a larger anion:

Table 27.3 Some compounds with perovskite structure

Name	Formula
Perovskite	$CaTiO_3$
Loparite	$CeNaTi_2O_6$
Bridgmanite	$MgSiO_3$
PZT (synthetic)	$Pb(Zr,Ti)O_3$
PST (synthetic)	$Pb(Sc,Ta)O_3$
YBCO (HTC) (synthetic)	$YBa_2Cu_3O_{7-x}$
BISCO (HTC) (synthetic)	$Bi_2Sr_2Ca_2Cu_3O_x$

Note: HTC, High-temperature superconductor.

cation ratio, are more open, and their polyhedra share only edges or corners. These are so-called polyhedral structures.

We have already discussed in Chapter 21 silica minerals with polyhedral structures of SiO_4^{4-} tetrahedra connected by corners, and we will return to related structures in Chapters 28–31 on silicates. As was pointed out in Chapter 2, if coordination tetrahedra shared faces, cations would be in too close a proximity for the structure to be stable.

Besides tetrahedral structures there are also octahedral polyhedral structures in oxides, most importantly in aluminum, magnesium, iron, titanium, and manganese oxides and hydroxides. They all have similar anion:cation ratios. As a first example, we consider the octahedral structure of rutile. If we look at the tetragonal unit cell, Ti^{4+} occupies corners and the center (Figure 27.7a). Six oxygen ions in the form of an octahedron surround each titanium. The TiO_6^{8-} octahedron shares two edges with adjacent octahedra, comprising a ribbon that extends along [001] (Figure 27.7b). There are two types of ribbon, oriented at right angles and shifted by half a unit cell along the z-axis. The two types of ribbon are linked over the remaining free octahedral corners. Between the ribbons there are infinite "channels" or "tunnels" that are visible when the structure is viewed along the c-axis (Figure 27.7c) or in a polyhedral representation (Figure 27.8a). Rutile often has a prismatic morphology (Plate 28b,c) and sometimes occurs as epitaxial overgrowth on hematite, with rutile ribbons aligning with octahedra in the hematite structure (Plate 3e).

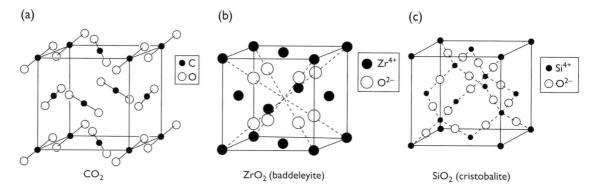

Fig. 27.6 Contrary to a classification with emphasis on anions, the structures of such different oxides as (a) CO_2, (b) baddeleyite (ZrO_2) (isostructural with fluorite), and (c) cristobalite (SiO_2) can all be visualized as cations in an fcc arrangement. (The oxide CO_2 is of course not an ionic structure and is simply added because of the structural analogy.)

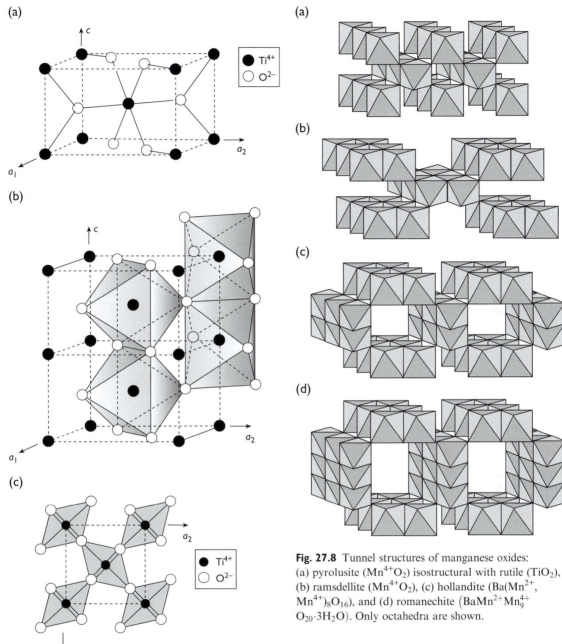

(a)

(b)

(c)

Fig. 27.7 (a) Structures of rutile, showing a body-centered tetragonal unit cell of Ti^{4+} ions. (b) Rutile can also be viewed as ribbons of edge-sharing TiO_6^{8-} octahedra that link at free corners of the octahedra. (c) View of the octahedral ribbons along the c-axis. The unit cell is indicated by dashed lines.

Fig. 27.8 Tunnel structures of manganese oxides: (a) pyrolusite ($Mn^{4+}O_2$) isostructural with rutile (TiO_2), (b) ramsdellite ($Mn^{4+}O_2$), (c) hollandite ($Ba(Mn^{2+}, Mn^{4+})_8O_{16}$), and (d) romanechite ($BaMn^{2+}Mn_9^{4+}O_{20}\cdot3H_2O$). Only octahedra are shown.

Cassiterite (SnO_2), pyrolusite (MnO_2), and stishovite (SiO_2) have structures similar to that of rutile.

There are numerous channel structures in manganese oxides that can be thought of as derivatives of the rutile structure. Pyrolusite ($Mn^{4+}O_2$) is isostructural with rutile (Figure 27.8a). In ramsdellite ($Mn^{4+}O_2$), two ribbons are joined by sharing edges to form a band and then stacked in a planar arrangement (Figure 27.8b). In hollandite ($Ba(Mn_2^{2+}, Mn_6^{4+})O_{16}$), bands of two octahedra alternate with a horizontal and vertical stacking, resulting in larger tunnels that can accommodate large cations such as barium (Figure 27.8c). In romanechite ($BaMn^{2+}Mn^{4+}_9 O_{20} \cdot 3H_2O$), bands of two and three ribbons alternate with a horizontal and vertical stacking (Figure 27.8d). This creates tunnels that are wide enough to accommodate water molecules. As you can imagine, there is some stacking disorder among these types, with intermediate or mixed arrangements in local regions. The structural variations are expressed in differences of cation–cation distances (within layers versus between adjacent layers) and can be used for identifying the minerals.

It is worth pointing out again that in stishovite silicon is in octahedral coordination, whereas in all other silica minerals silicon is in tetrahedral coordination. At extreme pressures O^{2-} deviates from a spherical shape and can be packed more closely around Si^{4+} with 6-fold coordination. Note that also in the high-pressure perovskite bridgmanite ($MgSiO_3$) Si^{4+} is in octahedral coordination.

In hydoxides brucite ($Mg(OH)_2$) and gibbsite ($Al(OH)_3$) octahedra, in which OH^- groups rather than O^{2-} form corners, are connected to infinite stacked sheets (Figure 27.9). The ionic sheets are electrostatically neutral and are held together with weak van der Waals bonds, giving these minerals a perfect planar cleavage. We will encounter brucite and gibbsite sheets again when we discuss sheet silicate structures (Chapter 29). In brucite the sheet is contiguous (Figure 27.9b) and is called a *trioctahedral sheet*. In gibbsite, owing to the trivalent Al^{3+}, one out of every three cations is missing, thus maintaining a charge balance and producing a *dioctahedral sheet* (Figure 27.9c).

Diaspore and boehmite, both with a composition AlOOH, have related octahedral structures (Figure 27.10). Anions, O^{2-} and OH^-, are nearly in a hexagonal close-packed arrangement. The structures contain double ribbons of distorted $Al(O,OH)_6$ octahedra. In diaspore they are stacked to form a framework with tunnels along the *c*-axis (Figure 27.10a, see also Figure 27.8b). In boehmite they are linked over corners to form a corrugated sheet (Figure 27.10b). The ribbon structure is expressed in an often fibrous morphology.

There are iron hydroxides FeOOH with structures analogous to these aluminum oxides. Goethite is isostructural with diaspore, and lepidocrocite with boehmite.

(a)

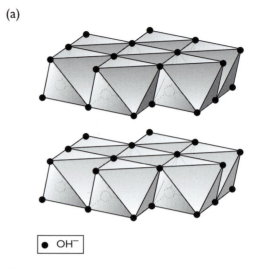

● OH^-

(b)

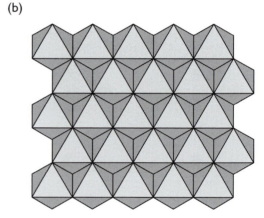

(c)

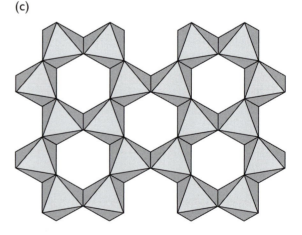

Fig. 27.9 Layered structures brucite ($Mg(OH)_2$) and gibbsite ($Al(OH)_3$), with octahedral sheets that are stacked (a). In brucite (b) all octahedra are occupied (trioctahedral), whereas in gibbsite (c) one out of three is vacant (dioctahedral) in order to achieve a charge balance.

(a)

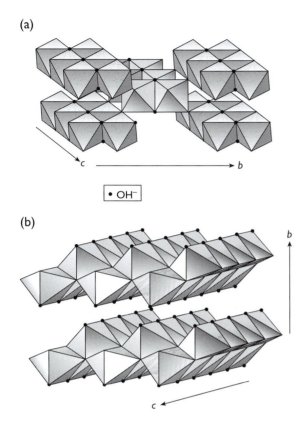

(b)

Fig. 27.10 Structures of (a) diaspore (isostructural with goethite and ramsdellite) with ribbons of two octahedra linking to a structure with tunnels parallel to the *c*-axis, and (b) boehmite (isostructural with lepidochrocite) with corrugated sheets of octahedra that link over corners. OH⁻ groups are indicated; all other octahedral corners are oxygens atoms.

Depending on temperature–pressure conditions, ice (H_2O) can exist in many different structures and a phase diagram was introduced earlier (Figure 19.1). The most common hexagonal ice Ih, stable at ambient pressure and occurring in glaciers and snow, has a structure that is closely related to tridymite (SiO_2), except that oxygen plays the role of silicon, forming hexagonal layers. Four hydrogen atoms tetrahedrally surround each oxygen, two are linked by covalent bonds at a distance of 1.0 Å and the other two by hydrogen bonds at a distance of 1.8 Å (Figure 27.11a). The weak hydrogen bond occurs if hydrogen is bonded covalently on one side to oxygen, and electrostatically to another oxygen of a H_2O group with a small electronegative charge. The different bond lengths result in distortions, giving rise to an open cage produced by linked layers (Figure 27.11b) and

this is the reason why ice Ih has a lower density than water.

There is another group of H_2O compounds called clathrates in which water molecules form a structure with large cages. The structure contains square, pentagonal, and hexagonal rings of water molecules that are linked to produce large three-dimensional cages. Five types of cages are known (Figure 27.12a): A is a cage with 12 5-fold rings, B a cage with 12 5-fold and 2 6-fold rings, C a cage with 12 5-fold and 4 6-fold rings, D a cage with 3 4-fold, 6 5-fold and 3 6-fold rings and E a large cage with 12 5-fold and 8 6-fold rings. These cages then combine to three even larger units (Figure 27.12b). Type I is a cubic body-centered structure with 2 A and 6 B cages, Type II is a diamond-type lattice with 16 A and 8 C cages and Type H has a hexagonal lattice with 3 A, 2 D and 1 E cages.

Clathrate cages contain large amounts of low molecular weight gases. Methane and carbon dioxide are most significant. Clathrates occur at higher pressures and lower temperatures relative to water (Figure 27.13, e.g., Sloan and Koh, 2007) and are geologically important components of sediments in continental shelves, especially in oceanic sediments at depths greater than 300 m, where the water temperature is around 2 °C. Some freshwater lakes also host gas hydrates, such as Lake Baikal (Siberia), and continental clathrate deposits in sandstone have been located in Alaska and Siberia. It is estimated that more than 6.5 trillion tons of methane is trapped in clathrates on the deep ocean floor. Interestingly, none of these naturally occurring clathrates have been approved as minerals in spite of their geological significance (e.g., Chakoumakos, 2004), but some analog structures with SiO_2 substituting for H_2O have recently been discovered (e.g., Momma *et al.*, 2011): melanophlogite is isostructural with type I and chibaite with type II structures.

27.4 Brief description of important oxide and hydroxide minerals

Some properties of important oxides and hydroxides are listed in Tables 27.4 and 27.5, respectively. In this section we will discuss some of them in more detail, in the same order as they are listed in the tables.

Oxides

Trigonal (rhombohedral) **corundum** (Al_2O_3), called "alumina" in the ceramics field, often contains minor

(a) 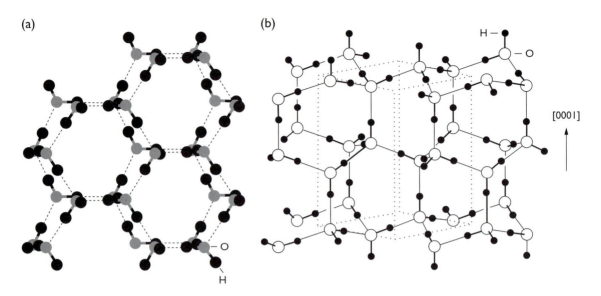 (b)

[0001]

H — ●

● — O

O

H

Fig. 27.11 Structure of hexagonal ice Ih. (a) Two layers of hexagonal nets of H_2O molecules formed by both covalent bonds (solid lines) and hydrogen bonds (dashed lines). (b) Three-dimensional stacking of these layers to form the hexagonal ice structure.

(a)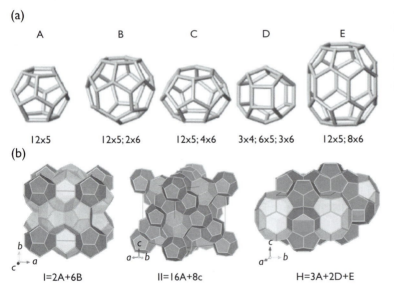

Fig. 27.12 Structure of clathrates. (a) H_2O creates 4-, 5-, 6-, and 8-membered rings that bond to form cages. (b) Combinations of these cages produce the three clathrate structures I, II, and H.

A   B   C   D   E

12x5   12x5; 2x6   12x5; 4x6   3x4; 6x5; 3x6   12x5; 8x6

(b)

I=2A+6B   II=16A+8c   H=3A+2D+E

amounts of chromium, titanium, and iron. Strong ionic–covalent bonds are expressed in its high hardness (Mohs scale 9) and the close-packed structure results in a high density (4.0 g/cm^3, close to the density of chalcopyrite ($CuFeS_2$), which contains much heavier elements). Euhedral crystals of corundum display a

combination of steep hexagonal bipyramid faces (which almost always have rough horizontal striations) (Plate 1e-g) and faces of a pinacoid (0001). The relative dominance of the two forms depends on the composition and the mineral-forming environment. The less silicon and the more alkali and

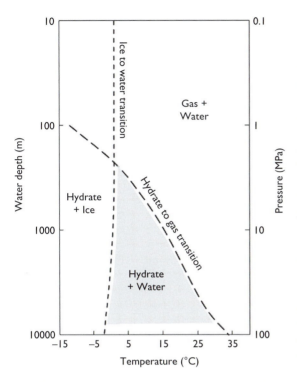

Fig. 27.13 Pressure–temperature phase diagram of the system methane and water.

alkali-earth metals the environment contains, the more elongate are corundum crystals. Besides being found as isolated crystals within a rock, corundum occurs also as massive granular aggregates known as emery. Corundum's color is white or gray if it is pure Al_2O_3, but it is *ruby*-red if some chromium is present (Plate 1f), *sapphire*-blue if it contains iron and titanium (Plates 1g, 28f, 32a), and yellow if only iron is present (Plate 1e). Corundum has a vitreous luster. In thin section, it displays a blue–white pleochroism. The high hardness, crystal habit, and color of the mineral serve as easy means of identification.

Corundum forms during hydrothermal alteration of volcanic and ultramafic rocks, under regional metamorphism of alumina-rich clays and bauxites, and in alkaline magmas that are alumina-supersaturated and silica-undersaturated. The latter mode is responsible for the formation of nepheline syenites and syenite pegmatites, with dispersions of sapphire-like corundum crystals. This mineral accumulates easily in placers. Corundum is used as a gemstone (ruby and sapphire) and as an industrial abrasive material.

Corundum and quartz never coexist owing to the following reaction:

$Al_2O_3 + SiO_2 \rightarrow Al_2SiO_5$ (kyanite, sillimanite, andalusite)

Hematite $\left(Fe_2^{3+}O_3\right)$ is isostructural with corundum. Crystals occur as hexagonal plates (Plate 3e) or as more complex tabular forms (Plate 6c, Plate 30c). Frequently it is found as earthy and irregular masses called red iron ore. The mineral has a black color and a semimetallic luster when found as crystals. The streak of hematite is remarkably cherry-red to brown-red, which is important for purposes of identification.

The most typical hematite deposits are in quartz veins; in skarns, where hematite is associated with epidote and quartz; in quartzites, schists, and other regional-metamorphic rocks; in oxidation zones of iron ores; and in laterites, where hematite is the most stable iron compound existing in an arid climate. Hematite is a major iron ore.

Ilmenite $(FeTiO_3)$ has the same structure as corundum (Figure 27.3) and hematite, but its cations are ordered in alternate layers parallel to the close-packed oxygen atoms. It occurs as tabular and platy crystals of black color with metallic luster. The streak is black. Sometimes ilmenite has weak magnetic properties. Its hardness is about 5.5. Ilmenite may be mistaken for magnetite, but differs in crystal morphology and has less pronounced magnetic properties.

Ilmenite occurs in commercial quantities as segregations, veins, and layers in pyroxenites and gabbros, as lenses in placers, and as thin beds in sandstones. Such ilmenite accumulations are an ore of titanium.

In **spinel** $(MgAl_2O_4)$ isomorphic substitutions of Fe^{2+}, Fe^{3+}, Mn^{2+}, Cr^{2+}, and Zn^{2+} cause a variety of colors. Compositions vary widely (Table 27.2). Ideally, pure magnesium aluminum oxide is colorless, but chromium produces pink or red color, and iron green, greenish-black, or blue color (depending on the total iron content and on the Fe^{2+}:Fe^{3+} ratio). The luster is vitreous. The crystal habit and high hardness (8) serve in the identification of the mineral.

Spinel forms at high temperature and pressure conditions. It occurs in metamorphic rocks rich in magnesium, such as marbles and calcsilicates, due to metamorphism of clay-rich carbonates. It also occurs in metasomatic calcsilicate rocks interbedded in Archean gneisses and schists. The mineral assemblage of all these deposits is very specific and includes calcite, diopside, forsterite, and phlogopite.

Magnetite $\left(Fe^{2+}Fe_2^{3+}O_4\right)$ has a very complex composition due to isomorphous substitutions of titanium (titanomagnetite), aluminum, magnesium, chromium,

Table 27.4 Anhydrous oxide minerals with some diagnostic properties; important minerals are given in italics. For silica minerals, see Table 19.1.

Mineral Formula	System	Morphology Cleavage	H	D	Color/luster Streak	n Pleochroism	Δ	$2V$ (Dispersion)
Corundum-type oxides								
Corundum Al_2O_3	Trig.	Pris. [0001]	9	4.0	Gray, blue, red	1.76–1.77	0.008	(−)
Hematite Fe_2O_3	Trig.	Platy (0001)	6.5	5.2	Black, red/(metallic) *Red*	2.80–3.04 *Yellow, red-brown, red*	0.245	(−)
Ilmenite $FeTiO_3$	Trig.	Platy (0001)	5–6	4.7	Black, brown/(metallic) *Brown*	opaque		
Spinel-type oxides								
Spinel $MgAl_2O_4$	Cubic	{111} {111}	8	3.7	Red, blue, green, yellow	1.72–2.05		
Magnetite $Fe^{2+}Fe_2^{3+}O_4$	Cubic	{111}, Gran. {111}	5.5–6	4.9	Black/metallic *Black*	opaque		
Chromite $FeCr_2O_4$	Cubic	{111}, Gran. {111}	5.5	4.8	Black, brown/metallic *Brown*	opaque		
Hausmannite $Mn^{4+}Mn_2^{2+}O_4$	Tetrag.	{111} (001)	5.5	4.7	Black, brown/metallic *Brown*	2.15–2.46	0.31	(−)
Chrysoberyl $BeAl_2O_4$	Ortho.	Tab. {011}	8.5	3.7	Green, green-yellow	1.75–1.76	0.009	+45–71
Ringwoodite Mg_2SiO_4	Cubic			3.9		clear 1.77		
Rutile-type structures								
Cassiterite SnO_2	Tetrag.	Equant {100}	6–7	7.0	Brown-black/(metallic) *Yellow-brown*	1.20–2.09	0.096	(+)
Pyrolusite MnO_2	Tetrag.	Pris. [001] {110}	5–6	5.0	Black/metallic *Black*	opaque		
Rutile TiO_2	Tetrag.	Pris. [001] {100}; {110}	6	4.2	Red-brown, black/(metallic) *Yellow-brown*	2.62–2.90	0.287	(+)

Mineral / Formula	System	Morphology	H	D	Color	n	Δ	2V
Stishovite SiO$_2$	Tetrag.	Pris. [001]		4.3	Clear	1.80–1.83	0.027	(+)
Other polyhedral oxides								
Bridgmanite MgSiO$_3$	Ortho			4.3				
Anatase TiO$_2$	Tetrag.	Pyr., Tab. (001), {111}	5.5–6	3.8	Blue-black, brown / *Red, brown*	2.49–2.56	0.022	(−)
Columbite (Fe,Mn)Nb$_2$O$_6$	Ortho.	Platy (010) (010)	6	5.3	Black/(metallic) / *Brown-black*	2.45		
Cuprite Cu$_2$O	Cubic	Oct. {111}	3.5–4	6.1	Red(metallic) / *Brown-red*	opaque		
Periclase MgO	Cubic	{111}{110} {001}	5.5–6	3.56	White, gray / *Yellow*	1.73		
Perovskite CaTiO$_3$	Ps. cubic	Cub., Oct. {100}	5.5–6	4.0	Black / *Gray-white*	2.38	0.017	90
Pyrochlore NaCaNb$_2$O$_6$(F,OH)	Cubic	Oct.	5–5.5	4.5	Brown, yellow, black / *Brown*	1.9–2.2		
Microlite NaCaTa$_2$O$_6$(F,OH)	Cubic	Oct.{111}	5.5	6.1	Black, brown / *Yellow, brown*	1.93–2.02		
Tantalite (Fe,Mn)Ta$_2$O$_6$	Ortho.	Pris. [001]	6	8.2	Black/(metallic) / *Brown-black*	2.26–2.43	0.17	+74 (r < v)
Tenorite CuO	Monocl.	Pris.	3.5	6.4	Black/(metallic) / *Black*	opaque		
Thorianite ThO$_2$	Cubic	{100}	6.5	9.8	Black, brown / *Gray*	2.3		
Uraninite (Pitchblende) UO$_2$	Cubic	Cub., Oct. {111}	4–6	10.6	Black, brown/(metallic) / *Green-black, brown*	opaque		

Notes: H, hardness; D, density (g/cm^3); n, range of refractive indices; Δ birefringence; 2V, axial angle for biaxial minerals. For uniaxial minerals (+) is positive and (−) is negative. Acute 2V is given in the table. If 2V is negative the mineral is biaxial negative and 2V is 2V_α, if it is positive, the mineral is biaxial positive and 2V is 2V_γ. Dispersion r < v means that acute 2V is larger for violet than for red.

System: Monocl., monoclinic; Ortho., orthorhombic; Ps., pseudo; Trig., trigonal; Tetrag., tetragonal.

Morphology: Cub., cubic; Fibr., fibrous; Gran., granular; Oct., octahedron; Pris., prismatic; Pyr., pyramidal; Tab., tabular; (hkl) is a single set of planes.

Color: Metallic luster is also included; parentheses indicate submetallic or occasionally metallic.

Table 27.5 Hydroxide minerals, with some diagnostic properties; important minerals are given in italics

Mineral Formula	System	Morphology Cleavage	H	D	Color/luster Streak	n Pleochroism	Δ	2V (Dispersion)
Brucite $Mg(OH)_2$	Trig.	Platy (0001) *(0001)*	2.5	2.4	White, green, brown	1.57–1.58	0.015	(+)
Boehmite γ-AlO(OH)	Ortho.	Platy (001) *(010)*	3.5	3.01	Clear, white	1.64–1.67	0.02	+74–88
Diaspore α-AlO(OH)	Ortho.	Platy (010) *(010)*	6–7	3.4	White, gray, brown	1.70–1.75	0.048	+84 (r < v)
Gibbsite $Al(OH)_3$	Monocl.	Platy (001) *(001)*	2.5–3	2.4	Clear, white	1.57–1.59	0.02	+(0)
Goethite $Fe^{3+}O(OH)$	Ortho.	Fibr. [001] *(010)*	5–5.5	4.3	Black-brown, red, yellow/(metallic) *Brown, yellow*	2.26–2.40 *Yellow-brown-orange, green*	0.140	–0–42 (r > v)
Lepidocrocite $Fe^{3+}O(OH)$	Ortho.	Platy (010) *(010)*	5	4.09	Red/(metallic) *Orange*	1.94–2.51 *Yellow-orange-brown*	0.57	–83
Manganite $Mn^{2+}Mn^{4+}O_2(OH)_2$	Monocl.	Pris. [001] *(010)*	4	4.4	Brown-black/(metallic) *Brown, red*	2.25–2.53	0.28	+0–5 (r > v)
Pyrochroite $Mn(OH)_2$	Hexag.	Tab. *(0001)*	2.5	3.3	Clear, white, brown *Brown-yellow, brown*	1.68–1.72	0.04	(–)
Romanechite $BaMn^{2+}Mn^{4+}_9 O_{20} \cdot 3H_2O$	Monocl.	Fibr., micr.	5–6	4.7	Black/(metallic) *Black, brown*	opaque		

Notes: H, hardness; D, density (g/cm^3); *n*, range of refractive indices; Δ, birefringence; 2V, axial angle for biaxial minerals. For uniaxial minerals (+) is positive and (–) is negative. Acute 2V is given in the table. If 2V is negative the mineral is biaxial negative and 2V is $2V_\alpha$; if it is positive, the mineral is biaxial positive and 2V is $2V_\gamma$. Dispersion r < v means that acute 2V is larger for violet than for red.

System: Hexag, hexagonal; Monocl., monoclinic; Ortho., orthorhombic; Trig., trigonal.

Morphology: Fibr., fibrous; Micr., microscopic; Pris., prismatic; Tab., tabular.

Color: Metallic luster is also included; parentheses indicate submetallic or occasionally metallic.

vanadium, manganese, and silicon, among others. Therefore the iron content may vary considerably. This compositional complexity and the distribution of the various cations among structurally nonequivalent positions are the causes of phase transformations at low temperature. Having crystallized at high temperatures as a homogeneous phase, magnetite undergoes exsolution into a mixture of mineral phases such as magnetite (chemically more pure), ilmenite, spinel, ulvite, and others.

Magnetite is found as euhedral octahedral crystals (Plate 3b) (rarely as dodecahedra), as granular aggregates, and as veinlets. The mineral is black, has a semimetallic or metallic luster, and exhibits pronounced magnetic properties that vary depending on its composition. The more magnesium and manganese that magnetite contains, the lower are its magnetic susceptibility and Curie point (the temperature above which magnetism disappears). Magnetism is a diagnostic property of magnetite, but it must be kept in mind that the degree of magnetization may vary within a crystal. Distinctive color, streak, luster, and strong magnetism provide for easy identification of this mineral.

Magnetite occurs in many geological environments. Economically important magnetite deposits are found as dispersions in olivinites, peridotites, and gabbros, or as skarns. In these deposits, magnetite has formed due partly to the addition of iron to solutions and partly to the release of iron from early formed iron-bearing skarn minerals such as hedenbergite and andradite, which in turn react with solutions. Other magnetite deposits are iron quartzites (metamorphic rocks), in which magnetite has formed by partial reduction of iron and by mineral transformations in primary volcanic and sedimentary rocks during high-grade metamorphism. When subjected to weathering, magnetite oxidizes and causes hematite and goethite to form. It is a major iron ore.

Chromite ($FeCr_2O_4$) is black or brownish-black in color with a greasy or metal-like luster. The brown, dark-brown, or greenish-brown streak is characteristic of chromite. The hardness ranges from 5.5 to 7, increasing with the chromium content. Weak magnetism is sometimes observed. Identifying chromite can be difficult because it has many similarities to magnetite. Chromite is almost always nonmagnetic or weakly magnetic; its streak is not black, but dirty brown or greenish-brown. The frequent association of the mineral with yellow and green masses of serpentine, with

emerald-green crystals and grains of uvarovite, and with pinkish-violet coatings of chromian chlorites helps in the identification.

Chromite is found in isolated grains and orbicular aggregates (nodules) located in massive layers within ultramafic rocks that are usually completely serpentinized. It is the only ore mineral of chromium. Important deposits are located in South Africa (Bushveld complex) and Zimbabwe (Great Dyke).

Ringwoodite (Mg_2SiO_4), with spinel structure, forms when olivine is subjected to high pressure. It occurs in some meteorites and is thought to be one of the major phases in the Earth's intermediate mantle ("transition zone"). At very high pressures its polyhedral sites are distorted and the cation:anion radius ratio rule discussed in Chapter 2 no longer applies. The phase diagram has been discussed in Chapter 20.

Tetragonal **cassiterite** (SnO_2) is isostructural with rutile (Figure 27.7) and crystals are often well developed, prismatic, columnar, or acicular in habit, and striated along their longest axis. Twins with an inclined axis are common. The luster is adamantine on faces and greasy in massive fine aggregates. Hardness ranges from 6 to 7. The mineral has a high density between 6.8 and 7.0 g/cm^3. Cassiterite's color is red, brown, or brownish-black, and the streak is generally colorless or yellow-white. Cassiterite can be recognized easily by its crystal morphology and by a combination of three extreme properties: very high hardness, very high density, and very strong luster. If the mineral is fine grained, identification is difficult and the so-called tin-mirror reaction may help. In this reaction cassiterite becomes covered with shiny metallic tin when placed on a heated zinc plate and treated with diluted hydrochloric acid.

Cassiterite occurs in granite pegmatites, greisens, high-temperature hydrothermal veins, and skarns (see Chapter 26). It accumulates also in marine placers. The mineral is mined from all these deposits as a tin ore.

Pyrolusite (MnO_2) also is isostructural with rutile. Crystals are extremely rare, but earthy masses and oolites are common. The mineral is black colored, has a matte luster, and the streak is black. It can be identified by its common occurrence as black oolites. When a drop of benzidine is put on the streak, its color turns to bluish-green.

Pyrolusite occurs in minable quantities as layers and lenses in sedimentary rocks and as earthy aggregates in oxidation zones. This is the most stable compound of manganese at surface conditions.

Rutile (TiO_2) usually forms prismatic and acicular crystals that often have well-developed faces (tetragonal-bipyramidal and prismatic) (Plate 28a–c). Rutile is black or reddish-brown in color and has a pale brownish streak and an adamantine luster. Prismatic cleavage is characteristic. The mineral is often found in quartz veins in metamorphic rocks. In such veins rutile forms spectacular acicular inclusions in quartz. The mineral accumulates in placers and is used as a titanium ore.

Anatase (TiO_2) is rare and found as isolated bluish-black tetragonal-bipyramidal crystals in quartz veins and fissures in metamorphic rocks (Plate 28d).

Columbite ($FeNb_2O_6$) together with **tantalite** ($FeTa_2O_6$) form isomorphous series. Iron may be substituted by manganese. These minerals have great economic importance as niobium and tantalum ores.

Cuprite (Cu_2O) is a cubic mineral that sometimes forms perfect octahedra with an adamantine luster on its faces (Plate 28e). More often, however, it occurs as compact granular aggregates with a metal-like and greasy luster. Typically it has a deep cherry-red color. Cleavage is seen in some cases. Cuprite's streak is brick-brown, but it becomes dark brown or green upon finer grinding. The latter feature helps to distinguish this mineral from iron oxides and hydroxides. Cuprite is often associated with native copper, malachite, and azurite. It is an intermediate oxidation product of chalcopyrite and other copper sulfide ores.

Periclase (MgO) is isostructural with halite ("magnesia"). It is only rarely found on the surface of the Earth but is likely to be a major component of the lower mantle, as we will discuss in Chapter 38.

Minerals of the "*perovskite group*" include **perovskite** ($CaTiO_3$) and its compositional varieties, including **loparite** that may contain up to 15% Nb_2O_5 and 0.7% Ta_2O_5. The perovskite group minerals show extensive isomorphic substitutions of the rare earth elements, thorium, and uranium for calcium by schemes such as the following:

(a) $Ca^{2+} + Ca^{2+} \rightleftharpoons Ce^{3+} + Na^+$
(b) $Ca^{2+} + Ti^{4+} \rightleftharpoons Ce^{3+} + Fe^{3+}$
(c) $Ca^{2+} + Ti^{4+} \rightleftharpoons Na^+ + (Nb,Ta)^{5+}$
(d) $Ca^{2+} + Ca^{2+} + Ti^{4+} \rightleftharpoons Th^{4+} + Na^+ + Fe^{3+}$
(e) $Ca^{2+} + Ca^{2+} + Ti^{4+} \rightleftharpoons U^{4+} + Na^+ + Fe^{3+}$

As a result, a general formula for perovskite is (Ca,Ce,Na,Th,U)(Ti,Nb,Ta,Fe)O_3.

Perovskite crystals are cubic or cube-octahedral in habit (Plate 27d); loparite forms star-like interpenetration twins of cubic crystals (Plate 27e). Perovskite has a resinous luster, a hardness of 5.5–6, and a weak-colored, gray-white streak. Neither perovskite nor loparite are easy to identify. Unlike magnetite, perovskite is not magnetic and has a pale streak.

In the crust these minerals occur almost exclusively in massifs of alkaline nepheline-bearing rocks, such as nepheline syenites, alkaline pyroxenites, and nepheline-bearing pegmatites, where they form scarce disseminated black grains. Alkaline pyroxenites of some of the massifs host large masses of fine-grained "fish-roe"-like perovskite, which constitutes up to 80–90% of the rock (an example is in the Afrikanda deposit on the Kola Peninsula, Russia). Perovskite minerals are also found in carbonatites. Loparite-bearing nepheline syenites are mined as a tantalum ore.

In the lower mantle of the Earth, silicate minerals with perovskite-like structures are the major components and will be discussed in Chapter 38. Most important is **bridgmanite** ($MgSiO_3$), also documented in meteorites, $CaSiO_3$ perovskite and "postperovskite" ($MgSiO_3$) at the core–mantle boundary.

In materials science the term "perovskite" is applied to a large number of synthetic compounds with a composition ABO_3 and a structure similar to that of perovskite. $CdTiO_3$, $CaSnO_3$, $BaThO_3$, $KMgF_3$, Pb$(Zr,Ti)O_3$ (PZT), Pb$(Sc,Ta)O_3$ (PST), $LaFeO_3$, and $BaPbBi_2O_6$ are just a few of these synthetic perovskites. Some have very useful ferroelectric properties that are applied in electronic devices. The perovskite structure has received a lot of attention because of its connection to high-temperature superconductivity in compounds such as $YBa_2Cu_3O_6$.

The *pyrochlore group* includes minerals of complex composition, among which are the three end members **pyrochlore** (Nb), **microlite** (Ta), and **betafite** (Ti). Similar isomorphic substitutions as in perovskite are found in the pyrochlore minerals. A general formula for pyrochlore is (Ca,Na,Ce,Th,U)$_2$(Nb,Ta,Ti)$_2O_6$(F,OH). Pyrochlore occurs as octahedral or cube-octahedral crystals and as rounded grains. As a rule, they have experienced metamictization and thus look like amorphous masses with a greasy luster and a shell-like fracture. Only a few pyrochlores exhibit a strong adamantine luster. Their color is yellow, brown, or black. This mineral is often radioactive.

Fig. 27.14 Aggregate of uraninite with the appearance of drooping pitch, hence the name pitchblende, from Schneeberg, Saxony, Germany (courtesy A. Massanek). Width 110 mm.

Pyrochlore forms in alkaline (nepheline syenite) pegmatites, in calcite veins within nepheline syenite intrusions, and in carbonatites. Pyrochlore ores are mined for niobium, tantalum, uranium, and rare earth elements.

Cubic **uraninite** (UO_2, approximately), also known as pitchblende, is always partly oxidized to the hexavalent state, which is present in the form of UO_2^{2+} uranyl ions. The hexavalent uranium occupies U^{4+} sites in the structure. The U^{4+} may be substituted by, among others, thorium, cerium, and lead, sometimes in significant amounts. It has black color, displays a resinous luster, and it is strongly radioactive. Uraninite is found in uranium-bearing skarns, in granites, in granite pegmatites as black cube-like crystals, and in uranium-bearing medium-temperature hydrothermal veins as kidney-shaped (Figure 27.14) and radiaxial aggregates. Uraninite is mined as an ore for uranium, thorium, radium, and rare earth elements. Red, orange, or bright-yellow fine-grained mixtures of secondary uranium minerals often replace uraninite. These have a glue-like appearance, consist of uranium hydroxides, carbonates, silicates, and phosphates; and are referred to as "gummites".

Hydroxides

Brucite ($Mg(OH)_2$) forms colorless platy crystals with hexagonal shape. This mineral is relatively rare, and it crystallizes as one of the products of hydrothermal alteration of ultramafic rocks and in some marbles, replacing olivine. Brucite is found in fine-grained masses, in crystals, and sometimes in parallel-acicular aggregates.

The most common *aluminum hydroxides* are **boehmite** ($AlOOH$), **diaspore** ($AlOOH$), and **gibbsite** ($Al(OH)_3$). The structural features with chains and tunnels parallel to the *c*-axis (see Figure 27.10) are expressed in the crystal morphology of boehmite and diaspore: crystals are elongate-platy in habit. Since hydrogen bonds play a role in the structures of boehmite and diaspore, these minerals are harder than gibbsite, which has no such bonds. Gibbsite and diaspore only rarely form crystals. Those of gibbsite are colorless hexagonal platelets with a perfect (001) cleavage and occur together with other products of hydrothermal alteration of nepheline. Diaspore is found as platy crystals with a rough striation and a strong pearly luster in metamorphic deposits. It has a hardness between 6 and 7.

Aluminum hydroxides rarely occur as individual grains. More often they associate with each other, with iron hydroxides and oxides, and with kaolinite to form *bauxites* (named after their occurrence at Les Baux, in Provence, France). The bauxites are compact fine-grained aggregates, loose or clay-like masses that often have oolitic texture. Their color is white, gray, red, or dark red. Bauxites form as a product of surficial weathering of granites in a tropical or subtropical climate. (Bauxites in France are of Eocene age, when the climate was very different from what it is today.) Genetically bauxites resemble laterites, but the latter form by weathering of ultramafic and mafic rocks and thus are poorer in aluminum hydroxides. The bauxites are a major source of aluminum and alumina, the latter a raw material used for manufacturing ceramics and cement.

Iron hydroxide minerals rarely occur as individual crystals. Goldish-brown needles of **goethite** ($Fe^{3+}OOH$) (Plate 27f) occur with quartz inside quartz-chalcedony amygdules in volcanic and sedimentary rocks. They also form botryoidal aggregates (Figure 10.19). **Lepidochrocite** ($Fe^{3+}OOH$) forms as goldish-brass-yellow micaceous coatings on hematite and other oxidized iron ores. More frequently, these minerals form part of mixtures known as *limonite*. Such mixtures are found as rusty-brown to black, earthy, kidney-shaped, fine-grained aggregates (often pseudomorphic after pyrite, hematite, and other minerals), and as oolites. The luster is dull or vitreous, and the streak is rusty-brown in color. The brown iron ores form due to surficial oxidation of different iron ores and iron minerals, and as colloid-chemical precipitates in sedimentary rocks. These minerals are an ore for iron, and powdery masses of them are used for paint manufacture.

Manganese hydroxides comprise about 20 mineral species that are combinations of manganese and elements such as barium, calcium, nickel, and zinc. All of them are morphologically alike. The manganese hydroxides commonly occur in association with pyrolusite (MnO_2) as earthy, colloform, oolitic aggregates. Such mixtures are often called *psilomelane*. They are black to brownish-black in color and have a brownish-black oily streak. The hardness of these minerals ranges from 2 to 6. Precise identification of these minerals is usually carried out by X-ray diffraction and chemical analysis. The mixtures that contain much water are brown in color and are called *wads*. They occur as dendrites on fracture surfaces (see Plate 4b), and as desert varnish, which is a red, brown, or black coating that forms on the surface of exposed rocks in desert areas. The manganese hydroxides form in some hydrothermal veins, due to weathering of manganese-bearing minerals, and as colloidal-chemical precipitates on the oceanic floor. They are an ore for manganese.

Manganite (MnOOH) is often associated with other manganese hydroxides formed by meteoric water and in low-temperature hydrothermal veins. It frequently alters to pyrolusite.

Romanechite ($BaMn^{2+}Mn_9^{4+}O_{20}·3H_2O$) and other minerals in this group constitute complex crystallohydrate manganese oxides with "tunnel" structures (similar to those shown in Figure 27.8b–d). The manganese in these minerals can display different valencies, and thus there is variability in the composition of the romanechite group. The romanechite group minerals occur as dark-brownish-black, brownish-black, or black powdery loose masses, or kidney-shaped aggregates, oolites, concretions, and dendrites. They have a powdery black or brownish-black streak. The romanechite minerals form in sedimentary rocks and in the oxidation zones of manganese-rich ores. Like most manganese oxides, they are used as a manganese ore.

Recently formed iron–manganese concretions from the ocean floor usually show very complex modal

and chemical compositions that may differ from one part of a concretion to another. Iron is concentrated in the form of hydroxides. The manganese has a different valency and is represented by oxides such as **pyrolusite** and **hollandite** ($BaMn_2^{3+}Mn_6^{4+}O_{16}$) with Mn^{3+} and Mn^{4+}, and by numerous poorly crystallized hydroxides such as **romanechite**, **todorokite** $((Mn^{2+}, Ca, Mg)Mn_3^{4+}O_7·H_2O)$, and **birnessite** $Na(Mn_3^{4+}Mn^{3+})O_8·3H_2O$. Nonferrous metals play a role as ion-exchange cations in some hydroxides of complex layered or tunnel structures. Frequently the concretions also include some fine-grained quartz, clay minerals, and zeolites. The iron–manganese concretions are a potential source for the extraction of nonferrous metals. Although the latter are present in small quantities (1–1.5% Cu; 1–1.5% Ni; about 0.2% Co) the economic importance is significant because the reserves are very large.

27.5 Summary

The chapter on oxide minerals is used to first review some basic ionic mineral structures. Many can be described by close-packing of larger anions and smaller cations in tetrahedral or octahedral interstices. Close-packing can be cubic or hexagonal. Of all known ionic A–X compounds 85% crystallize in one of those structures. The principle of close-packing also applies to more complex oxides such as "corundum" structures with hexagonal close-packing and aluminum in two-thirds of octahedral interstices or "spinel" structures with cubic close-packing and some cations in octahedral and some in tetrahedral interstices. Perovskite $CaTiO_3$ is a derivative with O^{2-} in an fcc arrangement (close-packing) but leaving one position vacant in the center of the cube, creating a large site with 12-fold coordination that is occupied by Ca^{2+} while Ti^{2+} is in octahedral sites. In other oxides and hydroxides, occupied octahedra form chains (e.g., rutile) and link to create complex tunnels (as in romanechite), or form sheets (as in brucite and gibbsite). In a section we also review H_2O, which has a variety of structures, the most common being hexagonal ice Ih with a similar structure as tridymite (SiO_2). In ice there are both covalent bonds and hydrogen bonds. A special type of structure in ice compounds is clathrates with large cages that contain methane and form large deposits in the deep ocean floor.

Important oxide minerals to remember

Name	Formula	System
Cuprite	Cu_2O	Cubic
Periclase	MgO	Cubic
Corundum	Al_2O_3	Trigonal
Hematite	$Fe_2^{3+}O_3$	Trigonal
Ilmenite	$Fe^{2+}TiO_3$	Trigonal
Spinel	$MgAl_2O_4$	Cubic
Magnetite	$Fe^{2+}Fe_2^{3+}O_4$	Cubic
Chromite	$Fe^{2+}Cr_2O_4$	Cubic
Rutile	TiO_2	Tetragonal
Cassiterite	SnO_2	Tetragonal
Pyrolusite	MnO_2	Tetragonal
Perovskite	$CaTiO_3$	Pseudocubic
Bridgmanite	$MgSiO_3$	Orthorhombic
Brucite	$Mg(OH)_2$	Trigonal
Gibbsite	$Al(OH)_3$	Monoclinic
Goethite	$Fe^{3+}O(OH)$	Orthorhombic
Manganite	$Mn^{2+}Mn^{4+}O_2(OH)_2$	Monoclinic
Ice	H_2O	Hexagonal

Test your knowledge

1. How can the principle of close-packing be used to explain the most common ionic structures? Think of examples for each group.
2. Give examples of structures in which the cation is (a) small, (b) intermediate, and (c) large. Give a mineral example for each and describe the coordination of the cation.
3. Some ionic structures are best described as polyhedra that are linked to form three-dimensional arrangements. Give examples of oxides with tetrahedra and with octahedra.
4. What structures does Mg_2SiO_4 (olivine at low pressure) have at high pressure in the Earth's mantle?
5. Oxides are important raw materials for iron, chromium, titanium, and manganese. Name the most important minerals used for the production of these metals and give their chemical composition.
6. Review the P–T phase diagram of SiO_2 and enter stability fields of the six most important silica polymorphs (see also Chapters 6 and 19).
7. The perovskite structure is of interest in geophysics and also in materials science. Why?
8. Which oxide minerals have tunnel structures?

Further reading

Lindsley, D. H. (ed.) (1991). *Oxide Minerals: Petrologic and Magnetic Significance*. Reviews in Mineralogy, vol. 25. Mineralogical Society of America, Washington, DC.
Navrotsky, A. and Weidner, D. J. (eds.) (1989). *Perovskite: A Structure of Great Interest to Geophysics and Materials Science*. Geophysical Monograph 45. American Geophysical Union, Washington, DC.
Rumble, D. (ed.) (1976). *Oxide Minerals*. Reviews in Mineralogy, vol. 3. Mineralogical Society of America, Washington, DC.

28 | Orthosilicates and ring silicates. Metamorphic mineral assemblages

Silicate minerals constitute over 90 vol.% of the Earth's crust and are thus the most common minerals that we encounter. Among the elements in the crust, oxygen (47 atomic %), silicon (28%), aluminum (8%), iron (5%), and calcium (4%) are the most abundant, and therefore minerals containing silica in combination with these other elements dominate. Feldspars, the most common silicate minerals, were discussed in Chapter 21. In the next four chapters we take a closer look at the large variety of silicates and their significance as rock-forming minerals. This chapter discusses first the general structural characteristics of silicates that are used for their classification into orthosilicates, sheet silicates, chain silicates, and framework silicates. Then focus will be on orthosilicates and ring silicates in which silicon tetrahedra are not linked to infinite structures, and we discuss their occurrence in metamorphic rocks.

28.1 General comments on silicates

Like other ionic compounds, silicate structures are built up of coordination polyhedra, mainly tetrahedra and octahedra. The Si–O bond is only about half ionic, while the remainder is covalent. The covalent Si $3p^3$–O $2p$ hybrid bonds (Figure 28.1) have *directional* properties, contrary to the spherical symmetry of electrostatic attraction in ionic bonding. Because of this, silicates in general have low crystal symmetry, a high anisotropy of physical properties, and a complex crystal structure

with large unit cells. The bonds define a coordination tetrahedron SiO_4^{4-}, which is the basic building unit of silicate minerals (see also Figure 21.2).

There are several important cation substitutions in silicates, on tetrahedral, octahedral, and larger structural sites (Table 28.1). It is significant that some of these substitutions are between ions of different charge. Because of the required overall electrostatic neutrality, it may be necessary to have coupled substitutions such as in plagioclase feldspar, as we have seen in Chapter 21, i.e.,

$$(Ca^{2+} \rightleftharpoons Na^+)(Al^{3+} \rightleftharpoons Si^{4+})\, AlSi_2O_8$$

Aluminum, with a radius ratio $r_{Al}:r_O = 0.43$, is close to the limit between octahedral and tetrahedral coordination (0.414) and can substitute either for tetrahedral silicon or for octahedral ions. This is one of the reasons for the tremendous variety of silicate structures. Silicates with all Al^{3+} in only tetrahedral coordination are called *alumosilicates*, and examples include all feldspars such as microcline ($KAl^{IV}Si_3O_8$; the roman superscript is used to indicate the coordination number), and the zeolites such as heulandite ($CaAl_2^{IV}Si_7O_{18} \cdot nH_2O$). Other silicates have at least some aluminum in octahedral coordination and those are called *aluminosilicates*. Among them are topaz

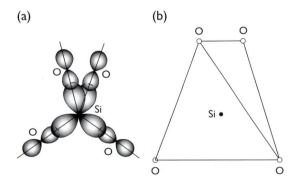

Fig. 28.1 Hybrid Si $3p^{3-}$ and O $2p$-orbitals define a SiO_4^{4-} tetrahedron.

Table 28.1 Important cations substituting in silicates, ordered in each group according to size

Coordination number	4 (tetrahedral)	6 (octahedral)	>6
Cations	Be^{2+}, Si^{4+}, Al^{3+}	Al^{3+}, Ti^{4+}, Fe^{3+}, Mg^{2+}, Fe^{2+}, Mn^{2+}	Ca^{2+}, Sr^{2+}, Na^+, K^+, Ba^{2+}

$(Al_2^{VI}(SiO_4)F_2)$, grossular garnet $(Ca_3Al_2^{VI}(SiO_4)_3)$, muscovite $(KAl_2^{VI}(Al^{IV}Si_3O_{10})(OH)_2)$, kyanite $(Al_2^{VI}SiO_5)$, and sillimanite $(Al^{VI}Al^{IV}SiO_5)$.

In order to classify the 150 different silicate structures that are known to date, various systems have been designed. Because of frequent substitutions, a rigorous *chemical classification* has failed. For example, take the magnesium silicates olivine (Mg_2SiO_4), enstatite ($MgSiO_3$), and talc ($Mg_3Si_4O_{10}(OH)_2$). All contain only Mg^{2+} and Si^{4+} as cations and one would expect similarities, yet the three minerals have entirely different morphologies and properties. The same is true for the magnesium–calcium silicates monticellite ($CaMgSiO_4$), akermanite ($Ca_2MgSi_2O_7$), and diopside ($CaMgSi_2O_6$), all with Ca^{2+}, Mg^{2+}, and Si^{4+} as cations. By contrast, minerals with different compositions are often very similar. It is difficult to distinguish olivine from monticellite, or enstatite from diopside. The reason is that the two pairs have very similar crystal structures. Therefore, a structural classification has been introduced that groups silicates based on structure rather than composition. For example, minerals such as enstatite ($MgSiO_3$), jadeite ($NaAlSi_2O_6$), and spodumene ($LiAlSi_2O_6$), with little chemical resemblance, all belong to the group chain silicates, or more specifically, pyroxenes. A structural classification is universally applied to silicates (e.g., Liebau, 1985).

For the structural classification of silicates, however, the system for ionic structures (Section 27.2), relying on the concept of close-packing of anions with cations in interstices, is not very appropriate. There are only a few silicate minerals with anions that are more or less close-packed. A natural grouping is based on silicon tetrahedra and their linkage. Each oxygen in a tetrahedron, such as those in Figure 28.1a, may share electrons with another tetrahedron, establishing linkages. This linking of tetrahedra is also referred to as *polymerization*, a term borrowed from organic chemistry.

A structural classification that primarily emphasizes the linkage of SiO_4^{4-} tetrahedra is used in most modern descriptions of silicates. Tetrahedra are either isolated

SiO_4^{4-} as in olivine (Figure 28.2a), or form groups of two with two tetrahedra sharing one oxygen $Si_2O_7^{6-}$ (e.g., lawsonite $(CaAl_2(Si_2O_7)(OH)_2 \cdot H_2O)$), Figure 28.2b). They also may form rings with three $Si_3O_9^{6-}$ (e.g., benitoite ($BaTiSi_3O_9$)), four $Si_4O_{12}^{8-}$ (e.g., axinite ($Ca_2Fe^{2+}Al_2OHBO_3(Si_4O_{12})$)), or six tetrahedra $Si_6O_{18}^{12-}$ (e.g., beryl ($Be_3Al_2(Si_6O_{18})$)), Figure 28.2c) where two corners of each tetrahedron are shared.

In *chain silicates* the tetrahedra are linked to form infinite chains, with either single chains with units $Si_2O_6^{4-}$, as in diopside ($CaMg(Si_2O_6)$) (Figure 28.2d), or double chains, with units $Si_4O_{11}^{6-}$, as in hornblende $((Na,K)_{0-1}(Ca,Na)_2 (Mg,Fe^{2+}) (Al,Fe^{3+}) (Si_7AlO_{22}) (OH)_2)$ (Figure 28.2e). In *sheet silicates*, such as mica, the tetrahedra form hexagonal sheets in which three corners of the tetrahedra are shared and the free corners all point in the same direction. The unit is $Si_2O_5^{2-}$ as in muscovite (Figure 28.2f). Finally, in *framework silicates*, all four corners of the tetrahedron are shared, resulting in a three-dimensional framework, SiO_2, such as that for tridymite (Figure 28.2g), quartz, and feldspars.

In silicates most oxygen atoms are linked to tetrahedra. The different groups, therefore, have characteristic Si:O ratios, which we can determine by counting all oxygen ions associated with tetrahedra. In orthosilicates with isolated tetrahedra all four oxygen ions belong to a single tetrahedron and the Si:O ratio is 1:4, as in Mg_2SiO_4. In a ring or a single-chain silicate, two oxygen ions of each tetrahedron are shared with adjacent tetrahedra and two are not $\left(2 \times \frac{1}{2} + 2 \times 1\right)$, resulting in a Si:O ratio of 1:3, as in $MgSiO_3$. In a framework silicate, all four oxygen ions are shared between adjacent tetrahedra $\left(4 \times \frac{1}{2}\right)$ and the Si:O ratio is 1:2 as in SiO_2. Note that tetrahedral aluminum is included with silicon in this count (e.g., in feldspars as in $K(AlSi_3O_8)$). In this case the tetrahedral ions (Al + Si) are sometimes expressed by the symbol T, for "tetrahedral". Notice that the tetrahedral component in silicates is often highlighted in the formula by putting it into parentheses or

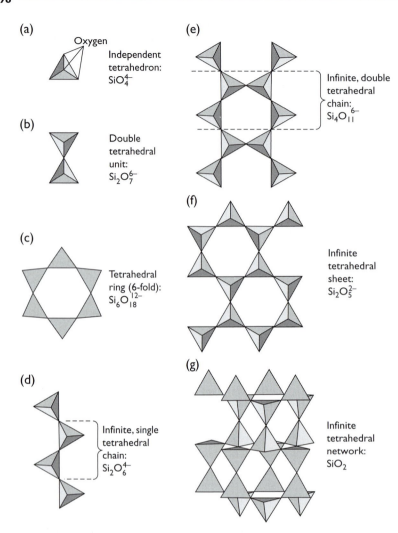

(a) Oxygen
Independent tetrahedron: SiO_4^{4-}

(b) Double tetrahedral unit: $Si_2O_7^{6-}$

(c) Tetrahedral ring (6-fold): $Si_6O_{18}^{12-}$

(d) Infinite, single tetrahedral chain: $Si_2O_6^{4-}$

(e) Infinite, double tetrahedral chain: $Si_4O_{11}^{6-}$

(f) Infinite tetrahedral sheet: $Si_2O_5^{2-}$

(g) Infinite tetrahedral network: SiO_2

Fig. 28.2 (a,b) Linkage of SiO_4^{4-} tetrahedra in *orthosilicates* with (a) isolated tetrahedra as in olivine and (b) groups of two linked tetrahedra as in lawsonite. (c) Linkage of SiO_4^{4-} tetrahedra in *ring silicates* with rings of six as in beryl. (d,e) Linkage of SiO_4^{4-} tetrahedra in *chain silicates* with (d) a single chain as in pyroxenes (translational repeat is indicated) and (e) a double chain as in amphiboles (translational repeat is indicated). (f) Linkage of SiO_4^{4-} tetrahedra in *sheet silicates* with an infinite two-dimensional sheet. (g) Linkage of SiO_4^{4-} tetrahedra in framework *silicates* as in tridymite (with the *c*-axis vertical). Oxygen atoms are in the corners of tetrahedra; silicon atom is in the center of tetrahedra.

brackets, e.g., $KAlSi_3O_8$ is written as $K(AlSi_3O_8)$ or K $[AlSi_3O_8]$. Here we generally use parentheses.

In this book we do not include silicates with exotic crystal structures such as those with broken (interrupted) frameworks (e.g., wenkite $(Ba,K)_4(Ca,Na)_6((SO_4)_3|(Si,Al)_{20}O_{39}(OH)_2)\cdot0.5H_2O)$, or with tube (tunnel) stuctures (e.g. charoite $(K,Sr,Ba)_5(Si_6O_{15})2$ $(Si_2O_7)(Si_4O_9)OH\cdot3H_2O$ and many others. They are numerous, but as rock-forming minerals they are insignificant.

Similar relationships exist for groups of two tetrahedra with Si:O ratios $2 : \left(2 \times \dfrac{1}{2} + 6 \times 1\right) = 2 : 7$ (Figure 28.2b), as in lawsonite $(CaAl_2(Si_2O_7)$

$(OH)_2 \cdot H_2O)$, for double chains with an Si:O ratio of $4 : \left(10 \times \dfrac{1}{2} + 6 \times 1\right) = 4 : 11$ (Figure 28.2e), as in tremolite $(Ca_2Mg_5(Si_8O_{22})(OH)_2)$, and for sheets $2 : \left(6 \times \dfrac{1}{2} + 2 \times 1\right) = 2 : 5$ (Figure 28.2f), as in kaolinite $(Al_2(Si_2O_5)(OH)_2)$. In these cases the ratios are best determined by counting Si and O within a repeat period. If we know the chemical formula and thus the Si:O or *T*:O ratio for a silicate mineral, we can assign it to a group. (There are a few exceptions to this simple rule. In orthosilicates in particular there are structures that have both isolated tetrahedra and groups. Also, there are examples where not all oxygen

Table 28.2 Classification of silicate structures according to polymerization of tetrahedra (cf. Figure 28.2). Parentheses are used in the formulas to emphasize the characteristic polymerization of Si tetrahedra.

Group	Structure	Si:O ratio	Example	Formula
Orthosilicates	Isolated tetrahedra	1:4	Olivine	$Mg_2(SiO_4)$
	Groups of two tetrahedra	1:3.5	Lawsonite	$CaAl_2(Si_2O_7)\,(OH)_2 \cdot H_2O$
Ring silicates	Rings of tetrahedra	1:3	Benitoite	$BaTi(Si_3O_9)$
Chain silicates	Single chains of tetrahedra	1:3	Enstatite	$Mg_2(Si_2O_6)$
	Double chains of tetrahedra	1:2.75	Tremolite	$Ca_2Mg_5(Si_8O_{22})(OH)_2$
Sheet silicates	Two-dimensional net	1:2.5	Kaolinite	$Al_2(Si_2O_5)(OH)_2$
Framework silicates	Three-dimensional network	1:2	Quartz	SiO_2
			Albite	$Na(Al^{IV}Si_3O_8)$

atoms are linked to silicon tetrahedra. Nevertheless, it is a good rule of thumb.) Table 28.2 summarizes the tetrahedral classification of silicates and gives an example for each group.

Except for pure SiO_2, which has a framework structure, the tetrahedra are not charge balanced and in silicates insertion of additional cations produces electrostatic neutrality. We will follow mainly this tetrahedral classification, first suggested by W. L. Bragg (1930), but we will keep in mind that, like all classifications, it is artificial and designed only to organize our thoughts. It is increasingly recognized that numerous intermediate structures exist between these groups. In many cases it is more useful to emphasize similarities and relationships between various structures in minerals of the different groups, rather than differences.

28.2 Orthosilicates

In this section we review those silicate minerals that have isolated tetrahedra and groups of tetrahedra. Minerals in this group are generally characterized by a low Si:O ratio, by fairly close-packing and a corresponding high density. Table 28.3 lists common mineral representatives and some of their diagnostic properties.

Olivine ($(Mg,Fe)_2(SiO_4)$) is typical of silicates with isolated tetrahedra. It forms a continuous solid solution between pure magnesium olivine *forsterite* ($Mg_2(SiO_4)$) and pure iron olivine *fayalite* ($Fe_2(SiO_4)$) (see Figure 20.4). The structure of olivine has orthorhombic symmetry, with the oxygen atoms forming nearly hexagonal close-packing and Figure 28.3a

shows two close-packed layers of oxygen atoms (large circles), parallel to (100). Magnesium and iron occupy half of the octahedral interstices (intermediate circles), and silicon is in one-eighth of the tetrahedral interstices (small circles). The sites are more easily visible in the representation of Figure 28.3b, which emphasizes coordination octahedra with different shadings indicating the levels along the a-axis. Note that the ribbons of edge-sharing octahedra parallel to the c-axis are linked over corners by tetrahedra. Figure 28.3c is an even more compact representation of the structure, showing only tetrahedra and octahedral cations (circles). In this figure it is obvious that tetrahedra are isolated in rows, pointing alternately up and down.

Figure 28.3 is an idealized view of the structure of olivine. In reality there are distortions. In the crystallographic representation of the unit cell in Figure 28.4 cations are labeled M (for metal) and T (for tetrahedral). As this figure illustrates, there are two octahedra of slightly different shapes, one of them called $M1$ and the other one $M2$. In particular, the $M1$ octahedra deviate considerably from an ideal octahedral geometry. Under most conditions iron and magnesium are distributed fairly randomly over the two structural sites, i.e., the Mg/Fe distribution is disordered.

In the tetragonal mineral *zircon* ($ZrSiO_4$), eight oxygen ions in an irregular dodecahedral coordination surround the large zirconium ion, while silicon is surrounded by four oxygen ions in a tetrahedron. The tetrahedra and dodecahedra share edges and extend as chains parallel to the c-axis (Figure 28.5a). Because of its size, the dodecahedron can accomodate larger ions such as uranium, thorium, and yttrium and those

Table 28.3 Common ortho- and ring silicate minerals with some diagnostic properties; most important minerals are given in italics

Mineral Formula	System a (Å)	Morphology Cleavage	H	D	Color	n Pleochroism	Δ	2V (Dispersion)
Orthosilicates								
Olivines								
Forsterite $Mg_2(SiO_4)$	Ortho.	Eq. *(010) poor*	6.5–7	3.2	Yellow-green	1.64–1.67	0.035	+86 (r < v)
Fayalite $Fe_2(SiO_4)$	Ortho.	Eq. *(010) poor*	6.5–7	4.3	Black	1.84–1.89	0.051	–47 (r < v)
Humite minerals								
Chondrodite $Mg_5(SiO_4)_2F_2$	Monocl.	Platy (010) *(100)*	6–6.5	3.2	Yellow, brown	1.60–1.66	0.03	72–90
Humite $Mg_7(SiO_4)_3F_2$	Ortho.	Platy (010) *(001)*	6–6.5	3.2	Yellow, brown	1.61–1.67	0.03	+65–84 (r > v)
Clinohumite $Mg_9(SiO_4)_4(OH)_2$	Monocl.		6–6.5	3.2	Yellow, brown	1.63–1.66	0.03	+74–90 (r > v)
Garnets								
Almandine $Fe_3^{2+}Al_2(Si_3O_{12})$	Cubic 11.53	Eq. {211}	6.5–7.5	4.2	Red, brown	1.76–1.83		
Andradite $Ca_3Fe_2^{3+}(Si_3O_{12})$	Cubic 12.05	Eq. {110}	6.5–7.5	3.8	Brown, black	1.89		
Grossular $Ca_3Al_2(Si_3O_{12})$	Cubic 11.85	Eq. {211}	6.5–7.5	3.5	White, pale-brown	1.74		
Pyrope $Mg_3Al_2(Si_3O_{12})$	Cubic 11.46	Eq. {211}	6.5–7.5	3.5	Crimson	1.76–1.83		
Spessartine $Mn_3Al_2(Si_3O_{12})$	Cubic 11.62	Eq. {211}	6.5–7.5	4.2	Pink	1.80		
Uvarovite $Ca_3Cr_2(Si_3O_{12})$	Cubic 12.00	Eq. {110}	6.5–7.5	3.9	Green	1.87		
Various orthosilicates								
Zircon $Zr(SiO_4)$	Tetrag.	Pris. [001] *{100} poor*	7–8	4.2	Brown, red	1.96–2.01	0.05	(+)

Mineral / Formula	System / Habit, cleavage	H	SG	Color	n		δ	2V (dispersion)
Topaz $Al_2(SiO_4)F_2$	Ortho. Pris. [001] (001)	8	3.6	Clear, yellow, blue	1.61–1.64		0.01	+48–65 (r > v)
Group silicates								
Epidote minerals								
Allanite $CaCeFe^{2+}Al_2(SiO_4)(Si_2O_7)O(OH)_2$	Monocl. Platy (100) (001)	5.5	3.6	Black	1.72–1.76	*Yellow-green-brown*	0.04 Anom.	90 (r > v)
Clinozoisite $Ca_2Al_3(SiO_4)(Si_2O_7)O(OH)_2$	Monocl. Pris. [010] (001)	6.5	3.4	Green-gray	1.72–1.73		0.01	+85 (r < v)
Epidote $Ca_2Fe^{3+}Al_2(SiO_4)(Si_2O_7)O(OH)_2$	Monocl. Pris. [010] (001)	6–7	3.4	Yellow-green, green	1.73–1.78	*(Clear-green-yellow)*	0.045 Anom.	−68–73
Piemontite $Ca_2Al_2Mn^{3+}(SiO_4)(Si_2O_7)O(OH)_2$	Monocl. Pris. [010] (001)	6–6.5	3.4	Red, brown	1.70–1.71	*Yellow/orange-pink-red*	0.04	+70–(−)70
Zoisite $Ca_2Al_3(SiO_4)(Si_2O_7)O(OH)_2$	Ortho. Pris. [010] (010)	6.5	3.3	Green-gray	1.70–1.71		0.004	+30–60
Various group silicates								
Vesuvianite $Ca_{10}Mg_2Al_4(SiO_4)_5(Si_2O_7)_2(OH)_4$	Tetrag. Pris. [001] (100) poor	6.5	3.3	Red-brown, green	1.70–1.73		0.04 Anom.	(−) (r < v)
Kyanite $Al_2(SiO_4)O$	Tricl. Pris. [001] (100) (010)	4–7	3.6	Blue	1.71–1.73	*(Clear-violet-blue)*	0.016	−82 (r > v)
Sillimanite $Al(SiAlO_4)O$	Ortho. Fibr. [001] (010)	6–7	3.2	White, gray	1.65–1.68		0.02	+25–30 (r > v)
Andalusite $AlAl(SiO_4)O$	Ortho. Pris. [001] {110}	7.5	3.2	Gray, red-brown	1.63–1.65	*(Rose)-clear-clear*	0.01	−83–85 (r < v)
Staurolite $(Fe^{2+},Mg)_2(Al,Fe^{3+})_9O_6(SiO_4)_4OOH$	Ortho. Pris. [001] (001)	7–7.5	3.7	Brown	1.74–1.76	*Clear-yellow-brown*	0.01	+79–88 (r > v)
Chloritoid $Fe_2^{2+}AlAl_3(SiO_4)_2O_2(OH)_4$	Monocl. Platy (001) (001)	6.5	3.5	Green, black	1.71–1.74	*Green-blue-yellow*	0.01	+36–68 (r > v)
Lawsonite $CaAl_2(OH)_2(Si_2O_7)·H_2O$	Ortho. Pris. (010) (010), (100)	6	3.1	White, gray, green	1.67–1.68		0.035	+84
Titanite $CaTi(SiO_4)O$	Monocl. {111} {110}, (100)	5–5.5	3.5	Yellow, green-brown	1.90–2.04		0.13	+23–34 (r ≫ v)

Table 28.3 (cont.)

Mineral Formula	System a (Å)	Morphology Cleavage	H	D	Color	n Pleochroism	Δ	$2V$ (Dispersion)
Ring silicates								
Tourmalines								
Schorl $NaFe_3Al_6(Si_6O_{18})(BO_3)_3(OH)_4$	Trig.	Pris. [0001]	7	3.2	Black	1.63–1.69 *Light–dark*	0.02	(−)
Rubellite $NaLi_{1.5}Al_{1.5}Al_6(Si_6O_{18})(BO_3)_3(OH)_4$	Trig.	Pris. [0001]	7	3.1	Pink	1.62–1.66 *Gray–pink*	0.02	(−)
Various ring silicates								
Axinite $Ca_2Fe^{2+}Al_2OHBO_3(Si_4O_{12})$	Tricl.	Platy (100)	6.5–7	3.3	Brown, gray	1.68–1.69	0.01	−63–76 (r < v)
Benitoite $BaTi(Si_3O_9)$	Hexag.	Pyr.	6.5	3.7	Black	1.76–1.80	0.047	(+)
Beryl $Be_3Al_2(Si_6O_{18})$	Hexag.	Pris. [0001]	7.0–8	2.7	White, green, blue	1.57–1.60	0.006	(−)
Cordierite $Al_3(Mg,Fe^{2+})_2(Si_5AlO_{18})$	Ortho.	Pris. [001] (010)	7–7.5	2.6	Blue, gray	1.54–1.55	0.007	−40–80 (r < v)
Dioptase $Cu_6(Si_6O_{18})\cdot 6H_2O$	Trig.	Pris. [0001] $\{01\bar{1}1\}$	5	3.3	Green	1.64–1.70	0.05	(+)

Notes: H, hardness; D, density (g/cm^3); n, range of refractive indices; Δ, birefringence; $2V$, axial angle for biaxial minerals. For uniaxial minerals (+) is positive and (−) is negative. Acute $2V$ is given in the table. If $2V$ is negative the mineral is biaxial negative and $2V$ is $2V_\alpha$; if it is positive, the mineral is biaxial positive and $2V$ is $2V\gamma$. Dispersion r < v means that acute $2V$ is larger for violet than for red; anom., anomalous dispersion or birefringence; a, lattice parameter for garnets.

System: Hexag., hexagonal; Monocl., monoclinic; Ortho., orthorhombic; Tricl., triclinic; Trig., trigonal; Tetrag., tetragonal.
Morphology: Eq., equiaxed; Fibr., fibrous; Pris., prismatic; Pyr., pyramidal.

(a) (b) (c)

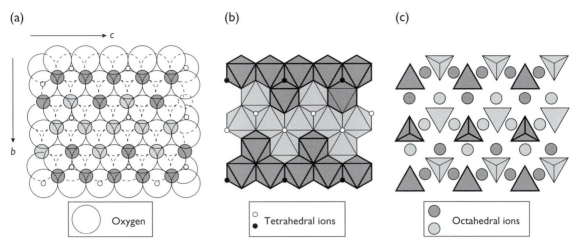

| | Oxygen |

| ○ ● | Tetrahedral ions |

| ● ● | Octahedral ions |

Fig. 28.3 Different representations of the idealized structure of olivine $Mg_2(SiO_4)$ in a (100) projection. (a) Close-packing of oxygen atoms with two layers shown (large circles, upper layer solid, lower layer dashed). Mg^{2+} (smaller circles) occupy octahedral interstices and Si^{4+} (small circles) occupy tetrahedral interstices. (b) Representation emphasizing octahedrally coordinated cations. The upper layer has darker shading. Occupied tetrahedral interstices are indicated with dots and circles. (c) Representation showing only tetrahedra and Mg^{2+} (circles).

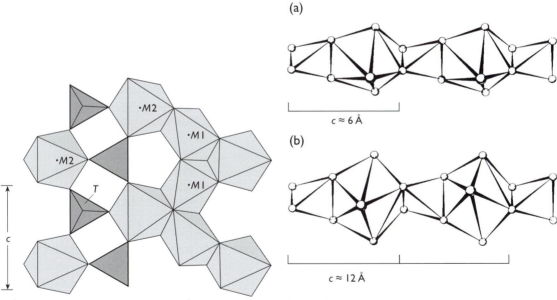

Fig. 28.4 Real structure of olivine in a (100) projection with octahedral and tetrahedral sites. Sites of atoms in the center of tetrahedra are marked with T, those in octahedra with M. There are two octahedral sites and $M1$ octahedra are more distorted than $M2$ octahedra.

(a)

$c \approx 6\,\text{Å}$

(b)

$c \approx 12\,\text{Å}$

Fig. 28.5 Structures of (a) zircon and (b) garnet, represented as chains with alternating tetrahedra containing silicon, and distorted dodecahedra. The chains extend along the c-axis. In zircon the motif repeats after one tetrahedon-dodecahedron unit ($c \approx 6$ Å); in garnet it repeats after two units ($c \approx 12$Å). In the case of garnet (b) two chains are linked by additional octahedra.

Table 28.4 Substitutions and physical properties of some natural garnet end members $X_3^{VIII}M_2^{VI}(Si^{IV}O_4)_3$

	X_3^{VIII}	M_2^{VI}	n	D (g/cm^3)	a (Å)
Almandine	Fe_3^{2+}	Al_2	1.830	4.32	11.53
Andradite	Ca_3	$(Fe^{3+}, Ti)_2$	1.887	3.86	12.05
Grossular	Ca_3	Al_2	1.734	3.59	11.85
Pyrope	Mg_3	Al_2	1.714	3.58	11.46
Spessartine	Mn_3	Al_2	1.800	4.19	11.62
Uvarovite	Ca_3	Cr_2	1.868	3.90	12.00

n, refractive index; D, density; a, cubic lattice parameter.

Table 28.5 Ionic substitutions in epidote minerals $X_2^{VIII}M_3^{VI}SiO_4Si_2O_7(O, OH, F)_2$

	X_2^{VIII} (2 sites)	M_3^{VI} (3 sites)
Zoisite, clinozoisite	Ca_2	Al_3
Epidote	Ca_2	$Fe^{3+}Al_2$
Piemontite	Ca_2	$(Mn^{3+},Fe^{3+},Al)_3$
Allanite	$Ca(La,Y)$	$(Fe^{3+},Mn^{3+},Al)_2Fe^{2+}$

frequently substitute for some of the zirconium in the zircon structure. Zircons are well suited for isotopic age determinations that depend on the radioactive decay of uranium and thorium.

The structure of the cubic mineral *garnet* has similar chains of edge-sharing tetrahedra and dodecahedra (Figure 28.5b), but they are linked in a manner slightly different from that of zircon and the structural motif repeats after two dodecahedral-tetrahedral units (~12 Å). Garnets contain additional cations: octahedrally coordinated aluminum link the dodecahedral-tetrahedral chains. Each oxygen has one tetrahedral (IV), one octahedral (VI), and two dodecahedral (VIII) cations as next neighbors, and the general formula of garnet can be expressed as $X_3^{VIII}M_2^{VI}Si_3^{IV}O_{12}$. This structure lends itself to many isomorphous substitutions. Table 28.4 shows some important substitutions in natural garnets, some corresponding physical properties (refractive index, density), and the a lattice parameter. These properties are used for purposes of identification, and determinative charts have been designed (see Figure 4.2).

The structure of tetragonal *vesuvianite* ($Ca_{10}Mg_2$ $Al_4(SiO_4)_5(Si_2O_7)_2(OH)_4$) is closely related to the garnet structure. The c lattice parameter is almost identical to the a lattice parameter of grossular

(11.85 Å), but contrary to garnet, vesuvianite contains both SiO_4^{4-} tetrahedra and $Si_2O_7^{6-}$ groups.

The *epidote* minerals, with the general formula $X_2^{VIII}M_3^{VI}SiO_4Si_2O_7(O, OH, F)_2$, also combine isolated SiO_4^{4-} tetrahedra and $Si_2O_7^{6-}$ groups. Table 28.5 lists some end-member compositions in this series of minerals in which ionic substitutions on the octahedral site M are ubiquitous.

Polymorphs of Al_2SiO_5, *kyanite, sillimanite*, and *andalusite*, have gained petrological significance because their occurrence in metamorphic pelitic schists can be related to temperature–pressure conditions during crystallization, as will be discussed later. In these minerals, the structures do not fit the simple classification described above and the Si:O ratio rule does not apply because one oxygen atom is not part of any tetrahedra. Thus the formula is sometimes written as Al_2SiO_4O.

In the triclinic high-pressure polymorph *kyanite*, oxygen atoms form cubic close-packing, all aluminum is in octahedral interstices, and silicon is in tetrahedral interstices. In the structure there are chains of edge-sharing octahedra parallel to the c-axis (Figure 28.6a). With its close-packed structure, kyanite has the highest density of the aluminosilicates (3.6 versus 3.2 g/cm^3 for sillimanite and andalusite).

Closely related to kyanite is *staurolite* ($FeAl_4(SiO_4)_2$ $O_2(OH)_2$) (the composition corresponding to 2 kyanite + $Fe(OH)_2$). In staurolite, aluminum substitutes for some of the tetrahedral silicon and additional aluminum atoms are inserted in the octahedral interstices that are unoccupied in kyanite.

In orthorhombic *sillimanite*, half of the aluminum is in octahedral sites and octahedra form chains parallel to the c-axis, as in kyanite. The other half of the aluminum is in tetrahedral sites. Aluminum and silicon tetrahedra alternate to form corner-sharing

(a)

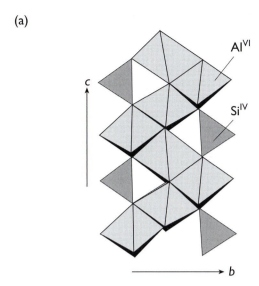

(b)

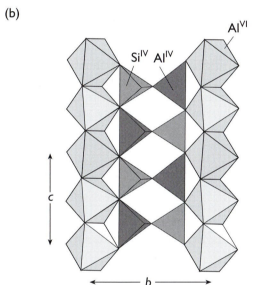

(c)

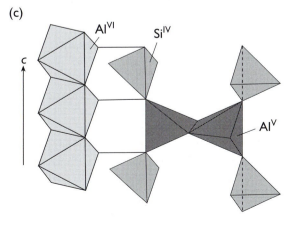

double chains that link the octahedral chains (Figure 28.6b). In this respect (tetrahedral chains), sillimanite has similarities with chain silicates with perfect cleavage and frequent fibrous growth.

In orthorhombic *andalusite*, aluminum not belonging to the octahedral chains occupies sites in a 5-fold coordinated polyhedron in the irregular shape of a trigonal bipyramid (Figure 28.6c).

28.3 Brief description of important orthosilicate minerals

Olivines are a group of minerals with a solid solution between **forsterite** (Mg_2SiO_4) and **fayalite** (Fe_2SiO_4). Olivine is rarely found as euhedral crystals with a prismatic habit (Plate 29a). The pure magnesian end member forsterite occurs in marbles and some types of skarn as greenish-yellow or sometimes colorless rounded grains, with anhedral crystals and having vitreous or greasy luster. In these rocks, forsterite forms under regional metamorphism of quartz-bearing dolomites with a reaction of the type

$$2CaMg(CO_3)_2 + SiO_2 \rightleftharpoons Mg_2SiO_4 + 2CaCO_3 + 2CO_2$$
Dolomite   Quartz Forsterite Calcite

Forsterite in these rocks is commonly accompanied by calcite, phlogopite, magnetite, black or pink spinel, and chondrodite.

Compositionally intermediate members of olivines are common in many igneous rocks. They occur in basalts and other extrusive rocks as isolated inclusions, translucent poorly developed crystals, and granular masses of green to brown color, with vitreous luster. They are also a major constituent of dunites and peridotites. A magnesium-rich olivine, though not pure forsterite, is the dominant mineral of the Earth's upper mantle and for this reason has received much attention from geophysicists. Fayalite is the only olivine that may occur together with quartz. This is a rare mineral, and it is found in some skarns, granite pegmatites, metamorphic rocks, and a few rhyolites.

Fig. 28.6 Structural units in aluminosilicates. (a) Kyanite with octahedral chains linked by silicon tetrahedra (projection on (100)). (b) Sillimanite with octahedral and tetrahedral double chains with alternating silicon and aluminum. (c) Andalusite with octahedral chains and silicon tetrahedra alternating with $(Al_2^VO_9)^{12-}$ groups.

Hydrothermal processes lead to the hydrolysis of forsterite and magnesium-rich olivines, replacing them by serpentine and talc. Schematically, these reactions may be represented in the following form:

$$4Mg_2SiO_4 + H_2O + 5CO_2 \rightarrow$$
Forsterite
$$Mg_3Si_4O_{10}(OH)_2 + 5MgCO_3$$
$$\quad\quad\text{Talc} \quad\quad\quad\quad\quad\text{Magnesite}$$

$$2Mg_2SiO_4 + 2H_2O + CO_2 \rightarrow$$
Forsterite
$$Mg_3Si_2O_5(OH)_4 + MgCO_3$$
$$\quad\quad\text{Serpentine} \quad\quad\quad\text{Magnesite}$$

Similar processes can take place when silicon interacts with the olivine-rich rocks in hydrothermal solutions. For example:

$$3Mg_2SiO_4 + 5SiO_2 + 2H_2O \rightleftharpoons 2Mg_3Si_4O_{10}(OH)_2$$
Forsterite   In solution   Talc

$$3Mg_2SiO_4 + SiO_2 + 2H_2O \rightleftharpoons 2Mg_3Si_2O_5(OH)_4$$
Forsterite   In solution   Serpentine

Transparent green olivines (peridot) are extracted from kimberlites and used as a gemstone.

One needs some experience to identify olivine with confidence. In basalts, olivine minerals look like fragments of green bottle-glass; in ultramafic rocks olivine is often partially replaced by talc and serpentine, both of which are easy to identify. Contrary to pyroxenes, olivine has conchoidal fracture, rather than a regular cleavage. In thin sections olivine is recognized by high birefringence (see Section 14.6 and Plate 14b).

Chondrodite, **humite**, and **clinohumite** are structurally related to olivine in the sense that layers of olivine (parallel to (100)) are intercalated between layers of brucite (Mg(F, OH)$_2$) (parallel to (0001)). As a result, the bulk formulas of these minerals can be written in the following way: for chondrodite $2Mg_2SiO_4 \cdot Mg(F,OH)_2$, for humite $3Mg_2SiO_4 \cdot Mg(F,OH)_2$, and for clinohumite $4Mg_2SiO_4 \cdot Mg(F,OH)_2$. Humite minerals occur in bright-orange and brownish-orange anhedral grains in marbles and are accompanied by forsterite, spinel, and dolomite. In thin section they have a yellow pleochroism. The color is diagnostic. It is difficult to distinguish between the humite varieties without a detailed analysis.

Garnets belong to a large group of minerals that have the general formula $X_3^{VIII}M_2^{VI}(SiO_4)_3$ (Table 28.4). All garnets form euhedral crystals in the form of rhombic dodecahedra and trapezohedra, or combinations of the two (Figure 28.7, Plates 29c and 31d). They have a vitreous luster and high hardness from 6.5 to 7.5. The most abundant garnets, their colors, and deposit types are listed in Table 28.6. On the whole, garnets are easy to identify because of their distinct crystal morphology and colors. At first it may be difficult to recognize green and brown garnets in massive aggregates of greasy luster, found in skarns and hornfelses. Observation of high hardness, lack of cleavage faces, and the presence of small well-developed crystals along calcite veinlets can be helpful. In thin sections, garnets are immediately

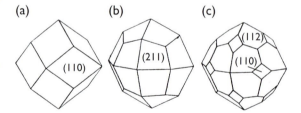

Fig. 28.7 Typical morphology of garnets: (a) rhombic dodecahedron {110}, (b) trapezohedron {211}, and (c) combination of dodecahedron and trapezohedron.

Table 28.6 Varieties of garnets, their color, and modes of occurrence (see also Table 28.4)

Mineral	Color	Deposits
Almandine	Red, brown	Schists, gneisses, pegmatites
Andradite	Dark brown, black	Skarns
Grossular	White, pale brown	Skarns, calcsilicate rocks
Pyrope	Crimson	Kimberlites, basalts, eclogites
Spessartine	Pink	Schists, gneisses, pegmatites
Uvarovite	Emerald green	Chromite ores in olivinites

recognized because the cubic mineral is isotropic (Plate 9), except for grossular, which may show slight birefringence. Almandine is used as an abrasive material. Clear pyrope and uvarovite are gemstones.

Zircon ($ZrSiO_4$) may contain high contents of uranium, thorium, cerium, hafnium, niobium, tantalum, and aluminum. It commonly forms perfect crystals with tetragonal symmetry (Plate 29d). The phenomenon of typomorphism can be well illustrated with zircon: silica-poor rocks usually contain flat, platy crystals, whereas elongated crystals are more common in siliceous rocks. Zircon has a brownish color in various shades and a diamantine luster. Its hardness ranges from 7 to 8.

Zircon is found in all igneous rocks, but it is especially abundant in nepheline syenites and related pegmatites. Radioactive varieties of zircon occur in granite pegmatites. Zircon is very stable and does not easily weather chemically or abrade physically. Therefore it accumulates in sands, which form commercial deposits of zircon placers in sandstones and conglomerates (zircon sands). Radioactive varieties are very prone to metamict decay, i.e., the periodic crystal structure is destroyed and they become amorphous. Since this process is associated with an increase in volume, the expanding zircon grains mechanically stress surrounding minerals and produce fracture patterns. Zircons are easily identified on the basis of their euhedral morphology, their diamantine luster, and their high hardness. In thin sections zircon has a high refractive index (2.0) and high birefringence (0.05). When included in biotite, zircon produces pleochroic halos (Plate 2e,f), due to structural damage from radiation.

Zircon is used as a casting material in metallurgy, and as a source of zirconium oxide ("zirconia") and hafnium (Hf substituting for Zr). Transparent reddish varieties ("hyacinths") are gemstones.

Topaz ($Al_2(SiO_4)F_2$) is orthorhombic with prismatic, lengthwise-striated crystals, elongated along the c-axis, and a perfect pinacoidal cleavage on (001). More rarely it occurs as columnar aggregates and irregularly shaped grains. Its luster is vitreous. Topaz crystals are transparent, colorless, blue, golden-yellow, or pink (Plate 29b), often with sector zoning and banded patterns of color distribution within a crystal. Sector zoning is due to the fact that different faces adsorb different amounts of trace elements (see Chapter 10). The prism faces and

corresponding growth sectors are usually pale blue (owing to traces of Fe^{2+}), growth sectors of inclining faces such as those of a rhombic bipyramid are golden-yellow (Fe^{3+}). The hardness of topaz is 8. The distinguishing features of topaz are its hardness, crystal morphology, striation, and cleavage perpendicular to its striation. Topaz differs from quartz in crystal morphology, in having a higher density (3.6 as compared to 2.6 g/cm^3), in its cleavage, and in its stronger luster. Unlike topaz, quartz crystals are striated perpendicular to their long axis.

Topaz occurs in granite pegmatites, where large crystals, weighing sometimes well over 10 kg, are concentrated in pockets. In greisens, topaz is often a major component and found either as granular masses or in veinlets and pockets as well-developed euhedral crystals and columnar aggregates. There are also hydrothermal topaz–quartz pockets in limestones. The formation of topaz in quartz–micaceous greisens results from the interaction between feldspars and high-temperature fluorine-rich hydrothermal supercritical solutions. This interaction may be schematically represented by the following reaction:

$$2KAlSi_3O_8 + 2F^- + 2H^+ \rightleftharpoons$$
$$K - feldspar$$
$$Al_2SiO_4F_2 + 5SiO_2 + 2K^+ + 2OH^-$$
$$Topaz \qquad Quartz$$

Topaz of good quality and intense color is a valued gemstone.

Vesuvianite ($Ca_{10}Al_4Mg_2(SiO_4)_5(Si_2O_7)_2(OH)_4$), formerly also called idocrase, is a typical mineral in skarns and calcsilicate rocks. In skarns, vesuvianite forms easily recognizable tetragonal-prismatic crystals and columnar aggregates of brown, grayish-brown, or red-brown color. If no crystals are observed, vesuvianite may be difficult to distinguish from andradite or grossular garnets because they have the same hardness and similar appearance (irregular masses of brown, greenish-yellow, and yellow color). It is easy to distinguish vesuvianite in thin section: cubic garnets are usually optically isotropic, whereas tetragonal vesuvianite shows anomalous birefringence (0.04).

The *epidote group* includes the isostructural minerals **clinozoisite**, **epidote**, and **allanite** with the general formula $X_2^{VIII}M_3^{VI}(SiO_4)(Si_2O_7)O(OH)_2$ (see Table 28.5). Clinozoisite is generally grayish-white, epidote forms prismatic striated crystals, elongated parallel to the *b*-axis, with a vitreous luster and a very characteristic pistachio or spinach-green color (Plate 29e). Allanite (also called orthite) contains rare earth elements (e.g., Ce, La) on the *X*-site. It occurs as elongated prismatic crystals of brown or black color. All epidote minerals have a poor cleavage. In thin section, epidote and allanite are easily recognized by brilliant anomalous interference colors (Plates 15f and 16f).

Clinozoisite is typical of low-grade metamorphic rocks. Epidote is very common in schists and skarns. Allanite is found in granites, in granite pegmatites, and in metamorphic rocks as an accessory. It is often metamict due to radiation damage from decay of substituting traces of radioactive elements.

Aluminosilicates (Al$_2$SiO$_5$) include the three poly-morphs **kyanite**, **andalusite**, and **sillimanite**. Kyanite occurs as colorless, cyan-blue (hence the name) elongate-tabular crystals with a pearly luster (Plate 29f). Kyanite crystals exhibit a pronounced anisotropy of hardness (about 7 perpendicular to the longest axis, and 3.5–4 parallel to it); for this reason, the mineral is sometimes called *disthen*, which means "double hard-ness" in Greek. Andalusite is found as prismatic pink, green, or gray crystals, often twinned (chiastolite) and altered to micaceous minerals. Sillimanite, in accord-ance with its chain-like crystal structure, occurs as colorless needles (Plate 17a) and fibrous masses (called *fibrolite*). The minerals form in aluminous meta-morphic rocks. Andalusite often forms as large por-phyroblasts (Plate 5c).

In thin section aluminosilicates are colorless and have fairly high refractive indices (1.65–1.73) with a birefringence higher than quartz and feldspars (0.015). Triclinic kyanite is distinguished from orthorhombic sillimanite by inclined extinction of the cleavage (Plate 17c,d).

The Al$_2$SiO$_5$ minerals are used for manufacturing refractory materials in metallurgy and the ceramics industry. The high-temperature mineral **mullite**, with a nonstoichiometric composition AlAl$_{1-2x}$Si$_{1-2x}$O$_{5-x}$, is structurally closely related to sillimanite. Mullite is rare in nature but is an import-ant ceramic product.

Staurolite ((Fe^{2+},Mg)$_2$(Al,Fe^{3+})$_9$O$_6$(SiO$_4$)$_4$OOH), similar to the aluminosilicates, occurs in micaceous schists. Staurolite is found as brown to black pris-matic, well-developed crystals (often poikiloblastic) and as cross-like twins of such crystals (Figure 28.8). Staurolite can occur in assemblages with micas

Fig. 28.8 Twinned staurolite from Keivy, Kola Peninsula, Russia (courtesy A. Massanek). Width 120 mm.

(Plate 17e,f), kyanite or sillimanite, quartz, and almandine. This mineral forms when goethite- and kaolinite-bearing shales undergo regional meta-morphism based on the following reaction:

$$10Al_2(Si_2O_5)OH + 2Fe^{3+}OOH \rightarrow Al_2SiO_5 + 11SiO_2$$

Kaolinite   Goethite   Kyanite   Quartz

$$+ 2Fe_2^{2+}Al_9O_6(SiO_4)_4OOH + 21H_2O + 14O_2$$

Staurolite   Water

Chloritoid ((Fe^{2+},Mg)$_2$(Al,Fe^{3+})Al$_3$(SiO$_4$)$_2$O$_2$(OH)$_4$) resembles chlorite and has a layered structure, but, contrary to sheet silicates, its layers of octahedral ions are tied together by isolated tetrahedra. Chloritoid occurs in some low-grade metamorphic rocks and has often been overlooked because of its resemblance to chlorite. In thin sections it is distinguished from chlorite by a higher refractive index and inclined extinction.

Titanite (CaTi(SiO$_4$)O), also called sphene, has monoclinic, wedge-shaped (diamond-like), brown/green-colored crystals (Plate 30a). Titanite has a strong vitreous to diamantine luster and imperfect cleavage that are used in the identification. Small yellow and honey-yellow radiaxial aggregates of tita-nite occur in granodiorites. Titanite has very high birefringence (2.0) and refractive index (0.13) and cannot be mistaken in thin section (Plate 9).

It is widespread in nature and occurs in various rocks as an accessory phase. Considerable concentra-tions of titanite (up to 50 vol.%) are found in nephe-line syenites and associated pegmatites.

Lawsonite (CaAl$_2$(OH)$_2$(Si$_2$O$_7$)·H$_2$O) is a relatively rare mineral in high-pressure–low-temperature

(a)   (b)

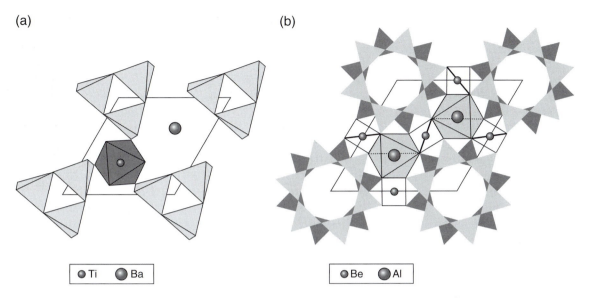

| ● Ti | ● Ba |

| ● Be | ● Al |

Fig. 28.9 Crystal structures of ring silicates. (a) Benitoite BaTi(Si$_3$O$_9$) with tetrahedral Si$_3$O$_9^{6-}$ rings that are linked over TiO$_6^{8-}$ octahedra. Ba^{2+} is in a larger site. (b) Beryl Be$_3$Al$_2$(Si$_6$O$_{18}$) with tetrahedral Si$_6$O$_{18}^{12-}$ rings connected by additional BeO$_4^{4-}$ tetrahedra and AlO$_6^{9-}$ octahedra. This structure contains channels.

metamorphic rocks. It occurs as veins of white, light green, or gray prismatic crystals. Lawsonite has good cleavage. It forms because of the breakdown of calcic plagioclase at high pressure, for example in subduction zones:

$$CaAl_2Si_2O_8 + 2H_2O \rightleftharpoons CaAl_2(OH)_2Si_2O_7 \cdot H_2O$$
Anorthite   Lawsonite

It is one of the high-pressure minerals that contain water and may be a significant hydrous phase in the upper mantle.

28.4 Ring silicates

In some silicates, tetrahedra are linked to form closed rings of three (benitoite, Figure 28.9a), four (axinite), or six tetrahedra (tourmaline, beryl (Figure 28.9b), cordierite, dioptase). In beryl, Be-tetrahedra link the Si$_6$O$_{18}$ rings, Al is octahedral. Depending on the multiplicity of the ring, ring silicates are often trigonal, tetragonal, or hexagonal, and form prismatic crystals elongated along the c-axis. The ring structures are characterized by relatively large channels that can accommodate water molecules, ions, and ion groups (e.g., OHBO$_3^{4-}$ in axinite, BO$_3^{3-}$ in tourmaline) that destroy the stoichiometry of the mineral composition and are relatively mobile. In a

ring, two oxygen atoms of each tetrahedron are shared, resulting in an Si:O ratio of 1:3.

28.5 Brief description of important ring silicate minerals

Tourmaline has a general formula $X_1Y_3Z_6(Si_6O_{18})$ $(BO_3)_3(OH)_4$, where X = Ca^{2+}, Na$^+$; Y = Mg^{2+}, Li^{2+}, Al^{2+}, Fe^{2+}, Mn; and Z = Al^{3+}, Fe^{3+}, Cr^{3+}. Chemical substitution of a pair Li$_{0.5}$Al$_{0.5}$ for Fe$_{1.0}^{3+}$ is typical in intermediate elbaite-shorl minerals. In the trigonal structure of tourmaline, major units are six-membered rings of silicon–oxygen tetrahedra and anionic groups BO$_3^{3-}$. Tourmaline crystals are often euhedral, ditrigonal-prismatic, and columnar in habit, with a rough lengthwise striation (see Plate 4c,d). Tourmalines cut perpendicular to the long axis [0001] have the shape of rounded triangles. The crystals of this noncentrosymmetrical mineral, in which the positive c-axis is distinct from the negative c-axis, are often terminated by pyramidal faces at one end and by basal faces at the opposite end. Radial aggregates are typical of some tourmalines (so-called "tourmaline suns"). Colors may be black (that of iron-rich *ilvaite* and *schorl* varieties) (Plate 30c), pink (owing to manganese in *elbaites* and *rubellites*), blue, green, brown, and white for compositionally different tourmalines. Multicolored crystals and crystals in which zones of different color alternate from the core outward are very typical (Plates 7d and 30d). Luster is vitreous on the faces and greasy on fractures. Tourmalines have a hardness of 7 and no cleavage. The crystals are very brittle and frequently have perpendicular fractures filled with quartz. In thin section, tourmaline is strongly pleochroic; however, contrary to the pleochroism in most other prismatic minerals, the darkest color (highest absorption) is observed when plane polarized light is vibrating perpendicular to the elongation direction.

Tourmalines are found in granites, pegmatites, high-temperature hydrothermal quartz veins, schists, and gneisses. In all these rocks, black ferriferous tourmalines are common, but in spodumene-bearing granite pegmatites, rubellites are found.

Beryl ($Be_3Al_2(Si_6O_{18})$) has a structure with six-membered tetrahedral rings that define large channels running parallel to the 6-fold axis of this hexagonal mineral. These channels can accommodate additional cations of alkali metals (Na^+) and water molecules. Some beryls contain small quantities of manganese, trivalent iron, and chromium that isomorphically substitute for aluminum. These trace substitutions are responsible for the varieties of color. There are blue *aquamarines* (iron at the beryllium sites) (Plates 18a, 30b); common green beryls (iron at the aluminum sites); colorless and milky-white, iron-free, alkali beryls; bright-green *emeralds* (chromium substituting for aluminum) (Plate 18b); pink *morganites* (manganese substituting for aluminum) (Plate 18c); and yellow *heliodores* (iron in the silicon sites). Luster is always vitreous on the faces and greasy on the fracture surfaces. Beryl crystals are hexagonal-prismatic in habit. The crystals are commonly striated lengthwise. Hardness is 7–8. Beryl can be recognized easily by its crystal morphology and by the typical green or blue color. White-colored alkali beryls, however, are often mistaken for quartz.

The beryls are a good illustration of how chemical composition and thus color depends on conditions of formation. Ferriferous beryls (green, pale blue) form in granite pegmatites, and high-temperature hydrothermal deposits. Colorless and milky-white alkali beryls are common accessories in granite pegmatites of the sodium–lithium type. Chromian beryls (emeralds) occur in fluorite-bearing micaceous metasomatic rocks, replacing ultramafics, and in quartz–albite

veins with rare earth carbonates in bituminous limestones. Beryl and its varieties are used for the extraction of beryllium and as gemstones.

Cordierite (($Mg,Fe^{2+})_2Al_3Si_5AlO_{18}$), though orthorhombic, has a structure that is closely related to that of hexagonal beryl. One aluminum ion substitutes for silicon in tetrahedral coordination. Generally, cordierite is ink-blue in color and translucent. Its hardness is 7–7.5. Translucent grains that change their color when rotated are used as gemstones (dichroism). Unless cordierite occurs as large crystals, it is difficult to identify, even in thin sections, and is easily mistaken for plagioclase. It displays characteristic pleochroic halos around monazite (not zircon) inclusions (Plate 2e,f) and often forms intergrowths of multiple twins with a pseudohexagonal symmetry. Rounded, small (2–3 mm) grains are not uncommon in aluminous metamorphic schists, gneisses, and hornfelses.

Dioptase ($Cu_6(Si_6O_{18}) \cdot 6H_2O$) is a rare rhombohedral ring silicate occurring as well-developed rhombohedral crystals of intense green color in calcite and dolomite veins in localities of DR Congo, Tsumeb (Namibia), and Pinal County in Arizona (USA). Dioptase is occasionally used as a gemstone.

28.6 Metamorphic minerals

Many orthosilicates – garnets, epidotes, and aluminosilicates among them – occur mainly in metamorphic rocks. They are the products of recrystallization in solid state from pre-existing mineral assemblages that are no longer stable because of changes in temperature or pressure. Under such conditions, chemical reactions take place to form new minerals. Changes in temperature and pressure can occur in different geological settings. If rocks are buried by accumulating sediments, and pressure and temperature increase progressively, the metamorphism is called *burial metamorphism* and is usually very extensive, covering large regions (*regional metamorphism*). Temperatures may also increase rapidly in the vicinity of an intruding pluton. In this case, heat has been transported to a shallow level by the magma, and pressures are relatively low as compared with those of a normal geothermal gradient. Such high-temperature–low-pressure metamorphism is limited to the direct vicinity of the igneous contact and is called *contact metamorphism*. In another setting, parts of the crust are subducted at continental margins; here, the pressure increases more rapidly than the temperature because, whereas pressure response is immediate, thermal conductivity of rocks is very slow. If such rocks are brought up to the surface before they have had time to heat up, either by rapid erosion or selective tectonic uplift, they display vestiges of high-pressure–low-temperature conditions, or *subduction metamorphism*.

During metamorphism, limestones transform into marbles, with minerals such as forsterite, humite, vesuvianite, and grossular; sandstones transform into quartzites; mudstones transform into slates and at higher temperature into pelitic schists, with such minerals as biotite, garnet, andalusite, kyanite, sillimanite, chloritoid, staurolite, and cordierite. Recrystallization during a temperature increase is almost necessarily accompanied by dehydration and decarbonation. With increasing temperature, muscovite, a hydrous sheet silicate in pelitic schists, transforms into anhydrous sillimanite and K-feldspar:

$$KAl_2AlSi_3O_{10}(OH)_2 + SiO_2 \rightarrow$$
Muscovite Quartz

$$Al_2SiO_5 + KAlSi_3O_8 + H_2O$$
Sillimanite K-feldspar Water

Plates 17a,b illustrate how a large grain of biotite breaks down to form needles of sillimanite, so-called fibrolite.

In quartz-bearing limestones, decarbonation occurs, as in the following reaction that produces wollastonite, a chain silicate:

$$CaCO_3 + SiO_2 \rightarrow CaSiO_3 + CO_2$$
Calcite Quartz Wollastonite

Reactions that occur with increasing temperature are called *prograde reactions*. You may wonder why metamorphic rocks do not revert to their original low-grade assemblages when they are uplifted and cooled? Why do we still observe sillimanite in metamorphic rocks, while muscovite and clay minerals would be stable at surface conditions? The prograde dehydration and decarbonation reactions such as those above are the main reason. During those transformations, volatile phases (i.e., H_2O and CO_2) are expelled and the reaction cannot be reversed. If water enters a metamorphic system during cooling, however, so-called *retrograde reactions* may occur and replace some of the high-grade metamorphic minerals. We have seen earlier in this chapter that high-temperature olivine may transform to talc and serpentine (e.g., Plate 14d).

Polymorphic minerals such as the aluminosilicates are excellent indicators of temperature–pressure conditions during metamorphism. Let us refer again to the phase diagram for these minerals (see Figure 19.2b) and recall that andalusite is stable at low pressure and intermediate temperature, kyanite at high pressure, and sillimanite at high temperature. The distribution of these minerals can be explained by collecting samples in the field and identifying the minerals in thin sections (Section 14.6, Plate 17c,d). An example is the Central Alps, where a granitic pluton has been emplaced in a stack of gneisses in the Miocene (Figure 28.10a). The distribution of the polymorphs is very regular and outlines regions of kyanite in the northwest, andalusite to the east (along the granite contact shown as a dotted line), and sillimanite in the central part. At the border between two regions, two aluminosilicate minerals coexist, and where all three regions meet, near A in the north and B in the south, the location corresponds to the "triple point". At this point all three polymorphs were in equilibrium at the time of metamorphism and in fact are observed in the same thin section (Plate 17c,d).

The same rocks that contain pure aluminosilicates also bear iron and magnesium aluminous silicates such as chloritoid, staurolite, and cordierite. The regional distribution of those minerals is equally regular (Figure 28.10b). The distribution of chloritoid is far to the north. Staurolite closely overlaps with kyanite, and cordierite with sillimanite. We can write reactions to illustrate the transformation of chloritoid to staurolite and cordierite with increasing temperature (assuming for simplicity pure iron end members): these reactions are very idealized and may not actually occur, in part because other phases are involved and because all minerals in these reactions are solid solutions. Nevertheless, they illustrate the general picture of metamorphic transformations.

$$2Fe_2AlAl_3(SiO_4)_2O_2(OH)_4 + 5Al_2SiO_5 \rightarrow$$
Chloritoid Andalusite

$$2Fe_2Al_9O_6(SiO_4)_4OOH + SiO_2 + H_2O$$
Staurolite Quartz Water

$$2Fe_2Al_9O_6(SiO_4)_4OOH + 13SiO_2 \rightarrow$$
Staurolite Quartz

$$4Fe_2Al_3Si_5AlO_{18} + Al_2SiO_5 + H_2O$$
Cordierite Sillimanite Water

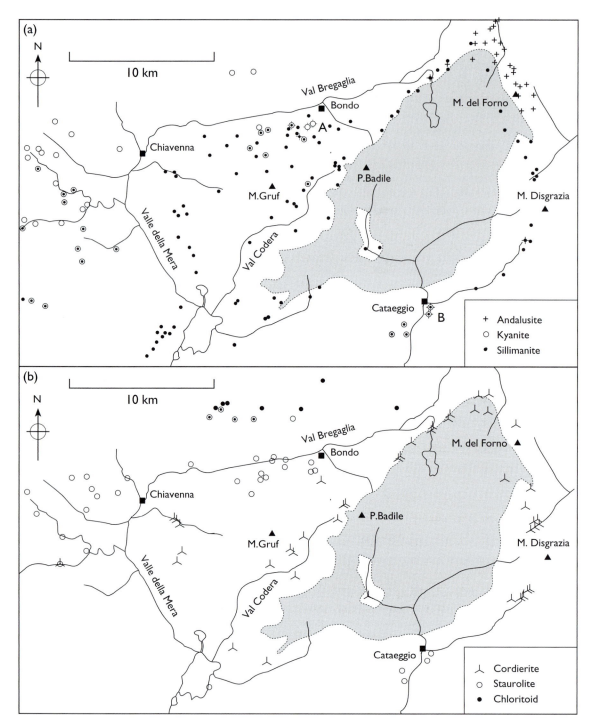

Fig. 28.10 Regular distribution of metamorphic minerals in pelitic schists from a region in the Central Alps. (a) Occurrence of andalusite, kyanite, and sillimanite. The extent of the Tertiary Bergell granite is shaded. Also indicated are locations A and B where all three polymorphs are present in the same rock. (b) Occurrence of cordierite, staurolite, and chloritoid (after Wenk *et al.*, 1974).

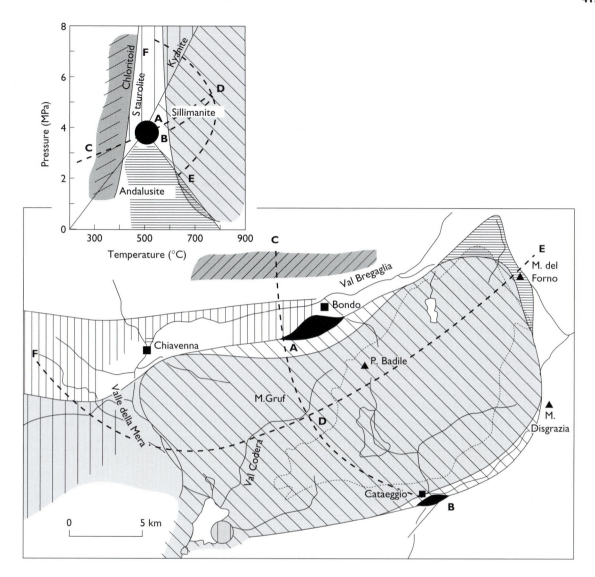

Fig. 28.11 (Top) The *P–T* phase diagram with stability fields of aluminosilicate minerals, the equilibrium curve for the breakdown of chloritoid to staurolite, magnetite, and quartz, and the minimum melting curve of granite as obtained experimentally. The light shading on the right side shows the field of melting of granitic rocks. The triple point region is black. (Bottom) Geological map, corresponding to Figure 28.10, on which the *P–T* shading patterns from the phase diagram have been transferred to visualize the temperature–pressure distribution during peak metamorphism.

The distribution of the minerals in the field (see Figure 28.10) can be compared with the phase diagram that displays the stability fields for aluminosilicates (see Figure 19.2b), the equilibrium curve for the reaction chloritoid transforming to staurolite, as well as the minimum melting curve of granite as obtained in experiments (see Figure 28.11 (top insert)). On the phase diagram we apply shadings to temperature–

pressure regions and transfer these shadings on to the geological map, on the basis of the distribution of minerals in Figure 28.10. We can now interpret the map in terms of temperature and pressure conditions. If we follow line C–A–D in the field and transpose it into the phase diagram, we start at low temperature and pressure, pass through the triple point at A (at 350 MPa and 550 °C), and end at high temperature and pressure at D. Line E–D–F starts at low pressure and moderate temperature at the northeastern granite contact at E, with a rapid increase in pressure towards D, ending in the high-pressure kyanite field at F. The field of sillimanite coincides more or less with the extent of melting, either as the complete melting of a granitic magma (as Bergell granite) or partial melting as in Alpine migmatites (light shading in Figure 28.11). Andalusite is present only in the low-pressure roof of the granite. At points A and B (Figures 28.10a and 28.11), andalusite, kyanite, and sillimanite coexist (Plate 17c,d). At this point temperature–pressure conditions are strictly defined.

Metamorphic rocks are generally classified according to the mineralogical composition of their major constituents. For example, gneisses contain quartz, feldspar, and mica; marbles contain carbonates (calcite or dolomite); and amphibolites contain hornblende. The mineralogical composition is in turn controlled by temperature and pressure conditions and by the chemical composition. An important consideration is the thermodynamic phase rule (Chapter 19), which is applied in Box 28.1 to metamorphic rocks.

On the basis of characteristic mineral assemblages, the temperature–pressure field has been divided into *metamorphic facies* as illustrated in Figure 28.13a,

---

Box 28.1 Additional information: Applying the phase rule to metamorphic rocks

In Chapter 19 we introduced the phase rule of Willard Gibbs, which states that, for a system in equilibrium, the following relation holds between the number of coexisting phases p, the number of components c, and the degrees of freedom f:

$$p = c + 2 - f \qquad (28.1)$$

During the process of mineral formation in metamorphic rocks, pressure P and temperature T are not constant but range over a large P–T interval, corresponding to a region in the P–T phase diagram. Thus both P and T are variable, providing two degrees of freedom. Under these conditions the phase rule reduces to

$$p \leq c \qquad (28.2)$$

which is known as the mineralogical phase rule and was introduced by V. M. Goldschmidt (1911). According to this rule, the number of different minerals in a rock should not exceed the number of components. (Because of solid solutions the number of minerals may be fewer.)

Let us take a ternary system with the components Al_2SiO_5–$CaSiO_3$–$(Mg,Fe)SiO_3$ (Figure 28.12). Such a system is representative of shales. In this triangle there are seven minerals at low-pressure–high-temperature conditions (corresponding to point E in Figure 28.11). Most of these minerals have been introduced in this chapter. We count Al_2SiO_5 polymorphs as one. Any combination of three minerals may coexist according to the phase rule. The particular combination depends on the overall chemical composition of the rock and the P–T conditions. (Only two minerals may coexist in two-component systems on the sides of the triangle.) We describe some of the combinations below, with each numbered combination referring to one of the labeled compositions in Figure 28.12:

1. In a calcium-free, aluminum- and (magnesium, iron)-bearing rock of composition 1 (corresponding to a pure shale), andalusite (or kyanite, or sillimanite, depending on P–T) and cordierite coexist.
2. If we add some calcium (composition 2), anorthite forms in addition to andalusite and cordierite. (If some sodium is present, a more sodium-rich plagioclase will form instead of anorthite.)
3. Decreasing the amount of aluminum shifts the composition to 3, and anorthite, cordierite, and hypersthene coexist. Aluminosilicates are no longer compatible.

Box 28.1 (cont.)

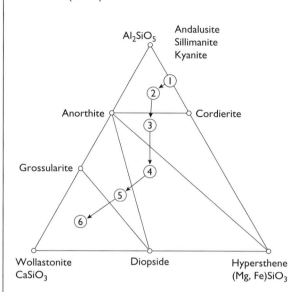

Fig. 28.12 Ternary diagram CaSiO$_3$–MgSiO$_3$–Al$_2$SiO$_5$ and stable minerals in high-temperature hornfelses.

4. Decreasing aluminum further, to composition 4, brings us into a field where anorthite, hypersthene, and diopside coexist. Conceivably anorthite, cordierite, and diopside could coexist, but this mineral assemblage is not observed under conditions of high-temperature metamorphism.
5. If we now once again increase calcium (e.g., in impure limestones) to arrive at composition 5, we are in a field with anorthite, diopside, and grossularite.
6. Increasing calcium even further (in more pure limestones) brings us into a field of coexisting grossularite, diopside, and wollastonite.

We mark coexisting phases with tielines. According to the phase rule, tielines always outline a triangle and can never cross. The system described above is idealized, but overall it is similar to nonfoliated metamorphic sedimentary rocks called *hornfelses*, which occur in the vicinity of an igneous contact. In those rocks there is generally an excess of SiO$_2$ that simply changes the discussion above by adding a fourth component (SiO$_2$) and a fourth phase (quartz) to any of the combinations above. Other variations to the diagram may occur if water and carbon dioxide are present in the system.

which also shows the normal geothermal gradient of 25 °C/km in the continental lithosphere. Some facies occur at higher pressure than a normal geothermal gradient (blueschist, eclogite facies), many follow a normal geothermal gradient (zeolite, greenschist, amphibolite, and granulite facies), and some occur at higher temperature (hornfels and sanidinite facies). Chlorite, epidote, and albite are typical minerals of the greenschist facies, which derives its name from the green minerals chlorite and epidote. The amphibolite facies is characterized by the presence of hornblende, plagioclase, and garnet (almandine), and by the absence of chlorite. Hypersthene, sillimanite, and cordierite occur in granulite facies, but compared with the amphibolite facies, hornblende is no longer present. Minerals typical of high-pressure metamorphism are lawsonite, jadeite, glaucophane, aragonite, and, in very extreme cases, coesite and diamond.

The temperature–pressure calibration of metamorphic facies relies on critical mineral reactions where equilibrium phase diagrams have been established either experimentally or by thermodynamic calculations. Some of these reactions are shown in Figure 28.13b.

(a)

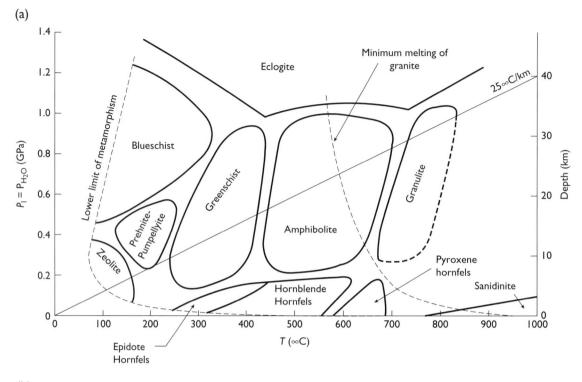

(b)

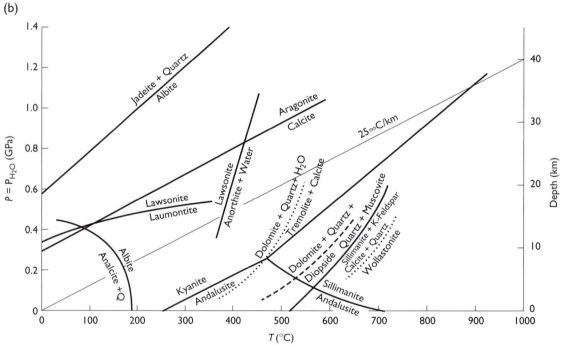

Fig. 28.13 (a) Temperature–pressure (depth) diagram with fields of important metamorphic facies. Minimum melting curve for the system quartz–orthoclase–albite–water is also shown (dashed line). The average geothermal gradient in continental crust (25 °C/km) is indicated. (b) Equilibrium curves for important reactions that can be used to establish the metamorphic grade (after Turner, 1981).

28.7 Summary

SiO_4^{4-} tetrahedra are the defining structural unit in silicates and they are classified according to the arrangement and linkage of these tetrahedra. Isolated or simple groups of tetrahedra are observed in ortho-silicates and group silicates; if tetrahedra form rings they constitute ring silicates and those are discussed in detail in this chapter. In chain silicates tetrahedra form chains, in sheet silicates they form sheets, and in frame-work silicates they are linked three-dimensionally to a framework. The Si:O ratio increases with increasing tetrahedral polymerization.

Orthosilicates such as olivine share many similar-ities to oxides, with oxygens in a close-packed struc-ture and cations in some of the octahedral and tetrahedral interstices. In garnets and zircon, chains occur with alternating tetrahedra and distorted dodecahedra. Chains are linked by octahedra. In ring silicates three-, four- and six-membered rings occur in benitoite, axinite, beryl and tourmaline, respectively.

Orthosilicates are important minerals in igneous and metamorphic rocks and their occurrence can be used to explore conditions. This is illustrated for the Al_2SiO_5 aluminosilicate polymorphs andalusite, kyan-ite, and sillimanite introduced in Chapter 19. The concept of metamorphic facies is used to relate min-eral assemblages and reactions to assess temperature–pressure conditions. Forsterite is one of the major minerals in the upper mantle of the Earth.

Important orthosilicates and ring silicates to remember

Mineral	Formula
Orthosilicates	
Forsterite (olivine)	$Mg_2(SiO_4)$
Fayalite (olivine)	$Fe_2(SiO_4)$
Zircon	$Zr(SiO_4)$
Titanite	$CaTi(SiO_4)O$
Almandine (garnet)	$Fe_3Al_2(Si_3O_{12})$
Pyrope (garnet)	$Mg_3Al_2(Si_3O_{12})$
Sillimanite, andalusite, kyanite	$Al_2(SiO_4)O$
Staurolite	$FeAl_4(SiO_4)_2O_2(OH)_2$

Epidote	$Ca_2Fe^{3+}Al_2(SiO_4)(Si_2O_7)$ $(O, OH, F)_2$
Ring silicates	
Beryl	$Be_3Al_2(Si_6O_{18})$
Schorl (tourmaline)	$NaFe_3^{2+}Al_6(Si_6O_{18})$ $(BO_3)_3(OH)_4$
Elbaite (tourmaline)	$NaLi_{1.5}Al_{1.5}Al_6(Si_6O_{18})$ $(BO_3)_3(OH)_4$

Test your knowledge

1. What are the main ionic substitutions in silicates? (Tetrahedral, octahedral, and large cations.)
2. Silicates are classified according to the linkage of tetrahedra. Describe the groups.
3. Derive the Si:O (T:O) ratios based on shared tetra-hedra for all groups.
4. Garnets are an important mineral group with vari-ous ionic substitutions. Mineral names have been assigned to the end members. Review these min-erals and their composition.
5. Aluminosilicates serve as indicators of meta-morphic conditions in pelitic rocks. What is their chemical formula, what are the mineral names? Describe their *P–T* phase diagram.
6. What is the main structural element of tourmaline, beryl, and cordierite?
7. Beryl occurs as a gemstone in a variety of colors and, depending on the color, different names are assigned. Review the names, colors, and reasons for the colors.

Further reading

Deer, W. A., Howie, R. A. and Zussman, D. J. (1982), *Rock-forming Minerals, Volume 1a, Orthosilicates*, 2nd edn. The Geological Society of London.

Kerrick, D. M. (1990). *The Al₂SiO₅ Polymorphs*. Reviews in Mineralogy, vol. 22. Mineralogical Society of America, Washington, DC.

Liebau, F. (1985). *Structural Chemistry of Silicates: Struc-ture, Bonding, Classification*. Springer-Verlag, Berlin.

Turner, F. J. (1968). *Metamorphic Petrology: Miner-alogical and Field Aspects*. McGraw-Hill, New York.

29 | Sheet silicates. Weathering of silicate rocks

In sheet silicates silicon tetrahedra are polymerized to form an infinite two-dimensional net with six-membered rings and hexagonal symmetry. Within a sheet silicate net, three corners of each tetrahedron are shared with another, resulting in a structural base of the tetrahedral network $Si_2O_5^{2-}$. Such a layer has a negative charge that is balanced by interstitial layers of cations and anions. Most significant are octahedral layers with Mg, Fe, and Al. There is a wide range of compositions and stacking. All sheet silicates contain some hydrogen, generally in the form of OH groups. Sheet silicates form in hydrous environments as clays such as montmorillonite and smectite. At higher temperature they transform to mica. Other sheet silicates form at late stages of crystallization of magmas, such as some micas, in hydrothermal ore deposits, and surrounding greisens, such as serpentine, which forms by hydration of olivine. We will introduce varieties of sheet silicates and then review the formation of clays during sedimentation, in soil, and during alteration.

29.1 Basic structural features

Sheet silicates are also called phyllosilicates (after the Greek *phyllon*, meaning "leaf"). Most of the minerals in this group have a flaky habit and an excellent single cleavage. The habit and cleavage are due to planar units in the crystal structure. All sheet silicates contain, in addition to O^{2-}, OH^- groups as anions and there is a large variety of compositions and structures. Many sheet silicates crystallize in the water of oceans and lakes as clays, others form by hydration of non-hydrous minerals.

We mentioned in Chapter 28 that silicon tetrahedra (SiO_4^{4-}) in sheet silicates are polymerized to form an infinite two-dimensional net with six-membered rings and hexagonal symmetry (Figure 29.1). Contrary to the tetrahedral nets of tridymite and cristobalite, where alternating apices point in opposite directions (and are linked to another net) (Figure 21.3a), in sheet silicates all free tetrahedral apices point in the same direction. Within a sheet silicate net, three corners of each tetrahedron are shared with another, resulting in a structural base of the tetrahedral network $Si_2O_5^{2-}$.

For more clarity, we will include this sheet network in parentheses when we write formulas of sheet silicates below.

The distance between free tetrahedral oxygen apices in a net is close to 1.3 Å, and this value is very similar to the O–O distance of a coordination octahedron (1.4 Å). Therefore, the tetrahedral layer fits almost perfectly on top of a layer of octahedra that lie on triangular sides and share edges (Figure 29.2a). As the tetrahedral layer is attached to the octahedral layer, corresponding oxygen atoms are shared. Those anions in octahedral layers that are shared with tetrahedra are O^{2-}; those that are not shared are hydroxyl groups (OH^-). We have seen structures with infinite octahedral layers in the discussion of hydroxides (see Chapter 27). The structure of brucite (see Figure 27.9a) consists of sheets of edge-sharing $Mg(OH)_6^{4-}$ octahedra (trioctahedral), and the structure of gibbsite (see Figure 27.9c) consists of sheets of $Al(OH)_6^{3-}$ octahedra with one-third of the octahedral sites vacant (dioctahedral). We can view the basic building unit of a sheet silicate as an octahedral layer (brucite or gibbsite type, depending on the charge of the cation) with an attached tetrahedral layer, and these units are periodically stacked. The main cations in sheet silicates are Mg^{2+}, Al^{3+}, Fe^{2+}, and Fe^{3+}. In the case of divalent ions, all octahedral interstices are occupied and the cation coverage is continuous (Figure 29.2a). For trivalent ions one out of three octahedral interstices is kept vacant to maintain charge balance (Figure 29.2b). The trioctahedral unit is more symmetrical than the dioctahedral unit. Indeed, the structure in Figure 29.2b is highly idealized. In a realistic structure of a dioctahedral sheet silicate, for example muscovite (Figure 29.3), the octahedra are distorted, the tetrahedra are rotated (as indicated by arrows), and the tetrahedral and octahedral rings are no longer ideally hexagonal.

There are two types of building units: in some sheet silicates, for example serpentine, tetrahedral sheets are attached only on one side of the octahedral layer (Figure 29.4a). This type is often referred to as a 1:1 structure. In others, for example talc, tetrahedral layers are attached to both sides (Figure 29.4b) and this type is referred to as a 2:1 structure. These tetrahedral-octahedral composite layers, which are charge balanced, are then stacked on top of each other. They are held together only by weak van der Waals bonds, which accounts for the excellent cleavage and low hardness (1 for talc on the Mohs scale).

The type of building unit, as well as the distinction between trioctahedral and dioctahedral structures accounts for the different chemical formulas of sheet silicates. Serpentine, with only one tetrahedral layer and OH^- only on one side of the octahedral layer has the formula $Mg_3(Si_2O_5)(OH)_4$. The corresponding dioctahedral kaolinite has the formula $Al_2(Si_2O_5)$

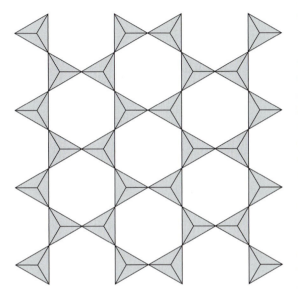

Fig. 29.1 Ideal hexagonal net of silicon tetrahedra in sheet silicates.

(a)

(b)

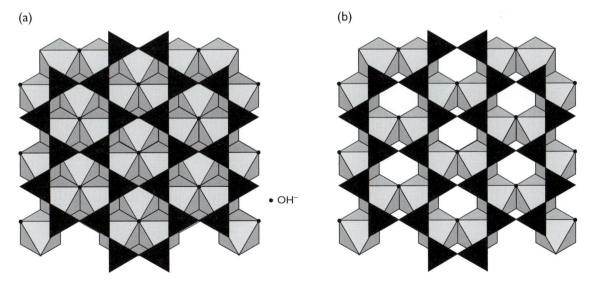

• OH^-

Fig. 29.2 Projection of the structure of (a) serpentine (trioctahedral) and (b) kaolinite (dioctahedral) on (001), illustrating how the tetrahedral layer (black) is attached to the octahedral layer (idealized). OH groups are indicated.

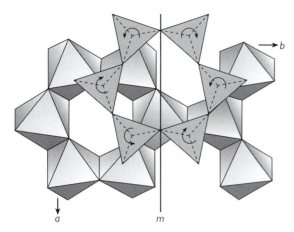

Fig. 29.3 Structure of muscovite. As compared to the idealized structure (Figure 29.2), in the real structure of a dioctahedral sheet silicate, octahedra are distorted and tetrahedra are rotated (indicated by arrows).

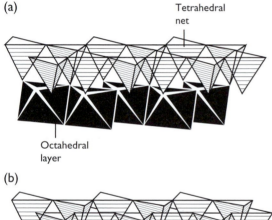

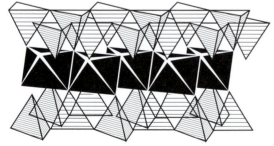

Fig. 29.4 Hexagonal net of silicon tetrahedra attached to an octahedral layer in (a) serpentine, a 1:1 structure, and (b) talc, a 2:1 structure.

$(OH)_4$. For 2:1 structures there are more Si^{4+} ions and fewer OH^- groups. The formula for trioctahedral talc is $Mg_3(Si_4O_{10})(OH)_2$ and for corresponding dioctahedral pyrophyllite $Al_2(Si_4O_{10})(OH)_2$. You can verify these

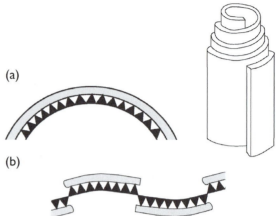

Fig. 29.5 Structure of (a) chrysotile with a continuous scroll of the tetrahedral-octahedral sheet and (b) antigorite with switching of units.

formulas by counting atoms in a repeat unit, see for example Figure 29.2.

The octahedral-tetrahedral sheet of talc is well balanced, with tetrahedral layers attached to both sides of the octahedral layer. In serpentine, with a tetrahedral net only on one side, and interatomic distances slightly shorter in tetrahedra than in octahedra, there are forces that bend the sheet. In the serpentine variety known as *chrysotile asbestos*, bending is continuous and the sheet becomes scroll-like (Figure 29.5a), giving rise to asbestos fibers with a textile-resembling fabric (Plate 31a). The scroll-like pattern can be seen in high-resolution electron micrographs (Figure 29.6). In the serpentine variety *antigorite*, bending occurs only over a few unit cells, after which the polarity of the $Si_4O_{10}^{12-}$ nets is reversed, producing a wavy structure (Figure 29.5b). This means that two-dimensional infinite nets break up into bands with free corners of tetrahedra pointing in alternate directions. In kaolinite this unbalanced structure is the cause for a very small crystal size.

In Figure 29.7 we illustrate the stacking of the octahedral-tetrahedral units in a more schematic way perpendicular to the plane of the layers. The structure of the hydroxides brucite–gibbsite (Figure 29.7a) is given for reference, followed by serpentine–kaolinite (Figure 29.7b) and talc–pyrophyllite (Figure 29.7c).

In *micas* part of the tetrahedral Si^{4+} is replaced by Al^{3+}. To maintain charge balance, large cations (Na^+, K^+, Ca^{2+}) are introduced between the sheets (Figure 29.7d). The large cations are in 12-fold oxygen coordination between six-membered rings of adjacent tetrahedral nets (see Figure 29.2). Since sheets are no longer electrostatically neutral, the bonding between sheets is partially ionic. Therefore, the hardness of

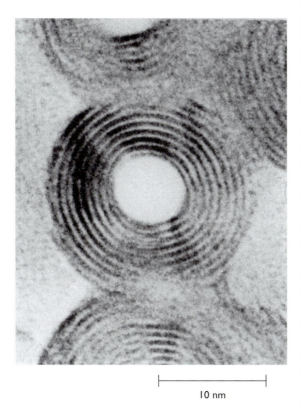

10 nm

Fig. 29.6 High-resolution TEM micrograph of chrysotile from Transvaal with concentric structure (from Yada, 1971).

micas is higher than that of talc, and the cleavage is less perfect. Micas can be subdivided according to chemical composition. Analogous to talc–pyrophyllite, there are trioctahedral and dioctahedral micas, depending on the charge of octahedral cations, mainly Mg^{2+} and Al^{3+}. Chemical substitutions of large cations (*X*-position) and octahedral cations (*M*-position) form many mineral species (Table 29.1). *Biotite* is a group name of trioctahedral micas enriched in Fe^{2+} and Fe^{3+}; their compositions are variable and lie between phlogopite, annite, and siderophyllite. *Zinnwaldites* are dark, lithium-rich micas intermediate in compositions between siderophyllite and polylithionite. *Lepidolite* is a group name for light (usually rose) lithium micas between polylithionite and trilithionite. Contrary to "ordinary micas" with K^+ and Na^+ in the interlayers, "brittle micas" such as *margarite* and *clintonite* contain Ca^{2+}. The bond strength of these divalent ions is larger and this is expressed in the brittle nature.

We noted earlier that trioctahedral sheets have almost hexagonal symmetry (see Figure 29.2a). Even though the stacking of the sheets makes most of the mica structures monoclinic, optical properties of trioctahedral micas such as biotite are almost uniaxial. The hexagonal symmetry is destroyed in the dioctahedral sheets (see Figures 29.2b and 29.3), and dioctahedral micas such as muscovite are optically biaxial with a large axial angle.

In *chlorites*, additional brucite layers, consisting of hydroxyl-coordinated octahedra (Figure 29.7e) replace the layers of large cations present in micas. There is a wide variety of chlorites having different chemical compositions, with Mg^{2+}, Fe^{2+}, Fe^{3+}, and

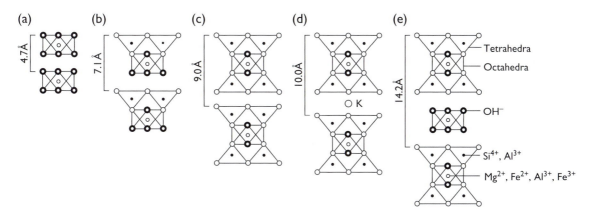

Fig. 29.7 General structural principles in sheet silicates with stacking of octahedral and tetrahedral layers: (a) brucite–gibbsite, (b) serpentine–kaolinite, (c) talc–pyrophyllite, (d) mica (biotite–muscovite), (e) chlorite. Repeat distance is indicated.

Table 29.1 Chemical substitutions and structure of micas $XM_{(2\ or\ 3)}(T_4O_{10})(OH,F)_2$

Mineral	X^{XII}	M^{VI}	T^{IV}	Principal polytype
Dioctahedral micas, ordinary				
Muscovite	K	$Al_2\square$	Si_3Al	$2M$
Paragonite	Na	$Al_2\square$	Si_3Al	$1M, 2M1$
Aluminoceladonite	K	$AlM_g\square$	Si_4	
Boromuscovite	K	$Al_2\square$	SiB	$1M, 2M1, 3T$
Dioctahedral micas, interlayer-deficient				
Glauconite	$K_{0.8}$	$Fe^{3+}_{1.33}Fe^{2+}{}_{0.67}\ \square$	$Si_{3.87}Al_{0.13}$	$1M$
Dioctahedral micas, brittle				
Margarite	Ca	$Al_2\square$	Si_2Al_2	$2M1$
Trioctahedral micas, ordinary				
Phlogopite	K	Mg_3	$Si3Al$	$1M, 2M$
Annite	K	Fe^{2+}_3	Si_3Al	$1M, 3T$
Siderophyllite	K	Fe^{2+}_2Al	Si_2Al_2	$1M$
"Biotite"[a]	K	$(Mg,Fe^{2+},Al,Fe^{3+})_3$	$Si_2(Si,Al,Fe^{3+})_2$	$1M, 2M$
"Zinnwaldite"[a]	K	$(Fe^{2+},Li,Al)_3$	Si_3Al	$1M, 3T$
Trilithionite[b]	K	$Li_{1.5}Al_{1.5}$	Si_3Al	$1M, 2M, 3T$
Polylithionite[b]	K	Li_2Al	Si_4	$1M$
Trioctahedral micas, brittle				
Clintonite[c]	Ca	Mg_2Al	Si_3Al	$1M, 2M$

Notes: The symbol X refers to the interlayer cation site, M to octahedral sites, and T to tetrahedral sites.
The symbol $\square$ signifies an octahedral vacancy. Polytype symbols are M for monoclinic and T for trigonal.
[a] Series name without standard formula.
[b] Older name: lepidolite.
[c] Older name: xanthophyllite.

Table 29.2 Compositional substitutions in chlorites

Mineral	Interlayer	Octahedra	Tetrahedra
Pennine	$Mg_3(OH)_6$	$(Mg_{2.5}Al_{0.5})(OH)_2$	$Si_{3.5}Al_{0.5}O_{10}$
Clinochlore	$Mg_3(OH)_6$	$(Mg_2Al)(OH)_2$	Si_3AlO_{10}
Daphnite	$Fe^{2+}_3(OH)_6$	$(Fe^{2+}Al)(OH)_2$	Si_3AlO_{10}
Chamosite	$Fe^{2+}_3(OH)_6$	$(Fe^{2+}_{2.5}Al_{0.5})(OH)_2$	$Si_{3.5}Al_{0.5}O_{10}$

Al^{3+} competing for octahedral positions. Table 29.2 gives some examples and the corresponding distribution of ions over the various structural sites.

As Figure 29.7 documents, a significant difference between the various sheet silicate structures is observed in the spacing of sheets. With increasing complexity, going from brucite to chlorite, the repeat distance increases from approximately 5 to 14 Å. The sheet plane is parallel to (001) and a diagnostic property is the interplanar spacing d_{001}, which can be measured easily with X-ray powder diffraction.

The magnesium minerals *paligorskite* and *sepiolite* can be viewed as degenerate talc structures with tetrahedral sheets switching directions to form three-dimensional connections (Figure 29.8). Octahedral layers (shaded) are no longer continuous, but stacked, producing large channels that contain molecular

(a)

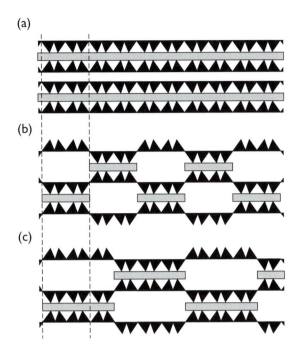

(b)

(c)

Fig. 29.8 Comparison of the structure of (a) talc with continuous octahedral sheets with those of (b) paligorskite and (c) sepiolite with stacked octahedral–tetrahedral units.

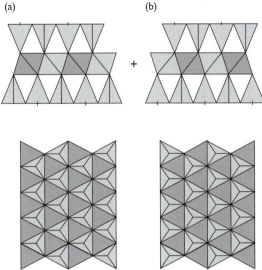

Fig. 29.9 (a and b) Octahedra in a sheet silicate may point in opposite directions relative to the bottom tetrahedral layer. Top: section perpendicular to the layers. Bottom: view on the octahedral layer parallel to (001).

water. These structures are reminiscent of manganese oxides (see Figure 27.8).

So far we have discussed stacking of layers with different compositions and various substitutions of cations. Another way to add diversity is to keep the composition constant but vary the geometrical relationship between adjacent layers. Such derivative structures that are variations of long-range stacking of layers are called *polytypes*. There is a great variety of polytypes in sheet silicates, and we will use them to discuss the general principle of polytypism.

29.2 Polytypism

In polymorphs and polytypes the chemical composition is the same but the structure differs. In polymorphs (see Chapter 3), short-range atomic arrangements differ, but in polytypes only the long-range stacking of structural units varies. Polymorphs have different mineral names whereas polytypes are structural varieties of the same mineral.

In sheet silicates, polytypism is due to stacking of tetrahedral-octahedral sheets. We confine our discussion to the basic principles of polytypism in ideal mica, where tetrahedral and octahedral layers have trigonal/hexagonal symmetry. There are two factors that contribute to polytypism: the orientation of the octahedra in the layer relative to a tetrahedral layer, and the disposition of the tetrahedral nets between subsequent layers.

Octahedra in a layer may be pointing in one direction relative to the top tetrahedral layer (Figure 29.9a, referred to as "+") or in the opposite direction (Figure 29.9b, referred to as "−"). If two tetrahedral-octahedral units are stacked on top of each other, this allows for two possibilities. In Figure 29.10a octahedra in all layers point in the same direction (+, +), and in Figure 29.10b octahedra in adjacent layers point in opposite directions (+, −). Figure 29.10a shows schematically that if octahedra in all layers point in the same direction, then the basic monoclinic symmetry is maintained and the structure pattern repeats after one layer. This polytype is called 1*M* (i.e., one-layer repeat, *M*onoclinic). If adjacent layers point in opposite directions, a mirror (*m*) plane parallel to the sheet plane is introduced, producing an

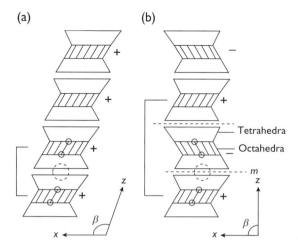

(a)   (b)

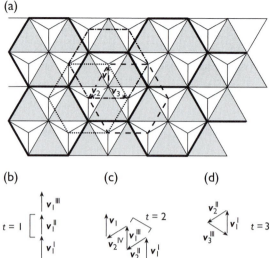

(a)

Fig. 29.10 Polytypism in sheet silicates. (a) Octahedra in adjacent layers point in the same direction (+, +, +, +), producing a $1M$ polytype. (b) Octahedra in adjacent layers point in opposite directions (+, −, +, −), producing a mirror symmetry and a 2-repeat in the $2O$ polytype.

Fig. 29.11 (a) (001) view of an octahedral layer. Top triangular surfaces of octahedra are shaded, depressions in unshaded triangles are oygen atoms on the lower layer. A tetrahedral net is attached to the top surface, connecting shared oxygen atoms at apices with solid lines. The three possibilities for tetrahedral nets attached to the lower surface of the octahedral sheet are indicated by dot-dashed, dotted, and dashed line hexagons. The three displacement vectors v_1, v_2, and v_3 between upper and lower hexagons are indicated. Sequence of displacements for (b) the $1M$ polytype after one translation, t, (c) the $2M$ polytype after two translations, and (d) the $3T$ polytype after three translations are also given.

orthorhombic structure called $2O$ (Figure 29.10b), and the repeat distance, and thus the unit cell, is doubled. We are using Arial font for polytype symbols (M for monoclinic, O for orthorhombic, T for trigonal) to distinguish them from atomic positions (M for octahedral and T for tetrahedral), e.g., in Table 29.1.

A second, more subtle, reason for polytypism in sheet silicates is the attachment of the tetrahedral layer to the octahedral layer. Figure 29.11a shows an octahedral layer with shaded triangles on the top surface. A tetrahedral net is attached to the top surface (solid lines connect those apices that are shared with octahedra). The net attached to the bottom surface (connected to oxygen atoms in the triangular depressions) is necessarily displaced (dotted, dot-dashed, and dashed hexagons). It is displaced in one of three directions, with vectors v_1, v_2, and v_3. As long as you have a single layer, the three possibilities are symmetrically equivalent and simply involve a 120° rotation. But if you stack layers, there are different options. In the simplest case all layers are in the same orientation (v_1 displacements: v_1^I, v_1^{II}, v_1^{III}, Roman superscripts indicating the layer) and this produces a $1M$ polytype with a one-layer repeat (Figure 29.11b). In the common $2M$ mica polytype, layers with tetrahedra displaced by v_1 and v_2 alternate (v_1^I, v_2^{II})

(Figure 29.11c). Displacements may form a triangular pattern (v_1^I, v_2^{II}, and v_3^{III}), and in that case a trigonal polytype with a three-layer repeat results, known as the $3T$ polytype (Figure 29.11d). Other stacking sequences are possible.

The two factors (orientation of octahedra, orientation of tetrahedral nets) may be combined, resulting in many possibilities. Also, regular stacking is often interrupted by stacking faults, as illustrated in a high-resolution electron micrograph with local domains of $2M$ and $3T$ polytypes in phengite (Figure 29.12). The favored polytype depends on conditions of formation and on chemical composition, although boundaries are not well defined and polytypes are generally not represented on equilibrium phase diagrams. Some common polytypes in mica minerals are listed in Table 29.1.

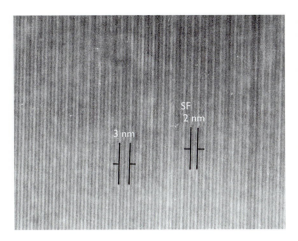

Fig. 29.12 High-resolution electron micrograph of a phengite. The polytypes are best visible in the upper thicker part with the $2M$ polytype on the right side (2 nm) and the $3T$ polytype on the left side (3 nm), with stacking faults (SF) interrupting the regular stacking sequence. The thinner (lower) part displays the structural 1 nm layer spacing (from Page and Wenk, 1979).

In the polytype symbol, numbers are used to classify the layer repeat, and capital italic letters are used to identify the symmetry. Subscripts are used if different possibilities exist. The nomenclature was originally introduced for SiC polytypes (Ramsdell, 1947) and has been adapted for sheet silicates (Smith and Yoder, 1956).

29.3 Structure of clay minerals

Clay minerals are essentially hydrous aluminous sheet silicates with variable composition and water content. They are important components of mudstones and claystones – sedimentary rocks that may also contain other minerals such as quartz and calcite. Clays are components of soils. Generally clay minerals are extremely fine grained and rarely exist as macroscopic crystals. The large surfaces of microcrystallites are electrostatically charged and adsorb ions and molecules. Clay minerals are the products of weathering and hydrothermal alteration of feldspars, mica, pyroxenes, and volcanic glasses. Clay minerals are classified in a similar way as other sheet silicates into major groups of 2:1 structures and 1:1 structures (Table 29.3). In 2:1 structures there are illite, smectite, vermiculite, and chlorite subgroups with di-octahedral

and tri-octahedral minerals, depending on composition, which can be quite variable.

The clay mineral *illite* is closely related to muscovite and has formerly also been called *hydromuscovite*. It contains less aluminum in the tetrahedral site than stoichiometric muscovite and, correspondingly, less potassium in the interlayer sites. As a result, illites have a general formula $K_xAl_2(Al_xSi_{4-x}O_{10}(OH)_2$ (e.g., $K_{0.6}Al_2(Al_{0.6}Si_{3.4}O_{10}(OH)_2)$. Illites form very fine-grained masses. Since illites adsorb H_2O, illite samples are water-enriched compared to the micas.

The *smectite* group includes a large number of clay minerals. Ideal end members in this group are saponite, beidellite, and nontronite. Unlike the structures of talc and pyrophyllite (see Figure 29.7c), smectites have layers of water molecules intercalated between octahedral-tetrahedral sheets (Figure 29.13b). Generally these clays do not have an ideal composition. Some Si^{4+} is substituted by Al^{3+}, and the charge is compensated by hydrated ionic complexes $(M^+ \cdot nH_2O)$ and $(M^{2+} \cdot nH_2O)$ entering the interlayer sites. Generally fewer than one-third of the tetrahedral sites are occupied by aluminum. The univalent interlayer cations or cation groups M^+ are Na^+, Li^+, NH_4^+, K^+, and Rb^+, and the divalent M^{2+} are Ca^{2+}, Mg^{2+}, and Co^{2+}. The interlayer ions can be exchanged with ions in surrounding aqueous solutions. Large organic molecules can be adsorbed. When water or molecules such as glycol are adsorbed, the structure swells. This process is reversible and, in a dry atmosphere or during heating, both water and hydrated cations leave the structure of smectites, causing the crystals to decrease in volume. In soils rich in smectite such swelling is often desirable to increase percolation. It can be achieved by adding fertilizers with Ca^{2+}, as in lime and gypsum. Smectite-rich soils are undesirable for construction because seasonal changes can induce expansion and shrinkage, sometimes resulting in landslides.

Illite, smectite and vermiculite have talc-type structures with two tetrahedral nets (2:1) (Figure 29.13b), whereas *kaolinite* minerals have serpentine-type structures with single tetrahedral nets (1:1) (Figure 29.13a). Dehydration of clay minerals by heating and ion exchange alter mainly the distance between layers, i.e., the lattice spacing (d_{001}). Variations of lattice spacings with heating, hydration, and glycolation are diagnostic characteristics and used to identify clay minerals in X-ray powder patterns (Table 29.3).

Table 29.3 Chemical composition and d_{001} spacings (in Å) of some clay minerals

Mineral Formula	d_{001} H_2O	d_{001} Glycol.	d_{001} Heat.	Conditions of formation (original material)
(2:1) structures				
Illite (hydromuscovite)	10	10	10	Alkaline conditions
$K_{0.6}Al_2(Al_{0.6}Si_{3.4}O_{10})(OH)_2$				(granitic rocks)
Smectite	14.2	17	10	Alkaline conditions
Montmorillonite $(Na,Ca)_{0.35}(Al,Mg)_2((Si,Al)_4O_{10})(OH)_2 \cdot nH_2O$				(mafic rocks)
Saponite $Ca_{0.25}(Mg,Fe)_3((Si,Al)_4O_{10})(OH)_2 \cdot nH_2O$				
Beidellite $(Al,Fe^{3+})_2((Si,Al)_4O_{10})(OH)_2 \cdot nH_2O)$				
Nontronite $Na_{0.3}Fe_2^{3+}((Si,Al)_4O_{10})(OH)_2 \cdot nH_2O)$				
Vermiculite	14.2	15.5	10–12	(Biotite, chlorite, hornblende)
$(Mg,Fe^{2+},Al)_3((Si,Al)_4O_{10})(OH)_2 \cdot 4H_2O$				
(1:1) structures				
Kaolinite (kandite)	7.1	7.1	7.1	Acid conditions
$Al_2(Si_2O_5)(OH)_4$				(granitic rocks)
Halloysite, dickite	10.2	~14	7.4	
$Al_2(Si_2O_5)(OH)_4 \cdot 2H_2O$				

H_2O, after water saturation; Glycol., after glycolation; Heat., after heating to 200 °C.

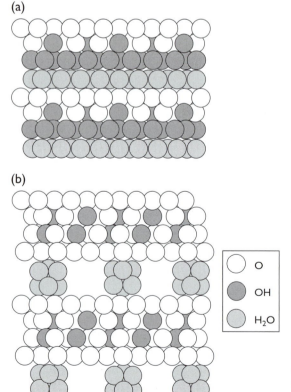

(a)

(b)

○ O

● OH

◐ H_2O

There are many minerals with more complicated sheet structures and chemical compositions. Paligorskite (Figure 29.8b), sepiolite (Figure 29.8c), and stilpnomelane are examples.

Properties of important sheet silicates are summarized in Table 29.4.

29.4 Brief description of important sheet silicate minerals (ordered as in Table 29.4)

In **serpentine** $(Mg_3(Si_2O_5)(OH)_4)$ some Fe^{2+} always substitutes for Mg^{2+}, but in limited amounts. There are three main polymorphs (antigorite, chrysotile, and lizardite), which differ in their long-range crystal structure. Antigorite is the most common of the three. Crystals are usually fine grained but occur in dark- to light-green aggregates with shiny curved surfaces. Chrysotile is fibrous and easily recognized (Plate 31a), although sometimes fibrous tremolite has a similar appearance and can be associated with chrysotile.

Fig. 29.13 Schematic view of clay mineral structures: (a) halloysite 1:1 (kandite), (b) montmorillonite 2:1 (smectite) (cf. Figure 29.7), with oxygen and hydroxyl ions and molecular water.

Table 29.4 Common sheet silicate minerals with some diagnostic properties; most important minerals are given in italics. (For clay minerals see Table 29.4)

Mineral / Formula	System	Morphology / Cleavage	H	D	Color / Streak	n / Pleochroism	Δ	2V (Dispersion)	d_{001}
Serpentine (antigorite) $Mg_3(Si_2O_5)(OH)_4$	Monocl.	Platy (001) *(001)*	3–4	2.6	Green	1.56–1.57	0.011	–40–60 (r > v)	7.3
Kaolinite $Al_2(Si_2O_5)(OH)_4$	Tricl.	Platy (001) *(001)*	2–2.5	2.6	White, yellow, green	1.55–1.57	0.007	–20–50 (r > v)	7.1
Talc $Mg_3(Si_4O_{10})(OH)_2$	Monocl.	Platy (001) *(001)*	1	2.7	White, green	1.54–1.60	0.05	–0–30 (r > v)	9.0
Pyrophyllite $Al_2(Si_4O_{10})(OH)_2$	Monocl.	Platy (001) *(001)*	1.5	2.8	White	1.55–1.60	0.05	–53–60 (r > v)	9.2
Paligorskite $Mg_5(Si_8O_{20})(OH)_2 \cdot 8H_2O$	Monocl.	Microcrystal	2–2.5	2.0	White, gray	1.51–1.53	0.01	(–)	
Sepiolite $Mg_4(Si_6O_{15})(OH)_2 \cdot 6H_2O$	Ortho.	Microcrystal	2–2.5	2.2	White, gray	1.51–1.53	0.01	(–)	
Mica minerals: trioctahedral									
Annite $KFe^{2+}_3(Si_3AlO_{10})(OH)_2$	Monocl.	Platy (001) *(001)*	2.5–3	3.3	Brown, black	1.62–1.67 *Yellow-brown-brown*	0.08	–0–5 (r < v)	10.1
Biotite $K(Mg,Fe)_3(Si_2(Si,Al,Fe^{3+})_2O_{10})(OH)_2$	Monocl.	Platy (001) *(001)*	2–3	2.9+	Brown, yellow, black	1.56–1.69 *Yellow-brown/green*	0.05	–0–9 (r < v)	10.1
Clintonite $CaMg_2Al(SiAl_3O_{10})(OH)_2$	Monocl.	Platy (001) *(001)*	3.5–6	3.1	Yellow, orange, brown	1.64–1.66 Clear-brown-brown	0.012	–2–40 (r < v)	10.1
Phlogopite $KMg_3(Si_3AlO_{10})(OH)_2$	Monocl.	Platy (001) *(001)*	2–3	2.8+	Brown, yellow	1.53–1.62 *Clear-yellow-yellow*	0.03	–0–10 (r < v)	10.1
Polylithionite $KLi_2Al(Si_4O_{10})F_2$	Monocl.	Platy (001) *(001)*	2–3	2.8	Pink, white	1.53–1.56 *Clear-green-green*	0.012	–0–40 (r > v)	10.1
Siderophyllite $KFe^{2+}_2Al(Si_2Al_2O_{10})(OH)_2$	Monocl.	Platy (001) *(001)*	2.5–3	3.1	Black	1.59–1.64 *Yellow brown-brown*	0.05	–0–5 (r < v)	10.1
Trilithionite (old: Lepidolite) $K(Li_{1.5}Al_{1.5})(Si_3Al)O_{10}(OH)_2$	Monocl.	Platy (001) *(001)*	2.5–4	2.8	Pink, violet	1.52–1.59 *Clear-pink-pink*	0.02	–0–50 (r > v)	10.1
Zinnwaldite $KLiFe^{2+}Al(AlSi_3O_{10})(OH)_2$	Monocl.	Platy (001) *(001)*	2.5–4	2.9	Brown, violet, gray	1.53–1.59 *Yellow-brown-brown*	0.03	–0–40 (r > v)	10.1

Table 29.4 (cont.)

Mineral Formula	System	Morphology Cleavage	H	D	Color Streak	n Pleochroism	Δ	$2V$ (Dispersion)	d_{001}
Mica minerals: dioctahedral									
Aluminoceladonite $KAlMg(Si_4O_{10})(OH)_2$	Monocl.	Massive (001)	2	3.0	Blue-gray	1.61–1.66	0.04	−5–8	10.0
Glauconite $(K,Na)(Mg,Fe^{2+},Fe^{3+},Al)_2(Si,Al)_4O_{10}(OH)_2$	Monocl.	Platy (001) (001)	2	2.8	Gray, yellow-gray	1.59–1.64 *Yellow-green-green*	0.025	−40–20 (r > v)	10.0
Margarite $CaAl_2(Si_2Al_2O_{10})(OH)_2$	Monocl.	Platy (001) (001)	3.5–4.5	3.1	Gray, yellow, green	1.63–1.65	0.012	−40–65 (r < v)	10.0
Muscovite $KAl_2(Si_3AlO_{10})(OH)_2$	Monocl.	Platy (001) (001)	2.5–3	2.9	White, gray	1.55–1.61	0.04	−30–45 (r > v)	10.0
Paragonite $NaAl_2(Si_3AlO_{10})(OH)_2$	Monocl.	Platy (001) (001)	2.5–3	2.9	White, gray	1.56–1.61	0.03	−0–40 (r < v)	10.0
Chlorite minerals									
Chamosite $Fe_3^{2+}(OH)_6\,Fe_2Al(Si_3AlO_{10})(OH)_2$	Monocl.	Oolite	2.5–3	3.2	Green-blue *Gray-green*	1.64–1.66 *Yellow-green-green*	0.005	−0 (r < v)	14.2
Clinochlore $Mg_3(OH)_6Mg_2Al(Si_3AlO_{10})(OH)_2$	Monocl.	Platy (001) (001)	2–2.5	2.6	Green *Gray-green*	1.57–1.59 *Clear-green-yellow*	0.005	Anom. +0–90 (r < v)	14.2
Other sheet silicates									
Stilpnomelane $KFe_8(Si_{11}AlO_{28})(OH)_8·2H_2O$	Monocl.	Platy (001) (001)	3–4	2.6–2.9	Black, brown	1.54–1.75 *Yellow-brown-black*	0.05	−0	
Prehnite $Ca_2Al(Si_3AlO_{10})(OH)_2$	Ortho.	Radiating (001)	6–6.5	2.9	Green, gray, white	1.61–1.66	0.03	65–69	

Notes: H, hardness; D, density(g/cm³); n, range of refractive indices; Δ, birefringence; $2V$, axial angle for biaxial minerals. For uniaxial minerals (+) is positive and (−) is negative. Acute $2V$ is given in the table. If $2V$ is negative the mineral is biaxial negative and $2V$ is $2V_{\alpha}$; if it is positive, the mineral is biaxial positive and $2V$ is $2V_{\gamma}$. Dispersion r < v means that acute $2V$ is larger for violet than for red; Anom., anomalous dispersion or birefringence. d_{001} is the lattice spacing for the basal planes.
System: Monocl., monoclinic; Ortho., orthorhombic; Tricl., triclinic.

Positive identification of lizardite must rely on X-ray diffraction.

Serpentine forms in ultramafic rocks as products of hydrothermal alteration of olivine and pyroxene (Plate 14d). Serpentine asbestos has been used as a refractory material, particularly for high-temperature applications. Today it is largely outlawed because of perceived carcinogenic health hazards (Chapter 36).

Kaolinite $(Al_2(Si_2O_5)(OH)_4)$ is often associated with, and is similar to, the related clay minerals dickite and halloysite. Kaolinite group minerals cannot be distinguished on the basis of their morphology. They usually occur as fine-grained aggregates in flour-like or clay-like white masses. Identification is done by X-ray diffraction (Table 29.4).

Talc $(Mg_3(Si_4O_{10})(OH)_2)$ may contain Fe^{2+} up to 1.5–2 weight%. Talc is found as light-green aggregates with a greasy touch in schists and marbles and as soft large plates with a perfect cleavage and pearly luster in altered dunites. It is easily distinguished from other sheet silicates such as chlorite and serpentine on the basis of its low hardness (1 on the Mohs scale; it can be scratched with fingernails). In thin sections talc is recognized by lower-order interference colors than muscovite, but higher-order interference colors than serpentine.

Similar to serpentine, talc is also frequently an alteration product of olivine and pyroxene. It occurs in low-temperature metamorphic rocks (schists and marbles). Talc is used as a refractory material, for brick-lining of blast furnaces, as a lubricant, and in the cosmetic industry.

Pyrophyllite $(Al_2(Si_4O_{10})(OH)_2)$, with an Fe^{3+} content of less than 0.5 weight%, occurs as cryptocrystalline solid masses of pink or greenish-gray color in metamorphic rocks and sometimes as radial aggregates (Plate 30e). It forms because of hydrolysis of aluminum silicates in granitic rocks. When present in significant concentrations, it is a raw material for manufacturing insulators and furnace brick-linings. Stonecarvers call translucent varieties of this mineral "agalmatolite". Pyrophyllite exfoliates on heating.

There is a continuous solid solution series of trioctahedral magnesium–iron–aluminum micas **phlogopite** $(KMg_3(Si_3AlO_{10})(OH)_2)$, **siderophyllite** $(KFe_2^{2+}Al(Si_2Al_2O_{10})(OH)_2)$, and **annite** $(KFe_3^{2+}(Si_3AlO_{10})(OH)_2)$. The intermediate compositions are generally called **biotite**, and also contain some

Fe^{3+}. Phlogopite and biotite are found as platelets and sometimes as very large crystals (e.g., Figure 10.12b). Their color varies from green to deep brown and black, depending on the iron content. Pure phlogopite is transparent and colorless and can be mistaken for muscovite but is recognized in thin sections by a small axial angle. In thin section colored biotites are pleochroic with brown, yellowish, and green colors. The strongest colors are observed when the vibration direction of the polarizer is parallel to the cleavage (Plate 11a).

Phlogopite crystals are used as electrical insulators. Commercially valuable concentrations of large phlogopite crystals are found in some alkali-rich ultramafic rocks (e.g., in Kovdor on the Kola Peninsula in Russia and Palabora in South Africa) and at contacts of marbles with gneisses and schists in crystalline shields.

Muscovite $(KAl_2(Si_3AlO_{10})(OH)_2)$ is an almost pure aluminous mica with very little iron (1–3 weight% Fe_2O_3). In some rare muscovites, aluminum is partly substituted by chromium, resulting in a bright-green color (*fuchsite* or *mariposite*). Other muscovites exist in which potassium is partly replaced by sodium. The isomorphic substitution $Al^{3+} + [Al^{3+}] \rightleftharpoons Mg^{2+} + [Si^{4+}]$ produces a series of micas with the common name *phengite* – in particular, $KAlMg(Si_4O_{10})(OH)_2$.

Muscovite usually occurs as tabular crystals of light brown or gray color (Figure 29.14a, Plate 30f). Thin cleavage platelets of muscovite are colorless. Large crystals and plates (e.g., Figure 10.12b) are found in granite pegmatites, while scaly aggregates are observed in granites, gneisses, and schists. *Sericite* is a yellow fine-grained or, sometimes, solid cryptocrystalline mass composed mainly of muscovite. It forms pseudomorphs after feldspar and aluminosilicates during hydrothermal alteration (Plate 7d,e).

Muscovite is used as a dielectric material in the electronics industry. It is mined from granite pegmatites.

Two lithium-bearing micas are noteworthy: **zinnwaldite** $(KLiFe^{2+}Al(AlSi_3O_{10})(OH)_2)$ is a magnesium–iron-bearing lithium mica (Figure 29.14b), while **lepidolite**, also known as trilithionite $(K(Li_{1.5}Al_{1.5})(Si_3Al)O_{10}(OH)_2)$, always contains some manganese in octahedral sites. The presence of manganese ions causes pink, silverish-pink, and lilac-pink colors that are diagnostic for the identification.

The micas (phlogopite, biotite, muscovite) form in igneous rocks as primary phases and as secondary postmagmatic minerals due to interaction between

(a)

(b)

Fig. 29.14 Morphology of sheet silicates. (a) A sheet of muscovite from Minas Gerais, Brazil (courtesy O. Medenbach). Width 55 mm. (b) Zinnwaldite on quartz from the Erzgebirge near Zinnwald (now Cinovec, Czech Republic) (courtesy A. Massanek). Width 90 mm.

solutions and olivines, pyroxenes, and hornblendes. Many volcanic rocks contain phenocrysts of biotite. In particular, large crystals of phlogopite, biotite, and muscovite are found in granite pegmatites. Spodumene-bearing pegmatites contain lithium micas. Metamorphic rocks (gneisses and schists) frequently bear muscovite and biotite as major mineral phases; phlogopites occur in dolomitic marbles.

Glauconite (K,Na)(Mg,Fe^{2+},Fe^{3+},Al)$_2$(Si,Al)$_4$O$_{10}$(OH)$_2$ is an example of an interlayer-deficient mica. It is blue or green in color, and it occurs as tiny roundish aggregates and colloform segregations in clays, marls, and dolomites.

A water-rich micaceous sheet silicate is **vermiculite** (Mg,Fe^{2+},Al)$_3$(Si,Al)$_4$O$_{10}$(OH)$_2$·4H$_2$O, which forms as a result of weathering of phlogopite and biotite in soils. Hydration of the mineral, and substitutions in its interlayer sites, bring about changes in its physical properties: compared to fresh phlogopite, vermiculite is more fragile, less shiny, and its cleavage platelets are not elastic and become crumpled, like wet paper. When vermiculite is heated, it loses the water and may shrink by 10% or more.

Clays are an important construction material and a raw material for the ceramic and porcelain industries. Pure montmorillonite clays, which exhibit pronounced adsorption properties, are known as bentonite.

Chlorites are a large group of compositionally complex sheet silicates (see Table 29.2). Two major members of the group are *clinochlore* (Mg$_3$(OH)$_6$Mg$_2$Al(Si$_3$AlO$_{10}$)(OH)$_2$), and *chamosite* (Fe$_3^{2+}$(OH)$_6$Fe$_2$Al(Si$_3$AlO$_{10}$)(OH)$_2$), which form a solid solution. Chlorites that are rich in magnesium are green in color and hence their name (Plate 31b,c). In thin sections they are recognized by low birefringence and anomalous brown or ink-blue interference colors owing to dispersion (Plate 15e,f). Iron-rich chlorites are often oxidized under weathering conditions and have a brown color. There are many other species among chlorites – nickel, zinc, manganese, and lithium substituting for magnesium in the clinochlore formula.

Chlorites are products of late-stage low-temperature alteration of olivine, pyroxenes, and hornblendes, as are serpentine, talc, and brucite (greenschist facies). They are one of the first metamorphic minerals to crystallize in slates.

Two sheet silicates with a more complex structure than micas and chlorite occur in low-grade metamorphic rocks. **Stilpnomelane** (KFe$_8$(Si$_{11}$AlO$_{28}$)(OH)$_8$·2H$_2$O is black, shows yellow and brown pleochroism, and can easily be mistaken for biotite (Plate 11b). However, cleavage of stilpnomelane is less perfect and it is more brittle and crystallizes often as prisms. In the extinction position (when viewed with a petrographic microscope), it does not show the mottling effect that is typical of biotite (Plate 16b). Stilpnomelane occurs mainly in low-grade iron-rich metamorphic rocks, whereas biotite is a high-grade mineral.

Prehnite (Ca$_2$Al(Si$_3$AlO$_{10}$)(OH)$_2$) is pale green, gray or white and generally occurs in botryoidal masses of tabular crystallites. It is found as a low-grade hydrothermal mineral in veins or cavities of mafic volcanic rocks and in altered metamorphic rocks, often associated with zeolites.

29.5 Formation conditions for sheet silicates and weathering of silicate rocks

Most sheet silicates are low-temperature and low-pressure minerals. Only muscovite, phlogopite, biotite, and lithium micas occur in igneous rocks. As a whole, endogenetic mineral deposits characteristically bear micas, talc, pyrophyllite, serpentines, and chlorites. Exogenetic conditions are appropriate for the formation of kaolinite, smectites, illites, some serpentines and chlorites. Clay minerals form directly by precipitation from seawater and by alteration of primary minerals. The six-membered tetrahedral $Si_6O_{18}^{12-}$ ring is one of the most stable polymers in solution and such rings combine to form either microcrystalline opal–cristobalite–tridymite-like structures in cherts, or clay minerals with sheet structures in clays, mudstones, and shales.

We will now take a closer look at alteration processes at surface conditions. These minerals are the main constituents of clays forming at surface and submarine conditions. That is why these minerals are sometimes called collectively *clay minerals*. Aluminous sheet silicates (kaolinite, etc.) develop in weathering crusts of granites and felsic (silicon-rich) volcanics, and they may later be transformed into bauxites.

Alteration of minerals and rocks on the Earth's surface, under the influence of physical, chemical, and biological factors, is described as a *weathering process*. Weathering is usually subdivided into four separate stages, according to the type of chemical processes involved. At first chemical decomposition of minerals is insignificant while mechanical weathering dominates. A coherent rock breaks down into fractured rock or loose sand. This is followed by the crystallization of clay minerals. The complete hydrolysis of the silicates causes oxides and hydroxides to form. Typically, aqueous solutions and colloidal solutions participate in all these processes.

Pervasive surface weathering takes place when several conditions work in combination. Lack of tectonic activity and weak erosion are important factors. Under surface conditions, zones of weathering form and cover unaltered rocks; these zones can extend to great depths, as much as 1 km, and in some cases their formation lasts over 15–20 million years.

In both ancient and recent zones of weathering, the intensity of the transformation of the initial rocks increases upwards (Figure 29.15), from the region of little alteration above the primary bedrock

towards zones where residual mineral deposits develop. There is often a regular zoning with a mechanical decomposition of pre-existing rocks into grain aggregates (or gruss), a zone of clay minerals, followed by kaolinite, and ultimately bauxite–*laterite* with oxides and hydroxides. In temperate zones with moderate rainfall, chemical weathering is usually confined to clay minerals. In arid regions transformations are minimal. The thickest weathering crusts form in a humid and warm climate such as in tropical rainforests. In those tropical regions with high precipitation, surface weathering of granites and other feldspathic rocks, as well as ultramafic rocks, causes laterite deposits to form. Alkalis, alkaline-earth elements, and silica are leached out of the original rock, while aluminum and iron oxides and hydroxides precipitate. The red-colored laterites are rich in iron oxides and hydroxides, whereas bauxites are rich in aluminum hydroxides. During alteration of some ultramafic rocks and serpentinites, residual nickel and magnesium deposits may appear.

The magnesium minerals paligorskite and sepiolite are important components of *calcrete* deposits in semi-arid climates. Calcretes develop by replacement of parent rocks by calcite and subsequent precipitation of paligorskite during evaporative episodes.

29.6 Clay minerals in soils

Having explored some aspects of weathering, it is natural that we extend this excursion by taking a brief look at soils and their mineralogical composition. In fact, of all minerals, clay minerals in soils have the most profound effect on life. Soils are the outermost thin layer of the solid Earth and support the majority of life, interfacing the lithosphere, the biosphere, and the atmosphere. Interactions between these systems occur mainly through plants and their root systems. Soils are very complicated systems, and we will address only a few issues directly related to mineralogy. Soil science, or *pedology*, investigates the composition, structure, and evolution of soils, and their relationship to the environment (e.g., Jenny, 2011; Troeh, 2005). In soils, most of the reactions take place on the surface of clays and organic matter.

Contrary to minerals in most rocks, soil minerals are extremely fine grained (<2 μm), creating a large surface area (per unit mass) on which reactions can

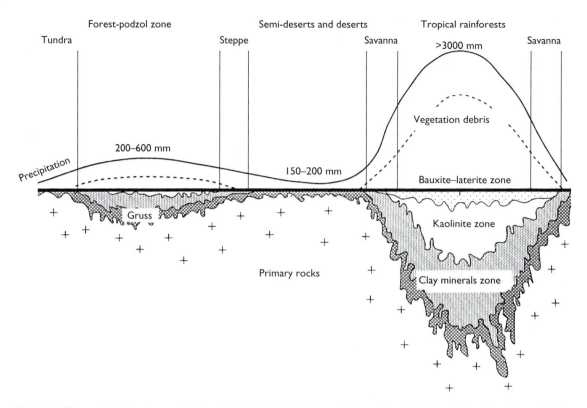

Fig. 29.15 Types of weathering, rainfall, and vegetation debris from polar regions (left) to equatorial regions (right) (based on Strakhov, 1967).

occur. An electron microscope is required to image individual crystals. Figure 29.16 is an SEM image of a clay sample composed largely of saponite. Thin lath- or blade-shaped crystals are visible, only a few micrometers in size. Most clay minerals have a negative charge within the tetrahedral-octahedral layers due to isomorphic substitutions. The charge is balanced by cations from the surrounding soil solution that attach to the surface of the crystallites. These cations exchange easily and are a major source of plant nutrients. Soil scientists define cation exchange capacity (CEC) as the amount of negative charge in the tetrahedral-octahedral layers per 100 g soil. Soil minerals with a higher CEC can hold on to more soil ions and are more reactive, benefiting plants. Since the surface area (per unit mass) increases enormously as the particle size decreases, clay minerals vary in their CEC, as is illustrated in Table 29.5. Some clays and organic humus have a variable surface charge that depends on the pH of the surrounding solution.

Minerals that are frequently found in soils are listed in Table 29.6. All clays adsorb water on their surface as discussed above, but some (such as smectite and halloysite) also allow water to enter between the layers in the crystal structure. Most of the soil minerals are aluminosilicates, i.e., they contain aluminum in octahedral layers. In kandites, with a 1:1 layer structure (see Figure 29.13a), there is little isomorphous substitution and octahedral-tetrahedral layers as well as interlayers are largely closed to water and cation exchange. As a result they are not very reactive. In contrast, smectites, with a 2:1 layer structure (see Figure 29.13b), have a significant amount of isomorphous substitution and the interlayers are open to water and cation exchange. Most of the substitution occurs within the octahedral layer. Therefore the negative charge is further from the crystal surface and cations are held less tightly. When soil water content increases, cation concentrations decrease through dilution, water enters the spaces between each 2:1 layer, and the clays swell. These soils are very reactive, shrinking when dry and expanding when wet.

There are also nonclay minerals in soils. Minerals such as halite, calcite, and gypsum are typical of

Table 29.5 Surface area and cation exchange capacity of some clay minerals and humus

Mineral	Surface area (10^3 m²/kg)	Cation exchange capacity (cmol charge/kg)
Kaolinite	10–20	1–10
Chlorite	70–150	20–40
Mica	70–120	20–40
Montmorillonite	600–800	80–120
Vermiculite	600–800	120–150
Humus	900	150–300

Source: From Singer and Munns, 2005.

Fig. 29.16 SEM image of a trioctahedral aggregate (saponite) from Redrock Canyon, California, USA, with minute lath-shaped crystallites typical of these phyllosilicates (courtesy T. Teague).

Table 29.6 Selected minerals that are frequent components of soils

Mineral	Idealized formula
Kandite (1:1)	
Kaolinite	$Al_2(Si_2O_5)(OH)_4$
Halloysite	$Al_2(Si_2O_5)(OH)_4 \cdot 2H_2O$
Smectite (2:1)	
Montmorillonite	$(Na,Ca)_{0.35}(Al,Mg)_{2\text{-}3}$ $(Si_{3.65}Al_{0.35}O_{10})(OH)_2 \cdot nH_2O$
Vermiculite	$(Mg,Fe^{2+},Al)_3((Si,Al)_4O_{10})$ $(OH)_2 \cdot 4H_2O$
Chlorite (clinochlore)	$Mg_5Al(Si_3AlO_{10})(OH)_8$
Gibbsite	$Al(OH)_3$
Goethite	$Fe^{3+}O(OH)_2$
Calcite	$CaCO_3$
Gypsum	$CaSO_4 \cdot 2H_2O$
Halite	$NaCl$
Apatite	$Ca_5(PO_4)_3OH$

pH. Minerals in the bedrock become unstable and are replaced by clays. Ions are lost to percolating aqueous solutions. Figure 29.17 illustrates a typical soil profile.

The lowest level is that of the unaltered initial bedrock (R) and microscopic fractures and fissures are pathways along which aqueous solutions penetrate and interact with the rock, altering initial minerals into mixtures of clays and hydroxides. The first alteration zone is called saprock (weathering horizon C). The initial minerals are more or less altered, but the petrographic structure of the rock is preserved. White areas may still outline euhedral feldspar crystals and dark areas amphiboles and biotite, though these minerals may have disappeared. Some components of the bedrock are recognized as partially unaltered fragments. While much of the alteration occurs *in situ*, some clay minerals are transported in the percolating solutions. Because of these processes the clay and mineral structure in zone C is highly heterogeneous, depending on the composition of the bedrock and the activity of the solutions. Geochemical processes govern this zone, with little biological activity.

Above zone C, the original petrographic structure has disappeared. In this so-called saprolite zone, clays have replaced most of the original minerals (horizon B). Some clays are still remnants of the original rock formed in local chemical reactions, but others are the result of mechanical transport, and clay minerals

soils in climates where evaporation is greater than precipitation. If they occur in significant concentrations, they have adverse effects on plant growth. Apatite occurs in some soils as a relict from the parent rock and is important because it is one of the few minerals that provide phosphorus, an important nutrient for plants.

Soils are a product of inorganic processes (such as physical and chemical weathering), as well as organic processes. Organic reactions are followed by the degradation of organisms and incorporation of organic matter into the soil. Soils are not a homogeneous system like most rocks, but display a vertical stratification. The soil profile is highly variable in content of clay, organic matter and water, as well as CEC and

434

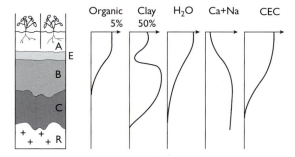

Fig. 29.17 Typical soil profile with (R) bedrock, (C) abiotic weathering zone, (B) clay horizon, (E) leached zone, and (A) biotic zone. Also shown are relative changes in organic matter, clay content, water content, calcium and sodium concentration, and cation exchange capacity (CEC).

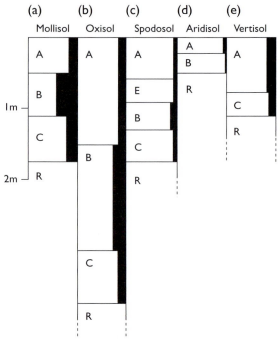

Fig. 29.18 Soil profiles for some typical soils: (a) mollisol, (b) oxisol, (c) spodosol, (d) aridisol, (e) vertisol. Horizons are labeled (cf. Figure 29.18 and text). In each layer, shading indicates the average content of clays (based on Singer and Munns, 2005).

present in this horizon may not have formed under the same chemical conditions. The composition of the B horizon depends on migration of cations in solution both from below and from above. Iron and aluminum are often concentrated in this zone. The B horizon may contain some organic matter, introduced from above, and occasionally carbonate and gypsum.

Above this B zone is sometimes a horizon (E) that has been leached of organic matter, silicate clay, as well as aluminum and iron, leaving a concentration of resistant sand and silt particles. The E horizon can also contain concentrations of calcium carbonate (caliche) in arid environments. The uppermost weathering horizon A contains much of what is commonly called "soil". The mineral horizon A is close to the surface and has accumulated organic matter such as humus. This accumulation is caused by high biological activity, resulting in acid conditions leading to rapid weathering of minerals and leaching of soluble products. The full stratification is observed in deep soils of tropical regions, where soil formation has been intense over long periods.

The characteristics and extent of each horizon are determined by numerous factors, among which composition of bedrock, climate (temperature and rainfall), organisms, time, and topography are the most important. Soil scientists consider how these factors differentially affect soil-forming processes and create different soil horizons. For instance, both time and precipitation affect hydrolytic weathering and soil mineral formation. Hydrolytic weathering results in some loss of silicon and alkalis from the initial rock,

whereas aluminum and iron remain. In total hydrolysis, all silicon and alkalis are lost, resulting in precipitation of gibbsite and goethite, typical components of laterites. If hydrolysis is less pervasive, only alkalis are lost, while some silicon remains. In such cases kaolinite is the typical mineral. If hydrolysis is even weaker, smectite-type clay minerals form.

Soil taxonomists have developed a system to classify soils based on their physical and chemical characteristics along the soil profile. For example, *mollisols* are soils with a high surface accumulation of organic matter in the A horizon and are typical of temperate grassland environments. The clay layer B consists of a variety of phyllosilicates (Figure 29.18a). *Oxisols* are most common in hot humid climates, where weathering and leaching are intense. The organic layer is reduced and the clay layer (B) consists of quartz, iron, and aluminum hydroxides, and kaolinite (Figure 29.18b). These soils are often acidic and infertile. Typical examples are the soils of the Amazon basin. *Spodosols* have a thin organic layer, very rich

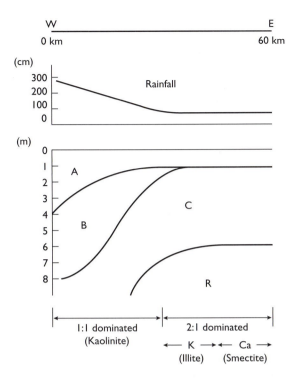

Fig. 29.19 West–east transect through the Gatt region in southwest India, illustrating the systematic change in soil profile and the transition from smectite to illite and kaolinite-dominated soils with increasing rainfall (after Pédro, 1997). For horizon symbols, see Figure 29.17.

in organic matter, immediately followed by a leached zone E, and have smectites below (B) (Figure 29.18c). These soils are usually found in cool, humid regions and on coarse-textured parent material where leaching occurs rapidly. *Aridisols* cover wide surfaces in arid regions and are depleted in organic matter. They are characterized by thin A horizons with low humus content. The clay horizon B consists of phyllosilicates and rests immediately on bedrock (R) (Figure 29.18d). The soil profile is shallow. In *vertisols* a stratified A horizon rests directly on the inorganic C horizon composed of smectite (Figure 29.18e). This horizon swells when wet and shrinks when dry, so that large cracks develop on the surface. Vertisols occur in climates sufficiently dry that soils dry regularly, as in central India and eastern Australia. They are easy to identify by their high shrink and swell capabilities.

A case study in southwestern India demonstrates the transition from a smectite-dominated soil to a kaolinite-dominated soil due to increasing precipitation in a 60 km transect (Figure 29.19). The bedrock is gneiss throughout and temperatures are high and constant all year (average 24 °C), but the rainfall increases from 76 cm in the dry zone (east) to 265 cm in the wet zone (west) because of the monsoon climate. In the dry zone (east) the soil profile is shallow, particularly the biotic A horizon (1 m). There is limited weathering and ion exchange. Plagioclase is retained, whereas biotite and amphiboles are replaced by low-charge 2:1 clays (smectites). With increasing weathering, plagioclase becomes affected and calcium is removed, yielding high-charge 2:1 K-phyllosilicates (sericite and illite). Finally, when weathering becomes intense in the region of high precipitation (west), the biotic horizon expands to 4 m. Alkalis are removed and the principal clay mineral is 1:1 kaolinite. On the basis of our knowledge of these different clays, we can assume that in the intermediate rainfall locations the soils have a high cation exchange capacity (CEC) and are productive, while at the high-rainfall location few cations remain and the soils are infertile.

The abundance and variety of clays in a soil profile determine the fertility and thus have a profound influence on the vegetation. In agricultural regions the composition of the natural soil, particularly in the A horizon, is often altered by adding fertilizer.

29.7 Summary

All sheet silicates share layers of octahedra that are linked to one or two layers of tetrahedra (1:1 and 2:1). Depending on the cation charge in octahedra either all octahedral interstices are occupied (trioctahedral for Mg^{2+}, Fe^{2+}) or one out of three is vacant (dioctahedral for Al^{3+}). These units are then stacked and bonded by van der Waals forces as in talc or by weak ionic bonds through large cations (K, Na, Ca) in the interlayers as in mica. There can be various types of stacking leading to polytypism. Clay minerals are hydrated sheet silicates with disordered structures.

Sheet silicates form mainly at moderate to low temperatures and are especially significant as weathering products of primary rocks. The composition of clay minerals in soils and shales is indicative of conditions under which they formed. Clay minerals are subject to ion exchange.

Important sheet silicate minerals to remember

Name	Formula
1:1	
Serpentine	$Mg_3(Si_2O_5)(OH)_4$
Kaolinite	$Al_2(Si_2O_5)(OH)_4$
2:1	
Talc	$Mg_3(Si_4O_{10})(OH)_2$
Pyrophyllite	$Al_2(Si_4O_{10})(OH)_2$
Micas	
Phlogopite (trioctahedral)	$KMg_3(Si_3AlO_{10})(OH)_2$
Biotite (trioctahedral)	$K(Mg,Fe)_3(Si_3AlO_{10})(OH)_2$
Muscovite (dioctahedral)	$KAl_2(Si_3AlO_{10})(OH)_2$
Paragonite (dioctahedral)	$NaAl_2(Si_3AlO_{10})(OH)_2$
Margarite (dioctahedral)	$CaAl_2(Si_2Al_2O_{10})(OH)_2$
Chlorite	
Clinochlore	$Mg_5Al(Si_3AlO_{10})(OH)_8$
Clay minerals	
Illite	$K_{0.6}Al_2(Al_{0.6}Si_{3.4}O_{10})(OH)_2$
Montmorillonite (smectite)	$(Na,Ca)_{0.35}(Al,Mg)_2(Si,Al)_4O_{10}(OH)_2 \cdot nH_2O$
Halloysite	$Al_2(Si_2O_5)(OH)_4 \cdot 2H_2O$

Test your knowledge

1. Describe the three principles by which sheet silicates can be classified. Give an example for each of the groups.
2. Some oxygen atoms in the structure of sheet silicates are not connected to tetrahedra. Where are they located and how do they influence the chemical formula of, for example, talc?
3. Describe the difference between chrysotile and antigorite and give a reason for their peculiar structures.
4. Write a reaction to form muscovite from the alteration of potassium-feldspar.
5. Clay minerals are a special group of sheet silicates. How are they best identified?
6. Describe clay minerals in typical soil profiles.

Further reading

Sheet silicates

Bailey, S. W. (ed.) (1984). *Micas.* Reviews in Mineralogy, vol. 13. Mineralogical Society of America, Washington, DC.

Deer, W. A., Howie, R. A. and Zussman, D. J. (2009). *Rock-forming Minerals, Volume 3B, Layered Silicates Excluding Micas and Clay Minerals*, 2nd edn. The Geological Society of London.

Fleet, M. E. (2013). *Rock-forming Minerals, Volume 3A, Micas*, 2nd edn. The Geological Society of London.

Moore, D. M. and Reynolds, R. C. (1997). *X-ray Diffraction and the Identification and Analysis of Clay Minerals*, 2nd edn. Oxford University Press, Oxford.

Wilson, M. J. (2013). *Rock-forming Minerals, Volume 3C, Sheet Silicates: Clay Minerals*, 2nd edn. The Geological Society of London.

Formation

Brady, N. C. and Weil, R. R. (2007). *The Nature and Properties of Soils*, 14th edn. Prentice Hall, Upper Saddle River, NJ.

Sposito, G. (2008). *The Chemistry of Soils*, 2nd edn. Oxford University Press, Oxford.

Troeh, F. R. (2005). *Soils and Soil Fertility*, 6th edn. Wiley-Blackwell, Oxford.

Velde, B. and Meunie, A. (2008). *The Origin of Clay: Minerals in Soils and Weathered Rocks*. Springer, Berlin.

30 | Chain silicates. Discussion of some igneous and metamorphic processes

In chain silicates the characteristic structural feature is an infinite chain of tetrahedra, like a sheet silicate cut into ribbons and then stacked. The two most common chain types are single chains in pyroxenes and double chains in amphiboles. Pyroxenes are anhydrous while amphiboles contain some hydrogen, in the form of OH^- groups. Both mineral groups are important in igneous and metamorphic rocks and the chapter will introduce some new aspects of these processes.

30.1 Structural and chemical features

Chain silicates are quite unique in that a few structure types can accommodate a wide range of compositions with mineral representatives in most igneous and metamorphic rocks. Enstatite ($Mg_2(Si_2O_6)$) occurs in the upper mantle, associated with olivine. Jadeite ($NaAl(Si_2O_6)$) forms by decomposition of albite at high pressure in subduction zones, and spodumene ($LiAl(Si_2O_6)$) is a mineral found in pegmatites.

In Chapter 29 we discussed the structures of sheet silicates with infinite two-dimensional layers of hexagonal tetrahedral nets. We noted that there are some deviations from infinite nets in the structures of paligorskite and sepiolite, where, instead of infinite sheets, finite strips are stacked in a brickwork fashion (see Figure 29.8). In chain silicates such strips are much narrower and extend only over one or two tetrahedra. From an infinite tetrahedral sheet, say of talc (Figure 30.1a), we can obtain simple strips in various ways. Figure 30.1b–d illustrates some of the possibilities for which there are mineral representatives. In Figure 30.1b a band consisting of four

(a)   (b)   (c)   (d)

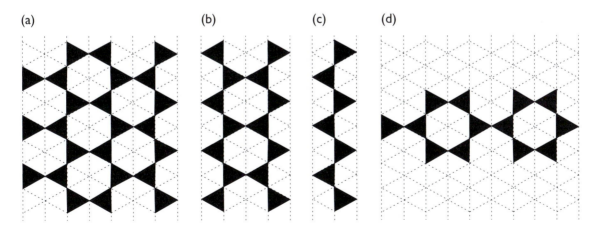

Fig. 30.1 Tetrahedral layer on a triangular reference grid. Apices are pointing down and only triangular surfaces are shown. (a) Infinite net of a sheet silicate. (b) Strip of four tetrahedra forming a double chain (amphibole). (c) Strip of two tetrahedra forming a single chain (pyroxene). (d) Strip of three tetrahedra, at right angles to (b) and (c) with a hybrid chain (howieite).

tetrahedra is cut from the sheet in the vertical direction. By doing so we obtain a *double chain*, which is present in *amphibole minerals*. In Figure 30.1c a narrow band of two tetrahedra forms a *single chain*, representative of *pyroxenes*. Finally, in Figure 30.1d a strip of three tetrahedra is cut from the sheet in the horizontal direction, at right angles to pyroxenes and amphiboles. Also this scheme is observed in nature in the rare iron-rich mineral *howieite* $\left(\mathrm{NaMg_{10}Fe_2^{3+}Si_{12}O_{31}(OH)_{13}}\right)$.

As in sheet silicates, tetrahedra in chain silicates are attached on both sides of octahedral units. This is illustrated for pyroxenes in Figure 30.2a,b and for amphiboles in Figure 30.2c,d. The chain units run along the *c*-axis of crystals and are parallel to (100). When viewed along *c*, the units are regularly

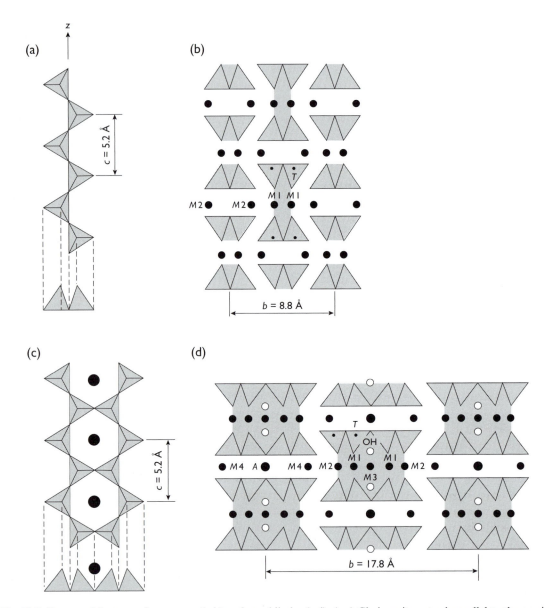

Fig. 30.2 Structural features of pyroxenes (a,b) and amphiboles (c,d). (a,c) Chain units extend parallel to the *c*-axis. (b,d) View in the chain direction, illustrating the stacking of the chain units. Tetrahedral chains are attached to each side of an octahedral layer. Structural cation sites *T*, *M*1, *M*2, *M*3, *M*4, and *A*, as well as $\mathrm{OH^-}$, are indicated. Tetrahedral–octahedral stack units are indicated by shading.

Table 30.1 Ionic substitutions in important pyroxenes and amphiboles and distribution of cations over structural sites. CN, coordination number; □, vacancy

A. Pyroxenes (formula based on six oxygen atoms)

Name	$M2$ CN 6–8	$M1$ CN 6	T CN 4	Symmetry
Enstatite	Mg	Mg	Si_2	Orthorhombic
Hypersthene[a]	(Mg,Fe^{2+})	(Mg,Fe^{2+})	Si_2	Orthorhombic
Clinoenstatite	Mg	Mg	Si_2	Monoclinic
Pigeonite[a]	(Mg,Ca,Fe^{2+})	(Mg,Fe^{2+})	Si_2	Monoclinic
Augite[a]	(Ca,Mg,Fe^{2+})	(Mg,Fe^{2+})	Si_2	Monoclinic
Diopside	Ca	Mg	Si_2	Monoclinic
Hedenbergite	Ca	Fe^{2+}	Si_2	Monoclinic
Essenite	Ca	Fe^{3+}	SiAl	Monoclinic
Jadeite	Na	Al	Si_2	Monoclinic
Aegirine	Na	Fe^{3+}	Si_2	Monoclinic
Omphacite[a]	(Ca,Na)	(Mg,Fe^{2+},Al)	Si_2	Monoclinic
Spodumene	Li	Al	Si_2	Monoclinic

B. Amphiboles (formula based on 22 oxygen atoms and 2(OH,F))

Name	A CN 12	$M4$ CN 6–8	$M1, M2, M3$ CN 6	T CN 4	Symmetry
Tremolite	□	Ca_2	Mg_5	Si_8	Monoclinic
Ferroactinolite	□	Ca_2	Fe_5^{2+}	Si_8	Monoclinic
Glaucophane	□	Na_2	Mg_3Al_2	Si_8	Monoclinic
Riebeckite	□	Na_2	$Fe_3^{2+}Fe_2^{3+}$	Si_8	Monoclinic
Arfvedsonite	Na	Na_2	$Fe_4^{2+}Fe^{3+}$	Si_8	Monoclinic
Ferrohornblende	□	Ca_2	$Fe_4^{2+}Al$	Si_7Al	Monoclinic
Hornblende[a]	$(Na,K)0–1$	$(Ca,Na)_2$	$(Mg,Fe^+)(Al,Fe^{3+})$	Si_7Al	Monoclinic
Edenite	Na	Ca_2	Mg_5	Si_7Al	Monoclinic
Pargasite	Na	Ca_2	Mg_4Al	Si_6Al_2	Monoclinic
Tschermakite	□	Ca_2	Mg_3Al_2	Si_6Al_2	Monoclinic
Cummingtonite	□	Mg_2	Mg_5	Si_8	Orthorhombic
Anthophyllite	□	Mg_2	Mg_5	Si_8	Orthorhombic
Grunerite	□	Fe_2^{2+}	Fe_5^{2+}	Si_8	Orthorhombic

[a] Series names widely used in petrology and therefore included in this book.

stacked (Figure 30.2b,d). In pyroxenes (Figure 30.2b) the structure is fairly compact with tetrahedral atoms (T) and octahedral atoms ($M1$ and $M2$), and the compositions are relatively simple (Table 30.2). In amphiboles (Figure 30.2d) the structure is more open, with several M sites and a larger space between opposite chains (A) which can accommodate large cations such as Na^+ and K^+. Also, there is space for OH^- in the octahedral units. Because of these ionic substitutions the chemical formulas of amphiboles are more complex (Table 30.1).

Figure 30.3 is a more schematic view of the stacking perpendicular to the chain direction. The preferred cleavage planes in chain silicates are those through which no Si–O bonds need to be broken. Since [001] is the chain direction and (100) the plane of octahedral and tetrahedral layers, cleavages traverse the structure diagonally parallel to (110) and $(1\bar{1}0)$ and, because of the width of the chains, are at an angle of about 88° in pyroxenes with narrow chains (Figure 30.3a) and a 125° angle in amphiboles with wider chains (Figure 30.3b). We noted in Chapter 14 that this cleavage

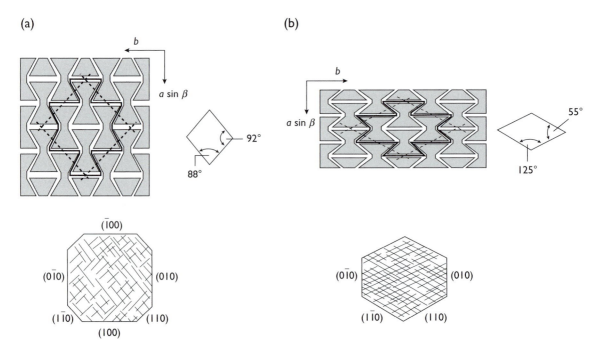

Fig. 30.3 Sketch illustrating the stacking of tetrahedral–octahedral chain units in (a) pyroxenes and (b) amphiboles, explaining the distinct cleavage angles.

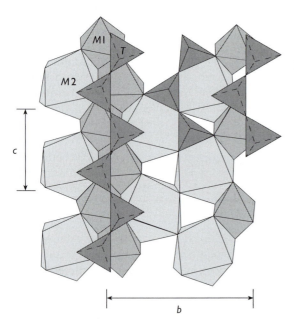

Fig. 30.4 Crystal structure of a monoclinic pyroxene such as diopside ($CaMgSi_2O_6$) in a (100) projection with tetrahedra T (Si), octahedra $M1$ (Mg), and larger cation sites $M2$ (Ca).

angle is diagnostic for identification of pyroxenes (Plate 14e) and amphiboles (Plate 11c) in thin sections.

In order to better understand the substitutions in pyroxenes we need to take a closer look at their crystal structure (represented in a (100) projection in Figure 30.4). There are three sites for cations, labeled T, $M1$, and $M2$. Silicon always occupies the tetrahedral (T) site. There are two octahedral sites: $M1$ is in the center of the octahedral unit (see also Figure 30.2b) in a fairly ideal octahedral coordination, and is occupied by Mg^{2+}, Fe^{2+}, Al^{3+}, and in some cases by Ti^{4+} (Table 30.1). The $M2$ site is on the outside and less confined. In fact its coordination is not strictly octahedral. In addition to cations such as Mg^{2+} and Fe^{2+}, it can accommodate larger Ca^{2+}, Na^+, and Li^+ ions. In diopside Mg^{2+} occupies the $M1$ site and Ca^{2+} the $M2$ site. Fe^{2+} can substitute for Mg^{2+} and there is a continuous solid solution between diopside (CaMg(Si_2O_6)) and hedenbergite (CaFe(Si_2O_6)), even at low temperature.

Calcic pyroxenes are very common in metamorphic and igneous rocks, and their composition is traditionally represented in a magnesium–iron–calcium quadrilateral, using only the lower (calcium-poor) part of

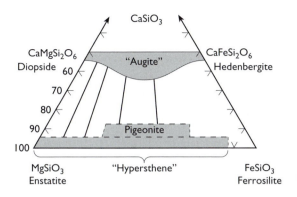

Fig. 30.5 Quadrilateral representation of chemical compositions (mol%) in the pyroxene system. Shaded areas give compositional ranges in igneous and metamorphic rocks. Individual minerals are indicated. Tielines give compositional pairs that are in equilibrium in igneous rocks or may form by exsolution.

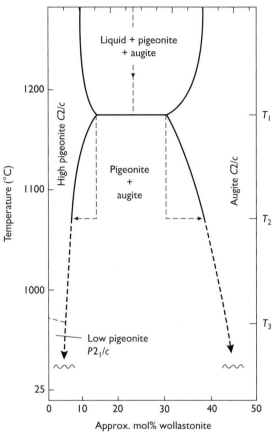

Fig. 30.6 Simplified phase diagram for the solid solution pigeonite ($Mg_{0.63}Fe_{0.32})_2Ca_{0.1}(Si_2O_6)$–diopside (augite) ($Mg_{0.65}Fe_{0.35})Ca(Si_2O_6)$ at low pressure (after Huebner, 1980). It illustrates the miscibility gap for intermediate compositions and the stability of phases. Indicated (with dashed lines) is the cooling history of a lunar pyroxene crystal with a microstructure shown in Figure 30.7. There are two exsolution events, at T_1 and T_2, as well as an ordering event at T_3.

the ternary diagram $CaSiO_3$–$MgSiO_3$–$FeSiO_3$ (Figure 30.5). Various minerals are indicated. Only a quadrilateral is used, because pyroxenes more calcic than diopside and hedenbergite do not exist, and pure $CaSiO_3$ has a different structure (wollastonite). This quadrilateral is particularly interesting with respect to phase transformations that occur during cooling of basaltic lava.

At high temperatures, all pyroxene compositions in the pyroxene quadrilateral are possible. During cooling, an initially homogeneous pyroxene may exsolve into a calcium-enriched diopside (also known as augite) and a calcium-poor enstatite or pigeonite, in a manner analogous to alkali feldspars decomposing into sodium-rich albite and potassium-rich orthoclase (see Chapters 17 and 18). Tielines in Figure 30.5 give the composition of coexisting pyroxenes. Figure 30.6 shows a simplified experimental temperature–composition phase diagram for a composition range pigeonite ($Mg_{0.65}Fe_{0.35})_2(Si_2O_6)$ to diopside ($Mg_{0.65}$ $Fe_{0.35})$ $Ca(Si_2O_6)$ under low-pressure conditions.

An additional complication is that enstatite, pigeonite, and augite (isostructural with diopside) have slightly different crystal structures and space-group symmetries. Enstatite is orthorhombic (space-group $Pbca$), diopside as well as augite and high-temperature pigeonite are monoclinic (space-group $C2/c$), low-temperature pigeonite is also monoclinic but with a more highly ordered structure

(space-group $P2_1/c$). Upon cooling, pigeonite undergoes a phase transformation (from $C2/c$ to $P2_1/c$), and since the symmetry changes, the phase transformation gives rise to antiphase boundaries (APBs; see Chapter 9).

Figure 30.7 is a transmission electron microscope image that illustrates a lamellar microstructure in a pyroxene from lunar basalt forming in two stages. In a first stage, at a higher temperature (schematically shown as T_1 in Figure 30.6), when the solidus is reached, first-generation coarse lamellae parallel to

Fig. 30.7 Clinopyroxene from lunar basalt 15058 with complex microstructure. The darkfield TEM image displays an earlier lamellar intergrowth of pigeonite and augite parallel to (100) (NW–SE). Later, at lower temperature, a second exsolution occurred on (001) (NE–SE). Finally, during a phase transformation in pigeonite, APBs developed that are visible as a mottled texture (from Wenk, 1976; photograph by W. F. Müller). Width 2 μm.

(100), consisting of alternating magnesium-rich augite and calcium-rich pigeonite, formed. At a lower temperature T_2, impure augite exsolved into purer augite and pigeonite, and impure pigeonite exsolved into purer pigeonite and augite. The second-generation finer lamellae are parallel to (001). In a last stage, at T_3, pigeonite underwent a phase transformation ($C2/c$ to $P2_1/c$), creating APBs. Dashed lines in Figure 30.6 indicate schematically the complicated cooling history of this crystal, which is recorded in the microstructure.

Whereas the microstructure in Figure 30.7 was observed with a transmission electron microscope, lamellar exsolutions in pyroxenes are often on a scale where they can be seen with an optical microscope (cf.

Plate 4b). Also, in many calcic igneous rocks such as andesites, a relatively calcic pyroxene (augite) and a relatively calcium-poor magnesium–iron pyroxene (pigeonite) coexist.

If a univalent ion such as Na^+ or Li^+ occupies an $M2$ site, charge balance is achieved by having trivalent ions such as Al^{3+} or Fe^{3+} on $M1$ sites. Examples include: the sodic pyroxene *jadeite* ($NaAlSi_2O_6$), which occurs in metamorphic rocks and is indicative of high pressures; *aegirine* ($NaFe^{3+}Si_2O_6$), which is a typical mineral in many alkaline rocks; and *spodumene* ($LiAlSi_2O_6$), which occurs in pegmatites, often in association with the lithium-mica lepidolite. There is miscibility between hedenbergite and aegirine, whereas the field of jadeite is isolated.

Often monoclinic pyroxenes contain some tetrahedral Al substituting for Si. This substitution is described with the hypothetical so-called *Tschermak pyroxene component* $CaAl(AlSiO_6)$ that is contained in most augites $Ca(Mg,Fe,Al)(Si,Al)_2O_6$. *Omphacite* is a diopside, containing about 25 mol% jadeite component $(Ca,Na)(Mg,Fe,Al)(Si_2O_6)$. It is typical of eclogites. Figure 30.8 shows compositions of 158 monoclinic pyroxenes in a range of igneous and metamorphic rocks. The plot illustrates that there is wide miscibility between diopside, hedenbergite, aegirine, and jadeite and assignment of mineral names is somewhat arbitrary. Omphacites in particular cluster halfway between diopside and jadeite.

We now look at compositional variations in amphiboles and relate them to structural features (Figure 30.9). Being somewhat intermediate between a single-chain and a sheet silicate (see Figure 30.1), the double-chain structure offers more compositional variety than the compact structure of pyroxenes. Contrary to pyroxenes, amphiboles contain some OH^- and can accommodate some Na^+ and K^+ in the often only partially occupied 12-fold coordinated A site (see also Figure 30.2c,d). As illustrated in Table 30.1B, the internal M sites $M1$, $M2$, and $M3$ with a strict octahedral coordination are occupied by ions such as Mg^{2+}, Fe^{2+}, and Al^{3+}, whereas the external $M4$ site is larger and can also accommodate Ca^{2+} and Na^+. Amphiboles are divided into a calcic and a sodic branch. As with the sodic pyroxene jadeite, sodic amphiboles such as glaucophane and riebeckite are typically in high-pressure metamorphic rocks. They are easily recognized in hand specimens by their

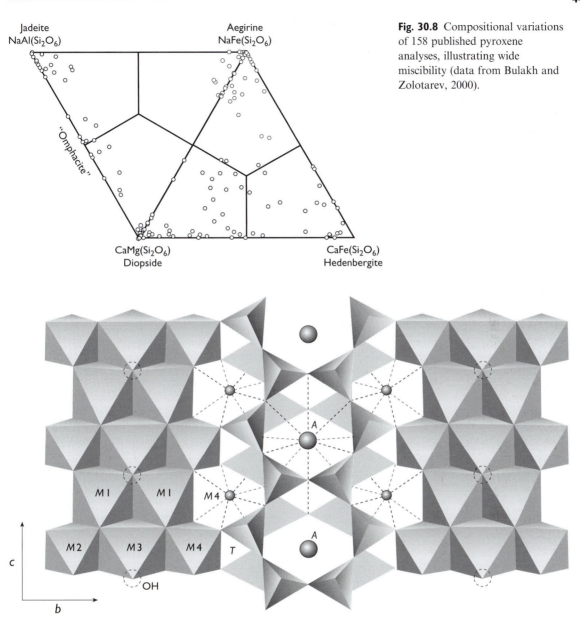

Jadeite
NaAl(Si₂O₆)

Aegirine
NaFe(Si₂O₆)

CaMg(Si₂O₆)
Diopside

CaFe(Si₂O₆)
Hedenbergite

Fig. 30.8 Compositional variations of 158 published pyroxene analyses, illustrating wide miscibility (data from Bulakh and Zolotarev, 2000).

Fig. 30.9 Crystal structure of a monoclinic amphibole such as tremolite $Ca_2Mg_5(Si_8O_{22})(OH)_2$ in a (100) projection with tetrahedra T, octahedra $M1$, $M2$, $M3$, and larger cation sites $M4$. Those anions that are not connected to chains are hydroxyl groups. Also shown is the large cation site A, between two six-membered rings. It is not occupied in the case of tremolite.

bluish color and in thin section by a blue pleochroism, which distinguishes sodic amphiboles from calcic amphiboles with green pleochroism. There is a wide range of cationic substitutions (Plate 11).

A quadrilateral representing minerals and compositions of calcic amphiboles is shown in Figure 30.10. It has some analogies with that for pyroxenes

(see Figure 30.5). Like the magnesium pyroxene enstatite, the magnesium amphibole anthophyllite is orthorhombic, while all others are monoclinic.

Hornblende contains some K^+ in the A site and has a very variable composition $(K,Na)_{0-1}(Ca,Na)_2$ $(Mg,Fe^+)(Al,Fe^{3+})(Si_7AlO_{22})(OH,F)_2$ that depends on rock composition and conditions of formation.

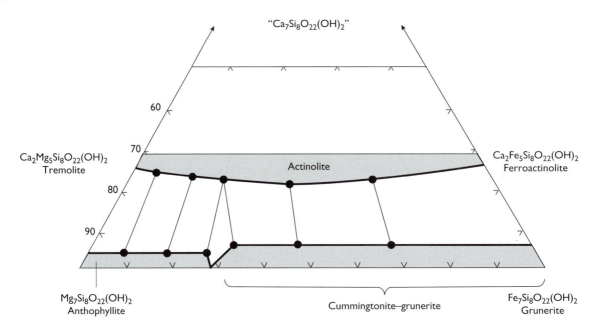

Fig. 30.10 Quadrilateral representation of compositions (in mol%) in calcic amphiboles (system $Mg_7Si_8O_{22}(OH)_2$–$Fe_7Si_8O_{22}(OH)_2$–$Ca_7Si_8O_{22}(OH)_2$). Compositional fields of minerals are shaded. Tielines are pairs of coexisting minerals in igneous rocks.

Calcic amphiboles such as tremolite, actinolite, and hornblende are characteristic of many regionally metamorphosed rocks of the amphibolite facies.

As you can imagine, the different structural units and combinations of such features as single, double, triple, and even quadruple chains, along with principles of polytypism similar to sheet silicates, open many possibilities for different structures as well as structural defects in amphiboles and pyroxenes. For example, the stacking of octahedra and their orientation that leads to polytypism in sheet silicates (see Section 29.2) creates monoclinic (clino-) and orthorhombic (ortho-) pyroxenes and amphiboles. Hybrids, intermediate between single, double, and triple chains and sheet silicates, have been found in metamorphic rocks using transmission electron microscopy (Figure 30.11) and have been labeled *biopyriboles* by J. Thompson (1978). Figure 30.12 shows the schematic stacking in the biopyriboles jimthompsonite and chesterite, as compared with that for enstatite and anthophyllite. Although rare, these minerals caution against a universal application of simple structural classifications.

Not all chain silicates can be classified as pyroxenes and amphiboles. There are more complicated types of chain silicates, and we have already mentioned

howieite earlier in this chapter (see Figure 30.1d). In pyroxenoids, chains are not linear but show kinks and are differently attached to octahedral layers (Figure 30.13). Friedrich Liebau (1962) classified chain silicates according to the number of connected

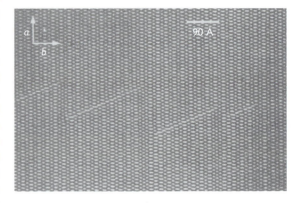

Fig. 30.11 High-resolution electron micrograph of a biopyribole, showing some stacking defects. At the top of the image is perfectly ordered chesterite with a regular 2–3 stacking. Below the diagonal faults the stacking is more irregular with regions of 2–2 stacking corresponding to anthophyllite (from Veblen and Buseck, 1980).

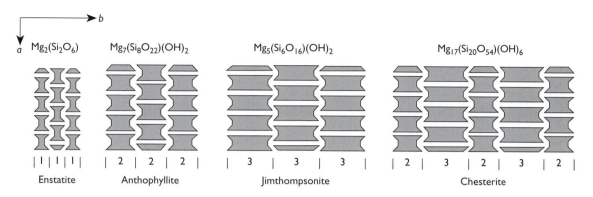

Fig. 30.12 Stacking patterns viewed along the chain direction in the pyroxene enstatite, the amphibole anthophyllite, and the biopyriboles jimthompsonite and chesterite.

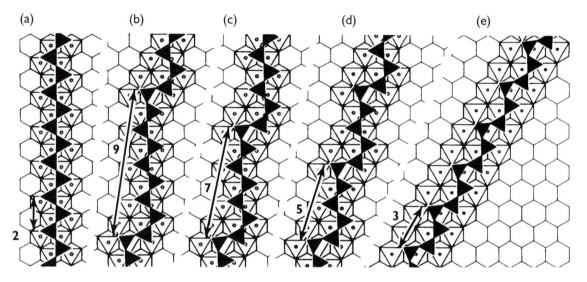

Fig. 30.13 Tetrahedral chains attached to an octahedral layer in single-chain silicates, plotted on a regular hexagonal grid. Octahedra marked with dots are part of the chain unit. Tetrahedra on the lower surface of the octahedral layer are not shown. Note that the chain is increasingly inclined going from (a) pyroxene (2-repeat) to (b) ferrosilite (9-repeat), (c) pyroxmangite (7-repeat), (d) rhodonite (5-repeat), and (e) wollastonite (3-repeat) (after Liebau, 1985).

tetrahedra in the chain with a distinct organizational pattern that repeats. Figure 30.13 shows different tetrahedral single chains attached to octahedral units. (Tetrahedral chains are attached on both sides of the octahedral layer; only the top side is shown in this figure.) The most common case is the 2-repeat in pyroxenes such as enstatite ($Mg_2(Si_2O_6)$) (Figure 30.13a). More rare is the 3-repeat in wollastonite ($Ca_3(Si_3O_9)$) (Figure 30.13e), the 5-repeat in rhodonite (Ca_4Mn_5 (Si_5O_{15})) (Figure 30.13d), the 7-repeat in pyroxmangite (($Fe, Mn)_7(Si_7O_{21})$) (Figure 30.13c), and the 9-repeat in ferrosilite ($Fe_9(Si_9O_{27})$) (Figure 30.13b). Multiples in

the chemical formulas are sometimes used to express this repeat (i.e., $Ca_3(Si_3O_9)$ rather than $Ca(SiO_3)$ for wollastonite).

Table 30.2 gives a list of the most important chain silicates, their compositions, and diagnostic properties.

30.2 Brief description of important chain silicate minerals

Pyroxene and **amphibole** crystals share some properties. They are elongated along [001] which is the direction of the tetrahedral chains. Amphibole

Table 30.2 Common chain silicates with some diagnostic properties; most important minerals are given in italics

Mineral Formula	System	Morphology Cleavage	H	D	Color	n Pleochroism	Δ	2V (Dispersion)
Pyroxenes								
Orthopyroxenes								
Enstatite	Ortho.	Pris. [001]	5–6	3.1	Gray, green	1.65–1.66	0.009	+55
$Mg_2(Si_2O_6)$		{210} 87°						(r < v)
Hypersthene	Ortho.	Pris. [001]	5–6	3.5	Brown, green	1.68–1.73	0.015	−40–90
$(Mg,Fe)_2(Si_2O_6)$		{210} 87°				*Pink-yellow-green*		(r < v)
Clinopyroxenes								
Aegirine	Monocl.	Pris. [001]	6–6.5	3.6	Gray, black	1.76–1.83	0.05	−60–70
$NaFe^{3+}(Si_2O_6)$		{110} 87°						(r < v)
Augite	Monocl.	Pris. [001]	6	3.4	Green, black	1.69–1.78	0.030	+25–85
$(Ca,Mg,Fe)_2(Si,Al)_2O_6)$		{110} 87°						(r > v)
Clinoenstatite	Monocl.	Pris. [001]	6	3.2	Green-yellow, yellow	1.65–1.66	0.009	+54
$Mg_2(Si_2O_6)$		{110} 87°						(r < v)
Diopside	Monocl.	Pris. [001]	5.5–6	3.3	Green, white	1.66–1.69	0.030	+59
$CaMg(Si_2O_6)$		{110} 87°						(r > v)
Essenite	Monocl.	Pris. [001]	6	3.5	Red-brown, yellow-green	1.80–1.82	0.030	−77
$CaFe^{3+}(SiAlO_6)$		{110} 87°						(r < v)
Hedenbergite	Monocl.	Platy (010)	5.5–6	3.5	Black, green	1.74–1.76	0.018	+60
$CaFe(Si_2O_6)$		{110} 87°						(r > v)
Jadeite	Monocl.	Fibr. [001]	6.5	3.4	White, green	1.64–1.67	0.009	+70–72
$NaAl(Si_2O_6)$		{110} 87°						(r < v)
Omphacite	Monocl.	Pris. [001]	5–6	3.3	Green	1.67–1.60	0.023	+60–70
$(Ca,Na)(Mg,Fe,Al)(Si_2O_6)$		{110} 87°						(r < v)
Pigeonite	Monocl.	Pris. [001]	6	3.4	Green, black	1.69–1.74	0.025	+0–50
$(Mg,Ca,Fe)_2(Si_2O_6)$		{110} 87°						(r < v)
Spodumene	Monocl.	Pris. [001]	6–7	3.2	White, yellow, violet	1.65–1.68	0.02	+54–56
$LiAl(Si_2O_6)$		{110} 87°						(r < v)
Amphiboles								
Calcic amphiboles								
Actinolite	Monocl.	Pris. [001]	5–6	3.4	Green	1.69–1.71	0.02	−10–20
$Ca_2Fe_5^{2+}(Si_8O_{22})(OH)_2$		{110} 124°				*(yellow)-blue green*		(r < v)
Ferrohornblende	Monocl.	Pris. [001]	5–6	3.2	Brown, black	1.67–1.72	0.015	−50–80
$Ca_2Fe_4^{2+}Al(Si_7AlO_{22})(OH)_2$		{110} 124°				*Green-violet-brown*		(r < v)

Mineral	System	Morphology	H	D	Colors	n / pleochroism	Δ	2V
Hornblende $(Na,K)_{0-1}(Ca,Na)_2(Mg,Fe^{2+})(Al,Fe^{3+})(Si_7AlO_{22})(OH,F)_2$	Monocl.	Pris. [001] {110}124°	5–6	3.2	Green, black	1.61–1.73 / *Yellow-green brown*	0.02	−60–88 (r < v)
Tremolite $Ca_2Mg_5(Si_8O_{22})(OH)_2$	Monocl.	Pris. [001] {110}124°	5–6	2.9	White, gray	1.60–1.63	0.02	−10–20 (r < v)
"Tschermakite" $Ca_2Mg_3Al_2(Si_6Al_2O_{22})(OH)_2$	Monocl.	Fibr. [001] {110}124°	5–6	3.3	Green	1.64–1.69 / *Yellow green-green*	0.02	−65–90
Sodic amphiboles								
Arfvedsonite $NaNa_2Fe_4^{2+}Fe^{3+}(Si_8O_{22})(OH)_2$	Monocl.	Pris. [001] {110}124°	5–6	3.4	Blue, black	1.67–1.71 / *Blue, green, violet*	0.015	−5–50
Glaucophane $Na_2Mg_3Al_2(Si_8O_{22})(OH)_2$	Monocl.	Pris. [001] {110}124°	5–6	3.1	Blue	1.61–1.67 / *(yellow)-violet-blue*	0.02	−50–0 (r ≫ v)
Edenite $NaCa_2Mg_5(Si_7AlO_{22})(OH)_2$	Monocl.	Pris. [001] {110}124°	5–6	3.1	White, gray	1.63–1.68 / *Brown, green-violet*	0.02	−50–80
Pargasite $NaCa_2Mg_4Al(Si_6Al_2O_{22})(OH)_2$	Monocl.	Pris. [001] {110}124°	5–6	3.1	White, green	1.61–1.68 / *Brown, green-violet*	0.02	+55–90 (r < v)
Riebeckite $NaFe_3^{2+}Fe_2^{3+}(Si_8O_{22})(OH)_2$	Monocl.	Pris. [001] {110}124°	5–6	3.4	Blue, black	1.65–1.71 / *Blue-brown-violet*	0.01	−40–90
Other amphiboles								
Anthophyllite $Mg_7(Si_8O_{22})(OH)_2$	Ortho.	Pris. [001] {210}124°	5.5	3.1	Brown, yellow-gray	1.60–1.67 / *brown, green-violet*	0.02	90 (r < v)
Cummingtonite $Mg_7(Si_8O_{22})(OH)_2$	Monocl.	Pris. [001] {110}124°	5–6	3.4	Green, gray, brown	1.61–1.64 / *yellow-brown-brown*	0.03	+80–90 (r < v)
Grunerite $Fe_7^{2+}(Si_8O_{22})(OH)_2$	Monocl.	Pris. [001] {110}124°	5–6	3.6	Gray, brown	1.69–1.71 / *yellow-brown-brown*	0.04	+84–90 (r < v)
Pyroxenoids								
Rhodonite $Ca_4Mn(Si_5O_{15})$	Tricl.	Pris. [010] (001), (100)	6	3.5	Red	1.72–1.74	0.02	+76 (r < v)
Wollastonite $Ca_3(Si_3O_9)$	Tricl.	Fibr. [010] (100), (001)	4.5–5	2.8	White	1.62–1.63	0.015	−35–40 (r < v)

Notes: H, hardness; D, density (g/cm³); *n*, range of refractive indices; Δ, birefringence; 2*V*, axial angle for biaxial minerals. Acute 2*V* is given in the table. If 2*V* is negative the mineral is biaxial negative and 2*V* is 2*V*$_\alpha$; if it is positive, the mineral is biaxial positive and 2*V* is 2*V*$_\gamma$.
Dispersion r < v means that acute 2*V* is larger for violet than for red; anom., anomalous dispersion or birefringence.

System: Monocl., monoclinic; Ortho., orthorhombic; Tricl., triclinic.

Morphology: Fibr., fibrous; Pris., prismatic.

Colors: Light colors are given in parentheses.

Table 30.3 Diagnostic features of pyroxenes and amphiboles

	Pyroxenes	Amphiboles
Aggregates	Granular (aegirine: columnar/radiaxial)	Columnar, fibrous, radiaxial
Color	Green, black, white, grayish	Green, blue, black, white, grayish
Pleochroism	Weak (pink-greenish in hypersthene)	Strong (green in calcic, blue in sodic Fe-containing amphiboles)
Luster	Weak vitreous, greasy	Strong vitreous
Cleavage	Poor (spodumene: perfect)	Perfect
Cleavage angle	$87°-89°$	$124°-126°$
Hardness	5.5–6 (spodumene: 7)	5.5–6
Streak	None	Green (black amphiboles)

crystals are more columnar and flattened, pyroxenes are more prismatic and elongated. Both have two distinct cleavages (Figure 14.21) with different angles based on the structure with single chains and double chains (Figure 30.3). The angle between {110} cleavages in pyroxenes is ~88° and in amphiboles ~125°, which is easily recognized in thin sections (e.g., Plates 14e, 11c, respectively). Some properties are compared in Table 30.3. In rocks undergoing prograde regional metamorphism, pyroxenes and amphiboles form in the following order with increasing temperature and pressure: actinolite (or tremolite) – hornblende – diopside – hypersthene. They easily become altered by hydrothermal solutions with decreasing temperature and pressure to micas, chlorite, sometimes serpentine, and talc.

Orthopyroxenes

Enstatite ($Mg_2(Si_2O_6)$) is orthorhombic, contrary to clinopyroxenes. It has a vitreous luster on cleavage surfaces. Hardness is 5.5–6 and density 3.2–3.6 for pure enstatite. Color ranges from yellowish to olive-green and brown. In thin sections it is recognized by symmetrical extinction with respect to the cleavage planes. At high temperatures there is a continuous substitution of Mg^{2+} by Fe^{2+} (Figure 30.5). The iron end member (>90% Fe) is called ferrosilite and has a different chain structure. Intermediate compositions are called "hypersthene". At very high pressure, enstatite transforms to monoclinic **clinoenstatite** ($Mg_2(Si_2O_6)$), which may also transform from orthorhombic enstatite under the influence of stress. It has been observed in meteorites and highly deformed mylonites. Enstatite occurs in

peridotites, gabbros, and basalts, generally associated with olivine.

Calcic clinopyroxenes

Similar to the enstatite–ferrosilite series, there is a solid solution between monoclinic **diopside** ($CaMg(Si_2O_6)$) and **hedenbergite** ($CaFe(Si_2O_6)$). These minerals are mainly observed in metamorphic rocks such as dolomitic marbles and skarns as prismatic crystals of white (Plates 31b,d), greenish to brown color. Darker members show some pleochroism. Prismatic crystals are frequently twinned on (001).

In igneous rocks, compositions and solid solutions in calcic pyroxenes are more complex, as discussed earlier (Figure 30.5). **Augite**, which is slightly depleted in Ca relative to diopside–hedenbergite, coexists with **pigeonite**, which has more Ca compared to enstatite–ferrosilite. We discussed an intermediate phase diagram earlier to interpret exsolution events in augite pigeonite mixtures during cooling of lunar basalt (Figures 30.6 and 30.7). The composition of clinopyroxenes can be estimated by measuring the extinction angle $n\gamma$ to [001] in (010) sections (Figure 14.22a). Large crystals are observed in some volcanic rocks (Figure 30.14a).

Sodic clinopyroxenes

At high pressure albite ($Na(AlSi_3O_8)$) breaks down to form **jadeite** ($NaAl(Si_2O_6)$) + quartz (SiO_2) and this mineral has been observed in subduction zones such as in California, USA. Jadeite may also form by a high-pressure–high-temperature reaction from albite ($Na(AlSi_3O_8)$) and nepheline ($KNa_3(AlSiO_4)_4$). It has

(a)

(b)

Fig. 30.14 Morphology of chain silicates. (a) Crystals of aluminous diopside (augite) from volcanic rocks at Mt. Kilimanjaro, Tanzania. Width 90 mm. (b) Radiating crystals of tremolite in metamorphic dolomite from Panamint Valley, California, USA. Width 60 mm.

green color, high hardness (6.5–7), and is extremely tough. Jadeite often occurs as aggregates of fibrous crystals, known as "jade" (note that there is another jade composed of the mineral nephrite which will be discussed below). It is highly prized in the Orient and worked into ornaments.

Omphacite $(Ca,Na)(Mg,Fe^{2+},Fe^{3+},Al)(Si_2O_6))$ has a complex composition and can be considered as a solid solution between augite and jadeite, with bright-green colors. It is a characteristic mineral of eclogites, mafic high-temperature–high-pressure rocks that occur in subduction zones (Plate 9) and mantle-derived ultramafic rocks.

Spodumene $(LiAl(Si_2O_6))$ is colorless, emerald-green (hiddenite) or lilac (kunzite). It is a characteristic mineral of lithium-rich pegmatites and typically associated with quartz, albite, lepidolite, and tourmaline. It can form giant crystals (Figure 10.12a) and is mined as a gemstone (especially the kunzite variety) and as a lithium ore. Major producers of spodumene ore are China and Zimbabwe.

Amphiboles

The composition of pyroxenes is relatively simple and students can derive chemical formulas. For amphiboles this is much more complex. We discussed some substitutions before (Table 30.1B and quadrilateral in Figure 30.10). **Hornblende** is the most common amphibole with a complicated formula that expresses many possible substitutions: $(Na,K)_{0-1}Ca_2(Mg,Fe^{2+},Fe^{3+},Al)_5(Si_{6-7}Al_{2-1}O_{22})(OH,F)_2$. It is relatively easy to recognize hornblende in thin sections of igneous and metamorphic rocks by prismatic habit, typical $125°$ cleavage, dark-green color, and pleochroism between yellow-green, green, and brown (Plate 11c).

Actinolite is a Mg-Fe solid solution between **tremolite** $Ca_2Mg_5(Si_8O_{22})(OH)_2$ and **ferroactinolite** $Ca_2Fe_5^{2+}(Si_8O_{22})(OH)_2$ (Figure 30.10). The color of tremolite is colorless or gray (Figure 30.14b), of actinolite pale to dark green, depending on iron content) and ferroactinolite is dark green to black. The latter is difficult to distinguish from hornblende in thin sections. Similarly there is a wide range of pleochroism from none in tremolite, and yellow to green in actinolite. Tremolite and actinolite occur largely in metamorphic rocks, e.g., from the reaction of dolomite and quartz:

$$5CaMg(CO_3)_2 + 8SiO_2 + H_2O \rightarrow$$
dolomite   quartz
$$Ca_2Mg_5(Si_8O_{22})(OH)_2 + 3CaCO_3 + 7CO_2$$
tremolite   calcite

We mentioned jade as a massive fibrous aggregate of the pyroxene jadeite. Similarly, *nephrite* consists of intimately interwoven actinolite needles. Both are gemstones that are especially valued in Myanmar (Burma) and China.

Anthophyllite $(Mg,Fe^{2+})_7(Si_8O_{22})(OH,F)_2)$ has a similar composition to actinolite, but without calcium. In contrast to all other amphiboles discussed here, anthophyllite is orthorhombic, corresponding to enstatite among pyroxenes but with a 125° cleavage. Anthophyllite is also a metamorphic mineral and attributed to metasomatism. It is often developed with an asbestiform habit, during regional metamorphism of ultramafic rocks, and is associated with talc.

Sodic amphiboles of the **glaucophane** $(Na_2Mg_3Al_2)$ $(Si_8O_{22})(OH)_2)$–**riebeckite** $(Na_2Fe_3^{2+}Al_2^{3+}Si_8O_{22})(OH)_2)$ series have a characteristic blue color and in thin section they display a distinct blue/yellow pleochroism (Plate 11d,e). Both color and pleochroism increase with iron content. These minerals are characteristic of the "blueschist metamorphic facies" and are indicative of high pressure and low temperature (see Figure 28.13a). The *crocidolite* variety of riebeckite is asbestiform and was mined in South Africa and Australia, e.g., for acid-resistant filters in chemical purification and as a refractory material in construction, but its application has been greatly reduced because of health risks (see Chapter 36).

Pyroxenoids

Rhodonite $(Ca_4Mn(Si_5O_{15}))$, from the Greek *rhodon*, meaning "rose", occurs in contact and regionally metamorphic rocks derived from oceanic crust. It is easily recognized by its pink (rose) color and solid granular masses. Black veinlets of manganese dioxide (MnO_2) often cross these masses. Rhodonite is used as a decorative stone.

Wollastonite $(Ca(SiO_3))$ is characteristic of igneous contacts with limestone, occurring in marbles and skarns. It forms fibrous and prismatic crystals; they are white with vitreous luster and perfect cleavage. Crystals are triclinic. Wollastonite is a valuable ceramic raw material.

30.3 Crystallization of igneous rocks

Pyroxenes and amphiboles are very common in igneous rocks, both volcanic and plutonic, and at this point it is useful to discuss some of the processes taking place during crystallization of magma.

While magma is cooling, minerals separate from the melt in a certain order that is determined by the physical and chemical rules of crystallization of compositionally complex multiphase melts. Mineral crystallization in plutonic rocks at depth proceeds from about 1300 to 700 °C, whereas most volcanic lavas crystallize between 1200 and 1000 °C. The mineral composition of a resultant igneous rock depends on the composition of the initial magma (and particularly its SiO_2 content). Contrary to metamorphic rocks, igneous rocks, on the whole, contain a relatively small number of major minerals (about 10 silicates, such as olivine, enstatite, hornblende, muscovite, biotite, alkali feldspar, plagioclase, nepheline, and quartz). This limited mineral diversity in igneous rocks arises because of the relatively limited range of initial magma compositions.

The first minerals to crystallize grow in free space in the melt and are generally euhedral (e.g., Figure 30.14a). Since a siliceous melt is very viscous, crystals remain suspended in the melt or sink only slowly. During growth of nonequiaxed crystals (platelets or needles) in a flowing magma, the crystals become aligned by the flow currents. With continued crystallization, free growth becomes limited. Minerals that crystallize later are only partially euhedral, or occur as irregular grains that occupy the interstices between the earlier formed minerals.

Plutonic rocks are classified according to the dominant rock-forming minerals in modal proportions. In Chapter 21 we discussed granitic rocks with various proportions of quartz, alkali feldspar, and plagioclase (see Figure 21.22). We now expand this system to consider rocks that are deficient in these minerals and have a high proportion of mafic minerals, such as olivine, pyroxene, and hornblende.

Mafic rocks are composed largely of plagioclase, pyroxene, and hornblende, and a corresponding triangle is used for their classification (Figure 30.15a). The mafic plutonic rocks are generally divided into anorthosite (>90% plagioclase), gabbro, and norite (10–90% plagioclase), and ultramafic rocks that include pyroxenite and hornblendite (<10% plagioclase). The distinction between gabbro and norite is made depending on whether clinopyroxene or orthopyroxene dominates.

In *ultramafic rocks*, where plagioclase constitutes less than 10%, the percentages of olivine, pyroxene, and hornblende are used for classification (Figure 30.15b). The three major ultramafic rock types are peridotite (>40% olivine), pyroxenite (<40% olivine and pyroxene > hornblende), and hornblendite (<40% olivine and hornblende > pyroxene).

The composition of volcanic rocks is more variable, and modal compositions in particular are more difficult to establish, since glass may be present and a

(a)

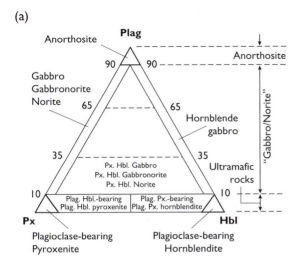

Plagioclase-bearing
Pyroxenite

Plagioclase-bearing
Hornblendite

(b)

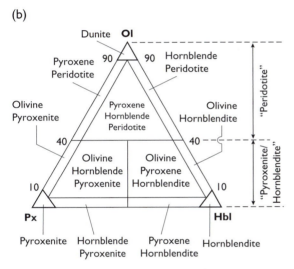

Pyroxenite   Hornblende   Pyroxene   Hornblendite
             Pyroxenite   Hornblendite

Fig. 30.15 Classification of (a) mafic and (b) ultramafic
rocks. Ol, olivine; Plag, plagioclase; Px, pyroxene; Hbl,
hornblende (after Le Bas and Streckeisen, 1991).

classification has to rely on modal as well as on chemical data. Without going into details, basalt
and andesite correspond roughly to gabbro, dacite to
granodiorite, and rhyolite to granite. The major
phases in mafic rocks, such as basalt and gabbro, are
plagioclase (solid solution anorthite–albite) and diopside containing aluminum and iron (also known as
augite). The ternary system diopside ($CaMgSi_2O_6$)–
albite ($NaAlSi_3O_8$)–anorthite ($CaAl_2Si_2O_8$) has
been studied in great detail and illustrates many features of the crystallization of mafic rocks (e.g., Morse,
1980).

At high temperature, the binary system albite–
anorthite forms a solid solution that was discussed in
Chapter 20 (see Box 20.1), where liquidus and solidus
curves were introduced, explaining the gradual
changes in plagioclase composition during cooling
(see Figure 20.2). Also, the binary system anorthite–
diopside has been discussed earlier (see Figure 20.1).
This system does not form a solid solution and,
depending on magma composition, either anorthite
or diopside will crystallize and grow. When the
magma cools and has reached a critical temperature
at the eutectic point, the residual anorthite and diopside components that remain in the liquid crystallize
simultaneously to form a groundmass composed of
the two minerals. The binary system albite–diopside
is similar.

For real mafic rocks we need to combine a binary
system without solid solution (diopside–plagioclase)
with a binary system of solid solution (albite–
anorthite). This ternary system albite–anorthite–
diopside is shown in Figure 30.16a, with temperature
as the third dimension (Bowen, 1915). The binary
phase diagrams that were discussed earlier are the
right side and the back side of this diagram. Contrary
to the binary systems, there is no eutectic point in the
ternary system, but rather a so-called *cotectic line* in
the valley between the diopside and anorthite slopes.

Some features are easier to see if we project this
three-dimensional diagram on to the compositional
triangle (Figure 30.16b). This diagram shows the liquidus surface with temperature contours and the cotectic
line. Also shown, with dashed lines, are plagioclase
compositions of the corresponding solidus. For
example, if the magma is pure plagioclase $An_{85}Ab_{15}$,
the plagioclase that crystallizes (around 1520 °C) has a
composition An_{95} (which we can also determine from
the binary phase diagram in Figure 20.2).

Let us now cool a magma of composition A,
i.e., $An_{50}Ab_{30}Di_{20}$ (Figure 30.16b). The first mineral
to crystallize, at 1400 °C, is a calcic plagioclase of
composition $An_{85}Ab_{15}$ (plagioclase compositions
indicated by dashed lines in Figure 30.16b). The composition of the melt descends down the liquidus surface, along the steepest temperature gradient, towards
the cotectic line, i.e., the boundary curve between
the plagioclase and diopside fields. The melt changes
composition (as we can read from the ternary diagram), and plagioclase that is in equilibrium with the
melt becomes progressively more sodic, as we can
read from the dashed lines. When the cooling magma

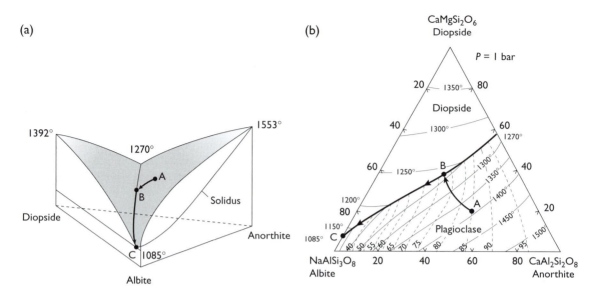

Fig. 30.16 Phase relations in the ternary system diopside ($CaMgSi_2O_6$)–albite ($NaAlSi_3O_8$)–anorthite ($CaAl_2Si_2O_8$) at atmospheric pressure. (a) Three-dimensional temperature–composition representation. Liquidus surface is illustrated. Line BC is the cotectic line. (b) Compositional triangle with contoured isotherms of the liquidus. In the field of plagioclase, dashed lines indicate the corresponding compositions of the plagioclase crystal in equilibrium with the melt (solidus) in mole % An. All temperatures are in °C.

reaches the cotectic line (at point B), plagioclase has a composition $An_{72}Ab_{28}$ and diopside begins to crystallize simultaneously. The crystallization of diopside and plagioclase continues down the cotectic line, towards point C near albite, until all the melt is used up. If there is still melt left when the cooling path reaches the eutectic point of the diopside–albite binary system (point C), then the remaining albite and diopside will crystallize simultaneously.

The microstructures that we are expecting to find are as follows. The first crystals to form in the melt are plagioclase and we expect them to grow as euhedral phenocrysts. As the magma cools, plagioclase becomes more sodic and we expect regular zoning, with a calcic core and a more sodic rim. At point B, diopside begins to crystallize. When point C is reached, the remaining plagioclase (albite) and diopside will crystallize simultaneously and instantaneously, resulting in an interwoven groundmass (Plate 7a).

A second system relevant for mafic rocks is forsterite (Mg_2SiO_4)–anorthite ($CaAl_2Si_2O_8$)–cristobalite (SiO_2) originally investigated by Andersen (1915). Figure 30.17a shows the ternary temperature–composition phase diagram at atmospheric pressure, where the surface represents the liquidus. Some reference temperatures are indicated. The liquidus surface is much

more complicated than in the system discussed previously. Before looking at the larger picture, let us explore the binary system forsterite–silica, on the front side of Figure 30.17a (Bowen and Anderson, 1914). A small sector of this system is enlarged in Figure 30.18. This system is different from a simple eutectic system such as diopside–anorthite (see Figure 20.1) because it contains an intermediate phase, enstatite ($Mg_2Si_2O_6$), at 30 weight% SiO_2. If the melt has a composition x_1 and cools, enstatite will start to crystallize at 1554 °C. Enstatite will continue to crystallize, thereby changing the composition of the melt, until the eutectic point E is reached at 1543 °C. At this point enstatite and silica, as cristobalite, crystallize simultaneously. Now consider a composition x_2 for the original melt. At temperature T_1 (1595 °C) forsterite begins to crystallize and continues to crystallize until point P is reached (1557 °C). At this point the liquid starts to react with forsterite, which is no longer stable, and enstatite is produced. The reaction point P is called the *peritectic point*. Box 30.1 explains this behavior, called *incongruent melting*, in terms of free energy of solid phases and melt. The reaction at point P consumes both liquid and forsterite, and, depending on the bulk composition, one or the other of these phases is depleted first, thus terminating the reaction.

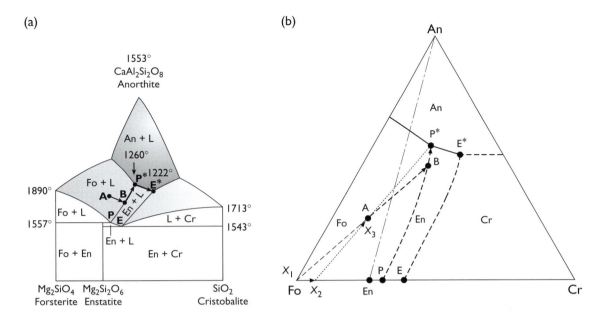

Fig. 30.17 Phase relations in the ternary system forsterite (Mg$_2$SiO$_4$)–anorthite (CaAl$_2$Si$_2$O$_8$)–silica (SiO$_2$) at atmospheric pressure. Liquidus surface is illustrated. (The phase diagram is simplified, and the two-liquids region for silica-rich compositions is omitted.) (a) Three-dimensional temperature–composition representation with peritectic point P* and eutectic point E*. The lines P–P* and E–E* are peritectic and eutectic lines, respectively. (b) Triangle of melt composition illustrating the cooling history of a melt of initial composition A. Fo, forsterite; An, anorthite; En, enstatite; Cr, cristobalite; L, liquid. All temperatures are in °C.

Composition x_2 is on the silica-poor side of the enstatite composition, and therefore cooling below the peritectic involves only forsterite and enstatite. Typically the earlier formed forsterite will show a mantle of enstatite produced during the reaction, as is observed in gabbros from Risör in Norway (Figure 30.19).

Let us return to the larger ternary system (see Figure 30.17a). There are four liquidus surfaces (for anorthite, forsterite, cristobalite, and enstatite) separated by boundaries. Consider a melt of composition A. Forsterite crystallizes and the melt changes composition towards B. At B, forsterite starts to react with the melt to form enstatite. Descending along the boundary P–P*, forsterite continuously reacts with the melt to form enstatite. At P* anorthite starts to crystallize. Further cooling will bring the system to the eutectic point E*, where a groundmass of enstatite, anorthite, and cristobalite would crystallize simultaneously if any melt were left. It is easier to quantify these changes if we project the system on a two-dimensional ternary diagram (see Figure 30.17b).

Consider the crystallization of a melt with composition A in Figure 30.18b. Since A is in the subtriangle forsterite–enstatite–anorthite (Fo, En, An), the melt must (according to the phase rule) ultimately crystallize to these three minerals, if equilibrium is maintained, but only at the ternary peritectic point P* do these three minerals coexist in equilibrium with the melt. Initially forsterite crystallizes from the melt, depleting the melt composition in the forsterite component and thus moving it in a straight line away from forsterite (indicated by dashed line A–B) until it reaches the peritectic reaction curve at B. Upon further cooling, the melt composition follows the peritectic line P–P* towards the peritectic point P*, and forsterite reacts with the melt to form enstatite. At point P* anorthite starts to crystallize.

We have followed changes in melt composition. There are corresponding changes in the bulk composition of the solid phases. As the melt moves from A to B, only forsterite crystallizes and the solid composition is therefore Fo (X_1). When the liquid reaches the peritectic line at B, enstatite starts to crystallize and this continues as the liquid moves towards P*. During this crystallization process the bulk solid composition moves from Fo towards En, as indicated by the arrow

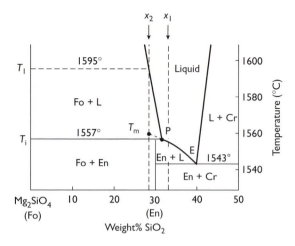

Fig. 30.18 Binary temperature–composition phase diagram for system forsterite (Mg_2SiO_4)–silica (SiO_2) (the silica-rich part is not shown), with peritectic point P and eutectic point E. Fo, forsterite; En, enstatite; Cr, cristobalite; L, liquid. T_m, congruent melting point; T_i incongruent melting point.

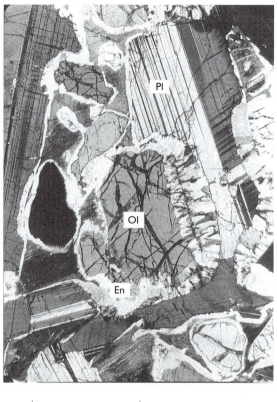

Fig. 30.19 Corona structure in gabbro from Risör in Norway. Olivines (Ol) are rimmed by enstatite (En). The large twinned crystals are plagioclase (Pl) (courtesy R. Joesten; see also Joesten, 1986).

at the bottom of Figure 30.17b, until it reaches X_2 (obtained by extrapolating the straight dotted line P*–A). At the peritectic P* the liquid does not change composition and anorthite crystallizes, in addition to forsterite and enstatite, moving the bulk solid from X_2 towards P* (dotted line). When the solids reach A (X_3), the composition of the initial melt, all liquid is used up and crystallization stops.

During this crystallization, there are also systematic changes in grain shapes and mineral intergrowths. The olivine crystals that form initially at point A are euhedral, but when the melt reaches the reaction curve at B, they become rounded owing to the reaction with the residual melt and are rimmed with enstatite that forms during this reaction. At point P*, plagioclase crystallizes in the spaces between enstatite (Figure 30.19).

In the example above, with an initial melt composition A, the eutectic point E* is never reached. This is different if the initial composition is in the subtriangle En–Cr–An. In that case enstatite, cristobalite, and anorthite are the equilibrium phases and crystallization proceeds to the eutectic point. Similar arguments as explained above can be used to follow the systematic evolution in composition of melt and solids.

The two systems described in Figures 30.16 and 30.17 were first studied experimentally by Norman L. Bowen and his coworkers at the Geophysical

Laboratory of the Carnegie Institution in Washington. They are the basis for a general model of magma evolution, which became known as Bowen's reaction principle. Bowen (1928) proposed that a primary mafic magma becomes increasingly siliceous through fractional crystallization. In the crystallization sequence he described two trends: continuous and discontinuous series.

There is a continuous change in plagioclase composition with crystallization of a melt, as we have seen in Figures 30.16 and 20.2. In a melt of basaltic composition, a calcium-rich plagioclase crystallizes first and becomes increasingly sodic, by continuously reacting with the magma, as crystallization proceeds. We have already noted that this process leads to zoning of crystals, with an anorthite-rich core and an albite-rich rim (Plate 7a). This *continuous series* is illustrated schematically on the right side of Figure 30.21.

Box 30.1 Additional information: Free energy and melting

Melting can be understood in terms of free energy. Consider, for example, what happens when we heat a crystal of diopside ($CaMgSi_2O_6$). For most solids the free energy decreases with increasing temperature but less than the free energy of the corresponding liquid does (see Figure 30.20a and review in Chapter 19). At some temperature the free energy of the solid crystal becomes larger than the free energy of the corresponding liquid. Where the two free energy curves cross, melting occurs. At that temperature ($T_m = 1392$ °C), crystal and melt coexist in equilibrium. This is called congruent melting.

The melting behavior is different for enstatite ($Mg_2Si_2O_6$) because another phase, forsterite (Mg_2SiO_4), enters the picture. For a certain temperature range it is energetically favorable to have solid forsterite coexist with a more siliceous melt according to the dissociation $Mg_2Si_2O_6 \rightleftharpoons Mg_2SiO_4(solid) + SiO_2(in melt)$. Figure 30.20b shows the free energy–temperature diagram for solid and liquid enstatite, as well as for a solid forsterite + liquid mixture. The point T_m, at the intersection of the free energy lines for solid and liquid enstatite, would be the equilibrium melting point of enstatite. However, the free energy curve for a compositionally equivalent solid forsterite + liquid mixture has a steeper slope than the solid enstatite curve and above a temperature T_i (1557 °C) melting begins, with simultaneous crystallization of forsterite because the free energy is lower than that for solid or liquid enstatite. With rising temperature the proportion and composition of the liquid that coexists with forsterite change and so do the corresponding free energy curves. The melt is first very siliceous but continuously approaches the composition of enstatite. The slopes of the free energy curves become steeper and define a curved minimum free energy envelope that converges on the enstatite liquid curve at T_1. At that temperature all forsterite has been consumed by the melt, above T_1 a homogeneous enstatite melt has the lowest free energy. Such melting behavior is called *incongruent melting*, as opposed to *congruent melting* as in diopside.

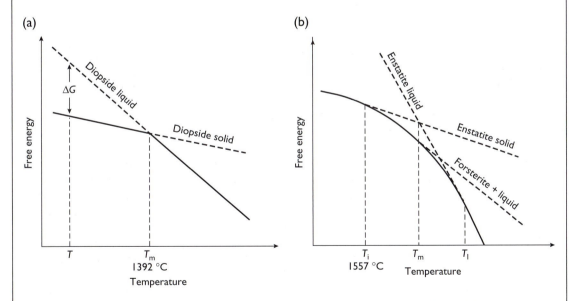

Fig. 30.20 Free energy–temperature diagrams, illustrating melting. (a) Congruent melting of diopside. (b) Incongruent melting of enstatite with intermediate formation of forsterite. T_m is the melting point; T_i is the incongruent melting point of enstatite, and T_1 is the temperature at which melting is complete.

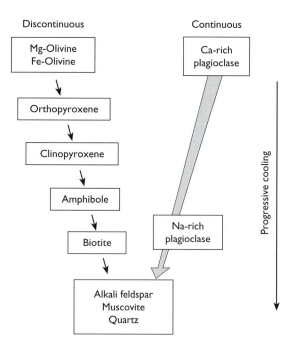

Fig. 30.21 Bowen's reaction series for mineralogical changes during cooling of a magma with the discontinuous branch on the left side and the continuous branch on the right side.

Parallel to this continuous crystallization, there is a *discontinuous series*, which applies to ferromagnesian minerals. With the ternary system in Figure 30.17 and the binary system in Figure 30.18, we have documented that in a mafic magma olivine crystallizes first, followed by orthopyroxene. In a more complete sequence olivine crystallizes at the highest temperature (first magnesium-rich, later more iron-rich; Figure 20.4), followed by orthopyroxene, clinopyroxene, amphibole, and biotite. This is illustrated on the left side of Figure 30.21 in Bowen's discontinuous branch.

We note that in both the discontinuous and the continuous series, the percentage of silicon in the melt increases progressively with decreasing temperature. In the discontinuous series, orthosilicates are followed by single-chain silicates, double-chain silicates, sheet silicates, and finally framework silicates. Some residual phases such as water and components corresponding to alkali feldspar, muscovite, and quartz remain in the melt, which increasingly becomes an aqueous hydrothermal solution.

Since the crystallization sequence goes more or less parallel with density of the minerals, there is a tendency for early forming minerals such as olivine to settle to the bottom of the magma chamber due to gravity, with the remaining melt becoming more siliceous as a result. This process is called *magmatic differentiation*. At lower temperatures, melts are generally too viscous for gravitational settling to occur. Bowen (1928) surmised that crystallization of a basaltic magma through differentiation could explain all igneous rocks, for example peridotites as accumulates of olivine and pyroxene that have been removed from the melt through gravitational settling, and granites as residual melts, depleted in ferromagnesian minerals. Today it is appreciated that the origin of igneous rocks is much more complicated and diverse than had previously been thought.

There are many complicating factors, such as mixing of magmas, melting of adjacent country rocks, removal or addition of volatile phases, and the fact that, for kinetic reasons, minerals that are lower in the series can crystallize at the same time as those higher in the series.

30.4 Metamorphic reactions in siliceous limestones

Bowen's reaction series highlights some systematic trends in the crystallization of mafic magmas. As we will see, there is an analog to this pattern in metamorphic rocks. Since we have now discussed most of the rock-forming minerals and since chain silicates also play an important role in metamorphic rocks, it is appropriate to return briefly to some reactions that occur in progressive metamorphism. In Chapter 28 we examined metamorphism in aluminous rocks, where minerals such as garnet, aluminosilicates, staurolite, and cordierite are commonly found. Here, we explore the metamorphism of siliceous limestones, rocks originally composed of calcite, dolomite, quartz, and occasionally magnesite, with or without water present. We use a ternary representation to show the stable combinations in this system at different metamorphic grades (Figure 30.22). At low temperature, for example during diagenesis, quartz, calcite, dolomite, and magnesite are the only stable phases (Figure 30.22a). Tielines show that dolomite may coexist with calcite and quartz, but does not coexist with both calcite and magnesite. With increasing temperature, under conditions of greenschist facies, dolomite is no longer stable with quartz and reacts to form talc and tremolite (in the presence of water).

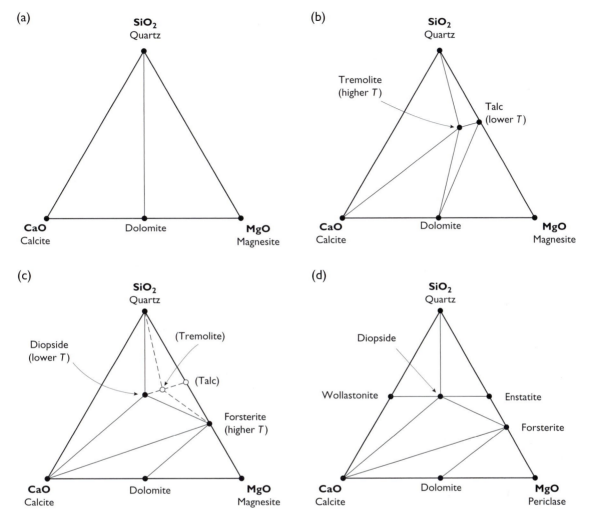

Fig. 30.22 Triangular representations of the system calcite–quartz–magnesite with the main minerals observed in metamorphic siliceous limestones and dolomites. Tielines give stable mineral assemblages for corresponding conditions. For the interpretation of these diagrams, see Box 28.1. (a) Unmetamorphic limestones and diagenesis. (b) Greenschist facies. (c) Lower amphibolite facies. (d) High amphibolite and granulite facies (after Winkler, 1979).

(The equilibrium curves for these reactions are shown in Figure 30.23 in a phase diagram with temperature and the CO_2 partial pressure as variables, reactions are referred to with numbers).

$3CaMg(CO_3)_2 + 4SiO_2 + H_2O \rightleftharpoons$
Dolomite   Quartz
    $Mg_3Si_4O_{10}(OH)_2 + 3CaCO_3 + 3CO_2$ #1
    Talc   Calcite

$5CaMg(CO_3)_2 + 8SiO_2 + H_2O \rightleftharpoons$
Dolomite   Quartz
    $Ca_2Mg_5Si_8O_{22}(OH)_2 + 3CaCO_3 + 7CO_2$ #2
    Tremolite   Calcite

Talc forms at a lower temperature than does tremolite. Note that for these silicate minerals water has to be present (reactions #1 and #2). Tielines in the triangle illustrate that, depending on rock composition, calcite–tremolite–dolomite, calcite–tremolite–quartz, tremolite–quartz–talc, tremolite–dolomite–talc, or dolomite–talc–magnesite coexist (Figure 30.22b). Quartz does not coexist with dolomite and magnesite at conditions of the greenschist facies.

With increasing metamorphic grade new minerals form – first diopside and then, at higher temperatures, forsterite:

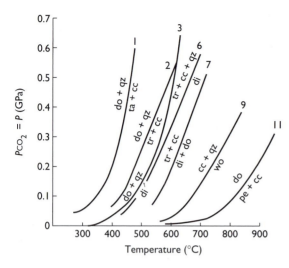

Fig. 30.23 Temperature–pressure phase diagram with experimentally determined equilibrium curves for reactions in siliceous limestones and dolomites: cc calcite; do, dolomite; qz, quartz; ta, talc; tr, tremolite; di, diopside; fo, forsterite; wo, wollastonite; pe, periclase. Numbers refer to reaction equations in the text. It is assumed that the CO_2 partial pressure equals the total pressure (after Turner, 1981).

$$CaMg(CO_3)_2 + 2SiO_2 \rightleftharpoons CaMgSi_2O_6 + 2CO_2$$
Dolomite   Quartz   Diopside   #3

$$2CaMg(CO_3)_2 + SiO_2 \rightleftharpoons 2CaCO_3 + Mg_2SiO_4 + 2CO_2$$
Dolomite   Quartz   Calcite   Forsterite

                                         #4

Figure 30.22c shows minerals stable in amphibolite facies conditions. Many combinations may exist.

In lower amphibolite facies, and for extremely calcium-poor rock compositions, talc and tremolite are still stable (indicated by open circles, parentheses and dashed lines), but at higher temperatures they break down to form diopside and forsterite (and secondary calcite, and dolomite) through reactions such as:

$$Mg_3Si_4O_{10}(OH)_2 + 3CaCO_3 + 2SiO_2 \rightleftharpoons$$
Talc   Calcite   Quartz
          $3CaMgSi_2O_6 + 3CO_2 + H_2O$ #5
          Diopside

$$Ca_2Mg_5Si_8O_{22}(OH)_2 + 3CaCO_3 + 2SiO_2 \rightleftharpoons$$
Tremolite   Calcite   Quartz
          $5CaMgSi_2O_6 + 3CO_2 + H_2O$ #6
          Diopside

$$Ca_2Mg_5Si_8O_{22}(OH)_2 + 3CaCO_3 \rightleftharpoons$$
Tremolite   Calcite
   $4CaMgSi_2O_6 + CaMg(CO_3)_2 + CO_2 + H_2O$ #7
   Diopside   Dolomite

$$Ca_2Mg_5Si_8O_{22}(OH)_2 + 11CaMg(CO_3)_2 \rightleftharpoons$$
Tremolite   Dolomite
   $8Mg_2Si_4 + 13CaCO_3 + 9CO_2 + H_2O$ #8
   Forsterite   Calcite

At the highest temperatures, in upper amphibolite and granulite facies, three new phases appear: wollastonite, enstatite, and periclase (Figure 30.22d):

$$CaCO_3 + SiO_2 \rightleftharpoons CaSiO_3 + CO_2$$
Calcite   Quartz   Wollastonite   #9

$$CaMg(CO_3)_2 + SiO_2 \rightleftharpoons CaCO_3 + MgSiO_3 + CO_2$$
Dolomite   Quartz   Calcite   Enstatite

                                        #10

$$CaMg(CO_3)_2 \rightleftharpoons CaCO_3 + MgO + CO_2$$
Dolomite   Calcite   Periclase   #11

Unless CO_2 pressures are very high, dolomite is no longer stable and breaks down to silicates and oxides, releasing CO_2. Phase relations become simpler again. Ultimately calcite would break down to CaO and CO_2, but in nature this reaction, which is significant in the production of cement, is rarely observed.

In Figure 30.23 it is assumed that the total pressure is equal to the CO_2 pressure. In a real metamorphic rock this may not be the case. For example, part of the gas may be water vapor. Or the system may not be closed and some CO_2 gas escapes. This has a profound effect on the phase diagram, as is illustrated for the wollastonite reaction #9 in Figure 30.24. If we assume a total pressure of 0.2 GPa, as is typical for amphibolite facies metamorphism, then the equilibrium temperature for the breakdown of calcite is reduced, as the partial pressure of CO_2 decreases from 100% (750 °C), to 50% (680 °C), and 0% (430 °C). The lowest temperature corresponds to heating calcite and quartz, with all CO_2 escaping. If the CO_2 gas pressure is 0%, then the system consists of three solid phases (calcite, quartz, and wollastonite), and the phase boundary is a straight line (cf. Chapter 19). Since the volumes of quartz and calcite are larger than that of wollastonite, the boundary has a negative slope (i.e., with increasing pressure, the stability field of wollastonite increases, contrary to the case where CO_2 gas is present in the system).

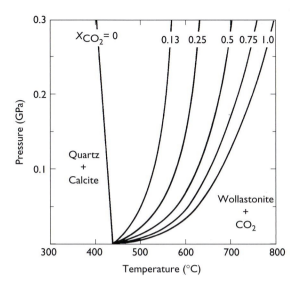

Fig. 30.24 Temperature–pressure phase diagram for the reaction quartz + calcite $\rightleftharpoons$ wollastonite + CO_2 for different partial pressures of CO_2, ranging from pure CO_2 ($X_{CO_2} = 1.0$) to pure water ($X_{CO_2} = 0$) (data from Greenwood, 1967).

If we look at these reactions in the temperature–pressure phase diagram (Figure 30.23), we notice that, by increasing temperature at a given pressure, progressively talc, tremolite, diopside, and forsterite become stable in these metamorphic marbles. In other words, the sequence is first sheet silicates, followed by amphiboles, pyroxenes, and finally orthosilicates, similar to the discontinuous reaction series in igneous rocks (see Figure 30.21). High-temperature metamorphic silicate minerals in marbles are not only depleted in volatile phases, such as H_2O and CO_2, they also have less silica.

In Chapter 28 we used the distribution of aluminosilicates in pelitic schists of the Central Alps to illustrate systematic changes in mineral assemblages with metamorphic grade. There is also a regular pattern in this region for metamorphic carbonate rocks. The map in Figure 30.25 extends over a much larger area than that in Figures 28.10 and 28.11, which was confined to the contact aureole of a granite on the eastern side. Metamorphic mineral assemblages in this wide region of the Central Alps are thought to be due to a Tertiary regional metamorphism that reached granulite facies conditions in the eastern part. The core of the metamorphic zone is defined by fields in which, progressively, tremolite, diopside, forsterite, and wollastonite occur, as indicated by different shadings.

We mentioned the analogy of metamorphic minerals with the discontinuous reaction series in igneous rocks. There is also a metamorphic equivalent of the continuous reaction series (Figure 30.21). As long as

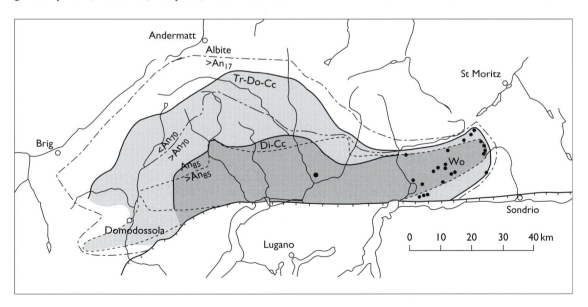

Fig. 30.25 Thermal metamorphism of carbonate rocks in the Central Alps. Solid lines and different shadings give distribution of calcsilicate minerals (after Trommsdorff, 1966), wollastonite localities are indicated by black dots; dashed lines are compositions (anorthite content) of plagioclase (after Wenk, 1970). Cc, calcite; Do, dolomite; Tr, tremolite; Di, diopside; Wo, wollastonite; An, anorthite content.

the overall composition of a rock is similar, the anorthite content of plagioclase increases with metamorphic grade. We use again the example of the metamorphic carbonate rocks of the Central Alps to illustrate this and show the contours of plagioclase compositions with dashed lines on the map of Figure 30.25. In the greenschist facies, outside the first contour, all plagioclase is albite. In amphibolite facies, plagioclase composition changes progressively to oligoclase, andesine, and labradorite (a contour at 70% An is shown). The field of bytownite (An_{70}–An_{90}) coincides closely with the stability field of tremolite. In the core of the metamorphic zone with the highest metamorphic grade, with diopside, forsterite, and wollastonite present, only very calcium-rich plagioclase is found (An > 85).

30.5 Summary

Chains can be cut from sheets. The simplest way is parallel to the line of single tetrahedra as in pyroxenes. In pyroxenoids the chain is not straight. In amphiboles two pyroxene chains join to form double chains. A hybrid is the mineral howieite with a chain of six-membered rings. The chains are then linked to a three-dimensional structure by stacking chains with displacements. This stacking results in {110} cleavages with angles of 88° in pyroxenes with single chains and 125° in amphiboles with double chains. Both pyroxenes and amphiboles show extensive compositional substitutions, especially Mg–Ca–Fe–Na–Al.

Chain silicates are important components of mafic igneous rocks and are used to explain the crystallization of basalts based on the classic experimental work of Bowen and theoretical concepts introduced in Chapters 19 and 20. Remember the discontinuous reaction series with progressive cooling beginning with olivine, then orthopyroxene, clinopyroxene, amphibole, biotite, and quartz, and the continuous series with calcium-rich plagioclase becoming progressively sodium-rich during cooling. A similar progression is observed in metamorphic rocks and documented by examples from the Central Alps.

Important chain silicate minerals to remember

Name	Formula
Pyroxenes	
Enstatite	$Mg_2Si_2O_6$
Diopside	$CaMgSi_2O_6$
Hedenbergite	$CaFeSi_2O_6$
Augite, pigeonite	$(Ca,Fe,Mg,Al)_2(Si_2O_6)$
Jadeite	$NaAl(Si_2O_6)$
Omphacite	$(Na,Fe,Mg,Al)_2(Si_2O_6)$
Spodumene	$LiAl(Si_2O_6)$
Amphiboles	
Tremolite	$Ca_2Mg_5(Si_8O_{22})(OH)_2$
Actinolite	$Ca_2Fe_5^{2+}(Si_8O_{22})(OH)_2$
Hornblende	$(Na,K)_{0-1}(Ca,Na)_2(Mg,Fe^{2+})$ $(Al,Fe^{3+})(Si_7AlO_{22})(OH, F)_2$
Glaucophane	$Na_2Mg_3Al_2(Si_8O_{22})(OH)_2$
Pyroxenoids	
Wollastonite	$Ca_3(Si_3O_9)$
Rhodonite	$Ca_4Mn(Si_5O_{15})$

Test your knowledge

1. Chain silicates are divided into two major groups. What distinguishes them in terms of structure and composition?
2. Give a structural explanation for the cleavage angle in pyroxenes and amphiboles.
3. Fill in mineral names in the quadrilateral of calcic pyroxenes and calcic amphiboles.
4. Compare the compositions of glaucophane and tremolite, and of riebeckite and actinolite.
5. Write a reaction to form wollastonite from calcite and diopside found in dolomite in impure limestone.
6. Enstatite and pigeonite are typical minerals in igneous rocks. If you cool a magma of basaltic composition, which minerals crystallize in which order (Bowen's reaction series)?
7. Sort the following minerals in marbles with increasing metamorphic grade: forsterite, talc, diopside, tremolite, periclase.

Further reading

Chain silicates

Deer, W. A., Howie, R. A. and Zussman, J. (1997). *Rock-forming Minerals, Volume 2A, Single-chain Silicates*, 2nd edn. The Geological Society of London.

Deer, W. A., Howie, R. A. and Zussman, D. J. (1997). *Rock-forming Minerals, Volume 2B, Double Chain Silicates*, 2nd edn. The Geological Society of London.

Prewitt, C. T. (ed.) (1980). *Pyroxenes.* Reviews in Mineralogy, vol. 7. Mineralogical Society of America, Washington, DC.

Igneous and metamorphic petrology

Best, M. G. and Christiansen, E. H. (2001). *Igneous Petrology.* Blackwell Science, Oxford.

Bowen, N. L. (1928). *The Evolution of Igneous Rocks.* Princeton University Press, Princeton, NJ.

Carmichael, I. S. E., Turner, F. J. and Verhoogen, J. (1974). *Igneous Petrology.* McGraw-Hill, New York.

La Maitre, R. W., Streckeisen, A., Zanettin, B., Le Bas, M. J., Bonin, B. and Bateman, P. (2005). *Igneous Rocks: A Classification and Glossary of Terms*, 2nd edn. Cambridge University Press, Cambridge.

Morse, S. A. (1980). *Basalts and Phase Diagrams: An Introduction to the Quantitative Use of Phase Diagrams in Igneous Petrology.* Springer.

Philpotts, A. R. and Ague, J. J. (2009). *Principles of Igneous and Metamorphic Petrology*, 2nd edn. Cambridge University Press, Cambridge.

Winter, J. D. (2001). *Introduction to Igneous and Metamorphic Petrology.* Prentice Hall, Upper Saddle River, NJ.

31 | Some framework silicates. Zeolites and ion exchange properties of minerals

Chapter 21 introduced two mineral groups with a tetrahedral framework structure that are most important for petrology and occur everywhere in the crust: silica minerals and feldspars. In this chapter some other important framework minerals will be discussed, especially zeolites. Zeolites have very open frameworks that can accommodate large ions and molecules, including water. They are of great technological importance because of ion exchange properties, as molecular sieves, as catalysts, and as nanopore materials.

31.1 The framework structure

Compared to all other silicates with some elements of close-packing of oxygens, framework silicates have much more open structures and thus low densities. Contrary to quartz and feldspars, zeolites and associated minerals have structures with open cages, a bit like clathrates introduced in Chapter 27 (Figure 27.12), or regular channels and thus unusual properties.

Framework silicates contain a three-dimensional framework of tetrahedra, in which all oxygen atoms are bonded to two tetrahedral cations, resulting in a general formula with a T:O ratio 1:2. Tetrahedral ions can be silicon or aluminum with $Al^{3+} \leq Si^{4+}$. Quartz, SiO_2, is an example with no aluminum, while anorthite ($CaSi_2Al_2O_8$) is an example where aluminum and silicon are equal. The aluminum-containing framework silicates are referred to as alumosilicates with all aluminum atoms tetrahedral, contrary to the aluminosilicates discussed in Chapter 28, where at least some aluminum is octahedral.

The composition of framework silicates is quite simple compared to pyroxenes and amphiboles: the major cations are silicon and aluminum; in addition, large cavities in the framework may contain large cations (e.g., K^+, Na^+, Ca^{2+}, Ba^{2+}, Sr^{2+}) as well as anions and anion groups (e.g., Cl^-, CO_3^{2-}, SO_4^{2-}, NH_4^+, and H_2O molecules). This is illustrated for *scapolite* in Figure 31.1. There are none of the octahedral building units that are an important part of all other silicate structures and therefore framework silicates do not contain octahedral cations (e.g., Mg^{2+}, Fe^{2+}). Among framework silicates, *feldspars* are most relevant and they have already been discussed in Chapter 21. (We will review some aspects of feldspar structures in this chapter.) *Feldspathoids* (e.g., nepheline, leucite) are a related group of anhydrous framework silicates, the main difference being that they are deficient in SiO_2, and the structure needs to be able to accommodate more alkali cations than

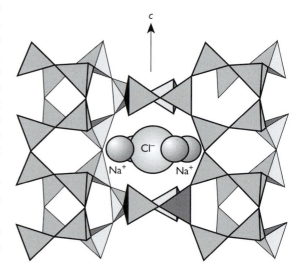

Fig. 31.1 Structure of the tetragonal mineral scapolite $Na_4(AlSi_3O_8)_3Cl$ projected on (100). The *c*-axis is vertical. It illustrates the open tetrahedral framework with large cages that contain cations (Na^+) and anions (Cl^-).

alkali feldspars. *Zeolites* are another important group. They are characterized by open frameworks with large cavities and open channels that contain variable amounts of H_2O which is easily lost upon heating (hence the name from the Greek *zein*, meaning "to boil" and *lithos*, meaning "stone").

The structures of framework silicates are complex and diverse because there are many ways to link tetrahedra in three-dimensional space. In the case of feldspars, the tetrahedra are linked together in rings of eight and four (Figure 31.2a). To simplify the structural representation of framework silicates and to better visualize the topology, often only tetrahedrally coordinated T cations are shown, leaving out tetrahedra and oxygen atoms (Figure 31.2b). In this figure a symbol U is used if the apex of a tetrahedron points up and D if it points down, connecting with another layer. By simply changing U and D, i.e., the connection to the next layer, we can derive from feldspars the structures of other framework silicates such as paracelsian (Figure 31.2c,d) and of the zeolite harmotome (Figure 31.2e,f) and recognize similarities and differences. Naturally, in such a compact two-dimensional

representation that highlights the topology, much information about the detailed geometry is lost.

The T-framework representation can be extended easily to three dimensions and this is illustrated in Figure 31.3 for zeolites. While tetrahedra are the primary building units in framework silicates, rings and "cages" composed of tetrahedra are secondary building units. The most common ring units are four-, five-, six-, and eight-membered rings (Figure 31.3a), although ten- and twelve-membered rings are also present. The smaller rings can be combined to form simple cages (Figure 31.3b). These cages, together with additional rings, can be combined to form larger units with different complexity as in gmelinite (Figure 31.4a), and chabazite (Figure 31.4b). In cubic sodalite cube-octahedral cages with four- and six-membered rings are linked together over four-membered rings (Figure 31.4c). In mazzite, twelve-membered rings delimit large channels (Figure 31.4d). Indeed, there is almost no limit to the different conceivable architectures, and many of these are observed in natural and synthetic zeolites.

The tetrahedral frameworks of silicates can not only accommodate large cations but also unusual

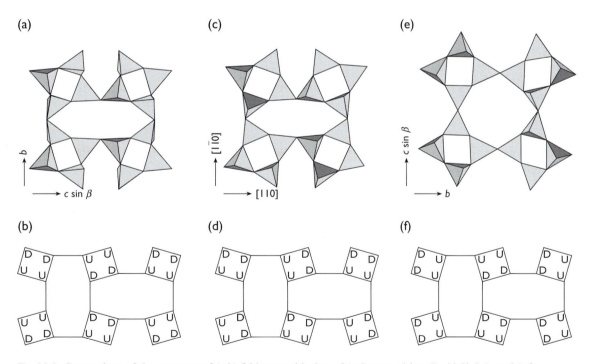

Fig. 31.2 Comparison of the structure of (a,b) feldspars with that of (c,d) paracelsian ($BaAl_2Si_2O_8$), and (e,f) monoclinic harmotome. Note the similarities with four- and eight-membered rings. The representations (a,c,e) show tetrahedra; (b,d,f) are idealized representations that show connections only between tetrahedral cations (Si^{4+}, Al^{3+}). Tetrahedral apices are pointing either up (U) or down (D).

anion groups, such as in scapolite minerals (see Figure 31.1), sodalite, cancrinite, and many others, mostly pertaining to the feldspathoid group of minerals. The cages are particularly large in zeolites (up to 9 Å in cross-section) and are also interconnected. These cavities contain groups and complexes of ions and water molecules. Natrolite is a typical example of a zeolite structure with an open framework and channels that contain water (Figure 31.5).

In zeolites, the intercavity cations, ion complexes, and molecules can exchange with surrounding solutions. A four-membered ring is too small to admit the passage of any atom. Six-membered rings have apertures 2.2–2.7 Å across, when the peripheries of the framework oxygen atoms are taken into account. This is still too small for any atom except hydrogen to pass through. The eight-membered rings are 3.7–4.1 Å across. Argon and methane are rapidly absorbed in cages with eight-membered rings; larger molecules are excluded. The structural feature that lets small molecules pass, while large ones are kept back, permits

zeolite to act as a molecular sieve, a property that has important technological applications.

The $T{:}O$ ratio for all alumosilicates (with Al^{3+} in the tetrahedral sites) is 1:2. However, tetrahedral ions can either be Si^{4+} or Al^{3+} and thus the composition of the framework differs. For example, the ratio Si:Al is 1:1 in nepheline (a silicon-deficient feldspathoid), 3:1 in albite, and 1:0 in silica minerals. There is never more aluminum than silicon because this situation would produce structures with adjacent AlO_4^{5-} tetrahedra, and Al–O–Al bonds are highly unstable (the "aluminum avoidance principle"; Loewenstein, 1954). In some framework silicates with solid solutions (for instance, plagioclase with the albite ($Na(AlSi_3O_8)$)–anorthite ($Ca(AlSi_3O_8)$) series, or the members of the scapolites with the marialite ($Na_4(AlSi_3O_8)_3Cl$)–meionite ($Ca_4(Al_2Si_2O_8)_3CO_3$) series), the ratio between aluminum and silicon changes continuously.

31.2 Morphology and physical properties

The diversity of crystal structures in the framework silicates is the cause of the widely different morphologies. Isometrically developed crystals are typical for leucite (Plate 16f), sodalite, lazurite, and analcime (Figure 31.6a) expressing the cubic or pseudocubic symmetry. In comparison, feldspar crystals, though often more or less equant, always have oblique angles between faces, consistent with monoclinic or triclinic symmetry. Some framework silicates are columnar or prismatic in habit such as scapolites and cancrinite, or fibrous such as natrolite (Figure 31.6b), scolecite, and mesolite. Tabular or platy habit is present in heulandite, stilbite, and clinoptilolite (Figure 31.6c).

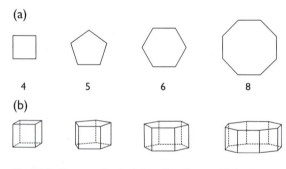

Fig. 31.3 Structural principles of zeolites, with basic building unit rings (a) assembling to cages (b).

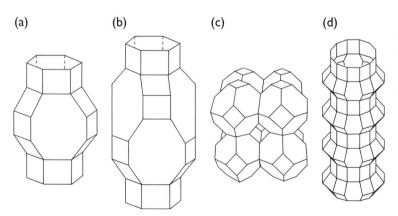

Fig. 31.4 The cages can combine to form larger units that are linked, such as in (a) gmelinite, (b) chabazite with eight-membered rings, (c) cubic sodalite, and (d) mazzite with large channels.

(a)

(b)

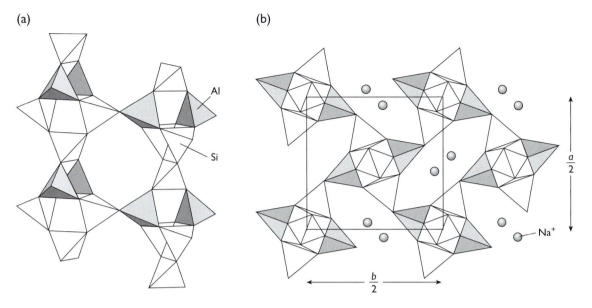

Fig. 31.5 Structure of natrolite $Na_2(Al_2Si_3O_{10})\cdot2H_2O$. (a) Three-dimensional view of two corner-linking chains that are parallel to the c-axis. (b) View along the c-axis, illustrating channels between the chains. The square outlines a quarter of the unit cell.

(a)

(b)

(c)

Fig. 31.6 Crystal structure is expressed in the morphology of framework silicates, such as in (a) cubic analcime from Ischia, Italy (width 60 μm), (b) fibrous natrolite from Mückenhalm, Germany (width 75 mm), and (c) tabular clinoptilolite from Creede, Colorado, USA (width 30 μm) ((b) is courtesy O. Medenbach; (a) and (c) are SEM images from Gottardi and Galli (1985)).

Table 31.1 Common framework silicates with some diagnostic properties; most important minerals are given in italics (For feldspars, see Table 19.1)

Mineral / Formula	Si:Al	System	Morphology / Cleavage	H	D	Color / Streak	n	Δ	$2V$ (Dispersion)
Feldspathoids and miscellaneous									
Cancrinite $Na_6Ca_2(AlSiO_4)_6(CO_3)_2 \cdot 3H_2O$	1:1	Hexag.	Pris. [0001]	5–6	2.4	Yellow, rose, blue	1.49–1.52	0.02	(−)
Lazurite (lapis lazuli) $Na_3Ca(AlSiO_4)_3(SO_4,S_2)$	1:1	Cubic	Microcryst.	5.5	2.4	Blue / *Blue*	1.5		
Leucite $K(AlSi_2O_6)$	2:1	Ps. Cubic	Xls. rare	5.5	2.5	White, gray	1.51	0.001	(+)
Nepheline $Na_3K(AlSiO_4)_4$	1:1	Hexag.	Pris. [0001] {10$\bar{1}$0}, (0001) poor	5.5–6	2.6	White, gray, red	1.53–1.55	0.004	(−)
Marialite (scapolite) $Na_4(AlSi_3O_8)_3Cl$	3:1	Tetrag.	Pris. [001] {100}, {110}	5–6.5	2.5	White, gray	1.54–1.55	0.002	(−)
Meionite (scapolite) $Ca_4(Al_2Si_2O_8)_3CO_3$	1:1	Tetrag.	Pris. [001] {100}, {110}	5–6.5	2.8	White, gray	1.56–1.59	0.04	(−)
Sodalite $Na_4(AlSiO_4)_3Cl$	1:1	Cubic	{110}	5.5–6	2.3	White, blue	1.48–1.49		
Zeolites									
Analcime $Na(AlSi_2O_6) \cdot H_2O$	2:1	Ps. Cubic	Trapez.{211}	5–5.5	2.3	White	1.48–1.49		
Chabazite-Ca $(Ca_{0.5},Na,K)_4(Al_4Si_8O_{24}) \cdot 12H_2O$	2:1	Trig.	Eq. {10$\bar{1}$1}	4.5	2.1	White	1.48	0.002 Anom.	−0–32
Clinoptilolite $(Na,K,Ca_{0.5})_6(Al_6Si_{30}O_{72}) \cdot 20H_2O$	5:1	Monocl.	Platy (010) (010)	3.5–4	2.1	White	1.48–1.59	0.003	−40 (r<v)
Erionite-K $(K_2,Ca,Na_{2,0.5})_2(Al_4Si_{13}O_{36}) \cdot 15H_2O$	13:5	Hexag.	Pris. Fibr.	3.5–4	2.1	White	1.46–1.48	0.002	(+)
Gismondine-Ca $Ca(Al_2Si_2O_8) \cdot 4H_2O$	1:1	Monocl.	Bipyr (010)	4.5	2.2	White	1.52–1.55	0.015	−85
Gmelinite-Na $(Na_2,Ca,K_2)_2(Al_4Si_8O_{24}) \cdot 11H_2O$	2:1	Hexag.	Eq. (10$\bar{1}$0)	4.5	2.1	White	1.48–1.47	0.015 Anom.	(−)

Mineral	Ratio	System	Morphology	H	D	Color	n	Δ	$2V$
Harmotome $(Ba_{0.5},Ca_{0.5},K,Na)_5(Al_5Si_{11}O_{32})\cdot12H_2O$	11:5	Monocl.	Platy (010) *(010)*	4.5	2.5	White	1.50–1.51	0.005	+43
Heulandite-Ca $(Ca_{0.5},Na,K)_5(Al_9Si_{27}O_{72})\cdot24H_2O$	3:1	Monocl.	Platy (010) *(010)*	3.5–4	2.2	White	1.49–1.50	0.005	+0–55 (r > v)
Laumontite $Ca(Al_2Si_4O_{12})\cdot4H_2O$	2:1	Monocl.	Pris. (010) *(010)(110)*	3–3.5	2.3	White	1.50–1.52	0.01	−26–47 (r≪v)
Mazzite-Mg $(Mg,K_2,Ca_5)_5(Al_{10}Si_{26}O_{72})\cdot28H_2O$	13:5	Hexag.	Fibr. [0001]	4	2.1	White	1.50–1.51	0.007	(−)
Mordenite $(Na_2,Ca,K_2)_4(Al_8Si_{40}O_{96})\cdot28H_2O$	5:1	Ortho.	Pris. (100)	3–4	2.1	White	1.47–1.48	0.005	−80–(+80)
Natrolite $Na_2(Al_2Si_3O_{10})\cdot2H_2O$	3:2	Ortho.	Fibr. [001] *(110)*	5–5.5	2.3	White	1.48–1.49	0.013	+60–63 (r < v)
Phillipsite-K $(K,Na,Ca_{0.5})_6(Al_6Si_{10}O_{32})\cdot12H_2O$	5:3	Monocl.	Platy (010) *(010),(100)*	4.5	2.2	White	1.48–1.50	0.005	+60–80 (r<v)
Stilbite $(Ca_{0.5},Na,K)_9(Al_9Si_{27}O_{72})\cdot28H_2O$	3:1	Monocl.	Platy (010) *(010)*	3.5–4	2.1	White	1.49–1.51	0.001	−33 (r < v)

Notes: H, hardness; D, density (g/cm^3); n, range of refractive indices; Δ, birefringence; $2V$, axial angle for biaxial minerals. For uniaxial minerals (+) is positive and (−) is negative. Acute $2V$ is given in the table. If $2V$ is negative the mineral is biaxial negative and $2V$ is $2V_\alpha$; if it is positive, the mineral is biaxial positive and $2V$ is $2V_\gamma$. Dispersion r < v means that acute $2V$ is larger for violet than for red; Anom., anomalous birefringence.

System: Hexag., hexagonal; Monocl., monoclinic; Ortho., orthorhombic; Ps, pseudo; Trig, trigonal; Tetrag., tetragonal.

Morphology: Eq., equiaxed; Fibr., fibrous; Microcryst., microcrystalline; Pris., prismatic; Bipyr., bypiramidal; Trapez., trapezohedral. Xls., crystals.

Some physical properties are fairly uniform among framework silicates. Strong covalent–ionic bonds and the open cell structure combine to give the framework silicates a medium hardness (4.5–6 on the Mohs scale) and a low density (2.1–2.6 g/cm^3). The bond type and composition of the framework silicates also determine the vitreous luster and transparence or translucence of these minerals. They are usually white because transition elements do not enter their structures. However, some framework alumosilicates do display colors related to the presence of color centers (see Chapter 15). For example, the ink-blue and dark-blue color of sodalite and lazurite is due to such color centers. The cellular structures can accommodate anions that are bigger than oxygen ions. For example, chloride (Cl$^-$), with an ionic radius of 1.81 Å, is present in sodalite and marialite (see Figure 31.1), and the persulfide ion S$_2^{2-}$, with an interatomic distance of 2.06 Å, occurs in lazurite. Inhomogeneities and cages of various shapes weaken the chemical bonds in framework silicates and result in good cleavages along two or more directions. Luster on cleavage planes is vitreous.

The above-described similarities in properties of framework alumosilicates are easily explained in terms of their structural similarity and the dominance of the same large cations (Na$^+$, K$^+$, Ca^{2+}), but these similarities also make their identification difficult.

Many minerals of the framework silicate group are important rock-forming minerals. *Feldspars* (already discussed in Chapter 21) are the main constituents of most igneous and metamorphic rocks. *Nepheline* is predominant in some alkaline rocks, and *leucite* occurs in some low-silica basalts. Important framework silicates and their properties are listed in Tables 31.1. For feldspars and silica minerals see Table 21.1.

31.3 Brief description of important framework silicate minerals

Minerals below are ordered as in Table 31.1; for feldspars, see Chapter 21.

Lazurite (Na$_3$Ca(AlSiO$_4$)$_3$(SO$_4$,S$_2$)) is characterized by its dark blue color. It occurs in contact metamorphic rocks, usually marble. Lapis lazuli is a mixture of lazurite, calcite, and pyroxenes and there are well-known deposits in Afghanistan. Lazurite was formerly used for blue paint pigment and is a highly valued decorative stone.

Leucite (KAlSi$_2$O$_6$) forms trapezohedral crystals of almost ideal shape (Plate 31e). These crystals are composed of thin twinned plates of tetragonal symmetry that form as a result of transformation from an original crystal of cubic symmetry. Such crystals form phenocrysts in some volcanic and plutonic rocks of basaltic composition. Leucite crystals are translucent, colorless, and have a vitreous luster.

Leucite is not very stable under many conditions. As basalt cools, leucite phenocrysts may react with the melt and leucite may be replaced by sanidine. Leucite crystals in already solidified rocks also may change into aggregates of microcline, muscovite, and zeolites under the influence of residual alkaline solutions. Finally, analcime, kaolinite, and carbonates can easily replace leucite during weathering. This process produces fertile soils, rich in the potassium that has been removed from leucite.

Leucite can easily be identified because of its crystal habit and its occurrence in volcanic rocks. Leucite never coexists with quartz because these two minerals would react to produce K-feldspar.

Nepheline (Na$_3$K(AlSiO$_4$)$_4$) is a typical mineral of alkaline rocks. Grains are often anhedral, but well-developed nepheline crystals with square or hexagonal sections are sometimes observed in alkalic volcanic rocks and syenites. Since nepheline is easily dissolved on weathering surfaces, it may leave pseudomorphic holes with a characteristic hexagonal shape. Deeper inside rocks, nepheline is frequently covered with a film of powder-like secondary minerals. Fresh nepheline surfaces have a flesh-red or greenish color and a greasy luster. Nepheline does not have a cleavage and this distinguishes it from feldspars (though sanidine may display only poor cleavage).

Nepheline may be mistaken for quartz, but quartz has a higher hardness and is resistant to weathering. As with leucite, nepheline never coexists with quartz because the two would react to form albite and K-feldspar:

$$Na_3K(AlSiO_4)_4 + 8SiO_2 \rightarrow 3NaAlSi_3O_8 + KAlSi_3O_8$$
Nepheline   Quartz   Albite   K-feldspar

Nepheline is a major constituent of nepheline syenites and associated pegmatites and in such rocks it may form deposits that are used for the extraction of aluminum.

Scapolites comprise a continuous isomorphic series **marialite** (Na$_4$(AlSi$_3$O$_8$)$_3$Cl)–**meionite** (Ca$_4$(Al$_2$Si$_2$O$_8$)$_3$CO$_3$). Compositionally the scapolites are analogs of plagioclases but contain additional anions Cl$^-$, CO$_3^{2-}$, and SO$_4^{2-}$; nonetheless, they have an entirely different structure (cf. Figures 31.1 and 21.10a). Scapolites are tetragonal with prismatic or columnar crystals. They are white, greenish, or pink, with vitreous luster. A weak {100} or {110} cleavage is sometimes observed. In thin section scapolite is distinguished from feldspars with interference figures (uniaxial). Scapolite minerals occur in skarns and in metamorphic carbonate rocks.

Sodalite ($Na_4(AlSiO_4)_3Cl$) occurs in grayish-blue and ink-blue veinlets and as clusters and massive aggregates. Macroscopic crystals are sometimes observed in nepheline syenites, trachytes, and phonolites. Exceptional transparent crystals have been found in the lavas of Mt. Vesuvius (Italy).

Zeolites are a large group of framework alumosilicates that include some 100 mineral species. The zeolites are low-temperature minerals that form under hydrothermal conditions – for instance, due to alteration of nepheline in alkaline rocks or crystallization from hydrothermal solutions in cavities and amygdules of basalts. However, zeolites also occur in sedimentary or tuffaceous–sedimentary rocks at a late-diagenetic early-metamorphic stage as a product of the transformation of feldspars and volcanic glass, or as an authigenic mineral in the matrix. The low-grade metamorphic conditions under which they form are known as the *zeolite facies*.

One can easily identify zeolites in alkaline igneous rocks, where they form euhedral crystals, often in radiaxial and plate-like aggregates. All zeolites are originally white in color, with a vitreous luster. They have low density (about 2.1–2.2 g/cm^3) and low hardness (3.5–5). Among the zeolites, **stilbite**, **heulandite**, and **clinoptilolite** form platy crystals (Figure 31.6c) with good cleavage parallel to (010); orthorhombic **natrolite** is characterized by prismatic crystals with a square cross-section. Natrolite, scolecite, and mesolite occur mostly as radiaxial fibrous aggregates (Figure 31.6b). For cubic **analcime**, trapezohedral crystals are typical (Figure 31.6a). Crystals of trigonal **chabazite** look like a cube squeezed along one of its axes (rhombohedron).

Zeolites in sedimentary rocks, tuffs, and soils occur as fine-grained aggregates and cannot be identified in hand specimens. X-ray diffraction techniques are required instead.

31.4 Ion exchange properties of some minerals

Compared to the majority of minerals, zeolites have an unusual property. They are able to lose water molecules and cations from their structure during heating, without changing the basic framework structure, and these dried (activated) zeolites can then reabsorb water, cations, and more complicated molecules from the environment. They are thus capable of *ion exchange* in a more extreme way than we have described for clays. We have already described typical structural features with large channels and cages through which atoms can move. In zeolites (e.g., chabazite), the diffusion is orders of magnitude faster, even at moderate temperature, than in olivine and feldspar (Table 31.2).

Zeolites have been classified according to their exchange properties (Table 31.3). Every zeolite has its own ion exchange capacity, but the experimental capacity is always less than its theoretical limit. The exchange capacity is usually given in milligram equivalents/gram (where "gram" is 1 g of a dried zeolite powder). For example, 1 mg equiv. of sodium is 22.99 mg, where 22.99 is the atomic weight of sodium, meaning that 22.99 mg of sodium can be absorbed in 1 g of a dried zeolite powder.

There are three main structural–chemical factors for ion exchange in zeolites. The first is the configuration and size of cavities (channels and cages) in their framework structures (Table 31.4). The monoclinic structures of heulandite and clinoptilolite (Figure 31.7), for example, are characterized by three types of open channel, defined by eight- and ten-membered tetrahedral rings. These channels are oriented in three directions: the larger channels with ten-membered rings extend along the *c*-axis (A in Figure 31.7a), while

Table 31.2 Diffusion in several minerals

Component	Mineral	T ($°C$)	Diffusivity (cm^2/s)
H_2O	Chabazite	45	1.3×10^{-7}
H_2O	Chabazite	500	4.4×10^{-4}
Ca	Chabasite	60	5.9×10^{-11}
Ca	Chabazite	500	7.3×10^{-7}
H_2O	Ice	−2	1.0×10^{-10}
Fe–Mg	Olivine	1100	8.9×10^{-18}
Fe–Mg	Olivine	1100	4×10^{-12}
Fe–Mg	Olivine	600	8.9×10^{-18}
O	Olivine	1400	$\sim 1 \times 10^{-14}$
Si	Olivine	1400	$\sim 1 \times 10^{-18}$
Na–K	Feldspar	900	1.0×10^{-13}

Source: After Kretz, 1994.

Table 31.3 Examples of the sorptive properties of some natural zeolites

Zeolite	Largest molecules absorbed	Stability at T (°C)	Exchange capacity (mg equiv./g)	Note
Phillipsite	H_2O	<200	3.3	Not used
Clinoptilolite	O_2	>700	2.2	
Mordenite	C_2H_4	>700	2.3	Have industrial importance
Erionite	C_3H_8	>700	3.1	

Source: After Helfferich, 1995.

Table 31.4 Types of channel in structures of some zeolites and their approximate dimensions (in Å)

Zeolite	Numbers of tetrahedra forming a ring			
	12	10	8	4
Chabazite	—	—	3.7×4.2	2.6
Phillipsite	—	—	4.0×4.2	
			2.8×4.8	
Clinoptilolite	—	4.3×7.1	3.9×5.4	
			3.9×5.2	
			4.0×4.6	—
Mordenite	6.7×7.0	—	2.3×5.2	—
Erionite	—	—	3.6×5.2	—

eight-membered rings form both channels extending along the a-axis (C in Figure 31.7b) and channels inclined at 50° to the a-axis (B in Figure 31.7a). It is obvious that each type of channel differs in its capacity for ion exchange and in the speed of diffusion that can take place through it.

The second factor affecting ion exchange in zeolites is the Si:Al ratio. We already have mentioned that every Si^{4+} replaced by Al^{3+} causes a charge imbalance in the framework, and this imbalance is compensated by cations in the channels. The lower the Si:Al ratio, the greater is the ion exchange capacity (mg equiv./g) of a zeolite. Among natural zeolites, mordenite (Si:Al ratio 5; exchange capacity 2.3), clinoptilolite (3.5; 2.2), and erionite (3.5; 3.1) are high-silica zeolites, whereas phillipsite (3.0; 3.3) and chabazite (2.0; 3.8) are considered low-silica zeolites. Note that the Si:Al ratio is not a fixed number but varies over a certain range. High-silica zeolites are more stable in most environments and particularly in acid solutions.

The third factor affecting ion exchange is the location of cations in the structures of zeolites. For example, an ideal heated and dried mordenite has a formula $Ca_4Al_8Si_{40}O_{96}$. Cations (Ca^{2+}) occupy four positions in its structure, with three of them placed in small channels (2.9 Å × 5.7 Å), and one in a wider channel (6.7 Å × 7.0 Å). The Ca^{2+} in the larger

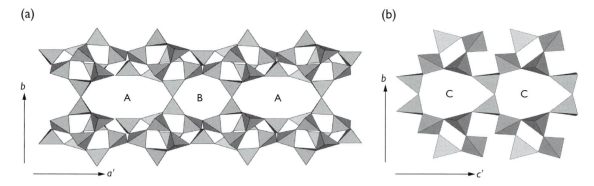

(a)

(b)

Fig. 31.7 Parts of the structure of monoclinic zeolite clinoptilolite $(Na,K,Ca_{0.5})_6$ $(Al_6Si_{30}O_{72}) \cdot 2OH_2O$ with large channels, A, B, and C. Representation of the tetrahedral framework with eight-membered (B and C) and ten-membered rings (A) outlining the channels. (a) View along the c-axis illustrating A and B channels. (b) View along the a-axis illustrating C channels. The channels are occupied by large cations that form hydrated complexes. a' and c' indicate that the corresponding axes are not in the projection plane.

channel can be easily replaced by, for example, Sr^{2+}, Ba^{2+}, K^+, Na^+, whereas the Ca^{2+} in the small channels can only be replaced by small cations (e.g., Na^+).

Heated and dried zeolite products differ in the speed at which an ion exchange process occurs. The speed increases with the size of cages and channels, the concentration of ions in the solution, and the temperature.

There is a wide range of uses for zeolites, both natural (from large sedimentary deposits) and synthetic (e.g., Mumpton, 1999). As a result, zeolites are among the most economically important minerals. Many of the industrial applications of zeolites were spearheaded in the USA and Japan, but they are now used worldwide. Both ion exchange and adsorption properties can be used. Noteworthy is a particular utilization in catalytic processes in the production of lead-free high-octane gasoline. Many other applications that have been developed are based on the ability of certain zeolites to exchange large cations selectively from aqueous solutions, such as the most common environmentally oriented applications found in the fields of phosphate-free detergents, wastewater treatments, and agricultural uses. Radioactive ^{137}Cs and ^{90}Sr can be removed from low-level waste streams of nuclear installations by extracting those ions with clinoptilolite filters. The same method can be used to extract ammonia and heavy metals from sewage and agricultural wastewaters. Zeolite filters, particularly mordenite and clinoptilolite, are also used to remove sulfur dioxide and other pollutants from stack gases of oil- and coal-burning power plants. Zeolite adsorption and catalytic properties can be used to enhance the oxygen content of water and air, because nitrogen is preferentially adsorbed. Zeolites are applied to clean drinking water or to decrease the hardness of industrial water.

The *molecular sieve* property of zeolites is widely applied in the petroleum industry to separate organic molecules during refinement of oil, retaining molecules larger than the width of the channel. For this application, synthetic zeolites, whose channel dimensions are tailored for particular molecular sizes, are used. Natural gas is cleaned in zeolite filters before it is transported in pipelines.

Increasingly, zeolites have become important in animal nutrition. Adding about 10% zeolite to the food of pigs, chickens, and ruminants results in significantly higher feed-conversion values and increased health of the animals. It appears that the main benefit

Table 31.5 Cation exchange capacity of some clay minerals (cf. Table 31.3)

Mineral	Exchange capacity (mg equiv./g)
Kaolinite	0.03–0.15
Halloysite·$2H_2O$	0.05–0.1
Halloysite·$4H_2O$	0.4–0.5
Montmorillonite	0.8–1.5
Illite	0.1–0.4
Vermiculite	1–1.5
Chlorite	0.1–0.4
Sepiolite–paligorskite	0.2–0.3

of this zeolite addition comes from the reduction of gases, particularly ammonia and methane, from the digestive system.

Clay minerals such as kaolinite, halloysite, montmorillonite, illite, and vermiculite also have exchange properties, although to a lesser extent (compare Table 31.3 with Table 31.5), and Ca^{2+}, Mg^{2+}, H^+, K^+, NH_4^+, and Na^+ are the most common exchange cations. There are several reasons for ion exchange phenomena in clay minerals. Broken bonds on the surfaces of clay particles enable adsorption. Due to substitutions of aluminum for silicon in montmorillonites, cations can be adsorbed at basal surfaces and between sheets to compensate charge. Also, hydrogen atoms of OH^- groups on surfaces of particles can be replaced by cations. As is the case in zeolites, the ion exchange capacity of clay minerals depends on temperature, concentration of solutions, and particle size. Montmorillonite and other clay minerals (e.g., in rocks known as bentonite) are widely applied in medicine, agriculture, and environmental hydrology.

31.5 Summary

The geologically most important framework silicates, feldspars and silica minerals, were discussed in Chapter 21. Here the focus is on zeolites and related minerals. They all have open tetrahedral cages with channels to accommodate large cations and water molecules. This is the reason for the unique applications of zeolites for ion exchange, molecular filters and, because of the large surfaces, excellent catalysts.

Important framework silicates to remember

Name	Formula
Feldspars	
Microcline, orthoclase, sanidine (alkali feldspar)	$KAlSi_3O_8$
Albite (plagioclase)	$NaAlSi_3O_8$
Anorthite (plagioclase)	$CaAl_2Si_2O_8$
Zeolites	
Chabazite	$(Ca_{0.5},Na,K)_4Al_4Si_8O_{24} \cdot 12H_2O$
Clinoptilolite	$(Na,K,Ca_{0.5})_6Al_6Si_{30}O_{72} \cdot 20H_2O$
Heulandite	$(Ca_{0.5},Na,K)_5Al_9Si_{27}O_{72} \cdot 24H_2O$
Laumontite	$Ca(Al_2Si_4O_{12}) \cdot 4H_2O$
Natrolite	$Na_2Al_2Si_3O_{10} \cdot 2H_2O$
Others	
Leucite	$KAlSi_2O_6$
Nepheline	$Na_3K(AlSiO_4)_4$
Scapolite	$Na_4(AlSi_3O_8)_3Cl-$
	$Ca_4(Al_2Si_2O_8)_3CO_3$

Test your knowledge

1. What factors determine the charge of the framework in the alumosilicate structures?
2. What is the difference between feldspar and feldspathoids?
3. What are the major features of zeolites (structure and properties)?
4. Name some technological and industrial applications of zeolites.
5. Zeolites occur at low temperature in the *zeolite facies* of metamorphism. Review metamorphism and give some typical minerals of (a) the blueschist facies, (b) the greenschist facies, (c) the amphibolite facies, and (d) the granulite facies.
6. Having arrived at the end of the systematic treatment of minerals, check how much you remember. Name rocks in which the following minerals occur (one rock example for each mineral): calcite, dolomite, olivine, garnet, epidote, biotite, antigorite, montmorillonite, enstatite, glaucophane, spodumene, microcline, scapolite, nepheline, and laumontite.

Further reading

Baerlocher, C., McCusker, L. B. and Olson, D. H. (eds.) (2007). *Atlas of Zeolite Framework Types*, 6th edn. Elsevier, Amsterdam.

Barrer, R. M. (1978). *Zeolites and Clay Minerals as Sorbents and Molecular Sieves.* Academic Press, London.

Bish, D. L. and Ming, D. W. (eds.) (2001). *Natural Zeolites: Occurrence, Properties and Applications.* Reviews in Mineralogy and Geochemistry, vol. 45. Mineralogical Society of America, Washington, DC.

Deer, W. A. and Howie, R. A. (2006). *Rock-forming Minerals, Volume 4B, Framework Silicates: Silica Minerals, Feldspathoids and Zeolites*, 2nd edn. The Geological Society of London.

Gottardi, G. and Galli, E. (1985). *Natural Zeolites.* Springer-Verlag, Berlin.

Tsitsishvili, G. V., Andronikashvili, T. G., Kirov, G. N. and Filizova, L. D. (1992). *Natural Zeolites.* Ellis Horwood, New York.

Van Bekkum, H., Flanigen, E. M. and Jansen, J. C. (eds.) (2001). *Introduction to Zeolite Science and Practice.* Elsevier, Amsterdam.

Vaughan, D. J. and Pattrick, R. A. D. (eds.) (1995). *Mineral Surfaces.* Chapman & Hall, London.

32 | Organic minerals

It is unlikely that you will find organic minerals. They are quite rare and often form only crusts on fractures. Nevertheless, there are about 50 organic minerals with fascinating structures. This chapter will briefly discuss some of these structures which are quite different from those introduced earlier, and also describe some of the minerals and their occurrence. First the chapter goes over some basic concepts of organic chemistry and then introduces some organic minerals that are not only significant in mineralogy but also in chemistry, such as formicaite=Ca-formate, idrialite=picene, kratochvilite=fluorene, and urea.

32.1 Organic compounds

Organic compounds are chemical compounds whose molecules contain carbon. For historic reasons a few types of carbon-containing compounds such as carbides, carbonates, simple oxides of carbon (such as CO and CO_2), cyanides (such as KCN), and all the polymorphs of elemental carbon (such as diamond and graphite) are considered inorganic. The distinction between organic and inorganic carbon compounds, while "useful in organizing the vast subject of chemistry... is somewhat arbitrary" (Seager and Slabaugh, 2013). Many of these compounds are directly related to biological processes, such as proteins, fats, sugars, and DNA, and many are synthesized. There are well over 100 000 organic compounds. A few form naturally with no biological control and are therefore minerals. About 50 organic minerals have been identified, which is less than 1% of all minerals; most of them occur in very small quantities and most mineralogists are not aware of them. Nevertheless, organic minerals are by no means new. For example, mellite was described in 1793, humboldtine in 1821, and idrialite in 1832. Mellite from the type locality in Thueringen, Germany, forms beautiful honey-colored crystals (named from the Latin *mel* for "honey") (Plate 31f). In this chapter we will discuss a few of them and explore their unique structures.

As an introduction we give a very brief summary of organic chemistry. Carbon has six electrons, two in 1*s*, two in 2*s* and one each in $2p_x$ and $2p_y$ orbitals (see Table 2.1). The energy of all second level orbitals is similar. With hydrogen present, carbon transforms to methane to reduce the energy. Carbon moves a 2*s* electron into the empty $2p_z$ orbital to create four orbitals with unpaired electrons, which then combine with hydrogens to form methane CH_4 (Figure 32.1a). Geometrically, orbitals arrange themselves to be as far apart as possible, producing a tetrahedral shape. We have seen similar bonding in silicon, which is just below carbon in the periodic table. Under different conditions, with less hydrogen, ethane C_2H_6 is formed. In this case two carbon atoms share orbitals $H_3C=CH_3$ (Figure 32.1b).

The carbon–carbon bond is very significant in organic chemistry. In benzene C_6H_6 carbons form hexagonal rings with alternating double and single bonds (Figure 32.2a). This diagram is often simplified by leaving out carbon and hydrogen atoms

(a)   (b)

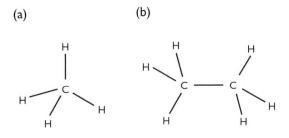

Fig. 32.1 (a) Bonding in methane CH_4 and (b) ethane C_2H_6.

(a)   (b)   (c)

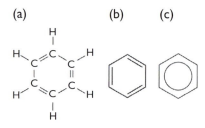

Figure 32.2 (a) Bonding in benzene C_6H_6 with a ring structure that is often abbreviated (b,c).

(a)   (b)

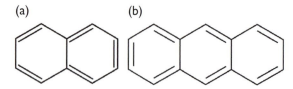

Figure 32.3 Benzene rings combine to form chains as in (a) naphthalene, (b) anthracene.

(a)   (b)

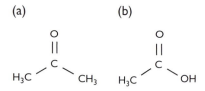

Fig. 32.4 Molecular structures of (a) acetone and (b) acetic acid.

(Figure 32.2b). Contrary to methane and ethane, benzene is a planar structure. In reality there is no alternation of single and double bonds in benzene rings but the three electrons from double bonds have an equal probability to be delocalized anywhere above or below the hexagonal ring (Figure 32.2c).

The benzene rings may combine to form pairs such as in naphthalene $C_{10}H_{12}$ (Figure 32.3a), or form combinations of three rings in anthracene $C_{14}H_{10}$ (Figure 32.3b), and even longer chains.

Another complexity arises if we combine carbon with oxygen and hydrogen. Two simple cases are acetone $(CH_3)_2CO$, where two groups CH_3 (as in ethane) are linked over a third carbon, bonded to oxygen (Figure 32.4a). A variation of this with OH is acetic acid CH_3COOH (Figure 32.4b). This establishes some of the most basic building units of organic chemistry.

32.2 Chemical classes and some structures of organic minerals

When it comes to organic minerals, compositions are usually much more complex and structures are intricate combinations of the basic molecules. Molecular construction of the lattice is typical for organic minerals. Several classes have been proposed:

- Carbohydrides (including carbocyclic compounds and paraffins)
- Salts of organic acids
- Oxyorganic and oxy-nitro-organic compounds

Carbohydrides are composed exclusively of carbon and hydrogen.

In **idrialite** ($C_{22}H_{14}$), with the chemical name picene, benzene rings are bonded to form chains (Figure 32.5a) and the chains are linked by Van der Waals forces (Figure 32.5b). Picene, doped with alkali metals, has superconducting properties.

Kratochvilite ($C_{14}H_{10}$) is a typical carbohydride with molecular stucture. In this case two hexagonal benzene rings are linked over a 5-fold ring to form the fluorene molecule (Figure 32.6a). These units are then, as in idrialite, linked in a chain-like fashion (Figure 32.6b).

For the structures of *salts of organic acids* there is an extensive review (Echigo and Kimata, 2010). They have the closest resemblance to some of the inorganic minerals. Salts of formic acid $H(COOH)$, acetic acid $CH_3(COOH)$, oxalic acid $(HCOOH)_2$, and benzoic acid $C_6H_5(COOH)$ exist.

We start with **formicaite** with the chemical name calcium formate $Ca(HCOO)_2$ (Figure 32.7a). Calcium is surrounded by six oxygens in an octahedral coordination and the oxygens are linked to carbon–hydrogen molecules (Figure 32.7b). These units then form a rather complex framework with tetragonal symmetry.

The mineral **humboldtine** has a formula $Fe^{2+}(C_2O_4)\cdot 2H_2O$. Fe^{2+} is again in octahedral coordination of oxygens. Two of the oxygens are actually H_2O and the other two are bonded to carbon. The carbons link the octahedra to form chains (Figure 32.8a). Parallel chains are linked over hydrogen atoms by hydrogen bonding (Figure 32.8b).

Oxyorganic and oxynitroorganic compounds have eight mineral representatives. **Urea** $CO(NH_2)_2$ consists of molecules with carbon linked to oxygen as well as to two H_2N groups (Figure 32.9a). Oxygen then

(a)

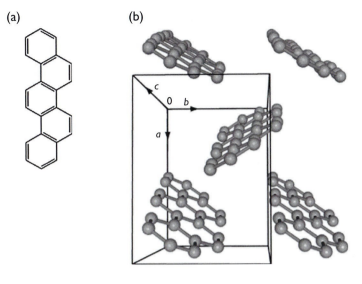

(b)

Fig. 32.5 Structure of idrialite ($C_{22}H_{14}$): (a) chains of benzene rings, (b) bonding with Van der Waals forces to create a three-dimensional structure (after Echigo and Kimata, 2010).

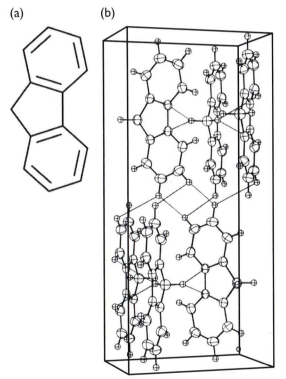

Fig. 32.6 Structure of kratochvilite ($C_{14}H_{10}$): (a) fluorene molecule, (b) linkage of fluorene molecules to form a three-dimensional structure (after Gerkin *et al.*, 1984).

(a)   (b)

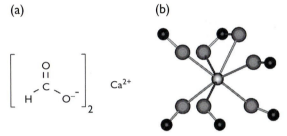

Fig. 32.7 Structure of formicaite (Ca(HCOO)2): (a) calcium formate molecule, (b) three-dimensional coordination.

32.3 Brief descriptions of some organic minerals

Idrialite $C_{22}H_{14}$ was originally discovered in the Idrjia region in Slovenia (1832). Since then it has been found at Skaggs Springs (Sonoma County, California, USA), and in Slovakia, Ukraine, and France. Idrialite probably formed by pyrolysis of organic material near hot springs and in bituminous clays. It often occurs with cinnabar, realgar, and opal. It is greenish-yellow or light brown with concoidal fracture. Hardness is 1.5.

Kratochvilite $C_{13}H_{10}$ or $C_6H_4 \cdot 2CH_2$ formed by burning of pyrite shale and occurs generally as drusy incrustations. It was first found in Kladno (Czech Republic). Color is white, hardness is 1.5. It has violet fluorescence, hence the chemical name fluorene. Polyfluorene polymers are electroluminescent and have been applied as organic light-emitting diodes.

bonds with two hydrogens to produce a chain-like structure (Figure 32.9b). Alternating chains are oriented perpendicular to each other and bonded via the other H–O bonds.

(a)

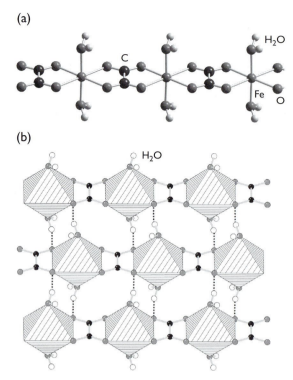

(b)

Fig. 32.8 Structure of humboldtine (Fe^{2+}(C$_2$O$_4$)□2H$_2$O): (a) molecular chain, (b) chains with Fe octahedra linked with hydrogen bonds to a three-dimensional structure (after Echigo and Kimata, 2010).

Fluorene is obtained industrially from coal tar, in a similar way to how kratochvilite forms.

Evenkite (CH$_3$)$_2$(CH$_2$)$_{22}$ is a natural paraffin mineral. It was first discovered in thin platy crystals in geodes lined with chalcedony and associated with a polymetallic vein cutting vesicular tuffs in the Evenki district (Krasnoyarsk district) in Siberia (Kotel'nikova *et al.*, 2007). The waxy hydrocarbon is colorless or slightly yellow with hardness of 1. Since then evenkite has been found in shales in the French Alps, probably produced by migration of hot fluids through the sedimentary rocks.

Formicaite Ca(HCOO)$_2$ is found in veinlets of copper deposits (e.g., Solongo, Russia) and in skarns (Novofrolovskoye, Russia). The chemical name is calcium formate. Color is white, hardness is 1.

Humboldtine Fe^{2+}(C$_2$O$_4$)·2H$_2$O was first found in lignite deposits in Korozluki (Bohemia, Czech Republic, 1821) and named after the German naturalist Alexander von Humboldt. It typically forms on fractures of coal deposits, sometimes associated

(a)

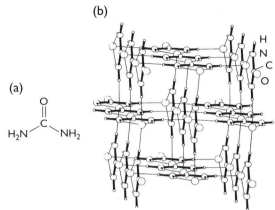

Fig. 32.9 Structure of urea: (a) urea molecule, (b) molecules link to form chains and then chains link with hydrogen bonds to form a three-dimensional structure (after Swaminathan *et al.*, 1984).

Fig. 32.10 SEM image of urea crystallites from Al Buqum, Saudi Arabia (courtesy Saudi Geological Survey, Jeddah).

with gypsum. There are quite a few occurrences in the Czech Republic, including the type locality at Bilin. In Germany it has been found in Hesse, Saxony, and Bavaria in coal mines. Other localities are in Cornwall (UK), Elba (Italy), Montreal (Canada), and Minas Gerais (Brazil). It forms small prismatic crystals, which are transparent, yellow, and of hardness 2.

Mellite Al$_2$[C$_6$(COO)$_6$]·16H$_2$O was the first organic mineral discovered in brown coal seams of lignite in Arten (Thueringen, Germany). It forms fine-grained aggregates, films, and more rarely large prismatic and pyramidal crystals (~1cm) of honey-like color

(Plate 31f). Since then mellite has been found in Tatabanya (Hungary) and in Tuscany (Italy). Hardness is 2, it is fluorescent and has a vitreous luster. The crystal structure is orthorhombic.

Urea $CO(NH_2)_2$ was found first around Kalgoorlie (Western Australia). It derives from bat guano and urine and is stable only under very arid conditions. Color is light brown. Small tetragonal crystals from Al Buqum, Saudi Arabia, are shown in Figure 32.10.

32.4 Summary

The 50 organic minerals are a small segment of over 100 000 organic compounds and remind us that some organic crystals form naturally without biological connections. They are quite rare and their conditions of formation are unusual. The chapter also provides a brief overview of organic bonding and basic structural principles.

Part VI | Applied mineralogy

33 | Metalliferous mineral deposits

So far the book has emphasized the fundamental principles of minerals and crystals, ranging from physical properties to chemical phase equilibria. This last part will highlight some practical applications, starting with ore minerals that are still the major source of metals. At one time mineralogy was applied largely to the field of mineral prospecting and the Latin name "mineralis" is directly linked to mining. Today the range of applications is much broader, with new branches of applied mineralogy, many of them making use of sophisticated instrumentation.

33.1 Applied mineralogy

Practical applications of mineralogical knowledge are the main employment opportunities for mineralogists and are of considerable economic significance. Mineralogical expertise is, of course, indispensable in geology and petrology. Other mineralogists work in gemology (see Chapter 34), in mineral extraction technology, in chemical plants, in the cement industry (see Chapter 35), and in ceramics and the manufacture of refractory materials. Some mineralogists are engaged in the fabrication of synthetic crystals, paints, enamels, and glazes, while others work in museums or become mineral dealers. Even in medicine there is a need for mineralogists, with minerals composing bones and teeth. Environmental mineralogy, dealing with hazardous minerals, has recently become an important new application. An example is the study and remediation of asbestos contamination (see Chapter 36). In this chapter we will start with still the most important aspect, metalliferous ore deposits.

Conventionally, the mineralogical features of ores have been the primary criteria for prospecting, and much has relied on the skills and accumulated wisdom of long-time miners, such as Pliny the Elder, Georg Bauer (Agricola), and Abraham Werner, who all had close links to the mining industry and contributed enormously to mineral science. For example, it has been common knowledge for centuries that malachite veinlets lead to copper ore; that a quartz–mica rock containing columnar aggregates of topaz indicates the presence of tin and tungsten ores; that garnet–diopside skarn relates to complex ores with lead, copper, and iron; and that quartz–microcline intergrowths (so-called "graphic granite", Plate 21c) are a sure sign of gem-bearing pegmatites.

Analytical techniques provide quantitative data about ore deposits and are the basis for a better understanding of ore-forming processes. Knowledge from mineralogy is integrated with that from geochemistry, petrology, and structural geology. Prospecting mineralogy relies on imaginative and quantitative field observations to localize mineral deposits. Samples retrieved from the field are the basis for laboratory investigations to quantify the mineralogical composition of the resources, classifying them by their geological and technological varieties. Later, the distribution of the minerals within the deposit is investigated and the compositionally different mineral resources within a deposit are mapped.

Traditionally, field mineralogical investigations have been carried out as a part of geological research. Detailed maps and sketches document the structure, position, age, morphology, and size of a deposit. Such notes not only identify the minerals that are present, but also their relative abundance. Field studies also address the extent of secondary processes of metamorphism, metasomatic alteration, surface weathering, and erosion. The main mineralogical prospecting techniques rely on visual identification of minerals in the field and require systematic sample collection for later laboratory investigations.

Today traditional mineralogical prospecting in the field with hammer and chisel has been largely replaced by sophisticated instrumentational methods. With remote-sensing techniques, potential deposits can be identified using a variety of spectral signatures and scales. Multispectral satellite imaging measures the electromagnetic spectrum reflected from the Earth's surface in relatively broad bands, with wide coverage but limited resolution. With hyperspectral imaging, often done from airplanes, a much higher resolution is achieved, although with a more limited coverage. These methods can be used to identify the spatial distribution of mineral types, such as those present after clay alteration and secondary iron oxide minerals. Once a potential ore deposit has been located, systematic drill cores are taken and analyzed for metal content to establish the extent and commercial feasibility of the deposit.

Some of the world's largest iron deposits, recently discovered in Brazil with advanced geophysical techniques, would not have been found using only traditional knowledge. Similarly, some of the largest new gold deposits in the western USA are so low grade that only a few decades ago they would have gone unnoticed.

33.2 Economically important minerals

In the next three chapters we will discuss the economically most important minerals. Most commonly we associate mineral resources with metalliferous ores, and indeed the world would be very different without metals. Table 33.1 lists important minerals, their use, and their economic significance. Metals are divided into ferrous, nonferrous (base and light), and precious. *Ferrous metals* are used to produce iron alloys, particularly steel. The name *base metals* was given by ancient alchemists who tried but failed to convert copper, zinc, lead, and tin to gold. The use of *light metals* is relatively new. They are characterized by low densities. *Precious metals* received their name because of their high value. More schematically, Figure 33.1 is a periodic table of elements illustrating which minerals are used mainly for the extraction of each element. The economic significance of various metal ores is summarized in Figure 33.2. Iron is by far the most important metal, followed by copper and aluminum. Iron is the key component of steel, and the world's annual steel production (1600 million tonnes) is similar in value to that of the world's annual oil production (30 billion barrels), i.e., around $US 1000 billion (in 2014).

In this chapter we concentrate on metalliferous ore deposits, reviewing both their occurrence and significance. The site of a mine depends primarily on the geological setting, but many other factors such as ease of transportation, local salaries, and political factors, including environmental laws, are becoming increasingly important. These other factors determine not only whether a mine will be operated at a given location but also the style of mining.

33.3 Geological setting of metal deposits

Ore-forming processes are diverse. Some are closely related to plate tectonic activity (e.g., Hutchinson, 1983; Sawkins, 1990). Along mid-oceanic ridges, seawater hydrothermal systems are formed. Continental hydrothermal activity is observed above subduction zones. Metamorphic waters are released during high-temperature metamorphism, generally along convergent margins. Even surface processes are influenced by plate tectonics. Placer deposits rely on mechanical erosion, which requires uplift. These examples illustrate the importance of the tectonic evolution of the Earth, which depends largely on changes in the rates and mechanisms for generation and dissipation of its internal heat. The igneous evolution in the crust, mountain building, and hydrothermal activity are all driven ultimately by thermal convection in the mantle. However, the chemical evolution of the surface and the hydrosphere is equally important. It depends largely on solar energy but is influenced by biological activity, including humans. Both magmatic as well as surface processes can produce large deposits of metal. Sometimes the two are related, as in the case of black smokers, where metal-bearing hydrothermal solutions interact with seawater. The wide range of geological processes that are active in the formation of mineral deposits is summarized in Table 33.2 (see, e.g., Lindgren, 1933).

Figure 33.3 shows four maps of the Earth, based on Kesler (1994). The first (Figure 33.3a) is a very simplified map of tectonic units displaying old Precambrian crystalline cratons, large overlying sedimentary basins, younger (Paleozoic and Cenozoic) orogenic belts at convergent margins, divergent mid-oceanic ridges, and active volcanoes. We are now going to explore how these tectonic units are related to ore deposits (Figure 33.3b–d).

Table 33.1 Economically most important minerals, their uses, and yearly production values that include processing (in billions of US dollars)

Products	Minerals	Production value
Ferrous metals		
Fe	Magnetite, hematite, goethite	700
Mn	Pyrolusite, manganite	9
Ni	Pentlandite, garnierite	40
Cr	Chromite, magnesiochromite	10
Si	Quartz	5
Mo	Molybdenite	7
Co	Co-pentlandite, linneite, asbolane	2
W	Scheelite, wolframite	2
Light metals		
Al	Bauxite (Al-hydroxides), nepheline	105
Mg	Magnesite, dolomite, also seawater	3
Ti	Ilmenite, rutile	3
Be	Bertrandite, phenakite, beryl	0.1
Nonferrous base metals		
Cu	Chalcopyrite, bornite, chalcocite	126
Zn	Sphalerite	28
Pb	Galena	12
Sn	Cassiterite	6
Hg	Cinnabar	0.1
Precious metals		
Au	Native gold, tellurides	125
Pt	Isoferroplatinum	93
Pd	Merenskyite, kotulskite, froodite, etc.	52
Ag	Acantite, native silver, etc.	20
Gemstones		
Diamond	Diamond, gem	19
Colored gems	Corundum, beryl, tourmaline, topaz, etc.	0.5
Abrasives		
Diamond	Diamond, technical	1
Garnets	Almandine, pyrope, andradite	1
Fertilizers		
P	Apatite	23
K	Sylvite, alunite	22
Chemical industry		
S	Sulfur, pyrite	9
Na, Cl	Halite	3
F	Fluorite	4
B	Kernite, borax	3
Portland cement	Calcite, clay, gypsum	800
Ceramic raw materials	Kaolinite, feldspars, baddeleyite	5
Clay minerals	Bentonite (smectite, nontronite, kaolinite)	8
Zeolites	Clinoptilolite, etc.	3
Glass	Quartz	12
Soda ash ($NaCO_3$)	Trona	0.2

Table 33.1 (*cont.*)

Products	Minerals	Production value
Energy minerals		
U	Uraninite, carnotite, torbernite, etc.	3
Energy resources (for comparison)		
Oil		2800
Gas		700
Coal		350

Source: Based on Petrov *et al.*, 2011.

1	**H** 1																	**He** 2
2	**Li** 3 Amblygonite Li-micas Spodumene	**Be** 4 Beryl Bertrandite Phenakite	**B** 5 Borates Datolite	**C** 6 Graphite	**N** 7	**O** 8	**F** 9 Apatite Cryolite Fluorite											**Ne** 10
3	**Na** 11 Halite	**Mg** 12 Dolomite Magnesite	**Al** 13 Alunite "Bauxites" Nepheline	**Si** 14 Quartz	**P** 15 Apatite	**S** 16 Sulfur	**Cl** 17 Halite											**Ar** 18
4	**K** 19 Carnallite Sylvite	**Ca** 20 Calcite	**Sc** 21	**Ti** 22 Ilmenite Rutile Ti-magnetite	**V** 23 Carnotite	**Cr** 24 Chromite	**Mn** 25 Braunite Psilomelane Pyrolusite Rhodochrosite	**Fe** 26 Goethite Hematite Magnetite Siderite	**Co** 27 Cobaltite Linnaeite Pyrrhotite	**Ni** 28 Garnierite Nickeline Pentlandite								
	Cu 29 Bornite Chalcopyrite Chalcocite Copper,Enargite	**Zn** 30 Smithsonite Sphalerite Willemite	**Ga** 31 "Bauxites" Sphalerite	**Ge** 32 Sphalerite	**As** 33 Arsenopyrite Enargite Orpiment Skyterrudite	**Se** 34 Copper ores Lollingite	**Br** 35	**Kr** 36										
5	**Rb** 37	**Sr** 38 Celestite Strontianite	**Y** 39 Xenotime	**Zr** 40 Zircon	**Nb** 41 Columbite Fergusonite Pyrochlore	**Mo** 42 Molybdendite	**Tc** 43	**Ru** 44 Laurite	**Rh** 45	**Pd** 46								
	Ag 47 Argenite Galena Silver Tennantite-tetrahedrite	**Cd** 48 Greenockite Sphalerite	**In** 49 Cassiterite Chalcopyrite Sphalerite Galena	**Sn** 50 Cassiterite Stannine	**Sb** 51 Stibnite Tetrahedrite	**Te** 52 Calaverite Copper sulfides	**I** 53	**Xe** 54										
6	**Cs** 55 Pollucite	**Ba** 56 Barite Witherite	**REE** 57-71 Apatite Bastnasite Monazite	**Hf** 72 Zircon	**Ta** 73 Loparite Microlite Tantalite	**W** 74 Scheelite Wolframite	**Re** 75 Molybdenite	**Os** 76 Iridosmine Osmiridium	**Ir** 77 Iridosmine Osmiridium Laurite	**Pt** 78 Platinum Sperrylite								
	Au 79 Calaverite Gold	**Hg** 80 Cinnabar Tennanite-tetrahedrite	**Tl** 81 Sphalerite Sulfide ores	**Pb** 82 Anglesite Cerussite Galena Sulfosalts	**Bi** 83 Bismuthinite Bismuth	**Po** 84	**At** 85	**Rn** 86										
7	**Fr** 87	**Ra** 88	**Ac** 89	**Th** 90 Monazite Thorite	**Pa** 91	**U** 92 Carnotite Coffinite Uraninite	**Np** 93											

Fig. 33.1 Periodic table of elements showing principal minerals used for their extraction (in alphabetical order).

Table 33.2 Important geological environments in which ore deposits form

Process	Example
Subsurface processes	
Magmatic precipitation	Immiscible magmas: Ni, Pt, Fe (Sudbury, Canada)
	Magma fractionation: Cr, V, Pt (Bushveld, South Africa)
Precipitation from aqueous solution	Hydrothermal brines: Pb, Zn (Mississippi Valley, USA)
	Magmatic water in porphyry copper deposits and skarn: Cu, Mo, W (Bingham, Utah, USA)
	Metamorphic water: Au, Cu (South Africa)
Surface processes	
Chemical sedimentation	Banded iron formations: Fe, Mn (Lake Superior, North America)
	Seawater in volcanogenic massive sulfides: Cu (Cyprus)
	Evaporites: Na, K, Cl, B (Saskatchewan, Canada)
Chemical weathering	Laterites: Al, Ni, Au, clays (North Australia)
Physical sedimentation	River placers: Au, Pt, diamond, gems (Witwatersrand, South Africa)
	Beach placers: Ti (Florida, USA)

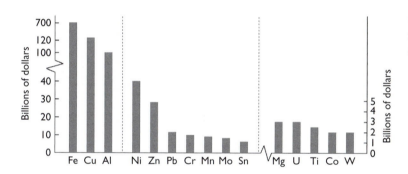

Fig. 33.2 World production of major metals (in US dollars). See also Table 33.1.

Convergent margins

Figure 33.3b is a world map identifying the distribution of deposits of copper and molybdenum. We observe an almost perfect match with young orogenic belts and volcanic activity at continental margins (Figure 33.3a). The location of important mercury deposits (Figure 33.3c) is related almost exclusively to the Paleozoic and Cenozoic orogenic belts. During subduction of oceanic crust under the continents, large volumes of sedimentary material, deposited along the continental shelf, also became subducted. At greater depths, and particularly with increasing temperature, mineral reactions start to occur, beginning with clay minerals transforming to metamorphic silicates and releasing water. Even with the low geothermal gradients typical of subduction zones, beyond depths greater than 50 km (1.5 GPa) temperatures are

reached where hydrous metamorphic minerals become unstable and dehydrate (Figure 33.4). The released water favors melting of the subducted sediments, and buoyant magmas rise to produce volcanism and, at deeper levels, batholithic intrusions. This igneous activity becomes the motor for driving hydrothermal processes. As is obvious from the isotherm in cross-section in Figure 33.4, temperatures to which source materials are subjected increase with increasing distance from the continental margin. This trend is the reason for the regular zonation of hydrothermal deposits in the western USA (Figure 33.5), with the lowest temperatures occurring in the Coast Ranges, where mercury deposits formed at temperatures below 200 °C at a shallow level. This mercury belt is followed by the gold belt in the Sierra Nevada foothills, including the Mother Lode deposit. Further west, in Nevada,

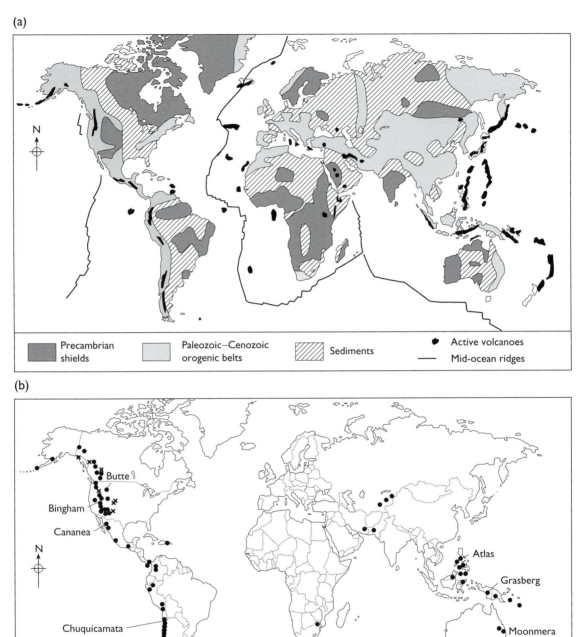

Fig. 33.3 Map of the world. (a) Major tectonic units with Precambrian cratons, sedimentary basins, Paleozoic–Cenozoic orogenic belts, and active volcanoes. Divergent oceanic margins are indicated. (b) Occurrence of porphyry copper and molybdenum deposits.

(c)

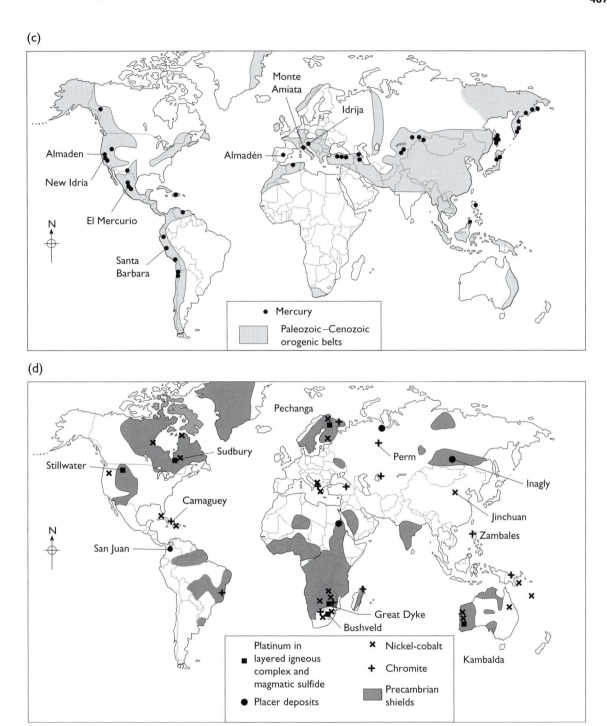

Fig. 33.3 (c) Mercury deposits. Cenozoic orogenic belts are indicated by shading. (d) Deposits of platinum, chromium, and nickel–cobalt. Precambrian shields are indicated by shading.

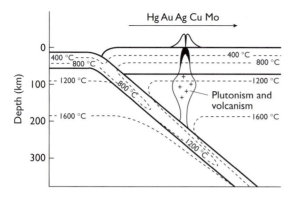

Fig. 33.4 Subduction of seafloor beneath a continent. Approximate isotherms are indicated. A batholith with a porphyritic intrusion is also shown. The vertical scale is exaggerated. The arrow illustrates the sequence of metal deposits.

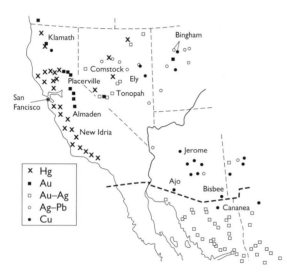

Fig. 33.5 Metal deposits (mercury, gold, silver, lead, and copper) in the western USA and northwestern Mexico. Some of the renowned mining districts are labeled (after Schneiderhöhn, 1941).

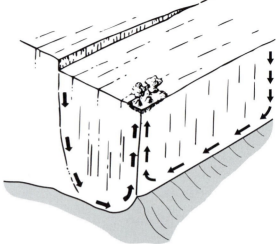

Fig. 33.6 Circulating fluids of seawater (arrows) at a mid-oceanic spreading ridge, giving rise to black smoker massive sulfide deposits (after Cann *et al.*, 1985).

dissolution of specific elements, which then precipitate at higher levels and lower temperature.

Divergent margins

The largest extrusions of magma occur along mid-oceanic ridges. In the highly fractured ocean floor, sea-water penetrates along fractures and faults, becomes heated, and circulates towards the upwelling ridge (Figure 33.6). At temperatures in excess of 350 °C, the hot waters react with basalt and precipitate dissolved calcium as epidote, titanite, and calcite, becoming increasingly acidic. The acidic waters dissolve trace metals such as copper, zinc, cobalt, and manganese in basalt. The metals are then precipitated as sulfides and oxides when the hydrothermal waters come into contact with cold seawater (see also Figure 26.13, and corresponding discussion in Chapter 26). These deposits are called volcanogenic massive sulfides and occasionally become attached to continents – as in Cyprus and Kuroko (Japan), where pillow lavas indicate their mid-oceanic origin.

Precambrian shields

While the previously discussed deposits are associated with relatively recent tectonic activity and make up the bulk of presently mined copper, lead, zinc, and

silver dominates, followed by copper, and finally copper–molybdenum, with hydrothermal solutions coming from the greatest depth. Clearly this regular pattern has many exceptions because mineralization depends not only on the temperature of the rising plutons and hydrothermal solutions but also on their composition and local conditions. Nevertheless, the general pattern is striking and indicates that the temperature of the hydrothermal waters influences the

mercury deposits, primary deposits of chromium, nickel, and platinum are old and occur mainly within Precambrian shields (Figure 33.3d). The rocks on these shields are primarily orthogneisses, derived from granodiorites and tonalites, and mafic to ultramafic volcanic rocks, known as *greenstone belts*. The Early Precambrian greenstone belts are host to some of the richest ores of chromium, nickel, cobalt, zinc, and gold of any tectonic unit. The ores are contained in ultramafic volcanic rocks, called komatiites. Komatiites have a characteristic microstructure of intergrown crisscrossing sheaves of bladed crystals, the so-called spinifex texture, named after Australian bunchgrasses. The ultramafic magmas formed at very high temperatures (>1500 °C), which are achieved only at great depth. Nickel sulfides form immiscible melts in komatiites, as in the Kambalda district of Western Australia. In greenstone belts, from South Africa, Canada, and Australia, gold-bearing veins occur uniformly, particularly at the contact zone between basaltic komatiites and surrounding granitic plutons. It is from such Early Precambrian greenstone gold mineralization that the giant placer deposits at Witwatersrand (South Africa) were derived (2.8–2.5 billion years ago).

Layered intrusions with stratiform layers of chromite occur in some of the oldest terrestrial rocks in West Greenland (3.8 billion years), the Stillwater Complex in Montana (2.9 billion years), and the mid-Tertiary Skaergaard complex of East Greenland (55 million years). However, the largest deposits of this type are the Great Dyke of Zimbabwe (2.5 billion years), and the Bushveld complex of South Africa (2.0 billion years).

The Bushveld complex near Pretoria, South Africa, is such a huge deposit that it deserves a brief discussion. Over an area extending more than 200 km, there are amazingly homogeneous layers of norite (Figure 33.7a) with a distinct stratigraphy. A lower zone contains alternating layers of almost pure chromite and mafic silicate rocks. The unique "critical zone" of the Merensky Reef also contains chromite, but in addition is very rich in native platinum. The main zone is not of great economic interest, but an upper zone contains chromium-bearing magnetite, rich in vanadium, and is mined mainly to extract that element. Only a very small portion of the deposit is presently mined. It is thought that the chromite layers do not represent immiscible melt, but rather that chromite crystals settled at the bottom of the magma

chamber and episodic pulses of new magma intrusion occurred (Figure 33.7b; see also Figure 26.7).

Sedimentary basins

Without a doubt, the sedimentary iron ores known as *banded iron formations* are the world's most important sedimentary metal deposits, providing about 60% of the world's iron ores. Algoma-type ores (named after the Algoma district in Ontario, Canada) formed when submarine hot springs, associated with volcanic activity, released iron-rich hydrothermal solutions into sedimentary basins (Figure 33.8). Deep ocean sediments display iron-rich layers, alternating with silica-rich layers on a scale of millimeters to centimeters. The younger, Superior-type banded iron formations (named after Lake Superior) are richer in iron, more uniformly bedded, and contain the bulk of the world's iron deposits. Examples of these formations are found in Michigan and Minnesota (Lake Superior, USA), Hammersley (Western Australia), Minas Gerais (Brazil), and the Kursk district (Russia). In Superior-type iron ores, there is no association with volcanism, and the occurrence with limestone suggests deposition in shallow water. Accumulated ferrous iron in the deep ocean is the most reasonable source for the iron as well as the silica. These accumulations were transported to shallow-water coastal environments by upwelling currents. The striking proliferation of Superior-type ores about 2600–1800 million years ago is one of the spectacular geochemical anomalies of geological history. In Box 33.1 we explore some aspects of the evolution of ore deposits with geological history.

33.4 Metal production around the world

The distribution of metals is of considerable political significance. Even if no wars have been fought recently over mineral deposits, minerals have greatly influenced the policies of colonial powers. Gold spurred the Spanish conquest of Mexico and the Inca empire, valuable minerals in southern Africa (gold and diamonds in South Africa, chromite in Zimbabwe) led to British and Dutch control, and copper and cobalt deposits in DR Congo were largely responsible for Belgium's colonization of that region. South Africa, now independent, as are the other countries mentioned above, plays a crucial role in the mineral industry. It is a major supplier of diamonds, gold,

(a)

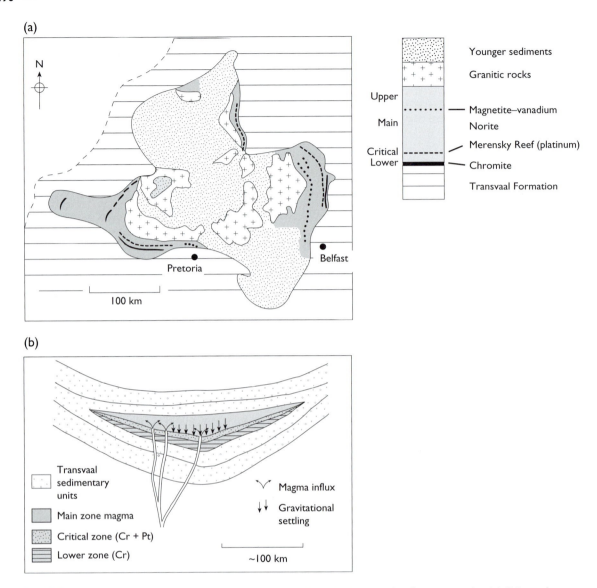

Fig. 33.7 Bushveld layered intrusion with mafic rocks intruding the Transvaal sedimentary units. (a) Schematic map showing main magnetite–vanadium ores in the upper zone, platinum mineralization (critical zone of Merensky Reef), and chromite in the lower zone. (b) Schematic cross-section illustrating periodic intrusion of magma and gravitational settling of chromite (after Carr *et al.*, 1994).

platinum, chromium, manganese, and other metallic raw materials.

In some mineral deposits only a single component is extracted (e.g., iron, chromium, manganese, aluminum, asbestos, graphite, sulfur, and diamond). Other deposits produce several valuable components. For example, nickel, copper, platinum, and cobalt are extracted simultaneously from copper–nickel sulfide deposits; muscovite, feldspar, quartz, and gemstones are obtained from pegmatites. The mines in the alkaline rock and carbonatite complexes of Kovdor (Russia) produce magnetite, apatite, and baddeleyite. The form in which a component occurs in a particular mineral may also be different: in some minerals single crystals with specific important properties are used (muscovite, fluorite, calcite, etc.). A mineral may be

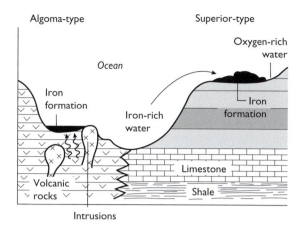

Fig. 33.8 Sedimentary iron deposits with (left side) Algoma-type deep ocean sedimentation that is associated with volcanism and hot springs and (right side) the shallow-water Superior-type banded iron formation (courtesy S. E. Kesler, see also Kesler, 1994).

mined because it contains major elements in its formula (copper in chalcopyrite, iron in magnetite, potassium in sylvite), or it may have some valuable minor isomorphic substitutions (cadmium in sphalerite, rhenium in molybdenite, hafnium in zircon).

In order to be profitable, a mineral must contain some component at a certain concentration. This concentration varies for different types of resource, from minor fractions of 1% to well over 10%. For example, in the USA (in the early 1980s) the "standard" (profitable) content of pure metal in ore had to exceed: 0.0005% for platinum; 0.001% for gold; 0.002% for cobalt; 0.05% for silver; 0.4% for uranium and lead; 0.7% for copper; 1% for titanium, tin, and tungsten; 1.5% for nickel and molybdenum; 4% for lead and zinc; 30% for aluminum, chromium, and iron; and 35% for manganese.

The standard content changes with time, since it depends on the current market value of a metal. It also varies with the type of deposit, with factors including the overall mineral composition of a processed ore, the structure and extent of an ore deposit, the presence or absence of undesirable impurities, and the general accessibility of a deposit. It also depends on local labor prices. For example, similar copper deposits may be highly profitable in DR Congo or China and worthless in France or the USA. As the mining technologies advance, lower grade deposits can be processed. Since the beginning of the nineteenth

century, the copper standard content has fallen from 10% to about 0.7%. Within the past 30 years very low-grade gold ore of a standard content of only 0.00003% (or about 0.3 g/tonne) has been mined profitably from disseminated epithermal deposits in Nevada, even with a depressed price of gold.

As we noted in the previous section, for geological and tectonic reasons, minerals are unevenly distributed around the world. These geological factors combine with political and economic factors to determine the geographical distribution of ore production, which is summarized in Table 33.3

Steel and ferrous metals

Iron is the main metal on which our civilization relies. The beginning of the Iron Age marked a significant advance over previous civilizations that relied on copper and bronze. Today iron is used in the form of steel, an alloy of iron with carbon and various metallic elements. Commercial carbon steels contain about 1% carbon and 0.5% manganese. High-strength and stainless steels are alloyed with up to 35% chromium, manganese, vanadium, and other elements. Overall, cobalt, chromium, molybdenum, tungsten, and vanadium add high-temperature hardness, and nickel provides low-temperature toughness; chromium, cobalt, manganese, and silicon reduce oxidation and are components of stainless steels. This list illustrates that the steel industry relies on several mineral resources.

The principal iron minerals are hematite and goethite, composed of ferric iron (Fe^{3+}), and forming in oxidizing environments. Secondary in importance are magnetite and siderite, which contain ferrous iron (Fe^{2+}) and form in reducing, oxygen-poor environments.

The extensive chemical sediments of the banded iron formations are mined as *taconite* in North America around the Great Lakes, *itabirite* in Brazil (Minas Gerais), *jaspilite* in Australia (Hamersley), and *banded ironstone* in South Africa (Thabazimbi), in India (Goa), and in China (Xishimen) (Figure 33.10). In magmatic iron deposits, magnetite forms in immiscible iron oxide melts, as in rhyolites in Kiruna (Sweden), Kovdor (Russia), Missouri (USA), and Cerro Mercado (Mexico).

Manganese, the most important secondary ingredient of steel, is mined mainly from chemical sediments in northern Australia, in China, and in South Africa,

Box 33.1 Additional information: Ore deposits and geological history

As we have seen, many ore deposits correlate with major tectonic settings. Charles Meyer (1988) explored ore deposits from the perspective of history and found surprising results. Ore formation is by no means uniform, as is evident from Figure 33.9. The banded iron formations date largely to between 2800 and 1800 million years ago, with practically no younger occurrences. Porphyry coppers and tungsten skarns, on the other hand, are mostly young, between 300 million years ago and Recent. Gold placers are either very old or very young. Chromite in layered intrusions occurs in spikes, at distant intervals. The patterns of ores over geological time seem to defy uniformitarianism. While much of the explanation for this sporadic evolution remains speculative, there is some rationale. For example, the scarcity of porphyry coppers in older rocks may be due partly to erosion of these rather shallow deposits. However, geochemical and tectonic conditions have changed over time.

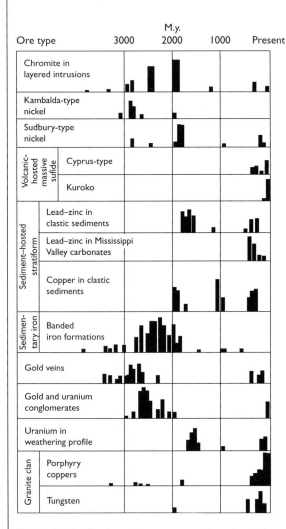

Fig. 33.9 Distribution of important ore deposits with geological time. The length of the bar is an estimate of the total quantity formed (in tonnage). M.y., million years ago.

The proportion of ore deposits that depend on volcanic activity is greater in the Early Precambrian than in any later segment of geological time. Komatiite flows are uniquely Early Precambrian. The high-temperature ultramafic komatiite extrusions require partial melts with a high percentage of melt. Such melts could have occurred by means of mantle penetration on major rifts. The deposits are very large, suggesting stable conditions over long periods of time, with a rather simple overall structure, which is also supported by the widespread and homogeneous distribution of gold in various greenstone belts.

At the end of the Early Precambrian there was a transition to gradual stabilization of continents, with development of large continental basins. These basins host the world's largest gold deposits in the quartz pebble conglomerate of Witwatersrand (South Africa), as well as the banded iron formations, the world's largest iron deposits. Both of these very different sediments were deposited in different basins over a period of more than 500 million years. The Late Precambrian banded iron formation coincides with the development of an oxygen-rich atmosphere and early biological life, with algal organisms producing oxygen by means of photosynthesis.

Around 1800 million years ago the stable conditions ended, and there is geological evidence for an increase in local tectonic and volcanic activity at this time. Unlike the Early Precambrian volcanism, ultramafic lavas were absent in this period of volcanism, with andesites and rhyolites dominating instead. Some of the volcanogenic deposits resemble porphyry coppers (Haib, Namibia), whereas others are more like mid-oceanic massive sulfides (as in Jerome, Arizona, USA). A new and

Box 33.1 (cont.)

very different type of deposit is seen in the clastic sediment-hosted stratiform lead–zinc ores, as in Broken Hill and Mount Isa (Australia), Gamsberg (South Africa), and Grenville Province (Ontario, Canada). The reason for the surge in lead deposits, starting around 1750 million years ago, is not clear. It correlates with increasing oxygen levels in the atmosphere and the evolution of oceans, but why should lead, among base metals, be affected to such a large degree? There is evidence for tectonic evolution: many rocks are heavily deformed, with numerous intrusions of anorthosites, as well as alkaline rocks and carbonatites (as in Kiruna, Sweden, and Kovdor, Russia).

Between 1000 and 500 million years ago, there was a period of minimum ore-forming activity, as is evident from most of the ore charts in Figure 33.9. After that period of geological time, major types of Precambrian deposits were rejuvenated in the Phanerozoic, with the notable exceptions of komatiitic nickel, anorthositic ilmenite, and banded iron. In the Paleozoic (300 million years ago) there was another spike of sedimentary lead–zinc deposits in clastic sediments, but the increase in lead–zinc deposits came mainly in carbonates at the edges of old basins (Mississippi Valley, USA). As we have seen earlier, Phanerozoic orogenic belts and igneous-related ores of the porphyry copper type closely coincide (see Figure 33.3a,b). These younger deposits show a much greater diversity and heterogeneity than do older ones, presumably because of the much more variable crustal composition and the local reworking of old ore deposits.

We can see that every geological period has some unique characteristics, and those characteristics are reflected in the types of metal ore deposit from that period. The complicated evolution of mineral formation with time becomes plausible in the context of general geological, tectonic, magmatic, and geochemical development of the Earth's crust. Since many deposits form at surface conditions or involve material that was at the surface, the importance of the geochemical evolution of the hydrosphere and atmosphere, including biological activity such as photosynthesis and sulfide oxidation, cannot be overemphasized. Obviously the geochemical laws controlling ore formation have remained the same, but the geological conditions of our planet have changed progressively, beginning with a fairly simple structure dominated by volcanic rocks (e.g., komatiites), followed by development of oceans and an atmosphere with organisms that produced oxygen through means such as photosynthesis. A fairly homogeneous supercontinent subsequently evolved, but this supercontinent, Pangea, later broke up into individual plates, with metamorphism, subduction, and mountain building occurring on and near plate boundaries, especially along convergent margins. The present period is very complex and diversified, but young ore deposits, such as porphyry coppers, show a pattern that is clearly related to plate tectonics.

and as a surface alteration in India and Brazil. Manganese nodules on the seafloor represent vast reserves but are not mined at present (Figure 33.10).

Nickel and *cobalt* deposits are mainly found in laterites of ultramafic rocks, representing mantle material that is juxtaposed with the crust. Nickel in olivine dissolves, percolates, and then precipitates at the base of the weathered zone as garnierite, a complex nickel silicate mineral. Nickel-rich laterite can contain up to 3% nickel and 0.2% cobalt. Large deposits are in New Caledonia, Cuba, the Dominican Republic, and Indonesia. Nickel is also mined as pentlandite and pyrrhotite in magmatic deposits of older parts of Precambrian cratons at Sudbury (Canada), in Western Australia, and in South Africa. An important

source of cobalt is linneite ($Co^{2+}Co_2^{3+}S_4$), which occurs in secondary copper deposits in DR Congo.

Chromium and *vanadium* are crucial components of high-performance steels, but their occurrence is very limited, restricted primarily to stratiform deposits in layered mafic rocks such as the Bushveld complex (South Africa) and the Great Dyke (Zimbabwe) (Figure 33.3d). There are some lenticular deposits in mafic and ultramafic igneous rocks at the base of the oceanic crust at convergent plate margins in Kazakhstan, Cuba, Turkey, and Albania. Southern Africa, Kazakhstan, and India account for over 70% of the world's chromium production. V^{3+} substitutes for Fe^{3+} in magnetite and is particularly rich in some horizons of the Bushveld.

Table 33.3 World and country production of some important extracted (but not refined) materials (t, tonnes)

Products	World production	Mine production for countries
Ferrous metals (10^6 t)		
Fe	2960	China (1400), Australia (550), Brazil (400), India (150), Russia (130)
Mn	17.5	South Africa (4.0), China (3.1), Australia (3.0), Gabon (2.0)
Ni	2.5	Indonesia (0.45), Philippines (0.44), Russia (0.25), Australia (0.24)
Cr	25.6	South Africa (11.0), Kazakhstan (4.1), India (3.0)
Mo	0.3	China (0.12), USA (0.06), Chile (0.04), Peru (0.02)
Co	0.12	DR Congo (0.06), Canada (0.01), Russia (0.007)
W	0.07	China (0.55), Canada (0.005), Russia (0.003)
Light metals (10^6 t)		
Bauxite	260	Australia (78), China (45), Brazil (35), Indonesia (30)
Al	47	China (21), Russia (4), Canada (3), USA (2)
Mg	6	China (4.0), Russia (0.5), Turkey (0.3)
Ti	7	South Africa (1.1), Australia (1.0), China (0.9), Canada (0.8)
Be	0.00025	USA, China, Mozambique, Kazakhstan
Nonferrous base metals (10^6 t)		
Cu	18	Chile (5.8), China (1.7), Peru (1.3), USA (1.2)
Zn	13.5	China (5.0), Australia (1.4), Peru (1.3), India (0.8), USA (0.7)
Pb	5.5	China (3.0), Australia (0.7), Peru (0.25), Mexico (0.25)
Sn	0.25	China (0.1), Indonesia (0.04), Peru (0.03), Bolivia (0.02)
Hg	0.002	China (0.0013), Kyrgyzstan (0.0003)
Precious metals (t)		
Au	2700	China (440), Australia (260), Russia (240), USA (230)
Pt	1900	South Africa (140), Russia (25), Zimbabwe (10)
Pd	2200	Russia (85), South Africa (80), Canada (15)
Ag	26 000	Mexico (6000), China (4000), Peru (3000), Russia (1800)
Gemstones (10^6 carats)		
Diamond	75	Russia (22), Botswana (16), Canada (12), Angola (8)
Colored gems	no data	Tanzania, Kenya, Madagascar, Brazil, India
Abrasives		
Diamond	90	Botswana (22), DR Congo (17), Russia (15), Australia (11)
Fertilizers (10^6 t)		
P	225	China (95), USA (38), Morocco (27), Russia (12)
K	37	Canada (11), Russia (5.5), Belarus (5), China (4)
Chemical industry (10^6 t)		
S	70	China (10), USA (9), Russia (7.5), Canada (6)
Na, Cl	265	China (72), USA (40), India (18), Germany (12)
F	7	China (4.4), Mexico (1.2), Mongolia (0.4)
B	5	Turkey (3), USA (0.7), Chile (0.4), Russia (0.3)
Portland cement	4000	China (2400), India (260), USA (80), Iran (75)

Source: Based on Petrov *et al.*, 2011.

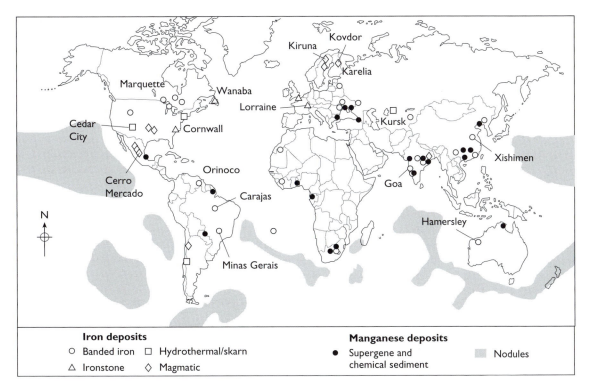

Fig. 33.10 Iron and manganese deposits.

Molybdenum occurs in molybdenite in porphyry copper. About 5–10% of mined molybdenum is used in the manufacturing of corrosion-resistant steel tools, 30% of its consumption is in chemicals such as orange pigments, and the remainder is used as lubricants and as a catalyst in oil refining.

Tungsten is added to steel for high-temperature and high-strength applications. Tungsten carbide is one of the hardest materials and has only recently been replaced for some applications by new materials such as synthetic diamonds, boron nitride, and titanium carbide (or coatings of these materials) to improve the durability of tools. Tungsten wires are used as filaments in light bulbs. Tungsten occurs in the minerals scheelite and wolframite, which form in skarns. Worldwide, China dominates the tungsten market and the Shizhu-yan mine in Hunan Province is the largest producer.

Light metals

Light metals are increasingly replacing both steel for structural applications and copper for transmission of electricity. While strong they have the advantage of a much lower density than iron and copper (1.7 for

magnesium, 2.7 for aluminum, and 4.5 for titanium, as compared to 7.9 for iron and 9.0 for copper). A disadvantage of aluminum, however, is its lower melting point (650 °C versus 1535 °C for iron). By means of alloying, the strength of aluminum can be improved at least at moderate temperatures. For high-temperature applications, titanium has received much interest because of its relatively low density and its much higher melting point (1678 °C) than aluminum. Over 30% of modern aircraft are now composed of titanium.

Aluminum is extracted from bauxite, an aluminum-rich laterite composed of a mixture of gibbsite, boeh-mite, and diaspore. It forms by intense weathering of felsic igneous rocks and shales in a climatic environ-ment of high rainfall and high temperature and extended geological stability because the weathering processes are very slow. The distribution of present-day laterites is superposed on the world map in Figure 33.11, and indeed the world's largest bauxite mines (northern Australia, China, Guinea, Jamaica, and Brazil) are all observed to be in lateritic regions. Aluminum is also extracted from nepheline in nepheline syenites (Khibini, Kola Peninsula, Russia) and alunite in altered volcanic rocks (Zardalek, Azerbaijan).

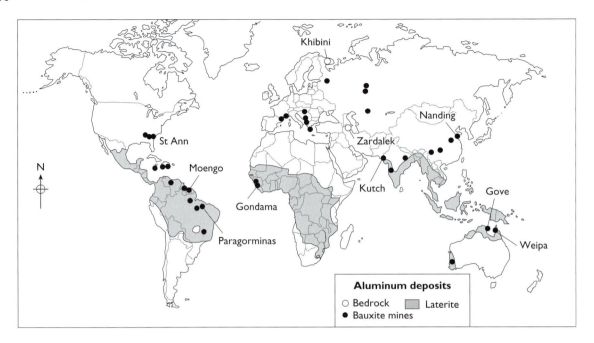

Fig. 33.11 Aluminum deposits. Shading indicates the extent of laterites.

Aluminum production requires high energy: first heat is required to produce alumina (Al_2O_3) from aluminum hydroxides, and then alumina reduction to aluminum requires considerable amounts of electricity. Thus, similar to iron, mining of the raw material does not occur at the same locations as the metal production. Aluminum is processed in countries with large hydroelectric power plants, such as Canada, New Zealand, and Brazil. Because of the high production cost, a significant amount of aluminum metal is recycled.

Titanium is mined as ilmenite and rutile. Titanium is used as a metal (5%), but by far its most important application (95%) is as a white paint pigment. It occurs in mafic intrusive rocks, particularly at the margins of anorthosites. Mining from these primary rocks is expensive (mines are in Australia, China, Norway, and Canada). Both ilmenite and rutile are hard refractory minerals and therefore accumulate as placers. About half of the world's production is from such deposits, which are found in uplifted beach dunes and beach sands in Florida (USA), on the Natal coast of South Africa, and in Queensland (Australia). These deposits are rapidly becoming exhausted.

Beryllium is mainly mined from betrandite ($Be_4(Si_2O_7)(OH)_2$), pheanakite ($Be_2(SiO_4)$), and beryl in pegmatites. The Spor Mountain mine in Utah (USA) is presently the largest source and occurs in altered rhyolites. Beryllium is very light (density 1.85 g/cm^3) with many applications.

Base metals

Copper appears to have been the first metal used by humans in the preparation of tools, the most likely reason being that copper occurs in metallic form at the Earth's surface. A copper axe was found along with the frozen remains of an Ice Age man (Oetzi) discovered in 1991 in northern Italy and dated to 3200 BC, and we know that copper was being smelted in the Middle East by 4000 BC. Where tin was available, as in Cornwall, UK, copper was alloyed with the tin to produce the harder alloy bronze, which initiated the Bronze Age. During the Roman Empire, zinc was alloyed with copper to produce brass. Today copper is used mainly because of its high electrical and thermal conductivity.

Copper occurs in many minerals, but chalcopyrite ($CuFeS_2$) is by far the most common and is found in hydrothermal deposits, the so-called *porphyry copper deposits* (see Figure 33.3b). As was noted in Chapter 26, these hydrothermal deposits formed around felsic intrusions that fed volcanoes, although the volcanoes are now largely eroded. Hydrothermal minerals crystallized in veins when magmatic waters were expelled along fractures. The largest deposits are found on convergent continental margins along the Pacific Rim in western Canada, the western USA, Mexico, Peru, Chile, and New Guinea/Philippines. Other hydrothermal copper deposits (the volcanogenic massive sulfides) formed as sediments where hydrothermal systems vented onto the seafloor along mid-oceanic ridges. Such deposits crop out on the island of Cyprus (named after the Greek word for copper), in Rio Tinto in Spain, and in Kuroko in Japan. Copper sulfide minerals are not stable at surface conditions and dissolve during weathering, producing secondary copper deposits.

Lead and *zinc* often occur together, in galena and sphalerite, respectively. Like chalcopyrite, galena and sphalerite occur almost exclusively in hydrothermal deposits, most of which are found along the western margins of the North and South American continents in Canada, Mexico, and Peru, and in northern Australia (Mount Isa) and China (e.g., Caijiaying mine and Hongtoushan mine in northern China). Currently the Red Dog mine in northwestern Alaska is the world's largest zinc producer, with a large deposit hosted in carboniferous black shales.

Tin is mainly retrieved from cassiterite (SnO_2), with ore deposits in only a few parts of the world such as China, Malaysia, Indonesia, and Thailand. We discussed cassiterite formation, which occurs as high-temperature hydrothermal alteration in the roof of granitic intrusions, producing rocks known as greisen (see Chapter 18). Cassiterite is highly resistant to erosion and accumulates in placer deposits. Most of the tin mining in Malaysia and Thailand relies on placers. Tin alloyed with copper or antimony produces pewter, which has been used since the beginning of the Bronze Age in the Near East.

Mercury and red cinnabar were used by ancient Egyptians, Greeks, and Romans in cosmetic applications and as medicine. Indeed, the mines at Monte Amiata (Italy) and Idrija (Slovenia) can be traced back to Roman times. In the sixteenth century the amalgamation method was invented to recover gold by dissolving gold from ore in mercury, and then evaporating the mercury. This discovery resulted in a large demand for mercury and the Almaden mine in Spain became a major supplier. In water, mercury can form the highly toxic compound methyl mercury. Because it is a potential health hazard, the use of mercury has been sharply reduced but it is still an important component of ammunition.

Cinnabar forms at the margins of larger hydrothermal systems, with a high solubility in alkaline low-temperature hydrothermal fluids. Most mercury deposits formed at temperatures lower than 200 °C. The distribution coincides with shallow subduction zones (see Figures 33.3c and 33.5).

Arsenic and mercury are found in similar environments. Arsenic occurs in arsenopyrite, enargite, and tennantite, but is produced mainly from realgar and orpiment, in low-temperature hydrothermal ore deposits. It is used as a wood preservative, herbicide, and insecticide because of its toxic properties.

Precious metals

Gold has been valued for its rich color and unique luster since ancient times and has long been used to make jewelry. Because gold is rare, it is very valuable. Throughout history, only about 110 000 tonnes of gold have been mined (corresponding to a cube that is ~21 m on each side), and most amazingly, about 80% of this total is preserved, almost half of it owned by central banks. Of today's production of approximately 2800 tonnes per year, more than 80% is used for jewelry, with about another 5% consumed by the electronics industry, where it is valued because of its extremely high conductivity and resistance to corrosion.

Gold occurs mainly in its native state or is alloyed with silver in electrum or, sometimes, with tellurium in tellurides. Primary deposits are hydrothermal (Figure 33.12). Epithermal gold deposits (<250 °C) are associated with felsic volcanism (e.g., the historic mines of Comstock in Nevada and Cripple Creek in Colorado, USA). Silver is often the dominant economic element. Mesothermal deposits (>250 °C) consist of gold-bearing quartz veins at deeper crustal levels. The veins are usually surrounded by calcite and other carbonates, suggesting that the hydrothermal solutions were rich in CO_2. Good examples in the USA are the Mother Lode district in California and the Homestake deposit in South Dakota. The Witwatersrand district in South Africa is a late Early Precambrian paleoplacer deposit which has long dominated global production but has now been surpassed by China, Australia, Russia, and the USA.

The principal use of *silver* has traditionally been in jewelry and tableware, but increasingly industrial applications are becoming important. Like gold, it is used in the electronics industries. Silver occurs in epithermal veins, forming as hot magmatic waters

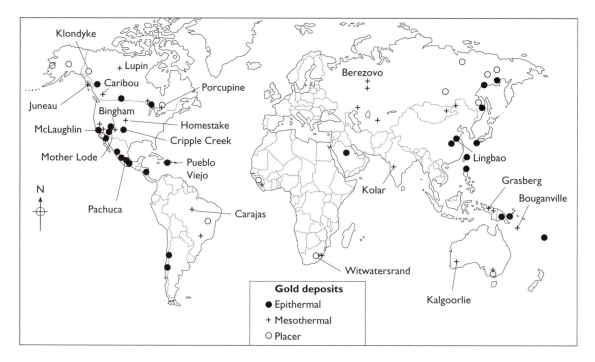

Fig. 33.12 Epithermal, mesothermal, and placer gold deposits. The most important mining districts are named.

from deeper levels cool to 100 °C to precipitate silver. The largest deposits are in Australia (Broken Hill, Mount Isa), Mexico, Peru, Bolivia, and the USA (Bingham, Utah).

Platinum production has increased dramatically over the past 50 years, while gold and silver production have remained relatively stable. The main market for platinum group elements (platinum, palladium, rhodium, ruthenium, iridium, osmium) is in the chemical industry, where it serves as a catalyst, much of it for catalytic converters in automobile exhaust systems. These devices speed oxidation reactions that convert hydrocarbons, carbon monoxide, and nitrous oxides to carbon dioxide, nitrogen, and water. However, even with all these industrial applications, platinum production is only a small fraction of that of gold.

We have already discussed the occurrence of platinum in layered ultramafic intrusions such as the Bushveld (Merensky Reef) and Great Dyke (see Figure 33.3d). Less important are magmatic sulfides such as those at Sudbury in Canada.

33.5 Reserves

The economy of a mineral resource depends on a variety of factors (e.g., David, 1977). Demand is determined by the applications and the price. In fact

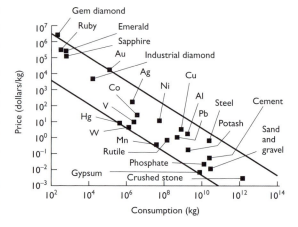

Fig. 33.13 The consumption versus price plot of a wide range of materials illustrates a very clear inverse relationship. Most commodities fall within a broad band. Gem diamond and gold are relatively expensive, whereas steel mercury, manganese, and gypsum are cheap (courtesy S. E. Kesler, see also Kesler, 1994).

there is an almost linearly inverse relation between consumption and price for all sorts of mineral commodities (Figure 33.13). The demand thus influences the production, and production from a finite reserve will determine how long a mineral resource lasts. It is not easy to give sensible estimates for reserves.

A geological determination of a metal reserve may rely on extrapolating the average crustal abundance of an element. By determining abundances for well-studied localities, we can statistically infer the abundance for the Earth, or at least for continental regions on the Earth. Another estimate is obtained by means

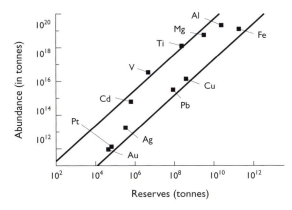

Fig. 33.14 Plot of estimated reserves versus average abundance in the crust for major metals (courtesy S. E. Kesler, see also Kesler, 1994).

of a detailed geological analysis of typical settings, and then extrapolating this analysis to regions with similar tectonic features but where no detailed prospecting work has been undertaken. There is quite a good correlation between geological reserve estimates and crustal abundance of metals (Figure 33.14).

One fundamental problem in formulating these estimates is the definition of a "reserve" itself. Take, for example, iron. Most rock-forming minerals contain iron in substantial amounts, and iron reserves interpreted in this context are practically infinite. However, there is no economic way to extract iron from silicates or, as we have seen earlier, from sulfides. Up to what minimum grade do we count the presence of a mineral as a potential reserve? Magnetite occurs in many plutonic and metamorphic rocks, but beyond a certain grade magnetite extraction is simply not feasible. In addition, the grade limit changes with time. As new technologies become available, lower grade ores can be processed and thus reserves will increase accordingly. This lower limit, however, is also influenced by demand and price. Large deposits of low-grade ore are available and therefore of great economic interest. It is

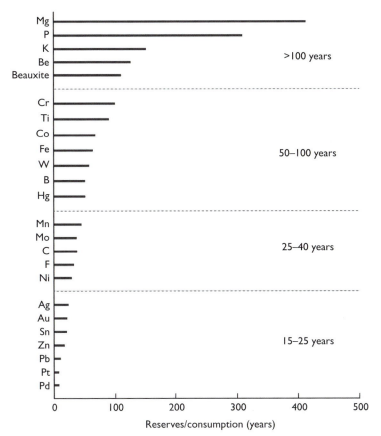

Fig. 33.15 Estimates of reserves (in years) for major mineral commodities, assuming a uniform consumption (after Petrov *et al.*, 2011).

often more useful to develop new extraction techniques and process large low-grade deposits, rather than prospect for small deposits of high-grade ores.

Demand for mineral resources is not easy to predict. For example, due to negative public opinion, very few nuclear power plants have been built in recent years, leading to a drop in the consumption and price of uranium. Iron as a structural material has been losing ground to concrete and plastics. Copper wires are being replaced by aluminum and fiberglass equivalents. Recycling, particularly of aluminum and iron, has reduced the demand for raw ores. Such factors certainly cause short-term fluctuations, but they can also have unpredictable long-lasting effects.

Nevertheless, if we divide the present-day consumption by the established reserve, we can get a general estimate of the lifespan for various mineral commodities. There has been much research into estimating reserves of oil and natural gas because these energy resources affect us directly in our daily life, but similar estimates also exist for minerals (Figure 33.15). For example, diamond, zinc, and gold reserves may be exhausted in a few decades, whereas sufficient iron and aluminum will be available for a long time.

33.6 Summary

Metalliferous ore deposits are of enormous economic importance, with a similar annual value as oil and gas. This chapter discusses the minerals that are most important for metal extraction such as (in order of value) hematite and goethite for iron, aluminum hydroxides for aluminum, native gold, chalcopyrite and bornite for copper, and pyrolusite for manganese. The ore deposits are closely linked to the geologic/tectonic history. Gold occurs largely in greenstones of Precambrian shields, the banded iron formation in Precambrian sedimentary basins is a main source for iron, mafic-ultramafic magmatism produced nickel, chromium, and platinum deposits. More recent Phanerozoic subduction zones developed porphyry coppers and mid-oceanic ridges contain massive sulfide deposits. Based on current reserves tentative estimates can be made on how long mineral resources may last.

Test your knowledge

1. Why are iron, aluminum, and copper economically the most important mineral resources?

2. Which minerals are used for the extraction of iron, manganese, aluminum, chromium, lead, and zinc?

3. Which mineral deposits are typical of convergent margins and of divergent margins? Give an example for each and explain how ores formed under both situations.

4. Large ore deposits are due to surface processes. Explain how the principal gold and iron deposits formed.

5. In the Precambrian cratons, there are some unique deposits of rare metals. Explain ore formation in the Bushveld complex (South Africa) and in Sudbury (Canada).

6. Give some reasons why conditions for metal ore deposition changed irreversibly over geological time.

7. Geographically, which countries are the main producers of iron, aluminum, chromium, gold, and copper?

8. The lifespans of ore deposits and of ore reserves depend on many factors. Discuss these factors and their relationships.

9. Of the following metals, which ones are likely to be exhausted first and which ones will be available for a long time: iron, chromium, molybdenum, silver, and titanium?

Further reading

Chang, L. L. Y. (2002). *Industrial Mineralogy*. Prentice Hall, Upper Saddle River, NJ.

Craig, J. R., Vaughan, D. J. and Skinner, B. J. (2010). *Earth Resources and the Environment*, 4th edn. Prentice Hall, Upper Saddle River, NJ.

Evans, A. M. (1993). *Ore Geology and Industrial-Minerals: An Introduction*, 3rd edn. Blackwell, Oxford.

Guilbert, J. M. and Park, C. F. (2007). *The Geology of Ore Deposits*. Waveland Press, Long Grove, IL.

Kesler, S. E. (1994). *Mineral Resources, Economics, and the Environment*. Macmillan, New York.

Ridley, J. (2013). *Ore Deposit Geology*. Cambridge University Press, Cambridge.

Robb, L. (2005). *Introduction to Ore-forming Processes*. Blackwell Science, Oxford.

Vanacek, M. (ed.) (1994). *Mineral Deposits of the World*. Elsevier, Amsterdam.

34 | Gemstones

The first mineral you ever examined conscientiously was probably a gemstone, and gems are the objects that most people associate directly with minerals. While mineralogical properties of most gemstones have been discussed earlier, we will summarize here the unique features that make them valuable, including how to prepare a gemstone out of a rough crystal and a review of testing instruments used by gemologists. There will be short sections about the most important gems and their occurrence and in conclusion methods of crystal synthesis will be described.

34.1 General comments about gems

Gemstones have been defined as minerals that are highly valued for their beauty, durability, and rarity; they may be worn for adornment or used to decorate art objects. Since they are rare, they have a high value. The value of gem diamond exceeds the price of gold (by weight unit corresponding to a Troy ounce) by a factor of about 200 (Table 34.1).

Note that in many cases the names used for gems are different from those of the regular minerals that they represent. For example, ruby and sapphire are varieties of corundum, emerald and aquamarine are varieties of beryl, and alexandrite is a variety of chrysoberyl. In addition, the weight units used in gemology are different from those used in ordinary science. The most common weight unit is the carat, which corresponds to 0.2 g. Table 34.2 lists some of the important minerals that are used as gemstones.

Table 34.1 Approximate 2015 prices for 1 carat (0.2 g) of some gems and gold

Gem	Price ($US)
Diamond	1000–1500
Emerald	100
Ruby	100
Sapphire	120
Gold	7

The visual appeal of different gemstones varies greatly. The majority are highly transparent to light; others, such as moonstone, opal, or tiger-eye, are translucent; and a few, such as lapis lazuli and jade, are opaque. Most gemstones are colored, and the color gives gems such as ruby, sapphire, emerald, and opal their visual appeal. Diamond is generally a colorless gemstone, and the attractiveness of a particular diamond depends on the interaction of white light with the crystal by internal reflection and dispersion. Many gemstones are single crystals. Exceptions include opal, which is amorphous, and jade, which is a polycrystalline aggregate.

Many gems are literally permanent and do not degrade in the atmosphere, in water, or by abrasion (although not all gems have the high hardness of diamond or corundum). Therefore, gems have been seen as a good investment for many generations. However, a cautionary note should be offered: some colored gems (such as amethyst, rose quartz, and yellow topaz) lose their coloring during prolonged exposure to sunlight. As another example, opal loses its high water content when moderately heated, and this dehydration destroys the opal's color pattern.

The worldwide production of gems is small. The yearly consumption of gem diamond, ruby, sapphire, and emerald amounts to only about 10 000 kg. However, their value is about the same as the yearly production of cement. De Beers, one of the largest diamond companies (e.g., van Zyl, 1988), sold over five billion US dollars of raw diamonds in 2013, about

Table 34.2 Most important gemstones, with mineral names, gem varieties, color, and principal occurrences (see Table 15.1 for origin of color)

Mineral name and formula	Gem variety	Color	Main source
Diamond C		Colorless, yellow, brown, green, blue, pink	Russia, Botswana, DR Congo, Australia, Canada, Zimbabwe, Angola, South Africa, Namibia, Sierra Leone
Corundum Al_2O_3	Ruby	Red	Myanmar, Thailand, Sri Lanka, Madagascar, Kenya, Tanzania, Vietnam, Thailand, Cambodia, Kenya, Pakistan, Madagascar, Nigeria, Myanmar, Sri Lanka, USA
	Sapphire	Blue	
Chrysoberyl $BeAl_2O_4$		Yellow, green	Brazil, Madagascar, Tanzania, India, Sri Lanka
	Alexandrite	Violet-red	Russia, Sri Lanka, Madagascar, Brazil
	Cat's-eye	Chatoyant	Sri Lanka
Beryl $Be_3Al_2Si_6O_{18}$	Emerald	Deep green	Colombia, Russia, Zambia, Madagascar, Zimbabwe, Afghanistan, Pakistan, Brazil
	Aquamarine	Blue-green	Brazil, Madagascar, Tanzania, Mozambique, Malawi, Zambia
	Morganite	Pink	Brazil, Pakistan
	Heliodore	Yellow	Brazil, Pakistan, Ukraine
Topaz Al_2SiO_4		Colorless, red, blue	Russia, Germany, Sri Lanka, Myanmar, Brazil, Ukraine
	Imperial topaz	Golden orange	Brazil, Pakistan
Tourmaline Na(Fe, Mg,Li,Al)$_3$(Al,Cr)$_6$ Si$_6$O$_{18}$(BO$_3$)$_3$ (OH,F)$_4$	Elbaite	Green, brown	Mozambique, Brazil, Nigeria, Zambia, Namibia
	Rubellite	Pink	Brazil,
	Indicolite	Blue	Namibia, Brazil
	Dravite	Yellow	Brazil, Austria
Quartz SiO_2	Amethyst	Violet	Russia, Zambia, Brazil, Canada, Uruguay
	Citrine	Yellow	Bolivia, Brazil
	Rose quartz	Pink	Brazil, Madagascar
	Smoky quartz	Brown	Switzerland, Russia, Brazil
Microcline $KAlSi_3O_8$	Amazonite	Green	India, Brazil, Russia, USA
	Moonstone	Iridescence	Sri Lanka, India
Spodumene $LiAlSi_2O_6$	Kunzite	Pink	Brazil, Afghanistan
	Hiddenite	Green	Brazil, Pakistan
Olivine Mg_2SiO_4	Peridot	Green	Egypt, Myanmar, USA, China, Pakistan
Turquoise $CuAl_6(PO_4)_4(OH)\cdot 4H_2O$		Blue	Iran, Chile, China, USA
Opal $SiO_2 \cdot nH_2O$		Iridescence	Australia, Mexico, Russia
Lazurite, sodalite, etc. $(Na,Ca)_8(Al,Si)_{12}$ $O_{24}(SO_4,S_2,Cl)$	Lapis lazuli	Blue	Afghanistan, Tajikistan, Russia

Table 34.2 (*cont.*)

Mineral name and formula	Gem variety	Color	Main source
Jadeite NaAlSi$_2$O$_6$	Jade	Green	Myanmar, Japan, Kazakhstan, China
Tremolite (nephrite) Ca$_2$Mg$_5$Si$_8$O$_{22}$(OH)$_2$	Jade	Green	China, Russia, New Zealand, Canada
Aragonite CaCO$_3$	Pearl (nacre)	Iridescence	

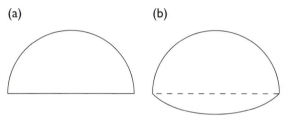

Fig. 34.1 The simplest cut of a gem is a cabochon with either (a) a single curved surface or (b) two curved surfaces.

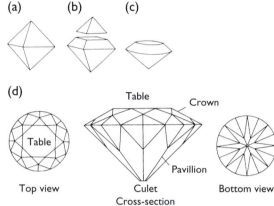

Fig. 34.2 Steps in converting a natural octahedral diamond crystal into a brilliant-cut gemstone with 58 facets. The top facet is called the table, the upper part the crown, the lower part the pavillion, and the bottom facet the culet.

half of it in the USA. This number is of considerable economic significance, particularly when one considers that cutting diamonds and setting them in jewelry increases their value about 10-fold.

Today many gems can be produced industrially, with properties similar to those of their natural counterparts. These artificial gems, however, sell at only a fraction of the price of the equivalent natural stones. Nevertheless, gem synthesis is a huge industry, for both jewelry manufacturing and industrial applications and some methods of synthesis will be described at the end of this chapter.

The aesthetic effect of gems is greatly enhanced by cutting and polishing. The simplest cut is a *cabochon* with either one flat surface (Figure 34.1a) or two spherical surfaces (Figure 34.1b). More sophisticated are faceted cuts with smooth planar surfaces. A classic cut, particularly for diamonds, is the "brilliant cut" with 58 facets. It is prepared from the frequently octahedral rough diamond (Figure 34.2a) by cutting off one corner (Figure 34.2b), polishing it to a conical shape by grinding away the corners of the crystal (Figure 34.2c), and finally polishing the facets in a certain sequence (Figure 34.2d). A brilliant cut consists of an upper part called the crown, topped by a table, and a lower part called the pavilion, with a

small facet called the culet at the bottom, parallel to the table. The reason for such a complicated cut is to optimize the reflection of light in the crystal. "Brilliance" depends on the ability of a crystal to reflect the light to the eye. The less the light that passes through the crystal and the more that is internally reflected upward through the crown of a faceted gemstone, the higher is the brilliance. Figure 34.3a illustrates how light arriving in different directions is reflected inside the crystal by total reflection towards the eye, making the cut gemstone act like a focusing device. "Fire", another desirable attribute, is the dispersion of light into the colors of the visible spectrum (Plate 23a, inset). Fire is due to refraction of the light that enters the crystal and it is most striking if the crystal has a high dispersion, such as diamond (dispersion, 0.044; refractive index, 2.417; see Figure 34.3b).

(a)

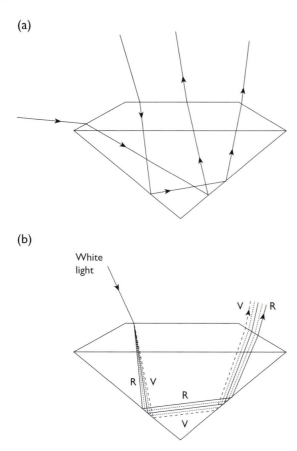

(b)

White
light

V   R

R   V

R

V

Fig. 34.3 (a) "Brilliance" is produced if light entering a crystal from all directions is focused towards the eye with total reflection on all faces. (b) Dispersion of white light entering the crystal into rainbow colors is perceived as "fire". R, red; V, violet.

Gemstones have been used since the dawn of history, and there are even examples from prehistoric times. In Bronze Age graves from Bohemia, pierced garnets were found that could be strung together to form a necklace. Early on, attractive stones were valued not merely for their beauty and used in jewelry; because of their rarity, they were made a part of local mythology and even had supernatural powers attributed to them. Some of these myths still persist today. As early as 5000 BC Assyrians and Hittites were using amulets made of hematite, lapis lazuli, amazonite, clear quartz, and chalcedony as talismans. In Egypt, the amulet took the form of the scarab beetle, with those made of emerald, ruby, amethyst, and turquoise being among the more precious. Around 2000 BC

stone beads were better rounded and polished and combined with gold to create sophisticated jewelry.

Very early on, a gem trade became established, with routes from Egypt to central Asia. Egyptians mined turquoise in the Sinai Desert and imported emerald and lapis lazuli from Afghanistan. At Gjebel Sikait between the Nile and the Red Sea are emerald mines that were worked in a systematic way at least 4000 years ago. The production reached a peak during the Roman occupation of Egypt, and sporadic mining in the so-called Cleopatra mines lasted until AD 1750. These gem mines are the oldest known. The lapis mines in Afghanistan have also had a long history.

Independently of the ancient Old World cultures, gemstones were used in the pre-Columbian American civilizations. Incan and Mayan cultures used emerald abundantly, and it is now established that Mexican emeralds were traded from mines in Colombia. The famous Muzo mine in Colombia has supplied emeralds for over 1000 years and is still one of the main producers of stones of unequalled quality (Plate 18b).

As far as we know, diamonds were first found in central India near the town of Golconda, whose name means "opulent wealth". In ancient epics they play an important role in the Hindu religion, and it was from India that diamonds were first introduced to Europe. One of the most famous, though not the largest, diamonds is known as the Koh-i-Noor, originally a 186 carat stone that is now part of the British Crown jewels (Box 34.1).

34.2 Instruments used by gemologists

Gemologists investigate gems using techniques similar to those applied by mineralogists, which were described in Chapters 13–15. There is one important difference. For most purposes the gem analysis must be done in a destruction-free manner. It is generally not possible to prepare a powder to conduct a standard X-ray diffraction (XRD) phase analysis or a quantitative X-ray fluorescence (XRF) chemical analysis, to cut a thin section for investigation of microstructure and optical properties, or even to prepare a grain mount to determine the refractive index with the immersion method. Gemologists use energy-dispersive X-ray fluorescence (EDX) on gem surfaces with a scanning electron microscope (see Chapter 16) and occasionally powder XRD on minute amounts of material scraped from surfaces. The most important instrument for gem identification is a binocular

---

Box 34.1 Background: Some famous Indian diamonds

The 186 carat *Koh-i-Noor* (mountain of light) was first described in the fourteenth century in central India in the family of the Rajah of Malwa, but tradition suggests that it was found thousands of years ago and was worn by heroes of the celebrated Hindu epic *Mahabharata*. It changed hands many times through conquests, was owned by the Mughal Emperors in India, and came into the possession of Nadir, Shah of Persia, after the conquest of India in 1739. It was later returned to India, and with the annexation of India by the British Government in 1849, the Koh-i-Noor was given to Queen Victoria, who had it recut to 109 carats because of its lack of brilliance. Unfortunately, the recutting diminished its historic value.

There are two other large diamonds from India that still have the traditional cutting: the *Orloff*, now in Russia, and the *Great Mogul*, at originally 793 carats the largest known Indian diamond, which was described by the French traveler Jean Baptiste Tavernier in 1665 but has since disappeared. Some sources claim that it may be the same stone as the Koh-i-Noor before it was cut.

More recent history surrounds the *Regent* (410 carats), which was found in 1701 near Golconda by an Indian slave. It was bought by Sir Thomas Pitt, then Governor of Madras. He sold it in 1717 to the Duke of Orléans, Regent of France for King Louis XV. The stone was used both in the King's crown and in the hat of Queen Marie Antoinette. After the French Revolution the Regent jewel, together with other diamonds, was stolen but later found. At one point it decorated the hilt of Napoleon's sword, but after Napoleon went into exile, his second wife, Marie Louise, gave it to her father, the Emperor of Austria. He returned it to France, and the Regent is presently exhibited in the Louvre in Paris.

---

gemological microscope ($10\times$ to $60\times$ magnification, with darkfield and brightfield illumination), used in reflected or transmitted mode. It serves to check inclusions and growth features, and for evidence of treatment and synthesis. In addition, three modified instruments are particularly important for gem analysis: a refractometer to determine the refractive index on a planar surface, a spectroscope to determine absorption spectra, and a polariscope to explore birefringence. A gem *refractometer* is similar to the Abbe refractometer described in Chapter 13 (see Figure 13.6), which is used to determine the refractive index of liquids. The refractometer makes use of the critical angle of total reflection. A face of the gem is mounted on a glass half-cylinder with an oil that has a higher refractive index than that of the gem (Figure 34.4). Light entering at various angles is shown. For rays A and B both reflection and refraction occurs. For ray C the angle of refraction is 90°, and in this case the incident angle α_c is called the critical angle. From Snell's law (Chapter 13) we obtain

$$n_c = n_g \sin \alpha_c \qquad (34.1)$$

where n_c is the unknown refractive index of the crystal, n_g is the refractive index of the glass cylinder

(typically 1.90), and α_c is the critical angle. For rays C and D (and any ray with a steeper incidence), no refraction occurs, because the angle of refraction is greater than 90° and all light is reflected. Viewing the brightness change as a function of α, one observes a relatively bright signal where no refraction occurs, and a darker signal where reflection is attenuated by refraction. The border between the two illuminated fields defines the critical angle and thus the refractive index of the crystal. In a gem refractometer a scale calibrated in refractive index units is superposed on the illuminated image and the refractive index is easily determined (e.g., 1.62 in Figure 34.4b). The refractive index of the glass cylinder (usually 1.90) sets an absolute upper limit for refractive index determination. If the refractive index of the oil is lower than that of the crystal, total reflection occurs on the oil rather than the crystal.

Because of dispersion, refractive indices vary with wavelength. In this case the boundary on the scale is not sharp, but consists of a band displaying spectral colors. The edges of the band (Figure 34.4c) can be used to determine refractive indices for red (e.g., 1.628) and violet (1.640), the difference between the two being the dispersion (0.012). If crystals are

(a)

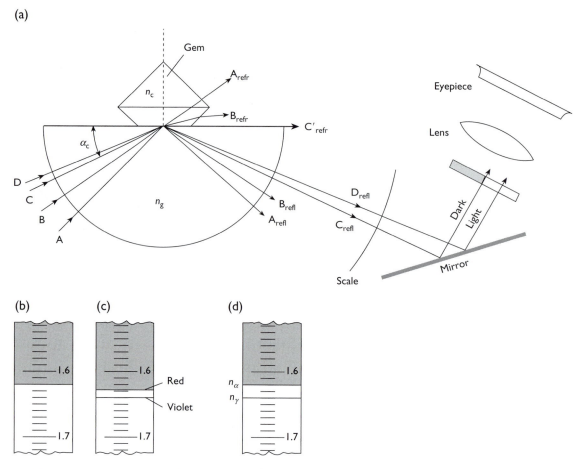

Fig. 34.4 Gem refractometer. (a) Illustration of the geometry for total reflection. The refractive index can be directly read on a scale for (b) simple case with a single refractive index, (c) crystal with dispersion and a color range of indices, and (d) a birefringent crystal with two refractive indices, n_α and n_γ, depending on crystal orientation. See the text for further explanation. Subscripts refr and refl indicate refracted and reflected rays, respectively.

birefringent, two shadow edges are observed on the scale (Figure 34.4d).

A gem *spectroscope* analyzes the absorption of light by the crystal. As white light travels through a colored crystal, one or more wavelengths are selectively absorbed. A gem is mounted on a black background and illuminated with a bright white light source. The white light produces a rainbow spectrum. The absorption spectrum displays dark bands superposed on the rainbow spectrum, and these bands are diagnostic of the coloring agent. The cause of these absorption bands was discussed in Chapter 15. In many gems color is due to chromatophore trace elements such as chromium, iron, and manganese. In others, particularly diamonds, the absorption spectrum is caused by defects in the crystal structure (color centers).

A *polariscope* is a simple version of a petrographic microscope, consisting of two polarizing plates with polarization directions at right angles (crossed polarizers) without an optical lens system. The crystal is held with tweezers between the two polarizers, with a light source underneath. With this instrument one can immediately distinguish colored glass beads from gems such as ruby or amethyst. Inclusions are often in good contrast.

Gemological laboratories routinely use several spectroscopy techniques to support their gem identification services. The main methods are visible spectroscopy for determining the cause of color, infrared spectroscopy for detecting foreign materials in treated gems, and Raman spectroscopy for mineral identification.

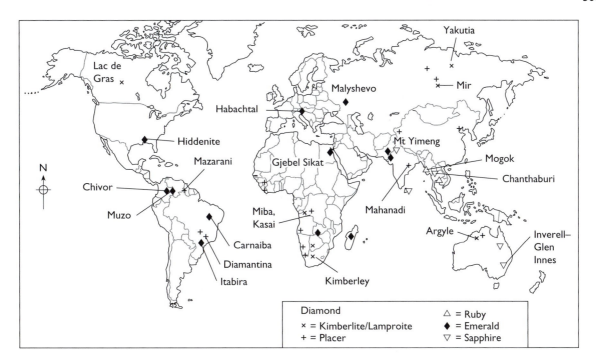

Fig. 34.5 Most important gem deposits.

Use of laser light to create a photoluminescence spectrum is also important for diamond identification.

34.3 Important gems

The most important and precious gemstones are diamond, emerald, ruby, and sapphire, but there are numerous others, some of which are listed in Table 34.2, and the most important gem deposits are displayed in Figure 34.5. We briefly review here some characteristics of these minerals, discussing both their occurrence and significance. Gem-quality minerals form in only a few primary environments. Diamonds, for example, form in the upper mantle and are brought up to the surface in *kimberlite pipes* (e.g., Harlow, 1998). Ruby and sapphire grow in high-temperature environments, both in alkaline magmas and in *aluminous metamorphic rocks*. Emerald is mainly a product of *hydrothermal* systems. Many other gemstones, including aquamarine, topaz, and tourmaline, are found in *pegmatites* (e.g., Shigley and Kampf, 1984).

Many commercial gem deposits are secondary and are mined in ancient and recent alluvial formations known as placers. Placers form as a result of water action with gradients in velocity. Where velocities are highest, lighter minerals are swept away while heavier ones collect in depressions. An extreme case of contrasting densities exists between precious metals such as gold and rock-forming minerals. The density of diamond, ruby, and sapphire is also higher than that of quartz, feldspar, and mica (around 4 versus 2.6–2.8 g/cm^3), though the contrast is much less pronounced. Another factor that contributes to the conservation of gems in placers is extreme hardness and chemical inertness, preventing mechanical abrasion and weathering. While most other minerals quickly degrade to fine sand, diamond, ruby, and sapphire resist breakdown and are preserved as large crystals. The alluvial action provides a selection in quality. Crystals with fractures, inclusions, and other flaws are destroyed in transit. For this reason placer gems have a much higher quality than gems mined from primary rocks. In the case of diamond, only 10% of stones mined from kimberlites are of gem quality, whereas placer gems make up 90%. Placer gem deposits are usually in the vicinity of the primary source, although this primary source may not be clearly established or may have been eroded in the course of the geological history, as in the case of the famous ruby deposits of Sri Lanka.

Diamond

We saw in Chapter 22 that most diamonds formed originally in the mantle at depths of up to 750 km (corresponding to a pressure of 35 GPa) and at temperatures of 2000 °C or more. From the original source they were brought up to the surface at great velocity and driven by expanding volatile gases. This transport took place in pipes, as inclusions in ultramafic rocks called kimberlite. Kimberlite pipes are carrot shaped, probably existing as narrow dikes at depth and expanding to a conical shape near the top (see Figure 22.7). Diamonds exist only in very small concentrations in kimberlites. South African kimberlites yield only about 0.1 carats (0.02 g) per tonne. Kimberlites in the Argyle region of western Australia yield up to 5 carats per tonne, although these diamonds are of lesser quality than those from Africa. Diamonds recently discovered in high-pressure metamorphic rocks are of no economic importance.

We have previously noted that all early diamonds used as gemstones originated from India. By the fifteenth century, Indian diamonds had become firmly established in Europe. With demand increasing, the Indian sources became depleted and exploration started in other parts of the world. In 1714 diamonds were discovered in Brazil. Like the Indian diamonds, they occur as placer deposits in gravels and conglomerates. In 1867 the first diamond was found in South Africa near the town of Kimberley, and that country became for many years the world's largest producer of gem-quality diamonds. Other diamond deposits have become important and are mined from kimberlites and related lamproites, all located in very old continental shields in Yakutia (Siberia, Russia), Botswana, Sierra Leone, Argyle (Western Australia), and the Northwestern Territories of Canada. Large placer mining operations are found in Namibia, Angola, Sierra Leone, Brazil (Diamantina), and Venezuela. Presently Botswana and DR Congo produce a larger proportion of diamonds than does South Africa, although most of these are industrial grade. The largest diamond ever recovered is the great Cullinan, originally at 3026 carats. It was found in 1906 in the Premier mine in South Africa. Later it was cut into over 100 polished stones, including the Greater Star of Africa (530 carats) and the Lesser Star of Africa (317 carats), which are the world's largest cut diamonds and belong to the British Crown. In November 2015, the world's second largest diamond (1111 carats) was

Fig. 34.6 1111 carat diamond from the Karowe mine in Botswana, width 65 mm (courtesy Lucara Diamond Corp.).

found in the Karowe mine in Botswana (Figure 34.6). Other famous diamonds are listed in Table 34.3. Only five flawless diamonds over 100 carats have sold in the last 25 years, most recently a gem of over 200 carats, mined in South Africa, cut and polished to 100 carats, and sold for $US 22 million in 2015.

Emerald

Emerald is the dark-green variety of beryl (Plate 18b). The color is due to traces of Cr^{3+} or V^{4+}, substituting for Al^{3+} in the crystal structure. Like ordinary beryl, most emerald deposits are in pegmatites and hydrothermal veins. However, these dikes and veins are associated with granitic intrusions into ultramafic rocks where the chromium and vanadium originate. The most important modern emerald deposits are at Muzo and Chivor in Colombia, which supply emeralds of the highest quality. As we discussed in Chapter 19, at elevated temperatures and pressures, water is capable of dissolving large amounts of elements. In Colombia hydrothermal calcite veins contain emerald. The host rock in which the hydrothermal veins occur is a chromium-bearing carbonaceous shale, which makes it relatively easy to extract the gemstones. In addition to clear crystals of emerald, there is another rare form of emerald that displays a star pattern, called *trapiche* (Plate 32 b,c).

Other emerald deposits occur in metamorphic rocks, such as biotite schists near Ekaterinburg in

Table 34.3 Large and famous diamonds

Name	Weight in carats		Origin	Present location
	Original	Cut		
Cullinan	3106	550, etc.[a]	Premier, South Africa	British Crown Jewels, London
Lesedi La Rona	1111		Karowe, Botswana	Lucara Corp.
Excelsior	995	21 stones	Jagersfontein, South Africa	Unknown
Star of Sierra Leone	969	770	Sierra Leone	British Crown Jewels, London
Great Mogul	793	280	India	Unknown
Vargas, brown	728		Brazil	Unknown
Jubilee	650	245	Jagersfontein, South Africa	Unknown
Regent	410	140	India	Louvre, Paris
Star of Yakutia	343	232	Yakutia, Russia	Treasury, Moscow
Orloff	787	190	India	Treasury, Moscow
Oppenheimer, yellow	254		Kimberley, South Africa	Smithsonian Institution, Washington, DC
Centenary	600	274	Premier, South Africa	Unknown
Tiffany, yellow	287	129	Kimberley, South Africa	Tiffany & Co., New York
Koh-i-Noor	>600	109	India	British Crown Jewels, London
Sancy	55		India	Louvre, Paris
Hope, blue	112	45	India	Smithsonian Institution, Washington, DC

[a] See the text.

the Ural Mountains (Russia), Habachtal (Austria), and Lake Manyara (Tanzania). During the seventeenth century, large Colombian emeralds were sought by the Mughal nobility of India, and some exceptional pieces are preserved. Most famous is the 218 carat Mogul (not to be mistaken for the Great Mogul diamond, mentioned earlier), inscribed with Islamic prayers and dated 1695. After the conquest of India by the Persians in 1739, many Colombian emeralds in India became part of the crown jewels of Iran, which still has the largest collection of emeralds in the world. Another exquisite collection is in the Topkapi museum in Istanbul (Turkey).

Because of the high value of emerald when cut for jewelry, only a few natural emerald crystals have survived. The largest known crystal is the 7025 carat Emilia crystal discovered in 1969 and owned by a private mining company. A superb 1759 carat crystal is in the collection of the Banco de la Republica in Bogotá, Colombia, and an 858 carat crystal is part of the Smithsonian Institution collection (Washington,

DC, USA). All these crystals have the typical beryl morphology, with a hexagonal prism topped by a basal plane.

Ruby and sapphire

Ruby (see Plate 1f) and sapphire (see Plate 1g) are varieties of the mineral corundum. The red color of ruby originates from Cr^{3+}, and the blue of sapphire is due to Fe^{2+}–Ti^{4+} charge transfer (see Chapter 15). Ruby and sapphire deposits are often closely associated, although one variety usually dominates.

Ruby never occurs as crystals of a size comparable to that of large diamonds and emeralds. Because of its rare natural occurrence and attractive properties such as brilliance in color and high hardness, ruby became the first candidate for modern gem synthesis. The highest quality ruby, with intense brilliance, is from limestone in Mogok, Myanmar; dark-purple crystals are mined in Thailand.

Ruby and sapphire form mainly in high-temperature metamorphic environments as a result of recrystallization from material rich in aluminum and poor in silicon and magnesium. These geochemical conditions are essential, because otherwise sillimanite and spinel would form. The Mogok deposit occurs in a low-pressure setting, with shallow granitic intrusions. Ruby occurs in calcite marbles, interbedded in pelitic schists and gneisses. Associated minerals are garnet, graphite, and diopside. Where the marble is dolomitic, forsterite and spinel crystallize, also with gem-quality crystals. Sapphires from Myanmar are valued for their dark-indigo-blue color.

Gem corundum occurs in a different environment in Thailand. In Chanthaburi, ruby and sapphire are found in weathered alkali basalt. However, the minerals did not crystallize in this basaltic magma but were simply transported in it to the surface, together with gem-quality olivine (peridot) in xenoliths.

In Myanmar and Thailand, ruby and sapphire are rarely found in fresh source rocks, and therefore the mineralogical conditions of formation are not very clear. The two gems are generally mined in highly weathered rocks and derived soils, as well as in associated alluvial placers. This is also true for the other economically important gem-quality corundum deposits in Sri Lanka. The hexagonal star pattern, occasionally observed in ruby and sapphire (Plate 32a), is attributed to oriented submicroscopic inclusions of rutile (see Figure 15.9a), which often grows in epitaxial patterns (see also Plate 3e).

Aquamarine, tourmaline, chrysoberyl, and topaz

While all the four major gemstones occur in very special geological settings, the largest variety of gem minerals (aquamarine, tourmaline, chrysoberyl, topaz, and others) are found in pegmatites. As discussed in Chapter 21, most pegmatites are of granitic composition and contain quartz, alkali feldspar, and sodic plagioclase as the principal minerals. Pegmatites form during the final stages of igneous activity, when most of the magma has crystallized except for a quartzo-feldspathic part. This uncrystallized remainder is enriched in volatiles and rises, injecting itself along fractures into pre-existing rocks. Pegmatites are enriched in noncompatible elements, i.e., elements that do not fit into the crystal structures of common rock-forming minerals. Such elements are beryllium, lithium, cesium, boron, manganese, phosphorus, and fluorine. Minerals containing these elements include tourmaline, beryl, chrysoberyl, kunzite (gem spodumene), and topaz. When pegmatites crystallize, large crystals form, with spodumene and beryl crystals measuring up to 20 m in length (Figure 10.12a).

Pegmatites that contain gem-quality crystals of rare minerals generally crystallize at shallow depth. Classic pegmatites that have been studied in great detail are those of Pala (in San Diego County, California, USA), the Black Hills (South Dakota, USA), Harding (New Mexico, USA), and the Ural Mountains (Russia), but by far the most famous gem pegmatites occur in Minas Gerais, Brazil. These deposits occurred around 500 million years ago when pegmatite dikes at the dome of large granitic batholiths intruded overlying metamorphic rocks in a large area of eastern Brazil, extending over hundreds of kilometers. Thousands of gem-bearing pegmatites are mined in Minas Gerais. The best-known mine locations are Cruzeiro (green and pink tourmaline), Golconda (green and bicolored tourmaline), and Virgem da Lapa (aquamarine, tourmaline, and blue topaz). Mines at these localities not only produce large quantities of samples for the international gem trade but also provide crystals that are on display in most major mineral museums around the world.

34.4 Gemstone enhancements

The principal use of gems is as adornments in jewelry. Therefore, great efforts are made not only to find attractive stones but also to improve their appearance. One way is to cut a natural crystal into a stone with either regular facets or smooth curved surfaces as in a cabochon. Since Egyptian times, much effort has also been spent on enhancing or altering natural colors and surface irregularities. Today gemstone enhancement is an important issue for the jewelry industry (e.g., Shigley and McClure, 2009). For example, a truly blue natural aquamarine is very rare and expensive, but by simple heat treatment, a greenish-blue color in aquamarine can be changed to pure blue. Often these changes are difficult to detect, particularly when one is using destruction-free methods. In this section we discuss briefly some of the methods used to improve the appearance of various types of gemstone.

Obvious gem defects that are easy to detect are fractures reaching the surface and producing unwanted reflection of light. Gem dealers use various methods to fill open fractures with materials that have a refractive index close to that of the crystal, which makes the fractures less noticeable. A traditional way has been to treat crystals with oils or natural resins such as Canada balsam. One problem with these media is that they wear off with time and are sensitive to heat, solvents, and detergents, obviating the desired durability property of a gem. On the other hand, an advantage of these natural oils is that they are relatively easy to remove or replace. More recently, synthetic resins related to epoxy have been applied; these are much more permanent and cannot be removed without destroying the gemstone. These fracture treatments can sometimes be detected by observing a gem under polarized light or by its ultra-violet fluorescence.

Dyeing is an enhancement technique that has been used for thousands of years. As an example, a beryl with microfractures can be treated with a dark-green resin to make it superficially resemble emerald. Obviously such dyes penetrate only the surface and are not very stable. Ruby dealers in Thailand treat almost all rough rubies with red "ruby oil" before selling them. There are also more permanent ways to change color, and many of these methods relate to the physical origin of color in crystals (see Chapter 15).

Heating can produce severe and permanent color changes. For example, brown, chromium-bearing topaz becomes pink at a temperature as low as 450–500 °C. Milky titanium-bearing corundum becomes blue when heated close to the melting point (2050 °C). In fact, the vast majority of blue sapphires in the jewelry trade today have been subjected to one or more high-temperature heat treatments. Such heat treatments may produce damage, such as fractures around inclusions or oxidation of included iron minerals, and therefore can sometimes be detected. Heating intensifies the blue color of corundum.

Heating of amethyst at low temperature can both reverse any radiation-induced damage that produced color centers and remove some of the darker smoky components from the spectrum. Heating at higher temperature produces yellow quartz that looks very similar to citrine. Almost all "citrine" on the market is heated amethyst. It can sometimes be distinguished from natural citrine by the presence of Dauphiné twinning, which is rare in true citrine.

Heating is often applied to topaz. Brown topaz of low value owes its color to a combination of a chromium-produced pink component and a color center-produced yellow component. Heating removes the color center defects and leaves only the pink color. There are numerous other applications of heating.

We have seen in Chapter 15 that ionizing radiation often produces color centers and that such radiation is responsible for the brown color of smoky quartz, the blue color of halite, and the purple color of amethyst. With artificial irradiation a whole range of color effects can be produced. This radiation may be in the form of alpha particles (helium nuclei), beta particles (accelerated electrons), γ-rays, or neutrons. There are basically three methods that are employed by the gem industry: γ-ray facilities, generally using radioactive ^{60}Co as a source; linear accelerators producing high-energy electrons; and nuclear reactors producing neutrons.

Neutron and electron irradiation are the preferred techniques to induce color in diamond. All clear quartz with some traces of aluminum becomes smoky when irradiated with γ-rays or neutrons, and irradiation also enhances the purple color of amethyst. Commercially, the most important radiation treatment is employed to produce the highly popular blue topaz, which is rare in nature. A first irradiation with γ-rays or electrons produces a brownish-green color. Subsequent heat treatment of the topaz removes the yellow component, resulting in the desired blue color.

Some of these radiation-induced colors are not permanent and disappear if crystals are exposed to ultra-violet light or are slightly heated. In addition, caution is advised with radiation-treated gemstones, as sometimes radioactivity is maintained for extended periods. Green diamonds, in particular, should be tested with a Geiger counter.

Thus, identifications of gemstone enhancements are big challenges for jewelers and gemologists. Some treatments have been historically accepted in the jewelry industry (e.g., heat treatment of aquamarine), whereas others are controversial (e.g., fracture fillings of diamonds and emeralds). Jewelers have the delicate task to meet legal and ethical disclosure requirements to consumers, and at the same time to present the information in a positive manner to maintain the interest of customers. There is a proliferation of new treatment techniques, some of which correspond closely to conditions in nature and are therefore difficult to detect.

34.5 Crystal synthesis

Because of the great value of gemstones, attempts have long been made to produce them synthetically. After many initial difficulties, fairly straightforward methods have been developed to produce large crystals of gem-quality ruby, emerald, and even diamond, and there is a large and profitable market for these synthetic gems. The methods are now used widely to produce crystals for many industrial applications as well (e.g., Elwell and Scheel, 1975). We discuss briefly some of the principal methods in this section.

Powder flame fusion: the Verneuil apparatus

In 1902 the first synthetic gem, a crystal of ruby 6 mm in diameter and 20 mm in length, was produced by Auguste Victor Louis Verneuil in Paris with a rather ingenious gadget. The idea was first to prepare a fine powder with the chemical composition of the desired final crystal (Al_2O_3 and about 2% Cr_2O_3 added for color). This powder was then dropped into a stream of hydrogen gas in a pipe. The hydrogen ignited when it came into contact with oxygen and produced a downward-pointing flame (Figure 34.7). In this flame, at a temperature of up to 2200 °C, the powder melted and accumulated first as a cone. At the tip of the cone a crystal nucleus formed that grew into the shape of an inverse droplet, which was called a *boule*. The success in producing a perfect ruby color hinged on using very pure starting materials, particularly materials without traces of iron. Verneuil ruby boules were widely produced, and the process became refined over time. The flame fusion-produced synthetic rubies always display curved growth bands due to slight fluctuations in chromium content and are therefore easily distinguished from natural rubies. There are large production facilities applying the Verneuil technique to produce gem-imitation rubies and sapphire (Plate 32d), as well as crystals used as instrument bearings. The cost of producing such crystals is low and the gem growth is

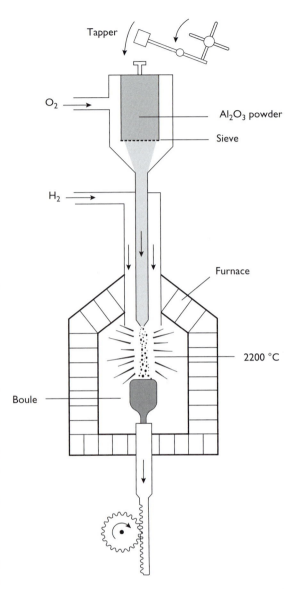

Fig. 34.7 Diagram of a Verneuil furnace to produce a ruby crystal. Fine-grained aluminum oxide powder with small amounts of chromium oxide is dropped into an oxygen stream which burns when combined with hydrogen, causing fusion of the powder. Droplets then crystallize on a seed crystal (boule) which is lowered slowly as it grows.

relatively rapid. Yet the growth heterogeneities proved detrimental for an important modern application of ruby in lasers, and thus a different method is preferred for such applications. It will be described in the next section.

Czochralski melt growth

In 1918, Jan Czochralski developed a method of growing crystals by pulling them from the melt. This method provides crystals of higher quality than those produced by flame fusion. The material to be grown is melted in a crucible with a radio-frequency induction heater. A power source feeds several kilowatts of electrical energy into a water-cooled copper coil. As the current through the coil changes at high frequency, power is induced in any electrically conducting materials near the coil. A preferred crucible material for a high-temperature Czochralski apparatus is iridium, with a melting point of 2442 °C (compared to a melting point of 2050 °C for Al_2O_3) (Figure 34.8). A control system is used to maintain the temperature just a few degrees above the melting point of ruby. A small seed crystal, a few millimeters in length, is now touched to the surface of the melt. The seed crystal is rotated and slowly pulled vertically from the melt at a rate of 5–25 mm per hour. The process is carried out in an inert atmosphere, almost pure nitrogen, to prevent oxidation of the iridium crucible.

The Czochralski method has become the preferred technique for producing silicon crystals used in the electronics industry and a large industrial Czochralski furnace with a silicon crystal is shown in Plate 32e. Silicon crystals up to 15 cm in diameter and 100 cm in length are grown routinely. The Czochralski method is also used for producing synthetic garnets such as YAG (*yttrium aluminum garnet*) and GGG (*gadolinium gallium garnet*). YAG, doped with neodymium, is used in neodymium lasers, with an output wavelength of 1.06 μm in the infrared range. GGG is applied in the electronics industry as a substrate for magnetic bubble domain memory units, which require perfect crystals without defects.

Flux growth

A method that relies on dissolving a material in a melt of a different composition, called a *flux*, at higher temperature and then precipitating the material again at lower temperature was developed mainly for producing synthetic emeralds. This process requires much lower temperatures than those needed to melt the actual gem material. A frequently used flux for emerald synthesis is lithium molybdate ($Li_2Mo_2O_7$) with a melting point of 705 °C. The constituents of beryl – Al_2O_3, SiO_2, and BeO – and a trace of Cr_2O_3 added as a coloring agent are dissolved in the flux at 800 °C. In order to have some convection in the flux, a temperature gradient is obtained, as for example in the flux cell illustrated in Figure 34.9, where the oxides dissolve in the hotter portion of the cell and the emerald grows from seeds in the cooler region. Emerald

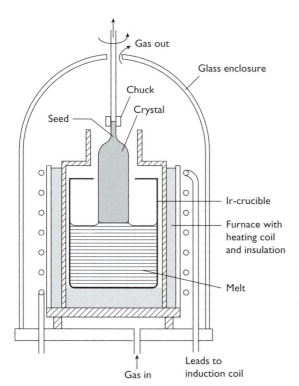

Fig. 34.8 Czochralski apparatus for crystal growth from a melt, with furnace and platinum crucible. The crystal is grown from a seed and slowly pulled out of the melt.

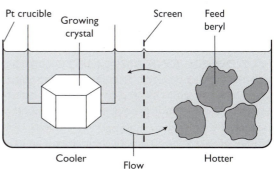

Fig. 34.9 Schematic illustrating growth of synthetic emerald from a flux.

crystals up to 2 cm in diameter have been grown from flux over periods of several months.

Hydrothermal growth of quartz

We discussed in Section 12.7 how quartz, with its noncentric crystal structure, has the property called piezoelectricity. If a stress is applied in certain directions, an electric field is induced and vice versa. For this reason, quartz is widely employed technologically in electronic filters and oscillators to control the frequency of electrical oscillations with superb precision. There is a large demand for high-purity and defect-free quartz for the radio and watch industries. Unfortunately, natural quartz is not only limited in supply but is also frequently internally twinned, which is not acceptable for piezoelectric applications. Thus, synthesis of large quartz crystals became a high priority.

Quartz is fairly insoluble in water, even at 100 °C. This insolubility changes considerably if water is heated to 400 °C at high pressure, corresponding to hydrothermal conditions (see Figure 18.2). At 300 °C and 140 MPa water may contain up to 0.1 weight% of SiO_2. This percentage is still not sufficient for good growth. To achieve a higher solubility in the range of several percentage points, a "mineralizer" such as NaOH is added to the solution. Experimentally, the conditions are achieved by filling a steel pressure vessel, also called an autoclave or "bomb", partially with water (e.g., 85%), some mineralizer, and nutrient (pure quartz), and then heating it externally from the bottom (Figure 34.10). In addition, some small millimeter-sized seed crystals are introduced at the top of the autoclave. Bottom heating introduces a temperature difference of 40−50 °C between the bottom and top of the autoclave. A saturated solution of quartz forms in the bottom hotter part, and the heating introduces thermal convection. In the upper, cooler region, the fluid can no longer hold all the silica in solution, and thus quartz crystallizes on the quartz seeds. The colder, denser solution sinks by convection and the cycle is repeated. It is desirable for quartz crystals not to grow too rapidly; otherwise, defects are introduced. A growth velocity of 1 mm per day is generally applied in industrial applications, and crystals of sizes up to 30 cm, weighing several pounds, are grown over the course of a few months (Figure 34.11). The morphology of the crystals is controlled by the shape of the seed crystals and the way that they are oriented in the convection current. The

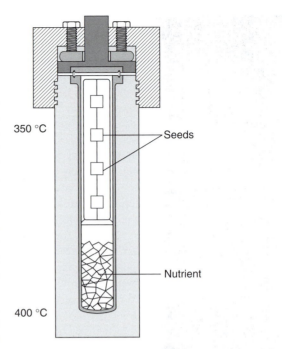

Fig. 34.10 Autoclave for hydrothermal growth of high-purity quartz crystals.

morphology is optimized for piezoelectric applications and often emphasizes a basal face (or close to it) or a minor rhombohedron $(01\bar{1}1) = z$ (Figure 34.12). The hydrothermal technique described above is also employed for industrial and gem manufacturing of ruby and emerald.

Ultra-high pressure

With diamond being the most precious of gemstones, it is not surprising that huge efforts have been made to synthesize it. Unlike the other gemstones discussed in this section whose synthesis relies on high temperature, diamond is the high-pressure polymorph of carbon that is stable at room temperature above 1.9 GPa. The pioneer of high-pressure synthesis was Percy Williams Bridgman, and his research earned him the Nobel Prize in Physics in 1946. He designed what became known as the piston cylinder apparatus. The idea behind this device is to exert a force with a relatively small piston on a sample that is shielded with a high-strength steel or tungsten carbide cylinder (Figure 34.13). The force can be applied hydraulically via a much larger piston. The sample is sealed within the cylinder, for example with a sleeve made of

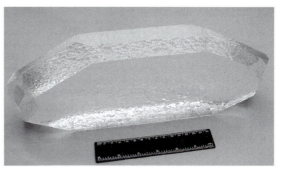

Fig. 34.12 Hydrothermally grown quartz crystal (courtesy D. Belakovskiy).

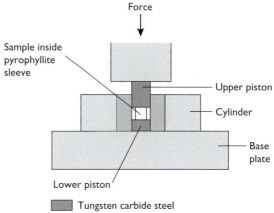

Fig. 34.11 Frame of quartz crystals after 1 year in the autoclave at the Institute for the Synthesis of Minerals, Aleksandrow, Russia (courtesy D. Belakovskiy).

Fig. 34.13 Bridgman-type piston cylinder apparatus.

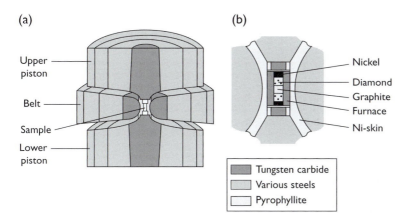

Fig. 34.14 High-pressure belt apparatus used for growth of diamonds. (a) Pressure vessel with pistons and belt. (b) Enlarged view of the sample with graphite powder, a skin of molten nickel, and precipitating diamond powder.

pyrophyllite, to prevent it from extruding. A furnace can also be applied to heat the sample. With this apparatus, Bridgman reached conditions in the diamond stability field (e.g., 3.5 GPa at 2000 °C), but at that temperature kinetics prevented diamond from crystallizing.

The original Bridgman design was modified to include a concentric set of cylinders of different high-strength materials and pistons (called the belt apparatus) for higher strength and a geometry with tapered pistons. With this device, pressures of 20 GPa and temperatures of 5000 °C could be reached simultaneously and maintained over many hours (Figure 34.14a). But physics alone was insufficient to produce diamonds – chemistry was also needed. In 1954, H. Tracy Hall, heading a team from the US firm General Electric, combined the high-pressure technology and the thermodynamics of the carbon system to produce the first significant synthetic diamonds.

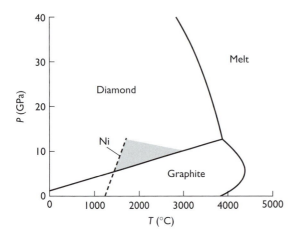

Fig. 34.15 Pressure–temperature phase diagram of carbon, showing the melting curve for nickel (dashed line). The region used for diamond growth is shaded.

The chemistry is best understood by returning to the phase diagram introduced in Figure 22.2. At low temperature, graphite does not convert to diamond for kinetic reasons. Hall proposed that a method similar to flux growth, but at high pressure, might solve the problem. The idea was to dissolve graphite in a molten metal and then recrystallize it from the melt as diamond, at a lower temperature (Figure 34.15).

In the Hall process, graphite is packed into a reaction unit (Figure 34.14b) contained in a pyrophyllite cylinder acting as a seal to prevent extrusion. Near the ends, nickel is introduced. The system is heated to 1800 °C, above the melting point of the metal. Graphite is dissolved in the nickel, and a film of metal moves across the sample, leaving behind carbon that crystallizes as diamond. In the first experiment, Hall observed diamond crystals 100–150 μm in diameter. Since then, the technique has been refined using diamond seeding, and several companies now produce large quantities of small diamonds that are used mainly for industrial applications, although many are of gem quality. Understandably there is much secrecy about diamond synthesis. General Electric produces about 100 million carats a year. A new split sphere high-pressure high-temperature apparatus developed in Russia is becoming very popular. The entire multi-anvil assembly is housed within two hemispherical steel castings (Figure 34.16). Pressure is applied hydraulically against the curved outer surfaces

Fig. 34.16 Split sphere diamond growth apparatus with hemispherical castings housing the eight anvils at Gemesis Corporation in Sarasota, Florida (photo by Tom Moses, GIA. Reprinted by permission).

of the eight anvils and a graphite furnace is used to heat the chamber. Typical diamond growth conditions are 5.0–6.5 GPa and 1350–1800 °C (e.g., Shigley *et al.*, 2002). Crystals up to 5 mm in diameter are used mainly in the semiconductor industry. Some of the largest synthetic diamonds weigh about 100 carats (20 g) (Plate 23b). The shape of the crystals depends on the temperature under which they form. Cube faces {100} form at 1300 °C, whereas at 1600 °C octahedra faces {111} dominate.

A more recent process for diamond synthesis is called chemical vapor deposition, where diamond is grown in a vacuum chamber at high temperatures but low pressures. A carbon-containing gas (such as methane) is introduced into the chamber, where it is subjected to a source of energy (such as a microwave beam) to form a plasma. Carbon atoms precipitate from the plasma and are deposited on flat seed plates to form tabular diamond crystals.

34.6 Summary

Gems are minerals that are highly valued for their beauty, color, durability, and rarity. Different names are assigned based on their color, generally due to traces of chromophore elements. Primary gemstones are diamond, ruby and sapphire (corundum), and emerald (beryl). Some of the history of famous gems is reviewed. Emeralds were mined at least 4000 years ago. Gemologists use special instruments to identify and verify gems to establish their value. Gem-quality minerals only occur under special geological conditions: most diamonds formed in the upper mantle and were brought up in kimberlite pipes. Ruby and sapphire grow in high-temperature aluminous magmas and metamorphic rocks. Emerald occurs in hydrothermal systems. Many gemstones (including aquamarine, spodumene, topaz, and tourmaline) are found in pegmatites. Methods of crystal synthesis have been developed to grow gem-quality crystals including Verneuil, Czochralski (e.g., ruby, silicon), flux growth (e.g., emerald), autoclave (e.g., quartz), Bridgman piston cylinder, and Hall process (diamonds).

Test your knowledge

1. Which properties make diamond and ruby the most valued gemstones?

2. What are the instruments with which every gemologist must be familiar?

3. Explore the history of some of the most famous diamonds. Where have they been found?

4. Why do secondary deposits produce higher quality gemstones than primary deposits?

5. In which environments are rubies and sapphires found? In which rocks should you not look for those minerals?

6. What produces the characteristic colors of emeralds, rubies, and sapphires?

7. List some gem minerals that are found in pegmatites. Which chemical elements do they contain?

8. Describe the methods used to produce synthetic gemstones.

9. How can one distinguish between natural gems and altered or synthetic gems?

Further reading

Arem, J. E. (1987). *Color Encyclopedia of Gemstones*, 2nd edn. Van Nostrand Reinhold, New York.

Bukanov, V. V. (2006). *Russian Gemstone Encyclopedia.* Granit, St. Petersburg, Prague.

Hughes, R. W. (1997). *Ruby and Sapphire.* RWH Publishing, Boulder, CO.

Hurlbut, C. S. and Kammerling, R. C. (1991). *Gemology*, 2nd edn. Wiley, New York.

Liddicoat, R. T. (1993). *Handbook of Gem Identification*, 12th edn. Gemological Institute of America, Santa Monica, CA.

Nassau, K. (1980). *Gems Made by Man.* Chilton Book Co., Radnor, PA.

Nassau, K. (1984). *Gemstone Enhancements.* Butterworth, London.

O'Donoghue, M. (2008). *Gems*, 6th edn. Robert Hale, London.

Read, P. G. (1999). *Gemmology*, 2nd edn. Butterworth, London.

Schumann, W. (2013). *Gemstones of the World.* 5th edn. Sterling, New York.

Sinkankas, J. and Read, P. G. (1986). *Beryl.* Butterworth, London.

Sofianides, A. S. and Harlow, G. E. (1990). *Gems and Crystals from the American Museum of Natural History.* Simon and Schuster, New York.

Webster, R. and Read, P. G. (1994). *Gems: Their Sources, Descriptions, and Identification*, 5th edn. Butterworth, Oxford.

35 | Cement minerals

Concrete is the closest industrial analog to a rock, forming a solid aggregation with what is called an aggregate (sand and gravel), cement, and water. Chemical reactions drive this process and the resulting product consists of compounds with many similarities to minerals. There are both hydraulic (water-resistant) and non-hydraulic varieties of cement. The latter are made with gypsum or lime and are not stable when exposed to water. Modern hydraulic cements, such as Portland cement, are much stronger, more stable and are currently universally used. We will look briefly at cement production, some of the cement compounds, and their mineral analogs. Some causes of concrete deterioration and cracking will be discussed.

35.1 Significance of cement

Concrete is the most widely used structural material in the world today. In 2013, 4 billion tonnes of Portland cement were converted into 40 billion tonnes of concrete, more than 6 tonnes for every human being. This is about five times the tonnage of steel consumption. Currently China is by far the largest producer of Portland cement. Even though concrete is considerably weaker than steel, it is preferred for several reasons. One reason is its excellent resistance to water. Today it is widely used in the construction of dams and offshore oil platforms, for example in the North Sea. A second reason is the ease with which concrete can be formed into almost any shape and size. Freshly made concrete is of a plastic consistency and can be poured into any prefabricated form. After a few hours it solidifies into a hardened and strong mass. Finally, at about $US 20 per tonne, concrete is the cheapest and most readily available building material, compared for example with steel ($US 500 per tonne).

Concrete is a composite material that essentially consists of a binding medium or *cement* that is combined with fragments of rocks, the so-called *aggregate* (Figure 35.1). The aggregate is granular material such as sand, gravel, or crushed rock. There are two types of cement, *hydraulic* and *nonhydraulic*. Hydraulic cements not only harden by reacting with water but also form a water-resistant product. Cements derived from calcination (i.e., obtaining calcium oxide by

heating) of gypsum or carbonates, such as limestone, are nonhydraulic, because their products are subject to dissolution and are not resistant to water.

Concrete has many similarities with rocks. The closest analogy is a cemented sandstone, in which sand grains are held together by a matrix with crystals of quartz or calcite. In concrete the minerals are far more complex and often not very well defined. In this chapter we will look briefly at "cement minerals" and the reactions by which they form, and by which they transform into new phases.

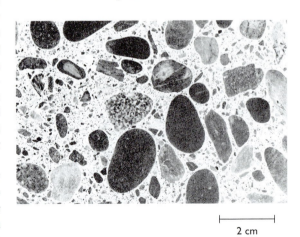

2 cm

Fig. 35.1 Polished section of a concrete specimen showing coarse and fine aggregate and a matrix consisting of hydrated cement paste (courtesy P. J. M. Monteiro).

There has been much recent interest in better understanding the processes that take place during the hardening of cement, and even more importantly the changes that occur owing to corrosion and deterioration. It is a large field of research that is generally pursued with the same methods as used in mineralogy: these methods include characterization using the techniques described in Part III, application of chemical thermodynamics and kinetics to understand transformations, and experiments under controlled conditions. Many mineralogists find employment in the cement industry.

35.2 Some features of nonhydraulic cements

Gypsum cement is still widely used as plaster of Paris for interior applications. The transformation of natural gypsum to anhydrite by heat treatment at 130–150 °C is described by the following chemical reaction:

$$2CaSO_4 \cdot 2H_2O \rightarrow CaSO_4 \cdot \frac{1}{2}H_2O + CaSO_4 + 3\frac{1}{2}H_2O$$
Gypsum   Hemihydrate   Anhydrite Vapor
(gypsum cement)

During curing, this reaction is reversed by adding water to gypsum cement and obtaining a hard and coherent gypsum aggregate. Unfortunately gypsum is very soluble in water.

A better nonhydraulic cement is *lime*, which was used extensively in Europe and the Middle East throughout antiquity, the Middle Ages, and well into the nineteenth century, and by the Mayas in Mexico. Lime is formed by decarbonation of calcite by heating at 900–1000 °C:

$$CaCO_3 \rightarrow CaO + CO_2$$
Calcite   Quick lime   Gas

Mixing the powder of quick lime with water produces a cement paste that quickly reacts and hardens by forming hydrated lime:

$$CaO + H_2O \rightarrow Ca(OH)_2$$
Lime   Water   Hydrated lime

Concrete with hydrated lime is not stable over long periods because $Ca(OH)_2$ is also soluble in water. However, if the exposure to water is not excessive, $Ca(OH)_2$ slowly carbonates in air to form stable calcite:

$$Ca(OH)_2 + CO_2 \rightarrow CaCO_3 + H_2O$$
Hydrated lime   Calcite   Vapor

Some of the earliest applications were by the Romans. Lime mortars that were used in the construction of aqueducts and retaining walls in harbors by the Ancient Greeks and Romans were rendered hydraulic by the addition of "pozzolanic" material, a volcanic ash that reacted with lime to produce a water-resistant cementitious product (e.g., Jackson *et al.*, 2014). (Pozzuoli is a town near Naples where volcanic ash was mined.) When silica-rich pozzolan is added to the system, a calcium silicate hydrate (C-S-H) is formed which is stable in water:

$$CaO + SiO_2 + H_2O \rightarrow CaSiO_2(OH)_2$$
Lime   Silica   Water   C-S-H

35.3 Portland cement

Portland cement is defined as a "hydraulic cement" and is produced by pulverizing a mixture of calcium silicates, calcium carbonates and also some aluminum and iron oxides. These raw materials are obtained by heating a mixture of limestone and clay at very high temperature. Portland cement was first produced in 1824 in the UK, although a very similar method was used by the Romans as mentioned above, but then forgotten. In a clever marketing strategy it was named after the Isle of Portland (UK), where a gray granite occurs which was highly popular at that time. Since then it has, with only minor changes, become the standard cement for most concrete applications. In a cement factory (Figure 35.2) limestone and clay are mixed and then milled for better reactivity and subsequently heated in a rotating kiln to a temperature of 1450–1550 °C. The reaction produces nodules 5–30 mm in diameter that are called clinker. About 5% gypsum is added to the clinker to control the early setting and hardening reactions of the cement. The composite is then ground to <75 μm in diameter.

This process releases large quantities of CO_2, through burning, as well as decomposition of carbonates. The cement industry is responsible for 8% of the world's industrial production of CO_2, and immense efforts are dedicated to reducing this problem.

The main minerals in clinker are tricalcium silicate (Ca_3SiO_5), also called *alite*, and dicalcium silicate (Ca_2SiO_4), called *belite*. Belite has a structure similar to that of olivine, with calcium substituting for magnesium. In detail the packing of oxygen ions around

Table 35.1 Chemical composition of cement minerals, with standard abbreviations used in the cement industry (Arial font is used for cement abbreviations in this chapter to avoid confusion)

Oxide	Abbreviation	Compound	Abbreviation
CaO	C	$3CaO \cdot SiO_2$	C₃S (alite)
SiO_2	S	$2CaO \cdot SiO_2$	C₂S (belite)
Al_2O_3	A	$3CaO \cdot Al_2O_3$	C₃A
Fe_2O_3	F	$4CaO \cdot Al_2O_3 \cdot Fe_2O_3$	C₄AF
SO_3	$\overline{S}$	$3CaO \cdot 3Al_2O_3 \cdot SO_3$	C₃A₃$\overline{S}$
H_2O	H	$CaO \cdot SO_3 \cdot 2H_2O$	C$\overline{S}$H₂ (gypsum)
		$CaO \cdot H_2O$	CH (portlandite)
		$6CaO \cdot Al_2O_3 \cdot 3SO_3 \cdot 32H_2O$	C₆A$\overline{S}$₃H₃₂ (ettringite)

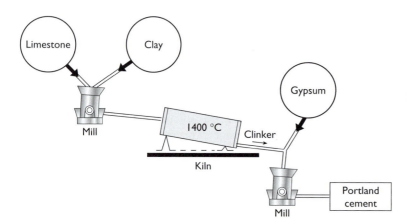

Fig. 35.2 Schematic representation of a cement plant. Limestone and clay are mixed and then heated in a kiln to form clinker. Gypsum is added to the clinker, and the mixture is then ground to the final product, Portland cement.

calcium is irregular, with O^- concentrated on one side of each Ca^{2+}. This leaves large structural holes that account for the high lattice energy and reactivity, particularly in the case of alite. If you read the cement literature you will come across a new nomenclature that is confusing at first. Each oxide is abbreviated with a letter, and compounds are given symbols according to the relative amounts of oxides present in them (Table 35.1).

In the language of cement research, Portland cement has an approximate composition of: 55% C₃S (Ca_3SiO_5), 25% C₂S (Ca_2SiO_4), 12% C₃A ($Ca_3Al_2O_6$), and 8% C₄AF ($Ca_4Al_2Fe_2O_{10}$). Clinker phases are easily recognized in an X-ray powder diffraction pattern (Figure 35.3).

When Portland cement is mixed with water, multiple reactions occur within minutes and continue over days and weeks. The systematic changes in mineral composition and microstructure with age of hydration are illustrated schematically in Figure 35.4. Corresponding to these changes is an increase in strength. At first the high-temperature compounds go into solution and the solution quickly becomes saturated with various ionic species. Within a few minutes, needle-shaped crystals of C₆A$\overline{S}$₃H₃₂ ($Ca_6Al_2S_3O_{18} \cdot 32H_2O$) called *ettringite* (which is also a rare, naturally occurring mineral) appear (Figure 35.5a). Also large prismatic or plate-like crystals of calcium hydroxide CH (Ca(OH)₂) (with the mineral name *portlandite*, derived from Portland cement, not the Isle of Portland, UK) form (Figure 35.5b). Portlandite is stoichiometric and forms large crystals constituting 20–25% of the volume. Its presence has an adverse effect on strength and durability.

Later, very small crystals of calcium silicate hydrates (C-S-H) begin to fill the empty space, formerly occupied by water. C-S-H is not a well-defined compound and its composition can vary considerably with temperature, age of hydration, and the water: cement ratio, therefore the notation C-S-H is used. On complete hydration, the approximate composition is C₃S₂H₃ ($Ca_3Si_2O_4(OH)_6$). C-S-H makes up

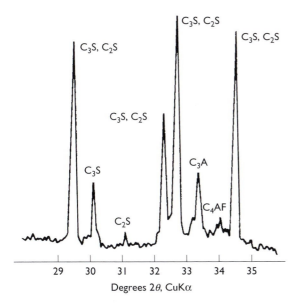

Fig. 35.3 X-ray diffraction pattern of Portland cement clinker with characteristic peaks of the main phases, particularly alite (C_3S) and belite (C_2S) and minor calcium aluminates. Many peaks overlap, yet those at $2\theta = 30.1°$ and $31.2°$ are diagnostic.

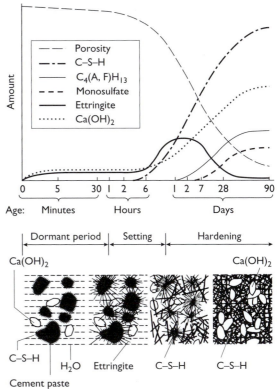

Fig. 35.4 Mineralogical and microstructural changes during hydration of cement (courtesy P. J. M. Monteiro).

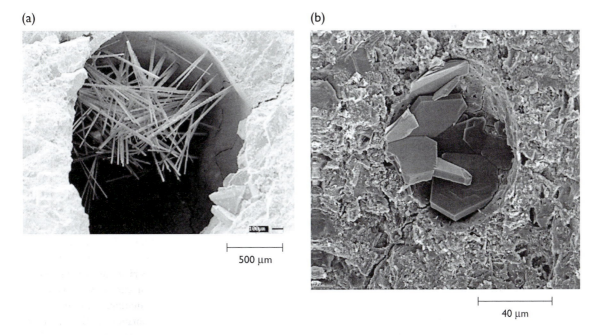

Fig. 35.5 SEM images of cement minerals in voids of concrete. (a) Cluster of ettringite needles. (b) Crystals of portlandite with platy morphology and excellent cleavage (courtesy P. J. M. Monteiro).

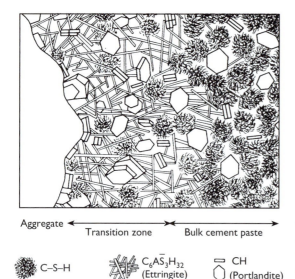

Aggregate ← Transition zone ↔ Bulk cement paste →

C–S–H $C_6A\bar{S}_3H_{32}$ (Ettringite)   CH (Portlandite)

Fig. 35.6 Schematic cross-section through aggregate and cement paste showing transition zone and bulk zone with different composition and microstructure (based on Mehta and Monteiro, 2006).

50–60% of the volume of the paste. The morphology varies from minute fibers to a network. The crystal structure is still not resolved in detail but it has similarities with the chain silicate minerals *tobermorite* $(Ca_5H_2(Si_3O_9)_2\cdot4H_2O)$ and *jennite* $(Ca_9(Si_3O_9)_2(OH)_6\cdot8H_2O)$ (e.g., Battocchio *et al.*, 2012), both related to wollastonite. C-S-H has an extremely high surface energy, providing strength through van der Waals forces (interlayer spaces are about 18 Å). After a few days of hydration, ettringite becomes unstable and decomposes to form monosulfate hydrate $(C_4A\bar{S}H_{12})$, which has hexagonal plate morphology.

Figure 35.6 is a sketch of a typical concrete microstructure. Heterogeneity is present at various scales. Pores in C-S-H are around 50 Å, and capillary voids occur in all sizes and shapes, from a few nanometers to micrometers. In concrete, the introduction of the rock aggregate produces an additional heterogeneity for the matrix and the interfacial transition zone between the aggregate and the cement is the weak link of the paste, because it is composed largely of low-strength portlandite and ettringite, while the bulk cement paste is composed largely of C-S-H, which provides the main strength.

Two important hydration reactions are exothermic, i.e., heat is released:

$$2C_3S + 6H \rightarrow$$
$$2Ca_3SiO_5\ 6H_2O$$
Alite   Water
$$C_3S_2H_3 + 3CH + heat\ (120cal/g)$$
$$Ca_3Si_2O_4(OH)_6\ 3Ca(OH)_2$$
(C-S-H)   Portlandite

$$2C_2S + 4H \rightarrow$$
$$2Ca_2SiO_4\ 4H_2O$$
Belite   Water
$$C_3S_2H_3 + CH + heat\ (70cal/g)$$
$$Ca_3Si_2O_4(OH)_6\ Ca(OH)_2$$
(C-S-H)   Portlandite

The heat released during these reactions can be detrimental for the curing process, and if concrete is applied in large quantities, it needs to be cooled. For example, in large concrete structures such as the Hoover Dam in the Arizona–Nevada Desert, thermal stresses could become very high and considerable efforts needed to be made to reduce the temperature rise inside the concrete mass. This can be done by using cements with low amounts of C_3S (alite) and CA, by using ice instead of water, or by installing cooling pipes inside the dam. Clinkers rich in C_3S are preferred for colder climates, whereas those rich in C_2S (belite) are more often used for hot weather conditions.

35.4 Some problems with concrete

There are several problems related to the deterioration of concrete and, since concrete is the major construction material for bridges, highways, and many buildings, the stability of concrete is of major economic importance. We are going to briefly discuss three issues: sulfate attack, the alkali silica reaction, and steel corrosion. Large research projects are dedicated to each and many mineralogists are engaged in the work. We will see that the reactions involved are similar to those that occur in mineral systems. To get an impression of the magnitude of the concrete problems, the US Federal Highway Administration recently estimated that the cost of correcting damage to highways, bridges, and buildings amounts to over 150 billion dollars in the USA alone.

Sulfate attack

Sulfate ions present in soil, groundwater, seawater, decaying organic matter, acid rain, and industrial effluents are known to have an adverse effect on the long-term durability of concrete. Sulfate attack on the hardened cement paste in concrete manifests itself in the form of cracking, spalling, increased permeability, and loss of strength. Therefore, concrete structures exposed to sulfate water must be designed for sulfate resistance.

Sulfate attack occurs when sulfate ions penetrate the concrete from the surrounding environment. As the sulfate ions permeate the concrete, they react with portlandite to form gypsum, which is accompanied by a large volume increase and causes expansion and cracking.

$$Ca(OH)_2 + SO_4^{2-} + 2H_2O \rightarrow$$

Portlandite   Water

$$CaSO_4 \cdot 2H_2O + 2OH^- \quad \Delta V = 123\%$$

Gypsum

where ΔV is the increase in volume.

For industrial Portland cements, monosulfate hydrate $C_4A\overline{S}H_{18}$ ($Ca_4Al_2SO_{10} \cdot 18H_2O$), which is a major component of fully hydrated hardened cement, reacts with gypsum to form secondary ettringite ($C_6A\overline{S}_3H_{32}$, trisulfate hydrate, $Ca_6Al_2S_3O_{18} \cdot 32H_2O$). This reaction also causes expansion of the solid components:

$$Ca_4Al_2SO_{10} \cdot 18H_2O + 2CaSO_4 \cdot 2H_2O + 10H_2O \rightarrow$$

Monosulfate hydrate   Gypsum   Water

$$Ca_6Al_2S_3O_{18} \cdot 32H_2O \quad \Delta V = 57\%$$

Ettringite

In the presence of portlandite (CH), the monosulfate hydrate ($C_4A\overline{S}H_{18}$) is converted to ettringite when the hydrated cement paste comes into contact with a sulfate ($\overline{S}$).

$$Ca_4Al_2SO_{10} \cdot 18H_2O + 2Ca(OH)_2 + 2SO_3 + 14H_2O \rightarrow$$

Monosulfate hydrate   Portlandite   Water

$$Ca_6Al_2S_3O_{18} \cdot 32H_2O \quad \Delta V = 91\%$$

Ettringite

All these reactions have a volume increase and the formation of gypsum and ettringite by sulfate attack causes internal stresses.

Details of these sulfate reactions are still not clear and are under much investigation. The reactions are dependent on the pH of the attacking sulfate solution, which is governed by the sulfate ion concentration. Weak sulfate solutions can react with the aluminate constituents of cement to form ettringite. Stronger solutions are somewhat acidic and capable of forming gypsum as a result of chemical reaction with portlandite. Highly acidic solutions can even decompose C-S-H.

Alkali–silica reaction

The second destructive problem is also of direct mineralogical significance. In the 1920s and 1930s, extensive cracking appeared within a few years of construction in a number of bridges and pavements along the US Californian coast from Monterey County to Los Angeles County. This could be attributed to a reaction of the opaline and cherty aggregate, which is a common rock in this region, and alkalis present in the cement paste. In theory, any aggregate containing silica has the potential to participate in the alkali–silica reaction. Yet it is siliceous minerals with disordered, amorphous, or defect-rich structures that are particularly susceptible to attack. This includes volcanic silica glass, opal, microcrystalline quartz, and highly deformed quartz in metamorphic rocks. If the aggregate contains significant amounts of such forms of silica, silica reacts with alkali ions that are present in the pore system of the cement paste, or enter from external sources (e.g., Na^+ from de-icing salts). This reaction produces a gel that transforms from an amorphous solid ultimately to a liquid phase, as water is taken into the structure. The initial hydrolysis of the siliceous fraction of the aggregate destroys the aggregate integrity and opens a pore system through which water can percolate. Swelling of the alkali silicate gel causes local stresses and cracking. This reaction occurs only if water is present and distress in structures is generally observed after 5–10 years:

$$2(K, Na)OH + SiO_2 + 2H_2O \rightarrow K_2O \cdot Na_2O \cdot SiO_2 \cdot 3H_2O$$

Alkali ions   Silica   Water   Gel

This reaction indicates that, in constructions with concrete, much attention ought to be paid to the mineralogical composition of the aggregate.

Corrosion

Structural concrete is usually designed in combination with reinforcing steel bars to increase the low tensile strength of the concrete. The alkaline environment of the concrete provides excellent protection for the steel, suppressing its tendency to corrode when exposed to the external natural environment. However, if solutions enter the concrete, for example in combination with sulfate attack or the alkali–silica reaction, corrosion does occur, producing serious structural damage. Corrosion of embedded steel reinforcing bars is a destructive electrochemical process that ultimately weakens concrete structures. Since corrosion products have a greater volume than the original steel reinforcing bars, internal stresses will develop in the cement mortar, at the steel/mortar interface. As a result, the surrounding concrete will crack and eventually spall away. The structural integrity of the concrete is increasingly compromised as cracking progresses. With the steel corroding away, the reinforcing bar cross-section is reduced, decreasing the member's tensile strength. Thus, corrosion of steel reinforcement and the associated concrete cracking lead to a loss of both tensile strength of the steel and compressive strength of the concrete.

Corrosion is essentially an electrochemical process. It involves the formation of a cathode and an anode, with an electrical current flowing between the two (cf. the discussion in Chapter 19). The anode is the steel bar. The cathode is located at a point where oxygen can enter by diffusion. At the anode, metallic iron is oxidized to Fe^{2+}, which dissolves and is transported to a cathode, where OH^- is produced. The Fe^{2+} and OH^-, which are moving in pore solutions, interact chemically to produce iron hydroxide at the anode, a reaction with a large volume increase:

$$Fe + \frac{1}{2}O_2 + H_2O \rightarrow Fe(OH)_2 \qquad \Delta V = 272\%$$

In general, the corrosion process occurs when metals revert back to lower energy states. Several conditions are required for corrosion to occur in an aerobic environment. The system must have an anode to produce electrons, and a cathode to accept electrons. This is usually satisfied by the metallic reinforcement and depends on microstructural defects in the metal and the presence of protective films. In addition, there has to be availability of oxygen and water at the cathode site as well as an electrical connection between the anode and cathode sites to transfer electrons and this depends on the properties of the cement, which can be viewed as a two-phase material composed of hydrated solid minerals and a pore fluid. The structure, size, distribution, and interconnection of the pores in the cement paste, in conjunction with the presence of cracks and microcracks, control the permeability of concrete. The more permeable the concrete, the greater the availability of oxygen and water at the cathode for electrochemical reactions. The corrosion process cannot occur without the availability of oxygen at the cathodic site. However, it has been established that even very dense concrete is fairly permeable to oxygen.

Techniques currently used to guard against corrosion include sealants, epoxy-coated reinforcing bars, galvanized steel, fiber-reinforced plastic reinforcement, cathodic protection, and protective overlays and membranes for bridge decks. A more traditional means of decreasing corrosion involves lowering the permeability of concrete. This can be accomplished by providing adequate concrete cover over the reinforcement, using water-reducing admixtures, and by proper curing of concrete.

35.5 Summary

Concrete can be considered an industrial clastic rock with fragments cemented together by a mineral microstructure. The earliest cements based on lime (produced from calcite) and calcium hemihydrate (produced from gypsum) were nonhydraulic and subject to dissolution. Modern Portland cements, with some similarities to Roman pozzolanic cements, are hydraulic, i.e., water resistant. Clinker produced by heating limestone and clay produces alite (Ca_3SiO_5) and belite (Ca_2SiO_4), with additional sulfur and aluminous phases which react if mixed with water within minutes to form portlandite ($Ca(OH)_2$) and ettringite ($Ca_6Al_2S_3O_{18}$ $32H_2O)_2$) providing a first coherent aggregation. This changes over days to form C-S-H, related to minerals tobermorite and jennite, which provides the main strength of concrete. Concrete produced by Portland cement can be subject to deterioration such as sulfate attack from groundwater, alkali–silica reactions with silica-rich aggregates, and corrosion of reinforcement steel. These processes are associated with volume expansion and cause fracturing and loss of strength.

Concrete is by far the most important building material, with tremendous significance for our society. It is also an example of how chemical reactions between mineral-like crystals cause transformations to produce a material of very high compressive strength.

Test your knowledge

1. Why is concrete increasingly replacing steel as a structural material?
2. Review the chemical reactions involved in the production and use of lime.
3. Which minerals are used as raw materials in the manufacturing of modern Portland cement?

4. Why is a cement rich in alite preferred for concrete in cold climates, whereas belite-rich cement is better in hot climates?

5. Describe some of the problems encountered with concrete.

Further reading

Brandon, C. J., Hohlfelder, R. L., Jackson, M. D. and Oleson, J. P. (2014). *Building for Eternity: The History and Technology of Roman Concrete Engineering in the Sea.* Oxbow Books, Oxford.

Hewlett, P. C. (2004). *Lea's Chemistry of Cement and Concrete*, 4th edn. Butterworth-Heinemann, Oxford.

Mehta, P. M. and Monteiro, P. J. M. (2006). *Concrete: Microstructure, Properties, and Materials*, 3rd edn. McGraw-Hill, New York.

Neville, A. M. (2012). *Concrete Technology*, 5th edn. Trans-Atlantic Publications, Philadelphia, PA.

Taylor, H. F. W. (1997). *Cement Chemistry*, 2nd edn. Thomas Telford Publishing, London.

36 | Minerals and human health

Mineralogy is finding increasing use in medicine and in environmental health applications. Biologists, physicians, pharmacists, and environmental health professionals rely on expertise provided by mineralogists. On the one hand, minerals may constitute health hazards. Exposure to asbestos, toxic waste from mining operations, or radiation due to radioactive decay may cause cancer or other diseases. On the other hand, compounds contained in minerals such as sodium chloride and calcium are essential nutritional components and both bones and teeth are composed of mineral-like hydroxyapatite.

36.1 Mineral-like materials in the human body

The principal mineral-like compounds in humans are phosphates, but other mineral-like crystals occur as well (Table 36.1). Bones of adults consist of approximately 70% calcium phosphate and 30% organic matter. Calcium phosphate forms tiny prismatic crystals less than 1000 Å in length, with a structure and composition similar to that of apatite (e.g., Ivanova *et al.*, 2001). Organic matter contains combinations of different collagens, fats, and proteins. The apatite crystallites line up in chains and, together with organic material, form fibers of bone tissue. A portion of the phosphate material of newborns is amorphous. As a child grows, the amorphous component disappears and the existing crystals increase in size. At the same time the portion of organic material decreases with age, with the result that the fibers lose their elasticity and the bone tissue becomes more brittle. We have discussed this already in Section 25.5 (e.g., Figure 25.8).

Biogenic apatite-like minerals have a rather variable composition, which can be expressed approximately by a formula such as $Ca_{10-x}(PO_4)_{6-y}$ $(CO_3)_z(OH)_{2+w} \cdot nH_2O$. A number of PO_4^{3-} tetrahedra in the structure are replaced by CO_3^{2-}, and Ca is substituted by other cations (Frank-Kamenetskaya *et al.*, 2011). The main crystalline phase in cartilage is also biogenetic phosphate, analogous to apatite and constituting about 5% of the volume. Modifications of

apatite form 96% of tooth enamel (the outer coating of teeth) and 70% of dentine (the material beneath the enamel), with the rest of tissue volume composed of proteins. In tooth enamel, some OH^- is replaced by F^-, which makes teeth more resistant to decay. In enamel, crystals are organized in a layered structure to improve mechanical properties (see Figures 25.7 and 25.8).

Crystals may also grow abnormally within the human body. Aggregations of biogenetic apatite up to 2 cm in size have been discovered in some malignant tumors. The lungs of patients with tuberculosis show calcification, with apatite and whitlockite having been observed. Similarly, in people with heart disease, heart tissue, including arteries and the aorta, can become covered with apatite-like calcium phosphate crystals (Plate 32f).

Abnormal "stones" form in the bladder, kidneys, liver, gall bladder, and trachea, and are composed of amorphous or very diverse crystalline phases of phosphates, carbonates, oxalates, or urates (Table 36.2). The morphology of the stones resembles inorganically formed concretions, with rhythmical zoning, geometrical sorting, and subgrain formation. In some cases a drusy growth has been documented.

36.2 Minerals in nutrition

Apart from table salt, known by mineralogists as halite (NaCl), minerals are rarely consciously ingested

Table 36.1 Mineral-like substances in the human body

Name	Formula	Place
Apatite	$Ca_5(PO_4,CO_3,OH)_3(OH)$	Bones, teeth, kidneys, urinary bladder, salivary glands, prostate, lungs, heart, blood vessels
Brushite	$CaHPO_4 \cdot 2H_2O$	Bones, teeth, kidneys, urinary bladder, prostate
Struvite	$MgNH_4PO_4 \cdot 6H_2O$	Kidneys, urinary bladder, teeth
Newberyite	$MgHPO_4 \cdot 3H_2O$	Kidneys, teeth
Whitlockite	$Ca_9Mg(PO_4)_6(PO_3OH)$	Bones, teeth, kidneys, urinary bladder, prostate
Calcite	$CaCO_3$	Gall bladder, teeth, salivary glands, tumors, kidneys, lungs
Whewellite	$CaC_2O_4 \cdot H_2O$	Urinary bladder
Weddellite	$CaC_2O_4 \cdot 2H_2O$	Urinary system
Urinary acid	$C_5H_4N_4O_3$	Urinary system

Source: From Katkova, 1996.

Table 36.2 Composition of urinary and gall stones

Medical name	Mineral name	Urinary	Gall	Formula
Oxalates	Whewellite	X		$CaC_2O_4 \cdot H_2O$
	Weddellite	X		$CaC_2O_4 \cdot 2H_2O$
Phosphates	Struvite	X		$MgNH_4PO_4 \cdot 6H_2O$
	Apatite	X	X	$\sim Ca_5(PO_4,CO_3,OH)_3(OH)$
	Newberryite	X		$MgHPO_4 \cdot 3H_2O$
	Brushite	X		$CaHPO_4 \cdot 2H_2O$
	Whitlockite	X		$Ca_9Mg(PO_4)_6(PO_3OH)$
Carbonates	Vaterite	X	X	$CaCO_3$ hexagonal
	Calcite	X	X	$CaCO_3$ trigonal
	Aragonite		X	$CaCO_3$ orthorhombic
Oxides	Magnetite	X		$FeFe_2O_4$
	Hematite	X		Fe_2O_3
	Goethite	X		$FeOOH$
	Lepidocrocite	X		$FeOOH$
Urates	Urea	X		$C_5H_4N_4O_3$
	—	X		$C_5H_4N_4O_3 \cdot 2H_2O$
	—	X		$C_5H_2O_3N_4(NH_4)_2$
	—	X		$C_5H_2O_3N_4Na_2 \cdot H_2O$
	—	X		$C_5H_2O_3N_4Ca \cdot 2H_2O$
Organic compounds	Holesterine		X	$C_{27}H_{46}O$
	Holesterine, hydrous		X	$C_{27}H_{46}O \cdot H_2O$
	Ca-palmaniate		X	$CH_3(CH_2)_{14}(COO)_2Ca$

Note: X indicates positive association.
Source: From Katkova, 1996; Korago, 1992.

by humans. Among the exceptions are barite ($BaSO_4$), called by the Russian mineralogist A. E. Fersman "the most edible mineral", which has been used as an inert filling for chocolate in Russia, and kaolinite, which is added to some ice creams to provide consistency when they start to melt. There are also other examples, less well known, where minerals are part of our food. Yet on shelves in supermarkets, "minerals" in the form of

Table 36.3 Essential nutritional elements and their physiological functions

"Macrominerals" (required in large amounts)	
Ca	Bones, teeth, neural transmission, muscle functions
Cl	Water and electrolyte balance, digestive acid
Mg	Regulating chemical reactions, nerve transmission, blood vessels
P	Bones, cell functions, and blood supply
K	Growth, body fluid, muscle contraction, neural transmissions
Na	Regulating acid–base balance, neural transmissions, blood pressure
S	Constituent of proteins, thiamine, structure of hair, skin
"Microminerals" (required in trace amounts)	
Cr	Glucose metabolism
Co	Vitamin B12, red blood cells
Cu	Red blood cells, prevents anemia, nervous system, metabolism
F	Tooth decay, strong bones
Fe	Hemoglobin, immune system
I	Thyroid hormones, reproduction
Mn	Tendon and bone development, central nervous system, enzymatic reactions
Mo	Growth, enzymes
Se	Prevents cardiovascular disease, cancer, detoxifies pollutants, antioxidant
Zn	Enzymes, red blood cells, sense of taste/smell, immune system, protects liver

Source: From Dunn, 1983.

nutritional additives play a role almost as important as that of vitamins, and in every modern book on nutrition there is a chapter on minerals. This popularity is in part due to the rather free use of the term "mineral" by physicians, pharmacists, and nutritionists. Traditionally they call any inorganic compound "mineral", following an old usage that divided chemistry into two branches: organic and mineral.

In nutrition, so-called minerals are divided into macrominerals (calcium, chlorine, magnesium, phosphorus, potassium, sodium, and sulfur) and microminerals (such as chromium, cobalt, fluorine, iron, manganese, molybdenum, and zinc). The former are required in rather large quantities in our daily diet, while the latter are also essential for physiological functions, but only in trace amounts. Table 36.3 lists some of the physiological functions of macro- and microminerals. Ultimately most of these elements are derived from "real minerals", but indirectly through a long chain of events, both natural and artificial. Primary minerals in rocks decompose to clay minerals that become part of soils. Plants growing on those soils accumulate the inorganic elements and store them in roots and leaves. Animals eat the plants and transfer the elements into their tissue, and finally humans acquire these elements largely by consuming either plants or animals, or by eating mineral substitutes.

The quantity of elements stored in plants is considerable, particularly in the green parts (e.g., average contents on a moisture-free basis in some legumes and grasses are Ca 1–4 weight%, P 0.1–0.5 weight%, Fe 100–200 ppm, Cu 5–15 ppm). These amounts vary greatly with the mineral content of the soil, but they are also affected by many other factors such as climate and elemental balance. The mineral content of plants can have a direct effect on the health of animals. It has been observed that cattle grazing in pastures with underlying limestone are less likely to develop bone diseases than those grazing on granitic soils. The trace element selenium, an essential antioxidant to preserve the cellular membrane, can become toxic if concentrations are too high, as in some sedimentary rocks. Conversely, a lack of selenium in the diet of Bighorn sheep was recently implicated in the low survival rate of newborn lambs (e.g., Hnilicka *et al.*, 2002). In human nutrition, calcium, magnesium, phosphorus, and copper are stored in legumes, whereas chromium, iron, manganese, and zinc are enriched in cereals.

A number of drugs used in the treatment of internal and external diseases contain minerals. The halide mineral bischofite ($MgCl_2 \cdot 6H_2O$), for example, is used for treating arthritis and rheumatic fever. Calcite, dolomite, and apatite are used as calcium, magnesium, and phosphorus supplements.

Direct ingestion of soils as a food supplement and medicine, known as *geophagy*, is common among some primates and is still practiced in some countries by humans (e.g., Aufreiter *et al.*, 1997). Ancient Greeks and Romans used tablets of soil as a remedy against poisoning. Traditionally, and until fairly recently, soils were consumed in China as famine food. In Europe, well into the eighteenth century, clay was mixed with flour in the preparation of bread. Pomo Indians in northern California, USA, mixed clay with ground acorns to neutralize the acidity. The most widespread incidence of geophagy is in Central Africa, as well as among some African Americans in the southern USA. In Africa geophagic clays are widely used by pregnant women as food supplements containing elements such as phosphorus, potassium, magnesium, copper, zinc, manganese, and iron, and as remedies against diarrhea. Interestingly the chemical composition of these soils and soil extracts is remarkably similar to modern commercial mineral-nutrient substitutes. The main clay mineral in geophagic soils is kaolinite. Soils rich in smectite are less desirable because of their swelling properties (see Chapter 29).

Minerals are also extensively used in beauty and grooming products. For example, talc is an important ingredient of many cosmetic products, baby powder being one of the better known ones. Minerals such as kaolinite, smectite, nontronite, biotonite, and hectorite clays are used in cosmetics, toothpaste, and pharmaceuticals, while mica provides the sheen in lipstick. Most consumers are generally unaware of these mineral ingredients.

36.3 Minerals as health hazards

Diseases caused by particulates

Minerals are ubiquitous in our daily environment. Along with their synthetic analogs, they are used in household products, as abrasives, pharmaceuticals, catalysts, fillers, anti-caking agents, building materials, insulation, and pigments. We are exposed to minerals daily, often without being aware of it. Many workers, including miners, quarry workers, sandblasters, stone masons, and agricultural workers, are exposed to dust from a variety of sources and inhale small mineral fragments. These workers have an increased probability of developing pulmonary diseases. Since workers are often exposed to dust from a mixture of minerals, it is difficult to establish whether it is the number of ingested particles or a specific mineral that causes a particular disease. Minerals for which a dose–response relationship between the amount of exposure and the degree of injury has been established with some confidence are fibrous forms of asbestos.

There is amphibole asbestos (riebeckite, trade name *crocidolite*; grunerite, trade name *amosite*; tremolite, actinolite, and anthophyllite) and serpentine asbestos (*chrysotile*). Recently several other amphibole minerals (winchite, richterite, and arvedsonite) have been implicated as causing cancer in workers at the Libby, Montana (USA), vermiculite mines. A brief review of these disease-causing minerals and the methods used to assess and monitor their presence is given below.

The first reported case of the lung disease *asbestosis* was in 1927 in a chrysotile textile worker. Ten years later asbestosis became generally accepted by the industry as an occupational disease with distinct characteristics. Stanton *et al.* (1981) demonstrated with a classic, though still controversial, epidemiological study that rats exposed to fibrous asbestos dust developed carcinogenic tumors (Figure 36.1). In that study, rats exposed to equivalent amounts of nonasbestos dust (such as talc) did not develop cancer. The researchers concluded that the fibrous morphology caused the disease. Since then, the fibrous morphology

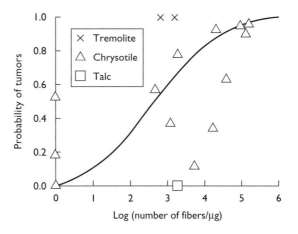

Fig. 36.1 Incidence of malignant tumors in rats as function of fiber concentration (data from Stanton *et al.*, 1981).

of asbestos has, in fact, been found to be only part of the reason that asbestos is harmful. Numerous other minerals exist that can occur in a fibrous morphology, such as talc, gypsum, and clays (i.e., kaolinite, halloysite, sepiolite) but they have not been associated with lung disease. In the early 1980s, the use of asbestos in the USA and in Europe was largely eliminated. However, much asbestos still exists in insulation, fireproofing, flooring, roofing, and surfacing materials of older buildings.

The detailed mechanisms of the lung diseases caused by inhaled dust are still unclear, but it has been established that sustained exposure to asbestos minerals can cause cancer of the lung, the trachea, and the bronchial walls. *Mesothelioma* is a rare malignant tumor, correlated with crocidolite exposure. It arises from the mesothelial membrane that lines the pleural cavity. Mesothelioma generally appears 20–40 years after asbestos exposure, but once it appears there is rapid growth, with the tumor spreading and invading adjacent organs such as the heart, liver, and lymph nodes. Death often occurs within one year after the first symptoms appear. *Asbestosis* is a nonmalignant disease that involves interstitial fibrosis with hardening of the lung tissue. It may lead to severe loss of lung function and ultimate respiratory or cardiac failure. The disease is often associated with pleural calcification and the appearance of asbestos bodies consisting of fibers coated with collagen (Figure 36.2). Asbestosis extends eventually to the walls of the alveoli (small air cavities where the oxygen exchange takes place) and leads to the destruction of alveolar spaces. The fibrous scar tissue narrows the airways, causing shortness of breath.

The harmful effect of asbestos dust is in part dependent on the physical shape of the particles, and the fibrous morphology of asbestos is particularly detrimental. However, it appears that the relative ability of the body to dissolve these materials is also of key importance. Asbestos is much less soluble in the body than are nondisease-causing fibrous minerals. Equally significant is the surface chemistry and reactivity of the particles. For example, fresh surfaces of minerals, exposed by fracture, are highly reactive due to the presence of under-coordinated surface atoms and broken bonds that accompany them. It has been observed that generation of free radicals by increased grinding of chrysotile fibers reduces the hemolytic activity because the particles become less crystalline. In chrysotile fibers enclosed in tissue, magnesium is preferentially leached from the fiber. If the

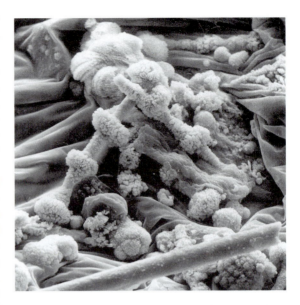

Fig. 36.2 SEM image of ferruginous bodies extracted from a human lung. Particles of asbestos are coated with an iron-rich material derived from proteins (see Guthrie and Mossman, 1993; photograph by L. Smith and A. Sorling).

surface chemistry of chrysotile is modified with polymers adsorbed to the particles, the toxic effect can be dramatically reduced.

As we have seen in Chapter 29, tetrahedral-octahedral sheets in chrysotile are rolled up similarly to a scroll (see Figure 29.5 and Plate 31a). The outside of the "scroll" is made up of the magnesium octahedral sheet, consisting of hydroxyl atoms on the surface which can be imaged with atomic force microscopy (AFM) (Figure 36.3). Each bright node in this image represents a hydroxyl on the surface, and each gray triangular region is a magnesium ion. It is on this surface that the chrysotile reacts with biological tissue. In the case of chrysotile, the surface is charge balanced and fairly regular.

In the amphibole crocidolite, the surface is dominated by {110} cleavages parallel to the silicate chains. The surface structure is much more irregular, containing not only OH^- but also tetrahedral Si^{4+}, octahedral Mg^{2+}/Fe^{2+}, and larger cations (Ca^{2+}, Na^+). During dissolution, amphiboles become depleted in iron, sodium, calcium, and magnesium. If iron is oxidized during the leaching process, it reprecipitates as ferric oxyhydroxide. Analyses of leached crocidolite fibers in human tissue display amorphous surface layers. Because of these layers, crocidolite fibers have a much

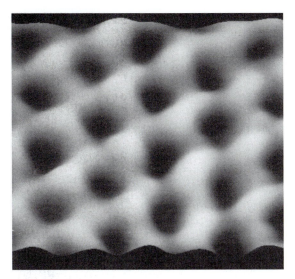

0.4 nm

Fig. 36.3 AFM image showing the atomic arrangement in the surface of lizardite, with a hexagonal pattern of hydroxyl ions (bright spots) and magnesium ions in depressions (gray) (from Wicks *et al.*, 1992).

longer lifetime than chrysotile, and for this reason crocidolite is more pathogenic.

Exposure to quartz dust leads to *silicosis*, a progressive lung disease characterized by the development of scar tissue. Inhalation of quartz particles 0.5–0.7 μm in size causes proteins to develop that surround the particles, stimulating fibroblast growth and producing collagen, an essential component of scar tissue. Fibrotic nodules develop in the region of small airways. As silicosis progresses, nodules coalesce and lesions develop that may involve one-third of the lung, leading ultimately to respiratory failure. Silicosis has symptoms similar to those of asbestosis. Unlike asbestos exposure, however, there is no clear evidence for a relationship between lung cancer and silicosis.

Coal workers' *pneumoconiosis* is caused by fine-grained coal dust composed of carbonaceous material. Dust-laden cells form a mantle around respiratory bronchioles, which dilate as the mantle enlarges, causing emphysema. Pneumoconiosis often takes many years to develop and, unlike silicosis, there is often no progression of this disease in the absence of further exposure.

These lung diseases illustrate that the interaction of fibers with human tissue is very complex. Fibers with

minor differences in composition and defect structure may have quite different biological activities. For example, glass fibers are not dangerous because they maintain their mechanical integrity and their dissolution rate is orders of magnitude faster than that of crystalline fibers.

Particle analysis

The assessment of hazardous concentrations of mineral particulates in the environment requires a combination of standardized industrial hygiene site-assessment techniques and mineralogical analytical procedures. We discuss these procedures in some detail because a fair number of mineralogists find employment in this field. The industrial hygiene assessment techniques involve a variety of sampling procedures (air drawn into membrane filters, wiping or vacuuming of known areas, direct sampling of building materials or rock/soil, etc.). The mineralogical procedures typically include the use of a polarized light microscope (PLM), phase contrast microscope (PCM), and transmission electron microscope (TEM) for asbestos, and X-ray diffraction (XRD) for quartz.

In the case of airborne particles, a known volume of air is collected onto a special type of membrane filter. Analysis of the particulate found on the filter is then performed using the appropriate technique. For asbestos air samples, a section of an air filter is either mounted on a glass slide and saturated with a special immersion oil for examination with a PLM or PCM, or is prepared to create a carbon film replica of the filter surface for examination with a TEM. Figure 36.4a is a PLM image of amosite. Figures 36.4b and c are TEM images of chrysotile and amosite, respectively, each with a characteristic morphology. The number of fibers in a given area is counted and, if the volume of air sampled is known, the number of fibers per unit volume of air (measured as fibers per cubic centimeter) can be calculated. For example, in the USA the Occupational Safety and Health Administration (OSHA) has established a permissible time-weighted exposure limit for workers at 0.1 fibers per cubic centimeter of air during an 8 hour work day. Many other countries have similar regulations.

In the case of asbestos bulk samples of building materials and rock/soil samples, a different technique called optical polarized light microscopy is generally used. It involves taking a small sample of the bulk

(a)

(b)

(c)

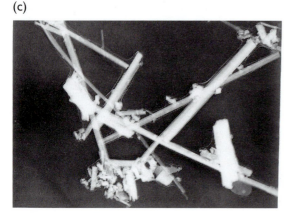

Fig. 36.4 Typical examples of asbestos found in building materials. (a) Polarized microscope image of amosite. Width 1.2 mm. (b) TEM image of chrysotile. Width 6 μm. (c) TEM image of amosite. Width 10 μm. (All micrographs are courtesy of Mark Bailey, Asbestos TEM Laboratories, Berkeley, California.)

material, mounting it in immersion oil (see Chapter 14), and identifying the minerals present, quantifying the amount of asbestos by area estimate or by point count. It is a fairly straightforward, although tedious, technique, which gives good general data on mass quantities of asbestos.

A major limitation in the performance of asbestos analysis is the fact that asbestos fibers are considered to be hazardous down to a length of 0.5 μm, which is well below the resolution limit of optical microscopes. For the precise identification of the extremely small asbestos fibers, the TEM is the analytical method of choice, as it can easily resolve particles much smaller than 0.5 μm (Figure 36.4b,c). Furthermore, the mineral identity of each individual asbestos fiber can be ascertained structurally by selected area electron diffraction (SAD) and chemically by energy dispersive X-ray analysis (EDXA).

Commercial testing laboratories that perform asbestos analysis must be certified by government agencies to perform each specific type of test that they offer. To become certified, they must pass a detailed inspection of their facilities and pass proficiency tests whereby blind samples are submitted to their facilities. Such laboratories employ a substantial number of mineralogists to perform both optical and electron microscopy analyses.

Chemical contamination from mining

With the increasing industrialization of society, the demand for both metallic and nonmetallic mineral products increases constantly. This causes irreparable damage to the environment. Open pit mines expand and deepen; underground mining causes subsidence; dumps of waste rocks grow and tailings of ore-dressing plants expand; the atmosphere becomes polluted with gases from smelters, often enriched in sulfur dioxide and carbon dioxide; and natural water systems are also polluted. Soils in the vicinity of Sudbury in Canada, which is one of the world's largest nickel producers, have a pH of only 3 and this causes extensive loss of vegetation. Metals such as nickel, lead, and copper vaporize during the high temperatures of smelting and are dispersed over extensive areas surrounding the smelters (Figure 36.5).

The most important pollutant of the hydrosphere is H^+, in the form of acid rain and acid mine

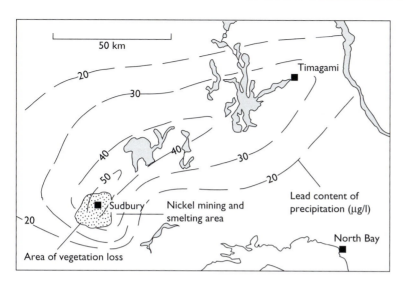

Fig. 36.5 Lead content of atmospheric precipitation around Sudbury (Canada) recorded 1970–4. The asymmetrical pattern is due to prevailing winds. Since then, the situation has much improved, but large concentrations of nickel and other heavy-metal contaminants remain in soils. A zone of vegetation loss near Sudbury is dotted (see Semkin and Kramer, 1976).

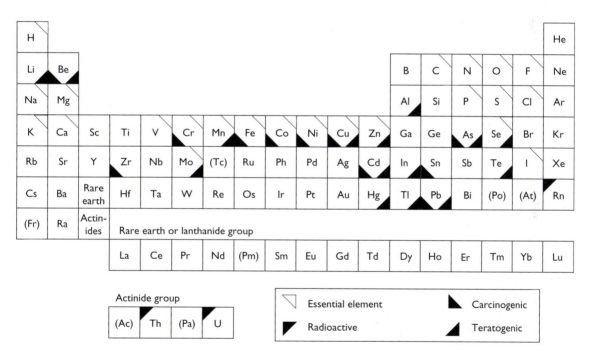

Fig. 36.6 Periodic table of elements, identifying those that are essential for human nutrition and those that are toxic and cause cancer (carcinogenic) or birth defects (teratogenic) if ingested in high doses. Also indicated are significant radioactive elements (data from Smith and Huyck, 1999).

drainage. Acid mine drainage results largely from the decomposition of pyrite to form iron hydroxide, H^+, and SO_4^{2-}. These reactions are often catalyzed by bacteria at low pH, increasing reaction rates by several orders of magnitude. Acid water produced by oxidation of sulfides can dissolve other metal sulfides and leach metals that are adsorbed in clays, thereby increasing the trace metal content in streams. Extremely low pH, even negative, never observed in natural systems, have been documented in mine waters in northern California, USA (e.g., Nordstrom and Alpers, 1999).

Many industrialized countries have put severe limitations on mining operations to maintain some environmental standards. One of the first environmental mining laws was the prohibition of hydraulic mining in California (1860) to prevent erosion and destruction of fertile farmland. Today in the USA, groundwater must be protected during mining operations, requiring elaborate schemes to ensure such protection. Only a few years ago, the majority of mining mineralogists were engaged in prospecting and extraction technologies. Today many are conducting research into remediation of environmental damage.

Mercury has been used extensively in gold extraction, in the process called amalgamation mentioned in Chapter 33. When this process was in extensive use, large amounts of mercury entered the atmosphere and rivers. Amalgamation was largely replaced by the environmentally more benevolent cyanide process in the early 1900s, but high concentrations of mercury are still present in soils around old mining districts, as well as in sediments of regions that receive the stream

and river drainage of these districts. An example is San Francisco Bay in California, which received deposits from streams draining Sierra Nevada mining districts, more than 160 km away, during the Gold Rush period of the 1850s.

Figure 36.6 is a periodic system of elements on which the essential human nutrients are marked, as well as toxic elements. The toxic elements are divided into those that are known to be carcinogenic, those that cause birth defects (teratogenic), and some that are radioactive. Among the radioactive elements, radon is most significant. It forms during radioactive decay of potassium, a major element in alkali feldspars, which are common in granitic rocks of continental shields.

This chapter has discussed some health aspects of minerals, both positive and negative, in nutrition and contamination. They rely on chemical properties and structural characteristics. In conclusion we want to briefly mention holistic healing powers, which have become a popular connection to minerals (Box 36.1).

Box 36.1 Additional information: Holistic healing attributes of crystals and minerals

The structure, chemistry, and conditions of formation of minerals have been the central topics of this book. They were deciphered by human investigators. There are other connections between humans and minerals and we discussed their impact on our health. Since antiquity they have also influenced the human mind. Ancient Egyptians used lapis lazuli, turquoise, and emerald for protection and health. Kings were buried with quartz on the forehead to help as a guide in the afterlife. Cleopatra used lapis lazuli for enlightenment and awareness. Ancient Greek sailors wore amulets of crystals to keep them safe at sea. Jade was highly valued in ancient China as a healing stone. Vedic Sanskrit texts of 1000 BC introduce chakra points in the body and use minerals associated with them, such as sapphire to bring mental balance or jasper to guarantee harmony. Though there is no scientific connection, the spiritual properties have seen a recent revival, which is exploited in the mineral trade. You might want to explore it on the Internet, without forgetting all the structural and chemical details you have acquired by studying this book. Here are some spiritual powers attributed to minerals:

- *Amethyst*: brings peace, provides spiritual insight, eases stress
- *Apatite*: strengthens intuition, suppresses food cravings, dissolves negativity
- *Diamonds*: powerful amplifiers, develop competence, bring purity, remove fog from the mind
- *Emerald*: detoxifies negative energies, emotional healer, improves memory
- *Hematite*: grounding tool for worldly tasks
- *Jadeite*: promotes prosperity, harmonizer, dream enhancer
- *Malachite*: protective stone, enhances psychic abilities, releases past traumas
- *Rose quartz*: eases heartache, relieves loneliness, promotes forgiveness
- *Serpentine*: enhances meditation, awakens spiritual center
- *Topaz*: promotes creativity, gives optimism

Some internet links that promote mineral healing:

http://www.crystalwellbeing.co.uk/introcrystalhealing.php
http://healing.about.com/od/crystaltherapy/

And others that remind you of its limitations:

http://rationalwiki.org/wiki/Crystal_healing
https://en.wikipedia.org/wiki/Crystal_healing

36.4 Summary

Minerals play an important part in human health. Apatite is the mineral structure that forms human bones and teeth and ingredients derived or extracted from minerals are essential for human nutrition. Since ancient times ingestion of soil has provided nutrients (geophagy). But minerals can also become serious health hazards. Breathing of mineral fibers can cause asbestosis, a form of lung cancer (mainly crocidolite and chrysotile) and exposure to quartz dust leads to silicosis, a progressive lung disease. Chemical contamination from mining in the form of acid rain and mine drainage is spreading carcinogenic elements such as Be, Ni, As, Zr, Cd, and Pb.

Test your knowledge

1. List some mineral-like crystals that are found in the human body.
2. In health and nutritional sciences the definition of *mineral* is somewhat different from that in mineralogy. Explain the difference.
3. List some ("true") minerals that are directly used in human nutrition.
4. Give examples of some elements that are essential for physiological functions but are toxic when used in larger doses.
5. Which minerals, if inhaled as particulate dust, are most hazardous to human health?
6. Review some of the dangers of acid mine drainage.

Further reading

Beeson, K. C. and Madrone, G. (1976). *The Soil Factor in Nutrition: Animal and Human.* M. Decker Inc., New York.

Dodson, R. F. and Hammar, S. P. (2011). *Asbestos: Risk Assessment, Epidemiology, and Health Effects*, 2nd edn. CRC Press, Boca Raton, FL.

Guthrie, G. D. and Mossman, B. T. (1993). *Health Effects of Mineral Dust.* Reviews in Mineralogy, vol. 28. Mineralogical Society of America, Washington, DC.

Jacobs, J. A., Lehr, J. H. and Testa, S. M. (2014). *Acid Mine Drainage, Rock Drainage, and Acid Sulfate Soils: Causes, Assessment, Prediction, Prevention, and Remediation.* Wiley, New York.

Plumlee, G. S. and Logsdon, M. J. (1999). *The Environmental Geochemistry of Mineral Deposits, Volume A.* Society of Economic Geologists, Littleton, CO.

Skinner, H. C. W., Ross, M. and Frondell, C. (1988). *Asbestos and Other Fibrous Materials: Mineralogy, Crystal Chemistry, and Health Effects.* Oxford University Press, New York.

37 | Mineral composition of the solar system

Minerals and mineral-like compounds occur in large parts of the universe. Using sophisticated telescopes and space missions, and with meteorites, the composition of the solar system and its planets is becoming better defined and the Earth emerges as just a small piece of a large ensemble with a lot of variety. This chapter first puts the solar system into the context of the universe, then discusses the mineral composition of the planets and the Moon. Some amazing progress has been made, such as a robotic rover performing X-ray diffraction experiments on Mars, closely analogous to what is done in mineralogy laboratories.

37.1 Elements in the universe

According to current theory, the universe began about 13.7 billion years ago during a primordial explosion referred to as the "Big Bang". Shortly after the Big Bang, the first elements were formed, primarily helium and hydrogen (minor amounts of other light elements, such as lithium, beryllium, and boron, also formed during this event). As the expanding universe continued to cool, areas of higher density matter, or protogalaxies, began to condense. As these early galaxies evolved, gravitational attraction between elements within the galaxies led to regions that collapsed under great pressure, triggering exothermic nuclear fusion reactions that resulted in the formation of the first stars. The nuclear reactions in the core of stars produce elements of light and intermediate weight (up to iron and nickel), with the mass of a particular star determining just what elements may form during the course of its lifetime. Our Sun is a relatively nondescript star of the "yellow dwarf" type; there are over one billion such dwarfs in our galaxy alone. In small stars such as the Sun, helium, carbon, oxygen, neon, and magnesium may form. In larger, more massive stars, however, heavier elements can synthesize, all the way up to iron. Magnesium, silicon, and iron are the main elements produced in these stars, making up less than 1% of the universe.

Everything heavier than iron and nickel formed during stellar explosions known as supernovae, as a result of endothermic nuclear reactions occurring at

high pressure. Figure 37.1 is an optical image of the supernova 1987A in the Large Magellanic Cloud, taken with the Hubble Space Telescope. How do we know the elemental composition of such remote objects? The answer is that these measurements are done with emission and absorption spectroscopic techniques very similar to those discussed in

Fig. 37.1 Three rings of glowing gas encircle the site of supernova 1987A. This supernova is 1.67×10^5 light-years away. The image was taken with the NASA Hubble Space Telescope in 1987.

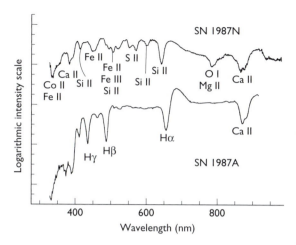

Fig. 37.2 Optical absorption spectra of supernovae, illustrating two types, one with heavier elements (SN 1987N, top) and one with lighter elements (SN 1987A, bottom). Logarithmic intensity scale with arbitrary origin. Roman numerals are used in spectroscopy to indicate the ionization state; I, elemental state; II, one electron removed; III, two electrons removed (courtesy A. Filippenko, U.C. Berkeley, see also Filippenko, 1997).

Chapter 16 and used to identify gems and other minerals. Figure 37.2 shows absorption spectra of several supernovae, including that shown in Figure 37.1. White light produced in the supernova explosion undergoes absorption when passing through the dust ejecta. Absorption bands in the resulting spectra can be attributed to hydrogen, helium, oxygen, sodium, magnesium, silicon, sulfur, calcium, cobalt, and iron. The heavier elements cannot be measured directly, and their abundance must be inferred from model calculations.

Thus, all the Earth elements heavier than iron were actually created in a supernova explosion somewhere else in our galaxy and then dispersed. The dispersed material eventually concentrated in nebulas, consisting mainly of hydrogen, helium, and minor amounts of lithium, beryllium, boron, magnesium, silicon, and iron. Dust particles of Fe(Ni), MgO, SiO_2, Mg_2SiO_4 (forsterite), and $MgSiO_3$ (enstatite) precipitated during this process. These nebulas also contain molecular groups such as H_2, CO, CN, CH_4, NH_3, H_2O, HCOOH, H_2CO, C_2H_6O, etc.

The solar nebula formed about 5 billion years ago. As part of the nebula cooled to close to absolute zero, almost all the gases condensed, resulting in the formation of helium–hydrogen icy planetesimals that later participated in the accretion of the planets. Some of these objects escaped later differentiation in the solar system because of their highly eccentric orbits (Figure 37.3). Comets and meteors are examples of such primitive material, and those that pass close to the Earth give scientists an opportunity to study the chemical as well as the mineralogical composition of the early solar system.

For example, *Comet Halley* passed close to the Earth in March 1986 and was investigated in detail. Compounds that were identified in it include gas (H_2O 80%, CO 10%, CO_2 3%, CH_4 2%, NH_3 1.5%, HCN 0.1%) and small particles of solids 0.1–10 μm in size. These particles turned out to be a mixture of H_2O ice and silicon–magnesium–iron–oxygen minerals. *Comet Hale–Bopp*, observed in March 1997, has a similar composition. However, improved techniques for compositional determinations allowed many more molecular compounds to be identified in the gas, as well as dust particles of olivine and enstatite. The tail of the comet, 50 million kilometers in length, consists largely of sodium atoms.

The ESA *Rosetta* spacecraft was launched in 2004 with a mission to research the composition and structure of comets. An encounter with comet 67P/ Churyumov–Gerasimenko in 2014 revealed the water–ice cycle, including oxygen gas (e.g., DeSanctis *et al.*, 2015).

Additional information about the solar system is obtained from interplanetary dust that the Earth accretes at roughly 40 000 tonnes per year, which equates to 1 g per 10 km^3. Dust samples are collected in high-altitude aircraft in the stratosphere and by satellites or space stations. Dust also enters the atmosphere, and indeed the Earth's atmosphere contains 1–2 million tonnes of such dust. The most suitable materials on the Earth's surface to study cosmic dust deposits are oceanic red clays (with an accumulation rate of 1 mm per 1000 years) and glaciers.

In these deposits, as well as in dust collected directly in space, glass, silicate minerals, metals, magnetite (Fe_3O_4), and cosmic globules containing such rare minerals as wüstite (FeO), kamacite (Fe,Ni), schreibersite (Fe_2NiP), and trevorite ($NiFe_2O_4$) have been identified. Most minerals in the dust, often as small as 100 Å in diameter or less, are also found in stony meteorites. Orthorhombic pyroxene and forsterite dominate in these dust particles, but they also contain serpentine, saponite

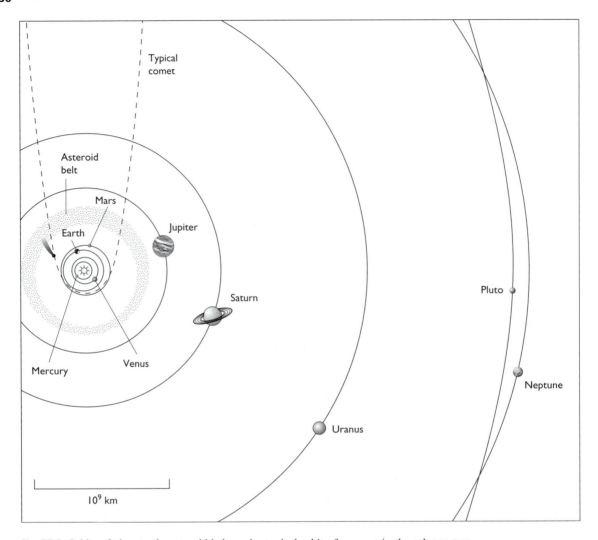

Fig. 37.3 Orbits of planets, the asteroid belt, and a typical orbit of a comet in the solar system.

$(Mg_3Si_4O_{10}(OH)_2 \cdot H_2O)$, chlorite, dolomite, calcite, graphite, diamond, and aromatic hydrocarbons. The complex organic molecules are highly enriched in deuterium over hydrogen as compared with terrestrial abundances.

37.2 Minerals of meteorites

Meteorites reach the Earth from outer space. Their presence on the Earth's surface has provided us very direct information about the mineralogical composition of the solar system, long before spacecraft could sample extraterrestrial bodies, spectroscopic methods became available to analyze comets, or cosmic dust

had even been detected. Most meteorites originate from the asteroid belt located between the inner ("terrestrial") and outer ("Jovian") planets (Figure 37.3). They show a range of compositions and textures that reflect their sources. For example, "primitive" meteorites tell us about the early history of the solar system. Other meteorites are more evolved and document various stages of the condensation and differentiation of the solar nebula. Meteorites are not all that rare. A total of 22 507 meteorites had been examined and classified by 1999 (Grady, 2000), and The Meteoritical Bulletin Database (www.lpi.usra.edu/meteor/meteorite-rss.php) lists over 50 000 as of 2015. These samples vary in their mineral, chemical, and isotopic

Table 37.1 Classification of meteorites (only the main groups are shown; the percentage of meteorite falls represented by each class is shown in parentheses)

Class	Subclass	Group	Principal minerals
Stony (92.8)	Chondrites (85.7)	Ordinary	Mg-rich olivine, Ca-poor pyroxene, kamacite
		Enstatite	Enstatite, kamacite, troilite
		Carbonaceous	
		Type 1	Serpentine, chlorite, olivine, pyroxene
		Type 2	Olivine, pyroxene
	Achondrites (7.1)	Ca-poor	Diopside, hypersthene, olivine, kamacite
		Ca-rich	Pigeonite and plagioclase, or hypersthene and plagioclase
		Primitive	Olivine, pyroxene, plagioclase
Stony iron (1.5)		Pallasites	Olivine, kamacite, taenite
		Mesosiderites	Pyroxene, plagioclase, kamacite
Iron (5.7)		Hexahedrites	Kamacite
		Octahedrites	Kamacite, taenite
		Ataxites	Taenite, kamacite

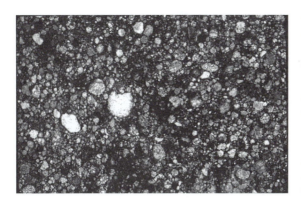

Fig. 37.4 Petrographic thin section of carbonaceous chondrite colony from Oklahoma, USA, with chondrules. Plane polarized light (courtesy O. Medenbach). Width 4.5 mm.

composition, and in their texture, structure, and size. On the basis of these investigations, meteorites have been divided into three major classes: stony (chondrites and achondrites), stony iron, and iron (Table 37.1).

Chondrites are agglomerated rocks containing spherules up to 5 mm in size; these spherules (chondrules) can make up to 70% of a chondrite's total mass (Figure 37.4). The structure of these so-called chondrules suggests that they formed during rapid cooling of droplets of molten material. The matrix typically consists of submicrometer-scale original condensate grains as well as broken fragments. Textures of chondrites indicate that they never underwent differentiation processes.

Ordinary chondrites are the most common variety of chondritic meteorites. High-temperature minerals such as olivine, enstatite, clinopyroxene, and sometimes plagioclase are major components of chondrules, and they are cemented with a matrix of the same fine-grained minerals, metals (Fe–Ni alloys), sulfides (troilite and others), and silicate glass. On average the composition of ordinary chondrites is (in vol.%): olivine 40–45, enstatite 30, Ni–Fe alloys 10–15, plagioclase 10, troilite 5–6. There is no consensus about the origin of chondrites. One group of theories assumes that chondrules formed directly out of the solar nebula gas. Another theory is that they formed during impacts on surfaces of meter-sized planetesimals.

Enstatite chondrites have a composition close to that of early condensed material at 1300 K. They are characterized by a very high extent of reduction consisting of iron-free enstatite, silicon-bearing kamacite (up to 3.5 weight% silicon), and troilite. These chondrites also contain some very rare and unusual minerals, such as alabandite ((Mn,Fe)S), niningerite ((Mg,Fe)S), oldhamite (CaS), and osbornite (TiN).

Carbonaceous chondrites are characterized both by the highest degree of oxidation among all meteorites and by high contents of water (up to 20 weight%, e.g., in serpentine and chlorite), carbon (up to 5 weight%, in graphite, amorphous carbon, organic matter such as polymers, and even traces of diamond), and sulfur (in gypsum and other sulfates). They also contain heavier elements in nearly their original cosmic proportions. The close correspondence between elemental abundances in the solar photosphere and in carbonaceous chondrites (Figure 37.5) suggests that these meteorites are representative of the primordial solar nebula. On the basis of their mineral compositions carbonaceous chondrites are viewed to be a combination of high-temperature minerals and other minerals (such as serpentine and chlorite) that were never heated above 500 K.

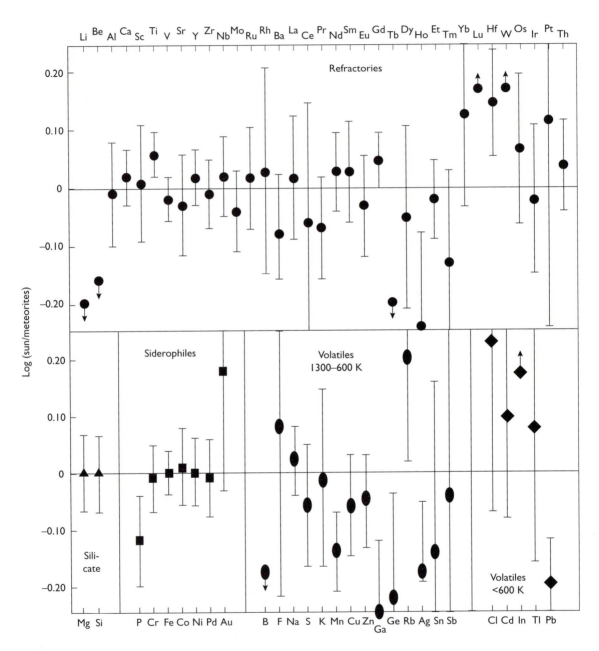

Fig. 37.5 Plot of the ratio of elemental abundances in the solar photosphere divided by those in carbonaceous chondrites. The close correspondence suggests that these chondrites are representative of the primordial solar nebula (data from Taylor, 2001).

Achondrites typically have igneous or brecciated texture with no chondrules. They were produced when some parent material melted and differentiated during cooling. On average, achondrites have a composition (in vol.%) of olivine 12–13, enstatite ~50, diopside 12, plagioclase 25, and Ni–Fe alloys ~1.

Stony-iron meteorites consist of metallic iron and silicates in approximately equal proportions. Pallasites contain olivine, and it has been suggested that they may have formed where planetary mantle material was in contact with an iron core (the core–mantle boundary). Mesosiderites contain plagioclase and pyroxenes, resembling those in crustal basalts.

Iron meteorites contain 95% metallic iron in the minerals kamacite and taenite. Typical accessory minerals are troilite and schreibersite. These meteorites have been classified mainly according to their nickel content (as well as by their iridium, gallium, and germanium contents). Kamacite (body-centered cubic) and taenite (face-centered cubic) are often intergrown in a geometry called the Widmanstätten pattern (see Plate 22f) and form large crystals, suggesting that they were buried deep inside parent bodies and cooled very slowly.

About 280 minerals are found in meteorites, and the majority of these are common on Earth. Because a major part of iron exists in meteorites as native metal, the formation conditions for the meteorites must have been highly reducing, except in the case of carbonaceous chondrites. Native elements and metallic alloys found in meteorites include kamacite, taenite, awaruite, lonsdaleite, diamond, graphite, gold, copper, and sulfur. There are also unusual carbides, silicides, nitrides, phosphides, and chlorides. Many sulfides that are rare on Earth are typical for meteorites, including troilite, alabandite, oldhamite, brezinaite, daubreelite, and djerfisherite. There are about 35 minerals that have been found so far only in meteorites but several are thought to be important components of the Earth's deep interior (Table 37.2).

The presence of diamonds in all types of meteorites provides information about conditions in the early solar system. The diamond content of meteorites is generally about 0.1 vol.% but can occasionally reach 2%. Usually meteoritic diamonds are very fine grained, but grains as large as 1–5 mm in diameter have also been observed. Figure 37.6 is a micrograph of such a diamond nanocrystal, only about 100 Å in diameter, which is thought to be older than the solar system and formed during circumstellar condensation. The diamonds in meteorites have a cubic, octahedral,

Table 37.2 Some minerals so far found only in meteorites but thought to be significant components of the deep Earth

Name	Formula
Kamacite (bcc)	$\gamma\text{-}(Fe_{0.9}Ni_{0.1})$
Taenite (fcc)	$\alpha\text{-}(Fe_{\sim 0.5}Ni_{\sim 0.5})$
Cohenite	Fe_3C
Moissanite	SiC
Bridgmanite (perovskite)	$(Mg,Fe)SiO_3$
Ringwoodite (spinel)	$\gamma\text{-}(Mg,Fe)_2SiO_4$
Majorite (garnet)	$Mg_3(MgSi)Si_3O_{12}$
Wadsleyite (spinelloid)	$\beta\text{-}Mg_2SiO_4$

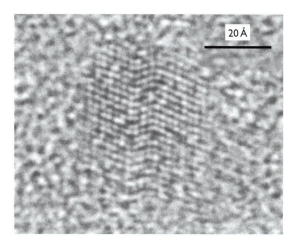

20 Å

Fig. 37.6 High-resolution TEM image of a twinned nanocrystalline diamond from the Allende meteorite. This diamond is supposed to have formed during presolar circumstellar condensation (from Daulton *et al.*, 1996).

or cube-octahedral habit. Graphite and lonsdaleite often form pseudomorphs after diamond.

There are several diamond–mineral associations in meteorites. The most common occurrence is as inclusions in metallic phases (kamacite and taenite). The presence of native silicon and aluminum indicates an extremely reducing environment with a high H_2 pressure, which was a major component of the original fluid envelopes of terrestrial planets. A second diamond association includes moissanite ("carborundum", SiC), troilite, cohenite $((Fe,Ni)_3C)$, schreibersite $((Fe,Ni)_3P)$, and spinel. These minerals also were formed under high H_2 pressures, and the following reactions could explain their formation:

$$SiO_2 + H_2 \rightleftharpoons SiO + H_2O$$

$$2MgO + 2SiO + CH_2 + CO \rightleftharpoons \\ Mg_2SiO_4 + SiC + C(diamond) + H_2O$$

The third association is typical for achondrites. As an example, in the meteorite "New Urei" graphite veinlets (40 μm thick) cut across olivine grains. Small droplets of kamacite are included in graphite veinlets. Cubic diamond crystals grow both on kamacite droplets and on olivine at the borders of the graphite veinlets. Crystals of diamond include droplets of kamacite and graphite and are full of fluid H_2 bubbles. One can suppose a reaction $CH_4 \rightarrow C(diamond) + 2H_2$ to explain the origin of the fluid inclusions and the active interaction of kamacite–graphite veinlets with the country rock. Olivine recrystallizes to form pure forsterite, enstatite, and native iron without nickel.

To find carbon, hydrogen, and deuterium compounds in solar dust is not surprising since these elements are ubiquitous in the large outer planets. What is perhaps surprising is that aromatic hydrocarbons are infrequently found in meteorites. The reason they are very rare in meteorites may be related to sampling. Both dust particles and meteorites approach the Earth at high velocities (>5 km/s). Larger meteorites must be strong to survive the entry into the atmosphere and only coherent silicate and oxide rock fragments survive. The dust particles, with a high surface:mass ratio, decelerate much more gently in the upper atmosphere, thus avoiding excessive heating and allowing less stable organic molecules to survive atmospheric entry.

Even though meteorites provide us with samples from the remote solar system, this sampling is very incomplete. The sources of the meteorite samples studied so far are only 70–80 parent bodies originating from the asteroid belt. Most of these meteorites formed before the oldest rocks of the Earth and Moon as a result of impacts and collisions between planetesimals. Isotopic studies indicate that most meteoritic material went from a condition of dispersed dust to incorporation within solid bodies between 4560 and 4571 million years ago, i.e., within 10–20 million years of the beginning of the solar system. Isotopic ages were modified by subsequent impacts, and most measured meteorite ages range between 4000 and 4500 million years, though a few are younger. Some carbonaceous chondrites may contain material that was formed earlier than the solar system, or that was carried from outside its boundaries.

In addition to the meteorites that originated from the asteroid belt, there are some unusual meteorites that are generally attributed to the Moon and Mars. The assumption is that they were ejected during asteroid impacts. These meteorites have textures typical of igneous rocks, are younger in age, and have a distinct composition of stable isotopes, particularly oxygen.

Meteorites not only give us information about the composition of the solar system, but can also be used to infer the composition of the Earth's deep interior, which cannot be sampled. It is assumed, for example, that the core of the Earth has a composition similar to that of iron meteorites and that the mantle is rather near in composition to some chondrites (Table 37.2).

37.3 Minerals of the planets

As the solar nebula condensed into early planetary bodies (called protoplanets), satellites, and asteroids, the first minerals started to form in the cooling plasma at temperatures below 2000 K. Figure 37.7 gives a simplified crystallization sequence (assuming that most of the gaseous components, mainly hydrogen

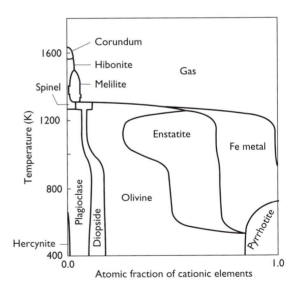

Fig. 37.7 Mineral stability in the solar nebula. Shown is the equilibrium condensation sequence of minerals as a function of temperature at a pressure of 1 Pa and solar system elemental abundances. The widths of the phase fields are proportional to the relative numbers of cations (Si^{4+}, Mg^{2+}, Fe^{2+}, Ca^{2+}, Al^{3+}, Na^+) incorporated into each (data from Grossman, 1972, and Wood and Hashimoto, 1993).

and helium, have been swept away). The first mineral to crystallize is corundum, followed by hibonite ($CaO·6Al_2O_3$), melilite (($Ca,Na)_2(Al,Mg)(Si,Al)_2O_7$), and spinel. Below 1300 K many minerals precipitate, including olivine (forsterite), diopside, plagioclase (anorthite and albite), enstatite, and metallic iron. Finally sulfides, H_2O, and CO_2 form. The minerals are those observed in most meteorites.

In the disk-shaped and rotating solar nebula, the protoplanets began to form about 5 billion years ago. The events that followed are poorly constrained by data, and models are highly speculative. One scenario proposes that originally all planets had relatively small cores with heavier elements, surrounded by a shell of light elements. These shells of the giant outer planets are to some extent still preserved, with those of Neptune and Uranus consisting mostly of H_2O and those of Saturn and Jupiter of helium–hydrogen. The protoplanets accumulated enormous masses, large enough for gravitational compression to produce thermal energy release and melting. This was followed by a stage of cooling and subsequent layering into liquid outer shells and iron-silicate-enriched (chondritic) interiors, stimulating an acceleration of the planetary rotation, development of centrifugal forces, and separation of satellites from fluid shells of planets. In the inner *terrestrial planets*, the outer shells were blown away by the solar wind because of their proximity to the Sun. The solar wind is an intense shower of protons and electrons originating from the Sun and moving with a great velocity. In this scenario the Earth, as well as Mars, Venus, and Mercury, lost most of their outer shells of light elements and with them over 95% of their mass at an early stage of aggregation.

There are many other models for the early evolution of the solar system, and we cannot possibly discuss them all here. We instead turn briefly to a description of the supposed mineralogical description of the planets and their satellites. The planets of the solar system can be divided into two subgroups: four inner terrestrial planets, resembling the Earth in size, density, composition, and surface temperature; and the remote giant gas planets, which are large in size, low in density, are composed mainly of light elements, and have low surface temperatures (see Figure 37.3). A summary is presented in Table 37.3.

The information about minerals of the outer planets has up to now been based only on indirect data, but amazing progress has been made in determining the elemental composition of some planets and their

Table 37.3 Information on composition of outer planets and their satellites

Jupiter	Silicate core, H mantle, liquid H
Io	*Surface*: Komatiite volcanism, sulfur, SO_2
Europe	*Surface*: H_2O ice, silicate fragments
Ganymede	*Surface*: H_2O ice, silicate fragments
Callisto	*Surface*: H_2O ice, silicate fragments
Saturn	*Surface and rings*: H_2O and NH_3 ice
Ariel	*Surface*: Ice
Titan	*Surface*: Ice, liquid CH_4
Uranus	NH_3 and CH_4 mantle
Ariel	*Surface*: H_2O and CO_2 ice
Miranda	H_2O ice
Neptune	H_2O, NH_3, and CH_4 mantle
Triton	N_2 ice
Pluto (dwarf planet)	*Surface*: CH_4 ice

satellites by analyzing their reflected light spectrum, just as is done for Earth by means of remote sensing.

Giant outer planets and their satellites

The spectacular NASA *Voyager* flyby missions to Jupiter, Saturn, Uranus, and Neptune provided considerable information about the outer planets and their satellites. The more recent *Galileo* mission added additional data on Jupiter and its moons, and the *Cassini* spacecraft provided a better understanding of the remote solar system.

The NASA *New Horizons* mission was launched in 2006 to explore the icy regions at the edge of the solar system. The flyby of Jupiter in 2007 provided amazing pictures of volcanic activity. An encounter with dwarf planet Pluto in 2015 recorded high-resolution images (Figure 37.8) of the shoreline of the Sputnik Planum basin with al-Idrisi mountains on the top left. The flat basin is enriched in methane (CH_4) and nitrogen (N_2). Since the surface temperature of Pluto is 35–45 K, these compounds, with melting points of 91 K for CH_4 and 63 K for N_2, are solid, but there is likely a liquid at depth. The mountainous region is probably composed of water–ice and the structures are indicative of crustal evolution and erosion (Stern *et al.*, 2015). The spacecraft is now heading towards the Kuiper Belt.

Table 37.4 Probable rocks and minerals on the surface of Mercury, Venus, and Mars based on reflection spectra, XRF, and XRD data and meteorites

Planets	*Rocks* and minerals
Mercury	*Fe, Ti-rich anorthosites*
Venus	*K-rich basalts, tholeiitic basalts, olivine gabbro–norite*
Mars	H_2O ice (polar region)
	Andesites and basalts with plagioclase, pigeonite, augite, enstatite, olivine, magnetite
	Secondary rocks with pyrrhotite, phyllosilicates (smectite), goethite, jarosite, gypsum, montmorillonite

Fig. 37.8 The mountainous shoreline of Sputnik Planum on the dwarf planet Pluto, recorded by the *New Horizons* spacecraft (courtesy NASA/JHUAPL/SwRI). Width ~70 km.

Since the outer planets are much further from the Sun than are the terrestrial planets, they were able to retain most of their volatile elements. The outer planets are on the whole very large, and they have a lower density and are richer in light chemical elements than the planets of the terrestrial group. While the internal structure of these planets is not known, it is conjectured that they consist of small silicate or metallic cores, surrounded by large regions of liquid and gas. In particular, it is thought that Jupiter has a small silicate core, covered by a zone composed of metallic hydrogen, followed by a liquid hydrogen zone and a gaseous atmosphere. The other outer planets may have similar structures, except that the outer part of Saturn is largely composed of water (H_2O) and ammonia (NH_3) ice, Uranus has an ammonia and methane (CH_4) ice mantle, and Neptune has a water, ammonia, and methane ice mantle.

The atmosphere of the giant planets, above their cloud tops, is roughly 75% H_2 and 25% He, with minor admixtures of CH_4, NH_3, and other gases. Compared to the solar nebula, the upper atmospheres of Jupiter and Saturn are greatly depleted in helium, presumably because heavier helium gravitationally settled into the interiors of these planets. The atmospheres of Uranus and Neptune are enriched in helium, possibly because hydrogen escaped from their atmospheres due to their lower gravitational fields. The clouds of Jupiter and Saturn are composed of ammonia ice particles, whereas those of Uranus and Neptune consist of methane ice. Methane absorbs red preferentially from sunlight and is responsible for the blue appearance of Uranus and Neptune.

In some respects, the moons of the giant planets are more comparable to the terrestrial planets than are their parent planets. For example, Jupiter's Io (3600 km in diameter), with an average surface temperature of 135 K, displays intense volcanism, with over 200 active hotspots, making this satellite the most volcanically active body in the solar system. Volcanism is very different from what it is on the Earth at present, where it is caused by tectonic activity. On Io heat is generated by tidal flexing owing to the gravitational pull of Jupiter and its other satellites. There are huge lava flows, over 250 km long (Figure 37.9a) and several eruptions on Io have been recorded by *Voyager* and *Galileo* spacecrafts and by the Hubble Space Telescope. Only a short time ago, it was assumed that unusual sulfur volcanism dominated on this satellite, in part because a large part of the surface is covered by sulfur dioxide frost. Very recently, extreme temperatures of 1750 K and higher have been recorded with infrared spectrometers (Lopez *et al.*, 2001). Such temperatures are incompatible with sulfur volcanism and even with basaltic melts and suggest ultramafic lavas (komatiites) that were common on Earth during the Precambrian era (see Box 33.1). The three other

(a)

(c)

(b)

(d)

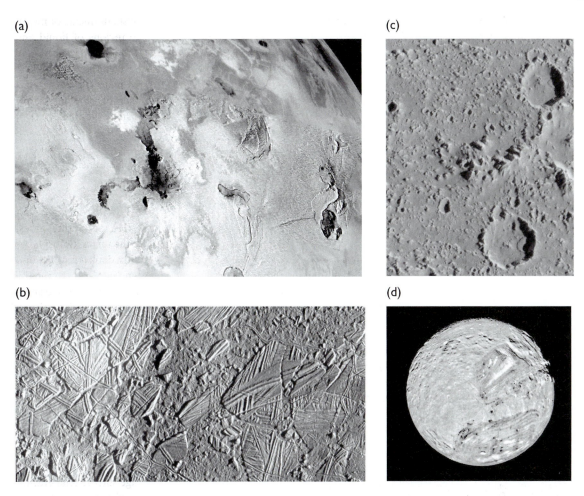

Fig. 37.9 *Voyager* and *Galileo* images of some of the moons of the outer planets. (a) Jupiter's Io displays intense volcanic activity with calderas and lava flows such as the 350 km Amirani flow in the center of the picture. (b) Jupiter's Europa, which is probably mainly composed of water ice, shows evidence of recent tectonic activity with ridges and rift valleys, resembling pack ice on polar seas. In this 30 km × 70 km *Galileo* image north is at the top. (c) Jupiter's moon Callisto has an ancient ice surface with numerous impact craters. The larger one on the image is over 15 km in diameter. (d) This *Voyager* image of Uranus' Miranda (500 km in diameter) reveals unusual mountain ranges made of water ice, with cliffs as high as 20 km. (All photos are courtesy NASA.)

large satellites of Jupiter – Europa (3100 km in diameter), Ganymede (5300 km), and Callisto (4800 km) – have surface temperatures between 125 and 170 K and are thought to be composed largely of water ice and silicate fragments. On Europa, there is good evidence for recent tectonic activity with large ridges and rifts visible in the ice that resemble the disruption of pack ice on polar seas (Figure 37.9b). Callisto, on the other hand, shows no evidence for recent tectonism. The surface is fairly flat, with numerous impact craters having accumulated over long periods of time

(Figure 37.9c). Saturn's largest satellite is Titan, which has a temperature of 95 K and a relatively thick hazy nitrogen atmosphere that covers large lakes of liquid methane below. The most mysterious and heterogeneous landforms exist on Uranus' innermost satellite, Miranda, only 500 km in diameter. The topography shows huge cliffs and canyons, as deep as 20 km (Figure 37.9d). Neptune's Triton, with a surface temperature of 38 K, has a nitrogen atmosphere and nitrogen ice caps and shows signs of volcanism resembling geyser-like plumes.

Table 37.5 Average chemical composition of stony meteorites (Mason, 1962), the lunar surface (Taylor, 2001), Mars (Rieder *et al.*, 1997 and Vaniman et *al.*, 2014), and the Earth's mantle (Ringwood, 1975) and crust (Ronov et *al.*, 1991), in weight% oxides

| | Chondrites | Moon (surface) | | Mars | | | Earth's mantle | Earth's crust | |
		Maria	Highlands	Barnacle Bill	Cumberland	Meteorites		Continental	Oceanic
SiO_2	36.6	45.4	45.5	55.0	43.0	8.2–52.7	43	60.2	48.7
Al_2O_3	2.3	14.9	24.0	12.4	8.6	0.7–12.0	3.9	15.2	16.5
FeO	9.7	14.1	5.9	12.7	22.4	17.6–27.1	9.3	6.05	9.9
MgO	23.7	9.2	7.5	3.1	9.4	9.3–31.6	38	3.1	6.8
CaO	1.8	11.8	15.9	5.3	6.3	0.6–15.8	3.7	5.5	12.3
Na_2O	0.9	0.6	0.6	—	3.0	—	1.8	3.0	2.6
K_2O	0.1	—	—	1.4	0.5	0.02–0.19	0.1	2.9	0.4

Terrestrial planets

Information about the composition of rocks and soils from terrestrial planets derives from reflection spectra. For Mars and Venus some data could be measured directly by instruments installed on landing vehicles by American and Russian space missions. In the case of Mars there is also evidence from meteorites. Table 37.5 summarizes these findings, which reveal no unusual minerals.

There is a class of meteorites (called shergottites and nakhlites) that are thought to have been ejected from Mars during an asteroid impact and provide material for direct analysis. These meteorites are characterized by igneous cumulative textures and have rather young crystallization ages (1600–650 million years). The proposed Martian origin of these meteorites is based on their isotopic compositions and abundances of argon (and other noble gases), nitrogen, and carbon dioxide that match characteristics of the Martian atmosphere. Of particular fame is the 1.9 kg Allan Hills meteorite ALH84001, discovered in Antarctica, where it landed about 13 000 years ago. This meteorite has fostered controversy as to whether morphological segmented features found on it actually represent bacterial fossils.

There have been a number of landing missions to Mars from Viking 1 in 1976 to the current Opportunity rover mission. The instruments on these missions include XRF for chemical characterization and XRD for mineral identification. A microfocus X-ray tube on the robotic rover irradiates the sample and diffraction images are recorded with a two-

dimensional detector (Figure 37.10a) comparable to those used at synchrotrons (compare Figure 11.12b). The data are then sent to Earth where they can be analyzed. The diffraction pattern of a soil sample from Cumberland on Mars (Figure 37.10b) reveals dominant plagioclase, phyllosilicate (smectite), and pyroxenes. Overall data suggest that the surface of Mars is dominated by igneous rocks of basaltic and andesitic composition. These rocks are partially covered by sandstone and clays suggesting a much wetter climate in the earlier history.

On the basis of the mineral composition of the surface of Mars, its density, and the moment of inertia, a model for the internal structure of Mars has been suggested which is fairly similar to that of the Earth, with a solid inner core, a liquid outer core, a lower and upper mantle, and a crust (Figure 37.10).

There has been much interest in extra-solar planets. About 2000 have been confirmed and there are about another 5000 unconfirmed candidates. Important in their discovery was the NASA *Kepler* space observatory, launched in 2009, which is monitoring brightness changes with a high-resolution photometer.

37.4 Minerals of the Moon

The year 1969 was one of great excitement for mineralogists. Not only did American astronauts first land on the Moon, but they also brought samples back for laboratory studies. Prior to the analysis of these samples, there was much speculation and uncertainty as to whether the Moon would contain compounds that were totally different from those that exist on

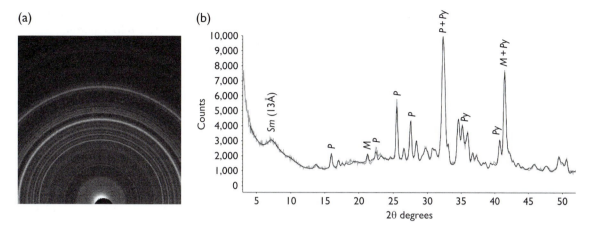

(a)

(b)

Fig. 37.10 (a) X-ray diffraction image of the Martian rock Cumberland collected by the Curiosity rover. (b) The diffraction pattern with CoKα radiation ($\lambda = 2.29$ Å) shows peaks of smectite (Sm), plagioclase (P), pyroxene (Py), and magnetite (M) (courtesy D. L. Bish, D. F. Blake, and NASA). Measured data (dotted line) and fit with the Rietveld method (solid line).

Earth. The answer was sobering: with the exception of three new minerals, all samples that were returned from the Moon were composed of the same minerals that are common in igneous rocks on Earth. Only then was it realized that the Moon was not primordial, but more like an evolved terrestrial planet, with all its complexities.

In total, 382 kg of rocks have been collected from the Moon's surface and brought to Earth by American astronauts during the six Apollo missions between 1969 and 1972, and by three unmanned Soviet robot Luna spacecrafts between 1970 and 1976. Results from the investigations of these lunar samples are published mainly in the *Proceedings* of the Lunar Science Conferences that are held each year in Houston, Texas, USA. The research on lunar samples has added greatly to the development of new analytical techniques, such as those described in Chapter 16.

About 90 minerals have been identified in these samples to date. Most of them are of magmatic origin, with a few having been formed or altered due to shock conditions during meteoritic impacts. Table 37.6 lists the important lunar minerals known today.

As Table 37.6 shows, nine minerals dominate the list of most abundant materials in samples from the Moon (+ symbols). All minerals and rocks have been collected in the so-called regolith, a thin cover of unconsolidated fragments on the surface of the Moon that was formed by meteorite impacts. The samples were collected at only eight locations, none deeper

Table 37.6 Important lunar minerals. Formulas are given only for minerals that have not been mentioned before. Minerals that are so far unique to the Moon are listed in italics

Silicates
*Alkali feldspar
+Olivine
+Plagioclase (mainly An_{60-100})
+Clinopyroxenes (augite and pigeonite)
+Orthopyroxenes (enstatite–hypersthene)
+*Pyroxferroite* $Fe_{0.9}Ca_{0.1}SiO_3$
**Tranquillityite* $Fe_8(Zr,Y)_2Ti_3Si_3O_{24}$
*Zircon

Sulfides
+Troilite

Metals
*Kamacite (ferrite)
*Taenite

Oxides and hydroxides
+*Armalcolite* $(Mg,Fe)Ti_2O_5$
+Ilmenite
*Perovskite
+Spinel
*Quartz

Phosphates
*Apatite
*Whitlockite $Ca_3(PO_4)_2$

Note: *accessory mineral; + rock-forming mineral.
Source: From Frondel, 1975.

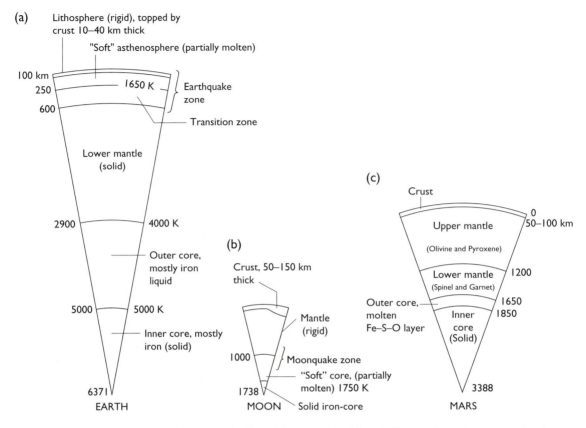

Fig. 37.11 Cross-section through (a) the Earth, (b) the Moon, and (c) Mars, indicating the major structural units.

than several centimeters, and therefore they do not necessarily provide a good average.

Nevertheless, from surface observations and indirect geophysical evidence, there is some consensus that the Moon, like the Earth, consists of a crust, a mantle, and a core (Figure 37.11b). A crustal layer 50–150 km thick is composed of anorthosites in the Lunar highlands and of basalts in the Maria regions, as is illustrated in the more detailed section in Figure 37.12. The crust covers a lithosphere that extends to a depth of 1000 km. The average composition of the mantle is thought to be analogous to the Earth's upper mantle, with dunite (an olivine rock) and olivine pyroxenite, which is the ultimate source of mare basalts (Ringwood, 1979). Unlike the Earth's mantle and crust, the lunar lithosphere is rigid, with no tectonic movements. Small moon-quakes have been recorded at a depth of about 1000 km, an indication of some tectonic activity, perhaps at the interface of the lithosphere and a partially molten core. A primitive lunar core also exists, probably composed of andesite, gabbro, and, less likely, metallic iron. There is no present-day

magnetic field associated with the Moon, but fossil magnetism is preserved in rocks.

Compared to that of the Earth, the geological history of the Moon is much simpler and shorter. Lunar rocks are ancient and preserve a record of the first billion years of the formation of the solar system. Overall, lunar minerals formed during four major episodes:

• Between 4.6 and 4.5 billion years ago the Moon formed. There are many different hypotheses about this event, which we will not discuss in this book.
• The second episode in lunar evolution (4.5–3.6 billion years ago) is characterized by intense meteorite bombardment. Extensive melting of the outer 100–200 km layer occurred as a result of the impact of meteorites, gravitational compression, and heating of the lunar surface by solar wind. During this period a large outer magma ocean formed. As this liquid layer subsequently cooled, minerals began to crystallize and gravitational fractionation took place. As a result, a crust of gabbro–anorthosite

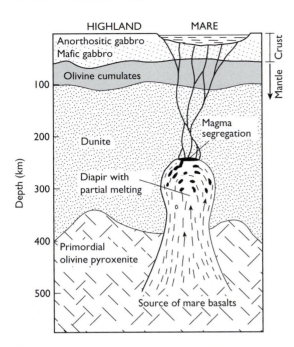

Fig. 37.12 Schematic cross-section through the crust and lithosphere of the Moon, illustrating a model for generation of mare basalts.

composition was formed. These oldest lunar rocks are called *ANT* (*a*northosite–*n*orite–*t*roctolite) and consist mainly of calcic plagioclase, pyroxenes, and olivine. This original crust is still present under the lunar highlands. It was broken up by large meteorite impacts and is generally brecciated.

- The third stage (3.8–3.0 billion years ago) comprises a period of volcanic eruptions that flooded impact-created basins with basaltic lava. These *mare basalts* formed as a result of radioactive heating and crystallized near the surface. The main minerals of mare basalts are pyroxenes, plagioclase, olivine, and ilmenite. All igneous activity on the lunar surface had ceased by about 3 billion years ago.

- In the final stage (3 billion years ago and later), slow erosion by means of meteoritic bombardments, the solar wind, and cosmic rays led to the destruction of the basalts and the formation of the regolith layer on the lunar surface. The regolith is a few tens of meters thick. In it are found *KREEP* rocks, which are metamorphosed, often glassy impact breccias. They are named after the first letters of the symbols or group names of the chemical elements in which they are slightly enriched (potassium, rare earth elements, phosphorus).

A striking contrast between the Earth and the Moon is the absence of water on the lunar surface and crust, except for small amounts of ice discovered recently at the poles of the Moon. This is the reason for the excellent preservation of old minerals on the surface and the lack of hydrothermal alteration and chemical weathering. The study of lunar pyroxenes has significantly contributed to our understanding of that mineral group and of the crystallization of basaltic melts. Unlike their lunar equivalents, pyroxenes in terrestrial rocks are often altered. The Moon is also depleted relative to the Earth in volatile elements such as sodium, potassium, and sulfur.

It is significant that the extraterrestrial materials in meteorites, from the Moon and from the planets that have been analyzed to date, contain many of the same minerals commonly found on Earth, including olivine, pyroxene, and plagioclase. However, if a mineralogy book were to concentrate on minerals in the universe, its emphasis would be different. Native iron–nickel would also figure among the prominent minerals, and compounds of hydrogen, nitrogen, oxygen, and carbon would figure as by far the most common minerals. Apart from water ice, none of them yet has a mineral name.

37.5 Summary

After the "Big Bang" ~14 billion years ago, elements started forming, first H and He. During subsequent cooling, gravitational attraction led to regions that collapsed under high pressure. Fusion reactions in these first stars produced elements of light to intermediate weight (up to Fe). Heavier elements were produced during supernova explosions somewhere else in our galaxy. The dispersed material concentrated in nebulas from dust particles of H, He, Mg, Si, and Fe about 5 billion years ago. Some comets captured material in this early disordered system and when meteorites reach the Earth they tell us about the ancient history. They are divided into chondrites and achondrites with compositions similar to the Earth's mantle and iron meteorites representative of the Earth's core. About 35 minerals are unique to meteorites and of those about six are thought to be major components of the deep Earth. While the inner planets (Mercury, Venus, and Mars) have a similar mineral composition as the Earth, the outer planets are very different, with H_2O, CO_2, CH_4, N_2, and S dominating. After the discussion of planets we focus on the

Moon. During intense meteorite bombardment 4.5–3.6 billion years ago, extensive melting occurred with a large magma ocean creating gabbros and anorthosites during cooling. Then local volcanic activity created mare basalts. All igneous activity ceased 3 billion years ago. Of 90 minerals identified in lunar samples, only three are unique to the Moon: armalcolite, pyroxferroite, and tranquillityite.

Test your knowledge

1. What are the most abundant elements in the universe?
2. Why are the inner planets depleted of light elements?
3. Why is the mineralogy of meteorites so important?
4. What are the principal classes of meteorites and which minerals do they contain?
5. What is the age of meteorites, lunar mare basalt, and the oldest rocks on Earth?
6. Which materials provide the best information about the original mineralogical and chemical composition of the solar nebula?
7. List the nine most abundant minerals on the surface of the Moon.
8. How do we know about the mineralogical composition of Mars? Give three lines of evidence.
9. Name some minerals that are the main constituents of the moons of Jupiter.

Further reading

Beatty, J. K., Peterson, C. C. and Chaikin, A. (eds.) (1999). *The New Solar System*, 4th edn, Cambridge University Press, New York.

Dodd, R. T. (1981). *Meteorites: A Petrologic-chemical Synthesis.* Cambridge University Press, Cambridge.

Hartmann, W. K. (2004). *Moons and Planets*, 5th edn. Cengage Learning, Andover, UK.

Hutchison, R. (2007). *Meteorites: A Petrologic, Chemical and Isotopic Synthesis.* Cambridge University Press, Cambridge.

Mason, B. (1962). *Meteorites.* Wiley, New York.

Papike, J. J. (1998). *Planetary Materials.* Reviews in Mineralogy, vol. 36. Mineralogical Society of America, Washington, DC.

Ringwood, A. E. (1979). *Origin of the Earth and Moon.* Springer, New York.

Rothery, D. A. (1999). *Satellites of the Outer Planets*, 2nd edn. Oxford University Press, Oxford.

Wasson, J. T. (1985), *Meteorites: Their Record of Early Solar-System History.* W. H. Freeman, New York.

38 | Mineral composition of the Earth

After an excursion into the solar system with minerals on the planets and on the Moon, we return back to Earth. With the background about minerals, their composition and structure, their stability and environments of formation, a review of minerals related to the Earth seems appropriate, starting with the crust, mantle, and core. The chapter concludes with mineralogy at the microscopic scale, which is increasingly becoming important in research.

38.1 Chemical composition of the Earth

On the Earth, only minerals located in the crust or suspended in the atmosphere and hydrosphere (oceans, lakes, and rivers) are accessible to direct investigation. Except for a few locations, where upper mantle material has been juxtaposed with the crust, the mineralogical compositions of the deeper zones can only be inferred from indirect evidence provided by studies of gravity, inertia, seismic wave propagation, magnetism, phase stability, and the general abundance of elements.

Particularly important for determining the structure of the Earth's interior are seismic waves. Their velocities increase with the elastic stiffness and the density of the material they pass through. Therefore one can draw conclusions about the material on the basis of travel times. Furthermore (as was discussed briefly in Chapter 12), there are two types of seismic waves. Longitudinal (or P waves) pass through solids as well as liquids, whereas transverse waves (S waves) pass only through solid material. This distinction can be used to identify regions of melt within the Earth.

Figure 38.1 shows a profile of the average longitudinal and transverse wave velocity from the surface to the center of the Earth, as well as the average increase in density and temperature, based on a model by Anderson and Hart (1976). The pressure is also indicated on the left. There is no linear relationship with depth because pressure depends on density which is a function of mineral composition. Note that the wave velocity graph shows both gradual changes (attributed to the general increase of velocity with pressure) and abrupt discontinuities, which are thought to be due to

changes in composition or phase transformations. Particularly striking is the discontinuity at 2900 km, where transverse waves vanish and the velocity of longitudinal waves drops from 14 to 8 km/s. This discontinuity is attributed to the transition from the largely solid silicate mantle to the liquid iron core.

It is generally believed that, except for the light elements such as hydrogen and helium, the bulk composition of the Earth is similar to the average composition of the solar system, probably best represented by the composition of stony meteorites (see Table 37.5). Note that the whole Earth is similar in composition to chondrites, whereas the Moon corresponds to the composition of the Earth's mantle.

In this concluding chapter, we review briefly the mineralogical composition of the different shells of the Earth: crust, upper and lower mantle, and core (see Figure 37.10a).

38.2 Composition of the crust

The bulk mineral composition of the *crust* can be estimated on the basis of the abundance of different types of rock in the crust as established by structural geology, combined with the actual (modal) mineral composition of those rocks. Such estimates have been done by many investigators, and an example is given in Table 38.1.

It may be useful to summarize, once again, in Table 38.2 the most important minerals in plutonic rocks, volcanic rocks, metamorphic rocks, and sediments. This table provides an absolute baseline of information about the mineralogical composition of

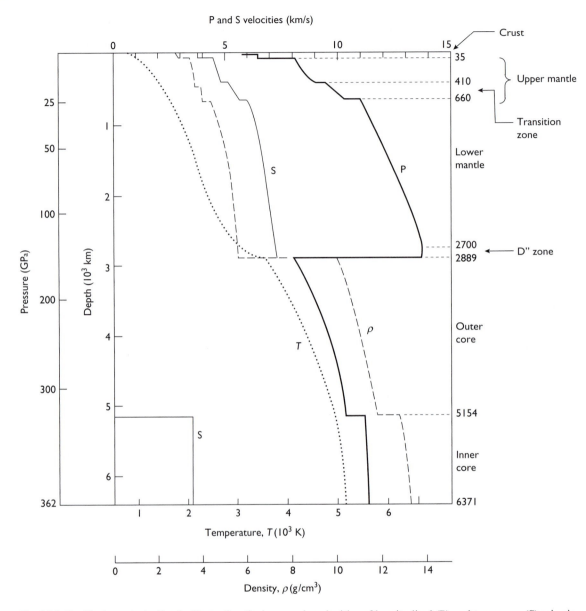

Fig. 38.1 Profile through the Earth, illustrating the increase in velocities of longitudinal (P) and transverse (S) seismic waves, density (ρ), and temperature (T) with depth (after Anderson and Hart, 1976). Pressure is also indicated. For D″, see the text.

rocks. As a further summary, we show again in Figure 38.2 the vol.% of 10 essential minerals in common igneous rocks. On the whole, the mineral composition of the Earth's crust is more diverse than the compositions of the mantle and core, of meteorites, of the Moon, and of other planets. This mineralogical diversity results from several factors. (1) The crust evolved during a complex process of melting and differentiation of primary mantle material in the Precambrian era. (2) As the atmosphere and hydrosphere evolved, material on the surface became subject to weathering and sedimentation, with the formation of numerous hydrous minerals. (3) Both igneous rocks and sediments in the crust became subjected to

Table 38.1 Mineral composition of the crust (vol.%)

Mineral	Average crust	Sediments	Granitic–metamorphic rocks	Mafic rocks	
				Oceanic	Continental
Feldspars	43.1	17.3	52.2	34.3	45.7
Pyroxenes	16.5	4.8	3.4	28.5	23.8
Quartz	11.9	18.4	22.5	—	11.6
Olivine	6.4	0.5	0.4	7.6	7.6
Amphiboles	5.1	<0.1	9.8	—	4.7
Micas	3.1	<0.1	5.6	—	3.3
Clay minerals	3.0	32.8	—	—	—
Other silicates	1.7	0.1	3.7	—	1.2
Carbonates	2.5	19.2	1.5	—	0.5
Ore minerals	1.5	0.6	0.7	3.0	1.6
Phosphates	0.4	0.2	0.2	—	—
Fe-hydroxides	0.2	2.0	—	—	—
Sulfates, chlorides	0.1	1.0	—	—	—
Organic material	—	0.4	—	—	—
Volcanic glass	—	2.3	—	26.6	—
Others	4.5	0.4	—	—	—

Source: From Bulakh, 1996; Yaroschevsky and Bulakh, 1994.

Table 38.2 The most important minerals in rocks

Mineral	Plutonic rocks	Volcanic rocks	Metamorphic rocks	Sediments
Quartz	×	o	×	×
Plagioclase	×	×	×	
Alkali feldspar	×	o	o	o
Mica	×	o	×	
Clay				×
Olivine	×	×	o	
Pyroxene	×	×	×	
Amphibole	×	o	×	
Garnet	o	o	×	
Calcite	o	o	×	×

Note: ×, common; o, subordinate.

tectonic activity, resulting in metamorphism and the creation of many new minerals.

38.3 Composition of the mantle

The mantle comprises the zone between the Mohorovicic discontinuity (the so-called Moho, located 30–50 km under continents and 5–10 km under oceanic crust) and the core–mantle boundary at 2900 km. The Moho defines the boundary between crust and mantle and is characterized by a rapid increase in seismic velocity below it, associated with a change in rock composition. Compared to the crust, the mantle is compositionally fairly uniform as far as we can tell. Largely based on seismic evidence, it is divided into four zones: the *upper mantle* (60–300 km), a *transition zone* (300–660 km) with numerous phase transformations, a more homogeneous *lower mantle* (660–2900 km), and the D″ zone at the bottom of the mantle and adjacent to the liquid outer core.

The mineral composition of the *upper mantle* is best known, since geophysical data constrain density, velocity, and temperature–pressure conditions. Some upper mantle minerals are present in samples of ultrabasic xenoliths in kimberlites, carbonatites, alkali basalts, and alpine peridotites juxtaposed with the crust. Furthermore, phase relations are fairly accessible to experimentation.

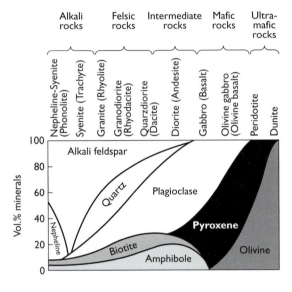

Fig. 38.2 Summary of the mineralogical compositions of major igneous rocks. Volcanic rocks are in parentheses.

Xenoliths in alkali basalts that originate from the upper mantle at depths of 60–90 km are mainly spinel peridotites (up to 80%), pyroxenites, and eclogites. Eclogites are more common in continental settings and along continental margins. Olivine, orthopyroxene (enstatite), clinopyroxene (chrome–diopside and augite), spinel minerals, amphiboles, titanium–phlogopite, garnet, plagioclase, and apatite are the main minerals in these xenoliths. Xenoliths in kimberlites are from greater depths (130 km and deeper; some xenoliths from pipes in Lesotho are attributed to depths of 250 km). Olivine, orthopyroxene (enstatite–hypersthene), clinopyroxene (omphacite, diopside), pyrope, chromite, phlogopite, and diamonds are characteristic minerals in kimberlite xenoliths. In diamonds there are microscopic inclusions of olivine, chrome pyrope, chromite, enstatite, diopside, omphacite, rutile, ilmenite, magnetite, corundum, pyrrhotite, zircon, sanidine, and phlogopite, suggesting that these minerals exist in the upper mantle (Plate 32g).

Figure 38.3 is a schematic picture of the crust and upper mantle with convection cells that induce plate collisions and subduction of continental crustal slabs, resulting in volcanism and mountain building, and high-pressure–low-temperature metamorphism in subduction zones. During upwelling, seafloor spreading is induced, with divergence and upwelling along mid-oceanic ridges and in rising diapirs, again with basaltic volcanism. The simple pattern is generally interrupted by transform faults. Most of this

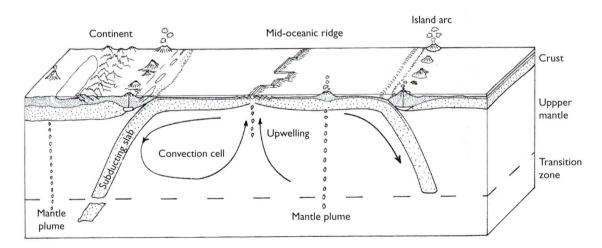

Fig. 38.3 Cross-section through the upper Earth with mantle convection, upwelling at mid-oceanic ridge, subduction of oceanic crust under continent and accumulation of sediments in a trench. Also shown are mantle plume volcanism and island arc volcanism (after Fowler, 2005).

upper mantle is solid, and only in the uppermost part is there evidence for partial melting. Nevertheless, the solid mantle behaves like a viscous liquid and is convecting in large cells driven by temperature gradients. This convection causes significant lateral and vertical heterogeneity.

A. E. Ringwood (1975, 1979) suggested that, on the whole, the upper mantle is composed of *pyrolite*, which is a type of peridotite enriched in aluminum and potassium, as compared with average alpine peridotites that make up the lithosphere and are occasionally observed in the crust. It is thought that the composition of most peridotites that reach the surface has been modified by extraction of basaltic melt. Pyrolite is thus a combination of three parts alpine peridotite (harzburgite) and one part oceanic basalt (tholeiite). During subduction along continental margins, basalts transform to eclogites at depths of 100–150 km.

In the *transition zone*, both the speed of longitudinal seismic waves and the material density increase with depth faster than they do in the upper mantle (see Figure 38.1). This change can be related to phase transitions (see discussion in Section 20.3). High-pressure experiments show that in this depth–temperature range, olivine (Mg_2SiO_4) transforms into wadsleyite and then into a material with spinel structure (*ringwoodite*, Mg_2SiO_4) (see Figure 20.2). The transition occurs at higher pressure for Mg-olivine than for Fe-olivine. Pyroxene transforms into a garnet structure (*majorite*) according to the reaction $2Mg_2^{VI}(Si_2^{IV}O_6) \rightarrow Mg_3^{VIII}(MgSi^{VI})(SiO_4)_3$ (superscripts indicate coordination).

At greater depths, around 660 km, there is again an abrupt increase in velocity (see Figure 38.1). This increase can be attributed to the breakdown of spinel (ringwoodite) into a mixture of MgO (*periclase*, or *ferropericlase* if some iron is present, with the structure of halite) and $MgSiO_3$ (*bridgmanite* with the structure of perovskite) via the reaction $Mg_2SiO_4 \rightarrow MgO + MgSiO_3$. Correspondingly majorite is transforming into bridgmanite. Compare the phase diagram (Figure 38.4) with Figure 38.1. Water is an important component at depths between 400 and 660 km, making up about 0.1 weight%, and supposedly contained mainly in magnesium- and alumino-silicates (e.g., Nishi *et al.*, 2014; Pamato *et al.*, 2014).

With increasing pressure, mineral transitions go from complicated and relatively open silicate structures to simple and highly symmetrical close-packed oxide

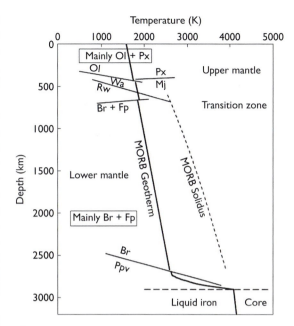

Fig. 38.4 Depth–temperature phase diagram for the lower mantle. Some phase boundaries and a MORB geotherm and solidus are indicated. Br, bridgmanite; Fp, ferropericlase; Mj, majorite; Ol, olivine; Ppv, post-perovskite; Px, pyroxene; Rw, ringwoodite; Wa wadsleyite.

structures. There is a large volume reduction during these reactions, which has been correlated with deep-focus earthquakes. Whereas silicon is always in tetrahedral oxygen coordination at low pressure, it goes into octahedral coordination at very high pressure, such as in bridgmanite with a perovskite structure ($MgSiO_3$) and stishovite with a rutile structure (SiO_2). The mineralogical composition also becomes simpler with increasing depth (Table 38.3 and Figure 38.4).

Above the core–mantle boundary, a thin, very heterogeneous layer exists, called D″, which is defined by its anomalous seismic properties with very small body-wave velocity gradients. It has been established experimentally and confirmed with first principle calculations that at those pressures bridgmanite with perovskite structure converts to "post-perovskite" with a different structure and different elastic properties, which can explain the seismic signature. The general phase diagram of the lower mantle (Figure 38.4) also shows a geotherm for MORB (mid-ocean ridge basalt) and the corresponding solidus. Clearly, most of the material in the lower mantle is in a solid state.

Table 38.3 Some important phase transformations in the mantle

Depth (km)	Phase transformation[a]
Transition zone	
410	olivine (α-$(Mg,Fe)_2SiO_4$) $\rightarrow$ wadsleyite (β-$(Mg,Fe)_2SiO_4$)
450	kyanite (Al_2SiO_5) $\rightarrow$ corundum (Al_2O_3) + stishovite (SiO_2)
520	wadsleyite (β-$(Mg,Fe)_2SiO_4$) $\rightarrow$ ringwoodite (γ-$(Mg,Fe)_2SiO_4$)
400–600	pyroxene ($(Mg,Fe)_2Si_2O_6$) $\rightarrow$ majorite ($Mg_3(MgSi)Si_3O_{12}$)
Lower mantle	
670	ringwoodite (γ-$(Mg,Fe)_2SiO_4$) $\rightarrow$ bridgmanite ($(Mg,Fe)SiO_3$) + ferropericlase $(Mg,Fe)O$
850–900	pyrope ($Mg_3Al_2Si_3O_{12}$) $\rightarrow$ bridgmanite $(Mg,Fe)SiO_3$ + solid solution of corundum (Al_2O_3) and "ilmenite" ($(Mg,Fe)SiO_3$)
1200	stishovite (SiO_2) $\rightarrow$ SiO_2 ("$CaCl_2$ structure")
1700	Metallization of chemical bonds in wüstite (FeO)
2000	SiO_2 ("$CaCl_2$ structure") $\rightarrow$ SiO_2 (structure intermediate between "PbO" and "ZrO_2")
2200–2300	corundum (Al_2O_3) $\rightarrow$ Al_2O_3 ("Rh_2O_3 structure")
D″ layer	
2800–2900	bridgmanite ($(Mg,Fe)SiO_3$) $\rightarrow$ "post-perovskite" ($(Mg,Fe)SiO_3$)

[a]Quote marks are used where names refer to crystal structure rather than minerals.
Source: From Hemley, 1998.

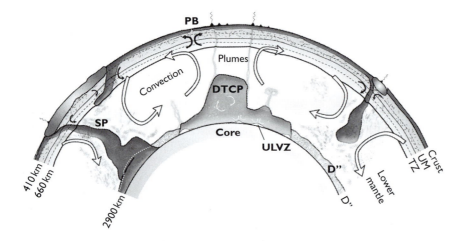

Fig. 38.5 Conceptual diagram illustrating the complex dynamics in the lower mantle with downwelling slabs (dark), convecting primordial material (arrows) and occasional upwelling of plumes. Also indicated is the D″ layer with an anomalous seismic signature. Dense thermochemical piles (DTCP) may be the hottest zones in the lower mantle and related to partially molten material in the ultra-low velocity zones (ULVZ). SP, subducting plate; PB, plate boundary; UM, upper mantle; TZ, transition zone (courtesy E. J. Garnero, see also Garnero *et al.*, 2005).

But seismic tomography has also revealed that the lower mantle is quite heterogeneous (e.g., Lekic *et al.*, 2012) as illustrated schematically in Figure 38.5. There are two large regions with low shear wave velocity (ULVZ for ultra-low velocity zones) that reach up over 1000 km from the core–mantle boundary and suggest compositional heterogeneity. One is under the Pacific and the second one under Africa. Associated with ULVZ are smaller regions of high density (DTCP for dense thermochemical piles) which may be the hottest zones in the lower mantle and related to partially molten material in the

Table 38.4 Important mantle minerals: estimates in vol.%

Mineral	vol.%
Upper mantle (from Moho down to 300 km)	
Olivine (~Fo$_{89}$)	57
Orthopyroxene	17
Omphacite	12
Garnet enriched in pyrope	14
First transition zone (300–450 km)	
Spinel phase (ringwoodite) (SiMg$_2$O$_4$)	57
Garnet phase (majorite) (Mg$_3$MgSi(SiO$_4$)$_3$)	43
Deeper transition zone (450–660 km)	
Periclase MgO with NaCl structure	29
Stishovite (SiO$_2$)	22
Phase of MgSiO$_3$–(Al,Cr,Fe)AlO$_3$ with the structure of ilmenite	24
Phase of (Ca,Fe)SiO$_3$ with the structure of perovskite	23
Phase of NaAlSiO$_4$ with the structure of spinel	2
Lower mantle (660–2700 km)	
Perovskite phase (bridgmanite) (Ca,Mg,Fe) (SiO$_3$)	70
Ferropericlase (Mg,Fe)O	25
Spinel phases "ferrite") NaAlSiO$_4$ + (Mg,Fe) (Al,Cr,Fe)$_2$O$_4$	5
	75
D" zone (2700–2900 km)	25
Post-perovskite phases (Ca,Mg,Fe)(SiO$_3$)	
Ferropericlase (Mg,Fe)O	

Source: Modified from Ringwood, 1979.

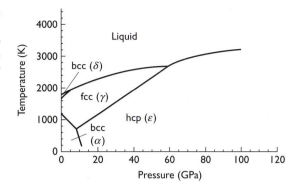

Fig. 38.6 Pressure–temperature phase diagram for iron (based on Shen *et al.*, 1998).

38.4 Composition of the inner core

To account for the overall density of the Earth, the core has to be composed of heavy elements. Of those, iron and nickel are of by far the greatest cosmic abundance. It is likely that the core has a composition similar to that of iron meteorites, with 95% iron and 5% nickel. Transverse seismic waves do not penetrate the core, and on this basis it is concluded that the outer core is liquid, but the inner core is solid. A phase diagram of Shen *et al.* (1998) suggests that iron in the inner core is present in the hexagonal close-packed form (ε) (Figure 38.6). Diamond anvil experiments have been able to reproduce the high pressures and even high temperatures in the center of the Earth (see, e.g., Tateno *et al.*, 2010, 2015 and Figure 16.12b).

Seismic measurements indicate that the density of the core is less than that of pure metallic iron (Allègre *et al.*, 1995). This problem could be solved if the core also contains some lighter elements such as sulfur, silicon, oxygen, or carbon. A recent model suggests that 3–6% oxygen may be the most likely substitution in the liquid outer core, or even hydrogen or helium (Badro *et al.*, 2014). In fact, hydrogen becomes metallic at core pressures, and experiments have shown that iron hydride is stable at pressures of more than 60 GPa (>1600 km depth). Hydrogen may have been incorporated into the core during the high pressures of an early Earth hydrogen atmosphere and may still be emanating from the liquid core. The water component lowers the melting temperature of rocks in the mantle and thus may initiate igneous activity.

ultra-low velocity zones. Relative to average MORB in regions of slab subduction, temperatures are depressed, and in upwelling plumes they are enhanced, causing different paths in the transition zone. Recent seismological observations indicate that some subducting slabs are transported to great depths. When this material heats up, it forms plumes that intrude towards the surface (Figure 38.5).

Thus, experimental and theoretical research in mineral physics, meteorite studies, and seismology make it possible to assess the mineralogical compositions in the deep Earth in regions that are far from direct observations. An estimate of the mineralogical composition of various parts of the mantle is given in Table 38.4.

Obviously, the further away we get from the surface, the more tentative conclusions become. Recall the famous statement of Francis Birch (1952), who was known as the father of high pressure research: "Unwary readers should take warning that ordinary language undergoes modification to a high pressure form, when applied to the interior of the Earth, e.g., *Certain* (high pressure form) – *Dubious* (Ordinary meaning); *Undoubtedly* – *Perhaps*; *Positive proof* – *Vague suggestion*; *Pure iron* – *Uncertain mixture of all the elements.*" Keep this in mind when you read about minerals in the deep Earth, as well as in the remote solar system.

38.5 Atmosphere and hydrosphere

There is also a wide variety of minerals in both the atmosphere and hydrosphere. Some of these are extraterrestrial in origin. For example, there are 1–2 million tonnes of cosmic dust at any given time in the Earth's atmosphere (see Chapter 37). However, less exotic terrestrially based minerals are more common in the atmosphere, ice being the predominant mineral. Other terrestrial minerals that are suspended in the atmosphere are minute particles of quartz, feldspars, calcite, chlorite, biotite, muscovite, hydromuscovite, kaolinite, montmorillonite, zircon, and apatite. Some minerals in the atmosphere are there because of volcanic eruptions or industrial pollution. The content of atmospheric aerosols varies from place to place and changes in time. Above oceans, a maximum concentration of mineral particles is observed at altitudes of 3–4 km.

Minerals of the hydrosphere correspond closely to those of the atmosphere. The sources are also similar. Minerals exist in suspensions, hydrosols, and gels. Other minerals are components of living organisms, such as apatite and other phosphates (in bones and teeth), aragonite, calcite (shells of mollusks, hard tissues of algae, and foraminifera), cristobalite, and tridymite (in diatoms).

38.6 Mineral evolution over Earth's history

In Chapter 37 and earlier in this chapter, we have described the present-day mineral composition of the Earth, the Moon, and the planets. We also presented general models on how the first minerals may have formed in the solar system and subsequently accreted in planets. Looking closer to home, how have

minerals evolved during the history of the Earth? Much of the interpretation of the early history of mineral development on Earth is based on characteristic isotope data of minerals found in various classes of meteorites, rocks from the Moon, and terrestrial rocks.

As we have seen, the hot solar nebula initially condensed into protoplanets, their satellites, and asteroids. Below 1500 K, in a turbulently convecting hydrogen silicate atmosphere, many minerals precipitate, including olivine, diopside, feldspar, enstatite, and metallic iron (Figure 38.7). In the differentiating Earth, the present structure, with core, mantle, and crust, developed during the first 500 million years. Differentiation occurred on the basis of density and melting point. Volatiles such as hydrogen, helium, sodium, potassium, lead, mercury, and zinc, with low melting points, accumulated in the outer parts of the Earth and were partially swept away by the solar wind. Iron and other ferrous elements condensed under the force of gravity and started to accrete in the core. At a later stage, high-temperature silicates and oxides crystallized and accreted as a primitive mantle. Earth minerals have been forming for the last 4.7 billion years in various stages. The oldest known mineral is a zircon from Western Australia that formed 4.4 billion years ago (Valley *et al.*, 2014). During an *early protoplanet stage* the list of minerals included about 40–50 species, corresponding largely to minerals in the oldest metallic meteorites and in primitive chondrites: enstatite, hypersthene, pigeonite, olivine, taenite, kamacite.

At the *basalt stage*, when the mantle was accreted and started to cool, minerals typical of basaltic magmas began to form in the Earth, as well as on the Moon and presumably other planets. The major new mineral species that appeared were feldspars. The mineralogical composition of this early Earth's mantle corresponded closely to rocks of the Moon, particularly:

- Major minerals (>10%): pyroxenes, plagioclase, olivine, ilmenite
- Secondary minerals (1–10%): cristobalite, tridymite, pyroxferroite

At the beginning of the development of the *crust* there were no more than 200–300 minerals occurring in the Earth. Over time, the environment became more complex; iron–nickel concentrated in the core under gravitational differentiation, the core and mantle degassed,

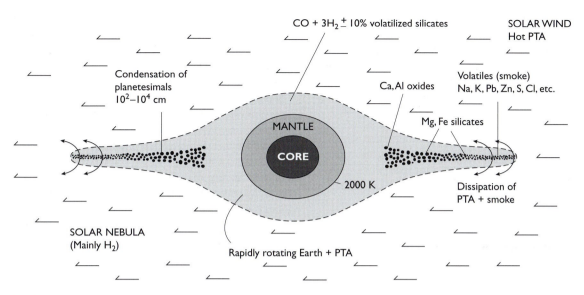

Fig. 38.7 During condensation of the solar nebula, the protoplanet Earth formed by accretion and differentiation of components, largely on the basis of their different densities. PTA, primitive terrestrial atmosphere.

and water appeared, first as freshwater and later accumulating in saline oceans. In the early stages there was an oxygen deficiency. Only after the formation of an oxygen-rich atmosphere by photosynthesis did new mineral species crystallize, most notably iron oxides and hydroxides of the Early Precambrian banded iron formations, as well as siliceous sediments. Crystallization of feldspars, micas, and quartz would later take place in granitic magmas. Surface minerals as well as diagenetic alterations added chlorites, serpentine, kaolinite, hematite, carbonates, and halides. Two tendencies are observed in this evolution. First, in similar geological conditions the number of minerals increases from older to younger rocks. Second, the chemical composition of minerals and their crystal structures becomes more complex with the evolution of a differentiated crust. Nevertheless, by about the Late Precambrian, most of the minerals that we know of today probably already existed.

Some of the important mineral-forming environments are summarized in Figure 38.3. The interaction of the convective mantle and the crust is the source of the major volcanic rocks, either during upwelling at ridges producing basalts, or during subduction with the formation of island arcs and remelting of andesitic material. Volcanism also occurs at hotspots. In the USA, alkali basalts in Hawaii and Yellowstone National Park are oceanic and continental examples, respectively. Plate divergence on continents often produces complex igneous activities, generally with alkaline rocks such as syenites and carbonatites, as in the Kola Peninsula (Russia), the Rhine Graben (Germany), and the East African Rift.

Mantle convection is also the driving force for tectonic activity and associated metamorphism: convergence of plates may produce subduction that results in high-pressure metamorphic rocks such as blueschists (containing glaucophane, jadeite, and aragonite) and ultimately eclogites (with omphacite and pyrope). Where granites intrude into country rock, contact metamorphism produces skarns and hydrothermal activity with typical minerals, garnet, vesuvianite, and epidote. During crustal shortening, overthrusting with regional metamorphism causes amphibolites, gneisses, and marbles to form.

Topographic elevation changes and the influence of water and ice cause original minerals to erode and dissolve; ultimately, they are transported to more stable settings. These dissolved and retransported minerals form the basis of sedimentary minerals such as clays, cristobalite–quartz, and calcite, which crystallize in lakes and oceans and are often associated with organisms. In humid tropical environments, supergene alteration of silicate rocks may occur and transform them to hydroxides (bauxite, goethite–limonite, manganese minerals) and clays (kaolinite). Recall James Hutton's geological cycle displayed in Figure 18.6.

38.7 Microscopic mineralogy

Having explored the universe, the solar system, and mineral-forming environments on the Earth, let us now turn in conclusion to the microscopic scale, where some very interesting mineralogy is happening right now. Materials science is rapidly changing as a result of large new initiatives in nanotechnology to take advantage of the unique properties of materials composed of crystals only a few unit cells in size, and mineralogy may play a critical part in this research. While there are many minute crystals in nature, until recently they have not received much attention, largely because they are difficult to study. Now, with new instrumental techniques, we are able to gain a wealth of information about these smallest crystals.

Such microscopic material calls into question the definition of a mineral. By now, it should be clear that our definition of a mineral as a structurally and chemically homogeneous and naturally occurring inorganic material, as introduced in Chapter 1, is highly idealistic, if not outright arbitrary. This was alluded to by Fritz Laves 50 years ago when he reviewed the status of mineralogy on the occasion of the fiftieth anniversary of the discovery of X-ray diffraction (Box 38.1).

There are numerous examples of minerals where seemingly homogeneous macroscopic crystals are composed of fine lamellae with different compositions and ordering states. Even in chemically homogeneous crystals, the periodic lattice structure is interrupted by dislocations. It is also interesting to note that among the new minerals discovered during the past 20 years, the majority are microscopic and have been found only in crystals of dimension less than 0.1 mm. However, one does not need to look at rare new minerals to find small crystals: most clay minerals occur as aggregates with crystals less than 0.01 mm in size. Devitrified glass is composed of very small particles. Ores formed by weathering – for example, iron, aluminum, and manganese hydroxides in laterites – are, as a rule, cryptocrystalline. Outside the Earth environment, minerals found in samples of the lunar regolith, the surface of Mars, chondritic meteorites, and cosmic dust are generally very fine grained, and electron microscopes are needed for their study.

Box 38.1 Background: What is a mineral? What is a crystal? (F. Laves, 1962)

The writer, as a professor of mineralogy, has frequently been asked "What is mineralogy?" An answer such as "it is the science that deals with minerals" can easily be given; some people, however, are not content with such an answer and want to know "What are minerals?" At this point it is better to leave the room, for there is no satisfactory answer to such a question.

Looking into textbooks of mineralogy one finds that a mineral is an inorganic substance produced by natural processes. Furthermore "homogeneity" is one of the properties that belongs to a mineral.

However, what does "homogeneous" mean? This question could not be answered in a satisfactory way before 1912, nor can it be answered today. Laue's discovery has not helped us to answer this basic question of mineralogy. However, in many special cases X-rays can tell us that substances previously thought to be homogeneous are actually mixtures of different substances; in other cases X-rays have revealed structural features in "minerals" which are fascinating and which add considerably to the liveliness of the present day research in structural mineralogy.

Much of Laves' conclusion, presented in 1962, still holds, even after more than half a century of new methods in mineralogy such as electron microscopy, microanalysis, and many spectroscopic techniques. Research in mineralogy answers some questions but also opens a whole range of new ones, which makes it a very dynamic branch of science. The classic definition of minerals as natural, inorganic, and homogeneous compounds with names and chemical formulas is increasingly blurred and of peripheral interest. Today, the concepts of genesis, heterogeneity, defects, and biological influences are in the center of modern mineralogical research, as foreseen by Laves.

Every natural mineral begins to grow as a nucleus, but then it generally advances to a macroscopic crystal to reduce the surface energy relative to the internal free energy of a stable crystal. It is thus reasonable to ask why many clay minerals never grow to a respectable size. All of them have a layered crystal structure with extreme heterogeneity in stacking and a lack of long-range order. They are composed internally of small units in which charges are locally balanced, so that the driving force for joining units to larger crystals is small compared, for example, with a crystal of sodium chloride containing many unbalanced ionic bonds on its surface.

The main reason that minerals remain fine grained is that chemical reactions stop and an equilibrium state is not achieved. This termination of chemical reactivity can be due to structural causes, as in the clays discussed above, or simply to kinetic reasons, particularly at low temperatures, as in heterogeneous carbonates in limestone.

On the other hand, reactivity may increase because of the small grain size and unusual reactions may take place because of the high surface energy. The grain size of quartz crystals has been found to have an influence on the temperature of polymorphic transformations. Small particles of pyrite are more enriched in gold and silver than larger crystals of the same mineral. Overall, microscopically dispersed ore is often richer in minerals of widely diverse and sometimes unusual composition as compared with simpler mineral compositions of the same coarse-grained ores. In this respect the *intergranular* substance (i.e., the material between grains and concentrated on grain boundaries as a thin film) becomes significant.

Because of the surface energy and the composition of the intergranular phase that must exist in metamorphic rocks in order for reactions to occur, very fine-grained materials have different reactivity. Locally, they also often have different thermodynamic properties. Surface reactions (e.g., ionic adsorption in clays and zeolites) are fields of current intense investigation in mineralogy. Methods such as EXAFS (see Chapter 16) are used to determine short-range order in clays. SEM and AFM have become indispensable in the characterization of crystal surfaces, as has TEM, a technique to investigate structural defects. In fact, TEM investigations (e.g., Buseck, 1992) have complicated and raised new questions in the area of systematic mineralogy. For example, how homogeneous does a compound have to be to deserve a mineral name? The results from TEM studies have shown that some of the most common minerals, such as feldspars, pyroxenes, amphiboles, and sheet silicates, are profusely heterogeneous.

Grain size not only influences the chemical aspects of mineral systems but also has profound effects on physical properties. An obvious influence is on the diffusivity because of the large free surface. In nanocrystalline materials, grain boundary diffusion is often an order of magnitude larger than lattice diffusion. Grain size also has an effect on mechanical properties. As was discussed in Chapter 17, large crystals deform primarily by dislocation glide and climb. For reduced grain size, surface diffusion becomes significant, and small particles may start to rotate independently, as with marbles, through a mechanism known as superplasticity where material can become very ductile. Grain size reduction takes place in heavily deformed mylonites through mechanical abrasion (cataclasis) and dynamic recrystallization (subgrain formation).

Nanocrystals are typically found in another class of minerals: biominerals. These minerals, which include apatite in bones, aragonite and calcite in mollusk shells, as well as minerals associated with bacteria (e.g., Banfield and Nealson, 1997), including minute magnetite crystals, are all extremely fine grained. It is this diversity of local structures, local defects, and surface structures that make minerals very relevant today for a wide range of sciences and has spawned much mineralogical research in areas unimagined several decades ago. This is the reason why mineralogy is suddenly claiming a central position between such science branches as geology, geophysics, materials science, physics, and biology, as illustrated in Figure 1.4 at the beginning of this book. While our text is now finished, for many of you this is just the beginning of a journey into further studies in areas that will provide you with deeper insight into the wonders of this world. Given mineralogy's broad range of relevancy to other fields, no doubt you will frequently find what you have learned in this book to be of great use to you in many different studies.

38.8 Summary

The chapter reviews the mineral composition of the Earth and corresponding mineral-forming environments. There is a brief discussion of the crust and then a more in-depth look at the mantle with mineral phase transformations that divide the mantle into upper mantle (mainly olivine), transition zone (wadsleyite,

ringwoodite-spinel, and majorite-garnet), lower
mantle (bridgmanite-perovskite and ferropericlase),
and D″ zone (postperovskite and ferropericlase), all
with characteristic mineral assemblages that can be
related to the seismic signature. While the outer core
is liquid iron, with about 5% nickel, it also is likely to
contain ~5% oxygen. The inner core is a solid Fe–Ni
alloy, corresponding in composition to iron meteorites
but with a hexagonal close-packed structure.

The book concludes with a discussion of microscopic mineralogy which is evolving into an advanced
science branch with broad applications.

Test your knowledge

1. Having finished this introduction to mineralogy,
   what are the most interesting aspects of minerals
   for you?
2. Review the major mineral-forming environments
   in the Earth.
3. What evidence do we have for the mineralogical and
   chemical compositions of the lower mantle and core?
4. How do we know that the inner core is solid
   hexagonal iron?
5. Name some environments where nanominerals
   form.

6. Give a general reason why nanomaterials have
   physical and chemical properties different from
   those of large crystals.

Further reading

Anderson, D. L. (2007). *New Theory of the Earth*, 2nd
  edn. Cambridge University Press, Cambridge.
Banfield, J. F. and Navrotsky, A. (eds.) (2001). *Nano-
  particles and the Environment.* Reviews in Mineral-
  ogy and Geochemistry, vol. 44. Mineralogical
  Society of America, Washington, DC.
Fowler, C. M. R. (2004). *The Solid Earth: An Intro-
  duction to Global Geophysics*, 2nd edn. Cambridge
  University Press, Cambridge.
Jackson, I. (2000). *The Earth's Mantle: Composition,
  Structure and Evolution.* Cambridge University
  Press, Cambridge.
Lowrie, W. (2007). *Fundamentals of Geophysics*, 2nd
  edn. Cambridge University Press, Cambridge.
Poirier, J. P. (2000). *Introduction to the Physics of the
  Earth's Interior*, 2nd edn. Cambridge University
  Press, Cambridge.
Turcotte, D. L. and Schubert, G. (2014). *Geodynamics*,
  3rd edn. Cambridge University Press, Cambridge.

Appendices

Abbreviations

Symmetry

Hexag.   Hexagonal
Monocl.   Monoclinic
Ortho.   Orthorhombic
Rhomb.   Rhombohedral
Tetrag.   Tetragonal
Tricl.   Triclinic
Trig.   Trigonal
Ps.   Pseudo

Colors

B.   Blue
B.ox   Blue (oxidized)
Bk.   Black
Br.   Brown
Cl.   Clear
G.   Green
Gy.   Gray
M.   Metallic
O.   Orange
R.   Red
V.   Violet
W.   White
Y.   Yellow

Morphology

Dodecah.   Dodecahedron
Octah.   Octahedron
Tetrah.   Tetrahedron
Trapezoh.   Trapezohedron

Mineral names in italic are common and important minerals.

Negative $2V$ indicates that the axial angle is $2V\alpha$, i.e., the mineral is biaxial negative.

Page number refers to table in text where mineral properties are listed, with a more detailed description nearby.

Appendix 1a.1 Metallic or submetallic luster, no cleavage or poor cleavage, sorted according to hardness

Mineral	Symmetry	Hardness	Density	Color	Streak	Morphology	Diagnostics
Mercury	Rhomb.	0	13.6	Y.		Liquid at 0 °C	Liquid droplets, often associated with cinnabar
Gold	Cubic	2.5	19.2	Y.	Y., M.	Cube, octah.	High density, unique color, soft and malleable
Copper	Cubic	3	8.7	R.	R., M.	Cube, octah.	Nodules or dendritic; color, often bluish oxidation; malleable
Silver	Cubic	3	10.5	W., Gy.	W., M.	Cube, octah.	Rare mineral, wire-like or skeletal crystals
Bornite	Ps. Cub. Tetrag.	3	5.1	Bronze, Box.	Gy.-B.	Cube, tetrah.	Fresh crystals are red but oxidize easily to brown or blue
Pentlandite	Cubic	3–4	4.8	Bronze-Br.	Bk.	Crystals rare	Strong metallic luster, octahedral cleavage
Tetrahedrite	Cubic	3–4	4.9	Gy., Bk.	R.-Gy., R.	Cube	Irregular gray-black aggregates, difficult to identify in hand specimens
Cuprite	Cubic	3.5–4	6.1	R.	Br.-R.	Octah., fibrous	Red crystals, streak brown to green, mostly adamantine luster, often with malachite
Chalcopyrite	Tetrag.	3.5–4	4.2	G.-Y.	G.-Bk.	Crystals rare	Brass color, often with blue or red oxidation films, rarely occurs as good crystals compared to pyrite
Platinum	Cubic	4	21.5	Gy.	Gy.	Cube, octah.	High density, in ultramafic rocks
Iron	Cubic	5–5.5	8.1	Silv., W.-Gy.	Gy.	Platy	Meteorites, Widmanstätten pattern
Ilmenite	Trig.	5–6	4.7	Bk., Br.	Bk.-Br.	Platy	Black crystals, streak black to brown, weakly magnetic
Nickeline	Hexag.	5.5	7.5	G.-red	Br.-Bk.	Crystals rare	Light copper-red kidney-shaped aggregates, association with annabergite color films and dust
Cobaltite	Cubic	5.5	6.2	W., Pink	Gy.-Bk.	Octah.	Similar to pyrite, but pinkish-gray, with cleavage, association with eritrine color films and dust
Arsenopyrite	Monocl.	5.5–6	6.0	Gy.-W.	Bk.	Prismatic	Elongated crystals with striations, garlic smell when hammered
Tantalite	Ortho.	6	8.2	Bk.	Br.-Bk.	Prismatic	Rectangular dark crystals of high density
Marcasite	Ortho.	6–6.5	4.8	Y.	G.-Bk.	Tabular	Radiating aggregates, straw-yellow color similar to pyrite
Pyrite	Cubic	6–6.5	5.1	Y.	G.-Bk.	Cube, pyritohedra	With striations, straw-yellow color, green streak, smells of sulfur when hammered, hardness higher than glass
Hematite	Trig.	6.5	5.2	Bk., R.	R., Br.	Platy or tabular	Also botryoidal, cherry-red streak

Appendix 1a.2 Metallic or submetallic luster, distinct cleavage, sorted according to hardness

Mineral	Symmetry	Hardness	Density	Color	Streak	Morphology	Diagnostics
Graphite	Hexag.	1	2.2	Bk.	Bk.-Gy.	Platy	Excellent cleavage, low hardness, black streak on paper
Molybdenite	Hexag.	1	4.7	Gy.	Gy.-Bk.	Platy	Perfect cleavage, low hardness and greasy touch, graphite has darker streak, adamantine luster
Cinnabar	Ps. Cub., Trig.	2–2.5	8.1	R.	R.	Equiaxed	Red crystals, high density, adamantine luster of crystals
Stibnite	Ortho.	2–2.5	4.6	Gy., Bk.	Gy.-Bk.	Prismatic	Dark-gray columnar crystals with striations, burns
Galena	Cubic	2.5	7.4	Gy.	Gy.-Bk.	Cube	Perfect cubic cleavage, high density
Chalcocite	Ortho.	2.5–3	5.7	Gy.-Bk.	Gy., M.	Tabular	Coating of blue azurite and green malachite
Bornite	Ps. Cub., Tetrag.	3.	5.1	Bronze, B.	Gy.-B.	Cube, tetrah.	Fresh crystals are red but oxidize easily to brown or blue
Sphalerite	Cubic	3–4	4.1	Y., Br., Bk.	W., Y., Br.	Tetrah.	Colorless, yellow to black, dodecahedral cleavage, hydrothermal, often associated with galena
Arsenic	Trig.	3.5	5.7	Gy.	Bk.	Equiaxed	Granular masses
Iron	Cubic	4	7.7	Gy.	Gy.		Meteorite, Widmanstätten pattern
Pyrrhotite	Monocl.	4	4.6	Bronze	Gy.-Bk.	Tabular	Bronze color, rather than golden-yellow of pyrite, often magnetic
Manganite	Monocl.	4	4.4	Br.-Bk.	Br.	Prismatic	Striated gray to black prisms, dark-brown streak
Uraninite	Cubic	4–6	10.6	Bk., Br.	G.-Bk., Br.	Cube, octah.	Black kidney-shaped aggregates, radioactive
Goethite	Ortho.	5–5.5	4.3	Bk.-Br., R., Y.	R., Br. Y.	Fibrous	Often in radiating rosettes, brown-yellow streak
Pyrolusite	Tetrag.	5–6	5.0	Bk.	Bk.	Prismatic	Often as oolites
Hausmannite	Tetrag.	5.5	4.7	Bk., Br.	Br.	Fine-grained	Brownish-black masses with submetallic luster
Arsenopyrite	Monocl.	5.5–6	6.0	Gy.-W.	Bk.	Prismatic, equant	Elongated crystals with striations, garlic smell when hammered
Magnetite	Cubic	5.5–6	4.9	Bk.	Bk.	Octah.	Black streak, strongly magnetic
Anatase	Tetrag.	5.5–6	3.8	B.-Bk., Br.	R., Br.	Pyramidal	Tetragonal–pyramidal morphology, mainly adamantine luster
Chromite	Cubic	5.5–7	4.8	Bk., Br.	Br.	Octah., granular	Mostly in nodules in ultramafic rocks, weakly magnetic, greenish-brown streak
Rutile	Tetrag.	6	4.2	R.-Br., Bk.	Y.-Br.	Prismatic	Black or red prismatic or acicular crystals, pale-brown streak, often adamantine luster
Columbite	Ortho.	6	5.3	Bk.	Br.-Bk.	Platy	Black to red brown, high gravity
Tantalite	Ortho.	6	8.2	Bk.	Br.-Bk.	Prismatic	Rectangular dark crystals of high density
Cassiterite	Tetrag.	6–7	7.0	R., Br.-Bk.	Y.-W.	Prismatic, equant	Striated crystals, light streak, high hardness and density, often adamantine luster

Appendix 1b.1 Nonmetallic luster, no cleavage or poor cleavage, sorted according to hardness

Mineral	Symmetry	Hardness	Density	Color	Morphology	Diagnostics
Tincalconite	Trig.	1	1.9	W.	Powder	Decomposition product of borax
Sulfur (α)	Ortho.	1.5-2	2.0	Y.	Prismatic, platy	Color, burns, adamantine luster of crystals
Realgar	Monocl.	1.5-2	3.5	R., O.	Prismatic	Orange streak to distinguish from cinnabar, low density
Carnallite	Ortho.	2.5	1.6	Cl., W., Y., R.	Granular, fibrous	Hygroscopic, bitter taste, evaporite mineral
Jarosite	Trig.	2.5-3.5	3.1	O.-Br.	Tabular	Ocher-yellow crusts, alteration product
Vanadinite	Hexag.	3	7.0	Y., Br., O.	Prismatic	Orange hexagonal prisms
Pyromorphite	Hexag.	3.5-4	6.8	W., G.	Prismatic	Green hexagonal prisms
Cuprite	Cubic	3.5-4	6.1	R.	Octah. or fibrous	Octahedral cherry-red crystals, streak brown to green, often adamantine luster
Bastnaesite	Trig.	4-4.5	5.1	Y., Br.	Tabular	Brown granular aggregates, mainly in carbonatites
Uraninite	Cubic	4-6	10.6	Bk., Br.	Cube, octah.	Black kidney-shaped aggregates, radioactive, black streak
Chabazite	Trig.	4.5	2.1	W.	Rhombs	Poorer cleavage than calcite, no HCl reaction
Apatite	Hexag.	5	3.2	Cl., G., Y., V.	Prismatic	Hexagonal prisms, various colors
Pyrochlore	Cubic	5-5.5	4.5	Br., Y., Bk.	Octah.	Often radioactive
Analcime	Cubic	5-5.5	2.3	W.	Trapezoh.	Free-growing crystals, leucite in rock matrix
Turquoise	Tricl.	5-6	2.7	B., B.-G.	Microcrystalline	Cryptocrystalline bluish-green aggregates
Ilmenite	Trig.	5-6	4.7	Bk., Br.	Platy	Tabular crystals, streak black to brown, weakly magnetic
Romanèchite	Monocl.	5-6	4.7	Bk.	Fibrous, massive	Botryoidal, black-brown streak ("psilomelane")
Microlite	Cubic	5.5	6.1	Bk., Br.	Octah.	Octahedral yellow to brown crystals, brown streak
Allanite	Monocl.	5.5	3.6	Bk.	Platy, prismatic	Prismatic striated brown to black crystals, no cleavage
Lazurite (Lapis)	Cubic	5.5	2.4	B.	Microcrystalline	Dark-blue color, marbles and contact metamorphic rocks; frequently associated with pyrite
Leucite	Ps. Cubic	5.5	2.5	W., Gy.	Crystals rare	Trapezohedral crystals in silica-deficient volcanic rocks
Opal	Amorphous	5.5-6	2.1	Many colors		Opalescence, chonchoidal fracture
Sodalite	Cubic	5.5-6	2.3	W., B.		Generally as bluish aggregates
Perovskite	Ps. Cubic	5.5-6	4.0	Bk.	Cube, octah.	Cube-octahedral habit, in alkaline rocks
Nepheline	Hexag.	5.5-6	2.6	W., Gy., R.	Prismatic	Often anhedral, no cleavage, in alkaline rocks, softer than quartz and feldspar, often fluorescent
Cryptomelane	Monocl.	6	4.3	Gy.	Fibrous, massive	Botryoidal, brown-black streak

Mineral	System	Hardness	Density	Colors	Habit	Description
Tantalite	Ortho.	6	8.2	Bk.	Prismatic	Rectangular dark crystals of high density, brown streak
Chondrodite	Monocl.	6–6.5	3.2	Y., Br.	Platy	Orange grains in marble and serpentine, yellow pleochroism
Humite	Ortho.	6–6.5	3.2	Y., Br.	Platy	Orange grains in marble, yellow pleochroism
Clinohumite	Monocl.	6–6.5	3.2	Y., Br.	Platy	Orange grains in marble, yellow pleochroism
Cristobalite	Tetrag.	6–7	2.32	Cl.	Octah.	Small pseudocubic crystals in volcanic rocks
Hematite	Trig.	6.5	5.2	Bk., R.	Platy, fibrous	Platy or tabular morphology, also botryoidal, berry-red streak
Benitoite	Hexag.	6.5	3.7	B.	Pyramidal	Blue crystals
Vesuvianite	Tetrag.	6.5	3.3	R.-Br., G.	Prismatic	Similar to grossular but not optically isotropic, crystals with square cross-section, poor cleavage
Forsterite	Ortho.	6.5–7	3.2	Y.-G.	Equant	Green, conchoidal fracture
Fayalite	Ortho.	6.5–7	4.3	Bk.	Equant	Black
Almandine	Cubic	6.5–7.5	4.2	R., Br.	Trapezoh., dodecah.	High hardness, in schists
Andradite	Cubic	6.5–7.5	3.8	Br., G., Bk.	Trapezoh., dodecah.	High hardness, in skarns
Grossular	Cubic	6.5–7.5	3.5	W., R.-Br.	Trapezoh., dodecah.	White, pale brown in calcsilicate rocks
Pyrope	Cubic	6.5–7.5	3.5	R.	Trapezoh., dodecah.	High hardness, crimson, in eclogites and kimberlites
Spessartine	Cubic	6.5–7.5	4.2	Y., R.-Br.	Trapezoh., dodecah.	High hardness, pink
Uvarovite	Cubic	6.5–7.5	3.9	G.	Trapezoh., dodecah.	High hardness, green, in ultramafic rocks
Cassiterite	Tetrag.	6–7	7.0	R., Br.-Bk.	Equant	Striated crystals, light streak, high hardness and density, adamantine luster
Quartz (α)	Trig.	7	2.65	Cl., V., Y.,Br.	Prismatic–pyramidal	Conchoidal fracture, botryoidal in microcrystalline chalcedony
Schorl (*tourmaline*)	Trig.	7	3.2	Bk.	Prismatic	Ditrigonal, striated, black, in pegmatites
Rubellite (*tourmaline*)	Trig.	7	3.1	Pink	Prismatic	Ditrigonal, striated, pink, in pegmatite
Cordierite	Ortho.	7–7.5	2.6	B., Gy.	Prismatic	Blue, dichroic prismatic morphology
Zircon	Tetrag.	7–8	4.2	Br., R.	Equant	Ideal tetragonal crystals, diamantine luster, high hardness
Beryl	Hexag.	7–8	2.7	G., B., W.	Prismatic	Hexagonal–prismatic, green or blue, mostly in pegmatites
Spinel	Cubic	8	3.7	R., B., G., Y.	Octah.	High hardness
Chrysoberyl	Ortho.	8.5	3.7	G., G.-Y.	Prismatic	Crystals of various color, high hardness, pegmatites
Corundum	Trig.	9	4.0	Gy., B., R.	Prismatic	High hardness, various colors

Appendix 1b.2 Nonmetallic luster, single cleavage (platy), sorted according to hardness

Mineral	Symmetry	Hardness	Density	Color	Morphology	Diagnostics
Graphite	Hexag.	1	2.2	Bk.	Platy	Excellent cleavage, low hardness, black-gray streak on paper
Talc	Monocl.	1	2.7	W., G.	Platy	Light green, perfect cleavage, greasy touch and low hardness
Pyrophyllite	Monocl.	1.5	2.8	W.	Platy	Often cryptocrystalline, sometimes radial-fibrous
Orpiment	Monocl.	1.5–2	3.5	Y.	Platy	Bright yellow with perfect cleavage, associated with realgar
Borax	Monocl.	2–2.5	1.8	Cl., W.	Prismatic	Clear prismatic crystals transform to white tincalconite
Autunite	Tetrag.	2–2.5	3.1	Y.-G., Y.	Platy	Crystals or powders with brilliant colors, radioactive
Kaolinite	Tricl.	2–2.5	2.6	W., Y., G.	Platy	Identify by X-ray diffraction
Chlorite	Monocl.	2–2.5	2.6	G.	Platy	Platelets with excellent cleavage, green, more brittle than mica, low birefringence
Thenardite	Ortho.	2–3	2.7	Cl., W., Br.	Tabular	White to brownish clusters, evaporate deposits
Biotite	Monocl.	2–3	2.9	Br., Y., Bk.	Platy	Platelets with excellent cleavage, brown
Phlogopite	Monocl.	2–3	2.8	Br., Y.	Platy	Platelets with excellent cleavage, light brown to colorless
Erythrite	Monocl.	2.5	2.9	Pink	Fibrous	Alteration product with striking pink color
Vivianite	Monocl.	2.5	2.7	Cl.-W. → B.	Tabular	Blue crystals, streak white but turns blue
Torbernite	Tetrag.	2.5	3.2	G.	Platy	Crystals or powders with brilliant colors, radioactive
Brucite	Trig.	2.5	2.4	W., G., Br.	Platy	Often in colorless fine-grained masses, harder than talc but softer than mica, hydrothermal alteration of ultramafic rocks
Annabergite	Monocl.	2.5–3	3.1	Y.-G.	Fibrous	Alteration product with pale-green color
Gibbsite	Monocl.	2.5–3	2.4	Cl., W.	Platy	Component of bauxite
Annite	Monocl.	2.5–3	3.3	Br., Bk.	Platy	Platelets with excellent cleavage, dark brown
Muscovite	Monocl.	2.5–3	2.9	W., Gy.	Platy	Flexible platelets with excellent cleavage (fuchsite is green)
Trilithionite	Monocl.	2.5–4	2.8	Pink, V.	Platy	Platelets with excellent cleavage, pink, in pegmatites
Atacamite	Ortho.	3–3.5	3.7	G.	Prismatic	Small emerald-green prisms, can be mistaken for malachite
Carnotite	Monocl.	3–4	4.5	Y., G.-Y.	Platy	Platy crystals or powders with brilliant colors, radioactive
Antigorite	Monocl.	3–4	2.6	G.	Platy	Dark- to light-green aggregates with shiny surfaces
Stilpnomelane	Monocl.	3–4	2.7	Bk., Br.	Platy	Low-grade metamorphic rocks (compare with biotite!)
Arsenic	Trig.	3.5	5.7	Gy.	Equant	Granular masses, black streak
Witherite	Ortho.	3.5	4.3	W., Gy., Y.	Equant, fibrous	Acid reaction, high density
Boehmite	Ortho.	3.5	3	Cl., W.	Platy	Component of bauxite
Aragonite	Ortho.	3.5–4	2.95	W., Y., Br.	Prismatic	Often pseudo-hexagonal twins, yellow-brown, poor cleavage

Mineral	System	Hardness	Density	Color	Habit	Description
Clinoptilolite	Monocl.	3.5–4	2.1	W.	Platy	Crystals with good cleavage
Heulandite	Monocl.	3.5–4	2.2	W.	Platy	Elongated crystals with good cleavage, pearly luster
Stilbite	Monocl.	3.5–4	2.1	W.	Platy	Platy crystals with good cleavage, often in radiating structures
Margarite	Monocl.	3.5–4.5	3.1	Gy.,Y., G.	Platy	Platelets, more brittle than muscovite
Malachite	Monocl.	4	4	G.	Fibrous	Botryoidal, occasionally fibrous, green
Manganite	Monocl.	4	4.4	Br.-Bk.	Prismatic	Striated gray to black prisms, dark-brown streak
Monazite	Monocl.	5–5.5	5.1	Y., Br.	Equant, tabular	Isometric crystals resembling garnet and zircon, pleochroic halos in cordierite
Wolframite	Monocl.	5–5.5	7	Br., Bk.	Platy	Tabular crystals, perfect cleavage, high density
Goethite	Ortho.	5–5.5	4.3	Bk.-Br., R., Y.	Fibrous	Often in radiating rosettes, brown-yellow streak
Natrolite	Ortho.	5–5.5	2.3	W.	Fibrous	Prismatic or fibrous with square cross-section
Anatase	Tetrag.	5.5–6	3.8	B.-Bk., Br.	Pyramidal	Tetragonal–pyramidal morphology, red streak
Rutile	Tetrag.	6	4.2	R.-Br., Bk.	Prismatic	Black or red prismatic or acicular crystals, pale-brown streak
Columbite	Ortho.	6	5.3	Bk.	Platy	Black to red-brown, high gravity, brown streak
Prehnite	Ortho.	6–6.5	2.9	G., Gy., W.	Platelets	Rosettes in cavities of igneous rocks
Clinozoisite	Monocl.	6.5	3.4	G.-Gy., W.	Prismatic	In metamorphic schists
Zoisite	Ortho.	6.5	3.3	G.-Gy.	Prismatic	Single cleavage (2 cleavages in amphiboles)
Vesuvianite	Tetrag.	6.5	3.3	R.-Br., G.	Prismatic	Similar to grossular but not optically isotropic, crystals with square cross-section, poor cleavage
Chloritoid	Monocl.	6.5	3.5	G., Bk.	Platy	Resembles chlorite but higher refractive index, in schists
Axinite	Tricl.	6.5–7	3.3	Br., Gy.	Platy	Tabular brown to gray crystals, compare with titanite
Diaspore	Ortho.	6–7	3.4	W., Gy., Br.	Platy	Component of bauxite
Epidote	Monocl.	6–7	3.4	Y.-G., G.	Prismatic	Striated bottle-green crystals
Staurolite	Ortho.	7–7.5	3.7	Br.	Prismatic	Brown to black crystals in schists
Topaz	Ortho.	8	3.6	Cl., Y., B.	Prismatic	Striated crystals, good cleavage, various colors, high hardness

Appendix 1b.3 Nonmetallic luster, polyhedral cleavage (three systems), sorted according to hardness

Mineral	Symmetry	Hardness	Density	Color	Morphology	Diagnostics
Salammoniac	Cubic	1–2	1.5	Cl., Y., Br.	Octah.	White crystals or crusts in volcanic fumaroles
Nitratite	Trig.	1.5–2	2.3	W., Y.	Equant	Water soluble
Gypsum	Monocl.	1.5–2	2.3	Cl., W.	Tabular, prismatic	Colorless or white, one perfect cleavage, low hardness
Halite	Cubic	2	2.1	Cl., Y., R., Bl.	Cube	Cubic cleavage, clear, salty taste, evaporate mineral
Sylvite	Cubic	2	1.9	W., Y., R.	Cube	Cubic cleavage, reddish color, bitter taste, evaporate mineral
Borax	Monocl.	2–2.5	1.8	Cl., W.	Prismatic	Clear crystals transform to white tincalconite
Cinnabar	Trig.	2–2.5	8.1	R.	Equant	Red crystals, high density, red streak
Cryolite	Monocl.	2.5	3.0	W., Pink, Br.	Equant	White pseudocubic crystals, pegmatites
Kernite	Monocl.	2.5	1.9	Cl., W.	Prismatic	Large clear crystals with two perfect cleavages
Calcite	Rhomb.	3	2.7	Cl., W., Y., etc.	Rhombs, etc.	Rhombohedral cleavage, acid test, composes limestone and marble
Polyhalite	Tricl.	3	2.8	R., W., Y.	Tabular	Bitter taste, evaporate mineral
Wulfenite	Tetrag.	3	6.8	Y.-O.	Platy	Thin brilliant orange-yellow crystals
Barite	Ortho.	3–3.5	4.5	W., Gy., Y., R.	Tabular	Often occurs as crested aggregates, high density
Celestite	Ortho.	3–3.5	4.0	Cl., W., B.	Prismatic	White-bluish to brownish crystals, similar to barite
Anhydrite	Ortho.	3–3.5	2.9	Cl., W.	Equant, prismatic	Excellent cleavage at right angles
Sphalerite	Cubic	3–4	4.1	Y., Br., Bk.	Tetrahedra	Colorless, yellow to black, dodecahedral cleavage, hydrothermal, often associated with galena
Dolomite	Rhomb.	3.5–4	2.9	Cl., W., Gy., Br.	Rhombs	Rhombohedral cleavage, concentrated acid test
Ankerite	Rhomb.	3.5–4	3.0	Br., Y., W.	Rhombs	Difficult to distinguish from dolomite
Azurite	Monocl.	3.5–4	3.8	B.	Tabular	Colloform aggregates, dark blue
Fluorite	Cubic	4	3.2	Cl., V., Y., G.	Cube, octah.	Perfect octahedral cleavage, fluorescence, variety of colors
Rhodochrosite	Rhomb.	4	3.5	Pink, R., W.	Rhombs	Pink color, mostly hydrothermal, rhodonite is harder
Magnesite	Rhomb.	4–4.5	3.0	W., Gy.	Rhombs	White-gray masses, e.g., veins in weathered serpentinites
Siderite	Rhomb.	4–4.5	3.9	Y., Br., W.	Rhombs	Saddle morphology
Bastnaesite	Trig.	4–4.5	5.1	Y., Br.	Tabular	Platy wax-yellow crystals with greasy luster
Kyanite	Tricl.	4–7	3.6	B.	Prismatic	Blue crystals, two good cleavages, schists

Scheelite	Tetrag.	4.5	6.0	Gy., W., Y.	Equant	Good cleavage, difficult to distinguish from calcite, fluorescence diagnostic, hydrothermal
Smithsonite	Rhomb.	5	4.4	Cl., Gy., G., Br.	Rhombs	Mostly in botryoidal masses
Dioptase	Trig.	5	3.3	G.	Rhombs	Green rhombohedral crystals, rare
Anatase	Tetrag.	5.5–6	3.8	B.-Bk., Br.	Pyramidal	Tetragonal–pyramidal morphology, red-brown streak, generally as bluish aggregates
Sodalite	Cubic	5.5–6	2.3	W., B.		Colorless, white to gray, frequently fluorescent
Scapolite	Tetrag.	5–6.5	2.5	W., Gy.	Prismatic	Often vitreous and with poor cleavage, in volcanic rocks
Sanidine	Monocl.	6	2.6	Cl., Y., Gy.	Tabular	Two inclined cleavages, often multiply twinned
Albite	Tricl.	6	2.6	Cl., W.	Tabular	Good cleavage
Lawsonite	Ortho.	6	3.1	W., Gy., G.	Prismatic	White or pink, two inclined cleavages, microscopic twinning (no regular twin striations)
Microcline	Tricl.	6–6.5	2.6	W., R., G.	Tabular	
Orthoclase	Monocl.	6–6.5	2.6	W., R., G.	Tabular	Two 90° blocky cleavages, Carlsbad twins
Anorthite	Tricl.	6–6.5	2.8	W., Gy.	Tabular	Microscope needed for identification
Diamond	Cubic	10	3.5	Cl., Y., B.	Octah.	Hardness, brilliance, equiaxed crystals

Appendix 1b.4 Nonmetallic luster, prismatic or fibrous cleavage (two systems), sorted according to hardness

Mineral	Symmetry	Hardness	Density	Color	Morphology	Diagnostics
Niter (saltpeter)	Ortho.	2	2.0	Cl., W., Gy.	Fibrous	Water soluble
Epsomite	Ortho.	2–2.5	1.7	Cl., W.	Fibrous	White botryoidal masses, bitter taste, deposited by salt springs
Strontianite	Ortho.	3.5	3.7	Cl., W., Y.	Prismatic	One good cleavage
Cerussite	Ortho.	3–3.5	6.5	W., Gy., B.	Prismatic, equant	High density
Laumontite	Monocl.	3–3.5	2.3	W.	Prismatic	Perfect cleavage
Azurite	Monocl.	3.5–4	3.8	B.	Tabular	Colloform aggregates, dark blue
Kyanite	Tricl.	4–7	3.6	B.	Prismatic	Blue crystals, two good cleavages, schists
Wollastonite	Tricl.	4.5–5	2.8	W.	Fibrous	White crystals, excellent cleavage, in marbles
Apatite	Hexag.	5	3.2	Cl., G., Y., V.	Prismatic	Hexagonal prisms, various colors
Titanite	Monocl.	5–5.5	3.5	Y., G.-Br.	Bladed	Wedge-shaped crystals, poor cleavage, high birefringence
Natrolite	Ortho.	5–5.5	2.3	W.	Fibrous	Square cross-section of fibers
Enstatite	Ortho.	5–6	3.1	Gy., G.	Prismatic	Green, 87° cleavage, mafic and ultramafic igneous rocks
Hypersthene	Ortho.	5–6	3.5	Br., G.	Prismatic	87° cleavage
Omphacite	Monocl.	5–6	3.3	G.	Prismatic	Blue-green, 87° cleavage, in eclogites
Ferroactinolite	Monocl.	5–6	3.4	G.	Prismatic	Green, 124° cleavage, metamorphic rocks
Ferrohornblende	Monocl.	5–6	3.2	Br., Bk.	Prismatic	Columnar, brown to black, 124° cleavage
Hornblende	Monocl.	5–6	3.2	G., Bk.	Prismatic	Columnar, 124° cleavage (tourmaline lacks cleavage), greenish streak
Tremolite	Monocl.	5–6	2.9	W., Gy.	Prismatic	Radiating, sometimes fibrous aggregates, 124° cleavage
Glaucophane	Monocl.	5–6	3.1	B.	Prismatic	Light blue, 124° cleavage
Pargasite	Monocl.	5–6	3.1	W., G.	Prismatic	White or light green, 124° cleavage
Riebeckite	Monocl.	5–6	3.4	B., Bk.	Prismatic, fibrous	Dark blue to black, 124° cleavage, often asbestiform (crocidolite)
Cummingtonite	Monocl.	5–6	3.4	G., Gy., Br.	Prismatic	124° cleavage
Grunerite	Monocl.	5–6	3.6	Gy., Br.	Prismatic	Gray to brown, 124° cleavage
Scapolite	Tetrag.	5–6.5	2.5	W., Gy.	Prismatic	Colorless, white to gray, frequently fluorescent

Mineral	System	Hardness	S.G.	Color	Habit	Description
Anthophyllite	Ortho.	5.5	3.1	Br, Y.-Gy.	Prismatic, fibrous	Gray or yellow, 124° cleavage
Diopside	Monocl.	5.5–6	3.3	G., W.	Prismatic	87° cleavage, generally green
Hedenbergite	Monocl.	5.5–6	3.5	Bk.	Prismatic, platy	Black, 87° cleavage, green streak
Rutile	Tetrag.	6	4.2	R.-Br., Bk.	Prismatic	Black or red sometimes acicular crystals, pale-brown streak
Lawsonite	Ortho.	6	3.1	W., Gy., G.	Prismatic	Good cleavage, subduction zones
Augite	Monocl.	6	3.4	G., Bk.	Prismatic	Green to black, 87° cleavage, common in basalt
Essenite	Monocl.	6	3.5	R.-Br.	Prismatic	Brown, 87° cleavage
Pigeonite	Monocl.	6	3.4	G., Bk.	Prismatic	Green to black, 87° cleavage
Rhodonite	Tricl.	6	3.5	R.	Prismatic	Pink color, often in masses, higher hardness than rhodochrosite
Aegirine	Monocl.	6–6.5	3.6	Gy., Bk.	Prismatic	Columnar or acicular crystals, black
Sillimanite	Ortho.	6–7	3.2	W., Gy.	Fibrous	White fibers in pelitic schists
Spodumene	Monocl.	6–7	3.2	W., Y., V.	Prismatic	Often large crystals, white to violet, 87 ° cleavage, in pegmatites
Jadeite	Monocl.	6.5	3.4	W., G.	Fibrous	Massive, high hardness, higher density than serpentine
Andalusite	Ortho.	7.5	3.2	Gy., R.-Br.	Prismatic	Pink, green or gray crystals, often twinned, schists

Appendix 2 Minerals that display some distinctive physical properties (in alphabetical order)

Magnetic minerals
Chromite
Franklinite
Ilmenite
Magnetite
Pyrrhotite

Water-soluble minerals
Borax
Carnallite
Epsomite
Halite
Kernite
Nitritite
Polyhalite
Sylvite

Minerals that fuse with a match
Carnallite
Enargite
Orpiment
Realgar
Stibnite
Sulfur

Minerals with excellent single cleavage
Brucite
Chlorite
Graphite
Mica
Molybdenite
Orpiment
Pyrophyllite
Talc

Minerals with high density (>5 g/cm^3)
Anglesite
Arsenic
Arsenopyrite
Bornite
Carnotite
Cassitterite
Cerussite
Chalcocite
Chromite
Cinnabar
Cobaltite
Columbite
Copper
Cuprite
Galena
Gold
Hematite
Magnetite
Monazite
Nickeline
Platinum
Pyrite
Pyrolusite
Scheelite
Silver
Tetrahedrite
Uraninite
Vanadinite
Wolframite
Wulfenite

Appendix 3 Rock-forming minerals that are colored in thin section (bold: common coloration)

Isotropic minerals

Red/orange	Pink/rose	Brown	Yellow	Green	Blue/violet	Gray
Sphalerite	Fluorite	**Garnet**	Fluorite	**Fluorite**	**Fluorite**	Sodalite
Spinel	**Garnet**	**Sphalerite**	Garnet	**Spinel**	**Hauyne**	
	Sodalite	**Spinel**	Sodalite	Garnet	**Sodalite**	
			Sphalerite		**Spinel**	
			Spinel			

Anisotropic minerals (pleochroism)

Red/orange	Pink/rose	Brown	Yellow	Green	Blue/violet	Gray
Iddingsite	**Andalusite**	**Aegirine**	**Actinolite**	**Actinolite**	**Chloritoid**	**Apatite**
Piedmontite	Corundum	**Augite**	Allanite	**Aegirine**	Cordierite	Chlorite
Rutile	**Hypersthene**	**Biotite**	Biotite	**Augite**	**Corundum**	Chloritoid
	Piedmontite	**Chondrodite**	**Chloritoid**	**Biotite**	**Glaucophane**	**Glaucophane**
	Staurolite	**Hornblende**	Chondrodite	**Chlorite**	Hauyne	Rutile
	Tourmaline	Hypersthene	Clinochlore	**Chloritoid**	**Piedmontite**	Titanite
		Iddingsite	Clinohumite	**Glauconite**	**Riebeckite**	Zircon
		Monazite	**Epidote**	**Hornblende**	**Tourmaline**	
		Phlogopite	Glauconite	**Hypersthene**		
		Rutile	**Glaucophane**	**Riebeckite**		
		Staurolite	**Humite**	**Stilpnomelane**		
		Stilpnomelane	Monazite	Titanite		
		Titanite	Phlogopite	**Tourmaline**		
		Tourmaline	**Piedmontite**			
		Zircon	Rutile			
			Staurolite			
			Stilpnomelane			
			Titanite			
			Tourmaline			
			Zircon			

Appendix 4a Optically isotropic minerals, sorted according to refractive index (common: italic)

Mineral	Symmetry	Refractive index
Opal	Amorphous	1.3–1.45
Fluorite	Cubic	1.434
Sodalite	Cubic	1.48–1.49
Analcime	Cubic	1.48–1.49
Lazurite (Lapis)	Cubic	1.5
Halite	Cubic	1.56
Spinel	Cubic	1.72–2.05
Periclase	Cubic	1.73
Grossular	Cubic	1.74
Almandine	Cubic	1.76–1.83
Pyrope	Cubic	1.76–1.83
Spessartine	Cubic	1.80
Uvarovite	Cubic	1.87
Andradite	Cubic	1.89
Pyrochlore	Cubic	1.9–2.2
Microlite	Cubic	1.93–2.02
Sphalerite	Cubic	2.37
Diamond	Cubic	2.411–2.447

Appendix 4b Minerals with very low birefringence (up to white interference colors in 30 μm thin sections), sorted according to birefringence (common: italic)

Mineral	Symmetry	Refractive index	Birefringence	2V
Uniaxial –				
Apatite	Hexag.	1.63–1.65	0.001	
Marialite (scapolite)	Tetrag.	1.54–1.55	0.002	
Cristobalite	Tetrag.	1.48–1.49	0.003	
Nepheline	Hexag.	1.53–1.55	0.004	
Beryl	Hexag.	1.57–1.60	0.006	
Uniaxial +				
Leucite	Ps. Cubic	1.51	0.001	
Erionite	Hexag.	1.46–1.48	0.002	
Biaxial –				
Stilbite	Monocl.	1.49–1.51	0.001	−33
Chabazite	Trig.	1.48	0.002	−0–32
Clinoptilolite	Monocl.	1.48–1.59	0.003	−40
Chamosite	Monocl.	1.64–1.66	0.005	−0
Microcline	Tricl.	1.518–1.526	0.006	−60–84
Orthoclase	Monocl.	1.518–1.530	0.006	−60–80
Sanidine	Monocl.	1.518–1.532	0.006	−0–20
Albite high	Tricl.	1.527–1.534	0.007	−50
Cordierite	Ortho.	1.54–1.55	0.007	−40–80
Kaolinite	Tricl.	1.55–1.57	0.007	−20–50
Biaxial ±				
Clinochlore	Monocl.	1.57–1.59	0.005	+0–90
Mordenite	Ortho.	1.47–1.48	0.005	−80–(+80)
Biaxial +				
Tridymite	Monocl.	1.47–1.48	0.004	+35
Zoisite	Ortho.	1.70–1.71	0.004	+30–60
Coesite	Monocl.	1.59–1.60	0.005	+64
Harmotome	Monocl.	1.50–1.51	0.005	+43
Heulandite	Monocl.	1.49–1.50	0.005	+0–55
Phillipsite	Monocl.	1.48–1.50	0.005	+60–80

Appendix 4c Minerals with low birefringence (up to first-order red interference colors in 30 μm thin sections), sorted according to birefringence (common: italic)

Mineral	Symmetry	Refractive index	Birefringence	2V
Uniaxial –				
Corundum	Trig.	1.76–1.77	0.008	
Torbernite	Tetrag.	1.58–1.59	0.01	
Paligorskite	Monocl.	1.51–1.53	0.01	
Sepiolite	Ortho.	1.51–1.53	0.01	
Pyromorphite	Hexag.	2.05–2.06	0.011	
Uniaxial +				
Quartz	Trig.	1.544–1.553	0.009	
Tincalconite	Trig.	1.46–1.47	0.01	
Biaxial –				
Andalusite	Ortho.	1.63–1.65	0.01	–83–85
Axinite	Tricl.	1.68–1.69	0.01	–63–76
Laumontite	Monocl.	1.50–1.52	0.01	–26–47
Antigorite	Monocl.	1.56–1.57	0.011	–40–60
Clintonite	Monocl.	1.64–1.66	0.012	–2–40
Polylithionite	Monocl.	1.53–1.56	0.012	–0–40
Margarite	Monocl.	1.63–1.65	0.012	–40–65
Anorthite	Tricl.	1.575–1.590	0.013	–77
Biaxial ±				
Celsian	Tricl.	1.585–1.595	0.01	–65–(+)95
Riebeckite	Monocl.	1.65–1.71	0.01	–40–90
Biaxial +				
Celestite	Ortho.	1.62–1.63	0.009	+51
Gypsum	Monocl.	1.52–1.53	0.009	+58
Chrysoberyl	Ortho.	1.75–1.76	0.009	+45–71
Enstatite	Ortho.	1.65–1.66	0.009	+55
Clinoenstatite	Monocl.	1.65–1.66	0.009	+54
Jadeite	Monocl.	1.64–1.67	0.009	+70–72
Albite	Tricl.	1.529–1.539	0.010	+77
Topaz	Ortho.	1.61–1.64	0.01	+48–65
Clinozoisite	Monocl.	1.72–1.73	0.01	+85
Staurolite	Ortho.	1.74–1.76	0.01	+79–88
Chloritoid	Monocl.	1.71–1.74	0.01	+36–68
Barite	Ortho.	1.64–1.65	0.012	+36
Natrolite	Ortho.	1.48–1.49	0.013	+60–63

Appendix 4d Minerals with high birefringence (second- to fourth-order interference colors in 30 μm thin sections), sorted according to birefringence (common: italic)

Mineral	Symmetry	Refractive index	Birefringence	2V
Uniaxial –				
Gmelinite	Hexag.	1.48–1.47	0.015	
Tourmaline	Trig.	1.63–1.69	0.02	
Cancrinite	Hexag.	1.49–1.52	0.02	
Vesuvianite	Tetrag.	1.70–1.73	0.04	
Meionite (scapolite)	Tetrag.	1.56–1.59	0.04	
Uniaxial +				
Brucite	Trig.	1.57–1.58	0.015	
Scheelite	Tetrag.	1.92–1.93	0.016	
Zircon	Tetrag.	1.96–2.01	0.05	
Biaxial –				
Realgar	Monocl.	2.46–2.61	0.015	–40
Ferrohornblende	Monocl.	1.67–1.72	0.015	–50–80
Arfvedsonite	Monocl.	1.67–1.71	0.015	–5–50
Wollastonite	Tricl.	1.62–1.63	0.015	–35–40
Gismondine	Monocl.	1.52–1.55	0.015	–85
Kyanite	Tricl.	1.71–1.73	0.016	–82
Trilithionite	Monocl.	1.52–1.59	0.02	–0–50
Actinolite	Monocl.	1.69–1.71	0.02	–10–20
Hornblende	Monocl.	1.61–1.73	0.02	–60–88
Tremolite	Monocl.	1.60–1.63	0.02	–10–20
Tschermakite	Monocl.	1.64–1.69	0.02	–65–90
Glaucophane	Monocl.	1.61–1.67	0.02	–0–50
Edenite	Monocl.	1.63–1.68	0.02	–50–80
Autunite	Tetrag.	1.55–1.58	0.024	–10–30
Glauconite	Monocl.	1.59–1.64	0.025	–0–20
Phlogopite	Monocl.	1.53–1.62	0.03	–0–10
Zinnwaldite	Monocl.	1.53–1.59	0.03	–0–40
Paragonite	Monocl.	1.56–1.61	0.03	–0–40
Essenite	Monocl.	1.80–1.82	0.03	–77
Aluminoceladonite	Monocl.	1.61–1.66	0.04	–5–8
Muscovite	Monocl.	1.55–1.61	0.04	–30–45
Epidote	Monocl.	1.73–1.78	0.045	–68–73
Talc	Monocl.	1.54–1.60	0.05	–0–30
Pyrophyllite	Monocl.	1.55–1.60	0.05	–53–60
Biotite	Monocl.	1.56–1.69	0.05	–0–9
Siderophyllite	Monocl.	1.59–1.64	0.05	–0–5
Aegirine	Monocl.	1.76–1.83	0.05	–60–70
Fayalite	Ortho.	1.84–1.89	0.051	–47
Biaxial ±				
Hypersthene	Ortho.	1.68–1.73	0.015	–40–90
Perovskite	Ps.cubic	2.38	0.017	90
Pargasite	Monocl.	1.61–1.68	0.02	+55–90
Anthophyllite	Ortho.	1.60–1.67	0.02	90
Clinohumite	Monocl.	1.63–1.66	0.03	+74–90

Appendix 4d (cont.)

Mineral	Symmetry	Refractive index	Birefringence	2V
Cummingtonite	Monocl.	1.61–1.64	0.03	+80–90
Chondrodite	Monocl.	1.60–1.66	0.03	+72–90
Forsterite	Ortho.	1.64–1.67	0.035	+86
Allanite	Monocl.	1.72–1.76	0.04	90
Piemontite	Monocl.	1.70–1.71	0.04	+70–(–)70
Grunerite	Monocl.	1.69–1.71	0.04	–84–90
Biaxial +				
Anglesite	Ortho.	1.88–1.89	0.017	+60–75
Hedenbergite	Monocl.	1.74–1.76	0.018	+60
Boehmite	Ortho.	1.64–1.67	0.02	+74–88
Gibbsite	Monocl.	1.57–1.59	0.02	+0
Sillimanite	Ortho.	1.65–1.68	0.02	+25–30
Spodumene	Monocl.	1.65–1.68	0.02	+54–56
Pyroxmangite	Tricl.	1.73–1.76	0.02	+37–46
Rhodonite	Tricl.	1.72–1.74	0.02	+76
Omphacite	Monocl.	1.67–1.60	0.023	+60–70
Pigeonite	Monocl.	1.69–1.74	0.025	+0–50
Humite	Ortho.	1.61–1.67	0.03	+65–84
Augite	Monocl.	1.69–1.78	0.030	+25–85
Diopside	Monocl.	1.66–1.69	0.030	+59
Lawsonite	Ortho.	1.67–1.68	0.035	+84
Zoisite	Ortho.	1.70–1.71	0.04	+85
Turquoise	Tricl.	1.61–1.65	0.04	+40
Anhydrite	Ortho.	1.57–1.61	0.044	+42
Monazite	Monocl.	1.80–1.84	0.045	+6–19
Diaspore	Ortho.	1.70–1.75	0.048	+84

Appendix 4e Minerals with very high birefringence (higher than third-order interference colors in 30 μm thin sections), sorted according to birefringence (common: italic)

Mineral	Symmetry	Refractive index	Birefringence	2V
Uniaxial –				
Jarosite	Trig.	1.72–1.82	0.10	
Wulfenite	Tetrag.	2.28–2.41	0.122	
Calcite	Rhomb.	1.47–1.66	0.172	
Dolomite	Rhomb.	1.59–1.68	0.179	
Ankerite	Rhomb.	1.52–1.75	0.19	
Kutnahorite	Rhomb.	1.54–1.73	0.19	
Magnesite	Rhomb.	1.60–1.70	0.191	
Rhodochrosite	Rhomb.	1.70–1.81	0.218	
Smithsonite	Rhomb.	1.73–1.85	0.228	
Siderite	Rhomb.	1.75–1.87	0.240	
Hematite	Trig.	2.80–3.04	0.245	
Hausmannite	Tetrag.	2.15–2.46	0.31	
Uniaxial +				
Xenotime	Tetrag.	1.72–1.82	0.095	
Cassiterite	Tetrag.	1.20–2.09	0.096	
Bastnaesite	Trig.	1.72–1.82	0.10	
Rutile	Tetrag.	2.62–2.90	0.287	
Cinnabar	Trig.	2.91–3.27	0.359	
Biaxial –				
Annite	Monocl.	1.62–1.67	0.08	–0–5
Goethite	Ortho.	2.26–2.40	0.140	–0–42
Witherite	Ortho.	1.53–1.68	0.148	–16
Strontianite	Ortho.	1.52–1.67	0.150	–8
Aragonite	Ortho.	1.53–1.69	0.156	–18
Malachite	Monocl.	1.66–1.91	0.254	–43
Cerussite	Ortho.	1.80–2.08	0.274	–8
Lepidocrocite	Ortho.	1.94–2.51	0.57	–83
Orpiment	Monocl.	2.4–3.0	0.6	–76
Biaxial +				
Azurite	Monocl.	1.73–1.84	0.108	+68
Titanite	Monocl.	1.90–2.04	0.13	+23–34
Wolframite	Monocl.	2.26–2.42	0.16	+76
Manganite	Monocl.	2.25–2.53	0.28	+0–5
Sulfur (α)	Ortho.	1.96–2.25	0.288	+70

Glossary

Acicular Greatly elongated, needle-shaped crystals.

Aggregate A composite of crystals, such as a rock.

Alkali rocks Rocks with more alkalis (Na, K, Rb) than are contained in feldspars.

Alkaline Having the quality of a base, rather than an acid.

Amorphous A substance without a regular crystal structure.

Amphibolite facies A metamorphic regime corresponding to intermediate temperature and pressure conditions (hornblende is a characteristic mineral).

Amygdale Cavity or vesicle in volcanic rocks that has become filled with secondary minerals (such as quartz).

Analyzer Component of a petrographic microscope producing plane polarized light.

Andesite A volcanic rock of intermediate composition with plagioclase as the major mineral.

Ångström Unit of measure, $1\text{Å} = 10^{-1}$ nm = 10^{-8} cm.

Anhedral A crystal that lacks well-developed faces.

Asbestiform Habit characterized by strong and flexible fibers.

Atomic coordinates Fractional coordinates specifying the position of an atom in the unit cell (x, y, z).

Atomic number The number of protons in the nucleus of an atom.

Aureole A contact metamorphic zone surrounding an igneous intrusion.

Authigenic minerals Minerals formed in place.

Basalt A volcanic rock of mafic composition with plagioclase, pyroxene, and/or olivine as major components.

Batholith A large body of intrusive rocks.

Bcc Body-centered cubic (mainly with reference to metal structures).

Becke line A band of light, visible under a microscope, that separates substances of different refractive indices, when slightly out of focus.

Biaxial crystals Crystals having two optic axes (orthorhombic, monoclinic, or triclinic).

Biogenic Resulting from physiological activities of organisms.

Birefringence The measure of difference in refractive indices in different directions.

Black smoker Columnar structure formed by hydrothermal processes along mid-oceanic ridges.

Blueschist facies A metamorphic regime corresponding to low-temperature and high-pressure conditions (glaucophane is a characteristic mineral).

Botryoidal Mineral aggregates having the shape of a bunch of grapes (smoothed curved surfaces).

Bragg's law Relates the diffraction angle of X-ray waves to the lattice spacing of crystals.

Bravais lattices Lattices of the 14 symmetrically distinct unit cells.

Brightfield A term used in transmission electron microscopy to indicate that the primary beam is used for image formation.

Calcsilicate minerals Calcium-rich silicates that generally form as a reaction of calcite, quartz, and other components in impure limestones during metamorphism.

Cartesian coordinate system An X–Y–Z coordinate system in which all axes are at right angles.

Chain silicates Silicate minerals characterized by chains of silicon tetrahedra (also known as inosilicates).

Chatoyant Having a luster, resembling the changing reflection from the eye of a cat. This property (e.g., of some chrysoberyl) is caused by fibrous inclusions that scatter light.

Clastic Being composed of fragments (clasts) of pre-existing rocks.

Clausius–Clapeyron equation An equation relating the slope of a reaction in a pressure–temperature phase diagram to the entropy and volume change of the reaction.

Clay minerals Minerals with particles of size less than 2 μm in clays and soils (mainly hydrous sheet silicates).

Cleavage Property that causes minerals to break along well-defined planes.

Close-packed structures The most efficient way to pack like atoms in three dimensions (ideally each atom has 12 like neighbors).

Colloidal In a state of fine dispersion with particle sizes less than 10^{-6} cm.

Conchoidal fracture Crystal fractures along curved surfaces (quartz is an example).

Conoscopic illumination Mode of microscope operation in which the light is focused to a point inside the crystal.

Contact metamorphism Metamorphism localized around an igneous intrusion.

Coordination number Number of closest neighbors of an atom or ion with which it forms chemical bonds.

Coordination polyhedron Polyhedron formed around an ion by connecting the centers of closest anion neighbors.

Cotectic line Univariant line that is the intersection of two stability fields in a ternary system.

Country rock Rocks surrounding, and penetrated by, mineral veins or igneous intrusions.

Covalent bond Ideal chemical bond that involves sharing of orbital electrons.

Craton A tectonically inactive large part of the Earth.

Cryptocrystalline Material with a grain size that is too small to be visible with an optical microscope.

Crystal A homogeneous substance with a regular lattice structure.

Crystal classes The 32 possible combinations of rotation and inversion (mirror reflection) symmetries (synonymous with point-groups).

Crystal field Electric field generated by the net negative charge of the anions bonded to a cation.

Crystal structure Spatial arrangement of atoms (or ions) in a unit cell.

Crystal systems The six symmetrically distinct crystallographic coordinate systems (cubic, tetragonal, hexagonal/rhombohedral, orthorhombic, monoclinic, triclinic).

Curie temperature Temperature above which a ferromagnetic substance becomes paramagnetic.

Cyclosilicates *See* Ring silicates.

Darkfield A term used in transmission electron microscopy to indicate that a diffracted beam is used for image formation.

Defects Imperfections in the crystal structure (planar, linear, or point).

Dendritic Mineral or mineral aggregate with a morphology resembling tree branches.

Density Mass of a unit volume (generally expressed in g/cm^3).

Diagenesis Mineralogical changes in sediments at conditions of low temperature and pressure.

Diamagnetic Lacking an internal magnetic field. An external field may induce a weak internal field that opposes the applied field.

Diapir Piercement fold, for example a salt dome that is injected into overlying rocks.

Diatreme General term for a volcanic vent drilled through enclosing rocks by the explosive gas-charged magma.

Diffraction Cooperative scattering of light or X-rays from a regular microstructure or lattice of a crystal.

Diffraction pattern A record of diffracted intensities, for example X-ray diffraction recorded on film. Also used as "powder diffraction pattern" which records intensity variations as functions of diffraction angle.

Dike A cross-cutting tabular body of igneous rocks.

Dislocation Linear defect, significant for ductile deformation of a crystal.

Disordered structure Random distribution of two or more different atoms or cations in the same type of coordination polyhedron.

Dolostone Sedimentary rock composed of dolomite.

Druse Crust of crystals lining the side of a cavity.

d-spacing Distance between adjacent lattice planes (also known as interplanar spacing).

Ductility Tenacity of a mineral that can be permanently deformed to a new shape without breaking (usually involving the movement of dislocations).

Eclogite A high-pressure–high-temperature metamorphic rock of a basaltic composition with characteristic minerals omphacite–pyroxene and pyrope–garnet.

Enantiomorphic repetition A symmetry operation that reverses the sense of a motif (left–right handedness). Inversion and mirror reflection are such repetitions.

End members A chemical formula representing one limit of a solid solution.

Entropy A thermodynamic variable representing the degree of randomness or disorder in a system.

Epigenetic A secondary mineral deposit that is emplaced after the host rock already exists.

Epitaxial growth A regular overgrowth of one crystal on the surface of another, with a plane of registry between the two structures.

Equant (or equidimensional) Crystals having approximately the same dimension in all directions.

Equilibrium A static state of a chemical system in which phases present do not undergo changes with time. There are stable and metastable equilibria.

Euhedral A crystal that is bounded by well-developed crystal faces.

Evaporite Mineral or rock that forms by precipitation from water due to evaporation.

Exhalative A hydrothermal deposit created on the surface, for example on the ocean floor from springs.

Exsolution A process by which a solid solution separates into two different phases, often in the form of a regular lamellar intergrowth.

Extinction (optical) A birefringent mineral grain is oriented such that it appears black when viewed under crossed polarizers.

Face A planar surface of a crystal, generally described by rational Miller indices: (*hkl*).

Facies (metamorphic) Specific mineral assemblages characterized by a range of pressure-temperature conditions.

Fcc Face-centered cubic (mainly with reference to metal structures).

Felsic Rocks characterized by light minerals and rich in silicon.

Ferromagnetic Type of magnetic order, characteristic of iron and magnetite, that causes a crystal to respond strongly to a magnetic field.

Fibrous Mineral composed of fibers (*see also* Acicular and Asbestiform).

Fluorescence Process by which crystals emit electromagnetic radiation of lower energy in response to irradiation at a higher energy level (e.g., ultraviolet light can produce fluorescence in the visible range).

Foliation A planar fabric of a rock that may be due to depositional layering in a sediment or to deformation in a metamorphic rock.

Form A set of crystal faces related by symmetry: {*hkl*}.

Fracture The different patterns and shapes of fragments and surface features produced when a mineral is crushed.

Framework silicates Silicate minerals characterized by silicon tetrahedra joining to form a three-dimensional network (also known as tectosilicates).

Fumarole A vent that emits steam or gaseous vapor in volcanic areas.

Gabbro A plutonic rock of mafic composition and corresponding compositionally to basalt.

Gem Mineral valued for its extraordinary aesthetic appearance due to color, light properties, and hardness. Used in jewelry or for ornamentation.

Geode A hollow rock cavity lined by euhedral minerals.

Geothermometer A mineral system with compositional variations that reflect the temperature under which it formed.

Gibbs free energy A thermodynamic variable that describes the change in energy of a substance brought about by independent changes in pressure and temperature.

Gibbs phase rule In any chemical system in equilibrium, the number of chemical components plus two is equal to the number of stable phases plus the number of degrees of freedom.

Glass An amorphous solid material without a long-range ordered crystal structure. In most cases, glasses are metastable supercooled liquids.

Glide plane Symmetry operation that repeats a motif by combined reflection and translation.

Gneiss A foliated metamorphic rock of granitic composition.

Graben A large tectonic block that has been displaced downward along faults, relative to adjacent rocks (German for ditch).

Grade (metamorphic) The degree of metamorphism, generally equivalent to the temperature of metamorphism.

Grain A particle (usually a discrete crystal) that composes a rock.

Granite A plutonic rock of felsic composition, with quartz, K-feldspar, and plagioclase as major components (may also contain muscovite, biotite, and some hornblende).

Granodiorite A plutonic rock similar to granite but with plagioclase dominating over K-feldspar.

Granulite facies   A metamorphic regime corresponding to the highest grade (temperature) of regional metamorphism.

Graywacke   Type of sandstone marked by quartz and feldspars in a clay matrix.

Greenschist facies   A metamorphic regime corresponding to fairly low temperature (300–500 °C) and pressure (0.2–0.8 GPa) conditions (chlorite is a characteristic mineral).

Greenstone belt   A metamorphic terrane with low-grade metamorphic volcanics and volcanogenic sediments.

Greisen   Hydrothermally altered rock of granitic composition. Often a host rock for tin ores.

Groundmass   Fine-grained material between phenocrysts in igneous rocks.

Habit   The characteristic appearance of a mineral due to crystal form or combination of forms, intergrowths, and aggregation.

Hardness   Relative resistance of a mineral surface to scratches.

Hcp   Hexagonal close-packed (with reference to metal structures).

Hornfels   A fine-grained contact-metamorphic rock lacking foliation or lineation.

Host rock   The rock that hosts ore minerals or an ore deposit.

Hydrothermal fluids   Hot water-rich fluids that circulate through rocks.

Igneous rocks   Rocks that form by solidification from a magma. They are divided into volcanic and plutonic rocks.

Indicatrix   Ellipsoidal surface used to describe the anisotropic propagation of light in crystals (main axes correspond to refractive indices of vibration directions).

Infrared light   Electromagnetic radiation with wavelengths longer than those of visible light.

Inosilicates   *See* Chain silicates.

Interference colors   The colors displayed by a birefringent crystal when viewed under crossed polarizers.

Inversion   Basic symmetry operation that repeats a motif by equidistant projection across a point (symbol i).

Ionic bonding   Ideal chemical bonding of electrostatic nature, between cations (+) and anions (−).

Ionic radius   Radius of the sphere effectively occupied by an ion in a particular structural environment.

Isochrome and isogyre   Components of an interference figure. Color bands and extinction locations, respectively.

Isograd   A boundary between minerals, mineral assemblages, or mineral compositions, which can be mapped in a metamorphic terrane.

Isomorphic crystals   Solid solution series in which the crystal structure is the same.

Isotope   Different isotopes of the same element have the same number of protons but vary in their number of neutrons.

Isotropic   Having the same properties in all directions.

Kimberlite   An alkalic peridotite associated with diatremes, typically containing phenocrysts of olivine and phlogopite. Some kimberlites host diamonds.

Komatiite   Volcanic rock of ultramafic composition.

Lamellar   Composed of thin layers, like leaves of a book (e.g., exsolution or twin lamellae).

Laterite   A highly weathered soil, rich in aluminum and iron hydroxides and variable amounts of quartz and clay minerals. Laterites form in high-rainfall tropical climates.

Lattice   Three-dimensional periodic array of points. It is used to represent the translational symmetry of a crystal structure.

Lattice parameters   Magnitudes and angles between unit translations of a lattice (a, b, c, α, β, γ).

Laue equation   Equation that describes the angular relationship between incident X-ray beam and diffracted X-ray beam, interacting with a lattice.

Lava   Molten extrusive magma.

Layer silicates   Silicate minerals characterized by layers of silicon tetrahedra as a primary frame (also known as sheet silicates and phyllosilicates).

Leucocratic   Term applied to light-colored rocks such as granite, containing less than 30% dark minerals.

Lineation   Any pervasive linear feature that may be observed in a rock. Examples are aligned prismatic minerals in a lava flow or striations in a metamorphic rock due to deformation.

Liquidus   The line on a temperature–composition diagram above which the system is completely liquid.

Luster Type and nature of reflection of light from mineral surfaces.

Mafic Rock composed largely of magnesian minerals that are generally dark.

Magma Naturally occurring molten rock.

Major elements An element that is a key part of a mineral's composition.

Martensitic A crystal structure that is produced by shear (from martensite, the metastable steel phase that forms from austenite).

Massif A large geological body composed of igneous and metamorphic rocks.

Metallic bonding Chemical bonding between atoms in which electrons are highly delocalized and free to move between atoms.

Metamict Structural alteration from exposure to radioactivity.

Metamorphism Mineralogical and textural change in a rock induced by changes in temperature and pressure. Metamorphism is often associated with deformation.

Metasomatism Process by which the bulk chemistry of a rock is changed (generally from movement of pore fluids during metamorphism).

Metastable equilibrium State of two or more phases that are at equilibrium but have not achieved the lowest free energy state.

Microcrystalline Crystals that can be seen only with a microscope.

Microtwins Twins that exist in small domains and can be seen only with a microscope (e.g., cross-hatched twinning in microcline).

Migmatite A mixed rock, generally with light and dark bands, due to partial melting at high-grade metamorphic conditions.

Miller indices Rational numbers used to describe the orientation of lattice planes with respect to crystal axes (hkl).

Mineral Naturally occurring chemically and structurally homogeneous substance with a lattice structure.

Mirror plane A basic symmetry operation across which a mirror image is created (symbol m).

Miscibility gap A range in temperature and composition of a solid solution series that is not stable as a single phase and exsolution (unmixing) occurs.

Monochromatic Electromagnetic radiation (X-rays or light) consisting of a single wavelength.

Nanometer SI unit of length generally used in crystal structure. 1 nm = 10 Å = 10^{-9} m.

Native element Element that occurs naturally as a mineral.

Neutron Elementary particle of neutral charge, generally found in atomic nuclei. A neutron has about the same mass as a proton. Neutrons are also produced by radioactive reactions in nuclear reactors.

Nucleus Central part of atoms consisting of protons or a combination of protons and neutrons.

Octahedron A regular polyhedron with six corners and eight faces composed of equilateral triangles.

Oölith Spherical body composed of generally radial crystals, often of concentric structure. Oölite rocks are composed of oöliths.

Opaque Mineral through which it is impossible to transmit light.

Optic angle Angle between the optic axes of biaxial crystals ($2V$).

Optic axis A direction in a crystal along which no birefringence occurs.

Optic sign Either + or −, depending on the relationship of refraction indices. In uniaxial crystals sign is + if $n\varepsilon > n\omega$, − if $n\varepsilon < n\omega$. Biaxial crystals are + if $n\beta$ is closer to $n\alpha$ than to $n\gamma$ and − if $n\beta$ is closer to $n\gamma$ than to $n\alpha$.

Ordered structure Structures with only one kind of atom or ion in a structurally distinct coordination polyhedron (*see also* Disordered structure).

Ore deposit An economic concentration of ore minerals.

Ore mineral Mineral, usually metallic, that is of economic value.

Orthoscopic illumination Illumination with light rays traveling parallel to the tube of a light microscope.

Oxidation Process of an atom losing one or more electrons through chemical reaction.

Parallelepiped Polyhedron defined by three pairs of parallel faces.

Paramagnetic (1) Crystals that have no net magnetic moment owing to random orientation of dipole moments of atoms. (2) Substances with a weak internal magnetic field and a weak attraction in an external field.

Path difference With reference to waves it signifies the distance by which two waves are displaced.

Pedology The science of soil investigation.

Pelite   Rock of composition corresponding to a clay-rich sediment (aluminum rich).

Penetration twin   A twin in which two crystals appear to penetrate or grow through each other, with an irregular contact interface (contrary to contact twins).

Peridotite   Igneous rock of ultramafic composition, composed mainly of olivine and minor pyroxene.

Petrology   Branch of science dealing with the formation, composition, description, and classification of rocks.

Phase diagram   Diagrams used to display the stability of minerals as functions of composition, temperature, pressure, or other variables.

Phase rule   In any chemical system in equilibrium, the number of chemical components plus two is equal to the number of stable phases plus the number of degrees of freedom.

Phase transformation   A mineral phase transforms to a new phase or phases with different structure and sometimes composition.

Phase transition   A mineral phase transforms to a new phase with different structure but the same composition.

Phenocryst   Large crystals that grow in a cooling magma (as opposed to fine groundmass).

Phosphorescence   Fluorescence in which emission of visible light persists after cessation of irradiation.

Phyllosilicates   *See* Sheet silicates.

Piezoelectric effect   Production of an electric field in acentric minerals owing to application of stress.

Placer   A deposit of gold or other heavy minerals in sand or gravel, produced by gravitational concentration during water flow.

Plane polarized light   Light constrained to vibrate only in one plane.

Plastic deformation   Deformation mode in which crystals do not break, but attain a new shape, generally through movements of dislocations or by diffusion.

Platy   Appearing like a plate (thin in one dimension compared with the other two).

Pleochroism   The property of anisotropic crystals to absorb different wavelengths of light in different directions (a color change is observed when the crystal is rotated in plane polarized light).

Plutonic rocks   Igneous rocks that crystallize at depth and cool slowly.

Poikiloblastic   A metamorphic growth texture in which large crystals contain numerous relics of original minerals.

Point defects   Defects that occur at points in a crystal structure (e.g., vacancies and interstitials).

Point-groups   Thirty-two different combinations of rotation, mirror reflection, and inversion in crystals (also called crystal classes).

Polarizer   Component of a petrographic microscope that produces plane polarized light.

Polymerization   Connection of polyhedra or other structural units into rings, chains, sheets, or three-dimensional networks by sharing atoms.

Polymorphism   Crystals of the same chemical composition but different structure.

Polytypism   Crystals of the same chemical composition with different types of stacking of the same structural unit (e.g., mica).

Porphyroblast   A large crystal that grows in a metamorphic rock by replacing pre-existing minerals.

Powder pattern (X-ray)   Collection of diffraction information from a powdered crystalline sample.

Prograde metamorphism   Metamorphism that proceeds from lower to higher grade.

Proton   Elementary particle in the nuclei of atoms with a positive charge.

Pseudomorph   One mineral replaces another and adopts its shape.

Pyroelectric effect   Production of an electric field in acentric minerals with a unique axis in response to heat (e.g., tourmaline).

Pyroelectricity   Property of some minerals to develop electric charges when heated.

Radioactivity   Spontaneous decay of atoms of certain isotopes into new isotopes, accompanied by emission of high-energy particles (neutrons, electrons, or γ-rays).

Radius ratio   The radius of the cation divided by the radius of the anion in a coordination polyhedron of an ionic structure.

Recrystallization   Formation, in solid state, of new crystals from pre-existing materials, generally under the influence of temperature and/or deformation.

Reflection   A basic symmetry operation with a mirror plane across which a mirror image is created (symbol *m*).

Refractive index The ratio of the velocity of light in vacuum to the velocity of light in a given material (symbol *n*).

Refractometer A device for determining the index of refraction.

Retardation Measure (in nm) of the lag of a slower light ray behind a faster light ray after passage through a birefringent crystal of arbitrary thickness.

Retrograde metamorphism Metamorphic transformations that take place during decreasing temperature (often involving water).

Ring silicates Silicate minerals characterized by rings of silicon tetrahedra (also known as cyclosilicates).

Rotation A basic symmetry operation that repeats a motif by rotation around an axis (*n*). Only 1-, 2-, 3-, 4-, and 6-fold rotations are possible in crystals with a lattice structure.

Scattering When an incident ray of electromagnetic radiation (e.g., X-rays) strikes an atom or other object and the radiation is emitted in different directions.

Scattering factor Expression of the scattering power of an atom. For X-ray scattering on electrons the scattering factor depends on atomic number, atom size, and diffraction angle.

Schiller effect A play of colors (as in opal and labradorite).

Schist A foliated metamorphic rock with parallel alignment of mica flakes and other tabular minerals.

Screw axis Symmetry operation repeating a motif after a combined rotation and translation.

Sedimentary rock Rock formed by the accumulation of sediment in water or from air.

Shale A foliated fine-grained sedimentary rock, rich in clay particles.

Sheet silicates Silicate minerals characterized by layers of silicon tetrahedra as the primary frame (also known as layer silicates and phyllosilicates).

Sialic Rock rich in silicon and aluminum (granitic rocks are sialic).

Silicate A compound with a crystal structure containing SiO_4 tetrahedra.

Skarn An altered carbonate rock in a contact-metamorphic zone. Skarns often contain important ore deposits.

Slaty A texture that is typical of slates with a parallel arrangement of platy minerals.

Slip plane A lattice plane in the crystal along which dislocations move, when subjected to a shear stress (also called glide plane).

Snell's law The law that relates the refractive indices of two media to the angles of incidence and refraction.

Soil Material composing the Earth's surface, produced by weathering and organic accumulations, and supporting rooting plants.

Solid solution A crystal with a continuous range of compositions for the same crystal structure.

Solidus The line on a temperature–composition diagram below which the system is completely solid.

Solvus The line on a temperature–composition phase diagram that outlines a miscibility gap.

Space-groups Possible and distinct combinations of rotations, mirror reflections, inversions, and translations into which crystal structures can be classified (230).

Specific gravity Density of a substance with respect to unit density of water.

Stability field Range of conditions over which a mineral or mineral assemblage is stable.

Stacking fault A type of planar defect caused by stacking irregularities.

Stock A medium-sized body of plutonic rock with steep contacts.

Stoichiometry The exact proportions of elements in a mineral, described by a formula with rational numbers.

Strain Deformation resulting from an applied force (may be elastic or plastic).

Stratovolcano A volcanic dome built of alternating layers of lava and pyroclastic rocks.

Streak Color of a mineral when powdered (generally determined by scratching a ceramic plate).

Stress Force per unit area on the surface to which the force is applied.

Structure factor Expression that describes the wave diffracted by a crystal structure. It is usually given as a complex number *F*. *F* depends on the diffracted plane *hkl* and on atomic positions.

Stylolytes Certain limestones develop grooved, curved surfaces owing to dissolution which are generally at high angles to the bedding planes.

Subduction Along collisional plate margins (produced by mantle convection), slabs of crust get subducted to great depth, maintaining relatively low temperatures.

Supercritical Beyond the critical point there is no distinction between liquid and gas (vapor). Water becomes supercritical above 374.4 °C.

Symmetry An object is symmetrical if a motif is repeated by rotation, mirror reflection, inversion, or translation.

Syngenetic Mineral deposit that formed at the same time as the host rock.

Tailings Portions of ore that are discarded after processing.

Tensor Mathematical quantity to describe, for example, physical properties of crystals that vary with direction.

Ternary system A system that can be characterized by three end members.

Terrigenous Produced from the earth.

Tetrahedron A regular polyhedron with four corners and four faces composed of equilateral triangles.

Thermal conductivity A physical property describing the heat flow through materials.

Thin section A thin slice of rock mounted on a glass slide and used for analyses with the petrographic microscope.

Tieline A line connecting two phases that are in equilibrium at a given temperature.

Trace element An element that is only present in very small quantities in a mineral.

Translation Basic symmetry operation that repeats a motif at regular intervals along a direction.

Tufa A chemical sedimentary rock composed of calcium carbonate or silica that is deposited from percolating water.

Twin Regular intergrowth of two crystals that share a lattice plane or a lattice direction but are in different orientations.

Typomorphism Minerals that display a conspicuous relationship between morphology, composition and properties, and conditions of formation are called typomorphic.

Ultramafic Rocks (igneous or metamorphic) rich in magnesium and depleted in silicon and containing primarily mafic minerals. Peridotite is an ultramafic rock with olivine as main constituent.

Ultraviolet light Electromagnetic radiation with wavelengths shorter than those of visible light but longer than those of X-rays.

Uniaxial Describing minerals with a single optic axis (tetragonal, trigonal, and hexagonal).

Unit cell Parallelepiped defined by three noncoplanar unit translations in a lattice.

Unmixing A process by which a solid solution separates into two different phases, often in the form of a regular lamellar intergrowth.

Vacancy A site in a crystal structure that is accidentally vacant in a particular unit cell.

Van der Waals bonds A weak type of electrostatic bond, created by brief fluctuations in the balance of positive and negative charges.

Vein Fracture in a rock filled with minerals that precipitated from aqueous solutions.

Vesicle A cavity in a volcanic rock formed by a gas bubble trapped during cooling of a lava.

Viscosity The internal resistance to flow (mainly used for liquids).

Vitreous A type of luster resembling that of a glass.

Wavelength Distance between wave crests of electromagnetic radiation expressed in nm or Å.

Weathering Alteration of rocks by surface agents, for example, water, wind, and sunlight (there is chemical and mechanical weathering).

Xenoliths A rock fragment picked up by a magma and preserved after the magma has solidified.

X-ray Electromagnetic radiation with wavelengths between 0.1 Å and 100 Å.

X-ray diffraction A change of direction of X-rays owing to interaction with the periodic crystal structure (used for structural characterization of minerals).

X-ray fluorescence A change in energy owing to interaction with X-ray photons and energy transitions in atoms (used for chemical characterization of minerals).

Zone axis A direction parallel to multiple faces in a crystal.

Zoning Variation in composition of a crystal, typically from the core to the margin.

References

See also "Further reading" at the end of individual chapters.

Ahrens, T. J. (ed.) (1995). *Mineral Physics and Crystallography. A Handbook of Physical Constants.* American Geophysical Union, Washington, DC.

Aines, R. D., Kirby, S. H. and Rossman, G. R. (1984). Hydrogen speciation in synthetic quartz. *Phys. Chem. Mineral.*, **11**, 204–212.

Allègre, C. J., Poirier, J. P., Humler, E. and Hofmann, A. W. (1995). The chemical composition of the Earth. *Earth Planet. Sci. Lett.*, **134**, 515–526.

Andersen, O. (1915). The system anorthite–forsterite–silica. *Am. J. Sci.*, **39**, 407–454.

Anderson, D. L. (1967). Phase changes in the upper mantle. *Science*, **157**, 1165–1173.

Anderson, D. L. and Hart, R. S. (1976). An earth model based on free oscillations and body waves. *J. Geophys. Res.*, **81**, 1461–1475.

Arnoth, J. (1986). *Achate, Bilder im Stein.* Birkhäuser, Basel, Switzerland.

Aufreiter, S., Hancock, R. G. V., Mahoney, W. C., Stambolic-Robb, A. and Sanmugadas, K. (1997). Geochemistry and mineralogy of soils eaten by humans. *Int. J. Food Sci. Nutr.*, **48**, 293–305.

Bacon, G. E. (1975). *Neutron Diffraction.* Oxford University Press, Oxford.

Badro, J., Cote, A. S. and Brodholt, J. P. (2014). A seismologically consistent compositional model of Earth's core. *Proc. Natl. Acad. Science*, **111**, 7542–7545.

Baikow, V. E. (1967). *Manufacture and Refining of Raw Cane Sugar.* Elsevier, Amsterdam.

Banerjee, S. K. (1991). Magnetic properties of Fe-Ti oxides. In *Oxide Minerals: Petrologic and Magnetic Significance*, Reviews in Mineralogy, vol. 25, pp. 107–128. Mineralogical Society of America, Washington, DC.

Banfield, J. F. and Nealson, K. H. (eds.) (1997). *Geomicrobiology: Interactions between Microbes and Minerals*, Reviews in Mineralogy, vol. 35. Mineralogical Society of America, Washington, DC.

Barber, A. H., Lu, D. and Pugno, N. M. (2015). Extreme strength observed in limpet teeth. *J. R. Soc. Interface*, **12**, 20141326, 1–6.

Barber, D. J. and Wenk, H.-R. (1979). On geological aspects of calcite microstructure. *Tectonophysics*, **54**, 45–60.

Barber, D. J., Heard, H. C. and Wenk, H.-R. (1981). Deformation of dolomite single crystals from 20–800 °C. *Phys. Chem. Miner.*, **7**, 271–286.

Barlow, W. (1897). A mechanical cause of homogeneity of structure and symmetry. *Proc. R. Dublin Soc.*, **8**, 527–690.

Barron, L. M. (1972). Thermodynamic multicomponent silicate equilibrium phase calculations. *Am. Mineral.*, **57**, 809–823.

Barth, T. F. W. (1962). *Theoretical Petrology*, 2nd edn. Wiley, New York.

Bartholinus, E. (1669). Experimenta crystalli Islandici disdiaclastici quibus mira et insolita refraction detegitur. Hafniae sumpt. Daniel Paulli Reg. Bibl. (English version, 1670: Experiments made on a crystal-like body sent from Iceland. *Phil. Trans. R. Soc. London*, **5**, 2039–2048.)

Battocchio, F., Monteiro, P. and Wenk, H.-R. (2012). Rietveld refinement of the structures of 1.0 C-S-H and 1.5 C-S-H. *Cement Concrete Res.*, **42**, 1534–1548.

Bauer, G. (Agricola) (1556). *De Re Metallica.* (English translation: Hoover, H. C. and Hoover, L. H. (1950), Dover, New York.)

Beck, A., Darbha, D. M. and Schloessin, H. H. (1978). Lattice conductivities of single-crystal and polycrystalline materials at mantle pressure and temperatures. *Phys. Earth Planet. Inter.*, **17**, 35–53.

Beck, R. (1909). *Lehre von den Erzlagerstätten*, 3rd edn. Bornträger, Berlin.

Becke, F. (1903). Über Mineralbestand und Struktur der kristallinen Schiefer. *Denkschr. Akad. Wiss., Vienna*, **75**, 1–53.

Bedogné, F., Maurizio, R., Montrasio, A. and Sciesa, E. (1995). *I Minerali della Provincia di Sondrio e della Bregaglia Grigionese.* Bettini, Sondrio.

Bell, K. (1989). *Carbonatites: Genesis and Evolution.* Springer.

Bentley, W. A. and Humphreys, W. J. (1962). *Snow Crystals*. Dover, New York. (Originally published by McGraw-Hill, 1931.)

Bergmann, T. (1773). Variae crystallorum formae a spata ortae. *Nov. Acta Reg. Soc. Sci. Upsala*, **1**.

Bindi, L., Steinhardt, P. J., Yao, N. and Lu, P. J. (2011). Icosahedrite, $Al_{63}Cu_{24}Fe_{13}$, the first natural quasicrystal. *Am. Mineral.*, **96**, 928–931.

Birch, F. (1952). Elasticity and constitution of the earth's interior. *J. Geophys. Res.*, **57**, 227–286.

Bischoff, W. D., Sharma, S. K. and MacKenzie, F. T. (1985). Carbonate ion disorder in synthetic and biogenic magnesian calcites: a Raman spectral study. *Am. Mineral.*, **70**, 581–589.

Bish, D. L., Blake, D. F., Vaniman, D. T. *et al.* and MSL Science Team (2013). X-ray diffraction results from Mars Science Laboratory: mineralogy of Rocknest and Gale Crater. *Science*, **341**. doi: 10.1126/science.1238932.

Bowen, N. L. (1913). The melting phenomena of the plagioclase feldspars. *Am. J. Sci.*, **35**, 577–599.

Bowen, N. L. (1915). The crystallization of haplobasaltic, haplodioritic, and related magmas. *Am. J. Sci.*, **40**, 161–185.

Bowen, N. L. (1928). *The Evolution of the Igneous Rocks*. Princeton University Press, Princeton, NJ.

Bowen, N. L. and Anderson, O. (1914). The binary system MgO-SiO_2. *Am. J. Sci.*, **37**, 487–500.

Bowen, N. L. and Tuttle, O. F. (1950). The system $NaAlSi_3O_8$–$KAlSi_3O_8$–H_2O. *J. Geol.*, **58**, 498–511.

Bragg, W. H. and Bragg, W. L. (1913). The reflection of X-rays by crystals. *Proc. R. Soc. London A*, **88**, 428–438.

Bragg, W. L. (1914). The structure of some crystals as indicated by their diffraction of X-rays. *Proc. R. Soc. London A*, **89**, 277–291.

Bragg, W. L. (1930). The structure of silicates. *Z. Kristallogr.*, **74**, 237–305.

Bravais, A. (1850). Les systèmes formés par des pointes distribués régulièrement sur un plan ou dans l'espace. *J. École Polytech.*, **19**, 1–128.

Breithaupt, A. (1849). *Die Paragenesis der Mineralien, mineralogisch, geognostisch und chemisch beleuchtet: mit besonderer Rücksicht auf Bergbau*. Engelhardt, Freiberg.

Brody, J. J. (1980). *Mimbres Painted Pottery*, 2nd edn. University of New Mexico Press, Albuquerque, NM.

Buerger M. J. (1951). Crystallographic aspects of phase transformations. In *Phase Transformations in Solids*, ed. R. Smoluchowski, J. E. Mayer and W. A. Weyls, pp. 183–221. Wiley, New York.

Buerger, M. J. (1978). *Elementary Crystallography: An Introduction to the Fundamental Geometric Features of Crystals*, revised edn. MIT Press, Cambridge, MA.

Bulakh, A. G. (1977). Thermodynamic properties of H_2O in region up to $1000°C$ and 100 kbar and specularities of phase transitions in the H_2O-system. *Int. Geol. Rev.*, **21**, 92–103.

Bulakh, A. G. (1989). *Mineralogy*. (In Russian.) Academia, Moscow.

Bulakh, A. G. (1996). Summary mineral composition of the earth's crust. (In Russian.) *Proc. Russ. Mineral. Soc.*, **4**, 23–28.

Bulakh, A. G. (2011). *Mineralogy*. (In Russian.) Academia, Moscow.

Bulakh, A. G. and Zolotarev, A. A. (2000). Composition of monoclinic Ca–Mg–Fe–Na pyroxenes of the C2/c space group and the 50% rule. *Proc. Russ. Mineral. Soc.*, **6**, 69–79.

Bulakh, A. G. and Zussman, J. (1994). Structural formulae. In *Advanced Mineralogy Volume 1, Composition, Structure and Properties of Mineral Matter*, ed. A. S. Marfunin, pp. 12–18. Springer-Verlag, Berlin.

Burchard, U. (1998). History of the development of the crystallographic goniometer. *Mineral. Record*, **29**, 517–577.

Burke, J. R. (1966). *Origins of the Science of Crystals*. University of California Press, Berkeley, CA.

Burri, C., Parker, R. L. and Wenk, E. (1967). *Die Optische Orientierung der Plagioklase*. Birkhäuser, Basel.

Buseck, P. R. (ed.) (1992). *Minerals and Reactions at the Atomic Scale: Transmission Electron Microscopy*. Reviews in Mineralogy, vol. 27. Mineralogical Society of America, Washington, DC.

Caley, E. R. and Richards, J. F. C. (1956). *Theophrastus on Stones. Introduction, Greek Text, English Translation, and Commentary*. Ohio State University Press, Columbus, OH.

Cameron, E. N., Jahns, R. H., McNair, A. H. and Page, L. R. (1949). *The Internal Structure of Granitic Pegmatites*. Economic Geology Monographs, vol. 2. American Geophysical Union, Washington, DC.

Cann, J. R., Strens, M. R. and Rice, A. (1985). A simple magma-driven thermal balance model for the formation of volcanogenic massive sulfides. *Earth Planet. Sci. Lett.*, **76**, 123–134.

Cappeller, M. A. (1723). *Prodromus Crystallographiae de Crystallis Improprie Sic Dictis Commentarium.* Wyssing, Lucerne. Translated by K. Mieleitner (1922). Piloty and Loehle, Munich.

Carr, H. W., Groves, D. I. and Cawthorne, R. G. (1994). The importance of synmagmatic deformation in the formation of Merensky Reef potholes in the Bushveld complex. *Econ. Geol.*, **89**, 1398–1410.

Cashman, K. V. and Marsh, B. D. (1988). Crystal size distribution (CSD) in rocks and the kinetics and dynamics of crystallization. II: Makaopuhi lava lake. *Contrib. Mineral. Petrol.*, **99**, 292–305.

Chai, M., Brown, J. M. and Slutsky, L. J. (1996). Thermal diffusivity of mantle minerals. *Phys. Chem. Mineral.*, **23**, 470–475.

Chakoumakos, B. C. (ed.) (2004). Clathrate Hydrates Special Issue. *Am. Mineral.*, **89**, 1153–1279.

Champness, P. and Lorimer, G. (1971). An electron microscopy study of lunar pyroxene. *Contrib. Mineral. Petrol.*, **33**, 171–183.

Champness, P. E. and Lorimer, G. W. (1976). Exsolution in silicates. In *Electron Microscopy in Mineralogy*, ed. H.-R. Wenk, pp. 174–204. Springer-Verlag, Berlin.

Chan, C. S., Fakra, S. C., Edwards, D. C., Merson, D. and Banfield, J.F. (2009). Iron oxyhydroxide mineralization on microbial extracellular polysaccharides. *Geochim. Cosmochim. Acta*, **73**, 3807–3818.

Chapman, C. H. (2010). *Fundamentals of Seismic Wave Propagation.* Cambridge University Press, Cambridge.

Christensen, J. N., Rosenfeld, J. L. and DePaolo, D. J. (1989). Rates of tectonometamorphic processes from rubidium and strontium isotopes in garnet. *Science*, **244**, 1465–1469.

Clark, S. P. (1966). *Handbook of Physical Constants.* GSA Memoir, 97, Geological Society of America, Washington, DC.

Craig, J. R. and Vaughan, D. J. (1994). *Ore Microscopy and Ore Petrography*, 2nd edn. Mineralogical Society of America, Washington, DC.

Curie, J. and Curie, P. (1880). Sur l'électricité polaire dans les cristaux hémidièdres à faces inclinés. *Comp. R. Acad. Sci.*, **91**, 383–389.

Dana, J. D. (1837). *System of Mineralogy, including extended Treatise on Crystallography: with an Appendix, containing the Application of Mathematics to Crystallographic Investigation, and a Mineralogical Bibliography.* Durrie and Peck, New Haven, CT.

Dana, J. D. and Brush, G. J. (1868). *A System of Mineralogy: Descriptive Mineralogy, Comprising the Most Recent Discoveries*, 5th edn. Wiley, New York.

Daulton, T. L., Eisenhour, D. D., Bernatowicz, T. J., Lewis, R. S. and Buseck, P. R. (1996). Genesis of presolar diamonds: comparative high-resolution transmission electron microscope study of meteoritic and terrestrial nano-diamonds. *Geochim. Cosmochim. Acta*, **60**, 4853–4872.

David, M. (1977). *Geostatistical Ore Reserve Estimation.* Elsevier, Amsterdam.

Davies, T. A. and Gorsline, D. S. (1976). Oceanic sediments and sedimentary processes. *Chem. Oceanogr.*, **5**, 1–80.

De Sanctis, M. C., Capaccioni, F., Ciarniello, M. *et al.* (2015). The diurnal cycle of water ice on comet 67P/Churyumov–Gerasimenko. *Nature*, **525**, 500–503.

Debye, P. P. and Scherrer, P. (1916). Interferenzen an regellos orientierten Teilchen im Röntgenlicht. I. Nachrichten von der Gesellschaft der Wissenschaften zu Göttingen. *Math. Physik. Klasse*, 1–15.

Deer, W. A., Howie, R. A. and Zussman, D. J. (1982). *Rock-forming Minerals, Volume 1a, Orthosilicates*, 2nd edn. The Geological Society of London.

Deer, W. A., Howie, R. A. and Zussman, J. (2013). *An Introduction to the Rock-forming Minerals*, 3rd edn. Mineralogical Society, Twickenham, UK.

Devouard, B., Posfai, M., Hua, X. *et al.* (1998). Magnetite from magnetotactic bacteria: size distributions and twinning. *Am. Mineral.*, **83**, 1387–1398.

Dillon, F. J. (1963). Domains and domain walls. In *Magnetism, Volume 3*, ed. G. T. Rado and H. Suhl, pp. 415–464. Academic Press, New York.

Donnay, J. D. H. (1947). Hexagonal four-index symbols. *Am. Mineral.*, **32**, 52–58.

Dove, H. W. (1860). Optische Notizen. *Ann. Phys.*, **110**, 286–290.

Downing, K. H., Meisheng Hu., Wenk, H.-R. and O'Keefe, A. O. (1990). Resolution of oxygen atoms in staurolite by three-dimensional transmission electron microscopy. *Nature*, **348**, 525–528.

Droop, G. T. R. (1987). A general equation for estimating Fe^{3+} concentrations in ferromagnesian silicates and oxides from microprobe analyses using stoichiometric criteria. *Mineral. Mag.*, **51**, 431–435.

Dunn, M. D. (1983). *Fundamentals of Nutrition.* CBI, Boston, MA.

Dziewonski, A. M. and Anderson, D. L. (1981). Preliminary reference Earth model. *Phys. Earth Planet. Interiors*, **25**, 297–356.

Echigo, T. and Kimata M. (2010). Crystal chemistry and genesis of organic minerals: a review of oxalate and polycyclic aromatic hydrocarbon minerals. *Can. Mineral.*, **48**, 1329–1358.

Elwell, D. and Scheel, H. J. (1975). *Crystal Growth from High-temperature Solutions.* Academic Press, London.

Eskola, P. (1946). *Kristalle und Gesteine, ein Lehrbuch der Kristallkunde und allgemeinen Mineralogie.* Springer-Verlag, Berlin.

Evans, A. M. (1993). *Ore Geology and Industrial Minerals: An Introduction*, 3rd edn. Blackwell, Oxford.

Evzikova, N. Z. (1984). *Prospecting Crystallomorphology of Minerals.* (In Russian.) Nedra, Moscow.

Ewald, P. P. (ed.) (1962). *Fifty Years of X-ray Diffraction.* International Union of Crystallographers and Oosthoek's, Utrecht.

Fairchild, I. J. and Baker, A. (2012). *Spaleothem Science: From Process to Past Environments.* Wiley.

Fedorow, E. S. von (1885). Elements of the rules of figures. (In Russian.) *Trans. R. Russ. Mineral. Soc. St. Petersburg*, **21**, 1–279. (See also 1890 German review by G. Wulff in *Z. Kristallogr*, **17**, 610–611.)

Fedorow, E. S. von (1892). Zusammenstellung der krystallographischen Resultate des Herrn Schoenflies und der meinigen. *Z. Kristallogr.*, **20**, 25–75.

Fei, Y., Van Orman, J., Li, J. *et al.* (2004). Experimentally determined postspinel transformation boundary in Mg_2SiO_4 using MgO as an internal pressure standard and its geophysical implications. *J. Geophys. Res.*, **109**, B02305, 1–8.

Feinberg, J., Scott, G. R., Renne, P. and Wenk, H.-R. (2005). Exsolved magnetite inclusions in silicates: features determining their remanence behavior. *Geology*, **33**, 513–516.

Fersman, A. (1939). *The Search for Mineral Deposits on the Basis of Geochemistry and Mineralogy* (in Russian). Press House of Academy of Sciences, Moscow.

Filippenko, A. V. (1997). Optical spectra of supernovae. *Annu. Rev. Astron. Astrophys.*, **35**, 309–355.

Fournier, R. O. (1985). The behavior of silica in hydrothermal solutions. In *Geology and Geochemistry of Epithermal Systems*, ed. B. R. Berger and P. M. Bethke. Reviews in Geology, vol. 2, pp. 63–79. Society of Economists and Geologists, Chelsea, MI.

Fowler, C. M. R. (2005). *The Solid Earth: An Introduction to Global Geophysics*, 2nd edn. Cambridge University Press, Cambridge.

Frank, F. C. (1949). The influence of dislocations on crystal growth. *Disc. Faraday Soc.*, **5**, 48–54.

Frank-Kamenetskaya, O., Kol'tzov A., Kuz'mina, M., Zorina, M. and Potitskaya, L. (2011). Ionsubstititions and non-stochiometry of carbonated apatite-(CaOH) synthesised by precipitation and hydrothermal methods. *J. Mol. Struct.*, **992**, 9–18.

Friedman, G. M. (1959). Identification of carbonate minerals by staining methods. *J. Sedim. Petrol.*, **29**, 87–97.

Friedrich, W., Knipping, P. and von Laue, M. (1912). *Interferenz-Erscheinungen bei Röntgenstrahlen.* Sitzungsberichte der mathematischnaturwissenschaftlichen Klasse der Königlich Bayerischen Akademie der Wissenschaften zu München, pp. 363–373.

Frondel, J. W. (1975). *Lunar Mineralogy.* Wiley, New York.

Gao Zhen-xi (1980). *Minerals in China.* Museum of Geology, Ministry of Geology, Beijing.

Garnero, E. J., Kennett, B. and Loper, D. E. (2005). Studies of the Earth's deep interior: Eighth Symposium. *Phys. Earth Planet. Inter*, **153**, 1–2.

Garrels, R. M. and Christ, C. L. (1990). *Solutions, Minerals and Equilibria*, 2nd edn. Jones and Bartlett, Boston, MA.

Garrels, R. M., Thompson, M. E. and Siever, R. (1960). Stability of some carbonates at 25 °C and one atmosphere total pressure. *Am. J. Sci.*, **258**, 402–418.

Gerkin, R. E., Lundstedt, A. P. and Reppart, W. J. (1984). Structure of fluorene $C_{13}H_{10}$, at 159 K. *Acta Cryst. C*, **40**, 1892–1894.

Goldschmidt. V. (1897). *Krystallographische Winkeltabellen.* Springer, Berlin.

Goldschmidt, V. M. (1911). Die Gesetze der Mineralassoziationen vom Standpunkt der Phasenregel. *Z. Anorgan. Chem.*, **71**, 313–322.

Goldschmidt, V. M. (1923–1927). *Geochemische Verteilungsgesetze der Elemente.* Norsk Videnskaps-akademi i Oslo. Skrifter. I.

Mathematisk-naturvidenskabelik Klasse. 1923, no. 3; 1924, nos. 4–5; 1925, nos. 5 and 7; 1926, nos. 1, 2 and 8; 1927, no. 4.

Goldsmith, J. R. and Heard, H. C. (1961). Subsolidus phase relations in the system $CaCO_3$–$MgCO_3$. *J. Geol.*, **69**, 45–74.

Goreva, J. S., Chi Ma and Rossman, G. R. (2001). Fibrous nanoinclusions in massive rose quartz: the origin of rose coloration. *Am. Mineral.*, **86**, 466–472.

Gottardi, G. and Galli, E. (1985). *Natural Zeolites*. Springer-Verlag, Berlin.

Gottshalk, M. (1997). Internally consistent thermodynamic data for rock forming minerals. *Eur. J. Mineral.*, **9**, 175–223.

Grady, M. M. (2000). *Catalogue of Meteorites*, 5th edn. Cambridge University Press, Cambridge.

Greenwood, H. J. (1967). Wollastonite: stability in H_2O–CO_2 mixtures and occurrence in a contact-metamorphic aureole near Salmo, British Columbia, Canada. *Am. Mineral.*, **52**, 1669–1680.

Grigor'ev, D. P. (1965). *Ontogeny of Minerals*. Israel Program for Scientific Translations, Jerusalem.

Grigoriev, I. S. and Meilikhov, E. Z. (eds.) (1997). *Handbook of Physical Quantities*. CRC Press, Boca Raton, FL.

Grossman, L. (1972). Condensation in the primitive solar nebula. *Geochim. Cosmochim. Acta*, **36**, 597–619.

Groth, P. H. (1904). *Einleitung in die Chemische Krystallographie*. Engelmann, Leipzig.

Groth, P. H. (1926). *Entwicklungsgeschichte der mineralogischen Wissenschaften*. Springer-Verlag, Berlin.

Guthrie, G. D. and Mossman, B. T. (1993). *Health Effects of Mineral Dust*. Reviews in Mineralogy, vol. 28. Mineralogical Society of America, Washington, DC.

Haeckel, E. (1904). *Kunstformen der Natur*. Bibliographische Institut, Leipzig, 204 pp. (English translation, 1974: *Art Forms in Nature*, Dover, New York.)

Hahn, T. (ed.) (2006). *International Tables for Crystallography. Volume A: Space-group Symmetry*. International Union of Crystallography.

Harding, W. D. (1944). *The Geology of the Mattawan-Olrig Area*. Ontario Department of Mines, 53rd Annual Report.

Harlow, G. (ed.) (1998). *Nature of Diamonds*. Cambridge University Press, Cambridge.

Haüy, R. J. (1784). *Essay d'une Théorie sur la Structure des Crystaux*. Gogué and Née de la Rochelle, Paris.

Haüy, R. J. (1801). *Traité de Minéralogie*, 4 vols. and atlas of 86 plates. Chez Louis, Paris.

Helfferich, F. (1995). *Ion Exchange*. Dover, New York.

Hemley, R. J. (ed.) (1998). *Ultra-high Pressure Mineralogy: Physics and Chemistry of the Earth's Deep Interior*. Reviews in Mineralogy, vol. 37. Mineralogical Society of America, Washington, DC.

Hessell, J. F. C. (1830). *Kristallonometrie oder Krystallometrie und Krystallographie*. In J. S. T. Gehler's *Physikalisches Wörterbuch*, vol. 8. (Separate printing Leipzig, 1931.)

Hibbard, M. J. (2002). *Mineralogy: A Geologist's Point of View*. Wiley, New York.

Hnilicka, P. A., Mionczynski, J., Mincher, B. J. *et al.* (2002). Bighorn sheep lamb survival, trace minerals, rainfall, and air pollution: are there any connections? *Proceedings of the Biennial Symposium of the Northern Wild Sheep and Goat Council*, **13**, 69–94.

Hoefs, J. (1987). *Stable Isotope Geochemistry*, 3rd edn. Springer-Verlag, Berlin.

Hoffmann, C. A. S. (1789). Mineralsystem des Herrn Inspektor Werners mit dessen Erlaubnis herausgegeben von C. A. S. Hoffmann. *Bergmännisches J.* **1**, 369–398.

Hofmann, F. and Massanek, A. (1998). *Die Mineralogische Sammlung der Bergakademie Freiberg*. Christian Weise Verlag, München.

Holden, P., Lanc, P., Ireland, T.R. *et al.* (2009). Mass-spectrometric mining of Hadean zircons by automated SHRIMP multi-collector and single-collector U/Pb zircon age dating: the first 100,000 grains. *Int. J. Mass Spectrom.*, **286**, 53–63.

Horai, K. (1971). Thermal conductivity of rock-forming minerals. *J. Geophys. Res.*, **76**, 1278–1308.

Hosking, K. F. G. (1951). Primary ore deposition in Cornwall. *Trans. R. Geol. Soc. Cornwall*, **18**, 309–356.

Hoszowska, J., Freund, A. K., Boller, E. *et al.* (2001). Characterization of synthetic diamond crystals by spatially resolved rocking curve measurements. *J. Phys. D*, **34**, A47–A51.

Hu, M., Wenk, H.-R. and Sinitsina, D. (1992). Microstructures in natural perovskites. *Am. Mineral.*, **77**, 359–373.

Huebner, J. S. (1980). Pyroxene phase equilibria at low pressure. In *Pyroxenes*, ed. C. T. Prewitt, Reviews in Mineralogy, vol. 7, pp. 213–288. Mineralogical Society of America, Washington, DC.

Hull, D. and Bacon, D. J. (2011). *Introduction to Dislocations*, 5th edn. Butterworth-Heinemann, Kidlington, UK.

Hutchinson, C. S. (1983). *Economic Deposits and their Tectonic Setting*. Wiley, New York.

Ivanova, T. I., Frank-Kamenetskaya, O. V., Kol'tsov, A. B. and Ugolkov, V. L. (2001). Crystal structure of calcium-deficient carbonated hydroxyapatite. Thermal decomposition. *J. Solid State Chem.*, **160**, 340–349.

Jackson, M. D., Landis, E. N., Brune, P. F. *et al.* (2014). Mechanical resilience and cementitious processes in Imperial Roman architectural mortar. *Proc. Natl. Acad. Sci.*, **111**, 18484–18489.

Jahns, R. H. and Burnham, C. W. (1969). Experimental studies of pegmatite genesis. I. A model for the derivation and crystallization of granitic pegmatites. *Econ. Geol.*, **64**, 843–864.

Jenny, H. (2011). *Factors of Soil Formation: A System of Quantitative Pedology*. Dover, New York.

Joesten, R. (1986). The role of magmatic reaction, diffusion and annealing in the evolution of coronitic microstructure in troctolitic gabbro from Risör, Norway. *Mineral. Mag.*, **50**, 441–467.

Kanamori, H., Fujii, N. and Mizutani, H. (1968). Thermal diffusivity measurement of rock-forming minerals from 400 to 1100 K. *J. Geophys. Res.*, **73**, 595–605.

Kanitpanyacharoen, W., Wenk, H.-R., Kets, F., Lehr, B. C. and Wirth, R. (2011). Texture and anisotropy analysis of Qusaiba shales. *Geophys. Prospecting*, **59**, 536–556.

Katkova, V. I. (1996). *Urinary Stones: Mineralogy and Origin*. (In Russian.) Russian Academy of Sciences, Syktyvkar.

Keffer, F. (1967). The magnetic properties of minerals. *Sci. Am.*, **217**, 222–238.

Keller, P. C. (1990). *Gemstones and Their Origins*. Van Nostrand Reinhold, New York.

Kelley, D. S., Delaney, J. R. and Yoerger, D. R. (2001). Geology and venting characteristics of the Mothra hydrothermal field, Endeavour segment, Juan de Fuca Ridge. *Geology*, **29**, 959–962.

Kelly, M. G. (1999). Effects of heavy metals on the aquatic biota. In *The Environmental Geochemistry of Mineral Deposits, Part A: Processes,* *Techniques, and Health Issues*, ed. G. S. Plumlee and M. J. Logsdon, pp. 363–371. Society of Economic Geologists, Littleton, CO.

Kesler, S. E. (1994). *Mineral Resources, Economics and the Environment*. Macmillan, New York.

Klassen-Neklyudova, M. V. (1964). *Mechanical Twinning of Crystals*. (Translated by J. E. S. Bradley.) Consultants Bureau, New York.

Kleber, W. (1970). *An Introduction to Crystallography* (English edition). VEB Verlag, Berlin.

Klein, C. (2002). *Manual of Mineral Science*, 22nd edn. Wiley, New York.

Klein, C. and Dutrow, B. (2007). *Manual of Mineral Science*, 23rd edn. Wiley, New York.

Klein, C. and Philpotts, A. (2012). *Earth Materials: Introduction to Mineralogy and Petrology*. Cambridge University Press, Cambridge.

Kocks, U. F., Tomé, C. N. and Wenk, H.-R. (2000). *Texture and Anisotropy: Preferred Orientation in Polycrystals and Their Effect on Materials Properties*, paperback edn. Cambridge University Press, Cambridge.

Kopylova, M. G., Gurney, J. J. and Daniels, L. R. M. (1997). Mineral inclusions in diamonds from the River Ranch kimberlite, Zimbabwe. *Contrib. Mineral. Petrol.*, **129**, 366–384.

Korago, A. A. (1992). *Introduction into Biomineralogy*. (In Russian.) Nedra, Leningrad.

Korzhinskii, D. S. (1959). *Physicochemical Basis of the Analysis of the Paragenesis of Minerals*. (English translation.) Consultants' Bureau, New York.

Korzhinskii, D. S. (1970). *Theory of Metasomatic Zoning*. Oxford University Press, Oxford.

Kotel'nikova, E. N., Platonova, N. V. and Filatov, S. K. (2007). Identification of biogenic paraffins and their thermal phase transitions. *Geol. Ore Deposits*, **49**, 697–709.

Kretz, R. (1994). *Metamorphic Crystallization*. Wiley, New York.

Krivovichev, V. G. and Charykova, M. V. (2014). Number of minerals of various chemical elements: statistics 2012 (a new approach to an old problem). *Geol. Ore Deposits*, **57**, 1–7.

Kundt, A. (1883). Über eine einfache Methode zur Untersuchung der Thermo-, Actinound Piezoelektrizität der Krystalle. *Annal. Physik*, **20**, 592–601.

Lally, J. S., Heuer, A. H. and Nord, G. L. (1976). Precipitation in the ilmenite–hematite system.

In *Electron Microscopy in Mineralogy*, ed. H.-R. Wenk, pp. 214–219. Springer-Verlag, Berlin.

Laves, F. (1962). The growing field of mineral structures. In *Fifty Years of X-ray Diffraction*, ed. P. P. Ewald, pp. 174–189. International Union of Crystallographers and Oosthoek's, Utrecht.

Laves, F. and Goldsmith, J. R. (1961). Polymorphism, order, disorder, diffusion and confusion in the feldspars. *Cursillos y Conferencias, Instituto di Lucas Mallado, CSIC, Spain*, **7**, pp. 71–80.

Le Bas, M. J. and Streckeisen, A. L. (1991). The IUGS systematics of igneous rocks. *J. Geol. Soc. London*, **148**, 825–833.

Lee, R. W. (1964). On the role of hydroxyl in the diffusion of hydrogen in fused silica. *Phys. Chem. Glasses*, **5**, 35–43.

Lekic, V., Cottaar, S., Dziewonski, A. and Romanowicz, B. (2012). Cluster analysis of global lower mantle tomography: a new class of structure and implications for chemical heterogeneity. *Earth Planet. Sci. Lett.*, **357–358**, 68–77.

Lenz, H. O. (1861). *Mineralogie der alten Griechen und Römer*. Thienemann, Gotha, 194 pp. (Reprinted Sandig Verlag, Wiesbaden, 1966.)

Liebau, F. (1962). Die Systematik der Silikate. *Naturwissenschaften*, **49**, 481–491.

Liebau, F. (1985). *Structural Chemistry of Silicates: Structure, Bonding, Classification*. Springer-Verlag, Berlin.

Lindgren, W. (1933). *Mineral Deposits*, 4th edn. McGraw Hill, New York.

Loeffler, B. M. and Burns, R. G. (1976). Shedding light on the color of gems and minerals. *Am. Sci.*, **64**, 636–649.

Loewenstein, W. (1954). The distribution of aluminum in the tetrahedra of silicates and aluminates. *Am. Mineral.*, **39**, 92–96.

Logvinenko, N. V. and Orlova, L. V. (1987). *Formation and Transformation of Sedimentary Rocks on Continents and in Oceans*. (In Russian.) Nedra, Leningrad.

London, D. (2008). *Pegmatites*. The Canadian Mineralogist Special Publication 10. Mineralogical Association of Canada, Québec.

Lopez, R. M. C., Kamp, L. W., Doute, S. *et al.* (2001). Io in the near infrared: near-infrared mapping spectrometer (NIMS) results from the Galileo flybys in 1999 and 2000. *J. Geophys. Res.*, **106**, 33053–33078.

MacGillavry, C. H. (1976). *Symmetry Aspects of M. C. Escher's Periodic Drawings*, 2nd edn. International Union of Crystallography and Bohn, Scheltma and Holkema, Utrecht.

Mallowan, M. E. L. and Cruikshank, R. J. (1933). Excavations at Tel Arpachiyah. *Iraq*, **2**, 1–178.

Manghnani, M. H. and Syono, Y. (eds.) (1987). *High Pressure Research in Mineral Physics*. Geophysical Monograph 39. American Geophysical Union, Washington, DC.

Mason, B. (1962). *Meteorites*. Wiley, New York.

Maucher, W. (1914). *Leitfaden für den Geologie-Unterricht*, 2nd edn. Craz und Gerlach, Freiberg.

McCammon, C. A. (1995). Mössbauer spectroscopy of minerals. In *Mineral Physics and Crystallography: A Handbook of Physical Constants*, pp. 332–347. American Geophysical Union, Washington, DC.

McCammon, C. A. (2000). Insights into phase transformations from Mössbauer spectroscopy. In *Transformation Processes in Minerals*, ed. S. A. T. Redfern and M. A. Carpenter, Reviews in Mineralogy, vol. 39. pp. 241–257. Mineralogical Society of America, Washington, DC.

McKeown, D. A. and Post, D. A. (2001). Characterization of manganese oxide mineralogy in rock varnish and dendrites using X-ray absorption spectroscopy. *Am. Mineral.*, **86**, 701–713.

McLaren, A. C. (1991). *Transmission Electron Microscopy of Minerals and Rocks*. Cambridge University Press, Cambridge.

Means, W. D. (1976). *Stress and Strain: Basic Concepts of Continuum Mechanics for Geologists*. Springer-Verlag, Berlin.

Medenbach, O. (2014). Putting the perspective on crystals. Historical polarizing microscopes made by the R. Fuess company in Berlin. In *The Munich Show: Mineralientage München*, pp. 190–214. Wachholtz Verlag, Kiel, Hamburg.

Medenbach, O. and Medenbach, U. (2001). *Mineralien, Erkennen und Bestimmen. Steinbach's Naturführer*. Mosaik Verlag, Niedernhausen, Germany.

Medenbach, O. and Wilk, H. (1986). *The Magic of Minerals*. Springer-Verlag, Berlin.

Mehta, P. K. and Monteiro, P. J. M. (1993). *Concrete: Structure, Properties, and Materials*. Prentice-Hall, Upper Saddle River, NJ.

Meyer, C. (1988). Ore deposits as guides to geologic history. *Annu. Rev. Earth Planet. Sci.*, **16**, 147–171.

Miller, W. H. (1839). *A Treatise on Crystallography.* Pitt Press, Cambridge.

Mills, S. J., Hatert, F., Nickel, E. H. and Ferraris, G. (2009). The standardization of mineral group hierarchies: application to recent nomenclature proposals. *Eur. J. Mineral.*, **21**, 1073–1080.

Mitchell, R. H. (1986). *Kimberlites. Mineralogy, Geochemistry and Petrology.* Plenum Press, New York.

Mitscherlich, E. (1820). Sur la relation que existe entre la forme cristalline et les proportions chimiques. *Ann. Chimie Phys.*, **14**, 172–190.

Mitscherlich, E. (1821). Sur la relation que existe entre la forme cristalline et les proportions chimiques. IIme. mémoire sur les arséniates et les phosphates. *Ann. Chimie Phys.*, **19**, 350–419.

Momma, K., Ikeda, T., Nishikubo, K. *et al.* (2011). New silica clathrate minerals that are isostructural with natural gas hydrates. *Nature Commun.*, **2**, 196.

Moradian-Oldak, J. (2012). Protein-mediated enamel mineralization. *Frontiers in Bioscience*, **17**, 1996–2023.

Morris, G. B., Raitt, R. W. and Shor, G. G. (1969). Velocity anisotropy and delay time maps of the mantle near Hawaii. *J. Geophys. Res.*, **74**, 4300–4316.

Morse, S. A. (1980). *Basalts and Phase Diagrams: An Introduction to the Quantitative Use of Phase Diagrams in Igneous Petrology.* Springer, New York.

Müller, W. F., Wenk, H. R. and Thomas, G. (1972). Structural variations in anorthites. *Contrib. Mineral. Petrol.*, **34**, 304–314.

Mullis, J. (1991). Bergkristall. *Schweizer Strahler*, **9**, 127–161.

Mullis, J., Dubessy, J., Poty, B. and O'Neil, J. (1994). Fluid regimes during late stages of a continental collision: physical, chemical, and stable isotope measurements of fluid inclusions in fissure quartz from a geotraverse through the Central Alps, Switzerland. *Geochim. Cosmochim. Acta*, **58**, 2239–2263.

Mumpton, F. A. (1999). La roca magica: uses of natural zeolites in agriculture and industry. *Proc. Natl. Acad. Sci. USA*, **96**, 3463–3470.

Nadeau, J. L., Davidson, M. W. and Connell, R. G. (2012). *Reflected Light Microscopy*, 2nd edn. Wiley, New York.

Nakamoto, K. (1997). *Infrared and Raman Spectra of Inorganic and Coordination Compounds: Part A Theory and Applications in Inorganic Chemistry*, 5th edn. Wiley, New York.

Nakaya, U. (1954). *Formation of snow crystals.* Snow, Ice and Permafrost Research Establishment, Research Paper 3. Corps of Engineers, US Army, Wilmette, IL.

Nassau, K. (1980). The causes of color. *Sci. Am.*, **243**, 124–156.

Neev, D. and Emery, K. O. (1967). The Dead Sea: depositional processes and environments of evaporites. *Israel Geol. Survey Bull.*, **41**.

Nesse, W. D. (2011). *Introduction to Mineralogy*, 2nd edn. Oxford University Press, New York.

Nesse, W. D. (2012). *Introduction to Optical Mineralogy*, 4th edn. Oxford University Press, Oxford.

Neumann, F. E. (1885). *Vorlesungen über die Theorie der Elastizität der festen Körper und des Lichtäthers, gehalten an der Universität Königsberg*, ed. O. E. Meyer. B. G. Teubner, Leipzig.

Newnham, R. E. (2005). *Properties of Materials: Anisotropy, Symmetry, Structure.* Oxford University Press, Oxford.

Niggli, P. (1920). *Lehrbuch der Mineralogie.* Borntraeger, Berlin.

Nishi, M., Irifune, T., Tsuchiya, J. *et al.* (2014). Stability of hydrous silicate at high pressures and water transport to the deep lower mantle. *Nature Geosci.*, **7**, 224–227.

Nomura, R., Hirose, K., Sata, N. and Ohishi, Y. (2010). Precise determination of the post-stishovite phase transition boundary and implications for seismic heterogeneities in the mid-lower mantle. *Phys. Earth Planet. Inter.*, **183**, 104–109.

Nordstrom, D. K. and Alpers, C. N. (1999). Negative pH, efflorescent mineralogy, and consequences for environmental restoration at the Iron Mountain Superfund site, California. *Proc. Natl. Acad. Sci.*, **96**, 3455–3462.

Nye, J. F. (1957). *Physical Properties of Crystals: Their Representation by Tensors and Matrices.* Oxford University Press, London. (Reprinted 1998.)

O'Keefe, M. and Hyde, B. G. (1985). An alternative approach to non-molecular crystal structures with emphasis on the arrangement of cations. *Struct. Bond.*, **61**, 77–144.

Orowan, E. (1934). Plasticity of crystals. *Z. Phys.*, **89**, 605–659.

Page, R. H. and Wenk, H.-R. (1979). Phyllosilicate alteration of plagioclase studied by transmission electron microscopy. *Geology*, **7**, 393–397.

Pamato, M. G., Myhill, R., Ballaran, T. B. *et al.* (2014). Lower-mantle water reservoir implied by the extreme stability of a hydrous aluminosilicate. *Nature Geosci.*, **8**, 75–79, doi:10.1038/ngeo2306.

Parsons, I. (ed.) (1994). *Feldspars and Their Reactions.* Kluwer Academic, Dordrecht, Netherlands.

Pauling, L. (1929). The principles determining the structure of complex ionic crystals. *J. Am. Chem. Soc.*, **51**, 1010–1026.

Pédro, G. (1997). Clay minerals in weathered rock materials and in soils. In *Soils and Sediments: Mineralogy and Geochemistry*, ed. H. Paquet and N. Clauer, pp. 1–20. Springer-Verlag, Berlin.

Peterson, M. N. A., Bien, G. S. and Berner, R. A. (1963). Radiocarbon studies of recent dolomite from Deep Spring Lake, California. *J. Geophys. Res.*, **68**, 6493–6505.

Petrov, S. V., Polekhovskiy, Yu. S., Borozdin, A. P. *et al.* (2011). *Mineral Resources of the World.* (In Russian.) St. Petersburg State University Publishing House, St. Petersburg.

Phillips, B. L. (2000). NMR spectroscopy of phase transitions in minerals. In *Transformation Processes in Minerals*, ed. S. A. T. Redfern and M. A. Carpenter, Reviews in Mineralogy, vol. 39, pp. 203–240. Mineralogical Society of America, Washington, DC.

Phillips, R. M. (1971). *Mineral Optics: Principles and Techniques.* W. H. Freeman and Co., San Francisco.

Poirier, J. P. (1985). *Creep of Crystals: High-temperature Deformation Processes in Metals, Ceramics and Minerals.* Cambridge University Press, Cambridge.

Polanyi, M. (1934). Lattice distortion which originates plastic flow. *Z. Phys.*, **89**, 660–604.

Pough, F. H. (1986). *A Field Guide to Rocks and Minerals*, 5th edn. Houghton Mifflin, New York.

Prechtl, J. J. (1810). Théorie de la cristallisation. *J. Mines*, **28**, 261–312.

Putnis, A. (1992). *Introduction to Mineral Sciences.* Cambridge University Press, Cambridge.

Ramsdell, L. S. (1947). Studies on silicon carbide. *Am. Mineral.*, **32**, 64–82.

Reed, S. J. B. (2010). *Electron Microprobe Analysis and Scanning Electron Microscopy in Geology*, 2nd edn. Cambridge University Press, Cambridge.

Reinhard, M. (1931). *Universal Drehtischmethoden.* Wepf, Basel.

Rickwood, P. C. (1981). The largest crystals. *Am. Mineral.*, **66**, 885–907.

Rieder, R., Economou, T., Wänke, H. *et al.* (1997). The chemical composition of Martian soil and rocks returned by the mobile alpha proton X-ray spectrometer: preliminary results from the X-ray mode. *Science*, **278**, 1771–1774.

Rietveld, H. M. (1969). A profile refinement method for nuclear and magnetic structures. *J. Appl. Cryst.*, **2**, 65–71.

Ringwood, A. E. (1975). *Composition and Petrology of the Earth's Mantle.* McGraw-Hill, New York.

Ringwood, A. E. (1979). *Origin of the Earth and Moon.* Springer-Verlag, Berlin.

Robie, R. A. and Hemingway, B. S. (1995). *Thermodynamic Properties of Minerals and Related Substances at 298.15 K and 1 Bar (105 Pascals) Pressure and at Higher Temperatures.* US Geological Survey Bulletin 2131.

Robinson, C., Kirkhamn, J. and Shore, R. C. (1995). *Dental Enamel: Formation to Destruction.* CRC Press, Bota Raton, FL.

Rogers, J. (1957). The distribution of marine carbonate sediments: a review. In *Regional Aspects of Carbonate Deposition*, ed. R. J. Le Blanc and J. G. Breeding, Society of Economic Paleontologists and Mineralogists, Special Publication 5, pp. 2–14. SEPM, Tulsa, OK.

Ronov, A. B. and Yaroshevsky, A. A. (1969). Chemical composition of the Earth's crust. In *The Earth's Crust and Upper Mantle*, ed. P. J. Hart, Geophysical Monograph 13, pp. 37–57. American Geophysical Union, Washington, DC.

Ronov, A. B., Yaroshevsky, A. A. and Migdisov, A. A. (1991). Chemical composition of the earth's crust and geochemical balance of elements. *Int. Geol. Rev.*, **17**, 941–1047.

Rosbaud, P. and Schmid, E. (1925). Über die Verfestigung von Einkristallen durch Legierung und Kaltreckung. *Z. Phys.*, **32**, 197–225.

Rosenbusch, H. (1876). Ein neues Mikroskop fuer mineralogische und petrographische Untersuchungen. *N. Jb. Mineral.*, 504–513.

Roth, W. L. (1958). Magnetic structures of MnO, FeO, CoO, and NiO. *Phys Rev.*, **110**, 1333–1341.

Sawkins, F. J. (1990). *Metal Deposits in Relation to Plate Tectonics*, 2nd edn. Springer-Verlag, Berlin.

Schaller, W. T. (1916). Gigantic crystals of spodumene. *US Geological Survey Bulletin*, **610**, 138.

Schmid, E. (1924). Zn-normal stress law. In *Proceedings of the International Congress of Applied Mechanics*, Delft, p. 342.

Schneider, S. L. (2007). *The World of Fluorescent Minerals*. Schiffer Publishing, Atglen, PA.

Schneiderhöhn, H. (1941). *Lehrbuch der Erzlagerstättenkunde, Volume 1*. Fischer Verlag, Jena.

Schoenflies, A. (1891). *Krystallsysteme und Krystallstructur*. B. G. Teubner, Leipzig.

Schubert, G., Turcotte, D. L. and Olson, P. (2001). *Mantle Convection in the Earth and Planets*. Cambridge University Press, Cambridge.

Scovil, J. A. (1996). *Photographing Minerals, Fossils and Lapidary Materials*. Geoscience Press, Missoula, MT.

Seager, S. L. and Slabaugh, M. R. (2013). *Chemistry for Today: General, Organic, and Biochemistry*, 8th edn. Brooks/2Cole, Pacific Grove, CA.

Seeber, L. A. (1824). Versuch einer Erklärung des inneren Baues der festen Körper. *Gilbert's Annal. Physik*, **76**, 229–248.

Semkin, R. G. and Kramer, J. R. (1976). Sediment geochemistry of Sudbury area lakes. *Can. Mineral.*, **14**, 73–90.

Shannon, R. D. and Prewitt, C. T. (1969). Effective ionic radii in oxides and fluorides. *Acta Crystallogr.*, **25**, 925–946.

Shen, G., Mao, H.-K., Hemley, R. J., Duffy, T. S. and Rivers, M. L. (1998). Melting and crystal structures of iron at high pressures and temperatures. *Geophys. Res. Lett.*, **25**, 373–376.

Shi, C. Y., Zhang, L., Yang, W. *et al.* (2013). Formation of iron melt network in silicate perovskite at Earth's lower mantle conditions. *Nature Geosci.*, **6**, 971–975.

Shigley, J. E. and Kampf, A. R. (1984). Gem-bearing pegmatites: a review. *Gems Gemol.*, **20**, 64–77.

Shigley, J. E. and McClure, S. F. (2009). Laboratory-treated gemstones. *Elements*, **5**, 175–178.

Shigley, J. E., Abbaschian, R. and Clarke, C. (2002). Gemesis laboratory-created diamonds. *Gems Gemol.*, **38**, 301–309.

Shtukenberg, A. G., Punin, Y. O., Gujral, A. and Kahr, B. (2014). Growth actuated bending and twisting of single crystals. *Angewandte Chemie Int. Edn.*, **53**, 672–699.

Shull, C. G. and Smart, J. S. (1949). Detection of antiferromagnetism by neutron diffraction. *Phys. Rev.*, **76**, 1256–1257.

Shull, C. G., Strauser, W. A. and Wollan, E. O. (1951). Neutron diffraction by paramagnetic and antiferromagnetic substances. *Phys. Rev.*, **83**, 333–345.

Sillitoe, R. H. (1973). The tops and bottoms of porphyry copper deposits. *Econ. Geol.*, **68**, 799–815.

Silver, P. G. (1996). Seismic anisotropy beneath the continents: probing the depths of geology. *Annu. Rev. Earth Planet. Sci.*, **24**, 385–432.

Simmons, G. and Wang, H. (1971). *Single Crystal Elastic Constants and Calculated Aggregate Average Properties: A Handbook*. MIT Press, Cambridge, MA.

Singer, M. J. and Munns, D. N. (2005). *Soils, an Introduction*, 6th edn. Prentice-Hall, Upper Saddle River, NJ.

Slemmons, D. B. (1962). *Determination of volcanic and plutonic plagioclases using a three or four-axis universal stage*. Geological Society of America Special Paper 69, GSA, Washington, DC.

Sloan, E. D. and Koh, C. A. (2007). *Clathrate Hydrates of Natural Gases*, 3rd edn. CRC Press, Boca Raton, FL.

Smith, G. I. (1979). *Subsurface stratigraphy and geochemistry of late Quaternary evaporites, Searles Lake, California*. US Geological Survey Professional Paper 1043, Washington, DC.

Smith, G. P. (1963). *Physical Geochemistry*. Addison-Wesley, Reading, MA.

Smith, J. V. and Brown, W. L. (1988). *Feldspar Mineralogy*. Springer-Verlag, Berlin.

Smith, J. V. and Yoder, H. S. (1956). Experimental and theoretical studies of the mica polymorphs. *Mineral. Mag.*, **31**, 209–235.

Smith, K. S. and Huyck, H. L. O. (1999). An overview of the abundance, relative mobility, bioavailability, and human toxicity of metals. In *The Environmental Geochemistry of Mineral Deposits, Volume A*, ed. G. S. Plumlee and M. J. Logsdon, pp. 29–70. Society of Economic Geologists, Littleton, CO.

Soda, Y. and Wenk, H.-R. (2014). Antigorite crystallographic preferred orientations in serpentinites from Japan. *Tectonophysics*, **615–616**, 199–212.

Sofianides, A. S. and Harlow, G. E. (1990). *Gems and Crystals from the American Museum of Natural History*. Simon and Schuster, New York.

Spry, A. (1969). *Metamorphic Textures*. Pergamon Press, Oxford.

Sriramadas, A. (1957). Diagrams for the correlation of unit cell edges and refractive indices with the chemical composition of garnets. *Am. Mineral.*, **42**, 294–298.

Stalder, H. A., de Quervain, F., Niggli, E. and Graeser, S. (1973). *Die Mineralfunde der Schweiz*. Wepf, Basel.

Stanton, M. F., Layard, M., Tegeris, A. *et al.* (1981). Relation of particle dimension to carcinogenicity in amphibole asbestoses and other fibrous materials. *J. Natl. Cancer Inst.*, **67**, 965–975.

Steiger, R. H. and Hart, S. R. (1967). The microcline–orthoclase transition within a contact aureole. *Am. Mineral.*, **52**, 87–116.

Steno, N. (1669). *Nicolai Stenonis de solido intra solidum naturaliter contento dissertationis prodromus ad serenissium Ferdinandum II*. Ex typographia sub signo Stellae, Florentiae.

Stern, S. A., Bagenal, F., Ennico, K. *et al.* (2015). The Pluto system: initial results from its exploration by New Horizons. *Science*, **350**, 6258.

Stolz, J. F. (1992). Magnetotactic bacteria: biomineralization, ecology, sediment magnetism, environmental indicator. In *Biomineralization Processes of Iron and Manganese: Modern and Ancient Environments*, ed. H. C. W. Skinner and R. W. Fitzpatrick, pp. 133–145. Catena Supplement 21. Catena Verlag, Cremlingen.

Strakhov, N. M. (1967). *Principles of Lithogenesis, Volume 1*. Oliver and Boyd, London.

Streckeisen, A. (1976). To each plutonic rock its proper name. *Earth Sci. Rev.*, **12**, 1–33.

Strunz, H. (1966). *Mineralogische Tabellen*, 4th edn. Akademische Verlagsgesellschaft, Leipzig.

Strunz, H. and Nickel, E. H. (2001). *Strunz Mineralogical Tables*, 9th edn. Schweizerbart, Stuttgart.

Sunagawa, I. (2007). *Crystals: Growth, Morphology and Perfection*. Cambridge University Press, Cambridge.

Swaminathan, S., Craven, B. M. and McMullan, R. K. (1984). The crystal structure and molecular thermal motion of urea at 12, 60 and 123 K from neutron diffraction. *Acta Cryst.*, **B40**, 300–306.

Tabor, D. (1954). Mohs's hardness scale – a physical interpretation. *Proc. Phys. Soc.*, **B67**, 249–257.

Takakura, M., Natoya, S. and Takahashi. H. (2001). Application of cathodoluminescence to EPMA. *JEOL News*, **36E**, 35–39.

Tateno, S., Hirose, K., Ohishi, Y. and Tatsumi, Y. (2010). The structure of iron in Earth's inner core. *Science*, **330**, 359–361.

Tateno, S., Kuwayana, Y., Hirose, K. and Ohishi, Y. (2015). The structure of Fe-Si alloy in Earth's inner core. *Earth Planet. Sci. Lett.*, **141**, 1–9.

Taylor, G. I. (1934). The mechanism of plastic deformation of crystals. *Proc. R. Soc. London A*, **145**, 362–387.

Taylor, S. R. (2001). *Solar System Evolution: A New Perspective*, 2nd edn. Cambridge University Press, Cambridge.

Taylor, W. H. (1933). The structure of sanidine and other feldspars. *Z. Kristallogr.*, **85**, 425–442.

Thompson, J. B. (1959). Local equilibrium in metasomatic processes. In *Researches in Geochemistry*, ed. P. H. Abelson, pp. 427–457. Wiley, New York.

Thompson, J. B. Jr. (1978). Biopyriboles and polysomatic series. *Am. Mineral.*, **63**, 239–249.

Troeh, F. R. (2005). *Soils and Soil Fertility*, 6th edn. Wiley-Blackwell, Oxford.

Trommsdorff, V. (1966). Progressive Metamorphose kieseliger Karbonatgesteine in den Zentralalpen zwischen Bernina und Simplon. *Schweiz. Mineral. Petrog. Mitt.*, **46**, 431–460.

Turner, F. J. (1981). *Metamorphic Petrology: Mineralogical, Field and Tectonic Aspects*, 2nd edn. McGraw-Hill, New York.

Valley, J. W., Cavosie, A. J., Ushikubo, T. *et al.* (2014). Hadean age for a post-magma-ocean zircon confirmed by atom-probe tomography. *Nature Geosci.*, **7**, 219–223.

Valyashko, M. G. (1962). *Geochemical Regularities of Formation of Deposits of Potassium Salts*. (In Russian.) Moscow University Press, Moscow.

van Zyl, A. A. (1988). De Beers' 100. *Geobulletin*, **31**, 24–28.

van't Hoff, J. H. (1905, 1909), *Zur Bildung der ozeanischen Salzablagerungen*. Viehweg, Braunschweig.

van't Hoff, J. H. (1912). *Untersuchungen über die Bildungsverhältnisse der Ozeanischen Salzablagerungen insbesondere des Stassfurter Salzlagers*. Akademische Verlagsgesellschaft, Leipzig.

Vaniman, D. T., Bish, D. L., Ming, D. W. *et al.* and MSL Science Team (2014). Mineralogy of a mudstone at Yellowknife Bay, Gale Crater, Mars. *Science*, 343, doi: 10.1126/science.1243480.

Vasin, R., Wenk, H.-R., Kanitpanyacharoen, W., Matthies, S. and Wirth, R. (2013). Anisotropy of Kimmeridge shale. *J. Geophys. Res.*, **118**, 1–26, doi:10.1002/jgrb.50259.

Vaughan, D. J. (ed.) (2006). *Sulfide Mineralogy and Geochemistry*. Reviews in Mineralogy, vol. 61. Mineralogical Society of America, Washington, DC.

Vaughan, D. J. and Craig, J. R. (1978). *Mineral Chemistry of Metal Sulfides*. Cambridge University Press, New York.

Veblen, D. R. and Buseck, P. (1980). Microstructures and reaction mechanisms in biopyriboles. *Am. Mineral.*, **65**, 599–623.

Verma, A. R. (1953). *Crystal Growth and Dislocations*. Academic Press, New York.

Verzilin, N. N. and Utsalu, K. R. (1990). New data on mineral composition of modern sediments of Lake Balkhash. *Dokl. Russ. Acad. Sci. USSR, Earth Sci.*, **T314**, 686–689.

Voigt, W. (1928). *Lehrbuch der Kristallphysik*. Teubner, Leipzig.

von Laue, M. (1913). Röntgenstrahlinterferenzen. *Phys. Z.*, **14**, 1075–1079.

Wagman, D. D., Evans, W. H., Parker, V. B. et al. (1982). *The NBS Tables of Chemical Thermodynamic Properties: Selected Values for Inorganic and C_1 and C_2 Organic Substances in SI Units*. American Chemical Society and Journal of Physical and Chemical Reference Data, vol. 11, Suppl. 2.

Wall, F. and Zaitsev, A. N. (eds.) (2004). *Phoscorites and Carbonatites from Mantle to Mine: The Key Example of the Kola Alkaline Province*. Mineralogical Society Book Series, vol. 10. Mineralogical Society, Twickenham, UK.

Wasastjerna, J. A. (1923). On the radii of ions. *Soc. Sci. Fenn.*, **1**(38), 1–25.

Weibel, M. (1973). *Die Mineralien der Schweiz, ein mineralogischer Führer*, 3rd edn. Birkhäuser, Basel.

Weiss, C. S. (1819). Über eine verbesserte Methode für die Bezeichnung der verschiedenen Flächen eines Krystallisationssystems nebst Bemerkungen über den Zustand von Polarisierung der Seiten in den Linien der krystallinischen Struktur. Abhandl.

Königl. *Akad. D. Wiss. Berlin* (1816–1817), pp. 287–336.

Wenk, E. (1970). Zur Regionalmetamorphose und Ultrametamorphose im Lepontin. *Fortschr. Mineral.*, **47**, 34–51.

Wenk, H.-R. (ed.) (1976). *Electron Microscopy in Mineralogy*. Springer-Verlag, Berlin.

Wenk, H.-R. (1979). An albite-anorthite assemblage in low-grade amphibolite facies rocks. *Am. Mineral.*, **64**, 1294–1299.

Wenk, H.-R. and Zenger, D. H. (1983). Sequential basal faults in Devonian dolomite, Nopah Range, Death Valley area, California. *Science*, **222**, 502–504.

Wenk, H.-R., Wenk, E. and Wallace, J. H. (1974). Metamorphic mineral assemblages in pelitic rocks of the Bergell Alps. *Schweiz. Mineral. Petrog. Mitt.*, **54**, 507–554.

Wenk, H.-R., Meisheng Hu, Lindsey, T. and Morris, W. (1991). Superstructures in ankerite and calcite. *Phys. Chem. Mineral.*, **17**, 527–539.

Westmacott, K. H., Barnes, R. S. and Smallman, R. E. (1962). The observation of dislocation "climb" source. *Phil. Mag.*, **7** (ser. 8), 1585–1613.

Wicks, F. J., Kjoller, K. and Henderson, G. S. (1992). Imaging the hydroxyl surface of lizardite at atomic resolution with the atomic force microscope. *Can. Mineral.*, **30**, 83–91.

Wilkinson, G. R. (1973). Raman spectra of ionic, covalent, and metallic crystals. In *The Raman Effect, Volume 2, Applications*, ed. A. Anderson, pp. 812–983. Marcel Decker, New York.

Winchell, A. N. (1929). Dispersion of minerals. *Am. Mineral.*, **14**, 125–149.

Winkler, H. G. F. (1979). *Petrogenesis of Metamorphic Rocks*, 5th edn. Springer-Verlag, Berlin.

Withers, P. J. (2007). X-ray nanotomography. *Materials Today*, **10**, 26–34.

Wollaston, W. H. (1813). On the elementary particles of certain crystals. *Phil. Trans. R. Soc. London*, **103**, 51–63.

Wood, J. A. and Hashimoto, A. (1993). Mineral equilibrium in fractionated nebular systems. *Geochim. Cosmochim. Acta*, **57**, 2377–2388.

Wulff, G. (1913). Über die Kristallröntgenogramme. *Phys. Z.*, **14**, 217–220.

Yada, K. (1971). Study of the microstructure of chrysotile asbestos by high resolution electron microscopy. *Acta Crystallogr. A*, **27**, 659–664.

Yaroschevsky, A. A. and Bulakh, A. G. (1994). The mineral composition of the Earth's crust, mantle, meteories, moon, planets. In *Advanced Mineralogy, Volume 1*, ed. A. Marfunin, pp. 27–36. Springer-Verlag, Berlin.

Yoder, H. S., Stewart, D. B. and Smith, J. R. (1956). Ternary feldspars. In *Annual Report of the Geophysics Laboratory*, pp. 206–214. Carnegie Institute, Washington, DC.

Yushkin, N. P. (1968). *Mineralogy and Paragenesis of Native Sulfur in Exogenic Deposits.* (In Russian.) Nauka, Leningrad.

Zholnerovich, V. A. (1990). Framboydal pyrite aggregates in sediments of modern lakes of humid zone. *Zapiski (Proceedings) of the Russian Mineralogical Society*, **119**, 4, 39–43.

Zorz, M. (2009). Quarz-Gwindel. *Mineralien der Welt*, **20**, 30–46.

Index

Bold entries are mineral names. Page numbers in bold refer to minerals with detailed descriptions. Page numbers in italics refer to pictures.